D1710520

Cellular Biophysics

Volume 1

Thomas Fischer Weiss

Cellular Biophysics

Volume 1: Transport

A Bradford Book
The MIT Press
Cambridge, Massachusetts
London, England

This book was set in Lucida Bright by Windfall Software using ZzTEX and was printed and bound in the United States of America.

Library of Congress Cataloging-in-Publication Data

Weiss, Thomas Fischer
Cellular biophysics / Thomas Fischer Weiss
v. <1- > ; cm.
Includes bibliographical references and index.
Contents: v. 1. Transport — v. 2. Electrical properties.
ISBN 0-262-23183-2 (v. 1). — ISBN 0-262-23184-0 (v. 2)
1. Cell physiology. 2. Biophysics. 3. Biological transport.
4. Electrophysiology. I. Title.
QH631.W44 1995
574.87'6041—dc20

95-9801
CIP

To Aurice B, Max, Elisa, and Eric

Contents

Contents in Detail

Preface

In scientific thought we adopt the simplest theory which will explain all the facts under consideration and enable us to predict new facts of the same kind. The catch in this criterion lies in the word 'simplest.' It is really an aesthetic canon such as we find implicit in our criticisms of poetry or painting. The layman finds such a law as $\partial x/\partial t = \kappa(\partial^2 x/\partial y^2)$ less simple than 'it oozes,' of which it is the mathematical statement. The physicist reverses this judgement, and his statement is certainly the more fruitful of the two, so far as prediction is concerned.
—Haldane, 1985

Subject and Orientation of the Book

This and the companion text (Weiss, 1996) consider two basic topics in cellular biophysics, which we pose here as questions:

- Which molecules are transported across cellular membranes, and what are the mechanisms of transport? How do cells maintain their compositions, volume, and membrane potential?
- How are potentials generated across the membranes of cells? What do these potentials do?

Although the questions posed are fundamentally biological questions, the methods for answering these questions are inherently multidisciplinary. For example, to understand the mechanism of transport of molecules across cellular membranes, it is essential to understand both the structure of membranes and the principles of mass transport through membranes. Since the transported matter may combine chemically with membrane-spanning macromolecules and/or carry an electrical charge, it is essential to understand the principles of chemical kinetics and of transport of charged molecules in an electric field.

Knowledge of transport through membranes is based on measurements. These measurements lead to physically and chemically based mathematical models that are used to test concepts based on measurements. The role of mathematical models is to express concepts precisely enough that precise conclusions can be drawn (see quote by Haldane). In connection with all the topics covered, we will consider both theory and experiment. For the student, the educational value of examining the interplay between theory and experiment transcends the value of the specific knowledge gained in the subject matter.

My aim has been to produce textbooks on cellular biophysics that clarify rather than mystify and that eschew glib development. Topics were chosen to emphasize well-established principles, but also to include some more recent and sometimes controversial material. Most topics are introduced with a brief historical perspective. Challenging problems are included to aid students in learning the material. Extensive references, much more extensive than the citations, are provided to aid readers to pursue topics of interest in further depth.

Expected Background of the Reader

It has been assumed that the background of the reader of these texts includes one year of college physics, a semester of chemistry, a semester of biology, one year of calculus, a semester of differential equations, plus a course in which differential equations are solved in some physical context (e.g., electric network theory, mechanical dynamics, chemical kinetics, etc.). Some background in probability theory is also desirable.

A Note to the Instructor

The material in this and the companion text (Weiss, 1996) has been used to teach an undergraduate course taken by juniors and seniors at the Massachusetts Institute of Technology, which is the first of a sequence of three courses in bioengineering offered in the School of Engineering. The course, called Quantitative Physiology: Cells and Tissues, consists of the following activities:

- Three lectures each week to introduce new material.
- Two recitations each week to review material, solve problems, and answer questions.
- One laboratory exercise, which requires a written report, which provides direct experience with experimental techniques and with communication of experimental results and interpretations.
- One theoretical research project on the Hodgkin-Huxley model, which requires a written report, to provide an opportunity for students to define a testable hypothesis, conduct a theoretical investigation, and communicate the results and conclusions.
- A homework assignment each week that enables students to actively assimilate the course material.
- Fifteen-minute weekly quizzes that encourage students to learn the material as it is presented and provide a regular assessment of their progress. Many of the exercises at the ends of each chapter were devised as quiz problems.
- One midterm examination and one final examination that give students an opportunity to integrate the course material and to obtain an objective evaluation of their understanding of the material. Many of the problems at the ends of each chapter were devised as examination problems.

Students are expected to devote twelve hours of time (contact hours plus preparation time) to the course each week for thirteen weeks. The two texts cover more material than is covered in the course. The list of lectures on the next page indicates the coverage of material in a one-semester course. In the course we use computers extensively in lectures, recitations, homework assignments, and projects, although the texts do not require the use of computers. The use of computers in teaching this course is discussed elsewhere (Weiss et al., 1992).

Preparation of the Manuscript

Typesetting was done in TEX with LATEX macros on a Macintosh computer using Textures. Spelling was checked with the LATEX spell checker Excalibur. Theoretical calculations were done with Mathematica, Macsyma, and MATLAB. Several tools were used to reproduce data from the literature. Figures from the literature were scanned at 300 dpi using either Ofoto or FotoLook. For figures consisting of a small number of data points, the coordinates of the points

List of Lectures

Lecture	Title	Volume	Sections
1	Introduction to transport	1	1.2–1.4, 2.1–2.5
2	Macroscopic laws of diffusion	1	3.1
3	Microscopic basis of diffusion	1	3.2
4	Diffusion through membranes; two-compartment diffusion	1	3.5.1, 3.6–3.7
5	Diffusion through cellular membranes	1	3.8
6	Solvent transport; van't Hoff's law	1	4.1–4.3, 4.5.1
7	Primary osmotic response of cells	1	4.7
8	Water channels; introduction to concurrent solute and solvent transport	1	4.8, 5.1
9	Concurrent solute and solvent transport	1	5.2, 5.3
10	Introduction to carrier transport, chemical kinetics	1	6.1, 6.2.1
11	Simple symmetric four-state carrier model, one solute	1	6.4.1
12	Simple symmetric six-state carrier model, two ligands	1	6.4.2
13	Glucose transport	1	6.5–6.7
14	Introduction to ion transport	1	7.1, 7.2.1, 7.2.2, 7.2.
15	Passive ion transport through membranes	1	7.4
16	Resting potential of cells	1	7.5
17	Active ion transport	1	7.6, 7.7
18	Active ion transport	1	7.7, 7.8
19	Cell homeostasis	1	8.3, 8.4.1, 8.4.2, 8.5
20	Introduction to cellular electric potentials	2	1.1–1.4
21	Lumped- and distributed-parameter models of cells	2	2.1–2.4.2
22	Core conductor model	2	2.4.3
23	Linear electric properties of cells, cable model	2	3.1–3.4.2
24	Time-independent and time-dependent solutions of the cable model	2	3.4.2–3.4.3
25	Spatial and temporal integration, introduction to action potentials	2	3.4.4–3.5, 4.1
26	Voltage-clamp currents	2	4.2.1–4.2.2
27	The Hodgkin-Huxley model of ionic conductances	2	4.2.3
28	Synthesis of the Hodgkin-Huxley model	2	4.3–4.4
29	Synthesis of the Hodgkin-Huxley model	2	4.4–4.5
30	Myelinated nerve fibers	2	5.1–5.2
31	Saltatory conduction	2	5.2–5.4, 5.6
32	Introduction to ion channels	2	6.1–6.2
33	Gating currents	2	6.3
34	Single-channel currents	2	6.4
35	Two-state gate model	2	6.5
36	Two-state gate model	2	6.5
37	Multiple-state gate models	2	6.6
38	Molecular biology of ion channels	2	6.7

were obtained from the scanned images by means of FlexiTrace. The data points were plotted by means of the charting program DeltaGraph Pro. More complex figures were processed with Adobe PhotoShop and/or Adobe Streamline. Most graphic files were imported to Adobe Illustrator for annotation and saved as encapsulated Postscript files that were included electronically in the text. Mathematical annotations were obtained by typesetting the mathematical expressions with Textures and saving the typeset version as a file that was read by Adobe Illustrator. Chemical formulas were done with ChemDraw Plus and three-dimensional models with Chem3D Plus. A database of genes of membrane proteins was obtained with DNAStar, which also allowed translation of nucleotide sequences into amino acids sequences from which properties of the protein could be displayed.

Personal Perspective

It was a shock to look over my records and to discover that, although I did not realize it at the time, I began to write these books in the spring of 1966, when I taught a course called Introduction to Neuroelectric Potentials. That spring, I started writing notes to clarify my thinking on the course matter and as a supplement to the lectures for students. At the time, there were no textbooks available on the topics I wished to cover—none that presented empirical findings as well as derivations of theoretical conceptions from first principles.

The notes grew annually and were reproduced for each fall's class. Changes in methods of reproduction of the notes reflect the changing technology. Originally, the notes were typed and reproduced, first using ditto masters—I think there is still some blue stuff under one fingernail from the last of these messy products—then by xerographic reproduction. With the availability of technical typesetting languages, the notes began to take on an improved appearance around 1984, when they approached several hundred pages of Troffed text. By 1986 the notes were converted to LaTeX, but the figures were still being pasted in with rubber cement—a process that left a succession of secretaries tired but surprisingly giddy. In order to save time, to improve the quality of the figures, and to avoid tripping out my secretary, in 1990 I began the long but rewarding process of replacing the figures pasted in with glue with figures included electronically in the notes. Experimental data were digitized from original sources, photographs were scanned, and theoretical calculations were done anew. It is remarkable what one can learn about a subject through this process. Frankly, this effort was only possible because I have an obsessive personality. A colleague once told me "you're too thorough" which I believe was meant as a mild put-down but which I always regarded as a compliment.

For over twenty years, the notes became more and more extensive and complete without any conscious inclination on my part to publish a text. I took seriously the admonition of a colleague who said, "Never publish a book." During this period, my children grew up—two of the three have married—and my wife and I have qualified for discounts as senior citizens. The transition from notes to textbook, which was clearly occurring subconsciously, took a final serendipitous turn into my consciousness. Sometime after 1984, I learned from a graduate student who came to MIT from mainland China that the notes had been translated into Chinese. Eventually it occurred to me that the notes were receiving a highly skewed international distribution with foci in China and Cambridge, Massachusetts. To achieve a more uniform global access, I decided to make the notes into a textbook. I sought and received a sabbatical leave in the calendar year 1991 to complete the textbook. Three and one half years later, the job is finally done—and then some. The notes have become two textbooks! The final task in the preparation of the manuscripts was spent producing indices, tasks for which I have new found respect. During this period I learned the meaning of the term, repetitive stress disorder. Apparently, one of my computers did as well. Four days before final completion, the Macintosh computer in my study simply quit. It just couldn't do any more. I am now off to a vacation in North Truro on Cape Cod in Massachusetts where many chapters of these books were contemplated in summers past. But not this summer!

Acknowledgments

I welcome the opportunity to acknowledge the contributions of the many colleagues, friends, and family members who have contributed directly or indirectly to this work. I have been influenced by many of my colleagues, but I would like to single out a few who have been particularly important in my intellectual growth—Nelson Kiang, Charlie Molnar, Bill Peake, Walter Rosenblith, and Bill Siebert. Members of the Department of Electrical Engineering and Computer Science and the School of Engineering have been very supportive of my efforts. These include department chairmen and deans, of whom I wish to particularly thank Dick Adler, Joel Moses, Paul Penfield, Jeff Shapiro, and Jerry Wilson.

During the latter portions of this project, I was the fortunate recipient of a chair donated to MIT by Gerd and Tom Perkins. This support has given me the freedom to pursue this writing project with additional vigor in the latter years.

A number of faculty and staff have shared in teaching the material that comprises these texts, including Dick Adler, Denny Freeman, Al Grodzinsky, Rafael Lee, Bill Peake, and Bill Siebert. All have made contributions to the development of these materials. In particular, Denny Freeman and Bill Peake have formulated problems that appear in this text. Many of the problems have been used for years, so I no longer remember who made them up. Denny Freeman and Bill Peake have also given me a great deal of critical feedback on the text. Weaknesses, from specious arguments to indefinite antecedents, have been reduced appreciably through their efforts. I also wish to thank my colleague John Wyatt, who told me that I could "continue to work on the material *after* the books are published." This remark made while waiting for an elevator relieved my angst in parting with the manuscripts, which are still somewhat short of what I have in mind. I thank my friend Helen Peake for her support and for translating a paper by J. Perrin for me. My secretaries have also been very important to the completion of the task. There have been three that I would like to single out: Sylvia Nelson, Susan Ross, and Janice Balzer. Janice, in particular, has helped in many ways.

Graduate student teaching assistants have also made important contributions to the text, particularly in troubleshooting the problems. In this regard, I would particularly like to acknowledge help from Eero Simoncelli. Many students have used the notes over the years. Their questions have been very important in helping me to determine the issues that they found difficult to comprehend. Often I found that these issues were ones on which my thinking was unclear. Students have also been a great source of proofreading over the years. A few students made heroic efforts to find errors in the notes. I especially thank Greg Allen, Andy Grumet, Chris Long, Karen Palmer, and Susan Voss for their efforts. Despite the efforts of many people, errors no doubt still exist. I would appreciate hearing from readers about errors—typographic or conceptual—either via mail or electronic mail (tfweiss@mit.edu).

I wish to thank Fiona Stevens of The MIT Press for her help throughout this process. After I had missed yet another deadline, Fiona told me that "the books that come in the latest are often the best books." That was a kind thing to say, and I appreciated it. I thank Katherine Arnoldi, senior editor at MIT Press, who always provided calm reassurance during the final stages of production of these books. The copy editors, Suzanne Schafer and Marilyn Martin, have allowed me to perpetuate the myth that I know how to use commas correctly and that I correctly distinguish between "that" and "which." Paul Anagnostopoulos of Windfall Software was my constant companion over the telephone and the internet as the electronic manuscripts were transformed into electronic books. I appreciate his constructive suggestions, expertise, and exuberance.

Finally, and most important, my family has been my greatest source of support. My children, Max, Elisa, and Eric, were always interested in this project and felt a sense of pride in me that helped me to keep going when I occasionally became despondent about ever finishing. My children-in-law, Kelly and Nico, were also very supportive. At times when I believed that the books were a total disaster and no one would ever read them, my wife, Aurice, always told me she was convinced they would turn out well. At times when I told her that I thought these books were the greatest texts written in any discipline, she told me she believed me. She always had a positive word to say, and she listened patiently to daily accountings of the state of the chapter of the day. It must have been terribly boring, but she always led me to believe that she was interested in the latest accounting.

References

Haldane, J. B. S. (1985). Science and theology as art-forms. In Smith, J. M., ed., *On Being the Right Size and Other Essays*, 32–44. Oxford University Press, New York.

Weiss, T. F. (1996). *Cellular Biophysics*, vol. 2, *Electrical Properties of Cells*. MIT Press, Cambridge, MA.

Weiss, T. F., Trevisan, G., Doering, E. B., Shah, D. M., Huang, D., and Berkenblit, S. I. (1992). Software for teaching physiology and biophysics. *J. Sci. Ed. Tech.*, 1:259–274.

Units, Physical Constants, and Symbols

Units

In this text we use the International Systems of Units, called the SI units (Mechtly, 1973).

Base SI Units

The SI units contain seven base units from which all other units are derived. The seven base units are presented in the following table.

Quantity	**Name**	**Symbol**
Amount of substance	mole	mol
Electric current	ampere	A
Length	meter	m
Luminous intensity	candela	cd
Mass	kilogram	kg
Thermodynamic temperature	kelvin	K
Time	second	s

Derived SI Units

Many units are derived from the base units. Some are obvious, e.g., the unit for volume is m^3; others involve new terms. The derived units that are not obvious and that are used in this text are as follows:

Quantity	Name	Symbol	Units
Capacitance	farad	F	$A \cdot s \cdot V^{-1}$
Electric resistance	ohm	Ω	$V \cdot A^{-1}$
Electric charge	coulomb	C	$A \cdot s$
Electric conductance	siemens	S	$A \cdot V^{-1}$
Electric potential difference	volt	V	$N \cdot m \cdot C^{-1}$
Energy	joule	J	$N \cdot m$
Force	newton	N	$kg \cdot m \cdot s^{-2}$
Frequency	hertz	Hz	s^{-1}
Inductance	henry	H	$V \cdot s/A$
Power	watt	W	$N \cdot m \cdot s^{-1}$
Pressure	pascal	Pa	$N \cdot m^{-2}$

Decimal Multiples and Submultiples of SI Units

The following prefixes are used to express multiples or submultiples of SI units.

Prefix	Symbol	Factor	Prefix	Symbol	Factor
exa	E	10^{18}	deci	d	10^{-1}
peta	P	10^{15}	centi	c	10^{-2}
tera	T	10^{12}	milli	m	10^{-3}
giga	G	10^{9}	micro	μ	10^{-6}
mega	M	10^{6}	nano	n	10^{-9}
kilo	k	10^{3}	pico	p	10^{-12}
hecto	h	10^{2}	femto	f	10^{-15}
deca	da	10^{1}	atto	a	10^{-18}

Commonly Used Non-SI Units and Conversion Factors

The following non-SI units are commonly used.

Quantity	Name	Symbol	Units
Energy	calorie	cal	4.184 J
Energy	erg	erg	10^{-7} J

Energy	electronvolt	eV	1.602×10^{-19} J
Force	dyne	dyne	10^{-5} N
Length	angstrom	Å	10^{-10} m
Molecular weight	dalton	D	$\mathrm{gm \cdot mol^{-1}}$
Pressure	atmosphere	atm	$1.013 \times 10^{5}\ \mathrm{N \cdot m^{-2}}$
Pressure	bar	bar	$1 \times 10^{5}\ \mathrm{N \cdot m^{-2}}$
Pressure	millimeter of mercury	mmHg	$133.3 \times 10^{5}\ \mathrm{N \cdot m^{-2}}$
Temperature	Celsius	°C	$T_C = T - 273.15$
Temperature	Fahrenheit	°F	$T_F = (9/5)T_C + 32$
Volume	liter	L	$10^{-3}\ \mathrm{m^3}$

T, T_C, and T_F are the temperatures in kelvins, degrees Celsius, and degrees Fahrenheit.

Physical Constants

Fundamental Physical Constants

The following fundamental physical constants (Mechtly, 1973) are used in the text.

Name	Symbol	Value
Avogadro's number	N_A	$6.022 \times 10^{23}\ \mathrm{mol^{-1}}$
Boltzmann's constant	$k = R/N_A$	$1.381 \times 10^{-23}\ \mathrm{J \cdot K^{-1}}$
Electronic charge	e	1.602×10^{-19} C
Faraday's constant	$F = N_A e$	$9.648 \times 10^{4}\ \mathrm{C \cdot mol^{-1}}$
Molar gas constant	R	$8.314\mathrm{J \cdot mol^{-1} \cdot K^{-1}}$
Permittivity of free space	ϵ_0	$8.854 \times 10^{-12}\ \mathrm{F \cdot m^{-1}}$
Planck's constant	h	$6.626 \times 10^{-34}\ \mathrm{J \cdot s}$

Physical Properties of Water

The following physical properties of water (Robinson and Stokes, 1959; Eisenberg and Crothers, 1979; Lide, 1990) are given at standard pressure (100 kPa).

Property	Temperature °C	Value
Density	20	$0.998\,g \cdot cm^{-3}$
Dielectric constant	20	$80.20\,cm^2 \cdot s^{-1}$
Diffusion coefficient	25	$2.4 \times 10^{-5}\,cm^2 \cdot s^{-1}$
Molar volume	20	$18.05\,cm^3 \cdot mol^{-1}$
Viscosity	20	$1.002\,mPa \cdot s$

Atomic Numbers and Weights

The atomic numbers and weights of the elements (Ebbing, 1984) are as follows (the values in parentheses are the mass numbers of the isotopes of longest half-life).

Element	Symbol	Atomic Number	Atomic Weight	Element	Symbol	Atomic Number	Atomic Weight
Actinium	Ac	89	227.0278	Molybdenum	Mo	42	95.94
Aluminum	Al	13	26.98154	Neodymium	Nd	60	144.24
Americium	Am	95	(243)	Neon	Ne	10	20.179
Antimony	Sb	51	121.75	Neptunium	Np	93	237.0482
Argon	Ar	18	39.948	Nickel	Ni	28	58.69
Arsenic	As	33	74.9216	Niobium	Nb	41	92.9064
Astatine	At	85	(210)	Nitrogen	N	7	14.0067
Barium	Ba	56	137.33	Nobelium	No	102	(259)
Berkelium	Bk	97	(247)	Osmium	Os	76	190.2
Beryllium	Be	4	9.01218	Oxygen	O	8	15.9994
Bismuth	Bi	83	208.9804	Palladium	Pd	46	106.42
Boron	B	5	10.81	Phosphorus	P	15	30.97376
Bromine	Br	35	79.909	Platinum	Pt	78	195.08±3
Cadmium	Cd	48	112.41	Plutonium	Pu	94	(244)
Cesium	Cs	55	132.9054	Polonium	Po	84	(209)
Calcium	Ca	20	40.08	Potassium	K	19	39.0983
Californium	Cf	98	(251)	Praseodymium	Pr	59	140.9077
Carbon	C	6	12.011	Promethium	Pm	61	(145)
Cerium	Ce	58	140.12	Protactinum	Pa	91	231.0359

Chlorine	Cl	17	35.453	Radium	Ra	88	226.0254
Chromium	Cr	24	51.996	Radon	Rn	86	(222)
Cobalt	Co	27	58.9332	Rhenium	Re	75	186.207
Copper	Cu	29	63.546	Rhodium	Rh	45	102.9055
Curium	Cm	96	(247)	Rubidium	Rb	37	85.4678
Dysprosium	Dy	66	162.50	Ruthenium	Ru	44	101.07
Einsteinium	Es	99	(252)	Samarium	Sm	62	150.36
Erbium	Er	68	167.26	Scandium	Sc	21	44.9559
Europium	Eu	63	151.96	Selenium	Se	34	78.96
Fermium	Fm	100	(257)	Silicon	Si	14	28.0855
Fluorine	F	9	18.998403	Silver	Ag	47	107.8682
Francium	Fr	87	(223)	Sodium	Na	11	22.98977
Gadolinium	Gd	64	157.25	Strontium	Sr	38	87.62
Gallium	Ga	31	69.72	Sulfur	S	16	32.06
Germanium	Ge	32	72.59	Tantalum	Ta	73	180.9479
Gold	Au	79	196.9665	Technetium	Tc	43	(98)
Hafnium	Hf	72	178.49	Tellurium	Te	52	127.60
Helium	He	2	4.00260	Terbium	Tb	65	158.9254
Holmium	Ho	67	164.9304	Thallium	Tl	81	204.383
Hydrogen	H	1	1.00794	Thorium	Th	90	232.0381
Indium	In	49	114.82	Thulium	Tm	69	168.9342
Iodine	I	53	126.9045	Tin	Sn	50	118.69
Iridium	Ir	77	192.22	Titanium	Ti	22	47.88
Iron	Fe	26	55.847	Tungsten	W	74	183.85
Krypton	Kr	36	83.80	Unnilhexium	Unh	106	(263)
Lanthanum	La	57	138.9055	Unnilpentium	Unp	105	(262)
Lawrencium	Lr	103	(260)	Unnilquadium	Unq	104	(261)
Lead	Pb	82	207.2	Uranium	U	92	238.0289
Lithium	Li	3	6.941	Vanadium	V	23	50.9415
Lutetium	Lu	71	174.967	Xenon	Xe	54	131.29
Magnesium	Mg	12	24.305	Ytterbium	Yb	70	173.04
Manganese	Mn	25	54.9380	Yttrium	Y	39	88.9059

Mendelevium	Md	101	(258)	Zinc	Zn	30	65.38
Mercury	Hg	80	200.59	Zirconium	Zr	40	91.22

Symbols

The principal symbols used in the text are given in the table below, which also indicates the chapters in which each symbol is used. The symbols are not all distinct, but the context should resolve ambiguities. The math accent (^) is used to indicate a normalized variable, and the tilde (˜) to distinguish between permeant and impermeant solutes. Vectors are indicated by boldface. An asterisk (*) in the Units column below indicates that the units depend on the context.

Name	**Symbold**	**Units**	**Chapters**
Name	Symbol	Units	Chapters
Absolute temperature	T	K	3–8
Area	A	m^2	3–8
Association constant	K_a	*	6
Avogadro's number	N_A	mol^{-1}	3
Boltzmann's constant	k	$J{\cdot}K^{-1}$	3, 6
Carrier density	$\mathfrak{N}$	$mol{\cdot}m^{-2}$	6
Charge density	ρ	$C{\cdot}m^{-3}$	7
Charge relaxation time	τ_r	s	7
Chemical potential	μ	$J{\cdot}mol^{-1}$	3–7
Conductance	$\mathcal{G}$	S	7
Convection velocity	ν	$m{\cdot}s^{-1}$	3, 7
Current	I	A	7
Current density	J	$A{\cdot}m^{-2}$	7, 8
Debye length	Λ_D	m	7
Diffusion coefficient	D	$m^2 \cdot s^{-1}$	3, 4, 5, 7
Diffusive permeability	P	$m{\cdot}s^{-1}$	3, 4, 5, 6, 8
Dissociation constant	K	*	6, 7, 8
Electric conductivity	σ_e	$S{\cdot}m^{-1}$	3, 7

Electric field intensity	$\mathcal{E}$	$V \cdot m^{-1}$	7
Electric potential	ψ	V	7
Electric potential difference	V	V	7, 8
Electrochemical potential	$\tilde{\mu}$	$J \cdot mol^{-1}$	7
Energy	E	J	6
Equilibrium time constant	τ_{eq}	s	3
Faraday's constant	F	$C \cdot mol^{-1}$	7
Force on a mole of particles	f	$N \cdot mol^{-1}$	3, 7
Force on a particle	f_p	N	3
Forward rate constant	α	*	6
Free energy	$\mathfrak{G}$	$J \cdot mol^{-1}$	7
Hydraulic conductivity	$\mathcal{L}_V$	$m \cdot Pa^{-1} \cdot s^{-1}$	4, 5, 8
Hydraulic permeability	κ	$m^2 \cdot Pa^{-1} \cdot s^{-1}$	4
Hydraulic pressure	p	Pa	4, 5, 8
Mass density	ρ_m	$kg \cdot m^{-3}$	3, 4
Mean free path	l	m	3
Mean free time	τ	s	3
Mean velocity	$\overline{v}$	$m \cdot s^{-1}$	3
Mean-squared velocity	$\overline{v^2}$	$m^2 \cdot s^{-2}$	3
Mean value	m	*	3
Membrane thickness	d	m	3, 4, 5
Membrane potential	V_m	V	7, 8
Molar concentration	c	$mol \cdot m^{-3}$	3–8
Molar electric mobility	$\hat{u}$	$m^2 \cdot s^{-1} \cdot V^{-1}$	7
Molar flux	ϕ	$mol \cdot m^{-2} \cdot s^{-1}$	3–8
Molar gas constant	R	$J \cdot mol^{-1} \cdot K^{-1}$	1, 3–8
Molar mechanical mobility	u	$m \cdot mol \cdot s^{-1} \cdot N^{-1}$	3, 7
Molar particle density	n	$mol \cdot m^{-2}$	3
Molar quantity	n	mol	5, 6, 8
Molecular weight	M	$g \cdot mol^{-1}$	3
Mole fraction	x	dimensionless	4
Osmotic coefficient	χ	dimensionless	4

Osmotic permeability	$\mathcal{P}$	$m \cdot s^{-1}$	4, 5
Osmotic pressure	π	Pa	4, 5
Partial molar volume	$\overline{\mathcal{V}}$	$m^3 \cdot mol^{-1}$	4, 5
Particle mass	m	kg	3
Particle mechanical mobility	u_p	$m \cdot s^{-1} \cdot N^{-1}$	3
Particle radius	a	m	3
Partition coefficient	k	dimensionless	3
Permittivity	ϵ	$C \cdot m^{-1} \cdot V^{-1}$	7
Pore density	$\mathcal{N}$	m^{-2}	3, 4, 5
Porosity	$\mathcal{P}$	dimensionless	3
Probability	W	dimensionless	3
Probability	x	dimensionless	7
Probability density	p	*	3
Q_{10}	Q_{10}	dimensionless	6
Radius — particle	a	m	3, 4
Radius — pore	r	m	3, 4, 5
Reflection coefficient	σ	dimensionless	5, 8
Resistance	$\mathcal{R}$	Ω	7
Resistivity	ρ_e	$\Omega \cdot m$	7
Resting membrane potential	V_m^o	V	7, 8
Reverse rate constant	β	*	6
Single-channel conductance	γ	S	7
Single-channel current	$\mathcal{I}$	A	7
Solute density	s	$mol \cdot m^{-2}$	6
Specific conductance	G	$S \cdot m^{-2}$	7, 8
Specific resistance	R	$\Omega \cdot m^2$	7
Standard deviation	σ	*	3
Steady-state time constant	τ_{ss}	s	3
Stoichiometric coefficient	ν	dimensionless	6, 7, 8
Surface charge density	Q_f	$C \cdot m^{-2}$	7
Tortuosity	$\mathcal{T}$	dimensionless	3
Total molar concentration	C_Σ	$mol \cdot m^{-3}$	4, 5, 6, 8

Total molar quantity	N_Σ	mol	4, 5, 8
Valence	z	dimensionless	7, 8
Viscosity	η	Pa·s	3, 4, 5
Volume	V	m^3	3, 4, 5, 6, 8
Volume flux	Φ_V	$m \cdot s^{-1}$	4, 5, 8

References

Ebbing, D. D. (1984). *General Chemistry*. Houghton Mifflin, Boston.

Eisenberg, D. and Crothers, D. (1979). *Physical Chemistry*. Benjamin-Cummings, Reading, MA.

Lide, R. R. (1990). *Handbook of Chemistry and Physics*, 71st ed. CRC Press, Boston.

Mechtly, E. A. (1973). *The International System of Units*. Technical Report NASA SP-7012, National Aeronautics and Space Administration, Washington, DC.

Robinson, R. A. and Stokes, R. H. (1959). *Electrolyte Solutions*. Butterworths, London.

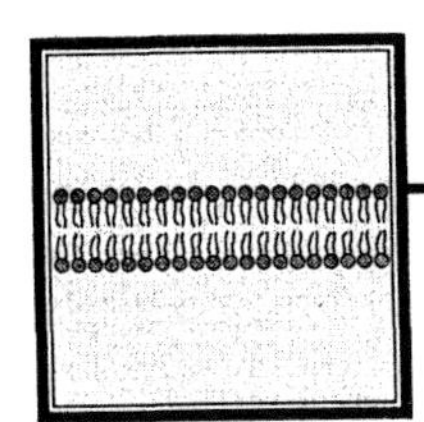

1 Introduction to Membranes

To stay alive, you have to be able to hold out against equilibrium, maintain imbalance, bank against entropy, and you can only transact this business with membranes in our kind of world.
—Thomas, 1974

1.1 Historical Perspective

1.1.1 Fundamental Concepts of Living Organisms

Modern concepts of living organisms have their origin in the nineteenth century. These concepts are so familiar that it is easy to overlook the fact that they are only about a century old.

▪ Perhaps the most profound concept in the life sciences is Darwin's *theory of evolution.* Evolutionary theory states that living organisms have evolved from their ancestors by a process in which randomly variable inheritance is passed on to descendants that are culled by natural selection.[1] By this mechanism, certain organismic traits favor the survival of a species in a given environment, and hence are passed on to descendants. In a slowly variable environment, changes in species occur relatively slowly on a geological time scale. However, when the environment changes rapidly, extinction and pro-

1. Discussions of evolutionary theory, compellingly told, can be found in the many books of S. J. Gould (Gould 1977, 1980, 1989) and of R. Dawkins (Dawkins 1976, 1987). Original works of Darwin document the voluminous observations that supported his ideas (Darwin 1859, 1871). Darwin's life, including the social pressures that delayed the publication of his heretical ideas and contributed to a mysterious illness that plagued him much of his life, is described in a comprehensive biography (Desmond and Moore, 1991).

liferation of species can occur rapidly. Fully extrapolated, this view of living organisms evolving naturally has enormous conceptual as well as pragmatic consequences. The theory gives order to the enormous range of living organisms and relates existing species to paleontological records of ancestral species. Thus, evolution relates all living organisms—past and present—through a common ancestral bond. Pragmatically, the theory suggests that understanding biological processes in any one organism may have much wider implications.

- The *cell doctrine* states that all living organisms consist of cells, which are the smallest units capable of sustaining life. Thus, to understand living organisms it is necessary to understand the structure and function of single cells. As a corollary to the cell doctrine, cells are formed from preexisting cells only; spontaneous generation of living cells does not occur. Quaintly put, cells beget cells. Together with the theory of evolution, this corollary implies that all life had a common ancestor—the first cell—to which, presumably, the general rule does not apply. The cell doctrine also implies the existence of some boundary of the cell that separates its interior from its exterior—the cell membrane.
- The intellectual ferment of nineteenth-century science led to the end of *vitalism*. It became clear that there was no need to find some special life force unique to living organisms and absent from the inanimate world. The operations of living organisms could be understood, in principle, by the same laws of physics and chemistry that apply to the inanimate world. No special chemistry and physics of living organisms is required, albeit that living organisms are generally more complex than are inanimate objects.

The twentieth century has produced major advances in the understanding of living systems at the molecular level.[2] A fundamental organizing principle of living organisms is that at the molecular level, the biological processes are universal. In all living organisms, the nucleic acids (DNA and RNA) contain the blueprints for proteins. Cells use these blueprints to produce proteins that control the biosynthesis of vital materials as well as myriad biochemical processes. The blueprint is passed on to descendants. As a consequence of a common heritage, certain biochemical processes are common to virtually all forms of life—from single-cell organisms such as bacteria to multicellular plants and animals. This attribute is of enormous practical importance. It implies that a great technical advantage may be attained if universal biochemical processes can be conveniently studied in simple organisms such as bacteria.

---

2. The history of molecular biology has been told in an engaging manner (Judson, 1979).

1.1.2 Emerging Concepts of Cell Membranes

Development of ideas on the structure of membranes went hand in hand with emerging notions of living organisms. Because cell membranes are central to all the topics in this text and the central issue in this chapter, we shall look at a brief chronology of the emergence of concepts of cell membranes. More detailed accounts are found elsewhere (Branton and Park, 1968; Cole, 1968; Dowben, 1969; Kepner, 1979; Tosteson, 1989).

1665 Hooke examined cork with a light microscope and described the compartments he saw as *cells*. These compartments were not living cells, but the remaining walls of plant cells that had long since died and disintegrated.

1675 Van Leeuwenhoek described the appearance of a large variety of single-cell organisms observed with a light microscope.

1839 Schleiden (a botanist) and Schwann (a zoologist) articulated the emerging cell doctrine: all living organisms consist of cells, which are the smallest structures that can be said to be alive.

1855 Nageli observed that living cells are selectively permeable to dyes and postulated the existence of a cellular membrane that separates the interior from the exterior of cells.

1858 von Virchow observed that new cells originate by cell division from preexisting cells.

1877 Pfeffer observed the swelling and shrinkage of cells in different aqueous solutions. He interpreted the change in volume as due to the flow of water and concluded that cells are more permeable to water than to most solutes. Many of these observations apparently influenced the work of van't Hoff, who formulated the osmotic pressure law in 1885.

1897 Overton firmly established the selective permeability of cells by extensive systematic observations of the rate of entry of a large variety of organic substances into several types of cells. He found that the permeability was correlated with lipid solubility and postulated a lipid structure for the cell membrane.

1902 Bernstein inferred that a negative electric potential exists between the inside and the outside of a cell, i.e., across the cell membrane, and that this resting potential is related to the selective permeability of the membrane to ions.

1910 Höber found that suspensions of intact erythrocytes have a high electrical resistance, while the cytoplasm has a conductivity that is similar to that of physiological saline. Höber inferred that the cellular membrane has a high electrical resistance.

1917 Langmuir showed that lipids form monolayers at air-water interfaces, with their polar ends in the water and their hydrocarbon tails in the air.

1925 Gorter and Grendel extracted the lipids from erythrocytes and measured the surface area of these lipids when they formed a monomolecular layer in a trough of water. They found that the surface area of the population of erythrocytes from which the lipids were extracted was half the surface area of the lipids in the monomolecular layer. They concluded that the lipids in the cellular membrane form a bimolecular layer. In the original study of Gorter and Grendel, the lipids were not totally extracted, but the surface area of the erythrocytes was underestimated. These errors canceled one another. The experiment was repeated (Bar et al., 1966) with more complete extraction procedures and with more accurate estimates of the surface area of erythrocytes, and the original conclusion of Gorter and Grendel was shown to be correct.

1925 Chambers and Plowe demonstrated that the membrane exists as a separate structural entity from the cytoplasm by microdissecting cells and by microinjecting dyes into cells. They observed that when cells were poked the cytoplasm came out and the cells lost their semipermeability. Dyes injected intracellularly into intact cells diffused in the cytoplasm, but either did not appear extracellularly or appeared there very slowly.

1925 Fricke measured the electrical capacitance of suspensions of cells, and from these measurements he estimated the capacitance of cell membranes. Based on the measured capacitance, Fricke estimated the thickness of the cell membrane and found it to be of molecular dimensions.

1925 Ruhland and Hoffman proposed the *pore hypothesis* to account for the dependence of solute permeability on molecular size. They found that certain molecules were transported through cellular membranes with a permeability that was inversely related to the molecular radius of the molecule. They proposed that the membrane contained pores of some fixed radius and acted as a molecular sieve, allowing small molecules to pass more readily than large molecules.

1933 Collander and his associates provided extensive quantitative measurements of the relation between the permeability of a solute through a cellular membrane and the lipid solubility of the solute. These measurements provided quantitative validation of Overton's observations and began an era of theoretical treatment of transport data.

1930s Harvey and Cole estimated the surface tension of cell membranes in aqueous solutions. They found that the surface tension was less than the surface tension of a lipid-water interface, but was closer to the surface tension of a lipid-protein interface with water. These results were interpreted to indicate that there was some protein in the membrane. Although this conclusion has turned out to be correct, it has also been found that phospholipids have low surface tension.

1935 Danielli and Davson summarized the existing state of knowledge of membranes and proposed a molecular model that consisted of a bimolecular lipid membrane with protein adsorbed to the surface.

1935–1941 Schmidt and Schmitt used polarized light and X-ray diffraction studies, respectively, to study the structure of nerve myelin. They found that myelin is made up of a repeating unit of lipoid material with a spacing of 170–185 Å. They inferred that lipid molecules are oriented in a radial direction with respect to the long axis of a nerve fiber. Since the gross structure of myelin was not known at that time, the significance of these results for the structure of membranes was unclear.

1950 Fernández-Morán observed the myelin sheaths of nerve fibers with the electron microscope and demonstrated the layered structure of myelin. The spacings of the layers were roughly consistent with the results of both polarized light and X-ray diffraction studies.

1952 Hodgkin and Huxley demonstrated that the electrically excitable properties of nerve membrane could be explained in terms of the kinetic properties of the selective permeability of the membrane to sodium and potassium ions. This work indicated that membrane properties were dynamic and accelerated interest in identifying the membrane structures responsible for the selective permeability.

1954 Geren showed that myelin consists of layers of membranes of Schwann cells that wrap themselves about neurons during development of the nervous system; hence, myelin is an almost pure membrane material. It subsequently became the focus of research on membrane structure. The information obtained by the previous workers on the structure of myelin indicated that the thickness of the membrane was about 75 Å.

1957 Fernández-Morán and Finean studied the same myelin specimens with both X-ray diffraction and electron microscopy and made a direct comparison between the periodicities seen by X-ray diffraction and by electron microscopy. These studies gave validity to measurements of dimensions of the cellular membrane based on these rather different techniques, which have quite different tissue preparation artifacts.

1959 Robertson summarized the extant data on membranes of many cells and coined the term *unit membrane* to suggest that the membranes of all cells and organelles have certain structural similarities. He viewed membranes as tripartite structures, lipid bilayers surrounded by protein coats. This conception was based largely on the emerging observations with the electron microscope, which allowed membranes to be "seen" for the first time.

1970s–1980s Techniques were developed to isolate cell fractions rich in membranes, to dissolve membranes in detergents, to isolate integral membrane proteins, to determine the amino acid sequence of membrane proteins, and to isolate individual genes that contain the code for a membrane protein.

1976 Neher and Sakmann recorded electric current from a small patch of membrane that contained a single ion channel. This development has allowed examination of the kinetic properties of single membrane-bound proteins that subserve ion transport.

1.2 Survey of Cell Structure

Cells are categorized into two types: *prokaryotes,* which contain no nucleus, and *eukaryotes,* which do. Bacteria are prokaryotes, whereas all plant and animal cells of multicellular organism as well as certain single-cell organisms are eukaryotes. When compared to eukaryotes, prokaryotes have a relatively simple, featureless cytoplasm (Figure 1.1). The cell interior contains ions, other small molecules, proteins, and nucleic acids. Eukaryotic cells are far more complex. Their linear dimensions are generally more than ten times as great as those of prokaryotes. The cell interior of a eukaryotic cell consists of the nucleus and the cytoplasm as well as the cell organelles (Figure 1.2).

Both prokaryotes and eukaryotes have a boundary that separates the cell exterior from its interior. Plant cells and certain bacteria contain an outer boundary, the *cell wall,* that is relatively rigid but highly permeable to solutes and water (Figure 1.3). In plant cells, the cell wall thickness varies among cells but can be several micrometers thick. The cell wall typically contains several layers: a secondary cell wall that can be several layers thick and is made of lignin (a relatively hard substance found in woody plants) and cellulose (a polysaccharide); an elastic primary wall consisting of cellulose and pectins (which are branched polysaccharides); and a middle lamella that is composed of pectins and is shared with neighboring cells. Between the cell wall, if a cell possesses one, and the cytoplasm is the cell membrane (also called the

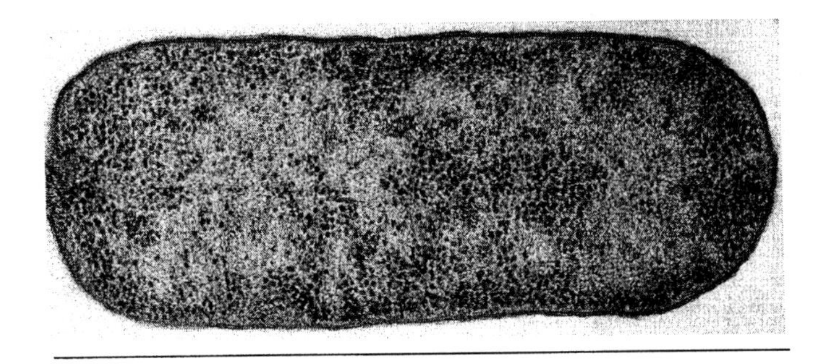

2.5 μm

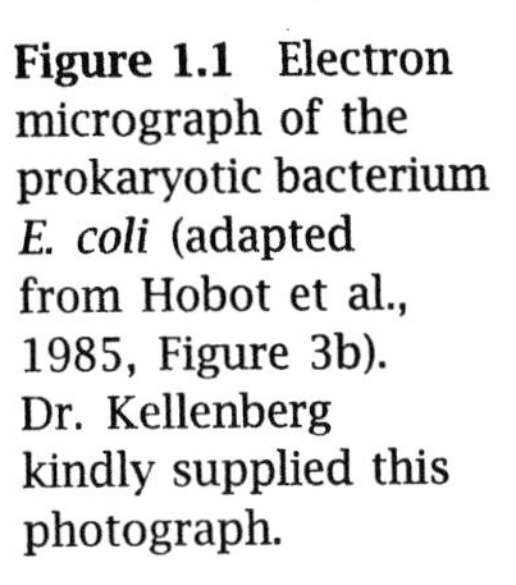

Figure 1.1 Electron micrograph of the prokaryotic bacterium *E. coli* (adapted from Hobot et al., 1985, Figure 3b). Dr. Kellenberg kindly supplied this photograph.

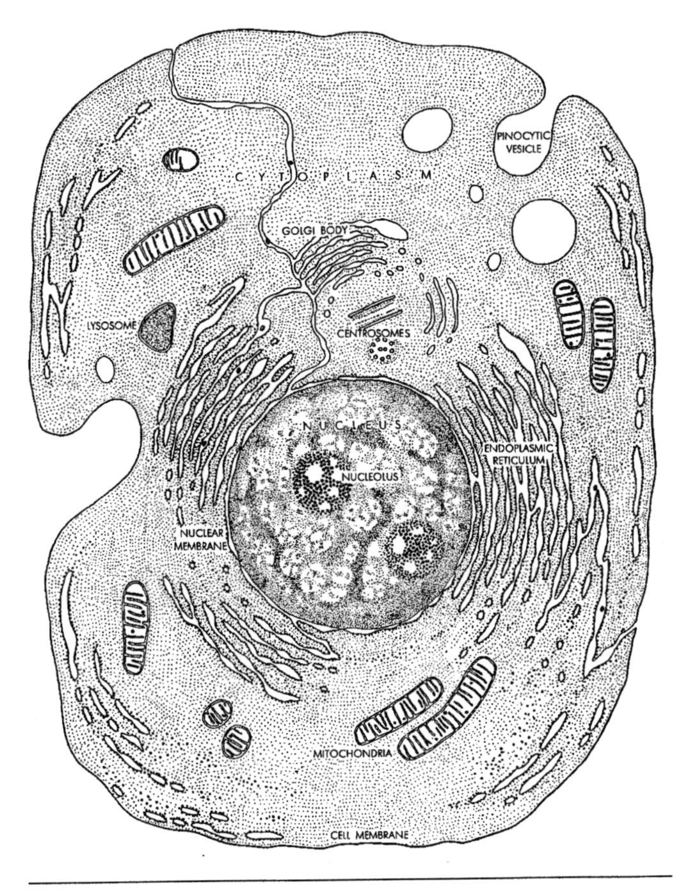

25 μm

Figure 1.2 Diagram of a generic eukaryotic animal cell (adapted from Brachet, 1965, page 9). Copyright ©(1965) by Scientific American, Inc. All rights reserved.

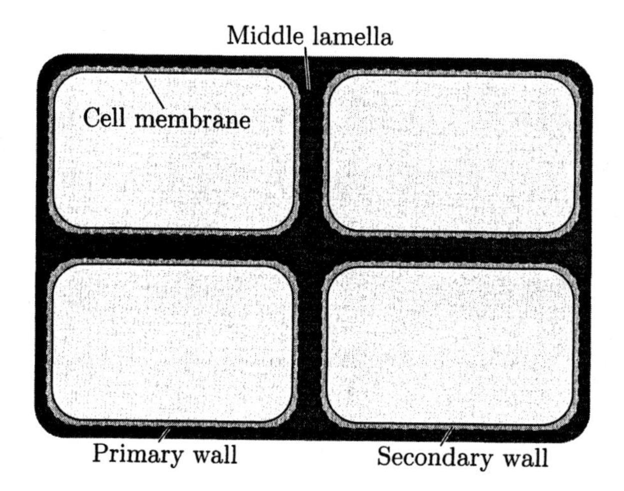

Figure 1.3 Schematic diagram of the cell walls of four contiguous plant cells. Only the cell walls and cell membranes are indicated; no other organelles are shown.

plasma membrane), which is the permeability barrier that separates the cell interior from its exterior. In cells without a cell wall, such as eukaryotic cells of animals, the cell membrane is the outer boundary of the cell.

Eukaryotic cells contain a prominent *nucleus,* which is the site of DNA replication and of RNA synthesis. The nucleus is surrounded by the *nuclear envelope,* which consists of two layers of membrane punctuated by nuclear pores 70 nm in diameter through which the nucleus communicates with the cytoplasm. The nucleus contains a *nucleolus,* which is the site of production of ribosomal RNA. The nucleus also contains the chromosomes made of DNA, which comprise the blueprint for the organism.

The *cytoplasm* is the portion of the cell interior outside the nucleus. It contains the *cytosol* (or cell water), which is the soluble portion of the cytoplasm and typically makes up more than 50% of the cell volume. The cytosol contains a variety of small and intermediate-size molecules. The free ribosomes in the cytosol are sites of protein synthesis. In addition, the cytosol contains enzymes that catalyze many intermediate metabolic reactions, including glycolysis. Coursing through the cytoplasm is the cytoskeleton, which comprises several distinct types of filaments. *Microtubules* are 25 nm diameter tubes made of the protein *tubulin. Microfilaments* are filaments 7 nm in diameter made of the protein *actin.* Both microtubules and microfilaments can be assembled and disassembled rapidly from building-block molecules present in the cytosol, *tubulin* for microtubules and *actin* for microfilaments. *Intermediate filaments* are a diverse group of protein filaments 10 nm in diameter. *Muscle thick filaments* are filaments 15 nm in diameter that are made of the protein *myosin,* which is found in a highly organized state in muscle cells. The cytoskeletal network of these filaments is responsible for the cell's shape, rigidity, organization, and motility, and for the intracellular transport of molecules. While it is present in all cells, the cytoskeleton is highly elaborated in motile cells such as muscle cells and ciliated epithelial cells.

The cytoplasm also contains an assortment of membrane-bound cell organelles that compartmentalize the cytoplasm into domains with different structures and functions. A prominent intracellular organelle is the *mitochondrion,* which is the cell's power plant. Mitochondria contain an outer membrane as well as an inner membrane consisting of invaginations of the outer membrane. The inner and outer membranes contain a number of enzymes, including those for the citric acid cycle and for the process of oxidative phosphorylation, by which the products of glycolysis are broken down to produce the energy storage molecule *adenosine triphosphate* (ATP). A metabolically active cell may contain a thousand mitochondria.

The *cytoplasmic vacuolar system* is a set of membrane-bound, interconnected tubes and flattened sacs that communicate with the nucleus. This system contains the *rough endoplasmic reticulum,* which contains ribosomes that are made of RNA and are the sites of synthesis of membrane-bound proteins as well as of proteins that are secreted by the cell, and the *smooth endoplasmic reticulum,* which is important in lipid metabolism. In addition, the cytoplasmic vacuolar system contains the *Golgi body,* which is involved in the sorting and packaging of substances into membrane-enclosed vesicles bound for export. The Golgi body is highly developed in secretory cells.

The cytoplasm also contains membrane-enclosed organelles that are used to degrade intracellular molecules meant for recycling. These include the *peroxisomes* found in all eukaryotes, the *lysosomes* found in animal cells, and the *glyoxysomes* found in plant cells. Plant cells contain *chloroplasts,* which have two layers of membrane enclosing the plastids, which contain the photosynthetic enzyme chlorophyll. This organelle is the site of photosynthesis, which generates ATP. Particularly prominent in plant cells are very large membrane-bound vesicles called *vacuoles.*

Thus, an examination of cell structure reveals that not only are all cells surrounded by a cell membrane, but cells are filled with organelles that themselves are delimited by membranes.

1.3 Molecules

In order to describe the structure of membranes in a meaningful manner, we shall first briefly describe the molecules that make up living organisms in

Table 1.1 Approximate element composition of the human body (adapted from Beck et al., 1991, Table 3-2). Estimates of calcium and phosphorus composition vary widely.

Element	Symbol	% weight
Oxygen	O	65
Carbon	C	18
Hydrogen	H	10
Nitrogen	N	3
Calcium	Ca	2
Phosphorus	P	1.1
Potassium	K	0.35
Sulfur	S	0.25
Chlorine	Cl	0.15
Sodium	Na	0.15
Magnesium	Mg	0.05
Iron	Fe	0.0006
Iodine	I	0.00006

general and cell membranes in particular. More detailed descriptions can be found elsewhere (Dickerson and Geis, 1969; Alberts et al., 1983; Darnell et al., 1990; Mathews and van Holde, 1990; Beck et al., 1991).

1.3.1 Atoms, Elements, and Bonds

1.3.1.1 Elements

Living organisms consist primarily of oxygen, carbon, hydrogen, nitrogen, calcium, and phosphorus, which collectively make up 99% of the element composition of the human body (Table 1.1). About 70% of the body consists of oxygen and hydrogen in the form of water. The remaining 30% consists primarily of compounds involving carbon, i.e., *organic compounds*.

1.3.1.2 Chemical Bonds

Atoms combine to produce molecules that are energetically more stable structures than are the individual atoms. A variety of chemical bonds can form between atoms, and these imbue the resultant molecules with distinct properties.

Table 1.2 Covalent bond strengths of biologically important bonds (adapted from Darnell et al., 1990, Table 1-1). C ≡ C is a triple carbon-carbon bond; C = C is a double carbon-carbon bond; C − C is a single carbon-carbon bond.

Bond	Energy (kJ/mol)	Bond	Energy (kJ/mol)
C ≡ C	821	N − H	391
C = O	716	C − O	351
C = N	619	C − C	348
C = C	615	S − H	339
P = O	505	C − N	292
O − H	463	C − S	259
H − H	436	N − O	223
P − O	421	S − S	215
C − H	413		

Primary Chemical Bonds

The primary chemical bonds, i.e., covalent bonds and ionic bonds, form to complete the valence shells of the participating atoms to yield a more stable structure. The strongest bonds, i.e., those that require the most energy to break, are the *covalent bonds* that form when two atoms share electrons in their outer shells (Table 1.2). In single covalent bonds, a single electron is shared. In double covalent bonds, two electrons are shared between the atoms. If the electronic cloud in a covalent bond is symmetrically disposed about the nuclei of the interacting atoms, the covalent bond is called *nonpolar*. Strictly speaking, a symmetric electronic cloud exists between two identical atoms only, e.g., a carbon-carbon bond. If the two atoms are not identical, one of the atoms in the bond will have a greater affinity for electrons than does the other atom. This property of an atom is called its *electronegativity*, and phenomenological scales have been defined to quantify electronegativity (Table 1.3). The electronic cloud moves closer to the nucleus of the more electronegative atom and moves farther from the nucleus of the less electronegative atom. While the bonded pair of atoms exhibits no net charge, it may exhibit an electric dipole moment. Such a bond is called *polar*. Table 1.3 indicates that the electronegativities of carbon and hydrogen are quite similar, so the carbon-hydrogen bond is relatively nonpolar, whereas the electronegativities of oxygen and hydrogen differ a great deal, so the oxygen-hydrogen bond is polar. Certain covalent chemical bonds are found predominantly in organic compounds (Figure 1.4). From an examination of the electronegativities of the atoms, we can see that

Table 1.3 Electronegativities of atoms important in living systems (adapted from Moore, 1972, Table 15.4).

Atom	Electronegativity
O	3.44
N	3.04
S	2.58
C	2.55
H	2.20
P	2.19

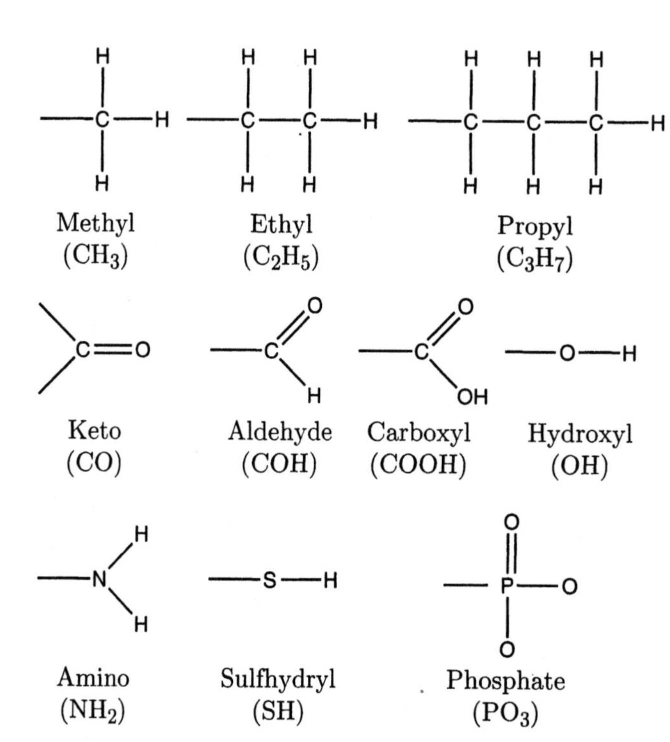

Figure 1.4 Common functional groups. The keto group is also called the *carbonyl group,* and the sulfhydryl group is also called the *thiol group.*

the keto, aldehyde, carboxyl, hydroxyl, and amino groups are polar, whereas the hydrocarbon methyl, ethyl, and propyl groups are nonpolar.

Ions are atoms that have either gained or lost one or more of their valence shell electrons. Since the original atoms are electrically neutral, the ions have a net electrostatic charge. Thus, oppositely charged ions have an electrostatic attraction and form *ionic bonds*. The ionic bond can be viewed as a covalent bond with a highly asymmetric electronic distribution for the valence electrons. The energy required to break ionic bonds is comparable to that of covalent bonds when the atoms are in a crystal, but in an aqueous solution the atoms readily dissociate into ions, a process called *ionization*. The high dielectric constant of water tends to shield the charges and to decrease the electrostatic attraction of the ions.

Secondary Chemical Bonds

Secondary chemical bonds involve electrostatic interactions between atoms. To understand these, imagine the spatial distribution of charge density about an atom. This distribution can be expanded in terms of its moments. The first term or first moment of the distribution reflects the net charge on the atom. The second term or second moment of the charge distribution reflects a separation of charge in the distribution and is called the *electric dipole moment*. Higher-order moments characterize the charge density distribution more precisely. The electrostatic potential energy of two atoms can be expanded in terms of the moments of the charge distribution. Different terms in the expansion depend differently upon interatomic distance. To explore this behavior, consider the electrostatic potential energy for simple charge distributions. The potential energy of two point charges is inversely proportional to the interatomic distance r (Table 1.4). The potential energy of a point charge and an electric dipole (which consists of two equal and opposite charges separated by an infinitesimal distance) decreases as $\propto 1/r^2$. The potential energy of two dipoles decreases as $\propto 1/r^3$. If the charge distribution is deformable, then bringing a point charge close to a neutral distribution of charges can induce a separation of charges, i.e., can induce a dipole moment. The interaction of a

Table 1.4 Electrostatic potential energy between two charges (q_1 and $-q_2$), a charge (q_1) and a dipole (p_2), two dipoles (p_1 and p_2), a charge (q) and an induced dipole, and a dipole (p) and an induced dipole spaced a distance r apart; ϵ is the permittivity of the medium, α is the polarizability, which is the induced dipole moment per unit electric field. The dipoles are aligned to maximize the potential energy.

Bond	Energy
Two charges (q_1 and $-q_2$)	$-\frac{q_1 q_2}{4\pi\epsilon r}$
Charge (q_1) and dipole (p_2)	$-\frac{q_1 p_2}{4\pi\epsilon r^2}$
Two dipoles (p_1 and p_2)	$-\frac{2 p_1 p_2}{4\pi\epsilon r^3}$
Charge (q) and induced dipole	$-\frac{2\alpha q^2}{(4\pi\epsilon)^2 r^4}$
Dipole (p) and induced dipole	$-\frac{2\alpha p^2}{(4\pi\epsilon)^2 r^5}$
Van der Waal's interaction	$-\frac{\kappa}{(4\pi\epsilon)^2 r^6}$

point charge with such an induced dipole moment has a potential energy that decreases as $\propto 1/r^4$. Similarly, a fixed dipole can induce a dipole moment in a deformable but previously electrostatically neutral charge distribution. The interaction of such a dipole with an induced dipole has a potential energy that decreases as $\propto 1/r^5$.

In general, the electrostatic potential energy of two atomic charge distributions can be expanded as a function of the internuclear distance. However, as the distance between the nuclei increases, the lowest-order term in the expansion dominates. So, for example, the potential energy of an ionic bond between two charged ions approaches that of two charges as the distance between atoms is increased, i.e., the potential energy becomes inversely proportional to the distance between the atoms. The higher-order terms can have a critical effect at small interatomic distances.

Among secondary chemical bonds, the *hydrogen bond* is one of the most important in organic molecules. Hydrogen covalently bonded to an electronegative atom such as oxygen or nitrogen forms a group that has a large dipole moment. This dipole moment can interact with other molecules that have a net charge or a large dipole moment to form a secondary chemical bond. However, this secondary chemical bond involving hydrogen is special. Because of the small size of the electronic shell of hydrogen, the two electronegative atoms bound by hydrogen are held so close together that they exclude other atoms. At room temperature (300 K), the thermal energy $RT \approx 2.5$ kJ/mol, whereas the energy required to break a hydrogen bond is typically 10–40 kJ/mol. This relatively small difference in energies implies that the rate at which hydrogen bonds are broken at room temperature is appreciable. Interactions of two dipoles depend upon the angle between the two dipoles and are maximal when the angles are aligned. The dependence of the energy on this angle makes the hydrogen bond highly directional. To summarize, hydrogen bonds are relatively weak and highly directional, and they exclude other atoms.

Water molecules readily make hydrogen bonds. Since oxygen is much more electronegative than hydrogen, the two electronic clouds of a water molecule have their centers of gravity much closer to the oxygen than to the hydrogen atoms. Thus, water has a dipole moment as shown in Figure 1.5. The dipole moments of neighboring water molecules result in the formation of a hydrogen bond that holds two water molecules together. Normally, water has a tetrahedral structure in which one water molecule forms hydrogen bonds with four other water molecules. The hydrogen bonds that hold water molecules together are much weaker than the covalent bonds that hold the two hydrogen and the oxygen atoms together in the water molecule. Hence,

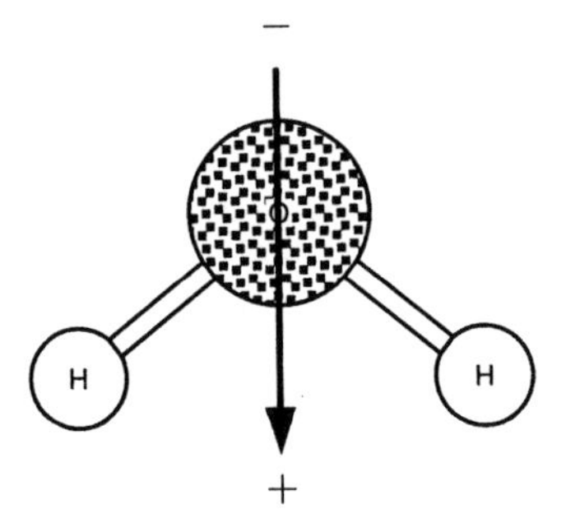

Figure 1.5 Model of a water molecule. The bond lengths are 0.94 Å, and the bond angle HOH is 104.5°. The direction of the dipole moment of a water molecule is shown superimposed on the model.

the covalent bonds are difficult to break at physiological temperatures, but the hydrogen bonds are relatively easily broken by thermal agitation of the molecules. Collections of water molecules are in continuous thermal agitation; the hydrogen bonds are continuously being broken and reformed as water molecules collide with their neighbors. Thus, water is made of clusters of water molecules, with the number of molecules in a cluster varying. This conception is called the *flickering cluster theory* of water structure. As the temperature of water is lowered, fewer hydrogen bonds are broken thermally, and water turns into solid ice. As the temperature is raised, hydrogen bonds are broken and water turns into a liquid. At still higher temperatures, water turns into a vapor in which the water molecules still exhibit hydrogen bonds.

In general, substances that have polar covalent bonds that can make hydrogen bonds with water are readily dissolved in water and are called *hydrophilic*. Such substances exist in water surrounded by clouds of water molecules. In particular, ions in water attract shells of water molecules. Such an ion is said to be *hydrated*. As we shall see in Chapter 7, when ions in water diffuse or drift (migrate) in an electric field, they drag their hydration shells with them. In contrast to polar substances, compounds with nonpolar bonds do not readily dissolve in water, and they are called *hydrophobic*. When these nonpolar substances are in water, they tend to cluster together to exclude water. This effect is called a *hydrophobic effect*.

Secondary chemical interactions can also occur between nonpolar substances. On average, the electronic distribution around a nonpolar molecule is symmetric. However, the electronic distribution fluctuates in time to produce a time-varying dipole moment. This time-varying dipole moment can either interact with another time-varying dipole or induce a time-varying dipole in a neighboring molecule. Thus, these nonpolar molecules can bind weakly with short-range binding energy. Such transient dipole interactions, which occur in both polar and nonpolar molecules, are called *van der Waal's* interactions.

1.3.2 Organic Molecules

In this section we shall describe the important organic molecules: carbohydrates, lipids, proteins, and nucleic acids. Each is described in terms of its constituent building block molecules: monosaccharides, fatty acids, amino acids, and nucleotides.

1.3.2.1 Carbohydrates

Carbohydrates are hydrates of carbon, i.e., they contain carbon and water. The water is in the form of hydrogen and hydroxyl groups bonded to carbon. Carbohydrates are formed as polymers of building block molecules, the simple sugars. The carbohydrates are major sources of energy for organisms. In addition, the carbohydrate cellulose is a structural molecule that makes up the cell walls of plant cells and certain bacteria.

Monosaccharides
Simple sugars or *monosaccharides* contain either an aldehyde or a keto group (Figure 1.4), have the molecular formula $C_n(H_2O)_n$ for n generally in the range of 3 to 9, and contain a number of hydroxyl groups. Monosaccharides that contain an aldehyde group are called *aldoses,* and those that contain a keto group are called *ketoses*. The monosaccharides that have three to nine carbon atoms are called *trioses, tetroses, pentoses, hexoses, heptoses, octoses,* and *nonoses,* respectively. Thus, a six-carbon aldose is called an *aldohexose,* etc. Since $C(H_2O)$ has a molecular weight of 30, an aldohexose such as glucose has a molecular weight that equals $6 \times 30 = 180$ daltons.

Figure 1.6 shows both a three-dimensional model and a chemical structural diagram of D-glucose in the ring form. D-glucose is found in solution both in a straight-chain form and in ring forms, but predominantly in ring forms such as the glucopyranose ring shown. D-glucose contains five OH groups. Rotation of any of the OH and H groups about the ring yields a different aldohexose with distinct chemical properties (Figure 1.7). Furthermore, the mirror-image molecule L-glucose is also distinct from D-glucose (Figure 1.8). As we shall see in Chapter 6, cell membrane transport mechanisms can distinguish between these two forms of glucose.

In summary, there are a number of variables that affect the chemical properties of monosaccharides: aldose versus ketose, number of carbon atoms, three-dimensional disposition of OH groups, and mirror-image symmetry. Therefore, many different monosaccharides are possible. However, a few monosaccharides are particularly important for living systems, including the

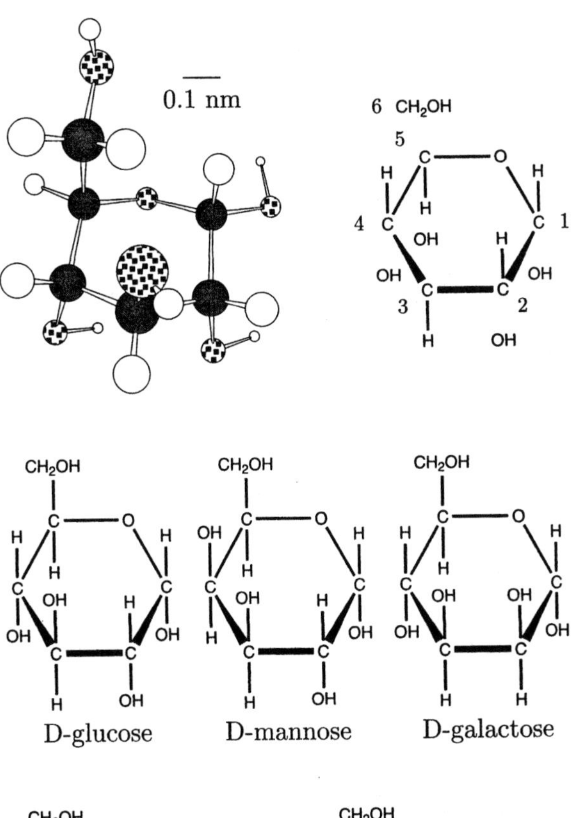

Figure 1.6 The structure of D-glucose ($C_6(H_2O)_6$) shown by means of a three-dimensional model and a chemical structural formula. The standard numbering of the carbon atoms is shown on the structural formula.

Figure 1.7 Chemical structural formulas of three hexoses. They differ from each other in the orientation of H and OH groups about the glucopyranose ring.

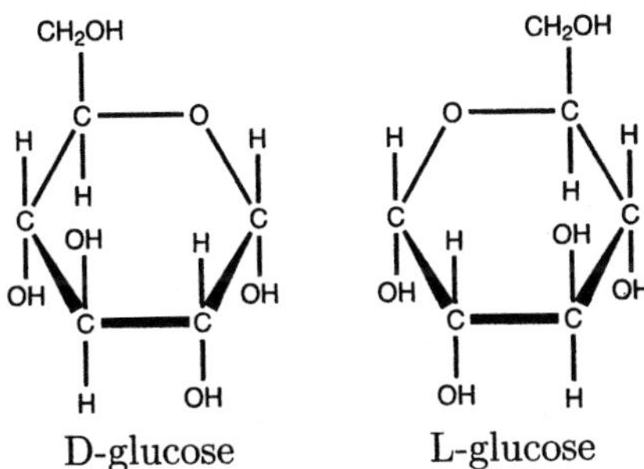

Figure 1.8 Chemical structural formulas of D-glucose and L-glucose.

aldotriose D-glyceraldehyde, the aldopentoses D-ribose and D-deoxyribose, the aldohexose D-glucose, and the ketohexose D-fructose.

Polysaccharides

Polysaccharides are polymers of monosaccharides. The simplest polysaccharide is obtained by linking together two monosaccharides to form a disaccharide by the removal of one molecule of water (Figure 1.9). This chemical

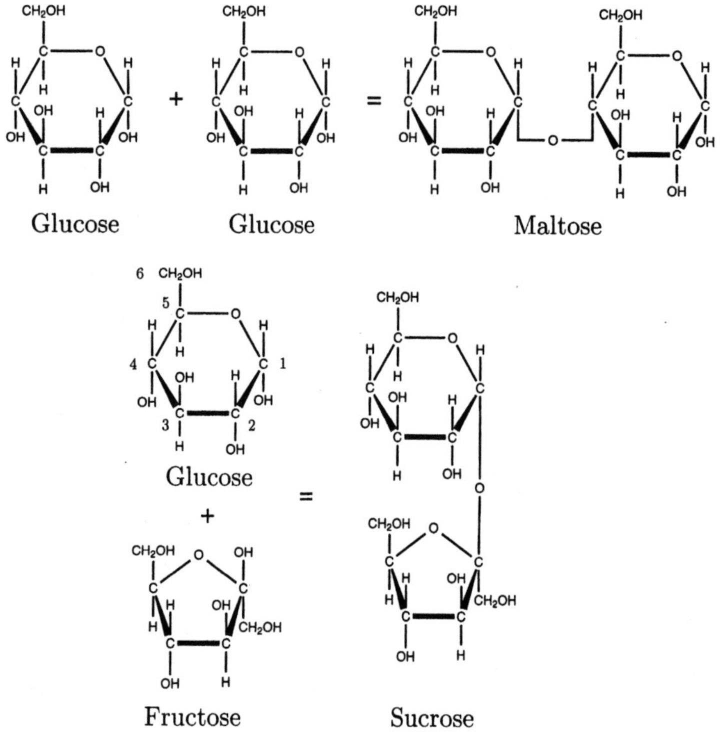

Glucose + Glucose = Maltose

Glucose + Fructose = Sucrose

Figure 1.9 Chemical structures of monosaccharides and disaccharides. Glucose and fructose are 6-carbon sugars (or hexoses). Two glucose molecules combine to produce the disaccharide maltose plus water (not shown). Similarly, one molecule of glucose combines with one molecule of fructose to produce the disaccharide sucrose plus water.

reaction is called a *condensation*. The reverse reaction, which occurs when the bond between monosaccharides is broken by the addition of a water molecule, is called a *hydrolysis*. Since condensations can occur between any two OH groups on two monosaccharides, a large number of disaccharides can be formed; two common ones are shown in Figure 1.9. The linkages are referred to by the numbers of the carbon atoms to which the OH groups on the two monosaccharides are attached as well as to the orientation of the linkages. For example, the linkage between glucose monomers in maltose links OH groups on carbons 1 and 4, and it is called the α1,4 linkage.

A polysaccharide is formed by linking together a large number of monosaccharides—as many as thousands. Since the number of monosaccharides is very large and since the number of condensation reactions between OH groups is large, there is an enormous number of possible polysaccharides. Polysaccharides have the molecular formula $C_n(H_2O)_m$ where $m \leq n$, because bonds between monosaccharides are formed by the removal of molecules of H_2O. Figure 1.10 shows examples of two types of important polysaccharides formed from glucose monomers. The unbranched polysaccharide represents

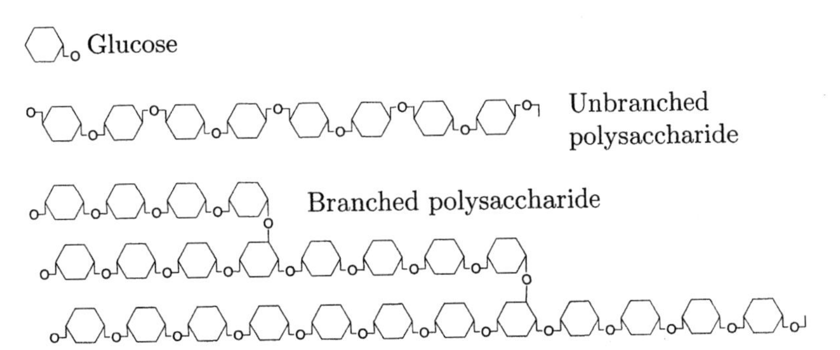

Figure 1.10 Chemical structures of unbranched and branched polysaccharides that are polymers of glucose. Note that there are three different bonds that link glucose monomers in these polysaccharides. The unbranched polysaccharide shown (which is typical of cellulose) has one type of bond, the β1,4 linkage. The branched polysaccharide shown (which is typical of starch and glycogen) has two types of bonds, the α1,4 linkages in the straight portions of the chain and the α1,6 linkages between chains.

the form found in cellulose. The branched polysaccharide represents the form found in both starch and glycogen. The bonds between the monomers differ in cellulose and in starch and glycogen. Since specific enzymes are required to break the bonds between monosaccharides in a polysaccharide, it is not surprising that some animals have enzymes that can break down some polysaccharides but not others. For example, the enzyme amylase, found in human saliva and pancreatic secretions, hydrolyzes the α1,4 linkage to produce mixtures of glucose and maltose, plus larger fragments from starch. Humans can digest starch but not cellulose since they do not synthesize the enzyme that breaks the β1,4 linkage. Grass-eating animals (e.g., cows and horses) harbor microorganisms that produce cellulases that hydrolyze the linkages in cellulose to produce glucose. Hence, these animals can digest the cellulose in grass, whereas humans cannot.

1.3.2.2 *Lipids*

A lipid molecule consists primarily of carbon and hydrogen, i.e., it is primarily a hydrocarbon. The hydrocarbons in lipids are relatively insoluble in water, but soluble in organic solvents including lipids. Lipids form important components of membranes and of compounds, such as the steroid hormones, that permeate membranes through their lipid domains. Lipids are formed from building block molecules called *fatty acids*.

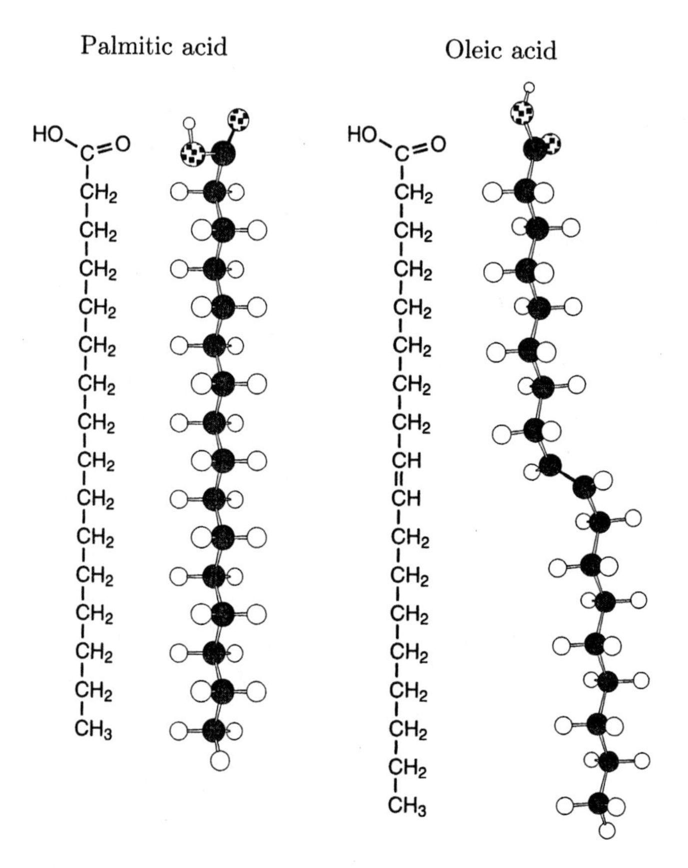

Figure 1.11 Chemical structural formulas and models of the three-dimensional structures of two fatty acids. Palmitic acid is a 16-carbon saturated fatty acid; oleic acid is an 18-carbon unsaturated fatty acid. The CH double bond causes a kink in the tail of oleic acid.

Fatty Acids

A fatty acid consists of a carboxyl group, which forms the head, connected to a hydrocarbon chain, which forms the tail of the molecule (Figure 1.11). When the fatty acid has the maximum number of possible hydrogen atoms attached to the hydrocarbon tail, it is called a *saturated fatty acid*. A saturated fatty acid has the molecular formula $CH_3(CH_2)_nCOOH$. When one or more carbon atoms has a double bond to an adjacent carbon atom and only a single bond to one hydrogen atom, so that the number of hydrogen atoms is less than the maximum possible on the hydrocarbon tails, the fatty acid is called an *unsaturated fatty acid*. If multiple carbons have double bonds, the fatty acid is called a *polyunsaturated fatty acid*. The molecular formula for an unsaturated fatty acid with m unsaturated carbon atoms is $CH_3(CH_2)_n(CH)_mCOOH$. Fatty acid hydrocarbon chains generally have an even number of carbon atoms, typically sixteen to eighteen.

Fatty acids have two functionally distinct groups. The carboxyl group is polar and water soluble and is called *hydrophilic,* whereas the hydrocarbon tail is nonpolar and water insoluble and is called *hydrophobic.* A molecule, like a fatty acid, that contains both hydrophilic and hydrophobic groups is called *amphipathic*. In solution, the carboxyl group is ionized and readily combines with other groups.

Triglycerides

Three fatty acids combine with glycerol to form a *triglyceride* (Figure 1.12). Triglycerides, also called fats, are insoluble in water and are stored in cells in large droplets of fat. Fat droplets are especially abundant in fat cells called *adipocytes*.

Phospholipids

Phospholipids are triglycerides with one fatty acid replaced by phosphoric acid linked to an alcohol (Figure 1.13). Thus, the head of the phospholipid is hydrophilic and the tail, which contains two fatty acids, is hydrophobic; therefore, phospholipids are amphipathic molecules. An important property of

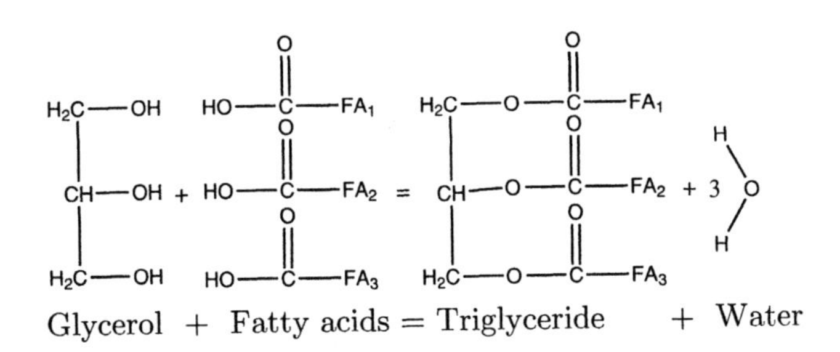

Figure 1.12 A triglyceride is formed by a condensation of three fatty acids (FA_1, FA_2, and FA_3) with glycerol.

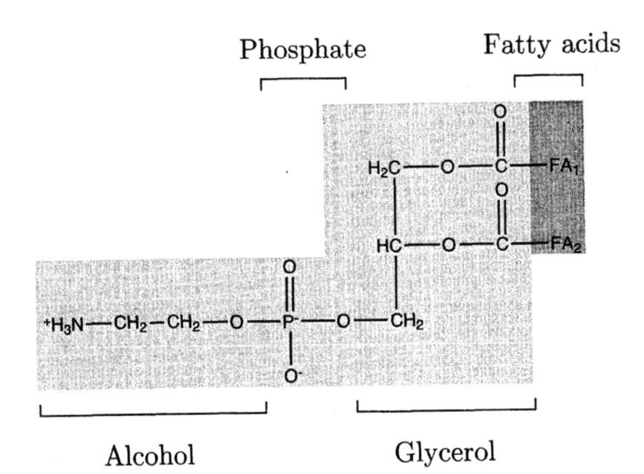

Figure 1.13 The chemical structure of a phospholipid consists of an alcohol linked to phosphoric acid, which replaces one fatty acid in a triglyceride. The phospholipid shown is phosphatidylethanolamine.

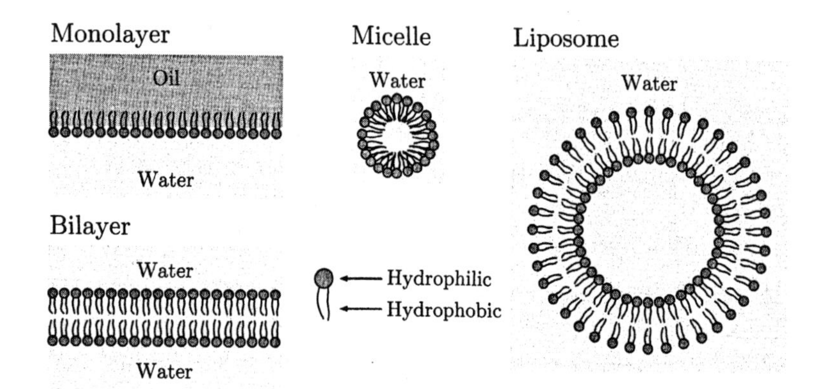

Figure 1.14 Phospholipids take on several types of stable structures in water.

phospholipids is that they self-assemble in a variety of structures as shown in Figure 1.14. At an oil-water interface, the hydrophobic (lipophilic) tails are oriented in the oil and the hydrophilic (lipophobic) heads are located in the water. In water, phospholipids form *micelles,* which are small spheres of molecules with the hydrophobic tails in the interior of the sphere and the hydrophilic heads forming the boundary with water. Liposomes are larger stable structures of phospholipids in water in which a lipid bilayer separates the water inside and outside the liposome. The hydrophilic heads point into the water, and the hydrophobic tails form the interior of the bilayer. Planar bilayers can be formed at an aperture in a partition that separates two aqueous solutions.

1.3.2.3 Proteins

Proteins serve diverse functions in cells. Those proteins known as *enzymes* catalyze a vast array of biochemical reactions. Proteins are also important structural elements of cells and of the extracellular matrix. In addition, proteins in cell membranes determine the passage of key molecules through the membrane. Proteins are also involved in cell motility and in the recognition of other molecules in the immune system. Hormonal proteins serve a signaling function between the secretory cell and its target cells. Proteins modulate gene expression by binding to nucleic acids.

Amino Acids

Proteins are unbranched polymers of building block molecules, the *amino acids,* which consist primarily of carbon, hydrogen, nitrogen, and oxygen. Amino acids consist of a central carbon atom, termed the α carbon, bonded to a carboxyl group, an amino group, a hydrogen, and another group, called the *side chain,* which varies among the amino acids (Figure 1.15). Twenty

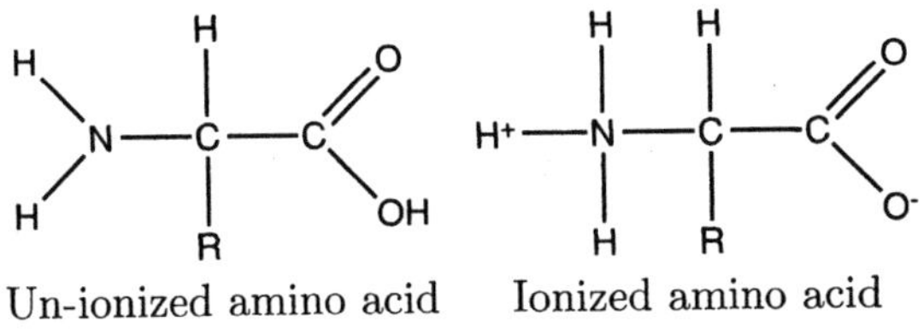

Un-ionized amino acid   Ionized amino acid

Figure 1.15 Un-ionized (left) and ionized (right) forms of an amino acid. An amino acid has an α carbon bonded with an amino, a carboxyl, a hydrogen, and a side group labeled R.

distinct amino acids are commonly found in proteins. Both the three-letter code (e.g., Lys, which is readily decoded) and the one-letter code (e.g., K, which is more convenient for specifying long sequences of amino acids) are given for each amino acid in Figure 1.16. Nineteen of the twenty amino acids have the structure shown in Figure 1.15; proline has a bond between its side group and the amino group. The state of ionization of the amino and carboxyl groups depends on the pH. The pKs of the carboxyl and amino groups vary somewhat among the amino acids, but for free amino acids in solution range from 1.8 to 2.6 for the carboxyl group and from 8.8 to 10.8 for the amino group.[3] Therefore, at a typical intracellular pH of about 7, both the carboxyl and the amino groups are ionized. The side chain, designated R for *residue,* defines the amino acid.

The differences in the chemical characteristics of the side chains are important in determining the properties of amino acids. Amino acids are categorized based on the properties of their side groups. For example, alanine, valine, leucine, isoleucine, and proline contain nonpolar side groups (e.g., CH_3) that do not react with water, but do react with other nonpolar substances. Therefore, these amino acids are categorized as *nonpolar*. Amino acids with an ionizable side group are called ionizable. If the ionizable side group is acidic, the amino acid is called acidic (e.g., aspartic and glutamic acid). If the side group is basic then the amino acid is basic (e.g., arginine and lysine). Both of these types of amino acids have side groups that react with water and are hydrophilic. To estimate the extent to which amino acids are hydrophobic, hydrophobicity indices have been developed (Kyte and Doolittle, 1982) based on the solubility of proteins in water and in organic solvents as well as on the tendency for amino acids to distribute themselves in the interior and exterior of globular proteins (Table 1.5).

The molecular formulas shown in Figure 1.16 do not indicate the three-dimensional arrangement of the atoms in an amino acid. Since the amino and carboxyl groups are asymmetrically located with respect to the α carbon, at

3. The pK of a reaction is described in Chapter 6.

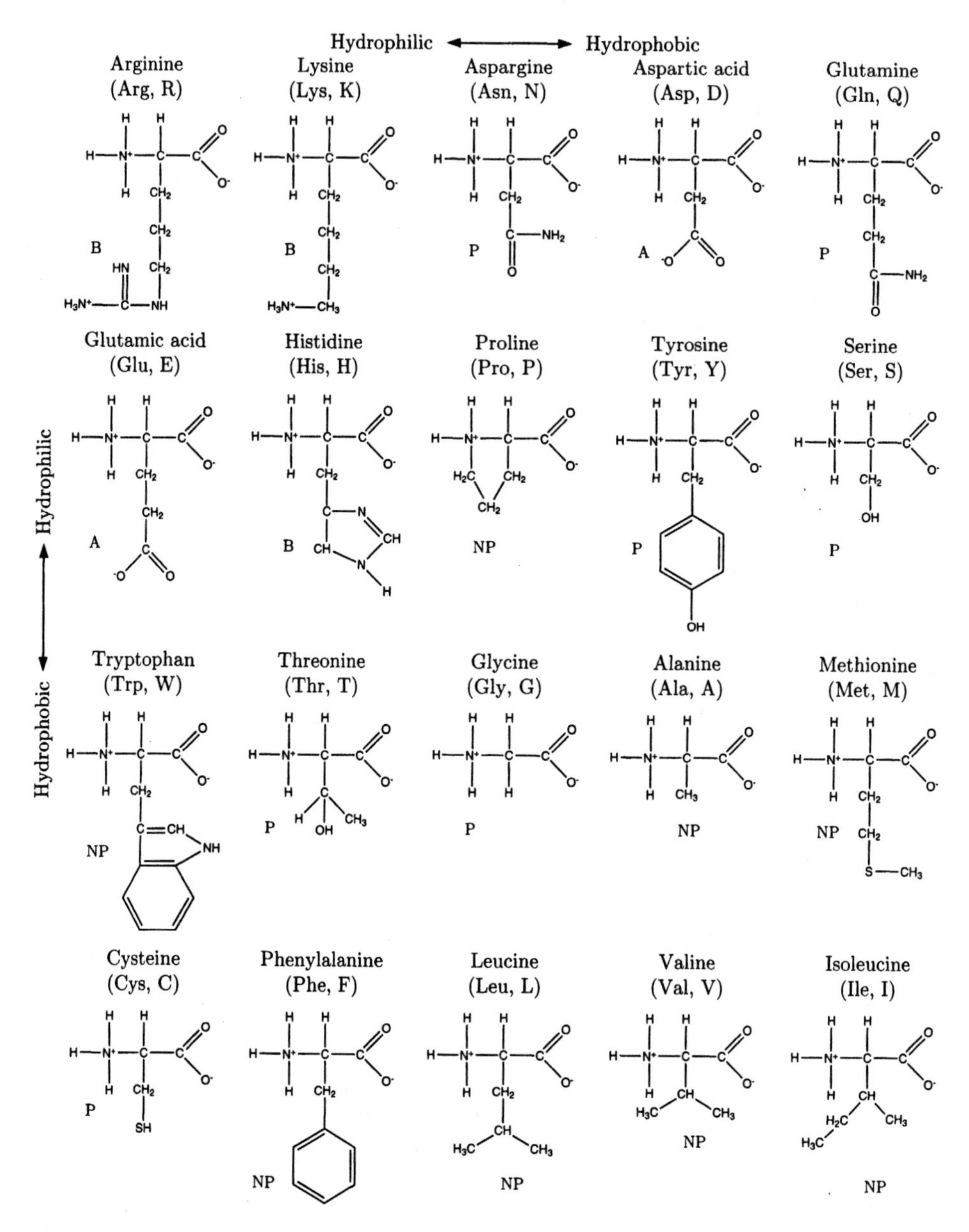

Figure 1.16 Chemical structural formulas for amino acids arranged according to their hydrophobicity, which increases from left to right and from top to bottom (see Table 1.5). The side chain of each amino acid is marked to indicate whether it is basic (B), acidic (A), or uncharged and polar (P) or nonpolar (NP). Both the amino and carboxyl groups are shown as ionized to correspond to a pH of about 7.

Table 1.5 Physical properties of amino acids. The molecular weight is for the nonionized amino acids minus water (Creighton, 1993). The pK of the side group is for the free amino acid (Mathews and van Holde, 1990). The amino acids are ordered according to their hydrophobicity index (Kyte and Doolittle, 1982). The more hydrophobic amino acids have the larger index.

Amino acid	Molecular weight (daltons)	pK of ionizing side group	Hydrophobicity index
Isoleucine	113.16	—	4.5
Valine	99.14	—	4.2
Leucine	113.16	—	3.8
Phenylalanine	147.18	—	2.8
Cysteine	103.15	8.3	2.5
Methionine	131.19	—	1.9
Alanine	71.09	—	1.8
Glycine	57.05	—	−0.4
Threonine	101.11	—	−0.7
Tryptophan	186.21	—	−0.9
Serine	87.08	—	−0.8
Tyrosine	163.18	10.1	−1.3
Proline	97.12	—	−1.6
Histidine	137.14	6.0	−3.2
Glutamic acid	129.12	4.2	−3.5
Glutamine	128.14	—	−3.5
Aspartic acid	115.09	3.9	−3.5
Asparagine	114.11	—	−3.5
Lysine	128.17	10.0	−3.9
Arginine	156.19	12.5	−4.5

least two different, mirror-image symmetric structures are possible for each amino acid. These are called D- and L-type amino acids, and solutions of these two types differ in their optical properties. The L-type amino acids are found predominantly in proteins.

Peptides

The carboxyl and amino groups of two amino acids bind in a condensation reaction to form a dipeptide (Figure 1.17) with the removal of one molecule of water. The bond between the amino acids is called a *peptide bond*. Chains of less than fifty amino acids linked by peptide bonds are generally called *polypeptides*. The peptide bond makes up the backbone of a polypeptide

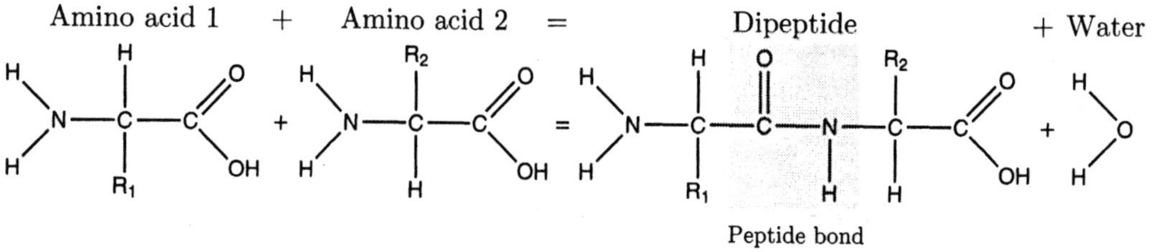

Figure 1.17 Two amino acids combine via a peptide bond to form a dipeptide plus a molecule of water.

chain, which contains the repeating sequence of carbon and nitrogen atoms in the form · · · NCCNCCNCCNCC · · ·. The side groups stick out from the backbone chain. The terminal amino acids of the chain have an amino group at one end, the "N-terminal" amino acid, and a carboxyl group at the other end, the "C-terminal" amino acid. Proteins are generally defined as either (1) polypeptides with unbranched chains that contain more than fifty amino acids and/or (2) molecules that contain multiple unbranched polypeptide chains that are cross linked.

Protein Structure

Proteins take on many different three-dimensional structures, from rodlike molecules with few folds to highly folded molecules. Protein structure and function are critically dependent on the exact sequence of amino acids. Since there are twenty amino acids, many different proteins are possible. Proteins may contain thousands of amino acids, so the possible structural variability among proteins is enormous. The structures of proteins are characterized at four different levels. The sequence of amino acids comprises the *primary structure* of a protein. Thus, the primary structure can be specified simply as a sequence of codes, one for each amino acid (Figure 1.16). Thus, for example, the protein insulin consists of two polypeptide chains linked by bonds as shown in Figure 1.18. The importance of the primary structure of the three-dimensional structure of the protein is illustrated for insulin in which the locations of the cysteine side chains determine the points of attachment of the two polypeptide chains. Thus, while the primary structure can give clues about the three-dimensional structure of the protein, the relation between the two structures is complex. For example, the three-dimensional structure of even a simple protein, such as insulin, can be quite complex, as indicated in Figure 1.19.

The *secondary structure* describes the local spatial relation of amino acids in the primary structure. Certain local structures are found repeatedly and

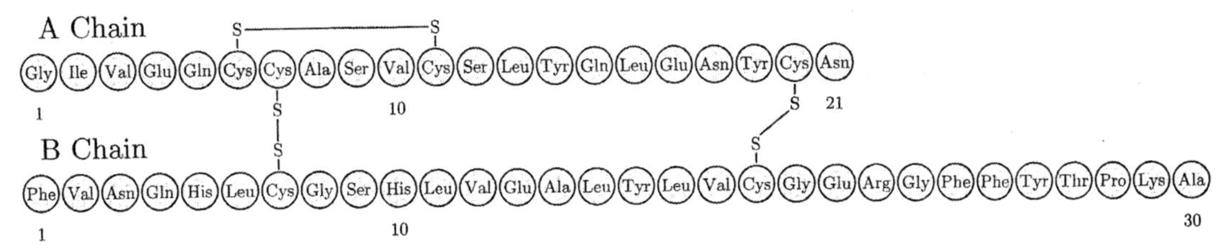

Figure 1.18 The primary structure of insulin consists of two polypeptide chains: the A chain, which consists of 21 amino acids, and the B chain, which consists of 30 amino acids. The two chains are linked by disulfide bonds on cysteine. Insulin is a relatively small protein with a molecular weight of 5733 daltons.

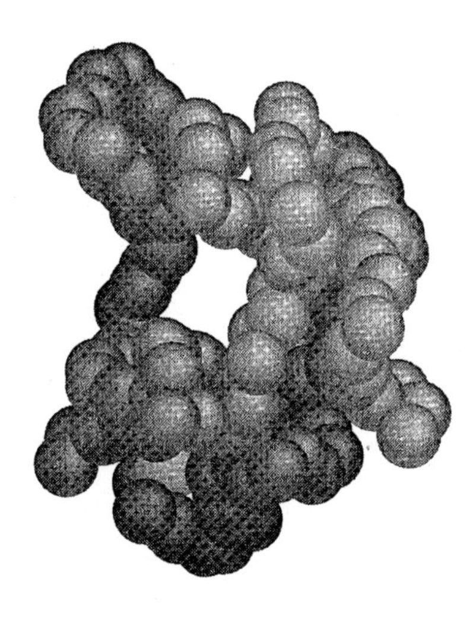

Figure 1.19 The three-dimensional structure of the backbone of insulin (from a pig) obtained from the Brookhaven Protein DataBank (9INS). The diagram shows the three-dimensional arrangement of the backbone of insulin. All the side chains are omitted so that the circuitous path of the backbone can be visualized.

are described below. These include the β pleated sheet and the α helix. The *tertiary structure* is the relation of segments of the secondary structure to each other. For example, a protein may have several sections whose secondary structure is α helical. The spatial relation of these α helices would describe the tertiary structure. The tertiary structure might be helical, in which case the α helical segments would themselves be part of a large helix. The *quaternary structure* describes the relation of several polypeptide chains to each other in a protein consisting of multiple chains. The three-dimensional structure or conformation of a protein is determined not only by the primary structure of the protein, but also by interactions of the amino acids with each other and with other molecules in their environment, e.g., with water when the protein is in solution, with lipids when the protein is embedded in membranes, or with ligands when the protein is an enzyme that catalyzes a reaction.

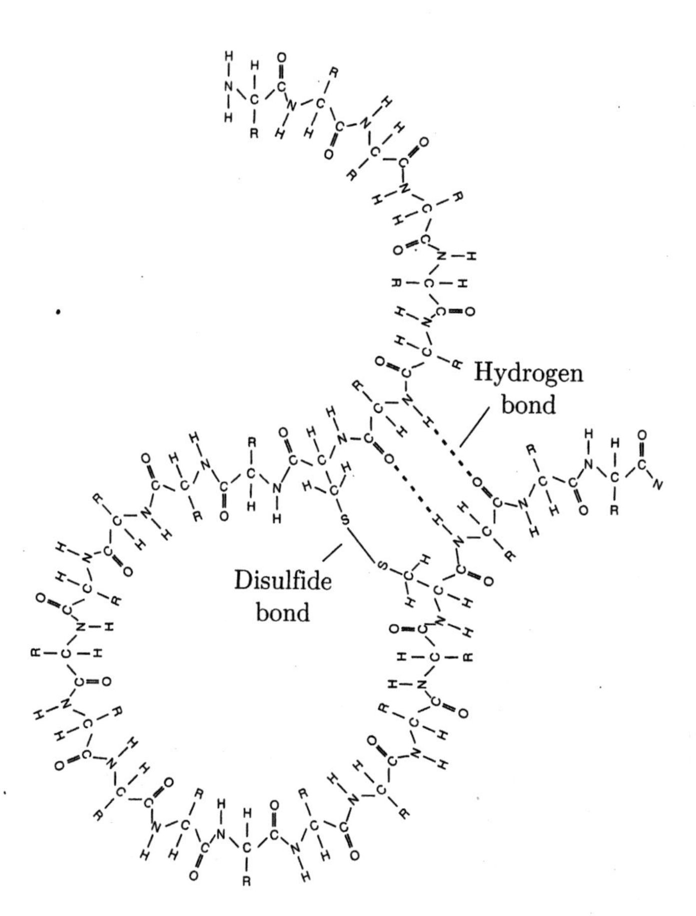

Figure 1.20 A peptide with a number of generic amino acids (with side chains indicated by *R*) and including two cysteine molecules. Two types of bonds between amino acids are illustrated. The disulfide bond links the side chains of two cysteine molecules. Hydrogen bonds link the backbones of two amino acids.

There are three types of bonds between amino acids in a polypeptide. One important interaction is the interaction of the atoms of two peptide bonds with each other (Figure 1.20). The carbonyl group of the peptide bond of one amino acid can form a hydrogen bond with the amide group (CONH) of the peptide bond of another amino acid in the polypeptide. These hydrogen bonds stabilize the conformation of a protein. Another important bond is between side groups of two amino acids. A common bond of this type is the disulfide bond that links cysteine side groups. A third type of bond is a hydrogen bond between the peptide bond of one amino acid and the side chain of another. These types of bonds are responsible for certain secondary structures that are widely found in proteins.

***β* Pleated Sheet** Globular proteins contain polypeptide segments that are hydrogen bonded to form *β* pleated sheets as shown in Figure 1.21. In

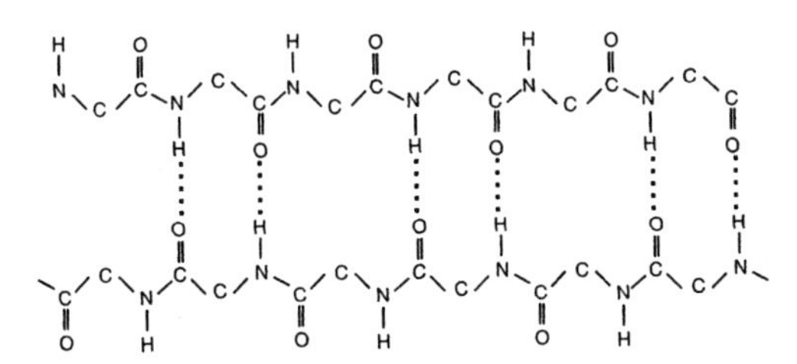

Figure 1.21 Chemical structure of a β pleated sheet. The carbon and nitrogen backbone is shown; the side chains are omitted for clarity. Hydrogen bonds link the amide hydrogen atoms with the carbonyl oxygen atoms.

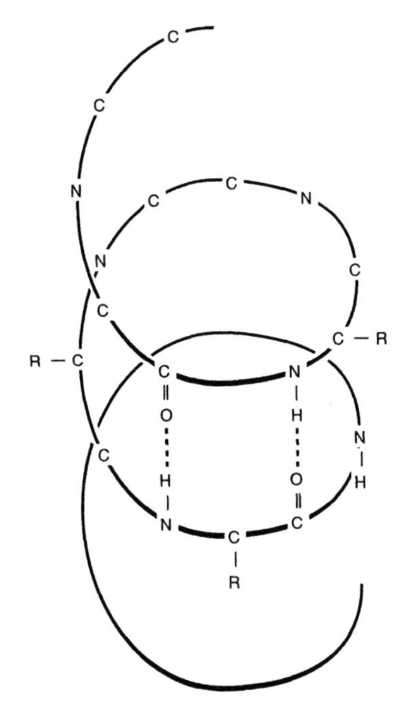

Figure 1.22 Schematic diagram of an α helix. The carbon and nitrogen backbone is shown along the helix; the side groups are shown for only a few of the amino acids. Two hydrogen bonds between amide hydrogen atoms and carbonyl oxygen atoms are shown; the rest are omitted for clarity. There are 3.6 amino acids in each turn of the α helix, which has a pitch, the axial distance between turns, of 5.4 Å.

the β pleated sheet a portion of an amino acid chain may fold back on itself so that two segments align themselves and are bound by hydrogen bonds. The polypeptide sequences may have parallel or antiparallel (as in Figure 1.21) symmetries. The peptide bonds of adjacent segments have hydrogen bonds linking amide hydrogen atoms with carbonyl oxygen atoms. Many such polypeptide segments can participate in β pleated sheets. The larger the number of segments that are hydrogen bonded, the more rigid the structure.

α Helix Another common secondary structure occurs when amino acid sequences arrange themselves in a helical configuration, called an α helix, as shown in Figure 1.22. The backbone of these amino acids makes up the helical core, and the side chains stick radially outward from the backbone. The

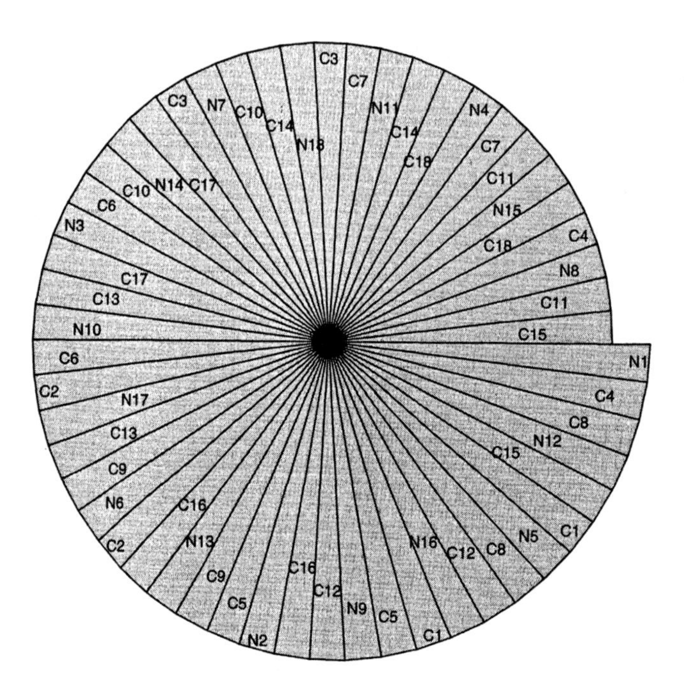

Figure 1.23 Schematic diagram of the three-dimensional arrangement of the amino acid backbone in an α helix looking down the axis of the helix. The sequence NjCjCj represents the nitrogen and two carbons of the jth amino acid in the sequence. The backbone of all 18 amino acids in five turns of the helix is shown. The amino acid backbone for subsequent turns is offset radially inward for clarity.

structure is stabilized by hydrogen bonds between the carbonyl oxygen of the peptide bond of each amino acid and the amide group of the peptide bond that occurs four amino acids farther along the chain. The sequence of backbone atoms is shown more completely in Figure 1.23. As an example of the hydrogen bonding, note that N5 is adjacent to the α carbon C1 in the helix. N5 is the nitrogen for the amino group in the peptide bond of amino acid 5, and C1 is from the carboxyl group of the peptide chain for amino acid 1. These two groups are adjacent in the α helix, and a hydrogen bond forms between the hydrogen of the amino group and the oxygen of the carboxyl group. The combination of the rigid peptide bond and the hydrogen bonding makes the α helix a stable structure.

Since all the hydrogens on amino groups and all the oxygens on carboxyl groups are hydrogen bonded in the core of the α helix, the hydrophobicity of the α helix is determined by the side chains. The sequence of side chains is shown in Figure 1.24. If all the side chains are hydrophilic, the α helical portion of the polypeptide is hydrophilic; if they are all hydrophobic, that portion is hydrophobic. The α helix can also be amphipathic if it contains both hydrophilic and hydrophobic side chains. As in indicated in Figure 1.24, a portion of the helix might be hydrophobic (lower left) and another portion (upper right) hydrophilic. Such a structure might be important as part of a

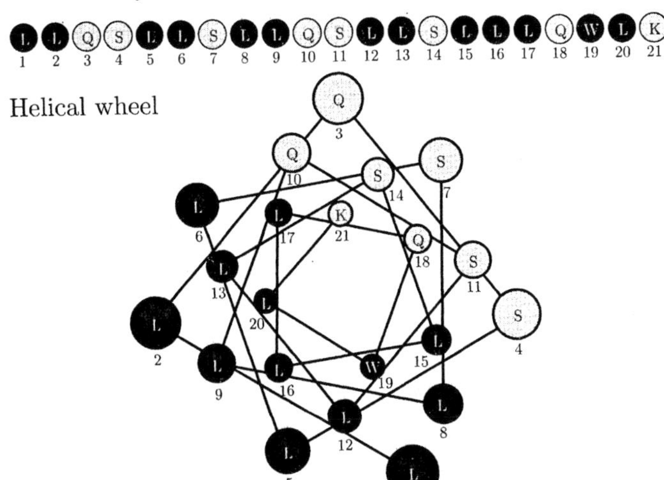

Figure 1.24 Schematic diagrams of the sequence of amino acid side chains (upper panel) and the sequence in an α helix looking down the axis of the helix (lower panel). The side chains of 21 amino acids comprising 5.5 turns of the helix are shown. Hydrophilic amino acids are indicated with grey filled circles, hydrophobic amino acids with black filled circles.

macromolecule making up a channel through a cellular membrane. Several α helical segments could be aligned so that their hydrophilic portions would form the channel interior. The hydrophobic portions would interface with the lipid bilayer or with other hydrophobic portions of a membrane-spanning protein.

1.3.2.4 Nucleic Acids

Nucleic acids store and transfer the genetic information in cells that is used to synthesize proteins. Cells contain two important nucleic acids, deoxyribonucleic acid (DNA) and ribonucleic acid (RNA). Both of these nucleic acids are linear polymers of building block molecules called *nucleotides*. RNA and DNA consist of long sequences of four types of nucleotides each.

Nucleosides and Nucleotides

A *nucleoside* consists of a pentose linked to an organic base molecule, whereas a nucleotide is a nucleoside linked to one or more phosphates (Figure 1.25). Nucleoside phosphates containing one, two, or three phosphates are called nucleoside monophosphates, nucleoside diphosphates, and nucleoside triphosphates, respectively. Nucleosides as well as nucleotides differ in their pentoses. For example, the nucleotides in RNA contain the pentose ribose, whereas those in DNA contain the pentose 2-deoxyribose (Figure 1.26).

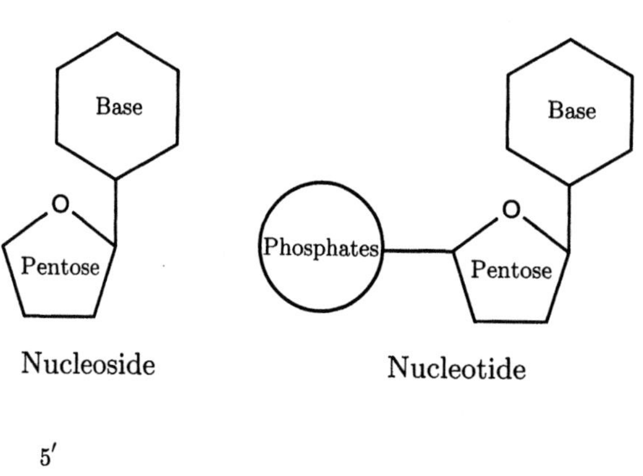

Figure 1.25 Schematic diagrams of a nucleoside and a nucleotide.

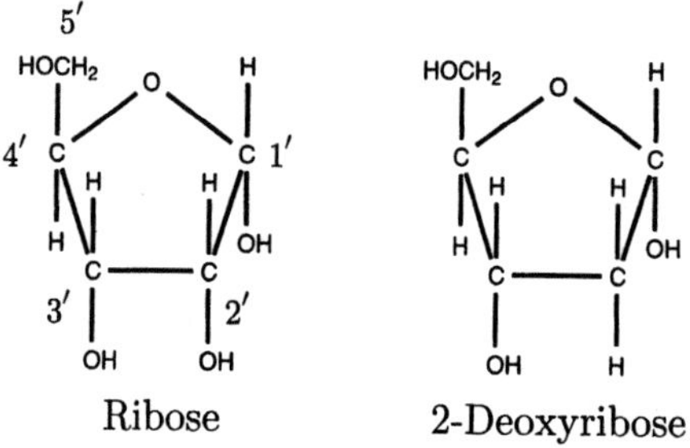

Figure 1.26 Chemical structures of ribose and 2-deoxyribose. The conventional primed numbers for carbon atoms are shown for ribose only. The same numbering is applicable to 2-deoxyribose.

Nucleosides and nucleotides also differ in their bases. There are five different base molecules in RNA and DNA: three different pyrimidines, each consisting of a ring structure with two nitrogens, and two different purines, each consisting of two fused rings (Figure 1.27). Adenine, guanine, and cytosine are found in both RNA and DNA. The fourth base is uracil in RNA and thymine in DNA. Thus, the nomenclature for nucleosides and nucleotides is determined by both the pentose and the base, as well as by the number of phosphates (Table 1.6).

Of particular importance in most energy-requiring reactions in cells are the nucleotides AMP, ADP, and ATP. AMP (Figure 1.28) contains the nucleoside adenosine, which is formed by the condensation of adenine with ribose. The ribose combines in another condensation reaction with a single phosphate group to yield the nucleotide *adenosine monophosphate* or AMP. Addition of a second phosphate yields *adenosine diphosphate* or ADP, and addition of a third phosphate yields the nucleotide *adenosine triphosphate* or ATP (Figure 1.29). ATP is the currency for energy in the body; when the phosphate bonds in ATP are broken, 31 kJ/mol of energy is released. The release of this energy is often coupled to another reaction that requires energy. Thus, the

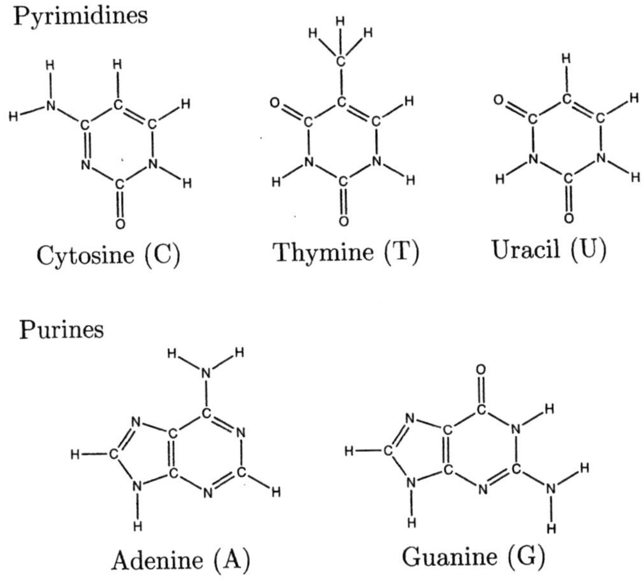

Pyrimidines

Cytosine (C) Thymine (T) Uracil (U)

Purines

Adenine (A) Guanine (G)

Figure 1.27 Chemical structures of pyrimidine and purine bases.

Table 1.6 Nomenclature for nucleosides and nucleotides.

Name	Bases			
	Purines		Pyrimidines	
	Adenine	Guanine	Cytosine	Thymine or Uracil
RNA nucleoside	Adenosine	Guanosine	Cytidine	Uridine
DNA nucleoside	Deoxyadenosine	Deoxyguanosine	Deoxycytidine	Deoxythymidine
RNA nucleotide	Adenylate	Guanylate	Cytidylate	Uridylate
DNA nucleotide	Deoxyadenylate	Deoxyguanylate	Deoxycytidylate	Thymidylate
Nucleoside monophosphate	AMP	GMP	CMP	UMP
Nucleoside diphosphate	ADP	GDP	CDP	UDP
Nucleoside triphosphate	ATP	GTP	CTP	UTP

energy in the phosphate bond of ATP is used to drive a variety of energy-requiring reactions.

Structure of Nucleic Acids

Two nucleotides can combine in a condensation reaction to form a dinucleotide as shown in Figure 1.30. The bond is called a *phosphodiester bond*. Nucleotides can polymerize further to form polynucleotides, or nucleic acids, each with a chain of thousands of nucleotides. The backbone of the molecule

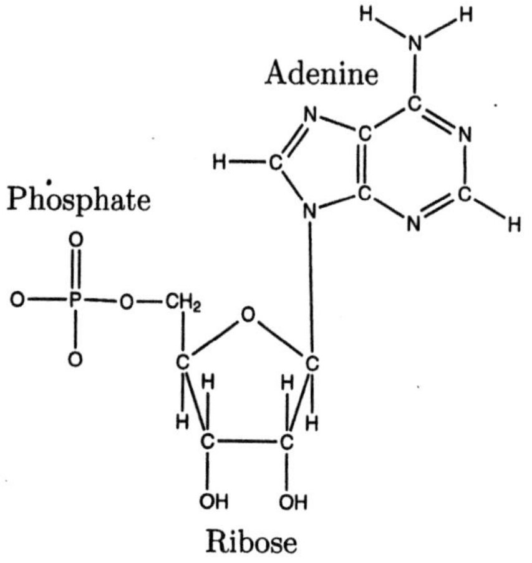

Figure 1.28 Chemical structure of adenosine monophosphate or AMP.

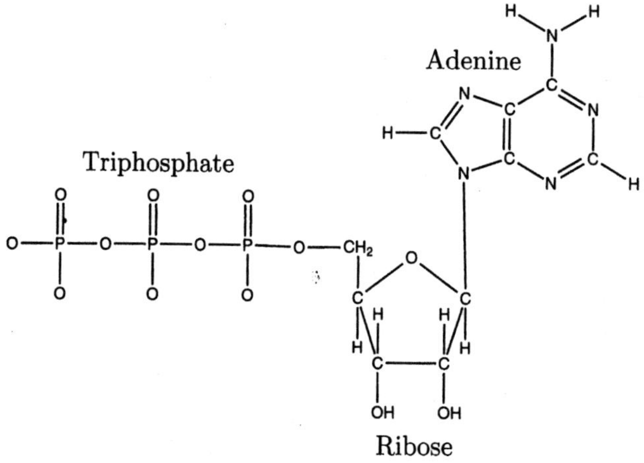

Figure 1.29 Chemical structure of adenosine triphosphate or ATP.

consists of the pentose and phosphate groups, with the bases as side chains. Thus, the primary structure of a nucleic acid is specified by the sequence of nucleotides, i.e., · · · TCTAATAGC · · ·. The nucleic acid has an orientation with a hydroxyl group attached to a phosphate group at one end, the 5′ terminal, and a hydroxyl group of a pentose at the other end, the 3′ terminal. The sequence is conventionally written starting at the 5′ terminal (at the left) and ending at the 3′ terminal (at the right). The secondary structures of DNA and RNA are similar. DNA normally consists of two strands of polynucleotides in a double helix bound together by hydrogen bonds and hydrophobic inter-

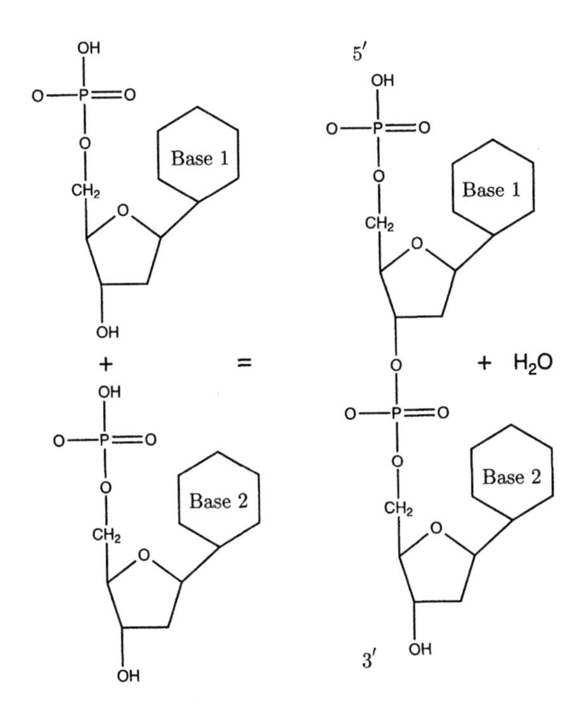

Figure 1.30 Chemical structure of a nucleic acid. The terminal nucleotide at one end of the nucleic acid is a phosphate group attached to the 5′ carbon of a pentose. This end of the nucleic acid is called the 5′ terminal. For similar reasons, the other end is called the 3′ terminal.

actions.[4] The two strands are complementary in that the bases on the two strands are paired in a regular arrangement, A with T and G with C.

Relation of Nucleic Acids to Proteins

The genetic code specifies the relation between a segment of DNA and the protein it encodes. However, the expression of that code into protein involves an intermediate RNA molecule. The processes that relate these three molecules are shown schematically in Figure 1.31. Briefly, genes that encode a protein are transcribed into messenger RNA (mRNA), a reaction that is catalyzed by RNA polymerase. The mRNA is translated into protein by ribosomal RNA. The relation between the nucleotide sequence of DNA and the amino acid sequence of the synthesized protein is a triplet code in which triplets of nucleotides in DNA specify an amino acid (Table 1.7). The triplet code for the amino acids is degenerate; several nucleotide triplets can encode a single amino acid.

4. The structure of DNA was worked out by Watson and Crick in 1953 based on X-ray diffraction of DNA crystals by Wilkins and on construction of molecular models. In 1962, the three workers were awarded the Nobel Prize for Physiology or Medicine. The story of the discovery of DNA has been told dramatically by one of the participants (Watson, 1968).

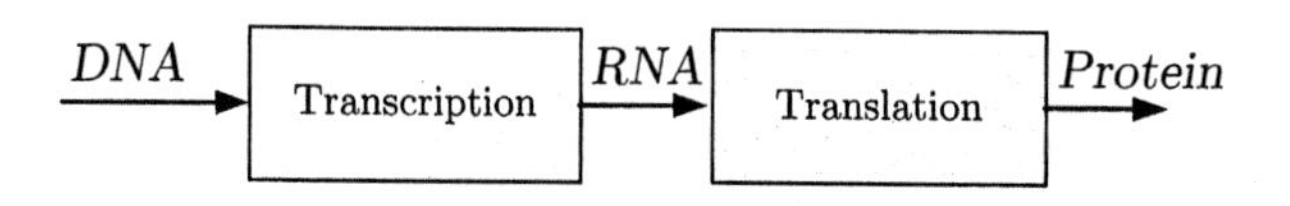

Figure 1.31 The direction of flow of genetic information from DNA to RNA to protein.

Table 1.7 The triplet code for encoding the amino acids. Three sequences do not code an amino acid but signal the termination of the protein—the stop codon. Another triplet, ATG, codes the beginning of the transcription of the protein.

Amino acid	DNA codon
Alanine	GCA, GCC, GCG, GCT
Arginine	AGA, AGG, CGA, CGC, CGG, CGT
Asparagine	AAC, AAT
Aspartic acid	GAC, GAT
Cysteine	TGC, TGT
Glutamic acid	GAA, GAG
Glutamine	CAA, CAG
Glycine	GGA, GGC, GGG, GGT
Histidine	CAC, CAT
Isoleucine	ATA, ATC, ATT
Leucine	CTA, CTC, CTG, CTT, TTA, TTG
Lysine	AAA, AAG
Methionine	ATG, *Start*
Phenylalanine	TTC, TTT
Proline	CCA, CCC, CCG, CCT
Serine	AGC, AGT, TCA, TCC, TCG, TCT
Threonine	ACA, ACC, ACG, ACT
Tryptophan	TGG
Tyrosine	TAC, TAT
Valine	GTA, GTC, GTG, GTT
Stop	TAA, TAG, TGA

1.4 Cell Membrane Structure

Membranes perform many functions. They compartmentalize both cells and their organelles; that is, membranes physically and chemically isolate cells and organelles from their environments. Membranes control the passage of solutes and water between the inside and outside of the enclosed cell or organelle. To do so, membranes house a variety of channels, carriers, and pumps whose transport properties are controlled by several classes of physicochemical variables and are modulated by signaling molecules such as hormones. Membranes also house signaling molecules by which intracellular sites can receive signals present in the extracellular environment without the signal-carrying molecules entering the cell. In addition, macromolecules tethered in membranes catalyze reactions that occur on membrane surfaces.

1.4.1 Contents of Membranes—Lipids, Proteins, and Carbohydrates

All membranes contain lipids and proteins. Lipids provide the basic compartmentalization function of membranes, whereas proteins invest membranes with their specialized functions—transport, signaling, catalysis, etc. In addition to lipids and proteins, some membranes contain carbohydrates as well. The carbohydrates are generally attached to proteins to form *glycoproteins* or attached to lipids to form *glycolipids*. It is thought that the carbohydrates play a critical role in cell recognition. The proportions of lipids, proteins, and carbohydrates vary among different types of membranes (Table 1.8) For example, the proportion of protein in myelin, which makes up the membranes of satellite cells of neurons, is much smaller than that of the protein in mitochondrial inner membrane. This difference in composition fits with the known functions of these two types of membranes. A major function of myelin is to provide electrical insulation of neurons from their environment. This function is consistent with the high proportion of lipid relative to protein in myelin. In contrast, mitochondrial inner membranes house several different enzyme

Table 1.8 Percent composition by weight of proteins, lipids, and carbohydrates in membranes (adapted from Darnell et al., 1990, Table 13-1). The compositions for myelin, mouse liver, human erythrocytes, amoeba, and *Halobacterium* purple membrane are for cellular membranes.

Membrane type	% protein	% lipid	% carbohydrate
Myelin	18	79	3
Mouse liver	44	52	4
Human erythrocyte	49	43	8
Bovine retinal rod	51	49	0
Mitochondrial outer membrane	52	48	0
Amoeba	54	42	4
Sarcoplasmic reticulum	67	33	0
Chloroplast spinach lamellae	70	30	0
Halobacterium purple membrane	75	25	0
Mitochondrial inner membrane	76	24	0

Table 1.9 Lipid composition (% lipid by weight) of membranes from different structures: CM, cellular membrane; RER, rough endoplasmic reticulum; SMR, smooth endoplasmic reticulum; MIM, mitochondrial inner membrane; MOM, mitochondrial outer membrane; NM, nuclear membrane; ROS, rod outer segment; CYTM, cytoplasmic membrane. The different lipid components are CH, cholesterol; PC, phosphatidylcholine; SM, sphingomyelin; PE, phosphatidylethanolamine; PI, phosphatidylinosotol; PS, phosphatidylserine; PG, phosphatidylglycerol; DPG, diphosphatidylinosotol; PA, phosphatidic acid; GL, glycolipids (adapted from Jain, 1988, Table 2-4).

Source	CH	PC	SM	PE	PI	PS	PG	DPG	PA	GL
Rat liver cell components										
CM	20	64	—	17	11	—	2	—	—	—
RER	6	55	3	16	8	3	—	—	—	—
SMR	10	55	12	21	6.7	—	—	1.9	—	—
MIM	< 3	45	2.5	25	6	1	2	18	0.7	—
MOM	< 5	50	5	23	13	2	2.5	3.5	1.3	—
NM	10	55	3	20	7	3	—	—	1	—
Golgi	7.5	40	10	15	6	3.5	—	—	—	—
Lysosomes	14	25	24	13	7	7	—	5	—	—
Other cells and cellular components										
Rat brain myelin	22	11	6	14	—	7	—	—	—	12
Rat brain synaptosome	20	24	3.5	20	2	8	—	—	1	21
Rat erythrocyte	24	31	8.5	15	2.2	7	—	—	< 0.1	—
Rat ROS	< 3	41	—	37	2	13	—	—	—	3
E. coli CYTM	0	—	—	80	—	—	15	5	—	—
B. subtilis	0	0	—	30	—	—	12	—	—	—
Chloroplast	0	4	—	—	1.5	—	6	—	—	55
Sindbis virus	0	26	18	35	—	20	—	—	—	—

systems, including those for oxidative phosphorylation, which accounts for the relatively high proportion of protein in this membrane. Results for several types of membranes (Table 1.9) indicate that the compositions of the lipids in membranes also vary with the membrane type.

1.4.2 The Ubiquitous Phospholipid Bilayer

Despite the diversity of lipid compositions of different membranes, the phospholipids in membranes are in the form of a bilayer as indicated schematically in Figure 1.14. This arrangement accounts for the images of membranes obtained with transmission electron microscopy. At the outset of the use of the electron microscope to examine cell structure in the 1950s, it was recognized that in tissue stained with a heavy metal such as osmium tetroxide, membranes appeared as distinct structural entities with characteristic profiles when cut and seen in cross section. Cells and organelles are seen to be surrounded by a tripartite structure whose thickness is about 75 Å (Figure 1.32). All cells and organelles are characteristically surrounded by what looks like a *railroad track,* because the electron-opaque heavy metals combine strongly with the polar heads of the phospholipids in the bilayer and much less so with the hydrocarbon interior. Thus, the outer leaflets of the membrane appear dark and the central portion appears light in an electron micrograph. The thickness of the membrane can be estimated conveniently by measuring the density of the image using a densitometer, which reflects the density of the electron opaque marker from an electron micrograph of a membrane (Figure 1.33).

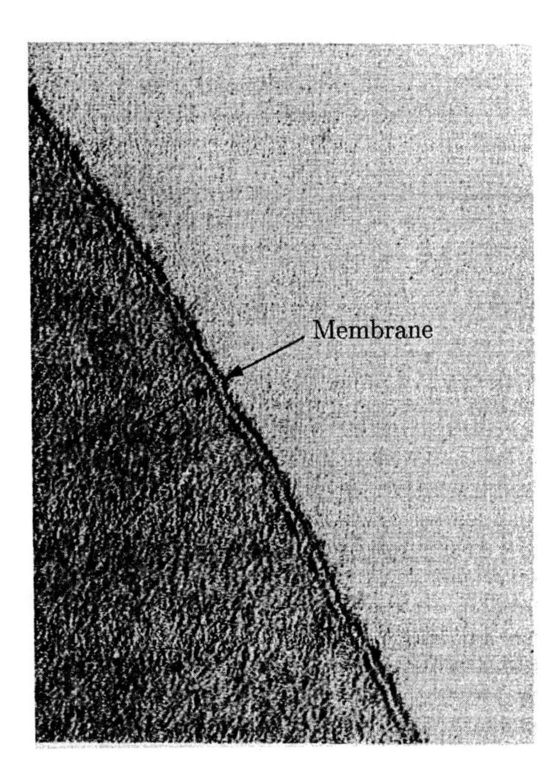

Figure 1.32 Electron micrograph of the plasma membrane of an erythrocyte (adapted from Weiss, 1983, Figure 1-12). The cytoplasm of the erythrocyte is on the left, and the extracellular space is on the right.

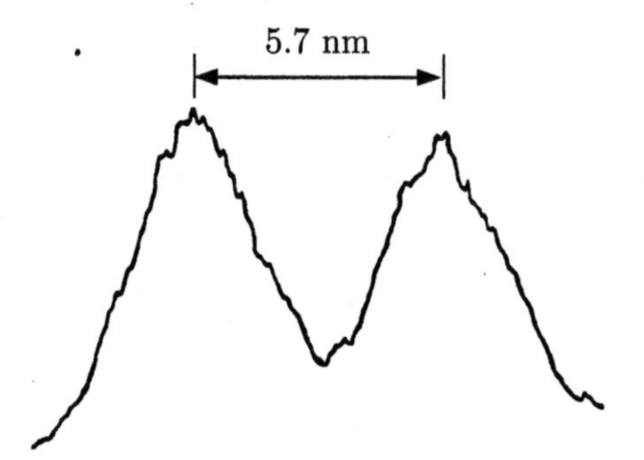

Figure 1.33 Opacity of an electron microscope image of a membrane of a bullfrog (ganglion cell) neuron (adapted from Yammamoto, 1963, Figure 3a). The measured optical density is plotted as a function of distance perpendicular to the membrane. The distance between the peaks in the density is 5.7 nm.

1.4.3 Membrane Fluidity

Cellular membranes are readily deformed by mechanical probes. Furthermore, markers of membrane components have been used to measure the diffusive movements of membrane components, both phospholipids and proteins. The diffusion coefficient for lateral diffusion in the membrane has been estimated to be 10^{-7}-10^{-9} cm^2/sec for some phospholipids and about $10^{-10}$$cm^2$/sec for some proteins (Jain, 1988). Both diffusion coefficients are less than those for the same components in water. However, diffusion of these same molecules from the inner to the outer membrane leaflet is many orders of magnitude slower. The conclusion to be drawn is that membranes can be described as two-dimensional fluid structures in which the diffusible membrane components can diffuse laterally (i.e., in the plane of the membrane), but diffusion in the orthogonal direction is much slower. Some membrane proteins appear to be relatively immobile in the membrane, presumably because they are tethered to the cytoskeleton.

1.4.4 Disposition of Membrane Proteins

The fact that membrane lipids are in the form of a bilayer and that membranes also contain appreciable quantities of protein was known before 1935 (see Section 1.1). However, it took some time to determine that there are two types of membrane proteins that differ in their disposition in the membrane: *peripheral* and *integral proteins*. Peripheral membrane proteins are relatively easily extracted, are water soluble, and are associated with hydrophilic surfaces of membranes. Integral membrane proteins are much harder to extract. They are generally not water soluble, but are soluble in detergents. The integral membrane proteins are anchored in the lipid bilayer.

As an example of a simple integral membrane protein, consider *glycophorin A,* whose primary sequence is shown schematically in Figure 1.34.

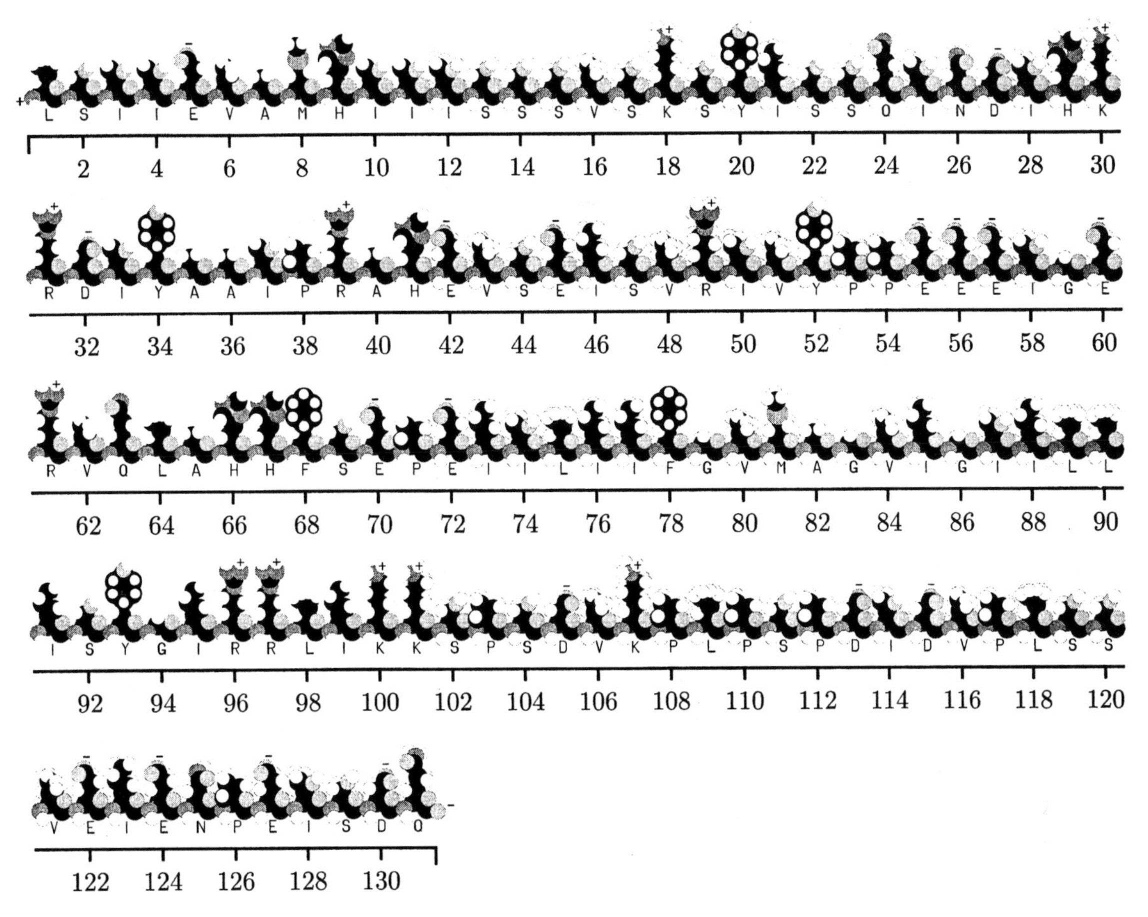

Figure 1.34 Primary structure of erythrocyte glycophorin A, indicated schematically with space-filling models of each amino acid. The amino end is shown at the upper left, and the carboxyl end is shown at the lower right. The amino acid is indicated below each space-filling model with a single-letter code, and the position of each amino acid is shown by the number below the code.

Glycophorin A, with its attached carbohydrates, is involved in cell recognition by the immune system. Glycophorin A contains 131 amino acids in a single strand and has a molecular weight of 14 kD. Hence, it is a relatively small integral membrane protein. Analysis of the amino acids shows that 38% are charged, 18% are acidic, 11% are basic, 42% are polar, and 38% are hydrophobic. The arrangement of the amino acids suggests the disposition of the protein in the membrane. Figure 1.35 shows the hydrophobicity index as a function of position along the protein. One segment (labeled *1*) is predom-

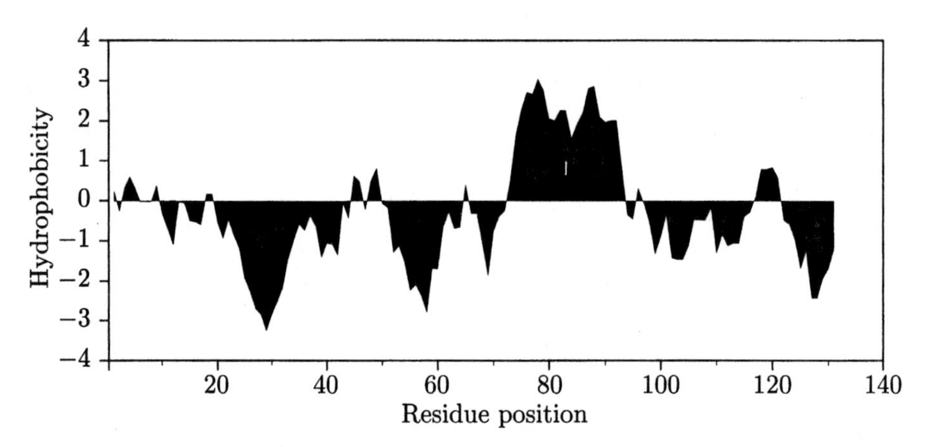

Figure 1.35 Hydrophobicity plot of human erythrocyte glycophorin A (Tomita and Marchesi, 1975). The amino acid sequence of glycophorin A was obtained from the Swiss and PIR and Translated data base (accession nos. A93801, A94584, and A03183). The hydrophobicity of each residue was assigned according to Table 1.5, and the resultant sequence of hydrophobicities was averaged with a window that was seven residues long to yield the averaged hydrophobicity shown.

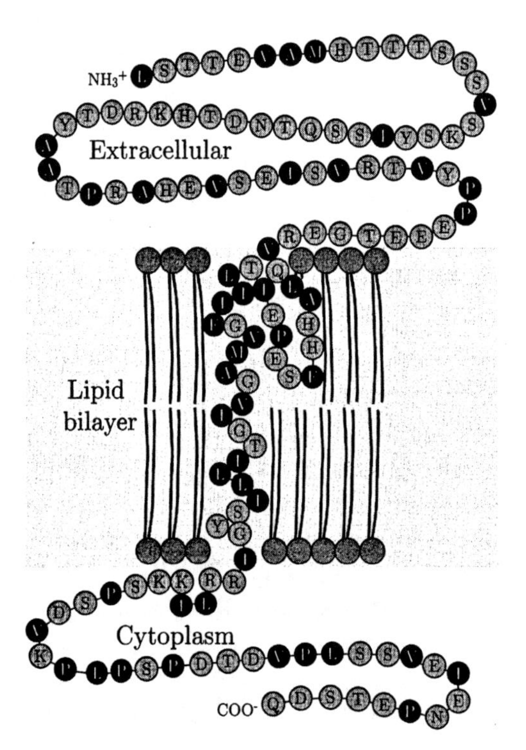

Figure 1.36 Model of glycophorin, an integral membrane protein in erythrocytes (adapted from Darnell et al., 1990, Figure 13-16). The hydrophilic amino acids are indicated with grey filled circles, the hydrophobic amino acids with black filled circles.

inantly composed of hydrophobic amino acids, i.e., 76% of the amino acids from position 75 to position 91 are nonpolar amino acids. This segment is contained in the hydrophobic lipid interior of the membrane. This arrangement of hydrophobic and hydrophilic amino acids in the membrane is shown schematically in Figure 1.36 by a two-dimensional model. In the model of glycophorin, 41% of the amino acids in the putative membrane-spanning portion are hydrophilic, and many of these occur in the polar regions of the bilayer. In the extracellular and cytoplasmic portions of the model of glycophorin, 71% of the amino acids are hydrophilic.

Analysis of the hydrophobicity of the amino acid sequences of integral membrane proteins has proved to be important in trying to determine their three-dimensional structures and their relation to the membrane. A key finding is that each such protein usually consists of predominantly hydrophobic amino acids separated by predominantly hydrophilic amino acid segments. Often the hydrophobic segments are of about the right dimensions to form an α helix that spans the membrane. On this basis it is possible to make models of the protein and its disposition in the membrane. A model of the relatively simple integral membrane protein glycophorin (Figure 1.36) contains only a single membrane-spanning segment. More typically, large integral membrane proteins that are responsible for transporting substances across the membrane may contain ten or more hydrophobic membrane-spanning segments. These regions can form a more complex structure in the membrane, such as a carrier or a channel.

The fluid mosaic model (Singer and Nicholson, 1972) of membranes is shown schematically in Figure 1.37. In this image, the lipid bilayer acts as

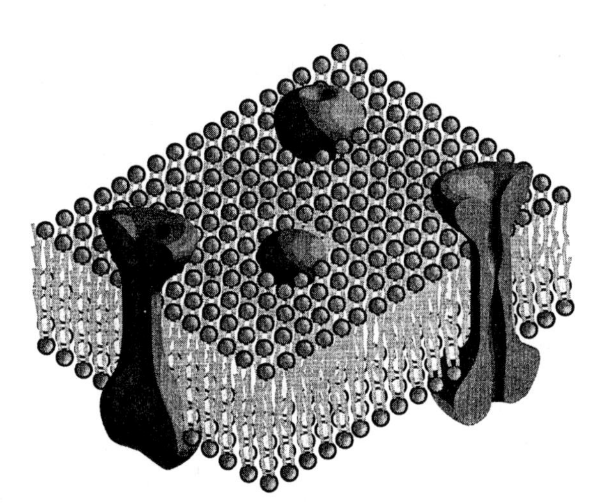

Figure 1.37 Model of a cellular membrane showing the lipid bilayer and four integral membrane proteins.

a two-dimensional lipid liquid to which the proteins are attached. Peripheral membrane proteins are found on the surface, while integral membrane proteins penetrate either into or through the membrane. These integral membrane proteins have different functions, including transmembrane transport and the reception and recognition of extracellular substances. Some of these proteins are free to diffuse in the plane of the membrane, while others are tethered to the cytoskeleton.

Exercises

1.1 Assume that the length of the carbon-carbon bond in a hydrocarbon is about 1.5 Å. Estimate the thickness of a cellular membrane that is due to hydrophobic tails of membrane lipids. Carefully state all your assumptions.

1.2 Estimate roughly how many different proteins might be formed from twenty amino acids in a chain that is one hundred amino acids long? Compare this number to the number of atoms in the universe, which has been estimated to be about 10^{79}.

1.3 One turn of an α helix of amino acids contains 3.6 amino acids and occupies an axial distance of about 5.4 Å.

a. For an integral membrane protein, how many amino acids would be contained in a membrane-spanning segment whose secondary structure was that of an α helix?

b. How many amino acids would be contained in a segment spanning the hydrophobic portion of the membrane only?

1.4 A polypeptide consists of twenty-five amino acids that form an amphipathic α helix such that the left half is hydrophobic and the right half is hydrophilic.

a. Using the single-letter code shown in Figure 1.16, specify the primary structure by an amino acid sequence.

b. Determine the axial length of the α helix.

1.5 Describe the differences in chemical structure of RNA and DNA.

1.6 Which chemical properties of lipids make them particularly suitable for cellular membranes?

1.7 What is the distinction between the cell wall and the cell membrane?

1.8 Portions of nucleotide sequences for two nucleic acids are · · · TCTAATA-GC · · · and · · · UCUAAUAGC · · ·. Which of these corresponds to RNA, and which to DNA?

1.9 Define the following terms:

a. Integral membrane protein
b. Prokaryote
c. Cytosol
d. Electronegativity
e. Hydrogen bond
f. Pentose
g. Peptide bond
h. Saturated fatty acid
i. Amphipathic

1.10 In this problem we consider that cells can be categorized as prokaryotes, eukaryote plant cells, and eukaryotic animal cells. Determine to which, if any, of these three cell types each of the following statements generally applies:

a. The cell cytoplasm is enclosed by a cell membrane.
b. The cell is enclosed by a cell wall that is enclosed by the cell membrane.
c. The cell contains a nucleus.
d. The cell contains mitochondria.
e. The cell contains cytoplasm.

1.11 Hydrophobicity plots have been found to be useful for studying membrane-bound proteins.

a. What is a hydrophobicity plot?
b. What useful information about a membrane protein is suggested by the hydrophobicity plot?

References

Books and Reviews

Alberts, B., Bray, D., Lewis, J., Raff, M., Roberts, K., and Watson, J. D. (1983). *Molecular Biology of the Cell.* Garland, Boston.

Beck, W. S., Liem, K. F., and Simpson, G. G. (1991). *Life: An Introduction to Biology.* HarperCollins, New York.

Berg, P. and Singer, M. (1992). *Dealing with Genes.* University Science Books, Mill Valley, CA.

Brachet, J. (1965). The living cell. In *The Living Cell*, 4–15. W. H. Freeman & Co., San Francisco.

Branton, D. and Park, R. B. (1968). *Papers on Biological Membrane Structure.* Little, Brown & Co., New York.

Bull, H. B. (1964). *An Introduction to Physical Biochemistry.* F. A. Davis, Philadelphia.

Cerdonio, M. and Noble, R. W. (1986). *Introductory Biophysics.* World Scientific, Singapore.

Cole, K. S. (1968). *Membranes, Ions, and Impulses.* University of California Press, Berkeley, CA.

Creighton, T. E. (1993). *Proteins.* W. H. Freeman & Co., New York.

Darnell, J., Lodish, H., and Baltimore, D. (1990). *Molecular Cell Biology.* Scientific American Books, New York.

Darwin, C. (1859). *On the Origin of Species by Means of Natural Selection; or, the Preservation of Favored Races in the Struggle for Life.* John Murray, London. Reprint, New American Library of World Literature, New York. 1958.

Darwin, C. (1871). *The Descent of Man, and Selection in Relation to Sex.* John Murray, London. Reprint, Princeton University Press, Princeton, NJ 1981.

Davson, H. and Danielli, J. F. (1952). *The Permeability of Natural Membranes.* Cambridge University Press, Cambridge, England.

Dawkins, R. (1976). *The Selfish Gene.* Oxford University Press, New York.

Dawkins, R. (1987). *The Blind Watchmaker.* W. W. Norton & Co., New York.

Desmond, A. and Moore, J. (1991). *Darwin.* Warner Books, New York.

Dickerson, R. E. and Geis, I. (1969). *The Structure and Action of Proteins.* Harper & Row, New York.

Dowben, R. M. (1969). *General Physiology. A molecular approach.* Harper & Row, New York.

Ebbing, D. D. (1984). *General Chemistry.* Houghton Mifflin, Boston.

Fawcett, D. W. (1966). *An Atlas of Fine Structure: The Cell.* W. B. Saunders, Philadelphia.

Giese, A. C. (1968). *Cell Physiology.* W. B. Saunders, Philadelphia.

Gould, S. J. (1977). *Ever Since Darwin.* W. W. Norton & Co., New York.

Gould, S. J. (1980). *The Panda's Thumb.* W. W. Norton & Co., New York.

Gould, S. J. (1989). *Wonderful Life.* W. W. Norton & Co., New York.

Hammel, H. T. and Scholander, P. F. (1976). *Osmosis and Tensile Solvent.* Springer-Verlag, New York.

Harrison, R. and Lunt, G. G. (1980). *Biological Membranes, Their Structure and Function.* John Wiley & Sons, New York.

Jain, M. K. (1988). *Introduction to Biological Membranes*. John Wiley & Sons, New York.
Judson, H. F. (1979). *The Eighth Day of Creation*. Simon & Schuster, New York.
Kepner, G. R. (1979). *Cell Membrane Permeability and Transport*. Dowden, Hutchinson and Ross, Stroudsburg, PA.
Kessel, R. G. and Shih, C. Y. (1974). *Scanning Electron Microscopy in Biology*. Springer-Verlag, New York.
Lehninger, A. H. (1970). *Biochemistry*. Worth, New York.
Lentz, T. L. (1971). *Cell Fine Structure*. W. B. Saunders, Philadelphia.
Mathews, C. K. and van Holde, K. E. (1990). *Biochemistry*. Benjamin-Cummings, Redwood City, CA.
Moore, W. J. (1972). *Physical Chemistry*. Prentice Hall, Englewood Cliffs, NJ.
Novikoff, A. B. and Holtzman, E. (1970). *Cells and Organelles*. Holt, Rinehart & Winston, New York.
Pfeffer, W. (1877). *Osmotic Investigations. Studies on Cell Mechanics*. Van Nostrand Reinholt, New York.
Porter, K. R. and Bonneville, M. A. (1968). *Fine Structure of Cells and Tissues*. Lea and Febiger, Philadelphia.
Snell, F. M., Shulman, S., and Moos, C. (1965). *Biophysical Principles of Structure and Function*. Addison-Wesley, New York.
Thomas, L. (1974). *The Lives of a Cell: Notes of A Biology Watcher*. Viking, New York.
Tosteson, D. C., editor (1989). *Membrane Transport: People and Ideas*. American Physiological Society, Bethesda, MD.
Vander, A. J., Sherman, J. H., and Luciano, D. S. (1990). *Human Physiology: The Mechanisms of Body Function*. McGraw-Hill, New York.
Watson, J. D. (1968). *The Double Helix*. Atheneum, New York.
Watson, J. D., Hopkins, N. H., Roberts, J. W., Steitz, J. A., and Weiner, A. M. (1987). *Molecular Biology of the Gene*, vol. 1, *General Principles*. Benjamin-Cummings, Menlo Park, CA.
Weiss, L. (1983). *Histology*. Elsevier, New York.
Weissmann, G. and Claiborne, R. (1975). *Cell Membranes, Biochemistry, Cell Biology and Pathology*. HP Publishing, New York.

Original Articles

Bar, R. S., Deamer, D. W., and Cornwall, D. G. (1966). Surface area of human erythrocyte lipids: Reinvestigation of experiments on plasma membrane. *Science*, 153:1010–1012.
Bernstein, J. (1902). Untersuchungen zur thermodynamik der bioelectrischen strome. *Pflügers Arch. Ges. Physiol.*, 92:521–562.
Collander, R. and Bärlund, H. (1933). Permeabilitätsstudien an *Chara ceratophylla*. II. Die permeabilität fur nichtelekrolyte. *Acta Bot. Fenn.*, 11:1–114.
Fick, A. (1855). On liquid diffusion. *Philos. Mag.*, 10:30–39.
Hobot, J. A., Villiger, W., Escaig, J., Maeder, M., Ryter, A., and Kellenberger, E. (1985). Shape and fine structure of nucleoids observed on sections of ultrarapidly frozen and cryosubstituted bacteria. *J. Bacteriol.*, 162:960–971.
Hodgkin, A. L. and Huxley, A. F. (1952). A quantitative description of membrane current and its application to conduction and excitation in nerve. *J. Physiol.*, 117:500–544.

Kyte, J. and Doolittle, R. F. (1982). A simple method for displaying the hydropathic character of a protein. *J. Mol. Biol.*, 157:105–132.

Neher, E. and Sakmann, B. (1976). Single-channel currents recorded from membrane of denervated frog muscle fibres. *Nature*, 260:799–802.

Robertson, J. D. (1960). The molecular structure and contact relationships of cell membranes. *Prog. Biophys. Biophys. Chem.*, 10:344–418.

Sigel, E. (1990). Use of *Xenopus* oocytes for the functional expression of plasma membrane proteins. *J. Membr. Biol.*, 117:201–221.

Singer, S. J. and Nicholson, G. L. (1972). The fluid mosaic model of the structure of cell membranes. *Science*, 175:720–731.

Tomita, M. and Marchesi, V. T. (1975). Amino-acid sequence and oligosaccharide attachment sites of human erythrocyte glycophorin. *Proc. Natl. Acad. Sci. U.S.A.*, 72:2964–2968.

Van't Hoff, J. (1887). The role of osmotic pressure in the analogy between solutions and gases. *Z. Physik*, 1:481–493.

Yammamoto, T. (1963). On the thickness of the unit membrane. *J. Cell Biol.*, 17:413–421.

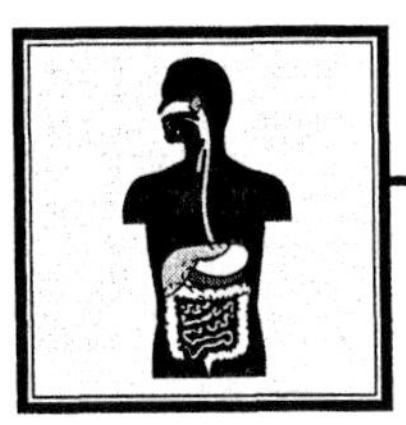

2 Introduction to Transport

[A]nimals have really two environments: a milieu extérieur in which the organism is situated, and a milieu intérieur in which the tissue elements live. The living organism does not really exist in the milieu extérieur (the atmosphere if it breathes, salt or fresh water if that is its element) but in the liquid milieu intérieur formed by the circulating organic liquid which surrounds and bathes all the tissue elements; this is the lymph or plasma, the liquid part of the blood which, in the higher animals, is diffused through the tissues and forms the ensemble of the intercellular liquids and is the basis of all local nutrition and the common factor of all elementary exchanges. A complex organism should be looked upon as an assemblage of simple organisms which are the anatomical elements that live in the liquid milieu intérieur.
—Bernard, 1898

2.1 Introduction

Living organisms—from simple, single-cell organisms such as bacteria to complex, multicell organisms such as humans—exchange matter with their external environments. Organisms take up nutrients and excrete waste products. Single-cell organisms extract nutrients directly from their external environments. However, most cells of complex, multicell organisms are not in direct contact with the organism's external environment, but rather are in contact with the internal environment. As indicated by the quotation at the beginning of this chapter, the internal environment consists of interstitial fluids that communicate with blood and lymph. Complex organisms have elaborate organ systems (Figure 2.1) that are in contact with both the internal and external environments. These organ systems control the internal environment of the body with which the cells communicate. For example, the food that enters a complex organism through the digestive system is not in a form that is digestible by the cells. The purpose of the digestive system is to break down the

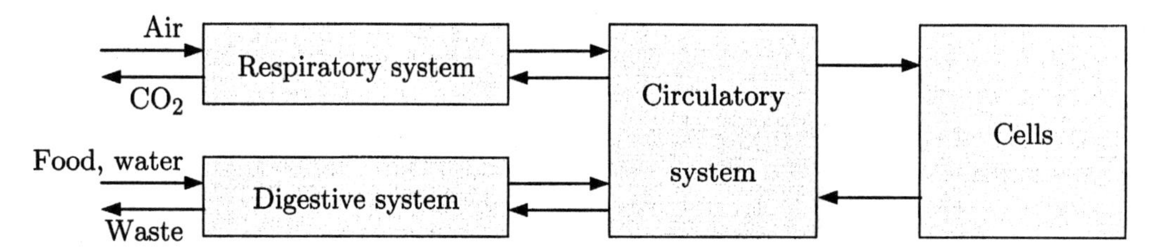

Figure 2.1 Schematic block diagram of the relations between the respiratory, digestive, and circulatory systems of a complex multicellular organism. Air, food, and water enter the respiratory and digestive systems and are broken down into building block molecules that are transported throughout the body via the circulatory system.

food into building block molecules—such as amino acids, monosaccharides, fatty acids, mononucleotides, etc.—that are usable by cells. In Chapters 3 through 8 we shall explore the transport of matter into and out of cells with a focus on membrane transport mechanisms. One purpose of this chapter is to point out the role of these membrane transport mechanisms in the overall transport of matter in an organism.

2.2 Cell Requirements

Which nutrients do cells require? Table 2.1 shows a culture medium in which a simple prokaryote, the bacterium *Escherichia coli* or *E. coli* (see Figure 1.1), can survive. The synthetic culture medium contains water, glucose, and some salts, which collectively serve as sources of carbon, hydrogen, nitrogen, oxygen, and phosphorus. Enriched media used to culture *E. coli* usually contain additional substances, including a mixture of amino acids, small peptides, lipids, and vitamins. In these simple media, *E. coli* synthesizes the required molecules, extracts energy for cellular functions, and replicates.

The eukaryotic cells (see Figure 1.2) of higher organisms are far more complex. Their linear dimensions are generally ten times as great as those of prokaryotes. Hence, their volumes are a thousand times greater than those of prokaryotes. The organelles of eukaryotes are about the dimensions of *E. coli*. Eukaryotes have more stringent needs because they cannot synthesize all the molecules they require from a medium as simple as that sufficient for *E. coli*. Typically, mammalian cells can be cultured in media that contain about a dozen essential amino acids (arginine, histidine, isoleucine, leucine,

Table 2.1 Culture medium for *E. coli* (adapted from Watson et al., 1987, Table 4-2).

Component	Quantity
NH_4Cl	1.0g
$MgSO_4$	0.13g
KH_2PO_4	3.0g
Na_2HPO_4	6.0g
Glucose	4.0g
Water	1.0L

lysine, methionine, phenylalanine, threonine, tryptophan, and valine), eight vitamins (choline, folic acid, nicotinamide, pantothenate, pyridoxal, thiamine, inositol, and riboflavin), salts (sodium, potassium, calcium, magnesium, chloride, phosphate, and bicarbonate), glucose, and blood serum. This mixture is still a relatively simple medium, yet cells can grow in such media by extracting all substances necessary to synthesize thousands of macromolecules. Cells require transport mechanisms to extract these substances from their environments, and higher organisms require organ systems to extract the nutrients and to deliver them to the cells.

2.3 Transport in the Body Illustrated Using a Potato

To illustrate some of the principles of transport in the body, we consider an ingested potato's breakdown into glucose molecules, which are transported by the digestive system into the circulatory system and then distributed to the cells in the body. For other food components, such as proteins and fats, the details of digestion, transport, and distribution differ from the illustration, but the broad outlines are similar.

2.3.1 Composition of a Potato

What is a potato? Although such a question has metaphysical overtones, we shall be satisfied with a description of the average composition of a potato (Table 2.2). The solid matter in a potato is predominantly carbohydrate in the form of starch. A potato has the biological function of storing the chemical energy contained in the chemical bonds of the starch molecule for the potato plant. Starch is a branched polysaccharide consisting of glucose monomers (Figures 1.9 and 1.10) linked together by condensations of OH groups.

Table 2.2 Average composition of a potato by percent weight of each component (Souci et al., 1986).

Component	Percentage
Water	78.7%
Protein	2.1%
Fat	0.1%
Available carbohydrate	15.6%
Total dietary fiber	2.5%
Minerals	1.0%

2.3.2 Digestion of a Potato

As indicated in Figure 2.2, the digestive system includes the gastrointestinal tract, consisting of the mouth, pharynx, esophagus, stomach, small intestine (duodenum, jejunum, and ileum), large intestine, rectum, and anus, plus associated secretory organs, such as the salivary glands, liver, gallbladder, and pancreas. Material inside the gastrointestinal tract is anatomically, topologically,[1] and physiologically outside the body. Just as the skin is lined with an epithelium that separates the outside of the body from its inside, so is the gastrointestinal tract. Though these epithelia have many common features, the epithelium of the gastrointestinal tract is specialized for digestion and transport. In addition, the tract houses microorganisms, some useful, that would be harmful if they "entered" the body.

A schematic diagram of the digestive process is indicated in Figure 2.3. Digestion begins with mastication (chewing) of food, in this case a potato, in the mouth, which divides the mass into smaller pieces. This process increases the potato's surface area, and hence, the access of the digestive enzymes to it. The salivary glands secret saliva into the mouth. Saliva contains both mucus, which lubricates the masticated potato, and amylase, which begins the process of digestion of the starch. When the bolus of masticated potato is swallowed, it passes from the mouth through the pharynx and esophagus and into the stomach. The gastric juices secreted by the gastric glands in the wall of the stomach contain, among other substances, HCl and pepsin. The HCl

1. Topologically, the gastrointestinal tract is outside the body, since a line drawn between arbitrary points on the inside of the gastrointestinal tract and at the outside of the body crosses an even number of epithelial layers.

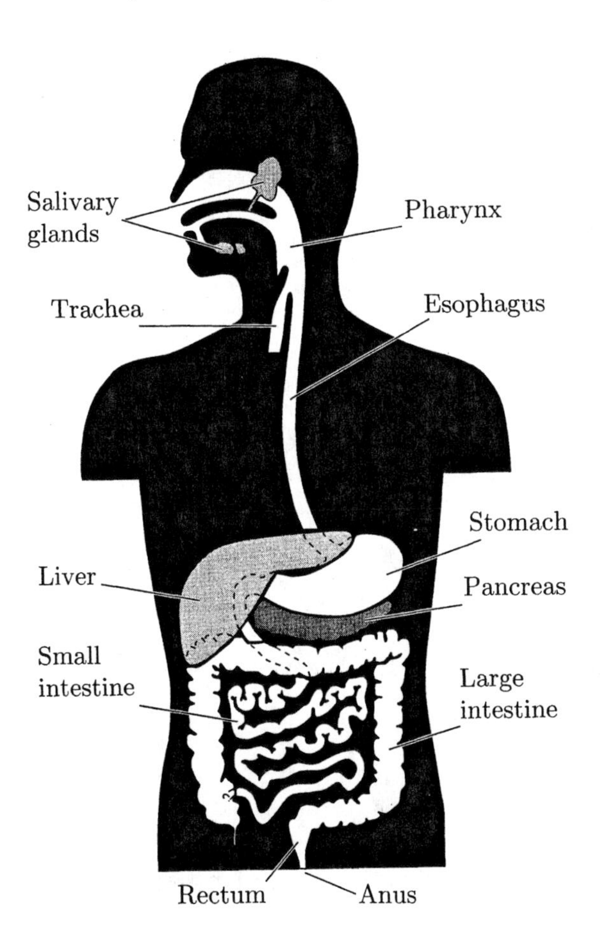

Figure 2.2 Schematic diagram of the digestive system (adapted from Vander et al., 1990, Figure 16-1). The gallbladder (which lies dorsal to the liver but is not shown), liver, and pancreas secrete digestive juices into the duodenum, which is the portion of the small intestine that connects to the stomach.

kills bacteria and breaks down the connective tissue, while the pepsin digests proteins.

The mixture of food and fluids, called *chyme,* proceeds to the small intestine, which is the most important site of digestion of food into building block molecules and is a major site of absorption of these molecules into the circulatory system. The pancreas, liver, and gallbladder secrete digestive enzymes into the duodenum. The pancreas secretes bicarbonate, which neutralizes the hydrochloric acid, and digestive enzymes that break down all of the major components of foodstuffs. These enzymes include trypsin, chymotrypsin, and carboxypeptidase, which digest proteins; lipase, which digests fat; amylase, which digests carbohydrates; and ribonuclease and deoxyribonuclease, which digest nucleic acids. Bile is produced by the liver, stored in the gallbladder, and secreted into the small intestine, where it is important in the digestion

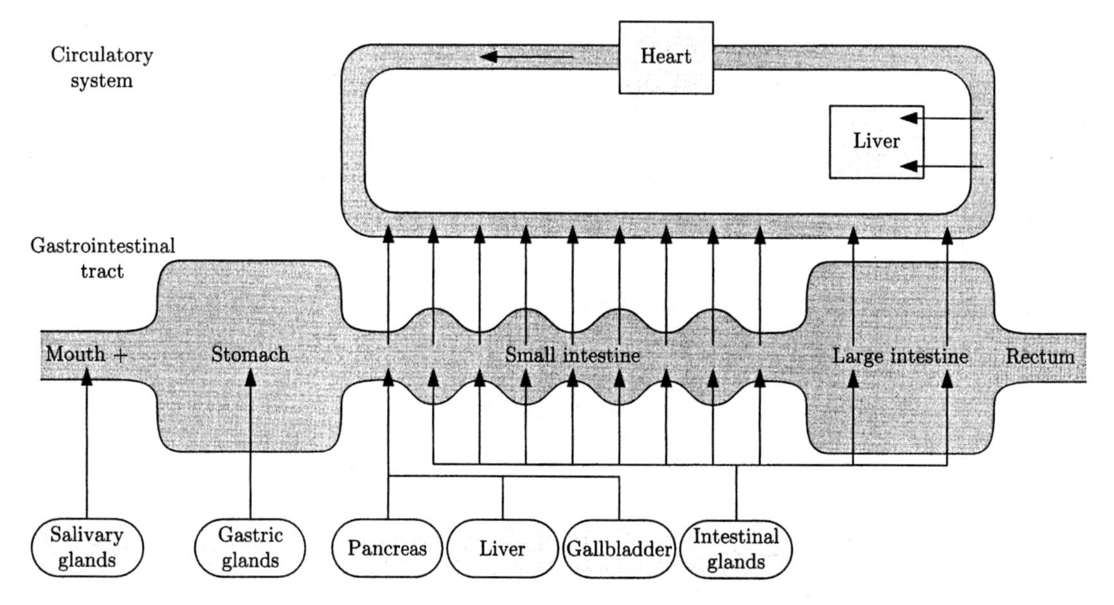

Figure 2.3 Summary of gastrointestinal activity (adapted from Vander et al., 1990, Figure 16-2).

of fats. The pancreatic amylase is most important in the breakdown of polysaccharides. In the small intestines, amylase hydrolyzes the α1,4 linkage (see Figure 1.10) to produce mixtures of glucose and maltose, plus larger polysaccharide fragments. In addition to these exocrine glands, the walls of the digestive tract contain secretory cells that secrete substances into the lumen of the digestive tract. The volume of all these secretions is appreciable (Vander et al., 1990). A typical human ingests 800 g of food and 1.2 L of water daily. To this is added 1.5 L of saliva, 2 L of gastric secretions, 0.5 L of bile, 1.5 L of pancreatic secretions, and 1.5 L of intestinal secretions. So about 2 L are ingested and another 7 L of digestive juices are added, but only a small fraction of this matter is excreted. Most of the digestive juices are themselves broken down and recycled.

2.3.3 Structure of the Small Intestine

In adult humans, the small intestine is a tube about 20 feet long and 1.5 inches in diameter extending from the pyloric valve of the stomach to the large intestine. It consists of three concatenated segments: the duodenum, which is attached to the dorsal abdominal wall, and the jejunum and ileum, which

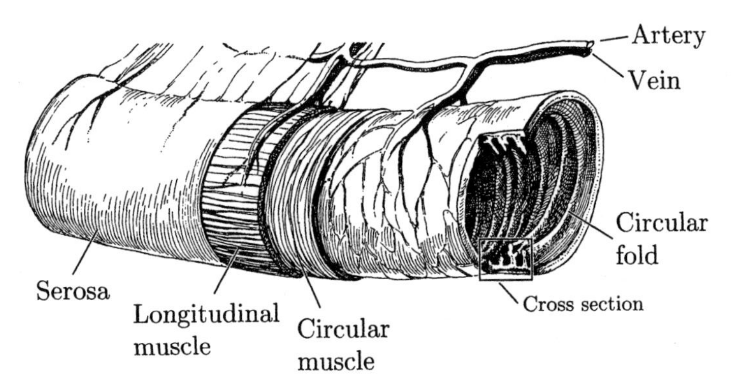

Figure 2.4 Diagram of a portion of the small intestine (adapted from Moog, 1981, page 156). Copyright © (1981) by Scientific American, Inc. All rights reserved. The diagram illustrates the concentric layers of tissue successively cut away to expose the lumen of the digestive tract.

are attached only loosely to the abdomen by a loose layer of connective tissue called the *mesentery*. The wall of the intestine consists of four layers of tissue (Figure 2.4)—the serosa, muscularis externa, submucosa, and mucosa—surrounding the lumen of the digestive tract. The serosa consists of layers of connective tissue that support the other tissue components and give the small intestine its structural integrity. The muscularis externa consists of a layer of longitudinal smooth muscle covering a layer of circular smooth muscle. Contractions of the longitudinal muscle shorten the intestinal tube, while contractions of the circular muscles constrict the intestinal tube. The involuntary smooth muscle cells in the intestine control the peristaltic movement of the intestine, which propels the digesting food down the gastrointestinal tract. The submucosa is a layer of connective tissue through which blood vessels and lymphatic vessels course to serve both the mucosa and the muscularis externa. Autonomic motor and sensory nerve fibers supply all the layers of tissue of the small intestine.

The mucosa is the innermost layer of tissue and is the site of the final stage of digestion and of transport in the small intestine. The mucosa contains circular folds 8 to 10 mm high that run circumferentially and project radially inward toward the lumen of the small intestine (Figure 2.5). The surface of the mucosal folds is covered with projections called *villi* that are 0.5 to 1.5 mm high and occur at a density of ten to forty per square mm. The villi have shapes that vary from fingerlike (Figure 2.6) to tonguelike in the different regions of the small intestine. The villi are separated by the *crypts of Lieberkuhn*. The villi give the intestinal inner surface a velvety appearance to the unaided eye.

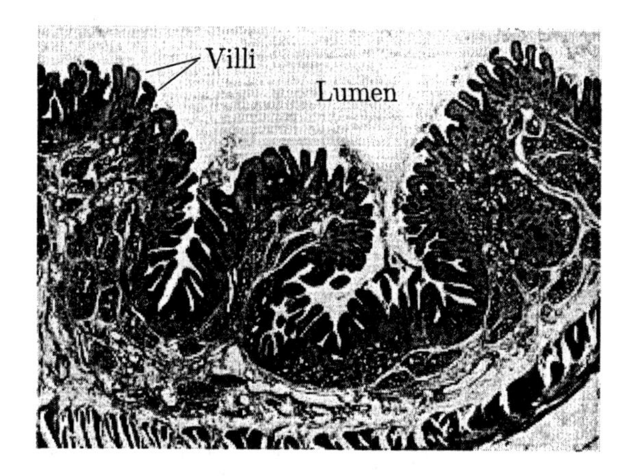

Figure 2.5 Photomicrograph of a section of a duodenum from a man who committed suicide by drinking formalin, thus preserving the structure of the intestine (adapted from Bloom and Fawcett, 1968, Figure 26-2). This cross section illustrates the circular folds covered by villi. The dark portion near the bottom is the muscularis externa.

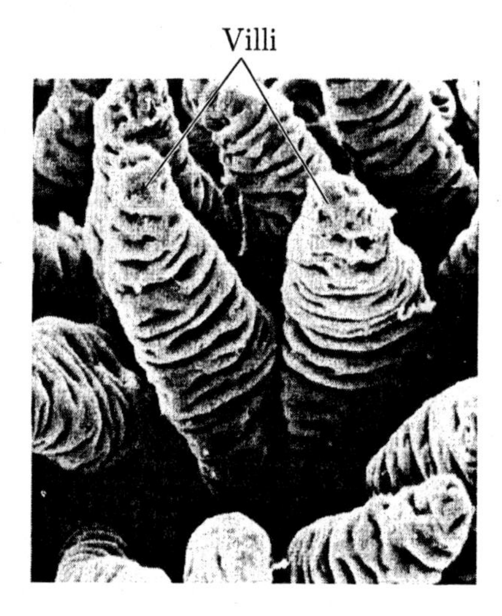

Figure 2.6 Scanning electron micrograph of the surface of the ileum of a monkey showing the villous surface (adapted from Weiss, 1983, Figure 19-25B). The wrinkling of the surface is caused by contraction of the smooth muscle in the villi.

Each villus is covered by a single layer of epithelial cells that surrounds a core of connective tissue that is traversed by blood and lymphatic vessels and contains smooth muscle cells (Figure 2.7). The epithelial cells consist of goblet cells, which secrete mucus, and absorptive epithelial cells, also called *enterocytes*, which are the sites of the transport of the end products of carbohydrate digestion. The blood vessels supply oxygen and other nutrients to the tissue and carry away both waste products and nutrients absorbed from the epithelium. Contraction of the smooth muscle cells results in motion of the villus. The epithelial cells that line the villus are formed by cell division in the crypts. They migrate up the villus, differentiating as they migrate, and reach the tip of the villus a few days later, when they are sloughed off into the lumen to

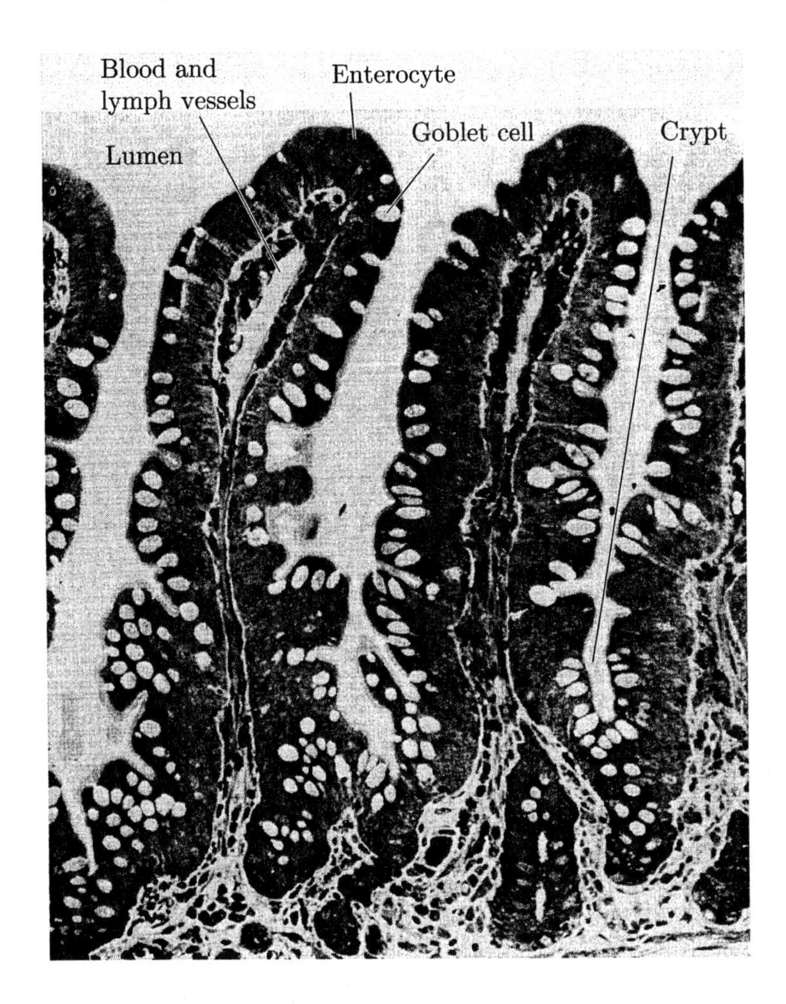

Figure 2.7 Photomicrograph of a section of the monkey jejunum showing four villi (adapted from Weiss, 1983, Figure 19-23). Both goblet cells, which are filled with mucus, and enterocytes are seen lining each villus. The core of each villus contains a large lymphatic vessel that ends blindly, plus blood capillaries.

be carried off with the lumen contents. It has been estimated that about 17 billion cells are discarded each day in the human small intestine.

Thus, the intestinal epithelium separates the lumen of the small intestine from the circulatory system. The part of the intestinal epithelial cell in contact with the lumen is called the *mucosal surface;* that in contact with the circulatory system is called the *serosal surface.*

2.3.4 Structure of Enterocytes

The lumenal surface of the intestinal epithelium, called the *brush border,* looks fuzzy under a light microscope. The structure of this fuzz as well as the fine structure of the rest of the epithelium is resolved in transmission electron

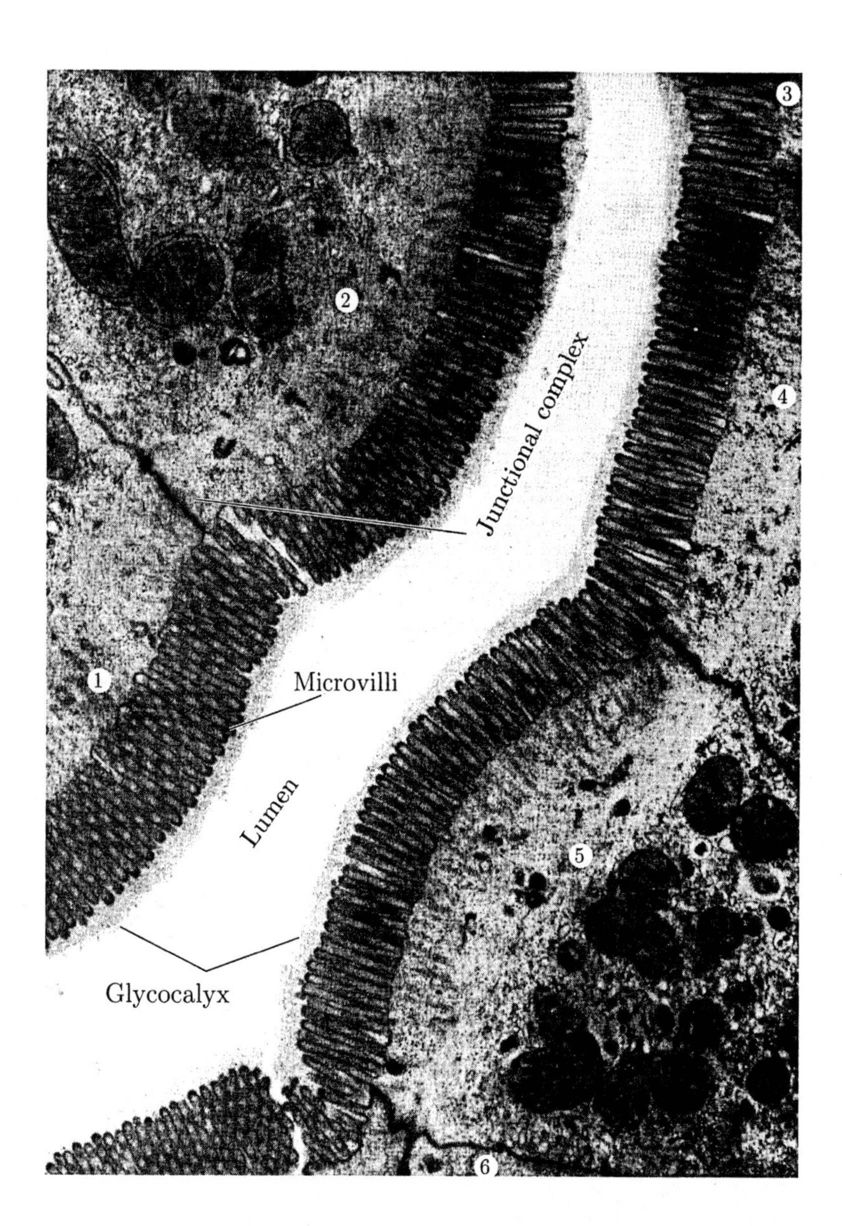

Figure 2.8 Low-power transmission electron micrograph of the wall of the intestine. The lumen of the intestine is shown lined by the microvilli of six (numbered) enterocytes (adapted from Fawcett, 1966, figure on page 347).

micrographs (Figure 2.8). The electron microscope reveals that the fuzz seen on the lumenal surface of the epithelium with the light microscope consists of the *microvilli* of enterocytes (Figure 2.9). Microvilli are cylindrical, membrane-enclosed protrusions from the lumenal surface of the enterocytes. Microvilli are about 1 μm long and 0.1 μm in diameter. The membrane of the enterocyte is continuous around the whole cell, including each of the microvilli. The

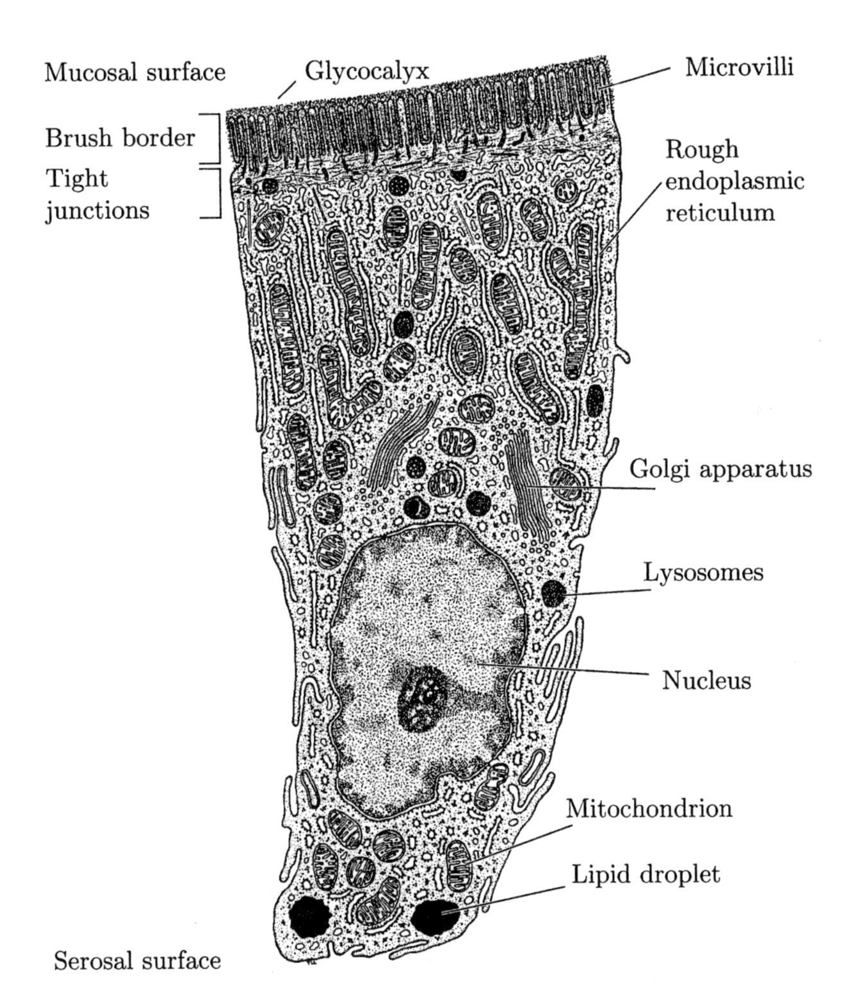

Figure 2.9 Drawing of ultrastructural features of an enterocyte (adapted from Lentz, 1971, Figure 76).

microvilli are one of three types of structural features of the intestinal wall that greatly extend the surface area of contact between the membranes of the enterocyte and the lumen. The three types of structures—circular folds, villi, and microvilli—are estimated to increase the surface area of the intestines by a factor of six hundred over that of a smooth cylinder of the same diameter.

The core of each microvillus is made up of a bundle of *actin* filaments. Actin is a structural protein, and these filamentary bundles endow the microvilli with their structural integrity. The actin filaments run down the core of each microvillus into a region of the cell called the *terminal web* that contains both actin and myosin filaments (Figure 2.10). The enterocyte is a columnar cell that is highly asymmetric. The microvilli are found on the lumenal side, also called the apical or mucosal side of the cell. The cell is rich in mitochondria, indicating the high rate of metabolic activity of enterocytes. The nucleus is located toward the opposite pole of the cell, also called the basal, basolateral or serosal portion of the cell.

Just below the brush border region of the cell begins a junctional complex that extends intermittently down the cell and links adjoining cells in all epithelia (Figures 2.8–2.10). This complex consists of three types of intercellular junctions that differ in morphology and in function. In order of occurrence from the mucosal to the serosal surface, the members of the junctional complex are *tight junctions, desmosomes,* and *gap junctions*. Desmosomes link cells in an epithelium mechanically and are important in intercellular adhesion. Gap junctions allow molecules of low molecular weight to pass between neighboring cells via channels that link the cytoplasms of adjacent cells. Of the three types of junctions, the tight junction is most important to our discussion of transport between body compartments.

In transmission electron micrographs, tight junctions are usually seen in cross section (i.e., a section perpendicular to the plane of the membrane) and appear as regions of fusion of the outer leaflets of the membranes of adjoining cells so that no appreciable intercellular space remains (Figure 2.11). These regions have been described quaintly as "kisses." Images obtained by the freeze-fracture technique give a view of these junctions (Figure 2.12) in the orthogonal direction (i.e., in the plane of the membrane). In freeze-fracture preparations these junctions appear as interweaving strands that cover the membrane surface of the cell. The presence of tight junctions greatly reduces the paracellular transport of matter (Figure 2.13), i.e., the transport that goes around the cells. Tight junctions appear in all epithelia and are the boundaries that separate body compartments. That is, tight junctions are the morphological sites of the seals between compartments. In the small intestine, the tight junctions separate the fluids on the mucosal surface from those on the serosal surface. The density and number of strands in tight junctions are correlated with the tightness of the seal of the epithelium; epithelia that have a small number of strands spaced far apart are leakier than those with more strands spaced closer together. The tight junctions divide the cell into two distinct regions: the apical and basolateral regions. The apical region is in contact with the mucosal surface, whereas the basolateral region is in contact with the serosal region. Thus, the two regions are in contact with extracellular media of quite different compositions.

2.3.5 Final Stage of Digestion

In low-power electron micrographs (Figure 2.8), the surface of the microvilli is seen to be covered by . . . more fuzz. This fuzz, called the *surface coat* or

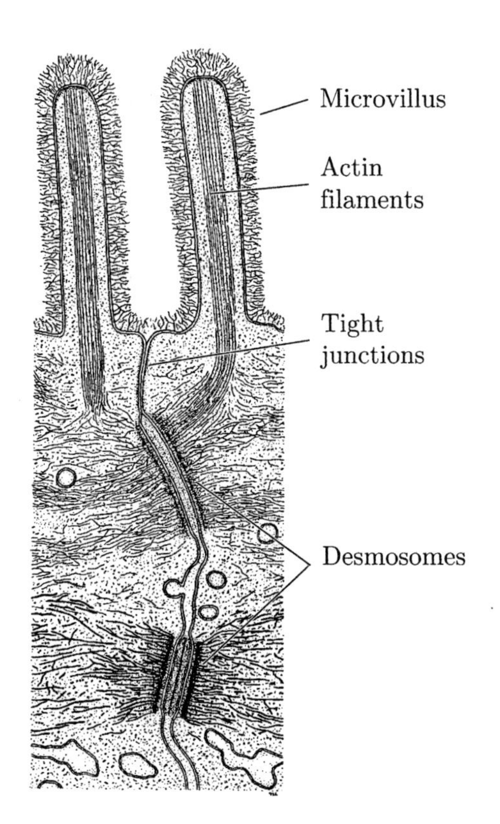

Figure 2.10 Schematic diagram of the apical regions of two adjacent enterocytes (adapted from Lentz, 1971, Figure 78). The glycocalyx is shown on the surface of each microvillus. Actin filaments in each microvillus terminate as rootlets in the terminal web region. Just below the microvilli is a tight junction that links the two enterocytes. Two desmosomes occur just below the tight junction.

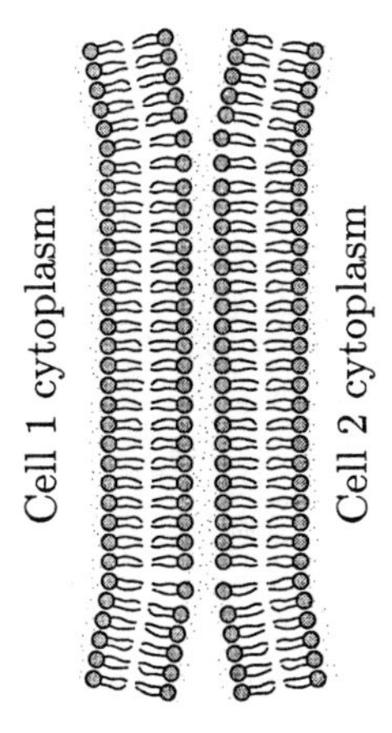

Figure 2.11 Schematic diagram of a cross section of a tight junction. At the tight junction, the outer leaflets of the plasma membrane are so close that they obliterate the intercellular space.

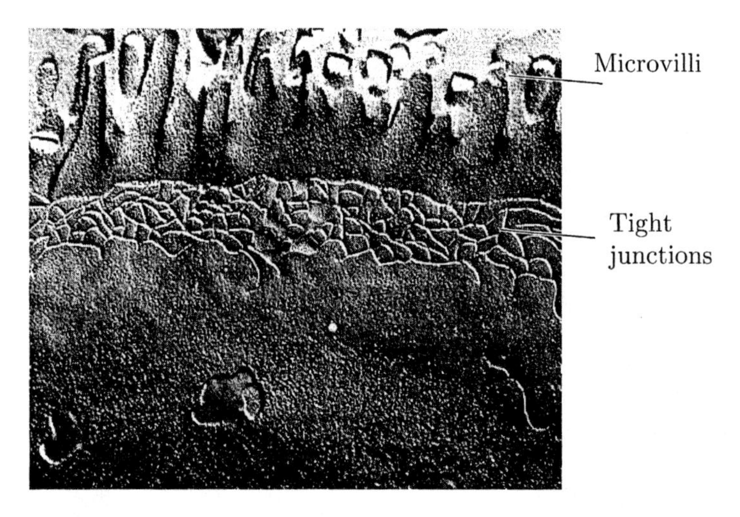

Figure 2.12 Appearance of tight junctions in a freeze-fracture preparation (adapted from Weiss, 1983, Figure 3-20b). In freeze-fracture preparations the tissue is frozen, subjected to a high vacuum, fractured with a microtome knife, replicated with a molecular layer of carbon, and shadowed with platinum, and the replica is examined with an electron microscope. Since the fracture planes are preferentially through the center of the lipid bilayer of the cell membrane, the electron micrographs reveal the plane of the membrane viewed either from the cytoplasmic or from the extracellular direction.

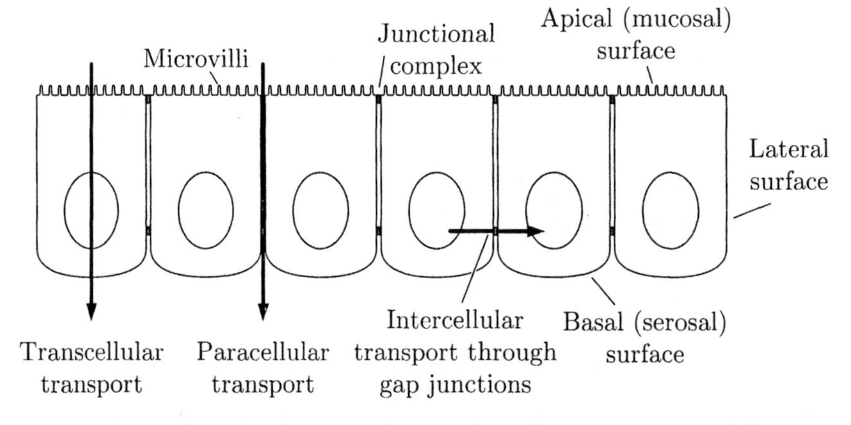

Figure 2.13 Schematic diagram of an epithelium illustrating the different transport pathways. The tight junctions reduce paracellular transport, which is transport from the mucosal side to the serosal side that does not go through the epithelial cells. The junctional complex at the apical surface of the epithelial cell consists of tight junctions and desmosomes. Gap junctions allow intercellular transport of small molecules between adjacent cells.

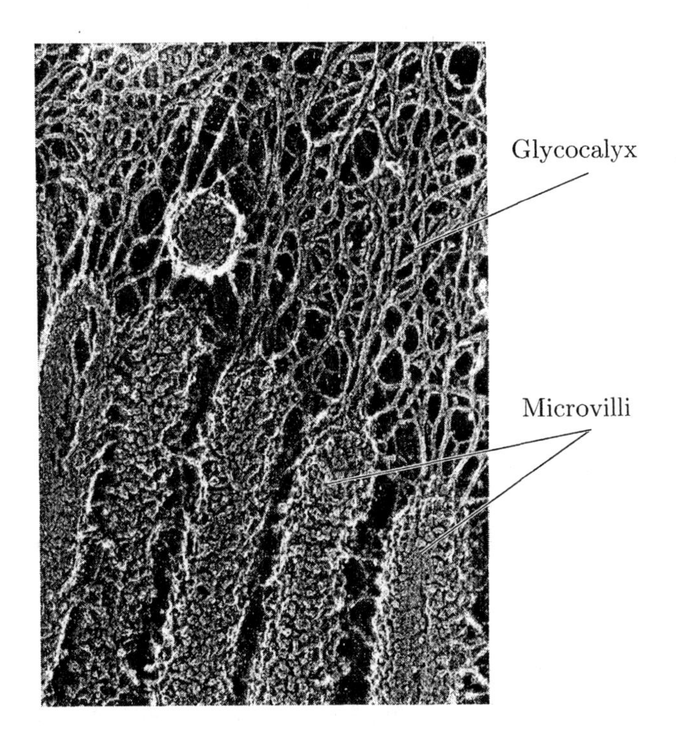

Figure 2.14 Electron micrograph of the surfaces of microvilli of an enterocyte prepared with the deep etching technique. The deep etching technique is similar to the freeze-fracture technique (Figure 2.12) except that after the frozen tissue has been fractured the surface ice is allowed to sublimate away. The glycocalyx is seen as a network of filaments attached to the microvilli (adapted from Hirokawa and Heuser, 1981, Figure 13).

glycocalyx, is much finer than that seen with the light microscope. In high-power electron micrographs, the glycocalyx can be seen to consist of filaments that are attached to the membrane that covers the microvilli (Figure 2.14). A number of important digestive enzymes are located in the glycocalyx region, including maltase, sucrase, lactase, aminopeptidase, etc. Maltase splits maltose into two glucose molecules, whereas sucrase splits sucrose into glucose plus fructose. These enzymes are members of a family of integral membrane proteins called *ectoenzymes.* Ectoenzymes contain a segment embedded in the membrane and a large extracellular segment that contains the catalytically active site. Ectoenzymes are the sites of the the final stage of digestion of the potato. Hence, for example, after carbohydrates have been digested into small fragments such as sucrose and maltose (see Figure 1.9), the ectoenzymes split these fragments into monosaccharides at sites strategically located near the surface of the microvilli.

2.3.6 Sugar Transport into and out of Enterocytes

The microvilli of enterocytes contain a special carrier-mediated transport system that transports monosaccharides in a highly stereospecific manner into the cytoplasm of the cell. This carrier mechanism is capable of concentrat-

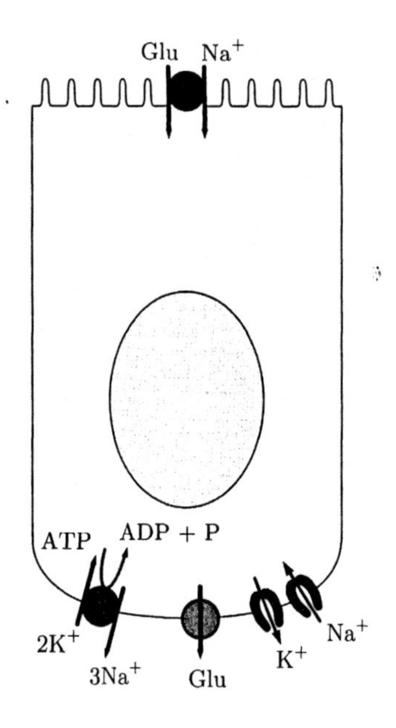

Figure 2.15 A schematic diagram of an enterocyte indicating the transport mechanisms that are involved in glucose transport from the mucosal surface to the serosal surface.

ing the monosaccharide in the cell. That is, it can transport glucose into the enterocyte even when the glucose concentration in the enterocyte exceeds that in the lumen of the digestive tract. This transport of glucose is linked to the transport of sodium; the carrier transports both glucose and sodium into the enterocyte. The energy to transport glucose against its concentration gradient comes from the potential energy stored in the difference of concentration of sodium. The concentration of cytoplasmic sodium is about an order of magnitude smaller than the concentration of extracellular sodium. Thus, the carrier transports sodium into the cell down its concentration gradient[2] and transports glucose with sodium. Therefore, this mechanism loads the cell with sodium as well as glucose. Additional transport mechanisms in enterocytes maintain the low intracellular sodium concentration and transport the glucose out on the serosal side (Figure 2.15).

The basolateral portion of the enterocyte contains another type of carrier mechanism that transports glucose outward through the membrane and into the serosal surface of the epithelium. This basolateral glucose carrier

2. In Chapter 7 we shall see that because there is a difference in electric potential across the membrane, it is the electrochemical potential difference for sodium, and not simply the difference in concentration of sodium, that is important in determining the sodium flux through the membrane.

(Chapter 6) differs from the sodium-linked glucose carrier resident in the microvilli. The basolateral glucose carrier transports glucose down its concentration gradient only, and the transport is not linked directly to the transport of any other solute. The basolateral membrane also contains a sodium-potassium pump that transports sodium out of and potassium into the enterocyte against both of their concentration gradients. This direction of transport requires energy that comes from the hydrolysis of ATP. Thus, the sodium-potassium pump operates through the presence of three ligands: sodium, potassium, and ATP (Chapter 7). Normally, for each molecule of ATP split to ADP and phosphate, three molecules of sodium are pumped out and two molecules of potassium are pumped in. There are also channels through which sodium and potassium leak down their concentration gradients. These transport mechanisms, which are only a few of those present in enterocytes, are most important in the transport of glucose. Glucose is brought into the cell on the mucosal side coupled to sodium and leaves the cell on the serosal side. The sodium accumulated in the cell by its influx on the serosal side and via other pathways is pumped out by the sodium-potassium pump. This pumping action maintains the low intracellular sodium concentration that powers the glucose flux. This discussion also indicates that the enterocytes are not just morphologically asymmetric along their apical to basal axis, but that their transport properties are also asymmetric.

2.3.7 Sugar Transport into and out of the Circulatory System

The glucose that is transported out of the enterocytes on the serosal side of the epithelium diffuses in the extracellular space, which is laced with blood capillaries. The walls of the capillaries are layered. The inner layer is formed from an endothelium one cell thick, which is surrounded by connective tissue elements. The capillaries of the intestinal villi contain fenestrated endothelial cells (Figure 2.16), which contain a large number of pores 500 to 800 Å in diameter. The glucose in the extracellular space diffuses through these pores into the capillaries, which empty into the portal vein, which in turn passes to the liver before connecting with the general circulation (Figure 2.3) via the inferior vena cava. Since the liver is a major site for glycogen storage, the effect of the routing of blood through the liver is that glucose is available to the liver before it is available to any other organ. The remaining glucose is then pumped by the heart through the rest of the circulatory system and leaks out of capillaries throughout the body, where it becomes available to cells.

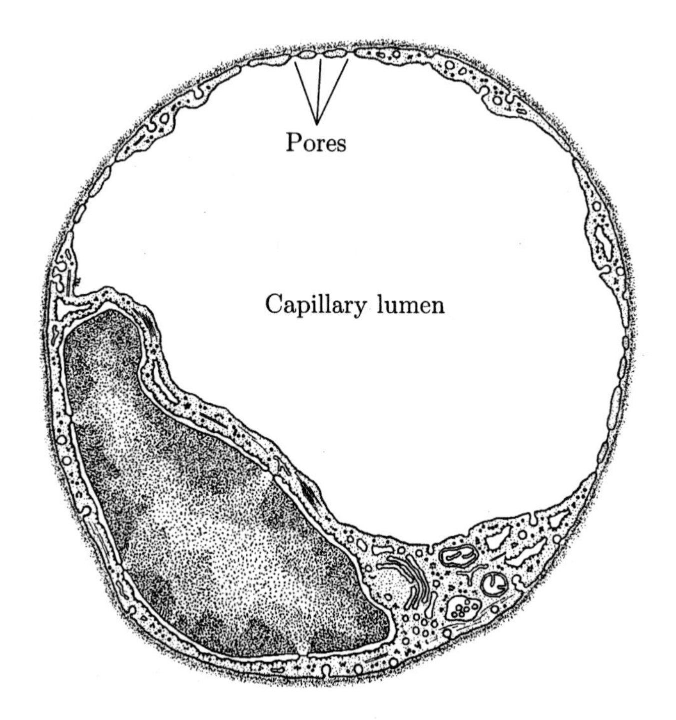

Figure 2.16 Cross section of a fenestrated capillary endothelial cell (adapted from Lentz, 1971, Figure 44).

2.3.8 Sugar Transport into Cells and Utilization of Sugars

All cells take up glucose from the circulatory systems via glucose transporters that are similar to those found in the *basolateral region* of enterocytes.[3] That is, glucose transport occurs in a direction that is down the glucose concentration gradient across the membrane. Glucose enters the cells through the glucose transporters and is metabolized in a complex sequence of reactions that can be divided into four steps as outlined in Figure 2.17. Glycolysis occurs in the cytosol and does not require oxygen. Glycolysis consists of ten reactions, each catalyzed by a distinct enzyme, by which the 6-carbon glucose molecule is broken down into two 3-carbon molecules of pyruvate. In addition, a net of two molecules of ADP are phosphorylated to ATP, and two molecules of nicotinamide adenine dinucleotide (NAD^+) are reduced (to form NADH). Glycolysis occurs in virtually all cells of all living organisms. Pyruvate is an important intermediary and the subsequent metabolism of pyruvate varies in different cell

3. As described in Chapter 6, there are several types of glucose transporters, and the different types are found in cells that serve distinct roles in the control, storage, and utilization of sugars.

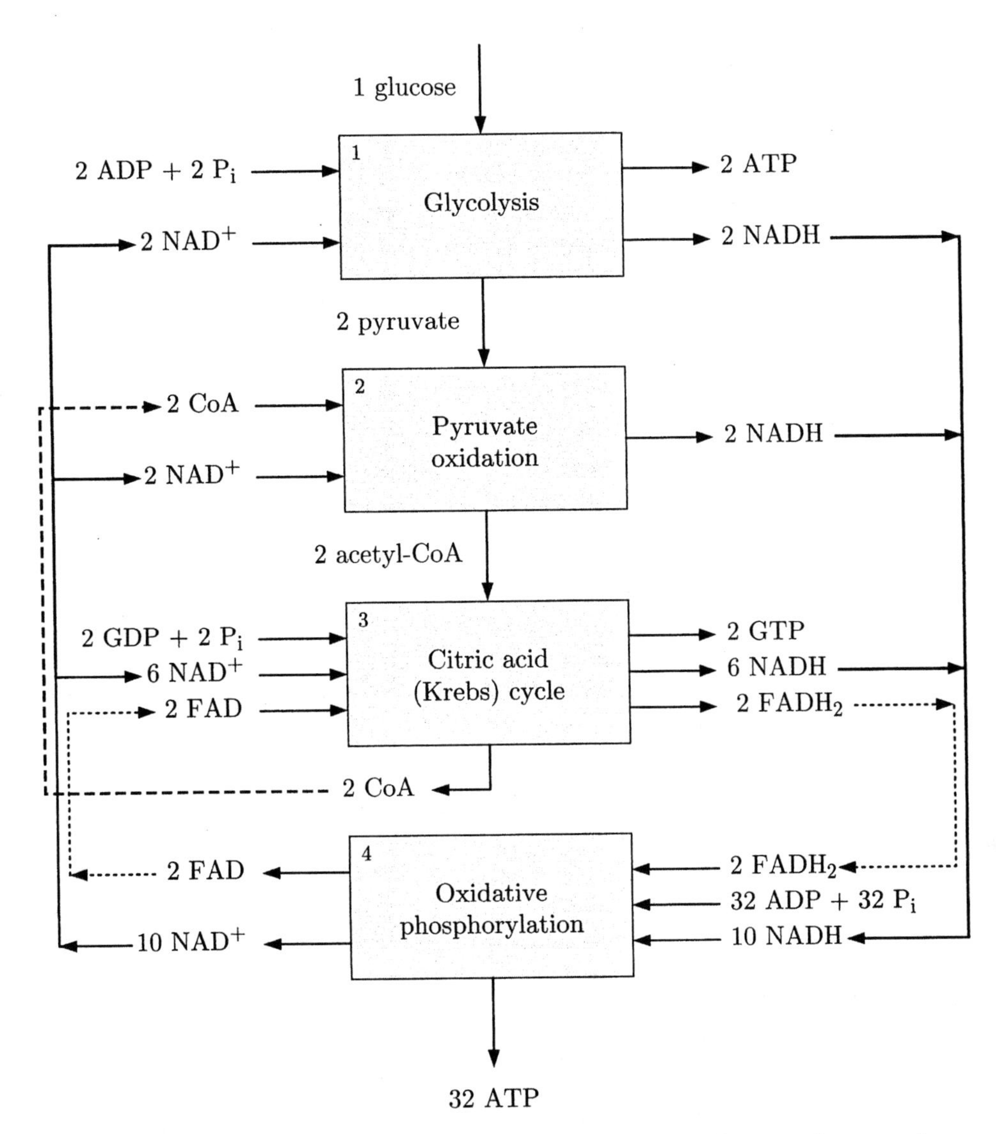

Figure 2.17 Schematic diagram of glucose metabolism. The diagram does not show hydrogen, carbon dioxide, or water in any of the reactions.

types. In many plant cells, pyruvate is metabolized to ethanol. In the absence of oxygen, many cells metabolize pyruvate to lactate.

The remaining stages of metabolism of glucose take place in the mitochondria. First, pyruvate is converted to acetyl-coenzymeA (acetyl-CoA) via the pyruvate dehydrogenase complex which consists of three enzymes plus coenzymes. Acetyl-CoA is also an important intermediary compound since it is involved in the metabolism of many other compounds including fatty acids, amino acids, etc. Acetyl-CoA is converted to CoA in the citric acid (Krebs) cycle which consists of eight enzymes. In both pyruvate oxidation and in the citric acid cycle, NAD^+ and FAD are reduced to NADH and $FADH_2$. These are then oxidized in the oxidative phosphorylation stage, in reactions that are linked to the phosphorylation of ADP to ATP.

In all these reactions, one molecule of glucose is metabolized to carbon dioxide and water,

$$C_6(H_2O)_6 + 6O_2 \rightleftharpoons 6CO_2 + 6H_2O.$$

Thus, the 6 carbons in glucose give rise to six molecules of carbon dioxide. Ideally, about 38% of the energy released in this reaction results in the phosphorylation of 36 molecules of ADP (GDP) to 36 molecules of ATP (GTP).

2.4 Cellular Transport Functions

The previous section has described the many roles of membrane transport in making available to cells the glucose molecules in the starch of a potato. But transmembrane transport is critical to many other cellular functions. We describe some of these briefly.

2.4.1 Maintenance of Intracellular Composition

The concentrations of solutes in cytoplasm differ appreciably from those in the extracellular space. These transmembrane differences in composition are both created and maintained by transport mechanisms resident in the membrane. Maintenance of these concentrations is critical for the control of a number of cellular processes. For example, the concentrations of Na^+, K^+, Cl^-, Ca^{++}, Mg^{++}, PO_4^-, and HCO_3^- are important for control of the electrical properties of cell membranes, especially in electrically excitable cells such as neurons and muscle cells. These ions are also involved in the electrochemical communication between cells. Ca^{++} and Mg^{++} are important in the control of muscle contraction, enzyme activation, and secretion. Phosphates are involved in the control of ATP production and in many intracellular enzymatic reactions. Monosaccharides, amino acids, fatty acids, and mononucleotides are the building block molecules used by cells for biosynthesis and energy.

2.4.2 Water Homeostasis

Sixty percent of the body weight is water, and two thirds of this water is found in cells. The relative constancy of body weight suggests that cellular water content, and hence of cellular volume, is controlled. Changes in extracellular and intracellular composition cause water transport across cellular membranes. The consequent change in cell volume results in a change in concentration of all intracellular solutes. Control of cell water is critical to the maintenance of cell vitality.

2.4.3 Secretion and Absorption

While all cells need to maintain the composition of their cytoplasms by transporting materials across their membranes, some cells, such as enterocytes, are specialized for secretion or absorption. For example, consider the parietal cells in the stomach. The walls of the stomach consist of many folds. The epithelium of the pits of these folds forms glands that secrete into the lumen of the stomach. The parietal cells (Figure 2.18) are part of the epithelium in these gastric glands. They transport HCl from the circulatory system (where the pH is about 7.4, which corresponds to a concentration of H^+ of about 0.0004 mmol/L) to the interior of the stomach (which has a pH as high as 0.8, which corresponds to a concentration of H^+ of about 150 mmol/L). Thus, the parietal cells transport HCl from a region of low HCL concentration to a region of high concentration, i.e., against a concentration ratio of about 10^6.

Interestingly, HCl is not transported directly, but rather the ions H^+ and Cl^- are transported separately. As with glucose transport in an enterocyte (Figure 2.15), the transport of HCl in a parietal cell results from the coordination of several different transport mechanisms, including a pump mechanism that transports hydrogen into the lumen of the stomach, a channel that transports chloride into the lumen of the stomach, and an anion exchanger that transports chloride in and bicarbonate out of the cell through the basolateral membrane.

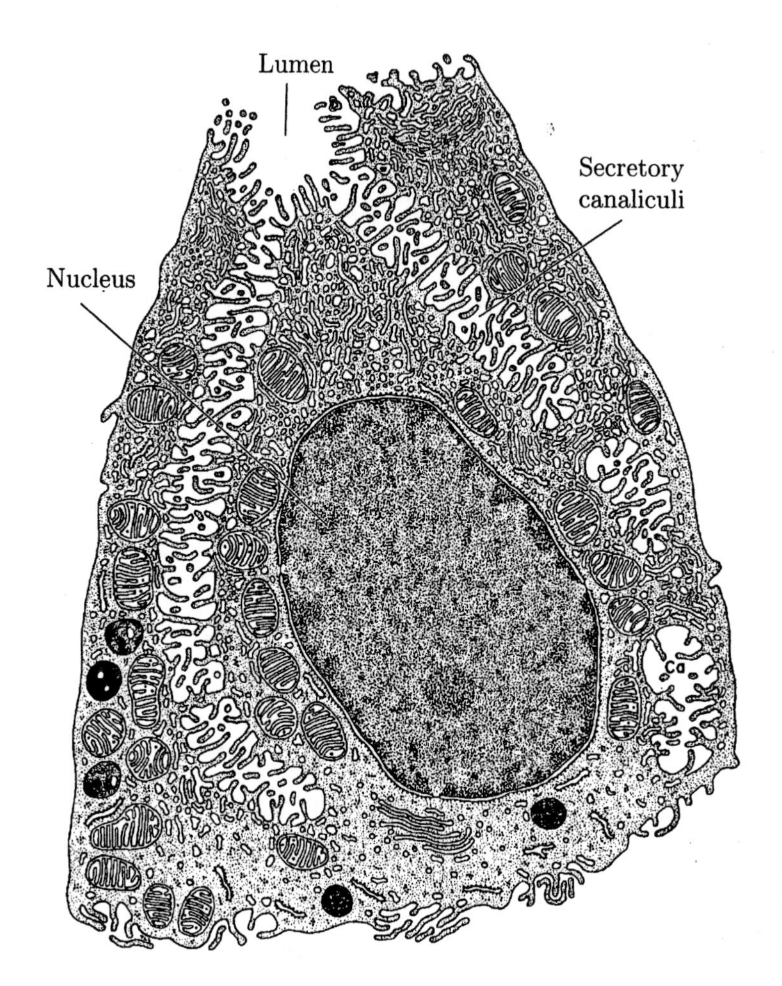

Figure 2.18 Drawing of ultrastructural features of a gastric parietal cell (adapted from Lentz, 1971, Figure 69). The cell has the usual cell organelles, but has a prominent set of intracellular canals, the *secretory canaliculi,* that communicate with the lumen of the gastric gland.

2.5 Survey of Transport Mechanisms

Chapters 3–7 are organized around the mechanisms that cells use to transport small molecules between the cytoplasm and the extracellular space. These mechanisms can be placed into the five categories shown schematically in Figure 2.19. The simplest mechanism is diffusion, which is the transport of solute resulting from a solute concentration gradient. As we shall discuss in Chapter 3, some solutes dissolve in and diffuse through the lipid bilayer portion of the membrane. Perhaps the most important molecule that diffuses through the lipid bilayer is water. Also important are molecules that are hydrophobic

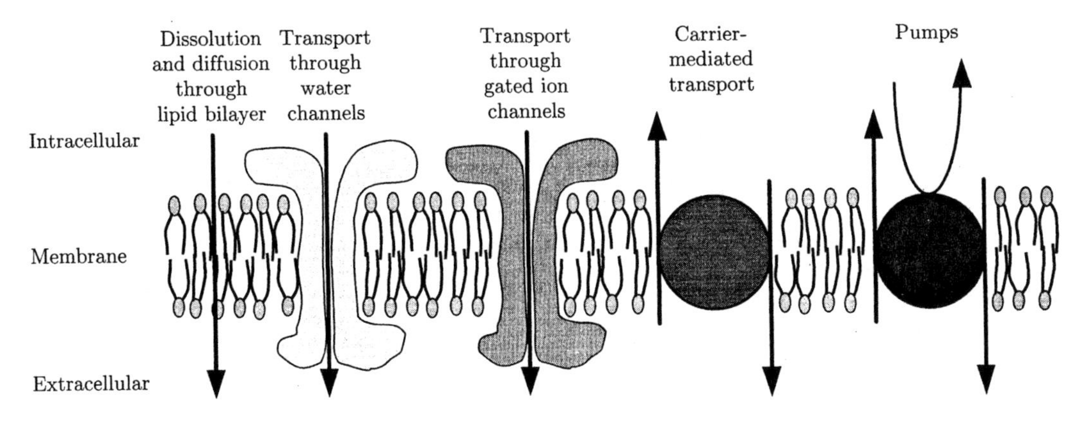

Figure 2.19 Schematic diagram of the mechanisms of transmembrane transport of small molecules.

or lipophilic, which include certain local anesthetics and the lipid-soluble hormones, the steroids, which act on intracellular targets.

Convective water transport is due to differences in hydraulic and osmotic pressure across a membrane (Chapter 4). Animal cells are highly deformable and do not support an appreciable difference in hydraulic pressure across the membrane, so water transport is due predominantly to the difference of osmotic pressure across the membrane. Osmosis is the *convective flow* of a *solvent* that results from a *solute concentration difference*. Convective water transport occurs through special channels that are highly selective for water.

When both solutes and water are transported simultaneously through a porous membrane, they may or may not interact, depending on whether they share the same conduit through the membrane (Chapter 5). For example, in transport through a porous membrane the transport of either solute or water can, in principle, drag the other through the membrane. The extent to which this occurs in water channels is presently unknown.

Carriers are membrane macromolecules that transport key building block molecules. In carrier-mediated transport (Chapter 6), the solute combines with the carrier at one face of the membrane, the complex translocates in the membrane, and the solute is released at the opposite face of the membrane. At each instant in time, the solute-binding site of a carrier is accessible to only one side of the membrane. There are a variety of such carriers. Some carriers transport a single type of solute, such as monosaccharides, amino acids, mononucleotides, phosphates, uric acid, choline, etc. Each type of carrier transports solutes in a highly structure-specific manner. For example, the glucose carrier transports D-glucose, but not its mirror image molecule, L-glucose. In addition, there are carriers that transport two solutes simultaneously (Chapter 8), either in the same direction through the membrane or in opposite directions through the membrane. Thus, the chemical energy stored in the difference of concentration of one solute can be used to transport the other solute in a direction opposite to its concentration gradient. An example of this type of mechanism is the sodium-coupled glucose carrier in the microvilli of enterocytes.

Ion channels are membrane macromolecules that transport ions through the membrane in a direction down the ionic electrochemical potential gradient (see Chapter 7 and Weiss, 1996, Chapter 6). There are many different types of ion channels that are characterized by the ion that permeates the channel and the manner in which the state of conduction of the channel is controlled. Gated ion channels are opened (or closed) by a gating variable. In contrast to carriers, solutes have simultaneous access to both sides of an open ion channel. Voltage-gated channels are found widely, but are especially prominent in electrically excitable cells. In voltage-gated channels, the potential across the cell membrane is the gating variable. For example, in voltage-gated sodium channels an increase in the membrane potential opens the channel transiently, and the open channel is preferentially permeable to sodium ions. There are many different types of voltage-gated ion channels even those that transport the same ion species. For example, there are many types of voltage-gated potassium channels. They differ in gating kinetics, in their permeation characteristics, and in their sensitivity to pharmacological agents. In addition, there are ligand-gated channels in which the state of the channel is dependent on the binding of a ligand to a receptor on the membrane at either an extracellular site or a cytoplasmic site. The ligand can be a small molecule such as calcium or ATP or a more complex molecule such as a neurotransmitter, neuromodulator, or hormone. Binding of the ligand may open or close the channel, depending on the characteristics of the channel macromolecule and the ligand. The permeation properties of open ligand-gated channels vary among the different channels. Some are highly permeable to only one type of ion; others are equally permeable to many small cations.

Ion pumps are variants of carriers in which ions are transported by a kinetic mechanism that is linked to the hydrolysis of ATP (Chapter 7). The energy liberated in the ATP reaction is used to transport ions up their concentration gradients. These pumps, including the ubiquitous sodium-potassium pump, are used to maintain the cytoplasmic concentrations of ions, which may be linked to the concentrations of other solutes.

While it is pedagogically helpful to examine these transport mechanisms in isolation (as is done in subsequent chapters), these mechanisms operate concurrently in cells. For example, a cell may transport sodium by several different mechanisms simultaneously, and these may be segregated spatially on the cell surface. Thus, to understand how cells can maintain their compositions, membrane potentials, and volumes, we must consider all transport mechanisms simultaneously (Chapter 8).

All the mechanisms described thus far are used by cells to transport relatively small solutes. Large molecules may be transported by entirely different mechanisms. In endocytosis, the membrane of a cell forms a pit and engulfs a portion of the extracellular medium. That medium may contain molecules of high molecular weight, viruses, or single-cell organisms. The engulfed material pinches off from the membrane to form a closed intracellular vesicle. In exocytosis, material is formed intracellularly in a membrane-enclosed vesicle that migrates to the cell membrane surface, combines with the membrane, and opens to the extracellular space to release its contents. An overview of these mechanisms can be found elsewhere (Alberts et al., 1983; Darnell et al., 1990).

2.6 Methods for Studying Membrane Transport

The experimental study of transport mechanism relies on a number of different methods and preparations. Here we shall describe some of these briefly.

2.6.1 Physicochemical Methods

The methods for measuring membrane transport fall into a small number of categories.

2.6.1.1 Chemical Methods

Chemical methods for measuring the quantity of solute transported are direct, but often tedious. Furthermore, these methods have relatively poor temporal resolution. The basic method is to load either the cell or the extracellular space with a solute and to collect the solution on either side of the membrane at regular time intervals and estimate the quantity of solute in the solution. The quantity of solute can be measured by flame photometry, in which the substance is burned and the color and intensity of the flame can be calibrated to yield the quantity of solute. Alternatively, the sample can be combined with a substance that reacts chemically with the solute to produce a detectable precipitate. The quantity of precipitate can be measured and used to indicate the quantity of solute present in the sample. Another method is activation analysis, in which the sample is exposed to an electron or proton beam. The energy distribution of the secondary emission emanating from the sample can be analyzed to indicate the quantity of any element in the sample. This method, which was initially developed to determine the element concentrations in liquid samples of body fluids, has been adapted to tissues. In this method, the tissue is freeze-dried and exposed to the beam, and the emissions are analyzed directly to obtain the distribution of elements in the tissue. The method can be used effectively in transport studies of intact organs by performing the analysis before and after exposing the tissue to a substance that affects transport, such as a hormone. However, it is difficult to calibrate the measurement of element concentration in tissue with this method.

2.6.1.2 Radioactive Tracers

Radioactive isotopes can be used in a manner that is similar to the use of chemical methods. First, a convenient (long half-life) isotope of the solute or of a constituent of the solute must be available. This requirement limits somewhat the class of solutes that can be studied with tracers. The radioactively labeled solute is collected, and the the amount of solute can be measured with a scintillation counter. The method needs to be calibrated, usually by counting a sample that is also analyzed by some other method. Radioactive tracer studies are also somewhat tedious and have a poor temporal resolution.

2.6.1.3 Optical Methods

Optical methods rely either on the changes in the optical properties of the solution caused by the transported substance or on tagging a substance whose transport is to be measured with a substance that is optically detectable. For example, the light transmitted through a sugar solution depends on the concentration of sugar. Thus, light passed through a suspension of cells transporting sugar into a solution can be analyzed to estimate the sugar concentration. As a second example, there are a variety of dyes that fluoresce in the presence of a particular substance. For example, aquorin fluoresces in the presence of calcium ions. The fluorescence can be calibrated to estimate the quantity of calcium present. As a third example, large molecules can be tagged with fluorescent tags, and the fluorescence can be traced in a system to reveal the transport of the fluorescent molecule.

Some of these methods are very indirect, and great care must be exercised in interpreting results. For example, changes in optical properties of the medium that are unrelated to the change in the concentration of the target solute can give misleading results. Furthermore, the quantifiable properties of the light, such as its intensity, may not be proportional to the concentration of the desired substance. For example, chemical products can quench fluorescence and lead to errant readings with fluorescent probes. On the other hand, the quenching of fluorescence by some solute can itself be used to measure solute concentration. For each method, careful calibration and control are required. Although more subject to artifacts, many of the optical methods have much better temporal resolution than either the chemical or the radioactive tracer methods. Some of the optical methods also have very fine spatial resolution and can reveal changes in concentration of substances within individual cell organelles.

2.6.1.4 Electrical Methods

Two types of electrical methods are widely used in transport studies. First, ion-selective electrodes are used to measure the concentration (activity) of an ion by means of an ion-selective material, e.g., a selective glass or resin. The ion-selective material is made part of a glass pipette that is filled with an ionic solution. When the pipette is placed into an ionic solution, a potential develops across the ion-selective material. The potential is approximately proportional to the logarithm of the ratio of target ion concentrations on the two sides of the ion-selective material. These electrodes can be made either large to measure extracellular ion concentrations or sufficiently small to penetrate single cells and to measure intracellular ion concentrations. Electrodes are available for a variety of ions, including H^+, K^+, Na^+, Ca^{++}, Mg^{++}, and Cl^-.

The different electrodes exhibit a variety of selectivities to the ions they are intended to measure. For example, a sodium electrode will exhibit a potential change in response to a change in potassium concentration. It is especially difficult to measure ion concentration in biological material, where a variety of organic substances may affect measurements of concentrations. Careful calibrations in a variety of solutions that mimic the biological situation are required to interpret measurements in tissue. Because of the large electrical capacitance of intracellular ion-selective electrodes, their temporal resolution is limited.

A second electrical method involves measuring the current carried by a specific ion. This highly successful method has very fine spatial and temporal resolution. The major difficulty is that many sources may contribute to the measurement of current, and these must be sorted out. Sorting can often be done by combining electrical measurements with more direct measurements using radioactive tracers or chemical methods, by changing concentrations of ions and determining the effect on the measured current, or by blocking various current pathways using pharmacological agents.

2.6.1.5 *Cell Volume Changes*

The passage of water across cellular membranes can be estimated by measuring the dimensions of single cells directly. Alternatively, the volume of a population of cells can be estimated by spinning the cells down in a centrifuge and estimating the volume of the cell suspension and that of supernatant fluid. Indirect measures of the cellular volume of a population of cells can also be obtained by optical methods, i.e. methods in which the opacity of a suspension of cells is a measure of the fraction of the suspension that contains cells. Since water is transported through cellular membranes more rapidly than most solutes and since a change in solute concentration is followed by an osmotic flow of water, measurements of cell volume can also be used to estimate the flow of solute. This method is indirect and relies on the validity of some theory to relate volume changes to solute concentration changes.

2.6.2 Preparations

The least direct measures of cellular transport mechanisms and technically the simplest to obtain are from whole tissues, such as whole muscles or kidneys, which may contain several different types of cells. It is usually difficult to discern membrane transport mechanisms from such experiments, although they can reveal the potential sensitivity of transport mechanisms to pharmacological variables in a crude manner. By successively dissecting parts of an organ, more homogeneous preparations of tissue can be obtained, as indicated in Figure 2.20. Somewhat more direct measurements of transport can be obtained from such tissues. For example, a tubule can be dissected from a nephron in the kidney and studied in isolation. This preparation is much more homogeneous than a whole kidney, but still contains a heterogeneous population of cells. Thus, results obtained from such preparations may be difficult to interpret in terms of cellular mechanisms, since measurements result from the superposition of transport from heterogeneous transport mechanisms.

Much more direct and easier to interpret than the preparations described above are preparations consisting of populations of one cell type. These can be obtained from tissues by enzymatic digestion and mechanical separation techniques. In some instances, such as in blood, no enzymatic digestion is necessary to obtain cells dissociated from tissues. By purifying these preparations, a subpopulation can be obtained that contains one cell type only. For

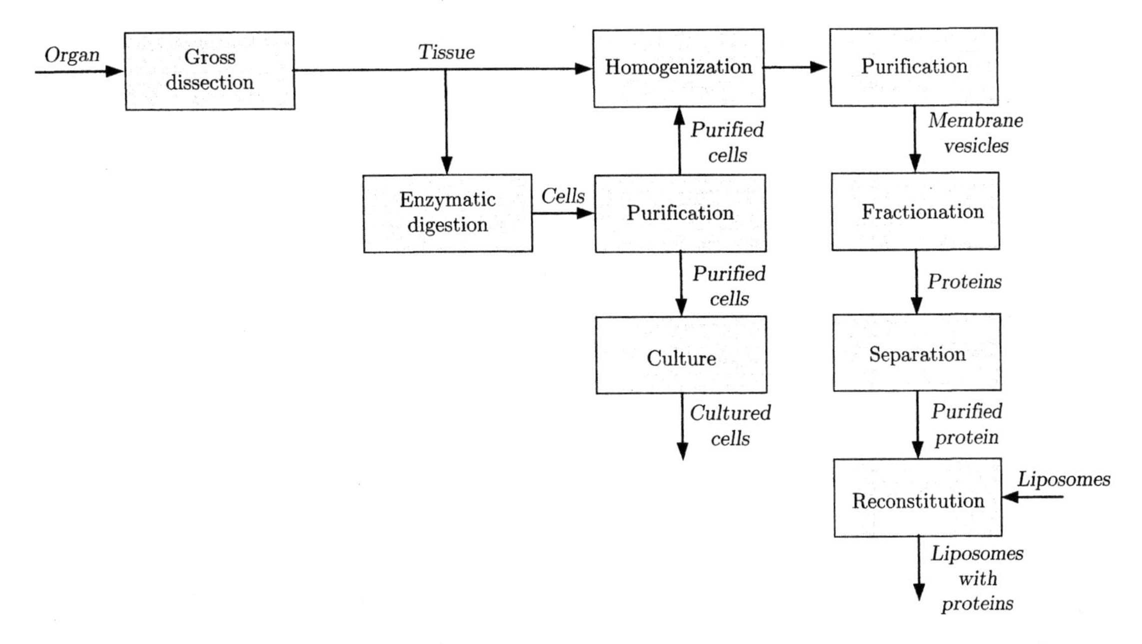

Figure 2.20 Illustration of methods for isolating cells or cell transport mechanisms to be used for studies of transport.

example, transport studies have been conducted on populations of erythrocytes. A population of such cells can be loaded with a particular molecule and the efflux of that molecule measured by one of the methods described above. Such preparations are more viable than single cell preparations, but the transport properties derived are averaged over the cell population. This procedure may hide important differences in the transport properties of different cells. Furthermore, it is often difficult to obtain a population of similar cell types without the contamination of other cell types; erythrocytes and other blood cells are the exception in this regard.

Measurement of transport from populations of cells is automated in flow cytometers, where the cells are mixed with some solution in a mixing chamber and then forced to flow at constant velocity down a tube and past a series of detectors (or collectors) spaced at known distances along the tube. Thus, each detector (or collector) measures the cell solution at a fixed time after the solution has left the mixing chamber. With this method, the time course of some detectable change can be measured rapidly. The detectors may be optical, chemical, or electrical, depending upon the method. A wide variety of cell properties have been studied with flow cytometers. Another method is to use a stop-flow apparatus in which the cells are mixed rapidly and then forced into a measurement chamber where sensors are used to measure transport.

Still more direct are preparations consisting of individual cells. Isolated cells tend to be more fragile, and only a few cell types that are fortuitously large or hardy lend themselves to studies of transport. Even these studies have some ambiguities that are common to all of the above methods as well. For example, transport may be affected by intracellular chemical processes that cannot easily be controlled by the experimenter. An example of this type occurs with glucose, which is metabolized in cells. Therefore, transport of glucose across the membrane of a cell, which depends upon the intracellular glucose concentration, will depend upon cellular metabolism, which consumes intracellular glucose. As with the study of transport by all methods described above, individual cells contain a number of transport mechanisms operating simultaneously. Thus, the study of a particular mechanism may be compromised by other mechanisms that may not yet be identified. For example, sodium is transported across cells by numerous mechanisms. Thus, the study of a particular sodium-transporting mechanism is difficult. Fortunately, pharmacological methods have been developed to block individual transport mechanisms. These can be used to eliminate unwanted transport mechanisms in particular cases.

In the 1970s, techniques were developed to remove a tiny (with dimensions on the order of 1 μm) patch of membrane from a variety of cells by applying suction to a micropipette that is in contact with the membrane of a cell. In this technique a patch of membrane seals the end of the pipette. Under control of the experimenter, the cytoplasmic and extracellular portions of the membrane face either the solution in the pipette or the bath solution in which the pipette is placed. With the patch recording method, transport can be studied in a controlled manner from a small area of membrane while both the intracellular and extracellular media and the electric potential across the membrane are under direct experimental control. Thus far, these studies have been restricted to the measurement of ion transport.

Although restricted to fortuitously large cells, methods have been developed to measure transport directly through an isolated membrane that has been freed of cytoplasmic constituents. Such preparations can give measurements that are most easily interpreted, but are often technically difficult. Among the most widely used animal cells for this purpose is the giant axon of the squid, a portion of a large nerve cell that is cylindrical (with a diameter on the order of 1 mm) and can be cannulated and perfused with various media. The cell can also be cut open and pinned out so that the membrane is isolated for transport studies. It is always essential to determine the effects of isolation of the membrane from the cytoplasm on the transport properties.

Methods have also been developed to study transport in highly purified membranes. Lipid bilayer membranes can be prepared from chemically synthesized lipids or from lipids extracted from cellular membranes. These lipids can be formed as membranes or as liposomes. Membrane transport proteins can be isolated from cells and incorporated into these lipid membranes, and transport properties of the reconstituted transport mechanism can be studied. Thus, it is possible to study transport in an artificial membrane whose lipid composition is known and which contains one type of transport protein only. With the methods of molecular biology, the transport protein can be manipulated to determine which portion of the protein has some key transport property.

2.7 Summary

The description of the digestion of a potato gives a glimpse of the processes involved in the digestion and distribution of nutrients and of the role of membrane transport in these processes. The potato, which consists primarily of starch, enters the digestive system and is broken down enzymatically into sugar fragments such as glucose. The enterocyte appears to have evolved to digest and absorb these sugar fragments. Through the circular folds of the wall of the intestine, the villous structure of the folds, and the microvillous structure of each enterocyte, a great profusion of enterocyte membrane occurs for the purpose of digestion and transport. This membrane contains tethered digestive enzymes on its surface, as well as transmembrane carrier molecules specialized for the intracellular transport of sugars.

The influx of sugars is linked to the influx of sodium. The presence of tight junctions between cells essentially minimizes intercellular transport between the lumen of the intestine and the circulatory system; matter is transported between these compartments primarily via cellular membranes. The basolateral membrane of the enterocyte contains specialized transport mechanisms to release the glucose on the serosal side of the epithelium. The glucose diffuses into the capillaries, is transported by the circulatory system throughout the body, and is released into the body tissues. Cells take up the glucose through specific sugar transport systems, some of which are under hormonal control. The glucose is used by cells for a variety of purposes. One important purpose is to serve as a substrate for glycolysis, in which process eventually one molecule of glucose can yield as many as thirty-six molecules of ATP (GTP). ATP is used as a source of energy that powers many cellular processes.

The transport of glucose into and out of enterocytes has been seen to depend upon the transport of other solutes, for example, sodium. Thus, understanding glucose transport in enterocytes requires understanding sodium transport in these cells, which itself occurs via more than one transport mechanism. In addition, since the transport of glucose is affected by the concentration of glucose, any factor that affects glucose concentration will affect glucose transport. The concentration of glucose depends upon the quantity of glucose and the quantity of water. But the transport of water is affected by the concentrations of all the solutes on the two sides of a membrane. Thus, it is clear that the transport of a solute across a cell membrane is linked, directly or indirectly, to the transport of other solutes. We will need to understand the transport mechanisms used by cells and the integration of these mechanisms in a cell in order to understand how cells function.

Exercises

2.1 Discuss the following statement: The main function of the digestive system is the elimination of the waste products of digestion from the body.

2.2 Estimate the surface area of membrane available for absorption of nutrients in the small intestine. Express your answer in units of tennis courts. A tennis court (for doubles) is 78 feet long and 36 feet wide.

2.3 Describe the differences between the villi and the microvilli.

2.4 Explain the importance of the sodium/potassium pump for the transport of glucose by enterocytes from the mucosal side to the serosal side of the small intestine.

2.5 Describe the structure and function of the following:

a. Enterocyte.

b. Glycocalyx.

c. Tight junction.

d. Lumen of the intestine.

e. Amylase.

2.6 Describe the role of the pancreas in the digestion of a potato.

2.7 Explain the role of tight junctions in epithelia.

2.8 In 100 words or less, describe the digestion of a potato focusing on the breakdown products of carbohydrate digestion. Do not explain digestion. Describe the structures—both the gross anatomical structures and the cellular structures—through which the breakdown products are transported.

2.9 It is desired to investigate the transport of an amino acid across the cellular membrane. The transport system is known to be the same in a species of bacteria (with a diameter of about 3 μm) and in giant nerve cells (with a diameter of about 50 μm) found in the nervous system of an invertebrate. Discuss the advantages and disadvantages of using each of these cell types for this study.

2.10 In 100 words or less, explain why it is important to understand membrane transport mechanisms in order to understand how carbohydrates are utilized in the body.

References

Books and Reviews

Alberts, B., Bray, D., Lewis, J., Raff, M., Roberts, K., and Watson, J. D. (1983). *Molecular Biology of the Cell*. Garland, Boston.

Beck, W. S., Liem, K. F., and Simpson, G. G. (1991). *Life: An Introduction to Biology*. HarperCollins, New York.

Bloom, W. and Fawcett, D. W. (1968). *A Textbook of Histology*. W. B. Saunders, Philadelphia.

Brachet, J. (1965). The living cell. In *Readings from Scientific American: The Living Cell*, 4–15. W. H. Freeman & Co., San Francisco.

Darnell, J., Lodish, H., and Baltimore, D. (1990). *Molecular Cell Biology*. Scientific American Books, New York.

Fawcett, D. W. (1966). *An Atlas of Fine Structure: The Cell*. W. B. Saunders, Philadelphia.

Friedman, M. H. (1986). *Principles and Models of Biological Transport*. Springer-Verlag, New York.

Fulton, J. F. and Wilson, L. G. (1966). *Selected Readings in the History of Physiology*. Charles C. Thomas, Springfield, IL.

Kenny, A. J. and Turner, A. J. (1987). *Mammalian Ectoenzymes*, vol. 14 of *Research Monographs in Cell and Tissue Physiology*. Elsevier, New York.

Kessel, R. G. and Shih, C. Y. (1974). *Scanning Electron Microscopy in Biology*. Springer-Verlag, New York.

Lehninger, A. L. (1970). *Biochemistry*. Worth, New York.
Lentz, T. L. (1971). *Cell Fine Structure*. W. B. Saunders, Philadelphia.
Moog, F. (1981). The lining of the small intestine. *Sci. Am.*, 245(5): 154–176.
Novikoff, A. B. and Holtzman, E. (1970). *Cells and Organelles*. Holt, Rinehart & Winston, New York.
Porter, K. R. and Bonneville, M. A. (1968). *Fine Structure of Cells and Tissues*. Lea and Febiger, Philadelphia.
Semenza, G. (1986). Anchoring and biosynthesis of stalked brush border membrane proteins: Glycosidases and peptidases of enterocytes and renal tubuli. *Ann. Rev. Cell Biol.*, 2:255–313.
Souci, S. W., Fachmann, W., and Kraut, H. (1986). *Food Composition and Nutrition Tables*. Wissenschafliche Verlagsgesellschaft mbH, Stuttgart, Germany.
Staehelin, L. A. and Hull, B. E. (1978). Junctions between living cells. *Sci. Am.*, 238(5): 140–152.
Stein, W. D. (1986). *Transport and Diffusion across Cell Membranes*. Academic Press, New York.
Vander, A. J., Sherman, J. H., and Luciano, D. S. (1990). *Human Physiology: The Mechanisms of Body Function*. McGraw-Hill, New York.
Watson, J. D., Hopkins, N. H., Roberts, J. W., Steitz, J. A., and Weiner, A. M. (1987). *Molecular Biology of the Gene*, vol. 1, *General Principles*. Benjamin-Cummings, Menlo Park, CA.
Weiss, L. (1983). *Histology*. Elsevier, New York.
Weiss, T. F. (1996). *Cellular Biophysics*, vol. 2, *Electrical Properties*. MIT Press, Cambridge, MA.
Weissmann, G. and Claiborne, R. (1975). *Cell Membranes, Biochemistry, Cell Biology and Pathology*. HP Publishing, New York.

Original Articles

Bernard, C. (1897–1898). *Leçons sur les Phénomènes de la Vie Communs aux Animaux et aux Végétaux*. J. B. Baillière et Fils, Paris, France. (Translation in Fulton and Wilson, 1966, p. 326
Hirokawa, N. and Heuser, J. E. (1981). Quick-freeze, deep-etch visualization of the cytoskeleton beneath surface differentiations of intestinal epithelial cells. *J. Cell Biol.*, 91:399–409.

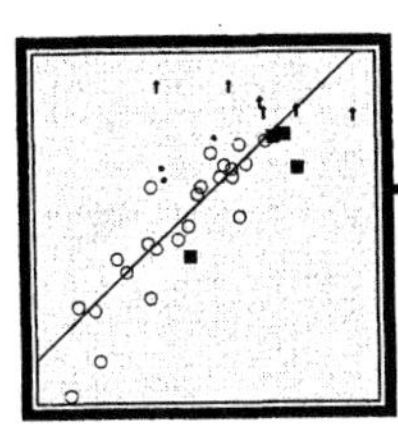

3 Diffusion

While examining the form of these particles immersed in water, I observed many of them very evidently in motion; their motion consisting not only of a change in place in the fluid, manifested by alterations in their relative positions, but also not unfrequently of a change of form in the particle itself; a contraction or curvature taking place repeatedly about the middle of one side, accompanied by a corresponding swelling or convexity on the opposite side of the particle.
—Brown, 1828

In this paper it will be shown that according to the molecular-kinetic theory of heat, bodies of microscopically-visible size suspended in a liquid will perform movements of such magnitude that they can be easily observed in a microscope, on account of the molecular motions of heat. It is possible that the movements to be discussed here are identical with the so-called "Brownian molecular motion"; however, the information available to me regarding the latter is so lacking in precision, that I can form no judgement in the matter.
—Einstein, 1905

3.1 Macroscopic Description

3.1.1 Background

A drop of water-soluble dye placed in a glass of water will spread out and its color will become less intense until the glass is filled with a solution of uniform color. The process by which a population of particles is transported from regions of high concentration to regions of low concentration so as to decrease the concentration gradient is called *diffusion*.

Diffusion phenomena must have been observed in antiquity, but they were not understood precisely until the nineteenth century (Cussler, 1984). The first systematic measurements were by Thomas Graham, a Scottish chemist, who studied diffusion of gases and liquids from 1828 to 1850. To study diffusion of substances in water, Graham filled two containers with different concentrations of acids or salts in water, allowed the contents of the containers to communicate for several days, and then analyzed their contents (Graham, 1850). Graham found that the diffusion of solutes in liquids was less than 10^{-3} as fast as that of the same substances in gases. Graham discovered that the quantity of solute transported decreased as a function of time. Most important, Graham found that the quantity of salt that diffused in a fixed time interval from a jar initially loaded with salt into an initially empty jar was proportional to the quantity originally present.

In 1855 Adolf Fick, a 25-year-old associate professor of physiology at the University of Zurich and a student of medicine and mathematical physics, proposed the laws of diffusion that now carry his name. Building on Graham's measurements, he proposed (Fick, 1855) these laws by analogy to the recently formulated laws of heat flow (Fourier's theory of heat conduction) and electrical conduction (Ohm's law):

A few years ago Graham published an extensive investigation on the diffusion of salts in water, in which he more especially compared the diffusibility of different salts. It appears to me a matter of regret, however, that in such an exceedingly valuable and extensive investigation, the development of a fundamental law, for the operation of diffusion in a single element of space, was neglected, and I have therefore endeavored to supply this omission.

It was quite natural to suppose, that this law for the diffusion of a salt in its solvent must be identical with that, according to which the diffusion of heat in a conducting body takes place; upon this law Fourier founded his celebrated theory of heat, and it is the same which Ohm applied with such extraordinary success, to the diffusion of electricity in a conductor. According to this law, the transfer of salt and water occurring in a unit of time, between two elements of space filled with differently concentrated solutions of the same salt, must be, cæteris paribus, *directly proportional to the difference of concentration, and inversely proportional to the distance of the elements from one another.*

3.1.2 Diffusion Variables

3.1.2.1 Concentration

Diffusion of particles is expressed in terms of the concentration and flux of particles. In one dimension, which is all we shall require for the most part, $c(x,t)$ is the concentration of particles at point x and at time t in units of mol/cm^3. To define the molar concentration of particles or the *molarity*, imagine an incremental volume element of volume $\Delta\mathcal{V}$ containing a number of particles Δn. The number of moles of particles in the volume element is $\Delta n / N_A$, and the number of moles per unit volume is $\Delta n/(N_A \Delta V)$. One mole of particles consists of $N_A = 6.022 \times 10^{23}$ particles, which is Avogadro's number. The concentration is defined as the limit of the number of moles per unit volume as the volume of the element approaches zero.

Because of the particulate nature of matter, there is a problem with this limiting operation. Namely, as the volume element is made arbitrarily small, much smaller than the size of the particle, the chance of finding a particle in the volume element also becomes arbitrarily small. However, as we shall see, the macroscopic laws of diffusion are valid for volume elements whose dimensions are small compared to the distance over which the concentration changes appreciably, but large enough to contain a large number of particles.

Units for the measurement of the concentration of a solute in a solution include the following:

$$\textit{Molarity} = \frac{\text{Moles of solute}}{\text{Liters of solution}} = c$$

$$\textit{Molality} = \frac{\text{Moles of solute}}{\text{Kilograms of solvent}} = m$$

$$\textit{Mole fraction} = \frac{\text{Moles of solute}}{\text{Moles of solution}}$$

$$\textit{Percent Mass} = \frac{\text{Kilograms of solute}}{\text{Kilograms of solution}} \times 100$$

The conversion from one unit of concentration to another usually requires knowledge of additional solution properties. We illustrate this by converting from molality, m, to molarity, c. The sum of the masses of solute and solvent in a kilogram of water is $m \cdot M + 10^3$ in grams, where M is the molecular weight of solute in g/mol. The volume of this solution in cm^3 is the mass divided by the density of the solution ρ in g/cm^3. Therefore, c, the number of moles of solute per liter of solution, is $c = \rho m/(10^{-3} m \cdot M + 1)$. Since the density of water is approximately 1 g/cm^3, the molarity and molality are numerically approximately equal at low solute concentration. However, differences between the two can be appreciable at high solute concentration.

3.1.2.2 Flux

The flux of particles $\phi(x,t)$ is the net number of moles of particles crossing per unit time at time t through a unit area perpendicular to the x-axis and located at x. The units of ϕ are in $mol/(cm^2 \cdot s)$.

3.1.3 Fick's First Law

Fick's first law states that the flux of particles in the positive x-direction, $\phi(x,t)$, is proportional to the spatial gradient of particle concentration, $c(x,t)$,

$$\phi(x,t) = -D\frac{\partial c(x,t)}{\partial x}, \tag{3.1}$$

where D is called the *diffusion coefficient* or *diffusivity*, and it is assumed that $D \geq 0$. With the units of flux and concentration specified above, the units of D equal $\left(\text{mol}/(\text{cm}^2 \cdot \text{s})\right) / \left(\text{mol}/\text{cm}^4\right) = \text{cm}^2/\text{s}$. Thus, the units of D are independent of the units used for the number of diffusing particles provided that consistent units are used for both the flux and the concentration. For example, both ϕ and c could be defined in terms of the number of particles rather than the number of moles of particles without changing the value of D. As we shall see later, the value of D does depend on the characteristics of the diffusing particles, as well as on the characteristics of the medium through which the particles diffuse.

Because $D \geq 0$, there is a strict relation between the sign of the concentration gradient and that of the flux, which is illustrated in Figure 3.1. A negative concentration gradient implies that the concentration is a decreasing function of x, and this leads to a flux of particles in the positive x-direction, i.e., a positive flux. More precisely, Fick's first law implies the following:

a. $\phi = 0$ if $\frac{\partial c}{\partial x} = 0$ and/or if $D = 0$;
b. $\phi \neq 0$ if $\frac{\partial c}{\partial x} \neq 0$ and $D \neq 0$; and
c. $\phi > 0$ if $\frac{\partial c}{\partial x} < 0$ and $\phi < 0$ if $\frac{\partial c}{\partial x} > 0$.

D governs the magnitude of the flux for a given concentration gradient. If $D = 0$, then the particles are fixed or indiffusible; they do not diffuse, no matter how large the concentration gradient. If D is large, the particles are highly diffusible.

The extension of Fick's first law from one dimension to three dimensions is

$$\boldsymbol{\phi} = -D\nabla c, \tag{3.2}$$

where $\boldsymbol{\phi}$ is the flux vector and ∇c is the spatial gradient of the concentration. In either one-dimensional or three-dimensional form, the analogies between the laws for diffusion, heat flow, electrical conduction, and fluid convection are apparent (Table 3.1).

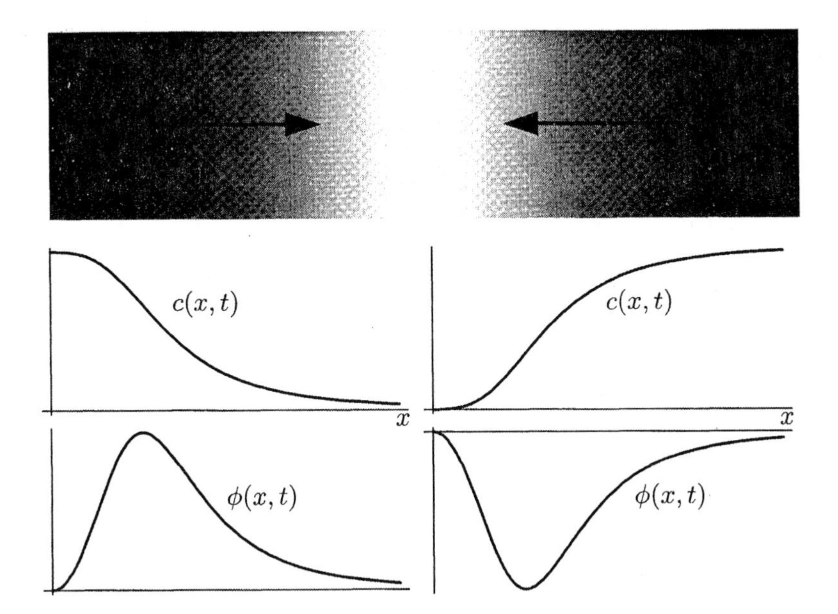

Figure 3.1 Two examples of the relation between flux and concentration gradient given by Fick's first law. The upper panels show schematic diagrams of concentration profiles; the gray scale is proportional to particle concentration, and each arrow indicates the direction of particle flux. The lower panels show the concentrations and the fluxes as functions of position at an instant in time t.

Table 3.1 Analogy between diffusion, heat flow, electrical conduction, and fluid convection. ϕ_H is heat flow, σ_H is the thermal conductivity, T is temperature, J is electric current density, σ_e is electrical conductivity, ψ is electric potential, Φ_V is fluid flow, p is hydraulic pressure, and κ is hydraulic permeability. The name of the person associated with the law for each process is given in parentheses. Vectors are indicated in boldface.

Flow process	One dimension	Three dimensions
Diffusion (Fick)	$\phi = -D\frac{\partial c}{\partial x}$	$\boldsymbol{\phi} = -D\nabla c$
Heat flow (Fourier)	$\phi_H = -\sigma_H\frac{\partial T}{\partial x}$	$\boldsymbol{\phi}_H = -\sigma_H\nabla T$
Electric conduction (Ohm)	$J = -\sigma_e\frac{\partial \psi}{\partial x}$	$\mathbf{J} = -\sigma_e\nabla\psi$
Convection (Darcy)	$\Phi_V = -\kappa\frac{\partial p}{\partial x}$	$\boldsymbol{\Phi}_V = -\kappa\nabla p$

In summary, Fick's first law states that there is a flux of diffusible particles from regions of high concentration to regions of low concentration and that the flux is largest where the concentration gradient is largest. Thus, the flux of particles is in a direction that reduces the particle concentration gradient.

3.1.4 The Continuity Equation

If particles are conserved, i.e., neither created nor destroyed at any point in space, then there is another relation between ϕ and c. Imagine a small rec-

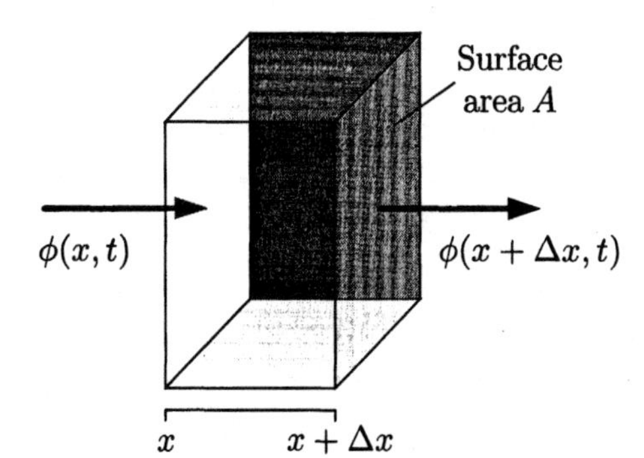

Figure 3.2 Conservation of particles in an incremental volume element of width Δx.

tangular volume element whose dimensions are small enough that the concentration does not vary appreciably with position in the element (Figure 3.2). The net influx of solute particles into the volume element in a time interval $(t, t + \Delta t)$ equals

$$(\phi(x,t) - \phi(x + \Delta x, t))\, A\Delta t, \tag{3.3}$$

where A is the surface area of the two faces of the volume element. The net increase in the number of particles in the volume element in this time interval can also be expressed in terms of the concentration as

$$(c(x + \Delta x/2, t + \Delta t) - c(x + \Delta x/2, t))\, A\Delta x, \tag{3.4}$$

where we assume that if the volume element's width is Δx, then the mean concentration in the volume is approximated by the concentration at the center of the volume element. If particles are conserved, the net influx must equal the increase in the number of particles in the volume element. Therefore, we can equate Equations 3.3 and 3.4 to obtain

$$\frac{\phi(x + \Delta x, t) - \phi(x,t)}{\Delta x} = -\frac{c(x + \Delta x/2, t + \Delta t) - c(x + \Delta x/2, t)}{\Delta t}. \tag{3.5}$$

If we take the limit of Equation 3.5 as $\Delta x \to 0$ and $\Delta t \to 0$, then we obtain

$$\frac{\partial \phi(x,t)}{\partial x} = -\frac{\partial c(x,t)}{\partial t}, \tag{3.6}$$

which is the equation for continuity or conservation of particles in one dimension. In three dimensions this takes the form

$$\nabla \cdot \boldsymbol{\phi} = -\frac{\partial c}{\partial t}. \tag{3.7}$$

The continuity equation links the rate of change of concentration of particles at a point in space with the divergence of the local flux.

The differential form of the continuity equation can be integrated over a closed volume $\mathcal{V}$ surrounded by the surface area A to yield the integral form of the continuity relation as follows:

$$\int_{\mathcal{V}} \nabla \cdot \boldsymbol{\phi} \, d\mathcal{V} = -\int_{\mathcal{V}} \frac{\partial c}{\partial t} \, d\mathcal{V}.$$

With the use of Gauss's integral theorem, the volume integral on the left-hand side can be expressed as a surface integral over the closed surface surrounding the volume

$$\oint_{A} \boldsymbol{\phi} \cdot d\mathbf{A} = -\int_{\mathcal{V}} \frac{\partial c}{\partial t} \, d\mathcal{V},$$

where $d\mathbf{A}$ is a vector whose magnitude equals the surface area of a differential element and whose direction is normal to the surface area element. Under the conditions that the flux is constant over the surface area and the concentration is constant over the volume, this equation can be expressed as

$$\phi = -\frac{\mathcal{V}}{A} \frac{dc}{dt}. \tag{3.8}$$

The integral form of the continuity equation states that the flux flowing out of a closed volume is proportional to the rate of decrease of concentration in the volume.

3.1.5 Fick's Second Law: The Diffusion Equation

Fick's first law (Equation 3.1) can be combined with the continuity equation (Equation 3.6) to eliminate $\phi(x, t)$ and thereby to derive an equation that $c(x, t)$ must satisfy. Differentiating Equation 3.1 with respect to x yields

$$\frac{\partial \phi}{\partial x} = -\frac{\partial}{\partial x}\left(D \frac{\partial c}{\partial x}\right),$$

from which

$$\frac{\partial \phi}{\partial x} = -\left(\frac{\partial D}{\partial x} \frac{\partial c}{\partial x} + D \frac{\partial^2 c}{\partial x^2}\right). \tag{3.9}$$

If the diffusion coefficient is independent of x, then $\partial D/\partial x = 0$, and if Equation 3.9 is substituted into Equation 3.6 then

$$\frac{\partial c}{\partial t} = D\frac{\partial^2 c}{\partial x^2}, \tag{3.10}$$

which is Fick's second law of diffusion. Equation 3.10 is also known as the one-dimensional diffusion equation or the heat equation. The three-dimensional diffusion equation is obtained by combining Equations 3.2 and 3.7 to obtain

$$\frac{\partial c}{\partial t} = D\nabla^2 c. \tag{3.11}$$

3.1.6 Diffusion with Convection and Chemical Reactions

Often diffusion of particles occurs in the presence of other physical processes. For example, the solute particles might be in a solution that is being transported, e.g., in response to a hydraulic pressure. If the particles are diffusing and being convected at a constant convection velocity v, then the solute flux, $\phi(x,t)$, is the sum of the diffusive flux, given by Fick's first law, and the convective flux, so that

$$\phi = -D\frac{\partial c}{\partial x} + vc. \tag{3.12}$$

Another mechanism for solute transport occurs if solute particles are subjected to a body force, e.g., due to gravity or to an electric field (for charged solutes). Motion of a solute caused by a body force is called *migration* or *drift*. The simplest description of the drift of particles in a medium is to assume that particles have a velocity proportional to the body force. If particles diffuse and drift due to a body force, then the flux can be expressed by an equation similar to Equation 3.12,

$$\phi = -D\frac{\partial c}{\partial x} + ucf, \tag{3.13}$$

where the drift velocity $v = uf$, u is the molar mechanical mobility, and f is the body force per mole of particles.

Another possibility is that the particles participate in a chemical reaction in the solution, e.g., by binding to some solution constituent.[1] In the simplest such possibility, if solute particles are binding to some constituent with a rate α, then the continuity equation is modified by the rate at which solute particles are being removed so that

1. For example, in semiconductors holes and electrons combine so that both "particles" are removed by recombination and neither is free to diffuse.

$$\frac{\partial \phi}{\partial x} = -\frac{\partial c}{\partial t} - \alpha c. \tag{3.14}$$

If both chemical binding and convection are present, the combination of Equations 3.12 and 3.14 yields a modified diffusion equation,

$$\frac{\partial c}{\partial t} = D\frac{\partial^2 c}{\partial x^2} - \nu\frac{\partial c}{\partial x} - \alpha c. \tag{3.15}$$

Thus, solute convection and a rate of chemical reaction between solute and solvent appear as additional terms in the diffusion equation. Of course, more complex reaction schemes will lead to more complex terms in Equation 3.15.

3.1.7 Postscript on Diffusion: The Second Law of Thermodynamics

Scientists and philosophers have been intrigued by Fick's laws since their formulation, because they embody the second law of thermodynamics in a manner that is more intuitive and less abstract than other formulations. The second law of thermodynamics states that the entropy of a closed system never decreases. The entropy of a system is a measure of its disorder. Highly disordered systems have greater entropy than do highly ordered systems. With this interpretation of entropy, the second law implies that closed systems spontaneously become increasingly disordered.

To see how Fick's laws tend to disorder an initially ordered distribution of particles, consider a vessel containing particles in solution with an initial concentration profile as shown in Figure 3.3. This concentration profile shows some initial order. By this we mean that the particles are not distributed uniformly in the vessel as would be the case for a maximally disordered distribution of particles. Nor does the initial distribution illustrated in Figure 3.3 exhibit the maximum order possible for which all the particles would be lo-

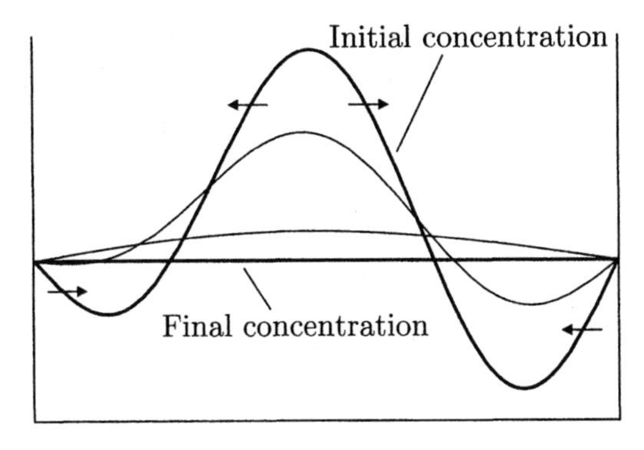

Figure 3.3 Illustration that diffusion tends to disorder an initially ordered spatial distribution of particles. The initial and final distributions of concentration versus position are shown along with two intermediate distributions. The arrows indicate the direction of the initial flux, which tends to fill in the valleys and decrease the peaks in the spatial distribution of particles.

cated at one position. Therefore, the initial concentration shown in Figure 3.3 shows some order, but not the maximum order that is possible. Recall that Fick's first law states that the flux of particles is down the particle concentration gradient, i.e., so as to decrease the existing concentration gradient. Thus, the flux will continue inexorably until the concentration gradients are eliminated and the system comes to diffusive equilibrium. At this point the concentration will have a uniform spatial distribution.

Diffusion processes in closed systems continue spontaneously until a completely disordered equilibrium state is reached. Furthermore, our experience tells us—and the second law states formally—that this process is irreversible. That is, a totally disordered system does not spontaneously order itself. If the particles are arranged uniformly in the vessel, we do not expect that at some later time all the particles will be found at one location. What is the microscopic physical mechanism that drives a set of diffusing particles inexorably, spontaneously, and irreversibly to a disordered equilibrium state? The answer is found by examining the microscopic basis for diffusion.

3.2 Microscopic Model

3.2.1 Introduction

The botanist Robert Brown (1828) observed with a simple microscope that small particles inside pollen grains exhibited erratic movements, subsequently called *Brownian motion*. Although he originally believed that this motion was somehow related to the vitality of the pollen grains, further observations indicated that small, inanimate particles displayed a similar motion. Brown convinced himself that this motion did not result from extraneous sources of convection such as thermal gradients. However, the physical basis of this Brownian motion eluded Brown. This basis was provided by Einstein (Einstein, 1905, 1906), who computed the expected magnitude of the motion of small solute particles in a liquid caused by thermal agitation and showed that this motion should be observable.[2] Quantitative experimental validation of Einstein's the-

2. Einstein was careful to indicate (see quote at the beginning of this chapter) that Brown's observations were not sufficiently quantitative to determine whether they could be explained by the newly developed theory of Brownian motion. His caution may have been appropriate. There has been some controversy about whether Brown observed "Brownian motion" (Cadée, 1991; Deutsch, 1991; Ford, 1991; Deutsch, 1992; Ford, 1992a, 1992b).

ory of Brownian motion was provided by Perrin (Perrin, 1909), who measured the motion and the equilibrium spatial distribution of mastic particles with a microscope (see Problem 3.19). This study reinforced the developing notion of the molecular nature of matter.

We now know that all matter is composed of molecules that are in continuous motion resulting from thermal energy. Thus, a particular particle in a medium (gas, liquid, or solid) is subject to a large number of collisions with molecules of the medium in any macroscopic observation interval. Brownian motion of the particle results from these collisions. The mean velocity of the particle can be shown to be directly proportional to the absolute temperature and inversely proportional to the mass of the particle. For example, oxygen molecules in air at 20°C move with a mean velocity between collisions of about 500 m/s. The mean free path, or mean distance between collisions, is about 0.1 μm. The mean free time, or average time between collisions, is about 0.2 ns. Hence, on a spatial scale of the mean free path and on a temporal scale of the mean free time, the particle trajectory is erratic and the particle is scattered in space. The details of the collisions, and hence of the erratic motion of the particle, depend upon the physical characteristics of the medium. The end result of these collisions—the erratic particle motion—is best described in probabilistic terms. The simplest probabilistic model that links the Brownian motion of particles to the macroscopic laws of diffusion is the one-dimensional random walk model. This model also reveals the microscopic basis of the irreversible nature of diffusion processes.

In a one-dimensional random walk, we assume a particle moves every τ seconds according to a simple probabilistic law. At each time increment, the particle moves a distance along the x-axis of either $+l$ with probability 1/2 or $-l$ with probability 1/2. Each time the particle is scheduled to move, we can imagine flipping a fair coin to decide whether the particle should move to the left or to the right. Such a process is called an *unbiased Bernoulli trial.* An unbiased one-dimensional random walk consists of a sequence of statistically identical and independent unbiased Bernoulli trials.

3.2.2 The Microscopic Basis for Fick's First Law

Fick's first law follows from this simple specification of a one-dimensional random walk (Einstein, 1908). To see this, consider a volume element whose linear dimension equals $2l$ and whose cross-sectional area is A (Figure 3.4). We shall derive an expression for the net number of particles that cross the plane (of area A) located at x in an interval of time τ. The derivation is greatly simplified by the particular choice of box dimensions and of the time

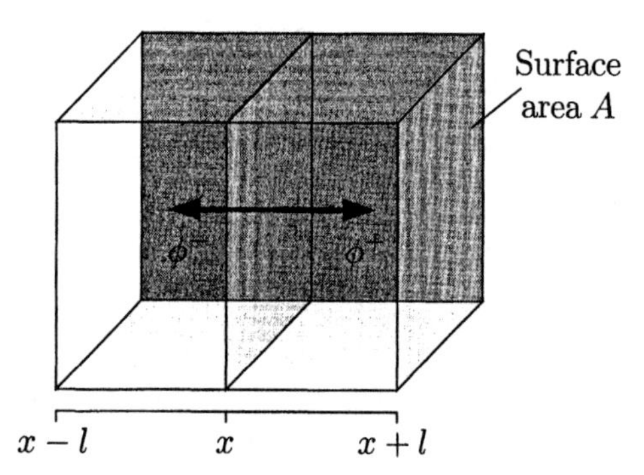

Figure 3.4 Volume elements used to prove Fick's first law for a population of particles undergoing random walks. A plane located at x separates two volume elements whose widths in the x-direction equal the step size of the random walk, l. ϕ^+ and ϕ^- are the fluxes of particles crossing the plane from left to right and from right to left, respectively.

scale. With these choices, the number of particles crossing the plane in the time increment τ depends only on the number of particles in the box at the beginning of the time interval and not on the particles outside the box. In the interval τ, 1/2 the particles in the left volume cross to the right, and 1/2 the particles in the right volume cross to the left. Let ϕ^+ equal the flux density of particles crossing from left to right and ϕ^- equal the flux density of particles from right to left. Then

$$\phi^+ = \frac{\frac{1}{2}c(x - l/2, t)Al}{A\tau}, \quad \text{and} \quad \phi^- = \frac{\frac{1}{2}c(x + l/2, t)Al}{A\tau}.$$

Hence, the net flux is

$$\phi = \phi^+ - \phi^- = \frac{l}{2\tau}\left(c(x - l/2, t) - c(x + l/2, t)\right). \tag{3.16}$$

Since for any macroscopic distance scale of interest $l \ll x$, we can expand the two expressions for c in Equation 3.16 in a Taylor's series and retain only the two leading terms to yield

$$\phi \approx \frac{l}{2\tau}\left(\left(c(x,t) - \frac{l}{2}\frac{\partial c(x,t)}{\partial x}\right) - \left(c(x,t) + \frac{l}{2}\frac{\partial c(x,t)}{\partial x}\right)\right), \tag{3.17}$$

which gives

$$\phi = -\frac{l^2}{2\tau}\frac{\partial c(x,t)}{\partial x}, \tag{3.18}$$

which is equivalent to Fick's first law (Equation 3.1). Comparison of Equation 3.1 with Equation 3.18 shows that $D = l^2/(2\tau)$, which relates the macroscopic diffusion coefficient with the microscopic quantities that define the random walk. This derivation shows, on a microscopic basis, why there can be a net flux of particles even though each particle has no preferred direction of motion. The net flux of particles results simply from the difference in the numbers of particles on the two sides of the partition.

3.2.3 The Microscopic Space-Time Evolution of Particle Location: The Binomial Distribution

Suppose a particle executes a one-dimensional random walk starting at $x = 0$ at $t = 0$. Therefore, it is located initially at $x = 0$ with probability 1. Since the location at any later time is determined probabilistically, we cannot say exactly where the particle is at some later time, but we can compute the probability of its location at any point in space at any time. We call this the *space-time evolution of particle location.* In this section, we shall compute this location at microscopic time and distance scales, i.e., those that have resolutions of τ and l, respectively. In the next section, we shall compute the space-time evolution with macroscopic resolutions.

A particular sample of a one-dimensional random walk might have the appearance shown in Figure 3.5. Every τ seconds the particle moves $+l$ or $-l$ with equal probability. How far does the particle get in time t? Let us define $W(m,n)$ as the probability that the particle is at position ml at time $n\tau$ assuming it started at $x = 0$ at $t = 0$, where m and n are integers. $W(m,n)$ is the probability that the particle has moved m units of distance in n units of time. In one unit of time, the particle moves $+1$ or -1 units of distance with equal probability. In two units of time, the particle moves $+2$ or -2 units of distance, each with probability 1/4, and moves a net distance of 0 with probability 1/2. This sequence can be displayed in a diagram (Figure 3.6). At $n = 0$, the particle starts at the top of the triangle. At $n = 1$, it moves $+1$ or -1 units of m with equal probability. If it is at $m = -1$ at $n = 1$, then it moves to either $m = -2$ or $m = 0$ at $n = 2$ with probability 1/2. If the particle is at

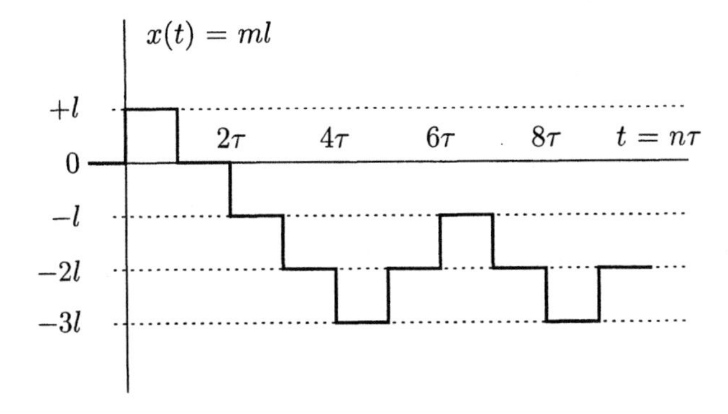

Figure 3.5 An example of displacement as a function of time for a particle undergoing a one-dimensional random walk with a step size l and a step interval τ.

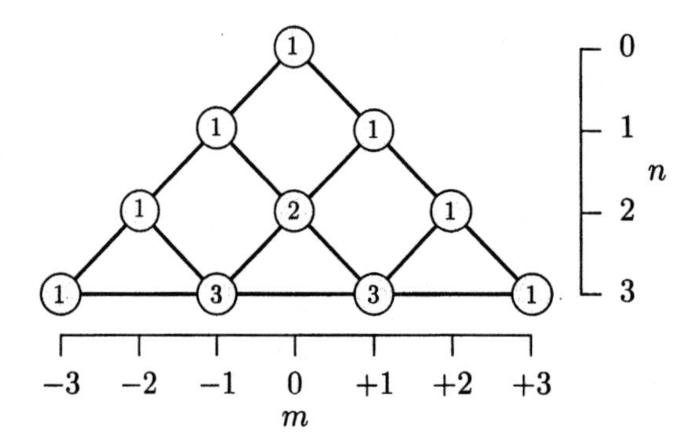

Figure 3.6 Pascal's triangle illustrates the different paths that are possible for a particle undergoing a random walk. The particle starts at the apex of the triangle ($m = 0$ at $n = 0$) and moves either one step to the left or one step to the right at each interval of time. The number in each circle is the number of different paths that are possible for the particle to reach that position; for example, there are three paths that result in a net displacement of $m = +1$ at $n = 3$. Note that the entry at a particular location at time n is obtained by adding the entries of all those locations at time $n - 1$ that converge upon that particular location.

$m = +1$ at $n = 1$, then it moves to $m = 0$ or $m = +2$ at $n = 2$, etc. This diagram allows the visualization of all possible random walks.

To evaluate $W(m, n)$, note that every sequence consisting of n steps has a probability of occurrence of $(1/2)^n$. Therefore, we need only determine the number of such sequences whose net displacement is m units. Suppose n_+ is the number of steps of length +1, and n_- is the number of steps of length −1. Then $n_+ + n_- = n$, and $n_+ - n_- = m$. Therefore, $n_+ = (n + m)/2$, and $n_- = (n - m)/2$. The number of different sequences of n steps for which $(n + m)/2$ is +1 and $(n - m)/2$ is −1 is given by the binomial coefficients. Therefore, the number of different sequences of net length m after n steps is

$$\frac{n!}{((n + m)/2)!\,((n - m)/2)!},$$

and each has probability of occurrence $(1/2)^n$. Hence,

$$W(m, n) = \frac{n!}{((n + m)/2)!\,((n - m)/2)!}\left(\frac{1}{2}\right)^n. \tag{3.19}$$

$W(m, n)$ is called the *binomial distribution*. Note that since n_+ and n_- are integers, $W(m, n)$ equals zero for noninteger values of $(n + m)/2$ and $(n - m)/2$. This result implies that when n is an even integer, m is an even integer and that when n is an odd integer, m is an odd integer. Equation 3.19 implies that

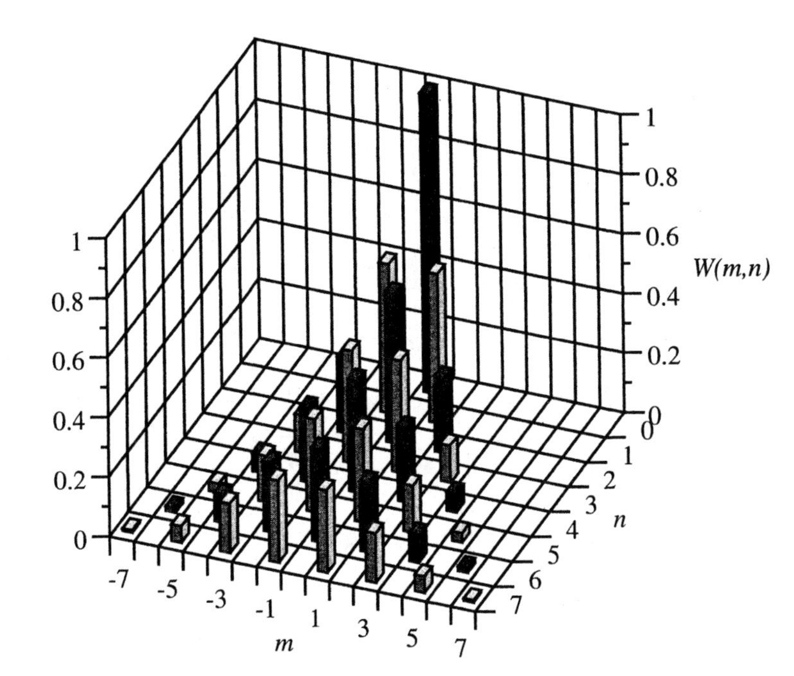

Figure 3.7 The binomial distribution, $W(m,n)$, is shown as a function of m and n for $0 \leq n \leq 7$.

$$W(m,n) = W(-m,n),$$

indicating that after n steps it is equally likely that the particle is found at position ml as at $-ml$.

We can get some insight into the random walk by plotting $W(m,n)$ as shown in Figure 3.7. As n increases, $W(m,n)$ gets broader, i.e., the probability that the particle wanders farther and farther from the origin increases. We can obtain quantitative estimates of the average displacements of the particle from the moments of the distribution $W(m,n)$. The first moment or center of gravity of the distribution is the average displacement of the particle, $\overline{m}$, where

$$\overline{m} = \sum_{m=-n}^{n} mW(m,n).$$

Since $W(m,n)$ is an even function of m, $\overline{m} = 0$ (also see Appendix 3.1). That is, the average displacement of the particle is zero! The particle is just as likely to move a positive distance from the origin as a negative distance, and "on the average" it does not go anywhere. However, it does tend to wander farther and farther away from the origin—in both directions.

This tendency to wander can be measured by computing the width of the $W(m,n)$ function. A convenient measure is the root-mean-squared value of m, σ_m, where σ_m^2 is the second central moment, or variance, of $W(m,n)$, which equals the second moment since $\overline{m} = 0$, i.e.,

$$\sigma_m^2 = \overline{m^2} = \sum_{m=-n}^{n} m^2 W(m,n).$$

The value of the second central moment is (see Appendix 3.1)

$$\sigma_m^2 = n,$$

so $\sigma_m = \sqrt{n}$. In n steps the distribution of probable locations of the particle has a width, or standard deviation, equal to $\sqrt{n}$. As we shall soon see, this result is characteristic of a diffusion process.

3.2.4 The Macroscopic Space-Time Evolution of Particle Location: The Gaussian Distribution

Next we shall examine the space-time evolution of particle position on macroscopic scales of time and distance. The relation between the microscopic and macroscopic scales is indicated in Figure 3.8. If the random walk is viewed at a resolution in time that is much coarser than τ and at a resolution in space that is much coarser than l, then the particle position appears to be a smooth function of time, i.e., the individual transitions in the random walk are not discernible. It is the wandering of the particle on this macroscopic scale that we shall now explore.

First we note that after a large number of steps n, the probability distribution $W(m,n)$ approaches a limiting form. To determine this limiting form, we use Stirling's approximation for factorials,

$$k! \approx \sqrt{2\pi k}\left(\frac{k}{e}\right)^k,$$

which is an exceedingly good approximation for large values of k. Therefore, assuming n is large and using Stirling's approximation in Equation 3.19, after some simplification we obtain

$$W(m,n) \approx \sqrt{2/\pi n}\left(1-(m/n)^2\right)^{-1/2}\left(1-(m/n)^2\right)^{-n/2}$$
$$(1+m/n)^{-m/2}(1-m/n)^{m/2}.$$

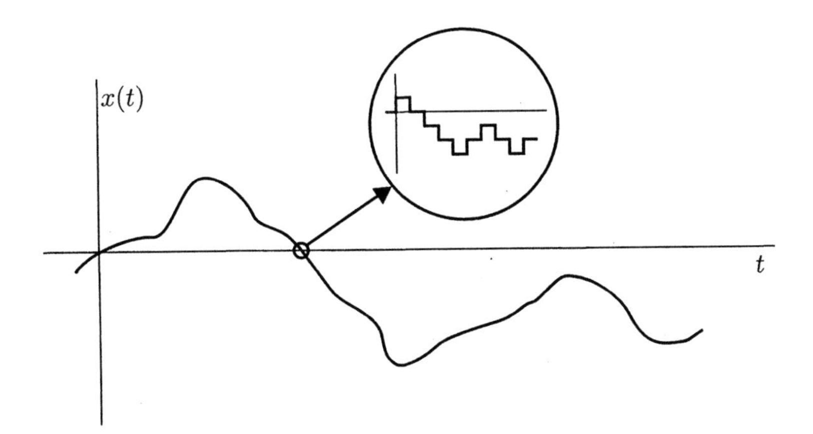

Figure 3.8 Relation between microscopic and macroscopic time and distance scales. The smooth trajectory represents the position of a particle as a function of time on macroscopic scales of distance and time. On the macroscopic scales, the individual transitions of the one-dimensional random walk are not discernible, and the trajectory appears smooth. The inset shows a magnification of a portion of the macroscopic trajectory to illustrate the underlying discontinuous microscopic trajectory.

Taking the logarithm of $W(m,n)$, we get

$$\ln W(m,n) \approx \frac{1}{2}\ln\left(\frac{2}{\pi n}\right) - \frac{1}{2}\ln\left(1-\left(\frac{m}{n}\right)^2\right) - \frac{n}{2}\ln\left(1-\left(\frac{m}{n}\right)^2\right)$$
$$- \frac{m}{2}\ln\left(1+\frac{m}{n}\right) + \frac{m}{2}\ln\left(1-\frac{m}{n}\right).$$

Since $W(m,n)$ is appreciable only for $m \ll n$, when n gets large (because the width of $W(m,n)$ is on the order of $\sqrt{n}$), we need only consider $W(m,n)$ for $m/n \ll 1$. If we use the approximation

$$\ln(1+x) \approx x \quad \text{for } x \ll 1,$$

we obtain

$$\ln W(m,n) \approx \frac{1}{2}\ln\left(\frac{2}{\pi n}\right) + \frac{1}{2}\left(\frac{m}{n}\right)^2 + \frac{n}{2}\left(\frac{m}{n}\right)^2 - \frac{m}{2}\left(\frac{m}{n}\right) - \frac{m}{2}\left(\frac{m}{n}\right),$$

and

$$\ln W(m,n) \approx \frac{1}{2}\ln\left(\frac{2}{\pi n}\right) + \frac{1}{2}\left(\frac{m}{n}\right)^2 - \frac{m^2}{2n}. \tag{3.20}$$

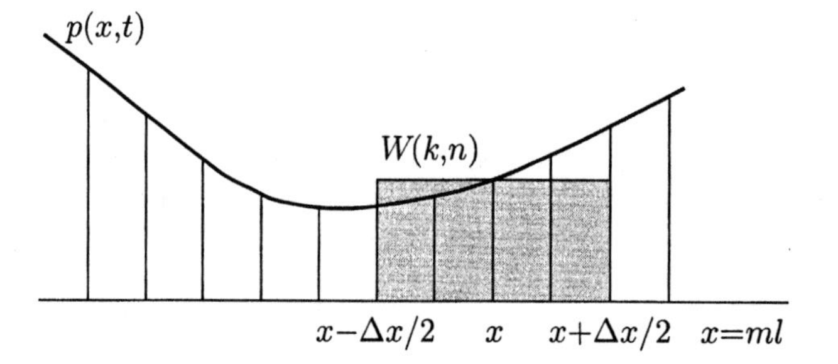

Figure 3.9 Illustration of the method for deriving the Gaussian probability density function $p(x,t)$ from the probability distribution $W(k,n)$.

For large n the middle term is negligible, and if we exponentiate Equation 3.20 we obtain

$$W(m,n) \approx \sqrt{\frac{2}{\pi n}} e^{-m^2/2n},$$

which is the Gaussian or normal distribution. Thus, the binomial distribution approaches the Gaussian distribution for large n.

Now we shall determine the probability that after n steps, which take $t = n\tau$ seconds, the particle lies in the interval $(x - \Delta x/2, x + \Delta x/2)$, where $x = ml$. We express this probability as $p(x,t)\Delta x$, where $p(x,t)$ is a probability density function. As indicated in Figure 3.9, the probability that the particle is in the indicated interval can be expressed in terms of both the continuous probability density function and the discrete probability distribution. These must be equal, so

$$p(x,t)\Delta x = \sum_k W(k,n), \quad \text{where } (ml - \Delta x/2) < kl < (ml + \Delta x/2).$$

If the interval Δx is sufficiently small and n sufficiently large,

$$p(x,t)\Delta x \approx W(m,n)\frac{\Delta x}{2l},$$

where $m = x/l$ and $n = t/\tau$. Therefore,

$$p(x,t) = \frac{1}{2l}\sqrt{\frac{2\tau}{\pi t}} e^{-(x/l)^2/(2t/\tau)},$$

and

$$p(x,t) = \frac{1}{\sqrt{4\pi D t}} e^{-x^2/4Dt},$$

where $D = l^2/2\tau$. This definition of D is consistent with the definition of D in Fick's first law, which we derived in Section 3.2.2.

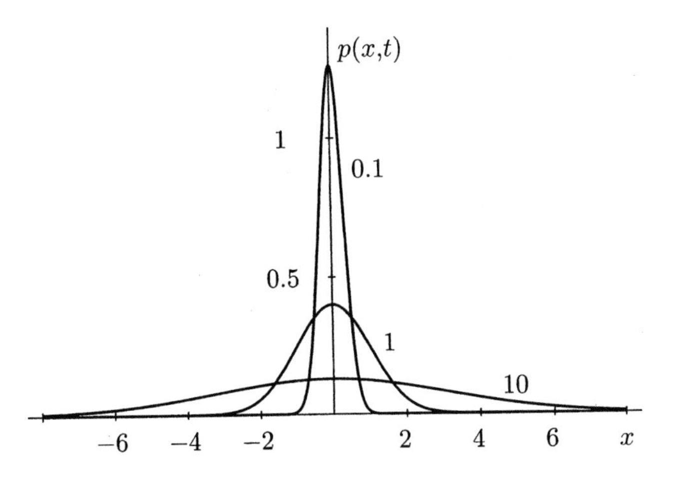

Figure 3.10 The Gaussian probability density function $p(x,t)$ plotted versus x for values of $2Dt = 0.1$, 1, and 10.

Table 3.2 Probability that the particle can be found in the indicated interval at time t.

Interval $-x$ to x	Probability $\Theta(x)$
$-0.67\sqrt{2Dt}$ to $+0.67\sqrt{2Dt}$	0.5
$-\sqrt{2Dt}$ to $+\sqrt{2Dt}$	0.68
$-2\sqrt{2Dt}$ to $+2\sqrt{2Dt}$	0.95
$-3\sqrt{2Dt}$ to $+3\sqrt{2Dt}$	0.997

A plot of $p(x,t)$, shown in Figure 3.10 as a function of x for three different times, shows that the spatial distribution of $p(x,t)$ gets broader and shallower as t increases. To determine the spread of this distribution quantitatively, we need a measure of the width of $p(x,t)$. A convenient measure is the root-mean-square value of x, σ_x, also called the standard deviation of $p(x,t)$, which is $\sigma_x = \sqrt{2Dt}$. Hence, the probable location of the particle spreads out as time increases, but only as fast as $\sqrt{2Dt}$. To see what this means quantitatively, we compute the probability that the particle is located a distance less than x from the origin, i.e.,

$$\Theta(x) = \int_{-x}^{+x} p(y,t)\,dy. \tag{3.21}$$

The table of values of Θ (Table 3.2) shows the probability of finding the particle within a given distance of the origin. Hence, $\sqrt{2Dt}$ is an effective measure of how far the particle is likely to wander in t seconds. Let us define $x_{1/2} = 0.67\sqrt{2Dt}$ as the distance such that the probability that the particle has wandered a distance greater than $x_{1/2}$ is 0.5.

To summarize, each particle starts out at the origin and wanders farther and farther as time goes on. Although the average displacement of the particle is zero, the distribution of its probable locations spreads out as $\sqrt{2Dt}$, so that the probability of finding the particle near the origin decreases and the probability of finding the particle far from the origin increases (Figure 3.10). Such a process is called a *Brownian process* or a *Wiener-Lévy process*.

3.2.5 Concentration as a Statistical Average of the Number of Particles per Unit Volume

Now we apply the results obtained for a single particle to a population of particles. Suppose that we start with n_0 moles of particles per unit area at the origin ($x = 0$) at $t = 0$ and that each particle undergoes an identically distributed, independent random walk. Then the mean number of moles of particles per unit area located in the interval $(x, x + \Delta x)$ at time t is $n_0 p(x,t)\Delta x$. Hence, $c(x,t) = n_0 p(x,t)$, where $c(x,t)$ is the concentration of particles (mol/cm^3) at x and t. Therefore, we see that the macroscopic concentration of particles is a statistical average over the number of particles undergoing independent random walks.

It follows directly that if we start with n_0 moles of particles per unit area at $x = 0$ and $t = 0$, then

$$c(x,t) = \frac{n_0}{\sqrt{4\pi Dt}} e^{-x^2/4Dt} \quad \text{for } t > 0.$$

Thus, $x_{1/2} = 0.67\sqrt{2Dt}$ is a distance such that at time t 1/2 the particles are located a distance $|x| > x_{1/2}$. How long does it take for $x = x_{1/2}$? If $t_{1/2}$ is the time it takes for 1/2 the particles to travel a net distance of at least $x_{1/2}$, then

$$t_{1/2} = \frac{1}{2D}\left(\frac{x_{1/2}}{0.67}\right)^2 \approx \frac{x_{1/2}^2}{D}.$$

This formula gives a rule of thumb for estimating the time it takes for diffusion to occur over a given distance. More precise estimates require solutions of the diffusion equation for the specified initial and boundary conditions.

3.3 The Diffusion Coefficient

The diffusion coefficient determines the time it takes a solute to diffuse a given distance in a medium. The diffusion coefficient depends not only on the physical characteristics of the solute, but also on those of the medium. We explore some of these dependencies in this section.

3.3.1 Solute in a Simple Fluid

3.3.1.1 Measurements

Measurements of the diffusion coefficient for over a hundred solutes are plotted versus the molecular weight of the solute in Figure 3.11 for two media, air and water; numerical values are given for selected substances in Table 3.3. Gas molecules of low molecular weight (e.g., hydrogen, oxygen, methane, etc.) diffusing in air have diffusion coefficients in the range of 0.1–1 cm^2/s, whereas these same molecules diffusing in water have diffusion coefficients in the range of 1–6 $\times 10^{-5}$ cm^2/s. Thus, the diffusion coefficient is four orders of magnitude smaller in water than in air. This difference reflects the much greater density of solvent molecules in water than in air.

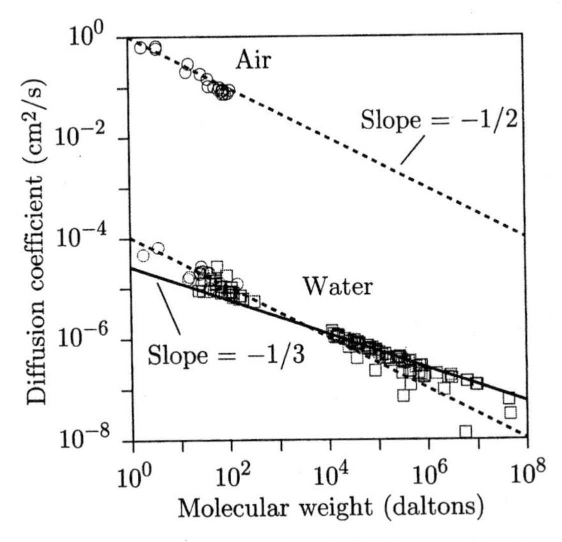

Figure 3.11 The measured diffusion coefficient D is plotted in logarithmic coordinates as a function of the molecular weight M for 19 gases diffusing in air and for 123 solutes diffusing in water. The data are from several sources (Tanford, 1961; Cohn and Edsall, 1965; Cussler, 1984; Lide, 1990). The measurements of solutes in water were made at temperatures in the range of 20–25°C and are extrapolated to infinite dilution. Those in air were made at atmospheric pressure and at temperatures in the range of 0–26.1°C. Circles represent gases diffusing either in air or in water; squares represent other solutes.

Table 3.3 Diffusion coefficients of selected molecules in air (above line) and water (below line). All these are included in Figure 3.11, which also gives citations. M is the molecular weight; D is the diffusion coefficient.

Molecule	Medium	Temp. (°C)	M (g/mol)	D (cm^2/s)
Hydrogen	Air	0	2	6.11×10^{-1}
Helium	Air	3	4	6.24×10^{-1}
Oxygen	Air	0	32	1.78×10^{-1}
Benzene	Air	25	78	9.60×10^{-2}
Hydrogen	Water	25	2	4.50×10^{-5}
Helium	Water	25	4	6.28×10^{-5}
Oxygen	Water	25	32	2.10×10^{-5}
Urea	Water	25	60	1.38×10^{-5}
Benzene	Water	25	78	1.02×10^{-5}
Sucrose	Water	25	342	5.23×10^{-6}
Ribonuclease	Water	20	13,683	1.19×10^{-6}
Hemoglobin	Water	20	68,000	6.90×10^{-7}
Catalase	Water	20	250,000	4.10×10^{-7}
Myosin	Water	20	493,000	1.16×10^{-7}
DNA	Water	20	6,000,000	1.30×10^{-8}
Tobacco mosaic virus	Water	20	50,000,000	3.00×10^{-8}

We are primarily concerned with diffusion of solutes in water and in cellular membranes. How long does it take for a solute molecule to diffuse over a biologically significant distance? A crude estimate of this diffusion time can be obtained as follows. Recall that it takes $t_{1/2}$ seconds for 1/2 the particles to diffuse a distance of at least $x_{1/2} = 0.67\sqrt{2Dt_{1/2}}$, where $t_{1/2} \approx x_{1/2}^2/D$. For a small particle (e.g., an oxygen molecule or a small ion) with $D \approx 10^{-5}\,cm^2/s$, Table 3.4 shows how long it takes. It takes a small molecule approximately 1 ms to go 1 μm and 1.2 days to go 1 cm. A large molecule (such as DNA) with $D \approx 10^{-8}\,cm^2/s$ takes 10^3 times longer or 1 second to go 1 μm and 3.17 years to go 1 cm. The dependence of time on the square of distance in diffusion has important consequences!

Figure 3.11 also indicates that D tends to decrease as the molecular weight (M) increases for particles in either air or water. In both these media, $D \propto M^{-1/2}$ for $M < 1000$. In water, as M increases above about 10^4, D varies as $M^{-1/3}$ for a selected group of solutes, but shows great scatter for many solutes of high molecular weight. The relation $D \propto M^{-1/3}$, which appears valid for some solutes, has a simple kinetic interpretation.

Table 3.4 Rough estimates of diffusion times for organelles, cells, and tissues of typical dimensions (Macey, 1980). The diffusion times were computed from the relation $t_{1/2} = x_{1/2}^2/D$ using a diffusion coefficient of $D = 10^{-5}$ cm^2/s.

$x_{1/2}$	$t_{1/2}$	Example
10 nm	100 ns	Thickness of cell membrane
1 μm	1 ms	"Size" of mitochondrion
10 μm	100 ms	Radius of a small mammalian cell
100 μm	10 s	Diameter of a large muscle fiber
250 μm	1 min	Radius of squid giant axon
1 mm	16.7 min	Half-thickness of frog sartorius muscle
2 mm	1.1 h	Half-thickness of lens in the eye
5 mm	6.9 h	Radius of mature ovarian follicle
2 cm	4.6 d	Thickness of ventricular myocardium
1 m	31.7 yrs	Length of a nerve or muscle cell

3.3.1.2 The Stokes-Einstein Relation

If we assume a spherical solute particle for which the molecular weight is proportional to the molecular volume, which is proportional to a^3, where a is the molecular radius, then $D \propto M^{-1/3}$ implies that $D \propto a^{-1}$. This relation is predicted by the *Stokes-Einstein relation.* A careful derivation of this relation is beyond our scope, but a plausible argument is given below.

The force, f_p, required to move a sphere of radius a through a viscous medium of viscosity η with a velocity of v is

$$f_p = 6\pi a \eta v, \tag{3.22}$$

which is known as Stokes's law (Stokes, 1851), whose derivation can be found elsewhere (Batchelor, 1967). The particle mobility, u_p, is defined as the ratio of the particle velocity to the force on the particle and is

$$u_p \equiv \frac{v}{f_p} = \frac{1}{6\pi a \eta}. \tag{3.23}$$

Stokes's law relates the particle mobility to the particle radius, and the following argument relates the mobility to the diffusion coefficient.

As we have seen, diffusion occurs because particles are scattered by collisions with molecules of the medium. In the absence of a body force, a particle bounces around in response to collisions with a mean free path of l, a mean free time τ, and a mean velocity of zero. That is, there is no preferred direction of motion for an individual particle. However, in the presence of a body force, the particle experiences an increment in velocity between collisions in the direction of the force. If the force on the particle is f_p, then the acceleration of the particle is f_p/m, where m is the particle mass. The increment in velocity called the *drift velocity* rises linearly from zero to a peak velocity. At one mean free time the particle drift velocity reaches a peak velocity of $v = (f_p/m)\tau$ and a mean velocity of $\overline{v} = f_p\tau/(2m)$. Thus, by definition the particle mobility is

$$u_p = \overline{v}/f_p = \tau/(2m).$$

From the equipartition theorem, the mean kinetic energy of a particle undergoing one-dimensional motion is

$$\frac{1}{2}m\overline{v^2} = \frac{1}{2}kT, \tag{3.24}$$

where k is Boltzmann's constant, T is absolute temperature, and $\overline{v^2}$ is the mean-squared particle velocity. Let the root-mean-squared particle velocity equal $\overline{v}$,[3] and assume that it is related to the parameters of the random walk by the relation $\overline{v} = l/\tau$. Then, by rearranging terms, we obtain

$$\frac{1}{2}m\frac{l^2}{\tau^2} = \frac{1}{2}kT, \text{ and } \left(\frac{2m}{\tau}\right)\left(\frac{l^2}{2\tau}\right) = kT.$$

Since $2m/\tau = 1/u_p$ and $l^2/2\tau = D$, we obtain the Einstein relation in terms of the particle mobility

$$D = u_p kT. \tag{3.25}$$

The Einstein relation can also be expressed in terms of molar quantities. Let

$$v = u_p f_p = \frac{u_p}{N_A} f_p N_A = uf,$$

where u is molar mechanical mobility, f is force per mole of particles, and N_A is Avogadro's number. Therefore,

$$D = u_p kT = uN_A kT = uRT, \tag{3.26}$$

3. The assumption that the root-mean-squared velocity equals the mean velocity is a weakness in this argument. In a more carefully constructed argument, the root-mean-squared velocity would be computed directly.

where R is molar gas constant. Finally, we can substitute Equation 3.23 into 3.25 to obtain the Stokes-Einstein relation,

$$D = \frac{kT}{6\pi a\eta}. \tag{3.27}$$

The derivation of the Stokes-Einstein relation between D and a assumes that the interaction between the spherical particle and the medium can be represented by continuum hydrodynamics. Clearly the validity of this assumption is suspect for solute particles whose dimensions are comparable to those of water molecules. This deviation from a sphere undoubtedly contributes to the deviation of the measurements from the Stokes-Einstein relation for solutes of small molecular weight (Figure 3.11). In addition, we do not expect that this law will apply to molecules whose geometries differ appreciably from spheres, as do those of many polymeric molecules with large molecular weights.[4]

3.3.2 Solute in a Polymer

In water, the dependence of the diffusion coefficient on solute dimensions (molecular weight) can be summarized for many solutes with the following approximation:

$$D \propto M^{-1/3} \quad \text{for } M > 10^3,$$

$$D \propto M^{-1/2} \quad \text{for } M < 10^3.$$

However, this dependence is quite different in other media. For example, consider a porous medium consisting of an impermeant matrix with permeable pores with diameter b. In such a medium the diffusion coefficient will be zero for all solutes whose smallest dimensions exceed b. Clearly the relation between diffusion coefficient and molecular weight seen in water and in air need not apply to other media.

Some investigators (Stein, 1986) have proposed that diffusion in membranes resembles that in polymers more than it does diffusion in simple liquids such as water. Therefore, it is of some interest to examine the dependence of the diffusion coefficient on molecular weight in polymers. Measurements in polymers (Figure 3.12) indicate that the dependence of the diffusion coefficient on molecular weight is much steeper than in water (Crank and Park,

4. Measurement of the diffusion coefficient together with Equation 3.27 is often used to compute the equivalent Stokes-Einstein radius, the radius of a spherical particle whose diffusion coefficient is equal to that of the non-spherical particle.

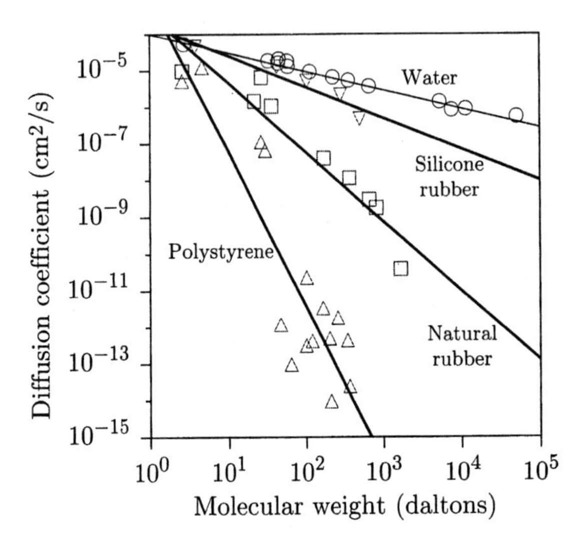

Figure 3.12 Diffusion coefficient as a function of solute molecular weight in water and in three polymers: silicone rubber, natural rubber, and polystyrene (adapted from Baker, 1987, Figure 2.5). The regression lines through the measurements had slopes and correlation coefficients of −0.51 and 0.99 (water), −0.86 and 0.96 (silicone rubber), −1.90 and 0.95 (natural rubber), and −4.20 and 0.89 (polystyrene).

1968; Comyn, 1985; Baker, 1987). The data shown in Figure 3.12 have been fitted with a power function relating the diffusion coefficient to the molecular weight

$$D \propto M^m,$$

where the exponent m varies from about −0.5 (for water) to less than −4 (for polystyrene).

3.4 Equivalent Diffusion "Force"

Diffusion results from the random motion of particles due to their thermal energy. If there is a difference in concentration with position, then there will be a net flow of particles down their concentration gradient. If the particles are solute particles in a solution, then we can readily imagine that the tendency to diffuse is balanced by a viscous force of the solvent on the solute particles. Hence, in the steady state the particles are not accelerated, but reach a terminal velocity. Therefore, we can find an equivalent body force that would give the particles the same terminal velocity as they acquire by diffusing. We shall find this force by first finding the relation between force and velocity for a population of particles subject to a body force.

If we assume that we have a population of particles subjected to a force f_e and that these particles collide with the medium so that the terminal velocity of the particles is proportional to the force, then $\phi = ucf_e$, where u is the molar mechanical mobility in units of (m/s)/N and f_e is the force per mole of particles which has units of N/mol. We can compute the equivalent force on particles moving as a result of a concentration gradient from Fick's first law as follows:

$$\phi = -D\frac{\partial c}{\partial x} = uc\left(-\frac{D}{uc}\frac{\partial c}{\partial x}\right) = uc\left(-\frac{D}{u}\frac{\partial}{\partial x}(\ln c)\right).$$

From the relation $D = RTu$, we can see that

$$\phi = uc\left(-\frac{\partial(RT\ln c)}{\partial x}\right). \tag{3.28}$$

Therefore,

$$f_e = -\frac{\partial}{\partial x}(RT\ln c) = -\frac{\partial \mu}{\partial x}. \tag{3.29}$$

Thus, the force is the negative gradient of a function, μ, which has the dimensions of energy per mole and is called the *chemical potential energy,*

$$\mu = \mu_o + RT\ln c,$$

where μ_o is the reference value of the chemical potential at $c = 1$ mol/L. Therefore, substitution of Equation 3.29 into Equation 3.28 yields

$$\phi = -uc\frac{\partial \mu}{\partial x}. \tag{3.30}$$

This result yields another interpretation of diffusion. A concentration gradient in a solution represents stored chemical energy. This stored chemical energy is dissipated spontaneously as the particles diffuse. The relation between flux and chemical potential gradient (Equation 3.30) is a common starting point for discussions of transport based on a thermodynamic point of view.

3.5 Diffusion Processes

In this section we shall consider diffusion processes[5] that are directly relevant to diffusion in cells and through cellular membranes.

5. A number of compendia of solutions to the diffusion equation exist (Carslaw and Jaeger, 1959; Crank, 1964; Jost, 1965; Jacobs, 1967; Widder, 1975; Cannon, 1984; Cussler, 1984).

3.5.1 Time-Invariant Diffusion Processes

First we shall consider cases in which the flux and the concentration are independent of time.

3.5.1.1 Equilibrium Diffusion Processes

By definition, at *equilibrium* the concentration is independent of time, and the flux is zero. For a diffusible particle ($D \neq 0$), Fick's first law implies that if $\phi = 0$, then $\partial c/\partial x = 0$. Thus, at diffusive equilibrium the concentration is constant in space and time, i.e., independent of x and t. A particle that is fixed, not diffusible ($D = 0$), is automatically at equilibrium, since $\phi = 0$ no matter how large the concentration gradient.

3.5.1.2 Steady-State Diffusion Processes

In *steady-state diffusion* we relax the requirement that the flux be zero and require that both ϕ and c be independent of time, t. That is, we allow a constant flow of particles. Under these conditions, the continuity equation,

$$\frac{\partial \phi}{\partial x} = -\frac{\partial c}{\partial t},$$

implies that

$$\frac{\partial \phi}{\partial x} = 0,$$

which implies that ϕ is constant (independent of x as well as t). Fick's first law can be simplified to

$$\phi = -D\frac{\partial c}{\partial x} = -D\frac{dc}{dx},$$

which for $D \neq 0$ can be integrated to yield

$$c(x) = c(x_o) - \frac{\phi}{D}(x - x_o), \tag{3.31}$$

where x_o is a reference location at which the concentration is known.

Hence, in steady-state diffusion the flux is constant and the concentration is a linear function of distance, x (Figure 3.13). Of course, in a physical system a constant flux in some region of space implies that there is a source of particles outside this region whose particle concentration must be changing. Therefore, the steady-state assumption cannot be valid for all time in a physi-

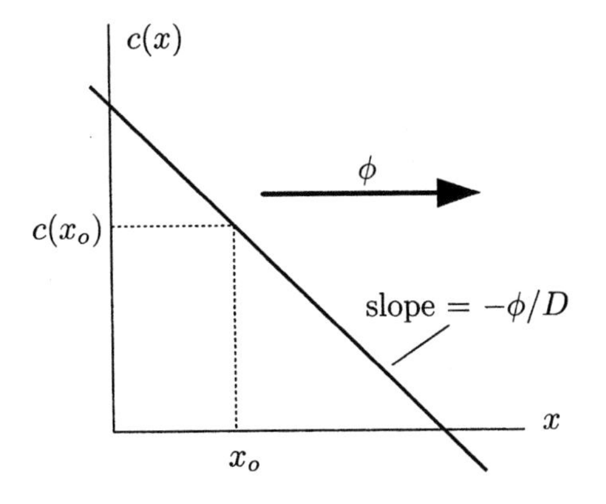

Figure 3.13 Under steady-state conditions, the spatial distribution of concentration is linear.

cal system. Nevertheless, it may be a very good assumption in cases in which there are large reservoirs of particles interacting diffusively with a small system. Note that if the flux is set to zero in Equation 3.31, then the concentration is constant, which defines the equilibrium diffusion process considered above.

3.5.2 Time-Varying Diffusion Processes

3.5.2.1 The Impulse Response (Green's Function)

Next we shall consider the solution to the diffusion equation for a point source of particles. This case corresponds to the physical situation of placing n_0 mol/cm^2 of particles at $t = 0$ at position $x = 0$. We shall find $c(x,t)$, which is the space-time evolution of concentration from a point source. The concentration $c(x,t)$ must satisfy the diffusion equation,

$$\frac{\partial c}{\partial t} = D\frac{\partial^2 c}{\partial x^2}, \text{ for } t > 0.$$

The solution can be found by a number of different methods. First, we note that if we have a candidate solution that satisfies the diffusion equation and matches all the initial and boundary conditions, then we have *the* solution. Second, we already know the solution to this problem on the basis of the statistical argument given in Section 3.2. Finally, we can use any one of several methods of solution of partial differential equations, such as the method of separation of variables described in Appendix 3.3.

The solution is

$$c(x,t) = \frac{n_0}{\sqrt{4\pi Dt}}e^{-x^2/4Dt}, \text{ for } t > 0.$$

as can be verified by substitution into the diffusion equation. Note also that conservation of particles guarantees that the integral

$$I = \int_{-\infty}^{\infty} c(x,t)\, dx = n_0,$$

which can be verified by evaluating the integral

$$I = n_0 \int_{-\infty}^{\infty} \frac{dx}{\sqrt{4\pi Dt}} e^{-x^2/4Dt}.$$

If $y = x/\sqrt{2Dt}$, then $dy = dx/\sqrt{2Dt}$ and

$$I = n_0 \int_{-\infty}^{\infty} \frac{dy}{\sqrt{2\pi}} e^{-y^2/2} = n_0,$$

which is evaluated in Appendix 3.2.

The concentration, $c(x,t)$, is sketched as a function of x and t in Figure 3.14. $c(x,t)$ has a Gaussian spatial distribution for any time $t > 0$ and has a maximum at $x = 0$. As time progresses, the distribution gets broader and shallower and the width increases as $\sqrt{2Dt}$, but the area remains constant. The temporal variation in concentration at any point $(x \neq 0)$ starts at $c(x,0) = 0$, increases to a maximum, and then decreases, approaching an asymptotic change in concentration that is proportional to $t^{-\frac{1}{2}}$. We shall see that the response to a *unit* point source is a particularly useful function called *Green's function*, $G(x,t)$, which is defined as

$$G(x,t) = \frac{1}{\sqrt{4\pi Dt}} e^{-x^2/4Dt} \quad \text{for } t > 0. \tag{3.32}$$

The Green's function describes the concentration that arises if 1 mol/cm^2 of particles are placed at $x = 0$ at $t = 0$.

3.5.2.2 The Space-Time Evolution of Concentration for an Arbitrary Initial Concentration

Because the diffusion equation is linear and has constant coefficients, it is possible to find the solution $c(x,t)$ for an arbitrary initial distribution $c(x,0)$ in terms of the Green's function for the diffusion equation. Consider the initial distribution shown in Figure 3.15. The differential element of thickness $\Delta\xi$ located at ξ represents $c(\xi,0)\Delta\xi$ mol/cm^2 of particles located at $x = \xi$ at $t = 0$. The contribution of these particles to the concentration of particles at any point x at time t is simply

$$c(\xi,0)\Delta\xi G(x-\xi,t).$$

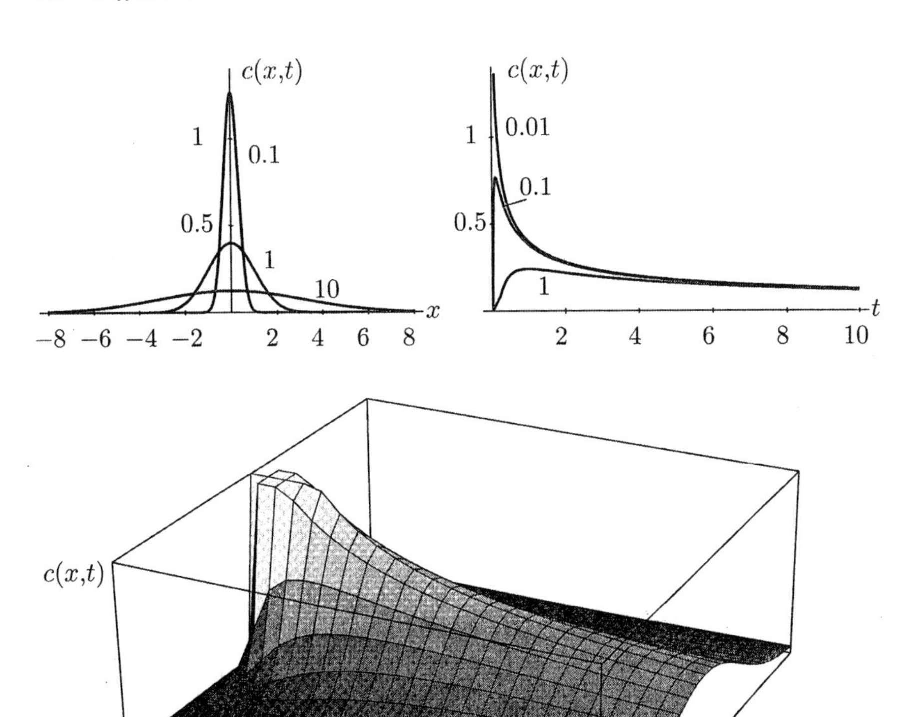

Figure 3.14 The space-time evolution of diffusion from a one-dimensional point source. The upper left panel shows the spatial distribution at three different times ($2Dt = 0.1$, 1, and 10), and the upper right panel shows the temporal distribution of concentration at three different positions ($x/\sqrt{2D} = 0.01$, 0.1, and 1). A three-dimensional display of concentration versus space and time is shown in the lower panel.

If we partition $c(x,0)$ into an infinite number of such differential elements and add up all the contributions of these elements to $c(x,t)$ then in the limit as $\Delta\xi \to 0$, we obtain

$$c(x,t) = \int_{-\infty}^{\infty} c(\xi,0)G(x-\xi,t)\,d\xi,$$

or

$$c(x,t) = \int_{-\infty}^{\infty} \frac{c(\xi,0)}{\sqrt{4\pi Dt}} e^{-(x-\xi)^2/4Dt}\,d\xi, \qquad t > 0. \tag{3.33}$$

This (integral) equation relates the concentration at some time, t, to the concentration at an earlier time (arbitrarily chosen to be zero).

As a simple example of the use of Equation 3.33, we shall consider the concentration that results from an initial concentration that represents an abrupt transition from 0 to C_0, occurring at $x = 0$, as shown in Figure 3.16.

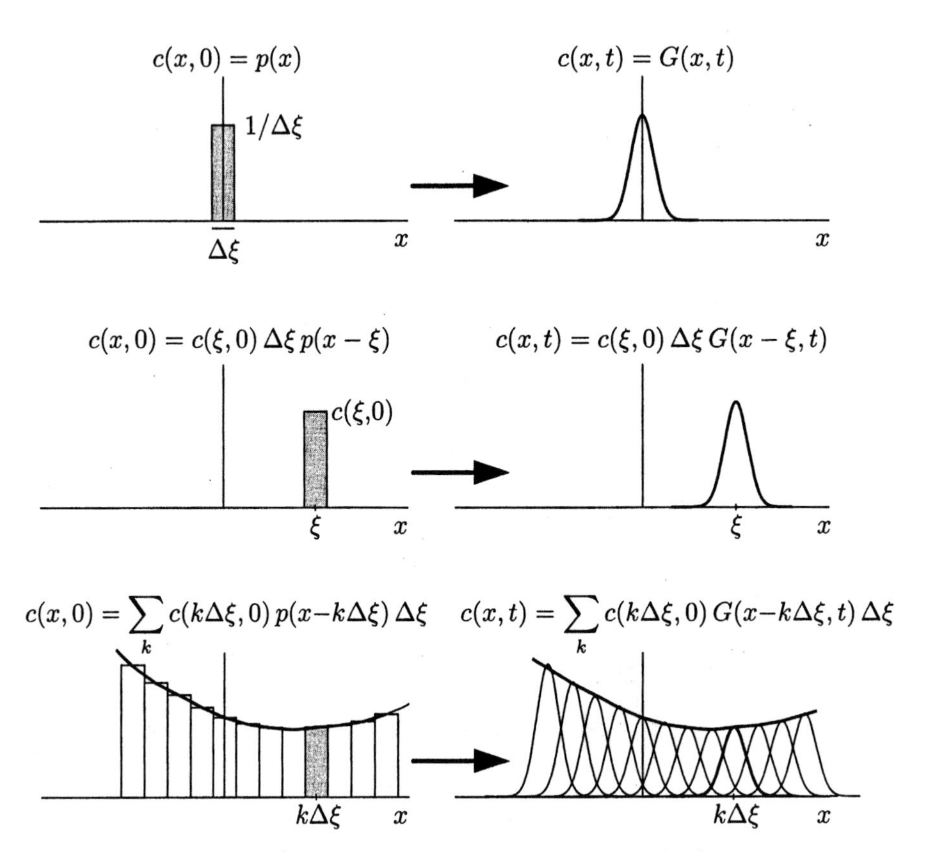

Figure 3.15 Sketch of proof of the superposition integral. In the left column are three initial concentrations, and in the right column are the concentrations at some later time t. The top row shows the results for an initial concentration consisting of a pulse of concentration, $p(x)$, of area one and of width $\Delta\xi$. For an arbitrarily narrow width, the concentration at time t is just $G(x,t)$. If the initial concentration is shifted to location ξ and has height $c(\xi,0)$ then this can be expressed by shifting and scaling $p(x)$. The concentration is simply $G(x,t)$ scaled and shifted by the identical factors used to shift and scale $p(x)$. With these results, an arbitrary initial distribution can be made up of a staircase approximation consisting of a sum of infinitesimal pulses (lowest panel). By superposition, the response to the concentration at time t resulting from the superposition of pulses is simply the superposition of Green's functions.

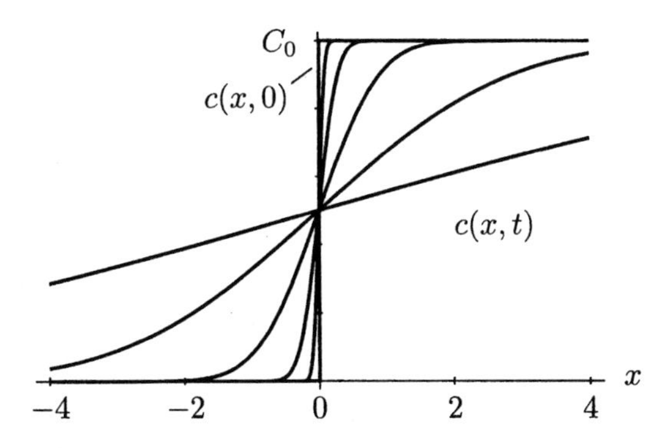

Figure 3.16 Solution for diffusion from an initial spatial step of concentration. The concentration is shown as a function of position for normalized times of $2Dt = 0$, 0.01, 0.1, 1, 10, and 100.

Hence,

$$c(x,t) = \int_0^\infty \frac{C_0}{\sqrt{4\pi Dt}} e^{-(x-\xi)^2/4Dt}\, d\xi \qquad \text{for } t > 0. \tag{3.34}$$

If $y = (x - \xi)/\sqrt{2Dt}$, then Equation 3.34 can be written as

$$c(x,t) = C_0 \int_{-\infty}^{x/\sqrt{2Dt}} \frac{1}{\sqrt{2\pi}} e^{-y^2/2}\, dy.$$

The integrand is a Gaussian function. The Gaussian function has the properties given in Equation 3.35,

$$\int_{-\infty}^{\infty} \frac{1}{\sqrt{2\pi}} e^{-y^2/2}\, dy = 1 \quad \text{and} \quad \int_0^\infty \frac{1}{\sqrt{2\pi}} e^{-y^2/2}\, dy = 0.5, \tag{3.35}$$

and the concentration is sketched in Figure 3.16 as a function of $x/\sqrt{2Dt}$. Note that $c(0,t)/C_0 = 0.5$ at $x = 0$ and that the slope of $c(x,t)$ at $x = 0$ is

$$\frac{\partial c(x,t)}{\partial x} = \frac{\partial}{\partial x}\left(C_0 \int_{-\infty}^{x/\sqrt{2Dt}} \frac{1}{\sqrt{2\pi}} e^{-y^2/2}\, dy \right),$$

$$\frac{\partial c(x,t)}{\partial x} = \frac{C_0}{\sqrt{2Dt}} \frac{1}{\sqrt{2\pi}} e^{-x^2/4Dt}.$$

Hence,

$$\left[\frac{\partial c(x,t)}{\partial x} \right]_{x=0} = \frac{C_0}{\sqrt{4\pi Dt}}.$$

Therefore, as $t \to \infty$, the slope goes to zero and $c(x,t) \to C_0/2$. Hence, at equilibrium the concentration is the average of the initial values for $x < 0$, which is 0, and that for $x > 0$, which is C_0. The average is $C_0/2$.

As a second example that illustrates the use of superposition, consider an initial concentration in the shape of a well that has value C_0 for $|x| > a$ and is zero for $|x| \le a$. This initial distribution represents the concentration profile for a slab of thickness $2a$ oriented perpendicular to the x-axis and immersed in a medium of concentration C_0. For example, this problem has great practical application in making textiles where the objective is to color a piece of cloth by dipping the cloth into a vat of dye. Initially the cloth contains no dye and the surrounding liquid has some concentration of dye. So the width of the well represents the cloth thickness. The solution to this problem can be found by using the superposition integral or by noting that the solution is a sum of solutions of the type we just found for an abrupt transition in concentration. The solution is

$$c(x,t) = \int_{-\infty}^{-a} C_0 G(x-\xi,t)\,d\xi + \int_{a}^{\infty} C_0 G(x-\xi,t)\,d\xi,$$

or

$$c(x,t) = C_0\left(1 - \int_{-a}^{a} G(x-\xi,t)\,d\xi\right). \tag{3.36}$$

Substitution of Equation 3.32 into Equation 3.36 yields

$$c(x,t) = C_0\left(1 - \int_{-a}^{a} \frac{1}{\sqrt{4\pi Dt}} e^{-(x-\xi)^2/4Dt}\,d\xi\right).$$

If $y = (x-\xi)/\sqrt{2Dt}$, then

$$c(x,t) = C_0\left(1 - \int_{(x-a)/\sqrt{2Dt}}^{(x+a)/\sqrt{2Dt}} \frac{1}{\sqrt{2\pi}} e^{-y^2/2}\,dy\right). \tag{3.37}$$

The concentration is plotted versus x at various values of t in Figure 3.17.

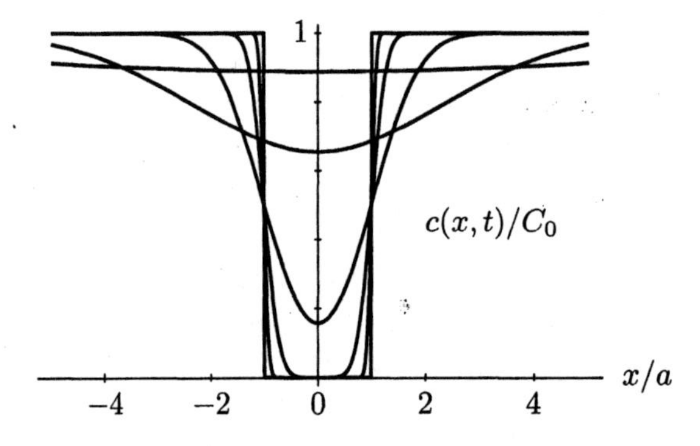

Figure 3.17 Concentration as a function of position at several instants in time for an initial concentration that is a rectangular well. The concentration is shown for normalized times $2Dt/a^2 = 0$, 0.01, 0.1, 1, 10, and 100.

3.5.2.3 Equilibration Times for Regular Geometric Volumes

In Section 3.3 we discussed a crude measure of the time it takes for particles to diffuse a fixed distance. We found that as a rough estimate the quantity $t_{1/2} \approx x_{1/2}^2/D$ gives an order of magnitude for the time it takes for particles to diffuse a distance of $x_{1/2}$. However, to accurately determine the time it takes for a given volume element to equilibrate with its surroundings, it is necessary to solve the diffusion equation under the appropriate boundary and initial conditions. This problem is beyond our scope, but to indicate the role of geometry in equilibration we shall examine the answer to the question, How long does it take for a volume element to equilibrate with a medium whose particle concentration is C_0 if the initial concentration of particles in the volume element is zero at $t = 0$? The solutions will not be derived, but the known solutions will be examined for three bodies of regular geometry: a rectangular slab of thickness $2a$ and both a cylinder and a sphere of radius a. We shall assume that the concentration of solute at the surface of the volume element is C_0.

To obtain a quantitative estimate of the degree of equilibration of the volume element, we shall compute the average concentration of particles in the volume element. Let $c(\mathbf{r}, t)$ be the concentration in the volume element at position $\mathbf{r}$. The spatial average of this concentration, $\overline{c(t)}$, is

$$\overline{c(t)} = \frac{1}{\mathcal{V}} \int_{\mathcal{V}} c(\mathbf{r}, t)\, dV,$$

where $\mathcal{V}$ is the volume of the element. The diffusion equation can be solved to obtain $\overline{c(t)}$ for volumes of simple geometry such as a slab, a cylinder, and a sphere. For each of these elements, the solution can be expressed in the form

$$\frac{\overline{c(t)}}{C_0} = 1 - e(t), \tag{3.38}$$

where $e(t)$ is the unequilibrated fraction for a volume element. The solutions for all three volume elements are (Crank, 1964)

$$e(u) = \begin{cases} \frac{8}{\pi^2} \sum_0^\infty \frac{1}{(2n+1)^2} e^{-(2n+1)^2\pi^2 u/4} & \text{for a slab,} \\ 4 \sum_1^\infty \frac{1}{\alpha_n^2} e^{-\alpha_n^2 u} & \text{for a cylinder, and} \\ \frac{6}{\pi^2} \sum_1^\infty \frac{1}{n^2} e^{-n^2\pi^2 u} & \text{for a sphere,} \end{cases}$$

where $u = Dt/a^2$ is normalized time, D is the diffusion coefficient inside the volume element, $2a$ is the linear dimension of the volume element, and α_n is the nth zero of the zeroth-order Bessel function of the first kind, i.e.,

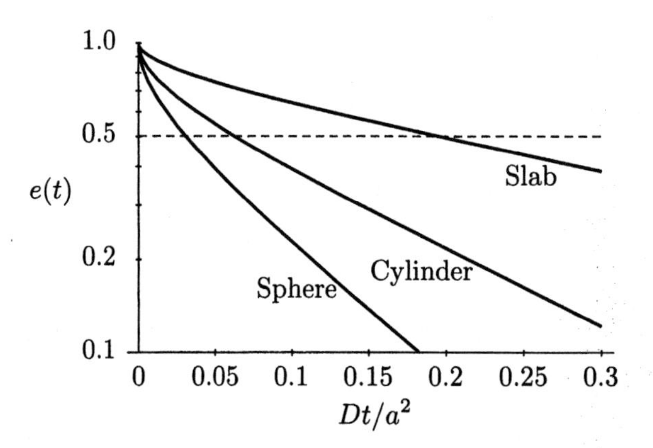

Figure 3.18 The time course of equilibration is shown for three regular geometric volume elements. The unequilibrated fraction, $e(t)$, is plotted versus normalized time Dt/a^2 in semilogarithmic coordinates.

Table 3.5 Equilibration times versus dimensions for regular geometric volumes. The times were obtained by finding the value of Dt/a^2 corresponding to $e(t) = 0.5$ from Figure 3.18 for each of the volume elements and using $D = 10^{-5}$ cm^2/s. The values were 0.21 (slab), 0.06 (cylinder), and 0.03 (sphere).

a	*Slab*	*Cylinder*	*Sphere*
10 nm	21 ns	6 ns	3 ns
1 μm	210 μs	60 μs	30 μs
10 μm	21 ms	6 ms	3 ms
1 mm	210 s	60 s	30 s
10 cm	24.3 d	6.9 d	3.5 d

$J_0(\alpha_n) = 0$. In Figure 3.18 $e(t)$ is plotted versus Dt/a^2. Initially, $\overline{c(t)} = 0$ and $e(0) = 1$; the volume is totally unequilibrated. Finally, as $t \to \infty$, $\overline{c(t)} \to C_0$ and $e(t) \to 0$; the volume element is totally equilibrated.

The results show that for 50% equilibration ($e(t) = 0.5$, i.e., the average concentration has reached half its final value), the value of Dt/a^2 increases from sphere to cylinder to slab. This order makes sense, since we expect the equilibration time to increase as the surface area to volume ratio decreases.

For $D = 10^{-5}$ cm^2/s, the equilibration time is given in Table 3.5. The table shows that it takes 21 ns for a slab the thickness of the plasma membrane (10 nm) to equilibrate (reach a concentration of 50% of its final value) for a small molecule ($D = 10^{-5}$ cm^2/s). However, it takes 3 ms for a spherical cell of radius 10 μm to equilibrate. A cylindrical cell of radius 1 mm, such as a neuron in an invertebrate (the largest animal cell), is equilibrated in 1 minute. We note again that the time course of diffusion is fast (on the scale of physiological phenomena) for small structures, but very slow for large structures such as organs.

3.6 Membrane Diffusion

We shall consider the theory of diffusion through membranes in this section. First we shall consider diffusion through a homogeneous membrane in which the solute dissolves and through which the solute diffuses. Then we shall consider diffusion through a porous membrane consisting of a matrix material that is impermeant to the solute and a set of pores through which the solute diffuses. These two mechanisms have been used to understand diffusion in cellular membranes, the topic we shall take up in Section 3.8.

3.6.1 Homogeneous Membranes

First we shall consider steady-state diffusion through a homogeneous membrane, i.e., diffusion when none of the diffusive variables is dependent on time. Then we shall consider the time it takes to reach this steady state.

3.6.1.1 Steady State

We shall consider diffusion through a membrane of thickness d that separates two solutions that contain solute n at concentrations c_n^i on the inside of the membrane and c_n^o on the outside of the membrane (Figure 3.19). The concentration of the solute in the membrane is $c_n(x)$. All the concentrations are time independent. We shall assume that the membrane is homogeneous and that the diffusion coefficient of the solute in the membrane is D_n. In the steady state, c_n is independent of t, and from the results of Section 3.5.1 we

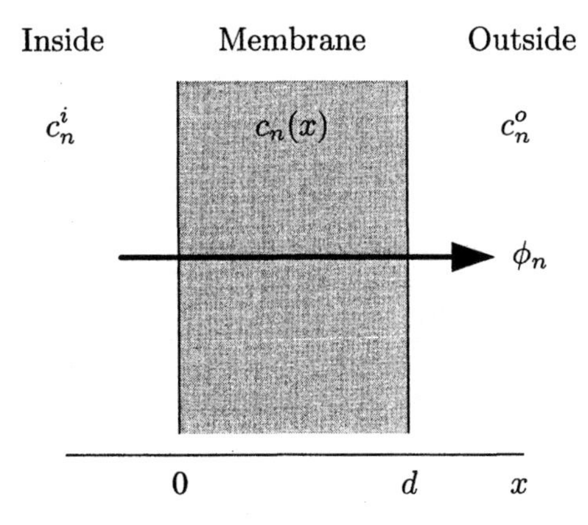

Figure 3.19 Definition of variables involved in deriving relations for steady-state diffusion in a membrane.

know that the flux of particles in the membrane ϕ_n is constant and that the concentration of n in the membrane depends linearly on x, i.e.,

$$c_n(x) = c_n(0) - \frac{\phi_n}{D_n}x, \tag{3.39}$$

which can be evaluated at $x = d$ to yield

$$\phi_n = \frac{D_n}{d}\left(c_n(0) - c_n(d)\right). \tag{3.40}$$

By combining Equations 3.39 and 3.40, we obtain

$$c_n(x) = c_n(0) - (c_n(0) - c_n(d))\,(x/d). \tag{3.41}$$

We need to formulate boundary conditions at the membrane: solution interfaces in order to relate the membrane variables to the external solution variables. We shall assume that at the interface between the membrane and the solution the solute will be partitioned according to its solubility in the membrane and in the solvent. Hence,

$$k_n = \frac{c_n(0)}{c_n^i} = \frac{c_n(d)}{c_n^o}, \tag{3.42}$$

where k_n is the membrane: solution *partition coefficient*. The notion of the partition coefficient becomes intuitively clear when we examine how it might be measured. Suppose we wish to determine the partition coefficient of some substance n in oil and in water, i.e., the oil: water partition coefficient. We might pour some oil and water into a container, add the solute, and then shake the solution. Oil and water are immiscible, and if we wait the oil will float on the water. The solute will distribute itself according to its partition coefficient. Therefore, at equilibrium the ratio of the concentration of the solute in the oil and in the water will equal the oil: water partition coefficient for this solute (see Figure 3.20).

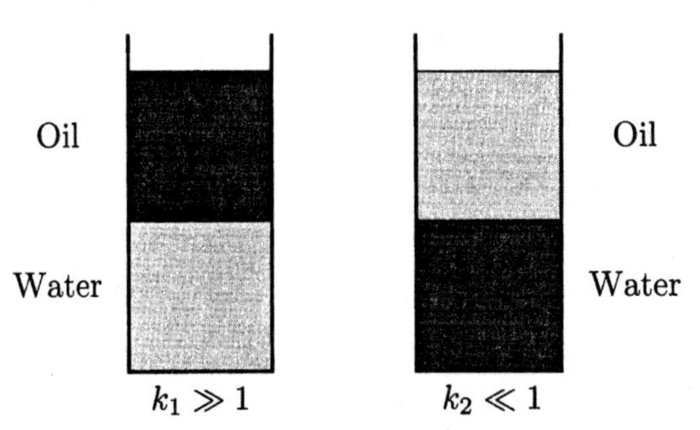

Figure 3.20 Illustration of the results obtained in measuring the oil: water partition coefficient for two solutes. One solute (left panel) is highly soluble in oil ($k_1 \gg 1$), while the other (right panel) is more soluble in water ($k_2 \ll 1$). Higher solute concentration is indicated by darker shading.

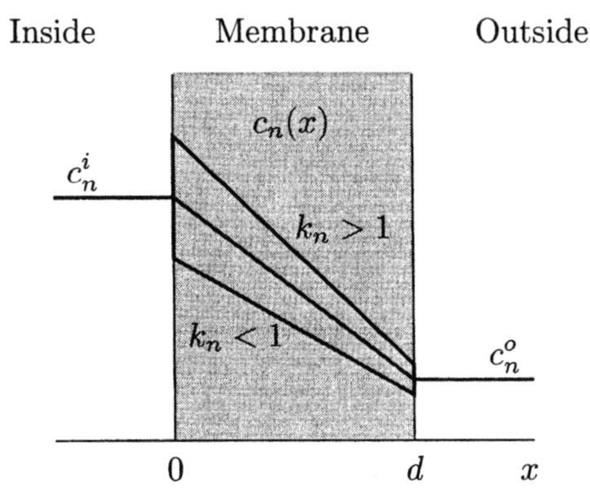

Figure 3.21 Illustration of effect of partition coefficient on steady-state concentration profile in membrane. When $k_n = 1$, the concentration profile is continuous; when the $k_n \neq 1$, the concentration profile is discontinuous at the solution membrane interface.

With the definition of the partition coefficient, Equation 3.41 can be expressed as

$$c_n(x) = k_n c_n^i - k_n(c_n^i - c_n^o)(x/d).$$

Hence, the profile of concentration in the membrane will depend on the partition coefficient as shown in Figure 3.21. Equation 3.40 can be expressed as

$$\phi_n = \frac{D_n k_n}{d}(c_n^i - c_n^o).$$

We define the permeability of the membrane for the solute, P_n, as

$$P_n = \frac{D_n k_n}{d}. \tag{3.43}$$

From the dimensions of D_n and d, we see that the permeability has dimensions of cm/s and is proportional to the diffusion coefficient and the partition coefficient and is inversely proportional to the membrane thickness.

Steady-state diffusion through a membrane can be characterized macroscopically by

$$\phi_n = P_n\left(c_n^i - c_n^o\right). \tag{3.44}$$

Equation 3.44 is sometimes called *Fick's first law for membranes*. This equation states that an outward flux of solute occurs ($\phi_n > 0$) if the inside concentration of solute exceeds the outside concentration ($c_n^i > c_n^o$). Thus, transport by diffusion is down the concentration gradient across the membrane. We call such transport *passive transport*. The magnitude of the flux is proportional to the product of the concentration difference and the permeability. If P_n is large, we say the membrane is highly *permeable* to the solute n. If $P_n = 0$, we say the membrane is *impermeable* to the solute n. Note that $P_n = 0$ either if $k_n = 0$

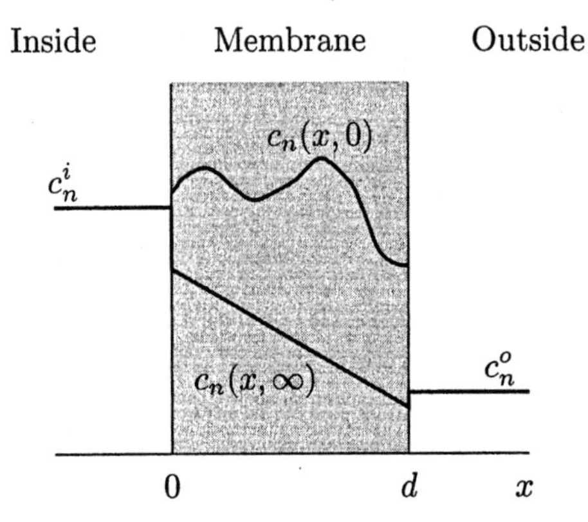

Figure 3.22 The concentration of solute n in a membrane is shown initially and in the steady state after the transient response has become negligible.

—i.e., if the solute is not soluble in the membrane—or if $D_n = 0$—i.e., the solute cannot diffuse in the membrane—or both. Although we have derived Fick's first law for membranes (Equation 3.44), which is a macroscopic relation between flux and concentration difference across a membrane, from a particular microscopic model of diffusion, we shall find that there are many microscopic models that are consistent with Fick's first law for membranes.

3.6.1.2 *Time to Reach Steady State*

We have shown that the steady-state concentration profile in a membrane is linear in distance. But how long does it take to establish this steady-state concentration profile? To determine this time, we shall assume that the membrane separates two baths whose concentrations remain constant at c_n^i and c_n^o and that the initial concentration profile in the membrane is arbitrary as shown in Figure 3.22. To solve this problem, we need to determine the concentration $c_n(x,t)$ that satisfies the diffusion equation and matches all the initial and boundary conditions. The initial condition is that $c_n(x,0)$ is an arbitrary function of x, call it $C(x)$. The boundary conditions are that for $t > 0$, $c_n(0,t) = k_n c_n^i$ at $x = 0$ and $c_n(d,t) = k_n c_n^o$ at $x = d$. But we know that the steady-state solution must be linear in space. Therefore, we can express the total solution as a sum of steady-state and transient components,

$$c_n(x,t) = c_n^s(x) + c_n^t(x,t),$$

where, from the previous section, the steady-state solution must be

$$c_n^s(x) = c_n(x,\infty) = k_n c_n^i - k_n(c_n^i - c_n^o)x/d.$$

Note that $c_n^s(x)$ already satisfies the boundary conditions at $x = 0$ and at $x = d$, which implies that $c_n^t(0,t) = c_n^t(d,t) = 0$ for $t > 0$. Thus, to solve the problem we need to find a function $c_n^t(x,t)$ that satisfies the diffusion equation, matches the boundary conditions, and has an arbitrary initial value of $c_n^t(x,0) = C(x) - k_n c_n^i + k_n(c_n^i - c_n^o)x/d$. As shown in Appendix 3.3, solutions of the diffusion equation can be written as the product of terms one of which has the form

$$e^{jpx} \quad \text{or} \quad \sin px \quad \text{or} \quad \cos px$$

and the other of which has the form

$$e^{-p^2 D_n t}.$$

In order to match the boundary conditions at 0 and d, we choose the $\sin px$ form of the solution with $p = l\pi/d$ and l an integer to obtain

$$c_n^t(x,t) = a_l \sin(l\pi x/d)e^{-t/\tau_l},$$

where

$$\tau_l = \frac{1}{p^2 D_n} = \frac{d^2}{l^2\pi^2 D_n}. \tag{3.45}$$

A more general solution to the diffusion equation that matches the boundary conditions is the superposition of such terms, which gives

$$c_n^t(x,t) = \sum_{l=1}^{\infty} a_l \sin(l\pi x/d)e^{-t/\tau_l}. \tag{3.46}$$

To match the initial condition, we must find the coefficients a_n such that

$$c_n^t(x,0) = \sum_{l=1}^{\infty} a_l \sin(l\pi x/d).$$

This equation is the Fourier series expansion of $c_n^t(x,0)$, and the coefficients are the Fourier series coefficients of $c_n^t(x,0)$, which can be found as

$$a_l = \frac{1}{2d}\int_0^d c_n^t(x,0)\sin(l\pi x/d)\,dx.$$

The total solution for the concentration profile is

$$c_n(x,t) = k_n c_n^i - k_n(c_n^i - c_n^o)x/d + \sum_{l=1}^{\infty} a_l \sin(l\pi x/d)e^{-t/\tau_l}, \tag{3.47}$$

which shows that the solution for the concentration profile in a membrane has a transient component whose components decrease in amplitude with time constants given by Equation 3.45. The components for higher l correspond to higher spatial frequencies and have shorter time constants. The longest time constant is the one for $l = 1$ and is defined as $\tau_{ss} = \tau_1$, so that

$$\tau_{ss} = \frac{d^2}{\pi^2 D_n}. \qquad (3.48)$$

We call τ_{ss} the *steady-state time constant,* since it limits the time taken to reach steady state in the membrane. The steady-state time constant for a membrane whose thickness is equal to that of the membrane of a cell can now be estimated. Suppose we take $d = 10$ nm and $D_n = 10^{-5}$ cm^2/s, which corresponds to the diffusion of a small molecule in water, then $\tau_{ss} \approx 10$ ns. Even if D_n were several orders of magnitude larger, the time to reach a steady-state concentration profile in the membrane would be very short for a membrane whose thickness was equal to that of cellular membranes.

3.6.2 Porous Membranes

In this section we shall consider a porous membrane composed of an impermeant matrix and pores filled with a solution in which the solute diffuses. First we shall discuss diffusion in porous membranes in which both the pores and the solute particles are large; then we shall describe diffusion in small pores. By a large solute particle we mean one whose dimensions are large compared to those of the solvent particles, so that the solvent can be treated as a continuum in which the macroscopic laws of continuum fluid dynamics hold. It appears that such theories remain valid approximations for solute diameters that are two to three times the diameters of solvent particles (Russel, 1981; Deen, 1987). By a large pore we mean a pore with dimensions much larger than those of the solute particle, so that viscous drag force on the solute particle caused by proximity to the pore walls is negligible.

3.6.2.1 Theory

Large Solutes, Large Pores

With both large solutes and large pores the permeability of the membrane can be expressed by a simple modification of Equation 3.43 as

$$P_n = \frac{D_n k_n \mathcal{P}}{d\mathcal{T}}, \qquad (3.49)$$

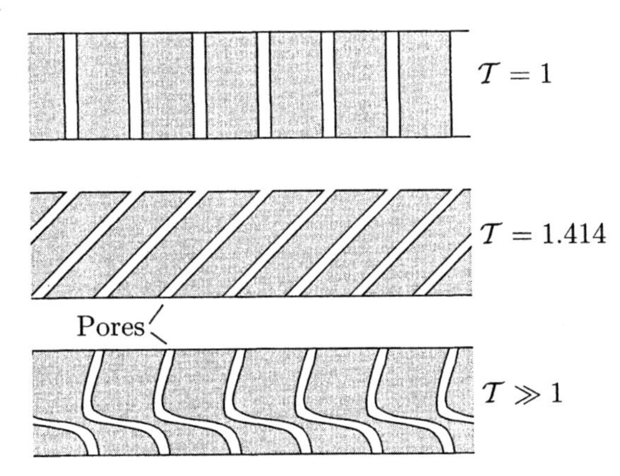

Figure 3.23 Examples of porous membranes of different tortuosities (adapted from Baker, 1987, Figure 2.1).

where $\mathcal{P}$ is called the *porosity* of the membrane and reflects the volume fraction of the membrane filled with pores, D_n is the diffusion coefficient of the solute in the free solution, and $\mathcal{T}$ is called the *tortuosity* of the membrane and reflects the fact that in a porous membrane the mean diffusion distance may exceed the thickness of the membrane, as is made clear in Figure 3.23. If the pores are water-filled right circular cylinders with radius r and the number of pores per unit area of membrane is $\mathcal{N}$, then the porosity is simply $\mathcal{N}\pi r^2$, the tortuosity is one, and the partition coefficient is one, so that the permeability is

$$P_{nl} = \frac{D_n \mathcal{N} \pi r^2}{d}. \tag{3.50}$$

For these conditions, the permeability depends on the total area of pores in the membrane, i.e., the total area available for diffusion of the solute. For a fixed pore density, the permeability is proportional to the pore radius squared.

Large Solutes, Small Pores

If the dimensions of pores become sufficiently small, the diffusion coefficient in a pore is less than that in free solution. Hence, diffusion in a porous membrane is called *hindered diffusion*. Derivation of the results for hindered diffusion in pores, which are reviewed elsewhere (Deen, 1987), is beyond our scope. To get some insight into diffusion in porous membranes, we shall examine some of the theoretical results obtained and compare these with measurements of diffusion through porous membranes with known pore geometries.

The diffusion of solutes in long cylindrical pores is derived in a manner analogous to that for diffusion in an infinite medium. Recall that Stokes's law

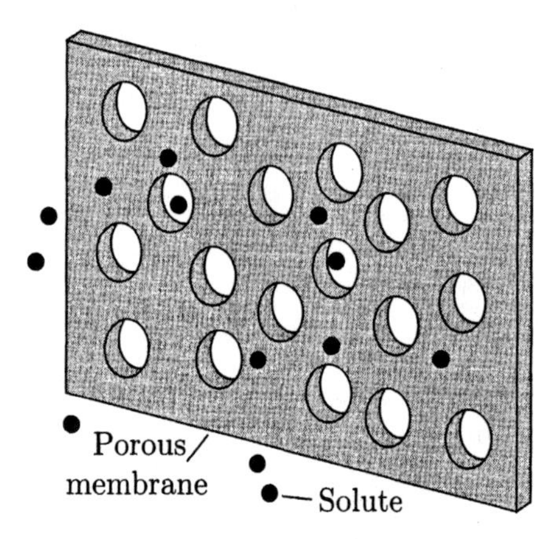

Figure 3.24 Schematic diagram of a porous membrane and solutes.

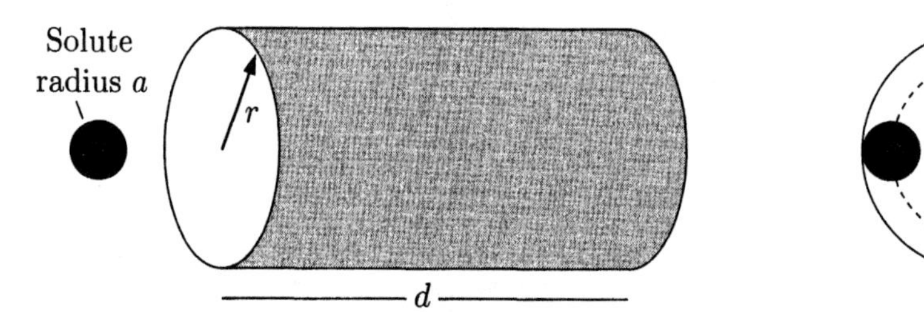

Figure 3.25 Schematic diagram of a membrane pore.

(Equation 3.22) for the force, f, required to move a mole of spherical particles at velocity v through an infinite extent of liquid with viscosity η is

$$f = (N_A 6\pi\eta a)v.$$

Now consider a membrane with cylindrical pores of radius r and length d and spherical solute particles with radius a (Figures 3.24 and 3.25). The force required to move a sphere through a pore of radius r is not known in general, although many approximate solutions exist (Deen, 1987). However, if the sphere moves down the center of the tube, the so-called *centerline approximation,* the force is (Renkin, 1954; Anderson and Quinn, 1974)

$$f = \frac{N_A 6\pi\eta a}{F(a/r)} v,$$

where

$$F(a/r) \approx 1 - 2.1044(a/r) + 2.089(a/r)^3 - 0.948(a/r)^5 \quad \text{for } a/r < 0.4,$$

and hence, by analogy with the Stokes-Einstein relation (Equation 3.27), the diffusion coefficient in a pore, D_n^*, can be defined as

$$D_n^* = F(a/r)D_n.$$

The quantity $F(a/r)$ is a decreasing function of a/r, so that for a fixed pore radius the diffusion coefficient decreases as the solute radius increases. This effect results because as the solute radius increases the solute is increasingly retarded in its motion by the viscous boundary layer of fluid at the pore walls. Thus, the term $F(a/r)$ results from the increase in friction between the solute particle and the solvent caused by the presence of the wall of the pore. Therefore, this reduction in diffusion coefficient is of hydrodynamic origin and is called the *hydrodynamic hindrance.*

The pores are assumed to be water-filled, but the concentration of solute in the pore differs from that in the bulk solution because the center of the solute particle cannot be located at a radius that exceeds $r - a$ (Figure 3.25). Thus, if the concentration of solutes is uniform in the pore at diffusive equilibrium, then the ratio of concentration in a pore to that in the surrounding medium is simply the ratio of the effective area of the pore to the total area which is $(r - a)^2/r^2 = (1 - a/r)^2$. This *steric hindrance* effect represents the partitioning of the solute between the pore and the solution at equilibrium.[6] Hence, this factor is a partition coefficient that is mathematically analogous to the partition coefficient of a homogeneous membrane although the physical mechanisms of partitioning for homogeneous and porous membranes differ.

Taking both the diffusion coefficient and the steric hindrance partition coefficient into account, the flux of solute n through a pore is

$$\phi_n^* = \frac{D_n^*(1 - a/r)^2}{d}(c_n^i - c_n^o).$$

The total flux through the membrane is

$$\phi_n = \frac{\mathcal{N}A_pD_n(1 - a/r)^2F(a/r)}{d}(c_n^i - c_n^o)$$

6. Of course, a water molecule also cannot be located at a distance of less than a_w from the pore wall, where a_w is the radius of the water molecule. Thus, a more accurate steric hindrance factor is $(1 - a/r)^2/(1 - a_w/r)^2$. This ratio has the desirable feature that the steric hindrance factor is 1 when the solute is (labeled) water so that the partition function of water in the pore and bulk solution is 1.

where $\mathcal{N}$ is the number of pores per unit area of membrane, and A_p is the area of a pore. Therefore,

$$\phi_n = P_{ns}(c_n^i - c_n^o)$$

where

$$P_{ns} = \frac{\mathcal{N}\pi r^2 D_n (1 - a/r)^2 F(a/r)}{d} = P_{nl}(1 - a/r)^2 F(a/r) \tag{3.51}$$

Thus, the factor $(1 - a/r)^2 F(a/r)$ is a measure of the hindrance, both steric and hydrodynamic, imposed by small pores.

3.6.2.2 Measurements on Porous Membranes

Now we shall examine diffusion of solutes through porous membranes made by first irradiating sheets of mica (3–5 μm thick) with a radioactive source and then etching them with hydrofluoric acid (Beck and Schultz, 1972). With these techniques Beck and Schultz generated pores that were straight with elliptical areas of eccentricity 1.5. Each membrane had a distribution of pore radii that varied about 13%, but different membranes were prepared with average pore radii in the range of 45–300 Å. Pore radii were determined directly from electron micrographs of the mica sheets and also by measuring the rate of water flow through the membrane caused by hydraulic pressure and using continuum hydrodynamics to compute the average pore radius from the rate of water flow. The two measures of average pore radius were consistent. Diffusion of seven solutes (urea, glucose, sucrose, raffinose, α-dextrin, β-dextrin, and ribonuclease) having radii in the range of 2.6–21.6 Å was measured. The results are compared with the predictions of Equation 3.51 in Figure 3.26.

The measurements and theoretical results all show that as the ratio of solute radius to pore radius increases, the normalized permeability decreases. Equation 3.51 gives a rough fit to the measurements and indicates that, to a first-order approximation, hindered permeation through porous membranes is consistent with Equation 3.51 down to pores of diameter 45.7 Å. Thus, it appears that for these dimensions continuum hydrodynamics can be used to account for hindered diffusion at least approximately. Since it has not been possible to produce simple porous membranes with pores of molecular dimensions, it has not been possible to determine whether the hydrodynamic results apply to pores of dimensions approaching 3–4 Å, which, as we shall see, may be the dimensions of putative pores in cellular membranes.

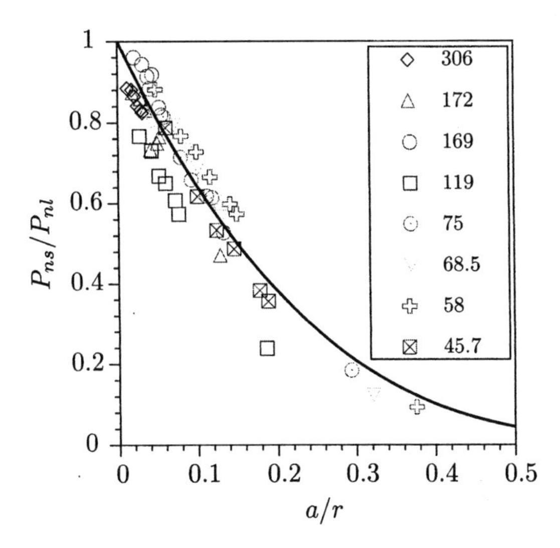

Figure 3.26 Permeability versus solute size for diffusion through mica membranes (based on Beck and Schultz, 1972, Table VII). The ordinate is the normalized permeability as defined in Equation 3.51, and the abscissa is the ratio of the radius of the solute to that of the pore. Results are shown for eight mica membranes, each with a different average pore radius, which is given in the legend in Å. The points are the measurements, and the solid line is plotted according to Equation 3.51.

3.7 Two-Compartment Diffusion

We have seen that the permeability defines the relation between flux and concentration difference across a membrane for a diffusing solute. How can we measure the permeability? If we wish to determine the permeability of a sheet of material, say mylar of a given thickness, to some solute, it is a relatively simple matter to mount the material so as to cover a hole in a partition that separates two compartments filled with solutions (Figure 3.27, left panel). Measurement of solute concentration can be accomplished using chemical or optical techniques or radiation detectors for radioactive solutes to measure both the flux of solute and the solute concentrations in the two baths. The flux can be estimated by taking the difference in solute quantity in one bath at two instants in time and dividing this difference by the time increment and the surface area of the membrane. The permeability can be measured as the ratio of the flux to the difference in concentrations between the baths. This measurement is valid if the concentrations of the baths change only incrementally over the time course of the measurements of both the flux and concentration difference. But how can we estimate the permeability of the membrane of a small cell such as an erythrocyte? The typical dimensions of an erythrocyte are those of a biconcave disc 7 μm in diameter and at most 2 μm thick. It is not presently feasible to dissect this membrane and mount it in a bath as shown in the left panel of Figure 3.27. Furthermore, the cell is so small that changes in

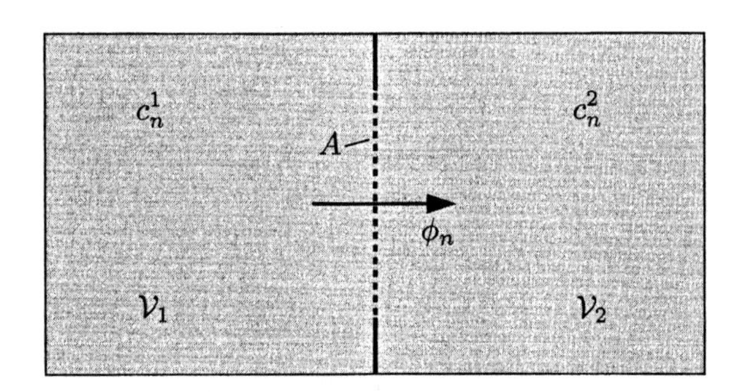

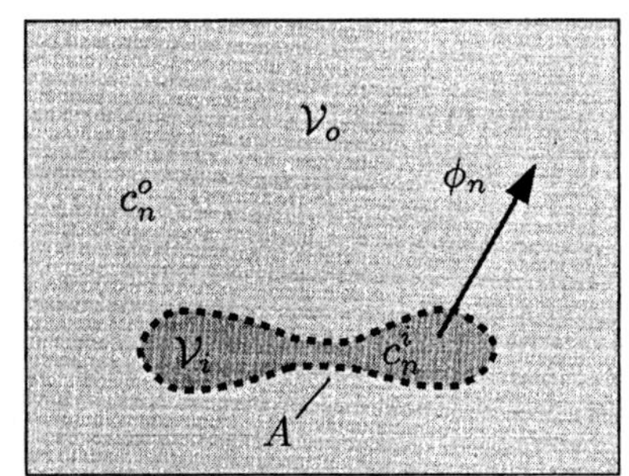

Figure 3.27 Geometry for two-compartment diffusion. The left panel shows two compartments, 1 and 2, separated by a membrane of surface area A. The flux of solute n from compartment 1 to compartment 2 is ϕ_n, and the solute concentrations and compartment volumes are c_n^1, c_n^2, $\mathcal{V}_1$, and $\mathcal{V}_2$. The right panel shows a schematic diagram of a cross section of a cell in a bath. The surface area of the cell is A, the outward flux of solute n is ϕ_n, and the volume of the cell interior and the bath are $\mathcal{V}_i$ and $\mathcal{V}_o$, respectively. The intracellular and extracellular concentrations of solute n are c_n^i and c_n^o, respectively.

concentration of diffusible solutes are very rapid, making it difficult to obtain estimates of the permeability from direct measurements of both solute concentrations and fluxes. As indicated in the right panel of Figure 3.27, diffusion between intracellular and extracellular compartments of a cell is equivalent to diffusion between two compartments. In this section we shall show that there is an alternative method of estimating the permeability of the membrane based on the time course of diffusion between two compartments.

3.7.1 Derivation for a Thin Membrane

We are considering two compartments of volume $\mathcal{V}_1$ and $\mathcal{V}_2$ separated by a membrane of area A, as shown in the left panel of Figure 3.27. We shall assume the following:

- The two compartments are well mixed, so that the concentration of solute n is uniform in space within each compartment and depends on time t. The concentrations are designated as $c_n^1(t)$ and $c_n^2(t)$.
- Solute particles are conserved, e.g., there is no chemical reaction present that either creates or destroys particles.
- The number of solute particles contained in the membrane at any time is negligibly small compared to the number in the two compartments.
- At each instant in time the relation between flux and concentration is given by Fick's law for membranes. As we shall see, this assumption is valid

if the time it takes to reach steady state in the membrane is significantly shorter than the time it takes for equilibration between the two compartments ($\tau_{ss} \ll \tau_{eq}$).

The last two assumptions constitute the *thin-membrane approximation.* After we derive the consequences of these assumptions, we will examine their validity.

To proceed further, we need to express the assumptions mathematically. If steady state applies at each instant in time in the membrane, then Equation 3.44 applies and

$$\phi_n = P_n \left(c_n^1(t) - c_n^2(t) \right), \tag{3.52}$$

where P_n is the permeability of the membrane for solute n. Conservation of solute implies that

$$\mathcal{V}_1 c_n^1(t) + \mathcal{V}_2 c_n^2(t) = N_n, \tag{3.53}$$

where N_n, the total number of moles of solute contained in the two compartments, is assumed to remain constant. The volumes of the two compartments are $\mathcal{V}_1$ and $\mathcal{V}_2$. Let $\phi_n(t)$ be the flux of solute from compartment 1 to compartment 2 at time t. The integral form of the continuity relation (Equation 3.8) is used to relate the rate of decrease of the total solute in compartment 1 to the rate at which solute leaves compartment 1. Therefore,

$$-\frac{1}{A}\frac{d\,(c_n^1(t)\mathcal{V}_1)}{dt} = \phi_n(t). \tag{3.54}$$

We assume that the volumes of the compartments remain constant and substitute Equation 3.52 into Equation 3.54 to give

$$-\frac{dc_n^1(t)}{dt} = \frac{AP_n}{\mathcal{V}_1} \left(c_n^1(t) - c_n^2(t) \right). \tag{3.55}$$

By solving for $c_n^2(t)$ from Equation 3.53 and substituting this expression into Equation 3.55, we obtain a differential equation that must be satisfied by $c_n^1(t)$,

$$\frac{dc_n^1(t)}{dt} + \frac{AP_n}{\mathcal{V}_e} c_n^1(t) = \frac{AP_n}{\mathcal{V}_1 \mathcal{V}_2} N_n, \tag{3.56}$$

where the equivalent volume of the two compartments, $\mathcal{V}_e$, is

$$\mathcal{V}_e = \frac{\mathcal{V}_1 \mathcal{V}_2}{\mathcal{V}_1 + \mathcal{V}_2}. \tag{3.57}$$

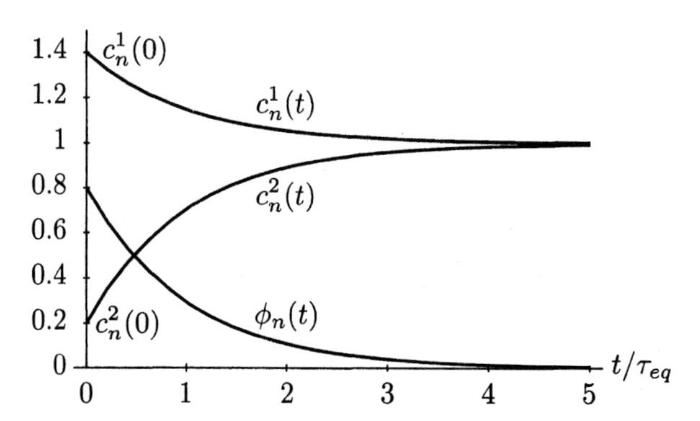

Figure 3.28 Time course of concentration in two compartments separated by a permeable membrane and the time course of the flux through the membrane. The scales have been normalized by the following choices of parameters: $N_n/(\mathcal{V}_1 + \mathcal{V}_2) = 1$, and $P_n = 2/3$. The plots show solutions for $c_n^1(0) = 1.4$ and $\mathcal{V}_1/\mathcal{V}_2 = 2$.

Equation 3.56 is a first-order ordinary linear differential equation with constant coefficients. Therefore, the solution varies exponentially in time between initial and final values (Figure 3.28), so that

$$c_n^1(t) = c_n^1(\infty) + \left(c_n^1(0) - c_n^1(\infty)\right) e^{-t/\tau_{eq}} \text{ for } t > 0, \tag{3.58}$$

where $c_n^1(0)$ is the initial concentration of solute in compartment 1, $c_n^1(\infty)$ is the final concentration, and τ_{eq} is the time constant. Using Equation 3.53 and Equation 3.54, we see that both $c_n^2(t)$ and $\phi_n(t)$ have an exponential time course with the same time constant as $c_n^1(t)$. Therefore, τ_{eq} is the time constant at which the two compartments come to equilibrium,

$$\tau_{eq} = \frac{\mathcal{V}_e}{AP_n}, \tag{3.59}$$

which can be seen directly by examining the second term on the left side of Equation 3.56, which must have units of concentration divided by time. Equation 3.59 indicates that the time constant is inversely proportional to the product of the permeability and the ratio of surface area to volume, i.e., the larger the permeability and the ratio of surface area to volume, the smaller the time constant. This result makes intuitive sense, since increasing the compartment volumes increases the total solute that needs to be equilibrated and, hence, increases the equilibration time. Furthermore, increasing either the permeability of the membrane or its surface area increases the flux through the membrane and, hence, decreases the equilibration time. Note also that τ_{eq} depends upon the *equivalent volume* $\mathcal{V}_e$, which is dominated by the smaller volume. In particular, Equation 3.57 indicates that if $\mathcal{V}_1 \gg \mathcal{V}_2$, then $\mathcal{V}_e \to \mathcal{V}_2$.

The final concentration of solute in compartment 1 can be determined by noting that at equilibrium the concentration is independent of time and, therefore, the derivative term in Equation 3.56 must be zero. Hence,

$$c_n^1(\infty) = \frac{\mathcal{V}_e}{\mathcal{V}_1\mathcal{V}_2}N_n = \frac{N_n}{\mathcal{V}_1 + \mathcal{V}_2} = \frac{\mathcal{V}_1 c_n^1(0) + \mathcal{V}_2 c_n^2(0)}{\mathcal{V}_1 + \mathcal{V}_2},$$

i.e., at equilibrium the total number of moles of solute are distributed uniformly in the total volume, $\mathcal{V}_1 + \mathcal{V}_2$. The fact that at equilibrium $c_n^1(\infty) = c_n^2(\infty) = N_n/(\mathcal{V}_1 + \mathcal{V}_2)$ can be verified by examining Equation 3.53.

3.7.2 Conditions for the Validity of the Thin-Membrane Approximation: A Specific Example

In the thin-membrane approximation, the distribution of concentration in the membrane is assumed to be the steady state distribution. In the steady state the concentration profile in the membrane is linear and Fick's law for membranes relates flux through the membrane to the concentration difference across the membrane. This condition occurs when $\tau_{eq} \gg \tau_{ss}$. That is, steady state will exist if the time it takes for the two volumes to equilibrate greatly exceeds the time it takes for the concentration profile in the membrane to reach its steady-state distribution.

To investigate the effect of the steady state and the equilibrium time constant in two-compartment diffusion, we shall examine the solution without making the thin-membrane approximation. The geometry of the problem is shown in Figure 3.29. We shall examine solutions for the concentration in the membrane and in the two compartments for a fixed membrane width of 100 μm, but for different compartment widths L. We shall also assume that the diffusion coefficient is $D = 1 \times 10^{-5}$ cm^2/s and the partition coefficient is 1. The two compartments are assumed to be well mixed, so that the concentration is uniform in each compartment.

Figure 3.30 shows the solutions at different instants in time for compartment widths of 1 cm, which are 100 times larger than the membrane thickness. The initial distribution in the membrane was arbitrarily chosen to be

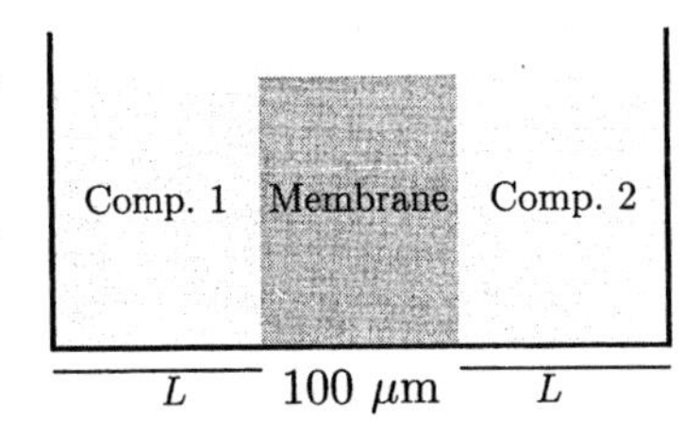

Figure 3.29 Geometry for a two-compartment diffusion problem without making the thin-membrane approximation. A membrane 100 μm wide separates two compartments of width L. The area of membrane and compartments is A.

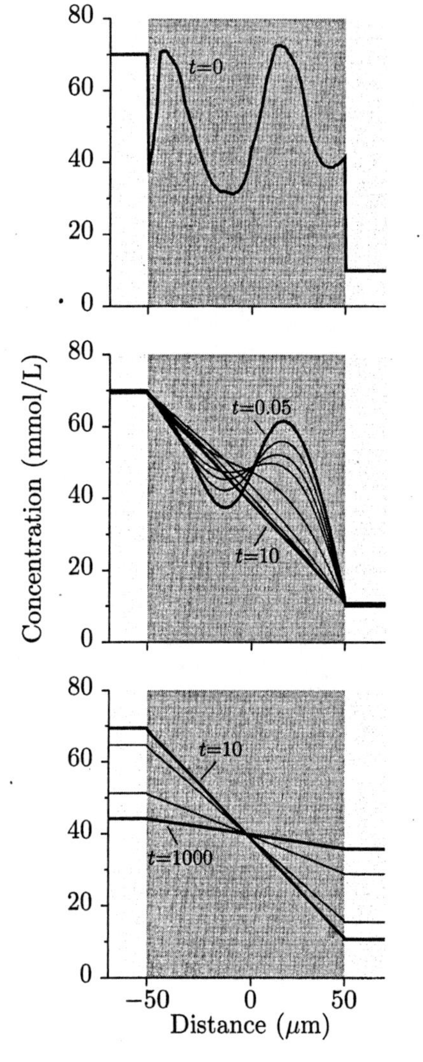

Figure 3.30 Solution for two-compartment diffusion for large compartments without making the thin-membrane approximation. The membrane is 100 μm wide and separates two compartments of width $L = 1$ cm. At time $t = 0$ the concentration in the left and right compartments are 70 and 10 mmol/L, respectively, and the concentration of solute in the membrane has been chosen arbitrarily as shown in the top panel. The diffusion equation was solved numerically with a special purpose software package developed for teaching purposes (Weiss et al., 1992). The solution is shown as a function of position in the membrane (dark shading) and for a portion of each compartment (light shading) at the times: 0.05, 0.1, 0.15, 0.2, 0.4, 1, 2, 10 s (middle panel); and at 10, 100, 500, 1000 s (bottom panel). The diffusion coefficient is $D = 1 \times 10^{-5}$ cm^2/s.

highly irregular. Nevertheless, the spatial dependence of concentration in the membrane changes in a matter of seconds to approach a linear profile. The time constant for this change is τ_{ss}. An estimate of the steady-state time constant can be computed from Equation 3.48 under the assumption that the compartment concentrations are constant. This assumption is valid for the first 10 s for the computations shown in Figure 3.30. This estimate of the steady-state time constant is $\tau_{ss} = (0.01)^2/(\pi^2 \times 10^{-5}) \approx 1$ s. Thus, numerical solutions for the two-compartment diffusion problem and the approximate estimate of the steady-state time constant are consistent. Both suggest that the profile in the membrane reaches steady state in a few seconds. The solution for $t > 10$ s exhibits a linear profile in the membrane, whose slope decreases systematically with time as the two compartments come to equilibrium. The time constant for this slower change is the equilibrium time constant. An estimate of the equilibrium time constant based on the thin-membrane approximation can be obtained from Equation 3.59. For the conditions in Figure 3.30, $\tau_{eq} = dL/(2D_n) = 500$ s. Thus, steady state in the membrane occurs 500 times faster than equilibration between the two compartments. Therefore, the time course of this solution is well described by the thin-membrane approximation. Note also that the equilibrium concentration of about 40 mmol/L is just the average concentration of the two compartments. This result occurs because with a thin membrane the quantity of solute in the membrane is negligible compared to the quantity of solute in the two compartments. Thus, the quantity of solute in the membrane does not contribute appreciably to the equilibrium concentration.

Figure 3.31 shows solutions to the diffusion equation for compartments with widths of 100 μm, which equal the membrane thickness. The concentration in the membrane reaches its steady-state linear profile in about 2 s, just as it did in the case with the large compartments shown in Figure 3.30. However, in contrast to the results with $L = 1$ cm, for $L = 100$ μm there is a significant change in the concentrations in the two compartments before the steady-state distribution is reached in the membrane. However, after about 2 s the profile of concentration in the membrane is linear, and it remains that way as the compartments equilibrate. The equilibrium time constant based on the thin-membrane approximation is $\tau_{eq} = dL/(2D_n) = 5$ s, which is only a factor of 5 greater than the estimate of the steady-state time constant. Thus, the thin-membrane approximation is only marginally successful in accounting for the diffusion between two compartments separated by a membrane when the widths of the compartments equal the membrane thickness. Note that the equilibrium concentration now exceeds the average compartment concentration, because the quantity of solute in the membrane is not negligible in this case.

As a final example, consider the case in which the compartments' widths are much smaller than the membrane thickness, as shown in Figure 3.32. The membrane still reaches steady state in about 2 s, but now there is a large change in the concentrations in the two compartments. For this case, the estimated equilibrium time constant based on the thin-membrane approximation is $\tau_{eq} = dL/(2D_n) = 0.5$, which is about half the steady-state time constant. Thus, the thin-membrane approximation is very poor for this case.

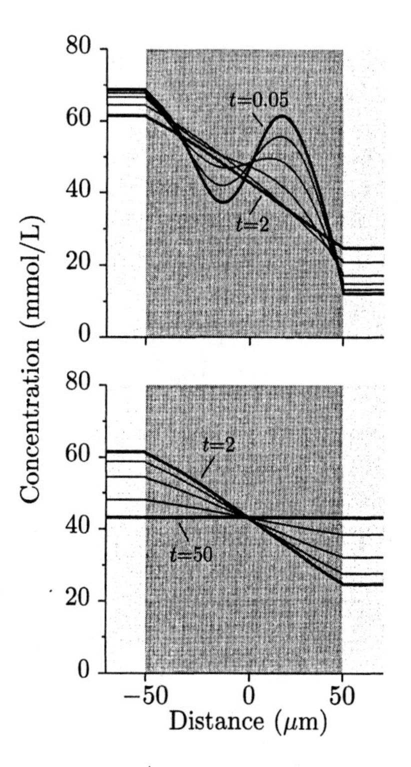

Figure 3.31 Solution for two-compartment diffusion for compartments whose widths are comparable to the membrane thickness. The membrane is 100 μm wide and separates two compartments of width $L = 100$ μm. At time $t = 0$ the concentrations in the left and right compartments are 70 and 10 mmol/L, respectively, and the concentration of solute in the membrane is the same as shown in Figure 3.30. The solution is shown as a function of position in the membrane (dark shading) and for a portion of each compartment (light shading) at the times 0.05, 0.1, 0.2, 0.4, 1, and 2 s (top panel) and at 2, 5, 10, 20, and 50 s (bottom panel).

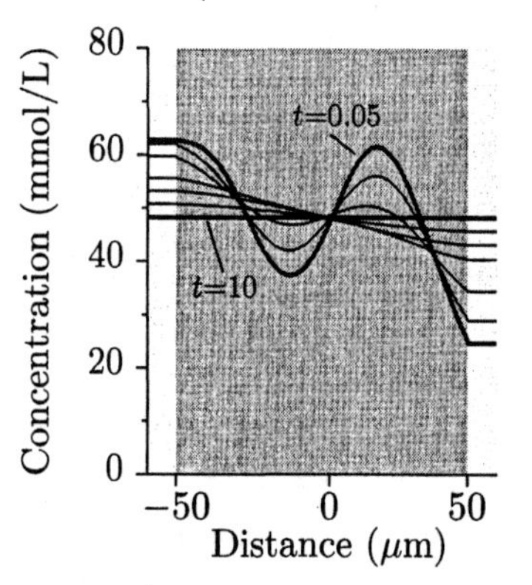

Figure 3.32 Solution for two-compartment diffusion for compartments whose widths are smaller than the membrane thickness. The membrane is 100 μm wide and separates two compartments of width $L = 10$ μm. At time $t = 0$ the concentrations in the left and right compartments are 70 and 10 mmol/L, respectively, and the concentration of solute in the membrane is the same as shown in Figure 3.30. The solution is shown as a function of position in the membrane (dark shading) and for a portion of each compartment (light shading) at the times 0.05, 0.1, 0.2, 0.5, 1, 2, and 10 s.

To determine the extent to which the thin-membrane approximation is accurate, we compute the ratio of the equilibrium to the steady-state time constants using Equations 3.59 and 3.48 as follows:

$$\frac{\tau_{eq}}{\tau_{ss}} = \frac{\mathcal{V}_e/(AP_n)}{d^2/(\pi^2 D_n)}.$$

If the permeability of the membrane is determined by simple diffusion in a homogeneous membrane, we can use Equation 3.43, which, after cancellation of common factors, yields

$$\frac{\tau_{eq}}{\tau_{ss}} = \frac{\pi^2 \mathcal{V}_e}{Adk_n}.$$

When this ratio is large, the steady-state assumption is accurate.

To see what this might mean for a typical cell, assume that the volume of external solution is large compared to that of the intracellular compartment, so that $\mathcal{V}_e \approx \mathcal{V}_i$. Also assume that the partition coefficient is one and that the cell is spherical with radius r. Then

$$\frac{\tau_{eq}}{\tau_{ss}} = \frac{\pi^2 (4/3)\pi r^3}{4\pi r^2 d} = \frac{\pi^2 r}{3d}.$$

This ratio[7] increases as r increases, but is large even for the radii of the smallest cells and even for the radii of cell organelles. Suppose that we have a 1 μm cell or organelle and a membrane whose thickness is 10 nm. Then $\tau_{eq}/\tau_{ss} \approx 300$. Therefore, the equilibrium time constant is 300 times the steady-state time constant. Under these conditions, the steady-state assumption is accurate for cellular membranes.

3.8 Measurements of Diffusion Through Cellular Membranes

3.8.1 Overton's Rules

Although living cells had been visualized by the early microscopists in the seventeenth century, the idea that living systems were made of cells was not clearly formulated until the middle of the nineteenth century. By that time, the general notion that cells were surrounded by a membrane had also been formulated, but the membrane was regarded as highly permeable to water and highly impermeable to other substances. By the end of the century, Charles Ernest Overton, an English biologist who worked primarily in Sweden, made extensive observations on the permeability of plant and animal cells to hundreds of substances and showed that cells exhibit a broad range of permeabilities to diverse substances (Collander, 1962–63). At the time, this was

7. The ratio is $\pi^2 r/2d$ for a cylindrical cell.

a revelation. Overton's motivation was to use different solutes as probes to investigate membranes. By discovering which substances permeated the membrane, Overton aimed to gain some insight into the structure of membranes.

Overton categorized substances according to their membrane permeabilities and developed a set of rules (Overton's rules) to predict the permeability of a given solute based on its physical-chemical properties. Overton's contributions to understanding transport through cellular membranes and the significance of these observations for understanding membrane structure include the following:

- Cell membranes are semipermeable, and the relative permeabilities to different solutes are similar for plant and animal cells.
- Membrane permeability is highly correlated with the solubility of the solute in organic solvents, e.g., ether, oil, etc.
- Based on the correlation between membrane permeability and solubility in organic solvents, Overton proposed that membranes were made of lipoid substances. He specifically suggested the substances cholesterol and phospholipids.
- Overton recognized that cells could concentrate certain substances at a concentration that was higher than in the extracellular environment. He understood that diffusion alone could not account for this phenomenon and clearly anticipated the discovery of processes now called *active transport*.
- Overton found that the potency of anesthetic substances was highly correlated with their lipid solubility. This finding, discovered independently by Meyer, is called the *Meyer-Overton theory of narcosis*.
- Overton found that while many cell types could continue to function in extracellular solutions lacking sodium ions (or analogs such as lithium), muscle cells did not contract in such solutions. Overton thus anticipated the sodium theory of electrically excitable cells (Hodgkin and Katz, 1949).

Overton's published results were largely qualitative. Although his plans to publish a magnum opus on cell transport and on membrane structure were not realized, his work stimulated the next generation of membrane biophysicists to test his ideas quantitatively.

3.8.2 Methods

Quantitative measurements of the permeability of membranes of plant and animal cells became more readily available in the 1920s and 1930s. Methods included some that depended on the osmotic responses of cells, which will be described more fully in Chapters 4 and 5. Direct and comprehensive measurements of the permeability of plant cell membranes to a large variety of solutes were obtained in the 1930s (Collander and Bärlund, 1933). The experiments were performed on algae *Chara ceratophylla* because they contain large cylindrical cells with diameters of about 1 mm and lengths of 1 cm. The geometry of these cells simplified estimation of their surface area and volume. The plant cells were immersed in solutions of various solutes for known intervals of time. Then the cell sap was squeezed out and its contents analyzed by microchemical methods. Measurements for different time intervals allowed the kinetics of concentration changes to be measured.

The permeability was computed from these kinetic measurements using the two-compartment diffusion model. The procedure follows directly from Equation 3.58, from which the intracellular concentration in the plant cell can be expressed in the form

$$\frac{c_n^i(t) - c_n^i(\infty)}{c_n^i(0) - c_n^i(\infty)} = e^{-t/\tau_{eq}}, \tag{3.60}$$

which can also be expressed as

$$\ln\left(\frac{c_n^i(t) - c_n^i(\infty)}{c_n^i(0) - c_n^i(\infty)}\right) = -t/\tau_{eq}, \tag{3.61}$$

where $c_n^i(0)$, $c_n^i(t)$, and $c_n^i(\infty)$ are the values of the intracellular concentration at time 0, t, and ∞. Thus, if the normalized concentration (seen on the left-hand side of Equation 3.60) is plotted in logarithmic coordinates versus time, the result is a straight line. The time constant, which determines the value of the permeability, can be estimated from the slope of the straight line. Measurements of the concentration of ethylene glycol in plant cells are shown in Figure 3.33 and compared with both Equation 3.60 and Equation 3.61.

An alternative experimental procedure (Collander, 1954) was to first load the cell with a solute by placing the cell in a solution containing the solute whose permeability was to be measured. After a time interval sufficient to allow the concentrations inside and outside the cell to come to equilibrium, the loaded cell was transferred to a glass vial filled with a solution that initially contained none of the solute. The cell was left in this vial for a fixed time interval, during which some of the solute diffused out of the cell and into the vial. The cell was then transferred to a fresh vial for another fixed time interval. The process was repeated with a succession of vials (Figure 3.34). The amount of solute in each vial was determined by chemical means. This se-

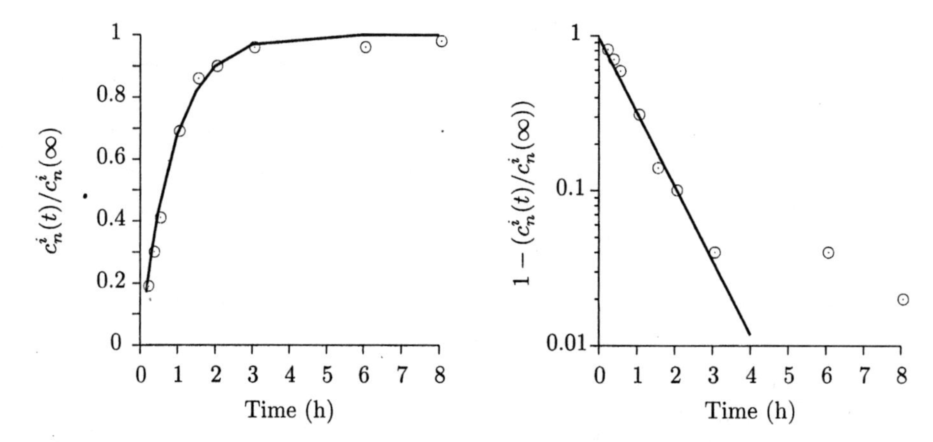

Figure 3.33 Measurements of diffusion of ethylene glycol through the membrane of a cell of the plant *Chara ceratophylla* (Collander and Bärlund, 1933). The left-hand side shows a plot of measurements (points) of the intracellular concentration of ethylene glycol normalized to the final concentration. The line is an exponential with a time constant of 0.99 h according to Equation 3.60 with the initial concentration of ethylene glycol assumed to be $c_n^i(0) = 0$. The right-hand side shows the same data plotted on semilogarithmic coordinates according to Equation 3.61. A straight line with a slope corresponding to $\tau_{eq} = 0.99$ h is superimposed on the measurements. This line was fitted to the first seven points using linear regression analysis (with a correlation coefficient of 0.99).

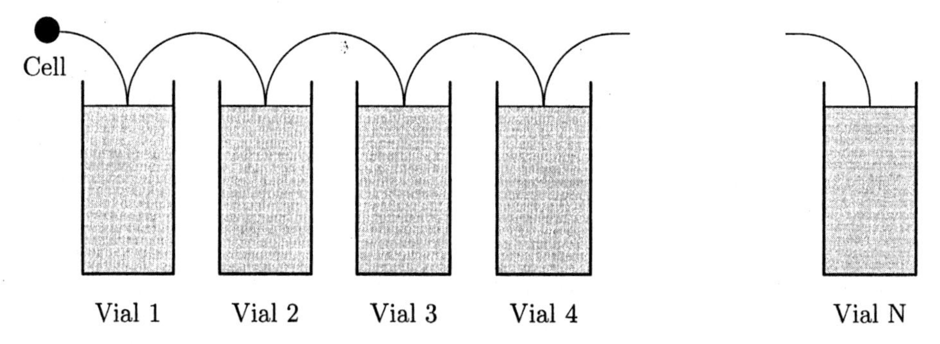

Figure 3.34 A method used to measure permeability of a cellular membrane to a solute. The cell, loaded with the solute, is placed for fixed intervals of time in a sequence of vials that initially do not contain the solute. Analysis of the contents of the vials is used to measure the rate of diffusion of solute out of the cell. The permeability is measured from the equilibrium time constant, which is estimated from the time sequence of vial contents.

quence of measurements gave the amount of solute transported out of the cell in successive time intervals. As in the experiments described above, exponential functions were fitted to these measurements, and the permeability was estimated from the measured time constant using an equation identical to Equation 3.60, but for concentration in the extracellular compartment c_n^o (see Problem 3.13).

The methods used by these workers were adequate for solutes and cells for which the time constant (Equation 3.59) was relatively large, but not for small (e.g., < 1 s) time constants. Small cells and highly permeant solutes yield small time constants. It was simply not feasible to sample intracellular concentrations of arbitrary solutes rapidly enough or to manipulate cells into a sequence of vials on such a time scale. Thus, new methods were devised to measure small time constants. In the 1950s and 1960s, devices called *flow cytometers* were developed to automate cytological measurements of several types. A flow cytometer is a device in which cells in solution flow through a tube while detectors measure some physical-chemical property of the cells or of the external solution as the flow proceeds (Shapiro, 1988). Many flow cytometers operate by detecting some optical signal (e.g., change in opacity, fluorescence, etc.) that indicates a change in solute concentration. In addition, flow cytometers have been adapted to perform analyses that are completely analogous to the method illustrated in Figure 3.34, but on a much faster time scale.

In the schematic diagram shown in Figure 3.35, the cells are loaded with the solute of interest, and the cells are mixed rapidly with a solution that does not contain the solute. The cells in solution are forced to flow rapidly and with uniform velocity down a tube whose fluid composition is sampled at known distances along the length of the tube. The key is that the cell suspension flows at constant velocity. If it does, then solution samples taken at fixed distances along the flow tube sample the solution at fixed times after it has exited the mixing chamber. Thus, if the solute is diffusing out of the cells, the samples represent the solute that has exited the cells in a given time interval after the onset of mixing. One method used to detect the solute is to use radioactively tagged solutes. The radioactivity of the fluid samples in the cytometer is then a measure of the efflux of the radioactive tracer. With such methods the permeability can be estimated even for solutes with very high permeabilities diffusing through the membranes of relatively small cells, as illustrated in Figure 3.36. Use of these methods, based on the two-compartment diffusion model, has allowed measurements of the permeability of a large number of solutes through cellular membranes. These measurements have led to certain generalizations, which we shall discuss next.

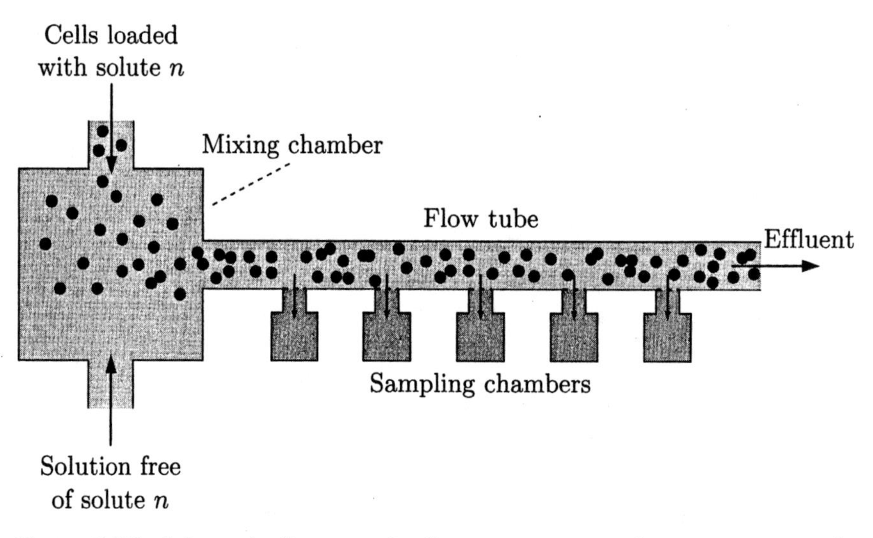

Figure 3.35 Schematic diagram of a flow cytometer used to measure membrane diffusion. Cells loaded with solute n are injected into one port of the mixing chamber, and solution free of solute n is injected into the other. The mixture flows down the flow tube, and the external solution is collected in the sampling chambers, which contain filters to exclude the cells.

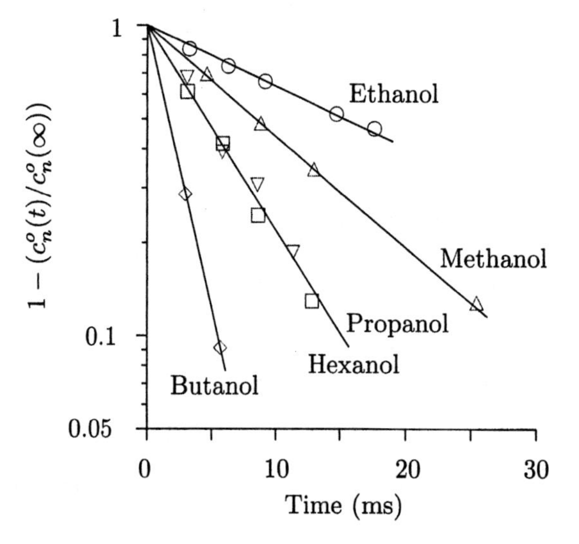

Figure 3.36 The dependence of extracellular concentration on time for human erythrocytes as determined by flow cytometry for five different alcohols plotted on semilogarithmic coordinates according to Equation 3.61 (adapted from Brahm, 1983, Figure 4). The cells were loaded with alcohol that was radioactively labeled. The concentration of alcohol for all these measurements was 1 mmol/L.

3.8.3 A Seminal Study

To test Overton's observations quantitatively, Collander and Bärlund (1933) measured the permeability of membranes of a species of plant algal cell to eighty-seven solutes and compared these permeabilities to the solutes' solubility in two organic solvents: olive oil and ether. The solutes had molecular weights that spanned the range of 19–480 and partition coefficients that spanned a range of more than four log units. This seminal study produced the most frequently reproduced data in the history of membrane transport studies (Figure 3.37). The type of plot shown in Figure 3.37, which plots a measure of the permeability of a solute through a cell membrane as a function of the partition coefficient between some organic solvent and water, is often called a *Collander plot*. The results clearly show a strong correlation between solute permeability in *Chara* membrane and solute olive oil:water partition coefficient, thus quantifying Overton's observations. A regression line is shown plotted through the data. The regression line is the line that best fits the data in the sense that the parameters of the line are chosen so as to minimize the

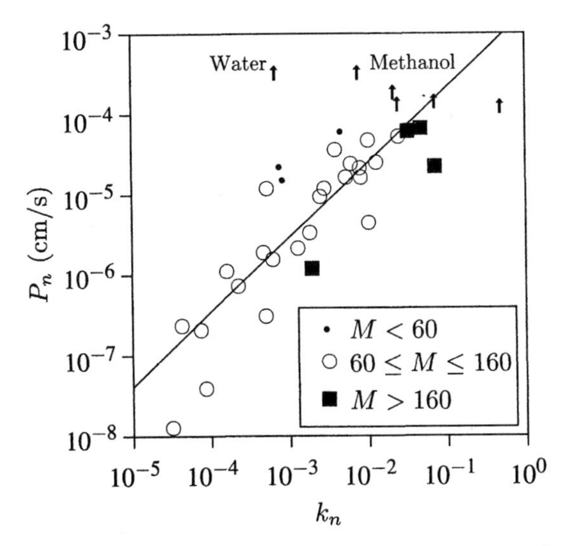

Figure 3.37 The permeability, P_n, of the membrane of the cell of the plant *Chara ceratophylla* versus the olive oil:water partition coefficient, k_n, for a number of solutes plotted on double logarithmic coordinates (adapted from Collander and Bärlund, 1933 Tables 6 and 10). The symbols are used to indicate the molecular weights of the solutes. The points with arrows indicate that the permeability of that solute was too large to be measured, but is above the value at the base of the arrow. The regression line has the equation $\log P_n = 0.930 \log k_n - 2.74$ with a correlation coefficient of 0.86.

mean-squared difference between the measurements and the line. To a first-order approximation, the regression analysis indicates that the permeability is proportional to the oil:water partition coefficient. The data also indicate that all the solutes of low molecular weight are above the regression line and all the solutes of high molecular weight are below the regression line. Thus, for solutes with the same partition coefficient, those with lower molecular weight are more permeant. A few points lie particularly far from the regression line, including the points for water and for methanol; the permeability of the membrane for these solutes far exceeds that which appears to fit most of the measurements.

In an effort to understand the mechanism(s) of membrane permeation, this basic experiment has been refined and repeated for more than half a century on several types of membranes, a large variety of solutes, and several organic solvents. In the process, a number of technical and conceptual hurdles had to be crossed. For example, when these studies were extended to cell-solute combinations for which equilibration of intra- and extracellular compartments was rapid, it was important to take into account the presence of stagnant or unstirred layers at the membrane surfaces (see Problem 3.8). Furthermore, some of the solutes used in the earlier studies are now known to be transported through the membranes of cells by mechanisms that differ fundamentally from diffusion; these will be discussed in Chapter 6.

From such measurements there evolved two principal hypotheses for the mechanisms of permeation of nonelectrolyte solutes through cellular membranes: the *dissolve-diffuse mechanism* and the *aqueous pore* or *water channel mechanism*. These two mechanisms are not mutually exclusive; we shall discuss them here briefly.

3.8.4 The Dissolve-Diffuse Mechanism

In the dissolve-diffuse mechanism, permeant solutes first dissolve in the membrane and then diffuse through it according to the theory of diffusion through a homogeneous membrane developed in Section 3.6. A direct test of this theory requires knowledge of both the membrane:water partition coefficients and the diffusion coefficients of solutes in the membrane. Neither of these quantities has been measured by independent methods. Thus, indirect tests of the hypothesis have been performed. The partitioning of solutes is measured in a solvent, and these partition coefficients are correlated with the permeabilities of the solutes in cellular membranes. To gain some appreciation of the reasoning involved, we shall analyze the most extensive study of the relation of permeability to the partition coefficient and molecular weight published to

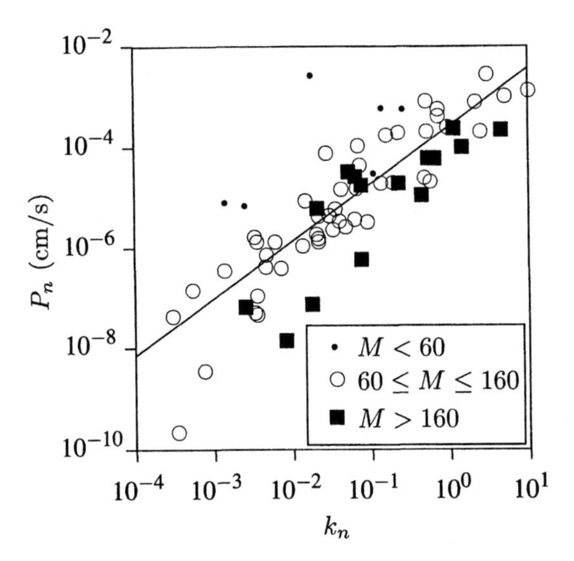

Figure 3.38 Collander plot of the permeability, P_n, of the membrane of a cell of the plant *Nitella mucronata* versus the ether : water partition coefficient, k_n (Collander, 1954). The regression line has the equation $\log P_n = 1.14 \log k_n - 3.58$ with a correlation coefficient of 0.83.

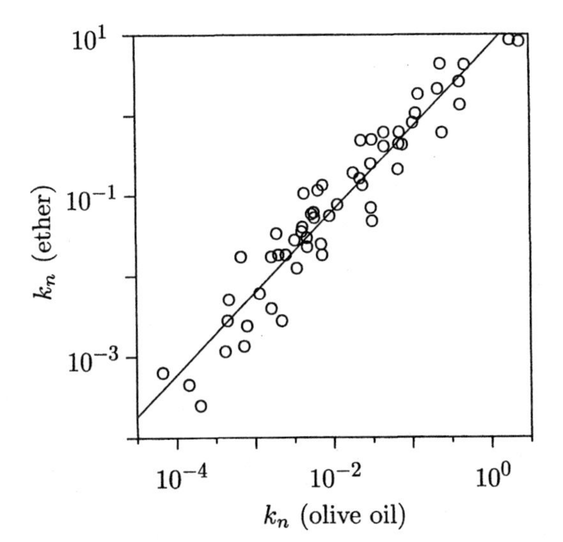

Figure 3.39 Relation between partition coefficients in ether and in olive oil (Collander, 1954). The regression line has the equation $\log k_n^{ether} = 1.02 \log k_n^{oil} - 0.86$ with a correlation coefficient of 0.96.

date (Collander, 1954). The results are illustrated in Figure 3.38. As with the earlier studies, these results show that the permeabilities of a large number of solutes are correlated with partitioning in an organic solvent, in this particular case ether. Because there is high correlation between partition coefficients in olive oil and in ether (Figure 3.39), the permeability of a solute is also highly correlated with the olive-oil : water partition coefficient.

Results such as those shown in Figure 3.38 and similar results obtained with different solvents raise the question, Which solvent best resembles partitioning in membranes? One criterion for deciding which solvent is best is to choose that solvent for which the relation of permeability to partition coefficient shows the least scatter. Organic solvents including olive oil, ether, hexadecane, and octanol have been studied, and some do lead to less scatter in the relation between permeability and partition coefficient. This result and the physical-chemical properties of the solvents have been used to suggest that some solvents better resemble cellular membranes than others, at least with respect to partitioning. In addition to the choice of model solvent, which other factors contribute to the scatter seen in the relation of permeability to partition coefficient? Can the dependence of the permeability on partition coefficient and molecular weight be factored in an insightful manner? Recall that $D_n \propto M^{-1/2}$ (Figure 3.11) for $M < 10^3$ daltons for solutes diffusing in water. If we assume that diffusion in membranes has the same dependence on molecular weight, substitution of this expression into Equation 3.43 gives $P_n M^{1/2} \propto k_n$. That is, the product of the permeability and the square root of the molecular weight should correlate with the membrane : solution partition coefficient for any solute that is transported through the membrane by simple diffusion in a fluid such as water. These size-normalized permeabilities (Figure 3.40) also show a dependence of permeability on the ether : water partition coefficient. However, a dependence of the permeability on the molecular weight remains. Although the scatter is measurably reduced, the reduction is not large (the correlation coefficient increases from 0.83 to 0.86).

Another approach to interpreting measurements of membrane permeability relies on the finding that the diffusion coefficient depends more steeply on molecular dimensions for diffusion in polymers than in water (Figure 3.12). This result has suggested (Lieb and Stein, 1986) that the permeability may be factored into a product of two terms, one of which represents the diffusion coefficient, which depends steeply on the solute's dimensions, and the other of which depends on its solubility in organic solvents. For example, suppose we try to fit the data in Figure 3.38 using a function of the form $\log P_n = A + \alpha \log k_n + \beta \log M$ and compute values of the parameters A, α, and β that minimize the mean-squared difference between the measurements and this function, i.e., compute a two-dimensional linear regression to the data in logarithmic coordinates. The results are that $A = 0.9 \pm 0.64$, $\alpha = 1.2 \pm 0.07$, and $\beta = -2.2 \pm 0.3$ (with a correlation coefficient of 0.91). These results suggest that the permeability varies approximately with the square of the molecular weight, which is consistent with diffusion in soft polymers (Figure 3.12). The quantity $P_n M^{2.2}$ is strongly correlated with the partition coefficient, but

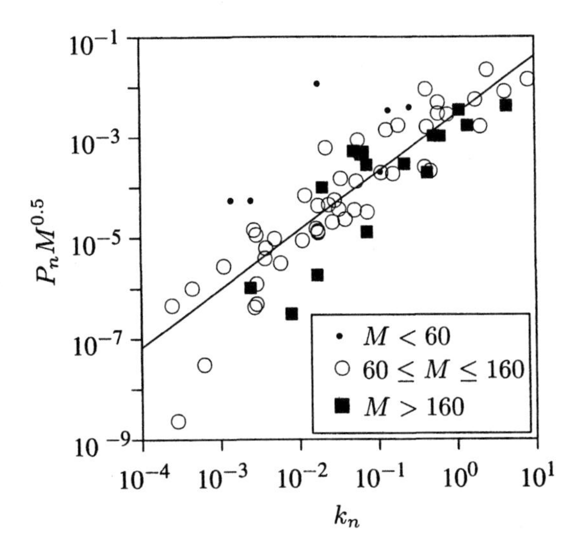

Figure 3.40 Collander plot of the permeability times the square root of the molecular weight ($P_n M^{0.5}$) versus the ether:water partition coefficient, k_n, for the same solutes shown in Figure 3.38. The regression line has the equation $\log(P_n M^{0.5}) = 1.16 \log k_n - 2.55$ with a correlation coefficient of 0.86.

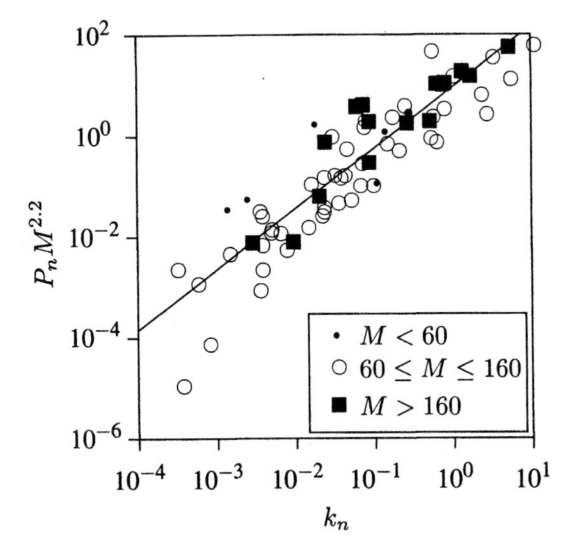

Figure 3.41 Collander plot of the permeability times the 2.2 power of the molecular weight ($P_n M^{2.2}$) versus the ether:water partition coefficient, k_n, for the same solutes shown in Figure 3.38. The regression line has the equation $\log(P_n M^{2.2}) = 1.20 \log k_n - 0.96$ with a correlation coefficient of 0.90.

no longer shows a strong variation with molecular weight. The scatter in the relation of $P_n M^{2.2}$ to k_n is smaller (Figure 3.41). These results suggest that the dependence of permeability on both partition coefficient and molecular weight can be accounted for by the dissolve-diffuse mechanism in a substance that is a lipoid polymer.

The types of measurements obtained in plant cells have also been obtained in erythrocytes using more modern techniques, but for a less extensive

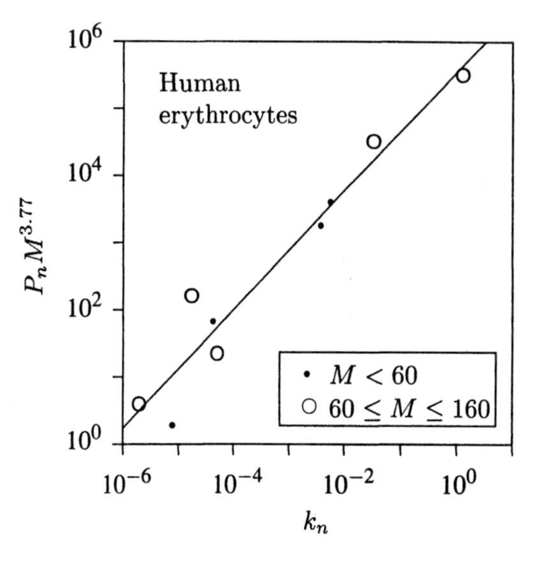

Figure 3.42 Collander plot of the permeability times the 3.77 power of the molecular weight versus the hexadecane:water partition coefficient, k_n, for erythrocytes (Lieb and Stein, 1986). The regression line has the equation $\log(P_nM^{3.77}) = 0.88 \log k_n + 5.53$ with a correlation coefficient of 0.97.

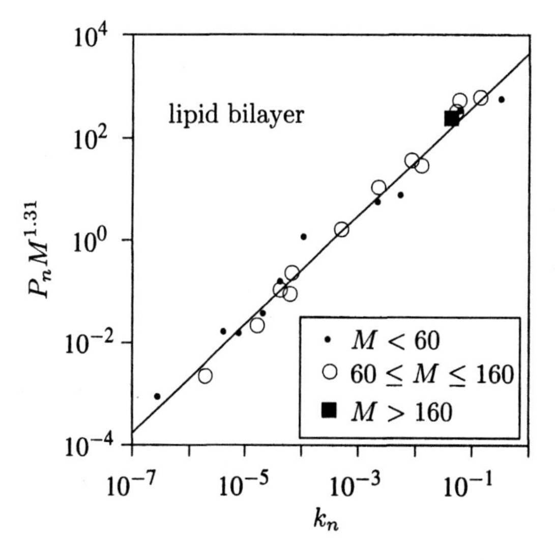

Figure 3.43 Collander plot of a lipid bilayer with the permeability multiplied by the 1.31 power of the molecular weight versus the hexadecane:water partition coefficient (Walter and Gutknecht, 1986). The regression line has the equation $\log(P_nM^{1.31}) = 1.06 \log k_n + 3.65$ with a correlation coefficient of 0.99.

set of solutes than is available for *Nitella*. The results are similar to those obtained for *Nitella*. The correlation of the permeability with the partition coefficient is strong, but shows a great deal of scatter. When normalized for the molecular weight of the solute, the correlation between the size-normalized permeability and the partition coefficient is very strong (Figure 3.42).

Similar measurements have also been made on artificial lipid bilayer membranes (Figure 3.43). These results show a very tight correlation between the

Table 3.6 Membrane permeabilities of selected solutes in *Chara, Nitella,* human erythrocyte, and artificial lipid membranes (Collander, 1954; Stein, 1990). M is the molecular weight, and k is the olive oil : water partition coefficient.

Solute characteristics			*Membrane permeability (cm/s)*			
Name	M	k	*Chara ceratophylla*	*Nitella mucronata*	Human erythrocytes	Artificial lipid
Water	18	1.3×10^{-3}	6.6×10^{-4}	2.5×10^{-3}	1.2×10^{-3}	2.2×10^{-3}
Formamide	45	1.1×10^{-6}	2.2×10^{-5}	7.6×10^{-6}	1.1×10^{-6}	1.0×10^{-4}
Ethanol	46	3.6×10^{-2}	1.6×10^{-4}	5.5×10^{-4}	2.1×10^{-3}	
Ethanediol	58	4.9×10^{-4}	1.1×10^{-5}		2.9×10^{-5}	8.8×10^{-5}
Butyramide	87	1.1×10^{-6}	5.0×10^{-5}	1.4×10^{-5}	1.1×10^{-6}	
Glycerol	92	7.0×10^{-5}	2.0×10^{-7}	3.2×10^{-9}	1.6×10^{-7}	5.4×10^{-6}
Erythritol	122	3.0×10^{-5}			6.7×10^{-9}	

size-normalized permeability and, in the case shown, the hexadecane partition coefficient. Ninety-nine percent of the variability in the size-normalized permeability can be accounted for by the variation in the partition coefficients of the solutes. Thus, other sources of variability, including those attributable to the measurement techniques, contribute only 1% of the variability. Furthermore, the slope of the relation is close to unity, suggesting that for these membranes the size-normalized permeability is proportional to the partition coefficient. The tight correlation between these variables in hexadecane, as opposed to other organic solvents, has been used to argue that diffusion in membranes resembles diffusion in hexadecane more than it does in other organic solvents.

Table 3.6 shows a comparison of measurements of permeability in four preparations for several solutes along with their molecular weights and olive oil : water partition coefficients. While values of the permeability for the same solute can vary a great deal among the different cell types, the dependence of the permeability on molecular weight and on partition coefficient is generally similar for all cell types. For example, the permeability is high for ethanol and low for glycerol in all four preparations. Glycerol with its 3 OH groups (Figure 3.44), each capable of making hydrogen bonds, is a polar molecule and is highly soluble in water and poorly soluble in organic solvents. This structure fits with its low permeability through membranes as shown in Table 3.6. Formamide and ethanol (Figure 3.45) have similar molecular weights, but very different olive oil : water partition coefficients. Ethanol is highly soluble in organic solvents, while formamide is poorly soluble in organic solvents. Thus,

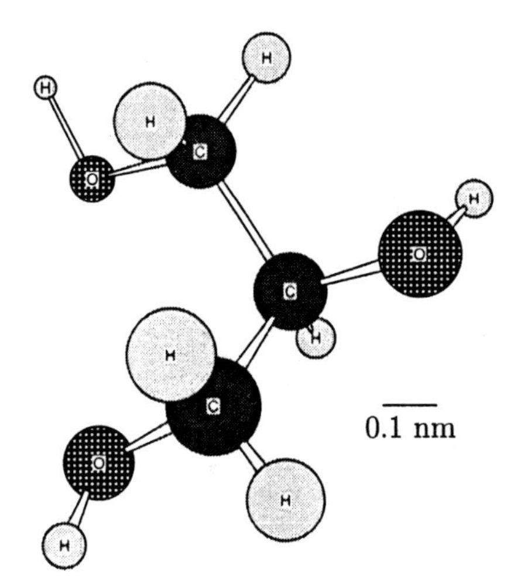

Figure 3.44 The structure of glycerol ($C_3H_8O_3$).

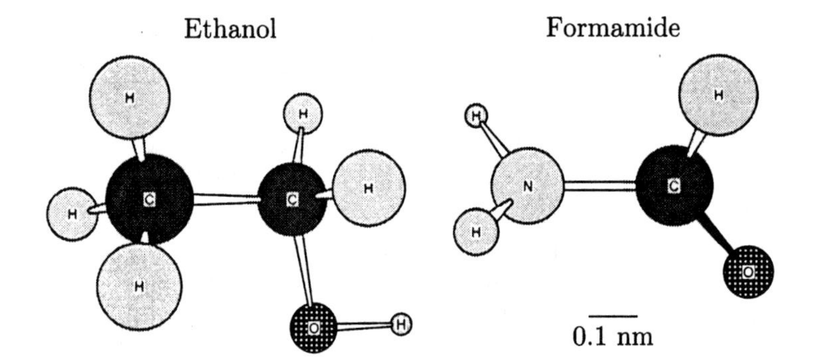

Figure 3.45 Comparison of the structures of ethanol (C_2H_6O) and formamide (NH_3O).

the olive oil:water partition coefficient of ethanol is more than 1000 times that of formamide. Consistent with the dissolve-diffuse hypothesis, membranes are much more permeable to ethanol than to formamide. In summary, the corpus of extant data suggests that there are a large number of substances whose transport through cellular membranes can be accounted for by the dissolve-diffuse mechanism.

3.8.5 The Water Channel Hypothesis

A different hypothesis arose in the 1920s and 1930s with the availability of the first direct measurements of permeability of solutes in plant and animal cells, and that hypothesis is still viable (Solomon, 1968, 1986). The idea was motivated by the observation that small, poorly lipid-soluble solutes, e.g., water, permeate membranes well. The hypothesis was that these molecules permeate membranes by a different mechanism than the dissolve-diffuse

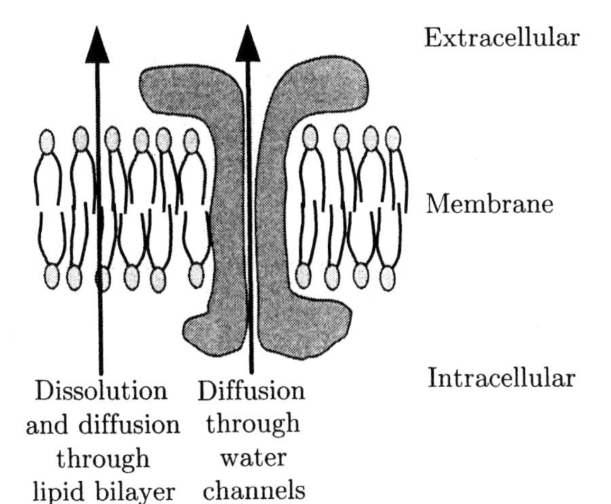

Figure 3.46 Schematic diagram illustrating two mechanisms for diffusion of solutes through membranes: one dissolution and diffusion through the lipid bilayer and the other diffusion through water channels in the membranes.

mechanism, namely by diffusion through aqueous pores or water channels in the membranes. Thus, in this hypothesis we imagine two pathways for diffusion through a membrane, one directly through the lipid bilayer and another through water channels that traverse the membrane, perhaps via proteinaceous macromolecules that insert into the membrane's matrix as shown schematically in Figure 3.46. The existence of water channels also has implications for the transport of water under a hydrostatic or osmotic pressure as well as for the coupling of water and solute transport. We shall return to this hypothesis in later chapters when we have developed the background required to understand the implication of such water channels on both diffusive and convective transport.

The dissolve-diffuse hypothesis and the water channel hypothesis are not in conflict. There is little doubt that the dissolve-diffuse hypothesis accounts for the transport of many solutes across membranes. The importance of a water channel pathway for transmembrane solute diffusion remains to be determined.

Appendix 3.1 Moments of the Binomial Distribution

To compute the moments of $W(m,n)$, where

$$W(m,n) = \frac{n!}{((n+m)/2)!\,((n-m)/2)!}\left(\frac{1}{2}\right)^n,$$

it is simpler to consider a more general case of the binomial distribution. First let $n_+ = (n+m)/2$, and let

$$F(n_+, n) = C_{n_+,n} p^{n_+} q^{n-n_+},$$

where

$$C_{n_+,n} = \frac{n!}{(n_+)!(n-n_+)!}$$

and $p + q = 1$. Note that for $p = q = 1/2$ and for $n_+ = (n+m)/2$, $F(n_+, n)$ equals $W(m, n)$. Hence, if we determine the moments of $F(n_+, n)$, we will have determined the moments of $W(m, n)$.

First we find the zeroth moment of $F(n_+, n)$, which is

$$\sum_{n_+=0}^{n} F(n_+, n). = \sum_{n_+=0}^{n} C_{n_+,n} p^{n_+} q^{n-n_+}.$$

By the binomial theorem,

$$\sum_{n_+=0}^{n} C_{n_+,n} p^{n_+} q^{n-n_+} = (p+q)^n = 1.$$

Hence,

$$\sum_{n_+=0}^{n} F(n_+, n) = 1.$$

The first moment or mean of $F(n_+, n)$ is

$$\overline{n_+} = \sum_{n_+=0}^{n} n_+ F(n_+, n) = \sum_{n_+=0}^{n} n_+ C_{n_+,n} p^{n_+} q^{n-n_+},$$

which can be written as

$$\overline{n_+} = \sum_{n_+=0}^{n} p \frac{d}{dp} (C_{n_+,n} p^{n_+} q^{n-n_+}) = p \frac{d}{dp} \left(\sum_{n_+=0}^{n} F(n_+, n) \right)$$

and yields

$$\overline{n_+} = p \frac{d}{dp} (p+q)^n = pn(p+q)^{n-1} = pn.$$

The second moment can be computed in a similar manner as follows

$$\overline{n_+^2} = \sum_{n_+=0}^{n} n_+^2 F(n_+, n).$$

Note that

$$p^2 \frac{d^2}{dp^2} F(n_+, n) = (n_+^2 - n_+) F(n_+, n),$$

and

$$\overline{n_+^2} = \overline{n_+} + p^2 \frac{d^2}{dp^2} ((p+q)^n) = pn + p^2 n(n-1).$$

For $p = 1/2$, $\overline{n_+} = n/2$, and $\overline{n_+^2} = n/2 + n(n-1)/4$. Therefore,

$$\overline{m} = 2\overline{n_+} - n = 0,$$

as was determined in Section 3.2.3, and

$$\overline{m^2} = \overline{(2n_+ - n)^2} = 4\overline{n_+^2} - 4n n_+ + n^2 = 4\,(n/2 + n(n-1)/4) - 4n(n/2) + n^2$$

$$\overline{m^2} = 2n + n^2 - n - 2n^2 + n^2 = n.$$

Appendix 3.2 Moments of the Gaussian Distribution

We shall first prove that the area under the Gaussian probability density function is unity and then find expressions for the moments of the function. Let

$$p(x) = \frac{1}{\sqrt{2\pi}\sigma} e^{-x^2/2\sigma^2}$$

and

$$I = \int_{-\infty}^{+\infty} \frac{1}{\sqrt{2\pi}\sigma} e^{-x^2/2\sigma^2}\, dx.$$

It is convenient to change variables as follows:

$$u = \frac{x}{\sigma}, \quad \text{and} \quad du = \frac{dx}{\sigma}.$$

Then

$$I = \int_{-\infty}^{+\infty} \frac{1}{\sqrt{2\pi}} e^{-u^2/2}\, du.$$

To evaluate I, we form the square of I and evaluate the volume integral under the Gaussian volume of revolution in cylindrical coordinates (Figure 3.47).

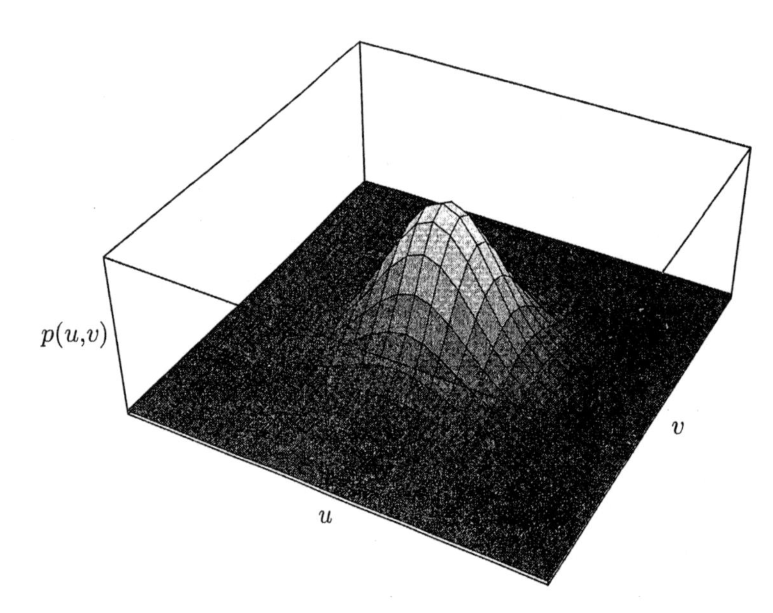

Figure 3.47 Two-dimensional Gaussian density function, $p(u,v) = 1/(2\pi)\exp(-(u^2 + v^2)/2)$.

Consider

$$\begin{aligned} I^2 &= \int_{-\infty}^{+\infty} \frac{1}{\sqrt{2\pi}} e^{-u^2/2}\,du \int_{-\infty}^{+\infty} \frac{1}{\sqrt{2\pi}} e^{-v^2/2}\,dv \\ &= \frac{1}{2\pi} \int_{-\infty}^{+\infty} \int_{-\infty}^{+\infty} e^{-(u^2+v^2)/2}\,du\,dv. \end{aligned} \tag{3.62}$$

To change variables from rectangular to cylindrical coordinates, let $u^2 + v^2 = r^2$. A differential element of area is expressed as $du\,dv$ in rectangular coordinates and as $r\,dr\,d\theta$ in cylindrical coordinates. In cylindrical coordinates, Equation 3.62 becomes

$$I^2 = \frac{1}{2\pi} \int_0^{2\pi} d\theta \int_0^{\infty} e^{-r^2/2} r\,dr\,d\theta = \int_0^{\infty} e^{-r^2/2} r\,dr. \tag{3.63}$$

Let $z = r^2/2$. Therefore, $dz = r\,dr$. Then Equation 3.63 becomes

$$I^2 = \int_0^{\infty} e^{-z}\,dz = 1.$$

The moments of a function are defined as

$$\mu_n = \int_{-\infty}^{\infty} x^n p(x)\,dx.$$

To find the moments of $p(x)$, it is convenient to first define the Fourier transform of $p(x)$, $P(\omega)$, which is

$$P(\omega) = \int_{-\infty}^{\infty} p(x)e^{-j\omega x}\,dx.$$

Then

$$\frac{d^n P(\omega)}{d\omega^n} = \int_{-\infty}^{\infty} (-jx)^n p(x)e^{-j\omega x}\,dx,$$

from which we find that

$$\mu_n = \frac{1}{(-j)^n}\left[\frac{d^n P(\omega)}{d\omega^n}\right]_{\omega=0}.$$

Thus, the nth moment of $p(x)$ can be found from the nth derivative of $P(\omega)$ evaluated at $\omega = 0$.

To find the moments of the Gaussian function, we first need to find its Fourier transform. Let the Gaussian function be denoted as

$$p(x) = \frac{1}{\sqrt{2\pi}\sigma}e^{-(x-m)^2/2\sigma^2}. \tag{3.64}$$

Hence,

$$P(\omega) = \int_{-\infty}^{\infty} \frac{1}{\sqrt{2\pi}\sigma}e^{-(x-m)^2/2\sigma^2}e^{-j\omega x}\,dx.$$

If we let $u = x - m$, then

$$P(\omega) = e^{-j\omega m}\int_{-\infty}^{\infty} \frac{1}{\sqrt{2\pi}\sigma}e^{-u^2/2\sigma^2}e^{-j\omega u}\,du. \tag{3.65}$$

By completing the square in the exponent of Equation 3.65, we obtain

$$P(\omega) = e^{-j\omega m}\int_{-\infty}^{\infty} \frac{1}{\sqrt{2\pi}\sigma}e^{-((u/\sqrt{2}\sigma)+(j\omega\sigma/\sqrt{2}))^2}e^{-\omega^2\sigma^2/2}\,du.$$

Let $z = u/\sigma + j\omega\sigma$. Then

$$P(\omega) = e^{-j\omega m}e^{-\omega^2\sigma^2/2}\int_{-\infty}^{\infty} \frac{1}{\sqrt{2\pi}}e^{-z^2/2}\,dz.$$

But the integral has value unity, and

$$P(\omega) = e^{-j\omega m}e^{-\omega^2\sigma^2/2}.$$

Using Equation 3.64, we can find the moments as follows:

$$\mu_0 = P(0) = 1,$$

$$\mu_1 = \frac{1}{-j}\left[\frac{dP(\omega)}{d\omega}\right]_{\omega=0} = \left[\frac{-(jm+\sigma^2\omega)}{-j}e^{-j\omega}e^{-\omega^2\sigma^2/2}\right]_{\omega=0} = m,$$

and

$$\mu_2 = \frac{1}{(-j)^2}\left[\frac{d^2P(\omega)}{d\omega^2}\right]_{\omega=0} = \left[\frac{-(m^2+\sigma^2)}{(-j)^2}e^{-j\omega}e^{-\omega^2\sigma^2/2}\right]_{\omega=0} = m^2+\sigma^2.$$

The nth central moment of $p(x)$ is defined as

$$s_n = \int_{-\infty}^{\infty}(x-m)^n p(x)\,dx,$$

from which we note that the second central moment of $p(x)$ is $\mu_2 - \mu_0^2 = \sigma^2$.

Appendix 3.3 Solution of the Homogeneous Diffusion Equation

We shall show that the solution of the homogeneous diffusion equation is a Gaussian function of position. We start with the one-dimensional diffusion equation

$$\frac{\partial c(x,t)}{\partial t} = D\frac{\partial^2 c(x,t)}{\partial x^2} \tag{3.66}$$

and seek a solution subject to the initial boundary condition $c(x,0)$. A solution to Equation 3.66 can be obtained by using the technique of separation of variables. Let

$$c(x,t) = X(x)T(t). \tag{3.67}$$

Substitution of Equation 3.67 into 3.66 and rearrangement of terms gives

$$\frac{1}{X(x)}\frac{d^2X(x)}{dx^2} = \frac{1}{DT(t)}\frac{dT(t)}{dt}. \tag{3.68}$$

Since the two sides of Equation 3.68 are functions of different independent variables, the two sides are equal only if each side equals the same constant value, which we will call $-p^2$ for convenience. Therefore,

$$\frac{1}{X(x)}\frac{d^2X(x)}{dx^2} = -p^2 \quad \text{and} \quad \frac{1}{DT(t)}\frac{dT(t)}{dt} = -p^2.$$

The solutions to these equations have the form

$$X(x) = e^{-jpx} \quad \text{or} \quad \sin px \quad \text{or} \quad \cos px \qquad \text{and} \qquad T(t) = e^{-p^2Dt}.$$

We assume that p is real, so the response will remain bounded as $t \to \infty$.

From Equation 3.67 we can write a general solution for $c(x,t)$. A more general solution is $C(p)X(x)T(t)$, where $C(p)$ is an arbitrary function of p. A still more general solution is obtained by a summation (integral) of terms of the form $C(p)X(x)T(t)$ as follows:

$$c(x,t) = \int_{-\infty}^{\infty} C(p)e^{-jpx}e^{-p^2Dt}\,dp. \tag{3.69}$$

Note that by substituting $t = 0$ into Equation 3.69, we obtain

$$c(x,0) = \int_{-\infty}^{\infty} C(p)e^{-jpx}\,dp. \tag{3.70}$$

From Equation 3.70 we can see that $C(p)$ is the Fourier transform of $c(x,0)$, i.e.,

$$C(p) = \frac{1}{2\pi}\int_{-\infty}^{\infty} c(x,0)e^{jpx}\,dx.$$

Now suppose that the initial distribution $c(x,0)$ has an infinitesimal spatial extent, but an area of unity, i.e., that $c(x,0)$ is a spatial impulse or a Dirac delta function. Then $C(p) = 1/2\pi$, and $c(x,t)$ is by definition the Green's function (Section 3.5.2.1),

$$c(x,t) = \frac{1}{2\pi}\int_{-\infty}^{\infty} e^{-p^2Dt}e^{-jpx}\,dp. \tag{3.71}$$

Note that Equation 3.71 shows that $c(x,t)$ is the Fourier transform of the function e^{-p^2Dt}, which can be evaluated by completing the square of the exponent in Equation 3.71 and then reducing the expression to a recognizable form. Note that

$$c(x,t) = \frac{1}{2\pi}\int_{-\infty}^{\infty} e^{-(p^2Dt+jpx)}\,dp = \frac{1}{2\pi}\int_{-\infty}^{\infty} e^{-(p\sqrt{Dt}+(jx)/2\sqrt{Dt})^2-x^2/4Dt}\,dp. \tag{3.72}$$

Equation 3.72 can be written as

$$c(x,t) = \frac{1}{2\pi}e^{-x^2/4Dt}\int_{-\infty}^{\infty} e^{-(p\sqrt{Dt}+(jx)/2\sqrt{Dt})^2}\,dp.$$

Now make the change of variable $y/2 = p\sqrt{Dt} + (jx)/2\sqrt{Dt}$, so that

$$c(x,t) = \frac{1}{\pi\sqrt{Dt}}e^{-x^2/4Dt}\int_{-\infty}^{\infty} e^{-y^2/2}\,dy.$$

From Appendix 3.2 we know that the integral has area $\sqrt{2\pi}$ and, therefore, that

$$c(x,t) = \frac{1}{\sqrt{4\pi Dt}}e^{-x^2/4Dt}.$$

Exercises

3.1 Define *permeability*.

3.2 Define *partition coefficient*.

3.3 Using no mathematical formulas or equations, describe the meaning of the continuity equation (Equation 3.6) in a few well-chosen English sentences.

3.4 Describe the distinction between Equation 3.1 and Equation 3.18.

3.5 The modern definition of the permeability coefficient was first formulated in the 1920s. Prior to that time, other definitions of the permeability coefficient were widely used. Consider the following definition based on measurements of the rate of change of concentration of a solute in a cell as a function of the intracellular and extracellular concentration:

$$-\frac{dc_n^i(t)}{dt} = P_n' \left(c_n^i(t) - c_n^o(t)\right),$$

where c_n^i and c_n^o are the concentrations inside and outside the cell and P_n' is an alternative definition of permeability. Contrast the definition of P_n' with that of the modern definition of permeability (Equation 3.44). Discuss the implications of using P_n' as a measure of the permeability of cells to a solute.

3.6 Discuss the distinction between equilibrium and steady-state diffusion.

3.7 Two time constants are involved in two-compartment diffusion through a membrane: the steady-state time constant in the membrane (τ_{ss}) and the equilibrium time constant for the two compartments (τ_{eq}). Without the use of equations, describe these two time constants.

3.8 Describe the dissolve-diffuse theory of diffusion through cellular membranes.

3.9 What is the purpose of displaying measurements in the form of Collander plots?

3.10 As shown in Figure 3.48, two solutions are separated by a membrane of thickness d, surface area A, and membrane : solution partition coefficient $k_n = 1.5$. Bath 1 has an infinite volume and has a concentration of solute n of C. The volume of bath 2 is $\mathcal{V}_2$. Assume that a steady state has been established.

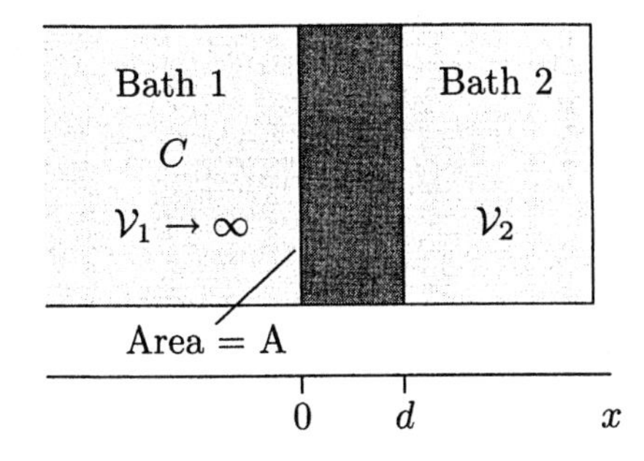

Figure 3.48 Arrangement of baths and membrane (Exercise 3.10).

a. Sketch the concentration of solute n, $c_n(x)$, for values of x that include the membrane and the two baths.

b. Is the system in equilibrium? Explain.

3.11 In the system of Exercise 3.10, assume that the membrane can be treated as a thin membrane with permeability P_n. The concentration of solute n in bath 2 is $c_n^2(t) = 0$ for $t < 0$. Sketch $c_n^2(t)$ for $t > 0$.

3.12 At a junction between two neurons, called a synapse, there is a 20 nm cleft that separates the cell membranes. A chemical transmitter substance is released by one cell (the presynaptic cell), diffuses across the cleft, and arrives at the membrane of the other (postsynaptic) cell. Assume that the diffusion coefficient of the chemical transmitter substance is $D = 5 \times 10^{-6}\,\text{cm}^2/\text{s}$. Make a rough estimate of the delay caused by diffusion of the transmitter substance across the cleft. What are the limitations of this estimate?

3.13 A tall cylinder is filled with a water solution with solute n. The cylinder is placed on a table until equilibrium is reached, and the concentration $c_n^E(y)$ is observed to vary with height as shown in Figure 3.49. The table with the cylinder is now moved to the planet Jupiter, where equilibrium is again established in the solution with the temperature maintained as it had been on Earth. Sketch the concentration $c_n^J(y)$ that you would expect on Jupiter on a scale that makes the relation between $c_n^J(y)$ and $c_n^E(y)$ clear. Explain your choice.

3.14 The time course of one-dimensional diffusion of a solute from a point source in space and time has the form

$$c_n(x,t) = \frac{n_o}{\sqrt{4\pi Dt}} e^{-x^2/4Dt},$$

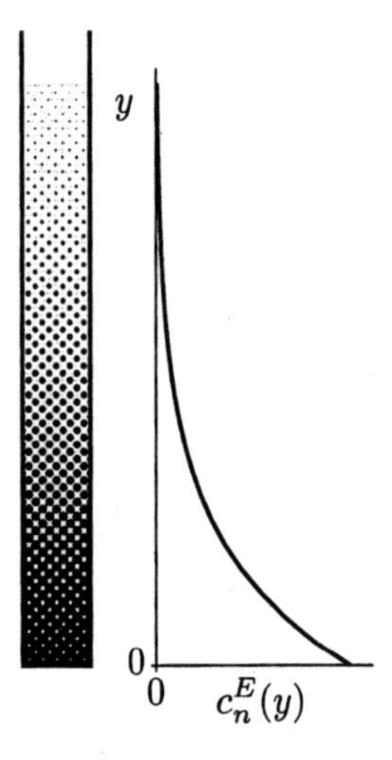

Figure 3.49 Distribution of solute n in a cylinder (Exercise 3.13). The cylinder is shown on the left, and the concentration is plotted versus height y on the right.

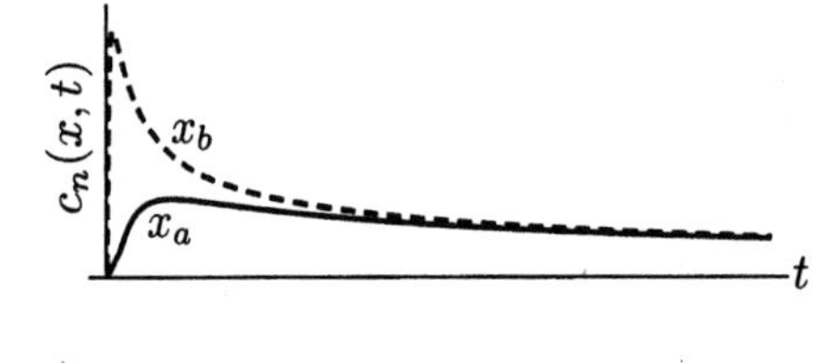

Figure 3.50 Concentration of solute n as a function of time for two locations, x_a and x_b (Exercise 3.14).

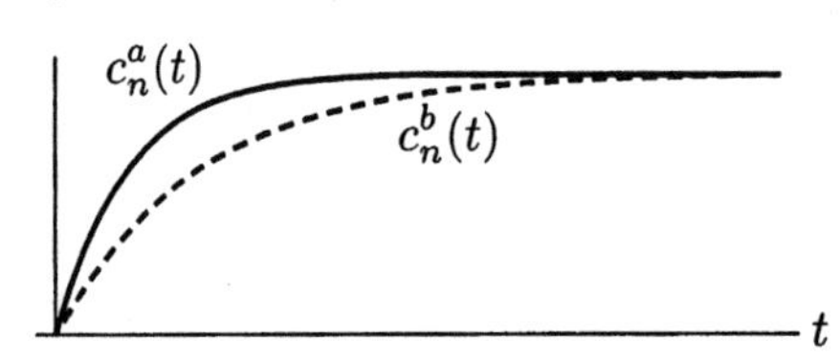

Figure 3.51 Concentration of solute n as a function of time in two cells, $c_n^a(t)$ and $c_n^b(t)$, that have different permeabilities to solute n (Exercise 3.15).

where n_o is the number of moles of solute per unit area placed at $x = 0$ at $t = 0$. As shown in Figure 3.50, $c_n(x,t)$ is computed for locations x_a and x_b. Is $x_a > x_b$, or is $x_a < x_b$? Explain.

3.15 Cell a and cell b have identical dimensions, but different permeabilities to solute n, P_n^a and P_n^b, respectively. The cells are placed in identical solutions that contain the permeant solute n. The intracellular concentrations for cell a and for cell b are shown in Figure 3.51. Is $P_n^a > P_n^b$, or is $P_n^a < P_n^b$? Explain.

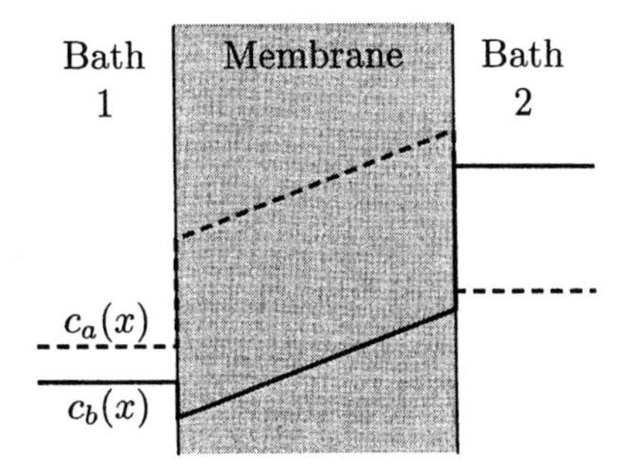

Figure 3.52 Concentration of solutes a and b as a function of position (Exercise 3.16).

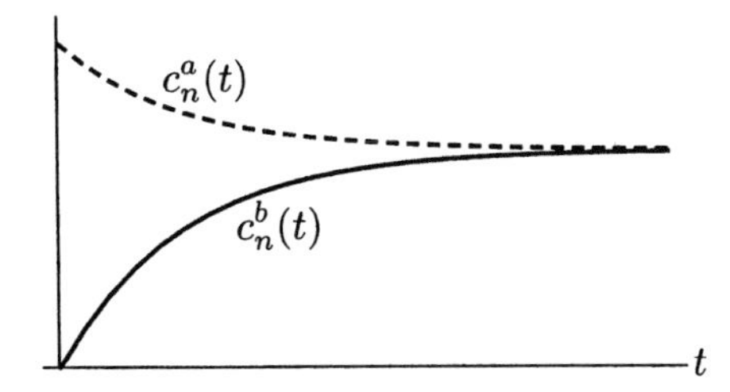

Figure 3.53 Concentration of solute n as a function of time in compartments a and b (Exercise 3.17).

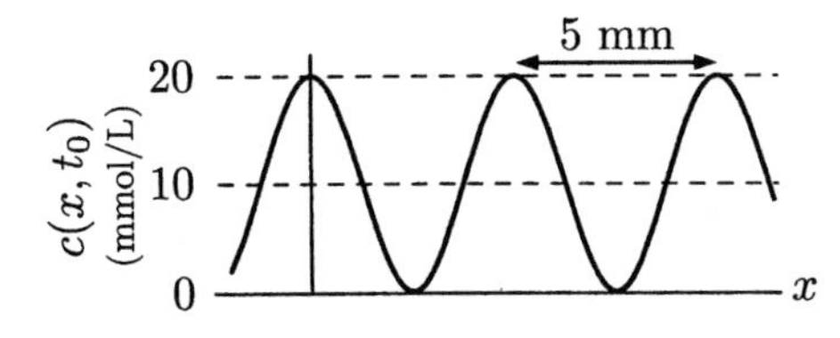

Figure 3.54 Concentration of solute as a function of position at time t_0 (Exercise 3.18).

3.16 Two solutes, a and b, diffuse in the steady state through a membrane with the concentration profiles shown in Figure 3.52. The membrane: solution partition coefficients for the two solutes, k_a and k_b, differ. Is $k_a > k_b$, or is $k_a < k_b$? Explain.

3.17 A solute n diffuses through a membrane that separates two compartments that have different initial concentrations. The concentrations in the two compartments as a function of time, $c_n^a(t)$ and $c_n^b(t)$, are shown in Figure 3.53. The volumes of the two compartments are $\mathcal{V}_a$ and $\mathcal{V}_b$. Is $\mathcal{V}_a > \mathcal{V}_b$, or is $\mathcal{V}_a < \mathcal{V}_b$? Explain.

3.18 The concentration of a solute at some time t_0 is shown as a function of x in Figure 3.54. Determine and sketch the flux of solute at time t_0.

Problems

3.1 Consider a cell composed of a cell body to which is attached a long, thin, tubular "process" (e.g., the axon of a neuron or the flagellum of a sperm cell) of length l cm (Figure 3.55). Suppose some substance n (e.g., a metabolite) is generated in the cell body and diffuses along the process, all along which the substance is consumed at a constant uniform rate, α_n mol/s per unit length. The continuity equation for the substance thus becomes

$$\frac{\partial c_n}{\partial t} = -\frac{\partial \phi_n}{\partial x} - \frac{\alpha_n}{A},$$

where A is the (assumed constant) cross-sectional area of the process.

a. Combine the modified continuity equation given above with Fick's law to obtain a modified form of the diffusion equation that must be satisfied by c_n.

b. Show that a solution of this equation in the steady state ($\partial c_n/\partial t = \partial \phi_n/\partial t = 0$) is

$$c_n(x) = \frac{\alpha_n}{2DA}x^2 + a_o x + b_o,$$

and find values of the constants a_o and b_o corresponding to the boundary conditions $c_n(o) = C_o$ and $\phi_n(l) = 0$.

c. Show that the requirement that α_n mol/(s·cm) be consumed uniformly along the process sets an upper limit (if C_o, D, A, and α_n are fixed) on the possible length l of the process, and find a formula for this upper limit.

3.2 Diffusion in the presence of both convection and a chemical reaction is described by Equations 3.12, 3.14, and 3.15.

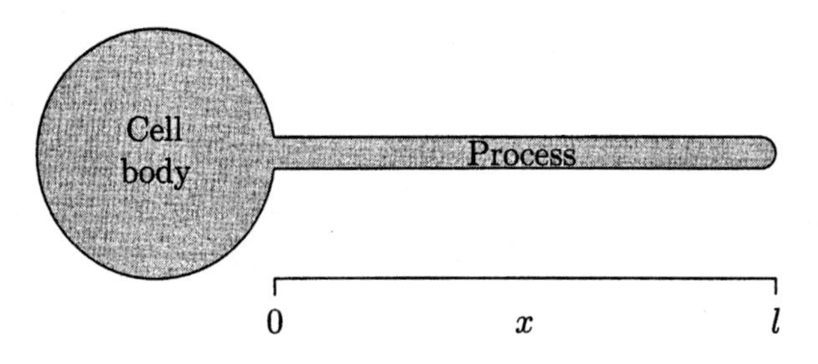

Figure 3.55 Cell body with a process (Problem 3.1).

a. Assume that there is no chemical reaction to eliminate particles, i.e., $\alpha = 0$.

   i. Determine the general form of the spatial distribution of concentration at equilibrium.

   ii. Determine the general form of the steady-state spatial distribution of concentration. Suppose that the steady-state solution satisfies the following boundary conditions: the concentration is 0 at $x = 0$ and 1 mol/cm^3 at $x = 1$. Find and sketch the solution for a diffusion coefficient of $D = 10^{-5}$ cm^2/s and a convection velocity of $v = 1$ cm/s. Repeat for a convection velocity of -1 cm/s.

b. Suppose there is a chemical reaction that removes particles at a rate $\alpha \neq 0$. Determine the distribution of particles at equilibrium.

3.3 Let $f(x,t)$ be the solution to the diffusion equation

$$\frac{\partial f(x,t)}{\partial t} = D\frac{\partial^2 f(x,t)}{\partial x^2}.$$

Show that the solution to the modified diffusion equation

$$\frac{\partial g(x,t)}{\partial t} = D\frac{\partial^2 g(x,t)}{\partial x^2} - \alpha g(x,t)$$

can be expressed as

$$g(x,t) = f(x,t)e^{-\alpha t}.$$

3.4 Let $f(z,t)$ be the solution to the diffusion equation

$$\frac{\partial f(z,t)}{\partial t} = D\frac{\partial^2 f(z,t)}{\partial z^2}.$$

Show that the solution to the modified diffusion equation

$$\frac{\partial g(x,t)}{\partial t} = D\frac{\partial^2 g(x,t)}{\partial x^2} - v\frac{\partial g(x,t)}{\partial x}$$

can be expressed as

$$g(x,t) = f(z,t),$$

where $z = x - vt$.

3.5 Figure 3.56 shows the concentration of a solute as a function of position 200 seconds after a point source of strength 1 mol/m^2 was applied at $t =$

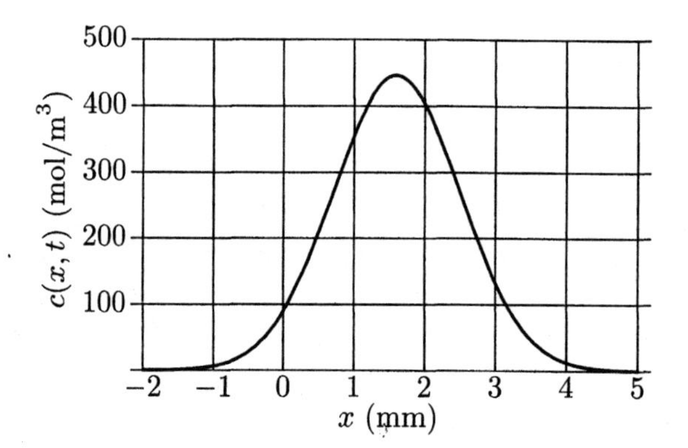

Figure 3.56 Concentration versus distance (Problem 3.5).

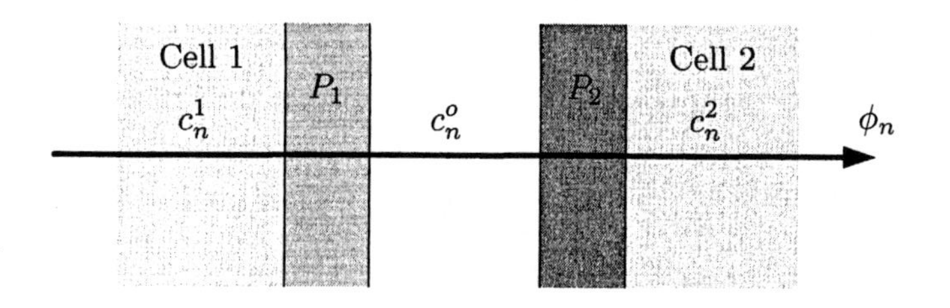

Figure 3.57 Two membranes in series (Problem 3.6).

0 at $x = 0$. The temperature is 300 K. The solute particles are subjected to a molar body force f. In the solution to this problem, use the result given in Problem 3.4 for the solution to the diffusion equation in the presence of solute drift caused by a body force.

a. Find a general expression for $c(x,t)$.

b. Determine the numerical value of the diffusion coefficient D.

c. Determine the numerical value of the molar mechanical mobility, u.

d. Determine the numerical value of the force on a mole of particles f.

3.6 Two adjoining cells have closely apposed membranes (Figure 3.57). The concentrations of uncharged solute n are c_n^1 and c_n^2 inside cells 1 and 2, respectively, and c_n^o in the intercellular space. The membrane permeabilities for this solute are P_1 and P_2 for the membranes of cells 1 and 2, respectively. Find the net permeability, P, between the inside of cell 1 and the inside of cell 2 in terms of P_1 and P_2, where

$$\phi_n = P\left(c_n^1 - c_n^2\right)$$

and ϕ_n is the steady-state flux of n in mol/(cm^2·s) across both membranes.

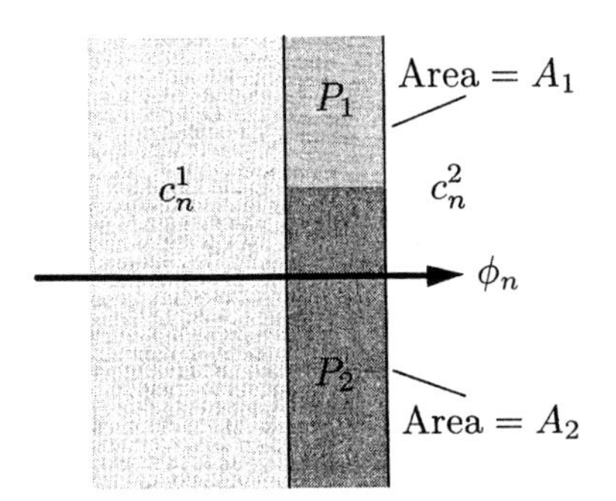

Figure 3.58 Two membranes in parallel (Problem 3.7).

3.7 A membrane separates two solutions containing the solute n as shown in Figure 3.58. The membrane consists of two contiguous patches. One has surface area A_1 and permeability P_1 for solute n, and the other has surface area A_2 and permeability P_2 for solute n. Assume that steady-state diffusion occurs through the membrane. Determine the permeability of the membrane for solute n where the permeability P is defined as

$$\phi_n = P\left(c_n^1 - c_n^2\right).$$

3.8 Measurements of the permeability of a membrane to a solute are often confounded by the presence of unstirred layers, also called stagnant layers, of solution near the membrane boundaries in which the concentration of solute differs from that in the bulk solution, i.e., far from the membrane. This problem deals with the effect of these unstirred layers on the measurement of permeability. Assume that in the steady state these unstirred layers can be represented as regions of fixed dimensions in which the concentration of solute varies linearly with distance from the membrane (Figure 3.59). Consider a membrane with true permeability, P, separating two solutions with bulk solute concentrations C_1 and C_2, respectively. The two unstirred layers have thickness d_1 and d_2, respectively. The diffusion coefficients are D_1 and D_2, respectively. Let the measured permeability be defined as $P_m = \phi/(C_1 - C_2)$, where ϕ is the solute flux through the membrane.

a. Find an expression for the measured permeability, P_m, in terms of the true permeability, P, the Ds, and the ds.

b. Is the difference between measured and true permeability largest for highly permeant solutes (which have a large value of P) or for relatively impermeant solutes (which have a small value of P)? Explain.

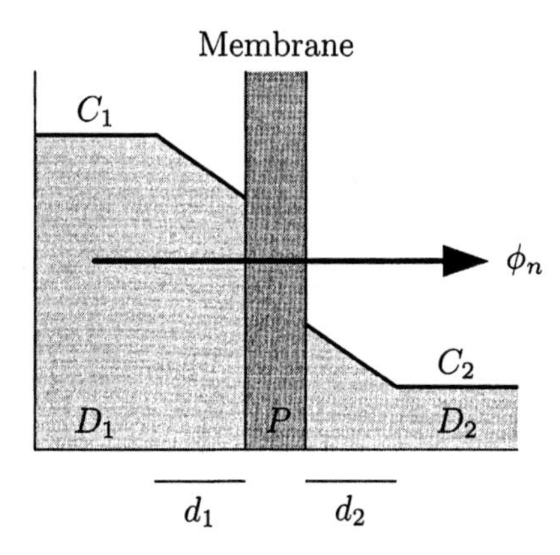

Figure 3.59 Unstirred layers (Problem 3.8).

3.9 A flow cytometer was used (Brahm, 1983) to measure the permeability of human erythrocytes (red blood cells) to several alcohols with the results shown in Figure 3.36. Direct measurements (Jay, 1975) indicate that for erythrocytes in physiological saline the surface area is $136.9 \pm 0.5\ \mu m^2$ and the volume is $104.2 \pm 0.6\ \mu m^3$. Using these results alone, estimate the permeability of erythrocytes to propanol.

3.10 Assume that n_o mol/cm^2 of sucrose (with a diffusion constant $D \approx 0.5 \times 10^{-5}$cm^2/s) are placed in a trough of water at a point $x = 0$ at time $t = 0$. Assume that the concentration of sucrose is a function of x and t only.

a. Show that for any fixed point x_p the maximum concentration occurs at time $t_m = x_p^2/2D$.

b. How long does it take for the concentration to reach a maximum at $x = \pm 1$ cm?

3.11 At time $t = 0$, a spherical cell that has a radius of 30 μm and a membrane that is 10 nm thick is placed in an external solution containing the permeant solute X at concentration c_X^o (which is maintained constant). The internal concentration of X, $c_X^i(t)$, which is assumed to be uniform inside the cell, is shown as a function of time in Figure 3.60 and has the form $c_X^i(t) = c_X^o(1 - \exp(-t/100)$, where t is in seconds. Assume that the cell's volume remains constant during this experiment.

a. Assume that X diffuses through the membrane so rapidly compared to the time course of $c_X^i(t)$ that steady-state conditions hold in the membrane at each instant of time and the membrane flux is independent of position in the membrane. Find the value of the permeability of the membrane to X, P_X.

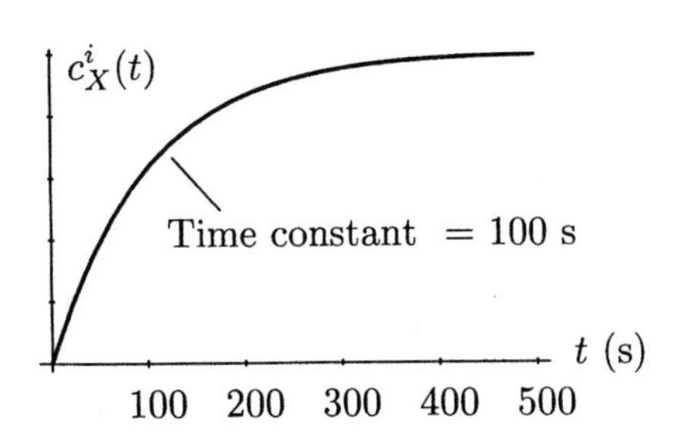

Figure 3.60 Time course of internal concentration (Problem 3.11).

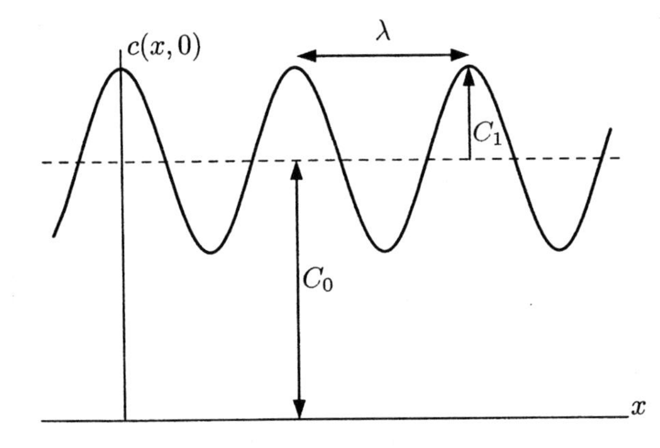

Figure 3.61 Sinusoidal concentration profile (Problem 3.12).

b. The membrane-solution partition coefficient of X is $k_X = 0.2$. Find the value of the diffusion coefficient, D_X, of X in the membrane.

c. Estimate the order of magnitude of the equilibration time of X in the membrane.

d. Is the steady-state assumption made in part a justified? Explain.

3.12 This problem deals with the use of Fourier analysis to obtain solutions to the diffusion equation.

a. Given the diffusion equation

$$D\frac{\partial^2 c(x,t)}{\partial x^2} = \frac{\partial c(x,t)}{\partial t},$$

prove that if $c_1(x,t)$ and $c_2(x,t)$ are solutions to the diffusion equation, then $ac_1(x,t) + bc_2(x,t)$ is also a solution to the diffusion equation.

b. Suppose that the initial concentration of solute is (Figure 3.61)

$$c(x,0) = C_0 + C_1 \cos(2\pi x/\lambda).$$

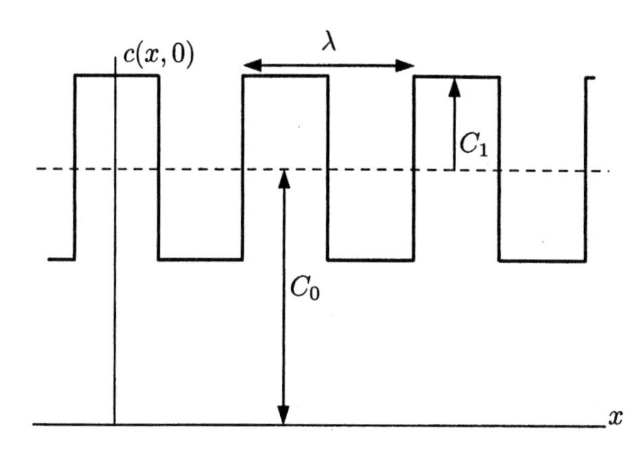

Figure 3.62 Rectangular distribution of concentration (Problem 3.12).

Show that the function

$$c(x,t) = C_0 + y(t)\cos(2\pi x/\lambda)$$

is a solution to the diffusion equation that matches the initial boundary condition, and show that

$$y(t) = C_1 e^{-t/\tau}.$$

Discuss the physical significance of the dependence of τ on λ and D.

c. Now assume that the initial distribution of the concentration, $c(x,0)$, is rectangular with spatial period, λ, as shown in Figure 3.62. Express $c(x,0)$ as a Fourier series in x, and use the results of parts a and b to find $c(x,t)$. Discuss qualitatively the changes in the spatial waveshape of $c(x,t)$ with increasing t.

3.13 An experiment is performed to determine the permeability, P_X, of the membrane of a cell to solute X. The cell, which is a sphere of radius 72 μm, is placed in a solution containing solute X for a sufficient time to load the cell with N_X moles of X. A set of identical vials containing identical solutions that do not contain the solute X are prepared. The cell is then immersed successively in a series of these vials for $T = 10$ minutes per vial (Figure 3.34), i.e., 10 minutes in vial 1, followed by 10 minutes in vial 2, followed by 10 minutes in vial 3, etc. The number of moles of solute X in vial k is $n_X(k)$. Assume that the volume of the cell is constant and negligible compared with the volume of a vial, and hence, the concentration of solute in a vial is always negligible compared to that in the cell. The solute permeates the membrane according to Fick's law for membranes.

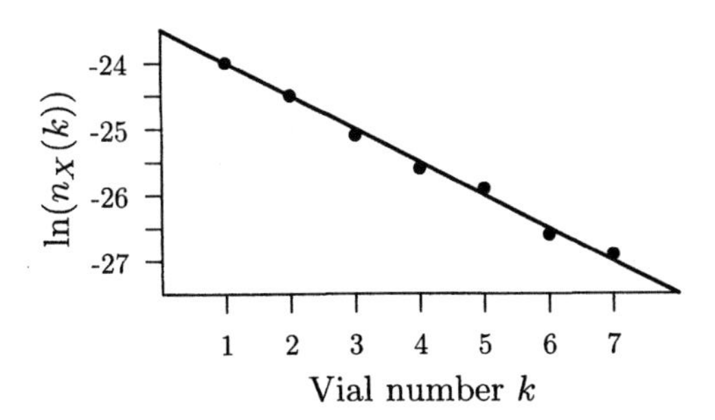

Figure 3.63 Vial measurements (Problem 3.13).

a. Determine an expression for the number of moles of X in the cell as a function of time, $n_X^i(t)$. You may write this expression in terms of literals such as N_X and a suitably defined time constant. Assume that the transfer of the cell from vial to vial takes no time.

b. Determine an expression for the total quantity of X in the kth vial, $n_X(k)$, in terms of literals such as N_X, T, and a suitably defined time constant.

c. For the measurements shown in Figure 3.63, determine the numerical values of P_X and N_X.

3.14 A basic assumption in the derivation of the results for two-compartment diffusion, described in Section 3.7, was that both compartments were well mixed, i.e., that the concentration of solute in each compartment was uniform. Applied to diffusion across the membrane of a cell, this assumption implies that both the extracellular space and the cytoplasm have uniform concentrations of solute. In this problem we explore the conditions for which this assumption is reasonable for the cytoplasm of a spherical cell. To explore this quantitatively, we compare the time it takes a spherical cell to equilibrate its concentration with an external solution with the time it would take in the absence of the semipermeable plasma membrane. Assume that the cell is a sphere with a radius of 10 μm. At $t = 0$ the cell is placed in a solution containing a concentration of urea, C, which can be assumed to remain constant. Urea has a diffusion coefficient in water of about $D = 1.4 \times 10^{-5}$cm^2/s. Urea is transported through the membrane of this cell by diffusion, and the permeability of the cell membrane to urea is $P = 1 \times 10^{-6}$cm/s. Assume that the volume of the cell remains constant and that the volume of the extracellular solution is much greater than the volume of the cell. The initial intracellular concentration of urea is zero.

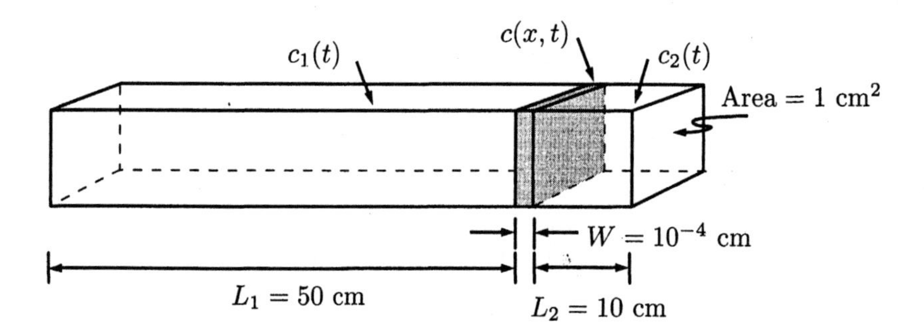

Figure 3.64 Two compartments separated by a membrane (Problem 3.15).

a. Compute the time, τ_{eq}, it takes for the intracellular concentration of urea to reach 63% of its final value assuming that the concentration of urea in the cell is uniform.

b. Determine the time, τ_d, it would take for the average concentration of urea to reach 63% of its final value for a sphere of water with a radius of 10 μm placed in a solution of urea.

c. How does the ratio τ_d/τ_{eq} depend on the radius of the sphere?

d. What do you conclude from these calculations?

3.15 Consider diffusion through a thin membrane that separates two otherwise closed compartments. As illustrated in Figure 3.64, the membrane and both compartments have cross-sectional areas $A = 1$ cm^2. Compartment 1 has width $L_1 = 50$ cm, compartment 2 has width $L_2 = 10$ cm, and the membrane width is $W = 10^{-4}$ cm. Assume that the compartments contain sugar solutions and that both compartments are well stirred, so that the concentration of sugar in compartment 1 can be written as $c_1(t)$ and that in compartment 2 can be written as $c_2(t)$; that the concentration of sugar in the membrane can be written as $c(x, t)$, where x represents distance through the membrane; that the diffusivity of sugar in the membrane is $D_{sugar} = 10^{-5}$ cm^2/s and that the membrane : water partition coefficient $k_{m:w}$ is 1; that the concentration of sugar in the membrane has reached steady state at time $t = 0$; and that $c_1(0) = 1$ mol/L and $c_2(0) = 0$ mol/L.

a. Compute the flux of sugar through the membrane at time $t = 0$, $\phi_s(0)$.

b. Compute the final value of the concentration of sugar in compartment 1, $c_1(\infty)$.

c. Let τ_{eq} characterize the amount of time required to reach equilibrium. What would happen to τ_{eq} if the diffusivity of sugar in the membrane were doubled? Explain your reasoning.

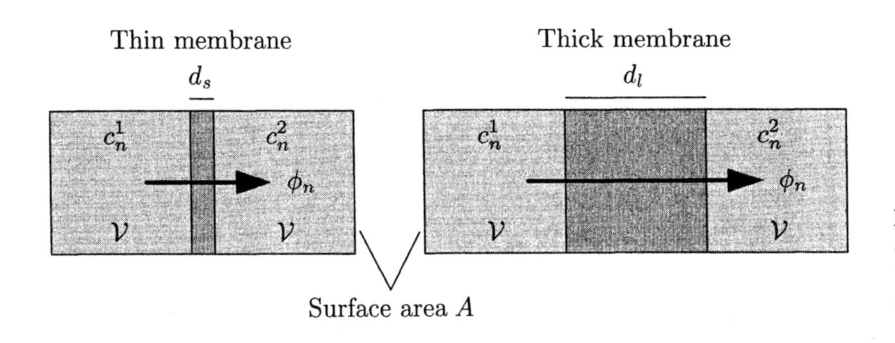

Figure 3.65 Schematic diagrams of thin and thick membranes (Problem 3.16).

3.16 A thin membrane and a thick membrane that are otherwise identical are used to separate identical solutions of volume $\mathcal{V} = 1\text{cm}^3$ (Figure 3.65). All the membrane surfaces facing the solutions have area $A = 1\text{cm}^2$. The thin membrane has thickness $d_s = 10^{-4}\text{cm}$; the thick membrane has thickness $d_l = 1\text{cm}$. For $t < 0$ the aqueous solutions on both sides of the membrane are identical and do not contain solute n. At $t = 0$ a small concentration of solute n is added to the solution on side 1 of the membrane. You may assume that the solutions are maintained in osmotic equilibrium so that there is no water flow across the membrane (see Chapter 4). The diffusion coefficient and membrane : solution partition coefficient of n in both membranes are $D_n = 10^{-5}\text{cm}^2/\text{s}$ and $k_n = 2$, respectively.

a. For each membrane, estimate the time, τ_{ss}, it takes for the concentration profile in the membrane to reach its steady-state spatial distribution.

b. For each membrane, find the time, τ_{eq}, it takes for the solutions on the two sides of the membrane to come to equilibrium assuming that the spatial distribution of solute in the membrane is at the steady-state distribution.

c. Is it reasonable to assume that Fick's law for membranes applies for the thin membrane, i.e., does

$$\phi_n = P_s(c_n^1 - c_n^2)\,?$$

d. Is it reasonable to assume that Fick's law for membranes applies for the thick membrane, i.e., does

$$\phi_n = P_l(c_n^1 - c_n^2)\,?$$

P_s and P_l are the permeabilities of the thin and thick membranes, respectively.

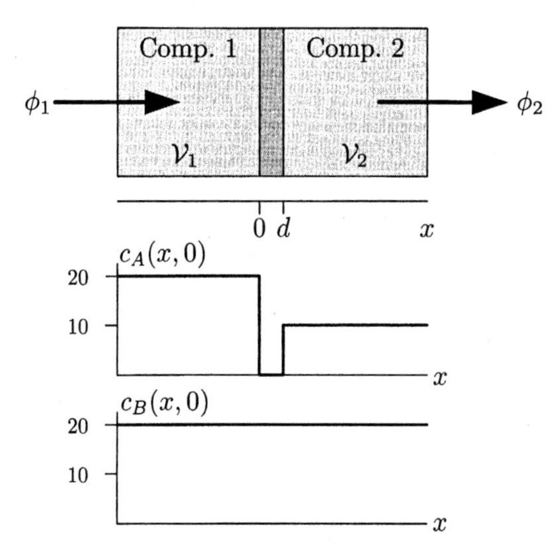

Figure 3.66 Arrangement of two compartments and two initial concentration profiles (Problem 3.17). Units for concentrations are mmol/L.

3.17 Two compartments, whose volumes are $\mathcal{V}_1$ and $\mathcal{V}_2$, are filled with aqueous solutions that contain a single solute (Figure 3.66). The concentration of solute within each compartment is kept uniform by stirring. The compartments are separated by a membrane of thickness d that has a solute diffusion coefficient of $D \geq 0$ and a membrane: solution partition coefficient of $k \geq 0$. Assume that the number of particles in the membrane is a negligible fraction of the number of particles in the two compartments. External sources determine the constant solute fluxes ϕ_1 and ϕ_2, where ϕ_1 is a flux into compartment 1 and ϕ_2 is a flux out of compartment 2. The solute concentration at location x and time t in this system is $c(x, t)$. Osmotic equilibrium is maintained, so that no water flows between the two compartments.

The goal of this problem is to determine the steady-state solute concentration profile $c(x, \infty)$ that results with particular initial solute concentration profiles $c_A(x, 0)$ and $c_B(x, 0)$ (as shown in Figure 3.66) for different parameters of this two-compartment diffusion system. For each of the conditions (a to j) shown in Table 3.7 determine *all* the steady-state concentration profiles shown in Figure 3.67 that apply and enter the profile number(s) into the right-hand column of Table 3.7. If none of the profiles in Figure 3.67 applies, enter *None* in Table 3.7. *In each case give a reason for your answer.*

3.18 A rigid homogeneous membrane separates two well-stirred compartments with rigid walls that contain aqueous solutions of a solute.

Table 3.7 Parameters for two-compartment diffusion through a membrane (Problem 3.17).

Part	$c(x,0)$	ϕ_1 & ϕ_2	$\mathcal{V}_1$ & $\mathcal{V}_2$	D	k	$c(x,\infty)$
a	$c_A(x,0)$	$\phi_1 = \phi_2 = 0$	Arbitrary	0	> 0	
b	$c_A(x,0)$	$\phi_1 = \phi_2 = 0$	Arbitrary	> 0	0	
c	$c_A(x,0)$	$\phi_1 = \phi_2 = 0$	Arbitrary	> 0	> 1	
d	$c_A(x,0)$	$\phi_1 = \phi_2 = 0$	Arbitrary	> 0	< 1	
e	$c_A(x,0)$	$\phi_1 = \phi_2 = 0$	$\mathcal{V}_1 \to \infty$, $\mathcal{V}_2$ finite	> 0	> 1	
f	$c_A(x,0)$	$\phi_1 = \phi_2 = 0$	$\mathcal{V}_1 \to \infty$, $\mathcal{V}_2$ finite	> 0	1	
g	$c_A(x,0)$	$\phi_1 > \phi_2 \neq 0$	Arbitrary	> 0	0	
h	$c_A(x,0)$	$\phi_1 > \phi_2 \neq 0$	Arbitrary	> 0	> 0	
i	$c_A(x,0)$	$\phi_1 = \phi_2 > 0$	Arbitrary	> 0	> 1	
j	$c_B(x,0)$	$\phi_1 = \phi_2 = 0$	Arbitrary	> 0	< 1	

The cross-sectional area of the membrane and the compartments is $A = 64\ \text{cm}^2$. The solute is transported through the membrane by diffusion only. The membrane:solution partition coefficient for the solute is 1.0; the value of the diffusion coefficient for the solute in the membrane is unknown, although it is known that $D \geq 0$. The widths are 10 cm for compartment 1, 1 cm for compartment 2, and 0.1 cm for the membrane. The geometry is shown in Figure 3.68. The concentrations of the solute in compartments 1 and 2 are $c^1(t)$ and $c^2(t)$, respectively. The concentration of the solute in the membrane depends on position in the membrane and is $c(x,t)$.

At $t = 0$ the concentration of the solute in compartment 1 is $c^1(0) = 100\ \text{mmol/cm}^3$, and the concentrations in the membrane and compartment 2 are zero, i.e., $c^2(0) = c(x,0) = 0$. The resulting initial spatial dependence of the concentrations is shown in Figure 3.69.

a. Sketch the solute concentrations in the membrane and in the two compartments at equilibrium. Make a sketch similar to Figure 3.69, and label important values on the curve.

b. For $t = 100$ seconds, the concentration $c(x,100)$, and flux, $\phi(x,100)$, in the membrane and the concentrations in the two compartments, $c^1(100)$ and $c^2(100)$, are shown in Figure 3.70.

   i. Is the concentration in the membrane at its steady-state distribution? Explain.

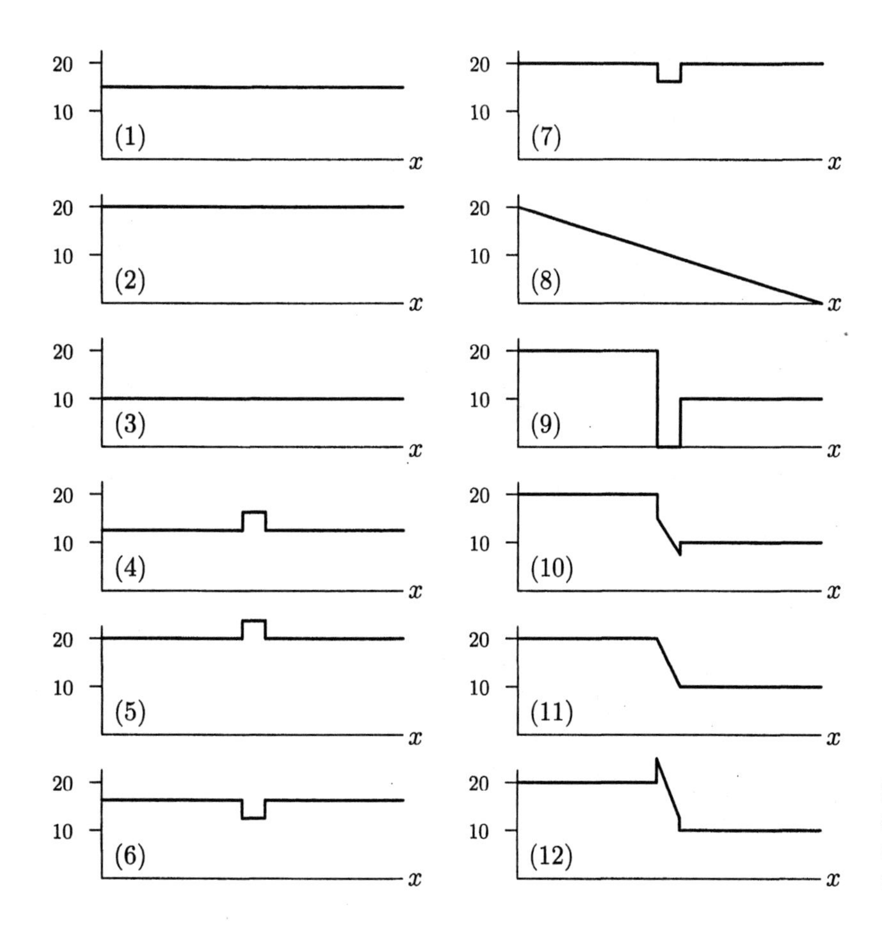

Figure 3.67 Twelve steady-state solute concentration profiles, $c(x, \infty)$ (Problem 3.17).

ii. Estimate the diffusion coefficient for the solute in the membrane, D, from the data in Figure 3.70. Clearly indicate your method.

c. For $t = 1000$ seconds, the concentration in the membrane, $c(x, 1000)$, and the concentrations in the compartments are shown in Figure 3.71. Figure 3.71 also shows the dependence of concentration on time at two locations: one in the membrane, $c(0.04, t)$, and the other in compartment 2, $c^2(t)$.

i. Is the concentration in the membrane at its steady-state distribution? Explain.

ii. Estimate the equilibrium time constant for the equilibration of the solute in the two compartments, τ_{eq}, from the data in Figure 3.71. Clearly indicate your method.

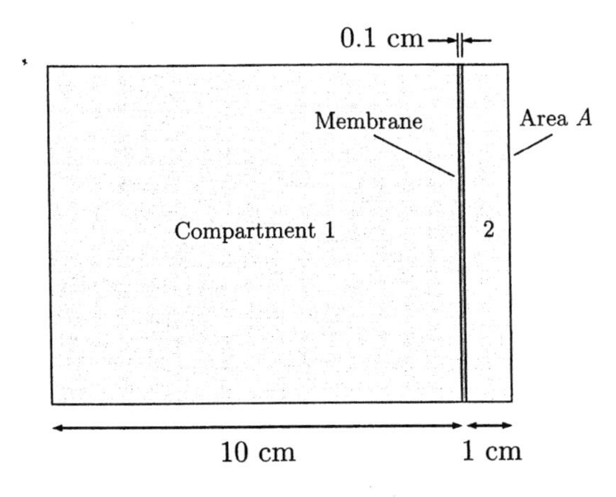

Figure 3.68 Geometry for two-compartment diffusion between compartment 1 and compartment 2 through a membrane (Problem 3.18).

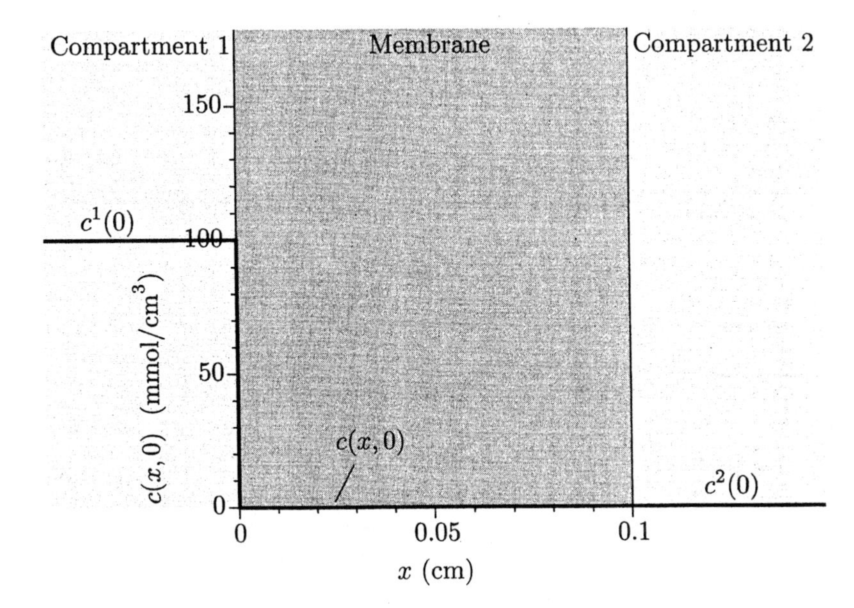

Figure 3.69 The initial concentrations of solute in the two compartments and in the membrane are shown at $t = 0$ (Problem 3.18). The figure illustrates the concentration in the membrane and in a portion of each compartment only. The concentrations in the compartments are uniform in space.

iii. Estimate the steady-state time constant for the solute in the membrane, τ_{ss}, from the data in Figure 3.71. Clearly indicate your method.

iv. Estimate the diffusion coefficient of the solute in the membrane D from the data in Figure 3.71. Clearly indicate your method.

v. Estimate the quantity $\int_0^{1000} \phi(0.1, t)\, dt$ from the data in Figure 3.71. Clearly indicate your method.

vi. Sketch $c^2(t)$ and $c(0.04, t)$ versus t from $t = 0$ to $t \approx 5\tau_{eq}$.

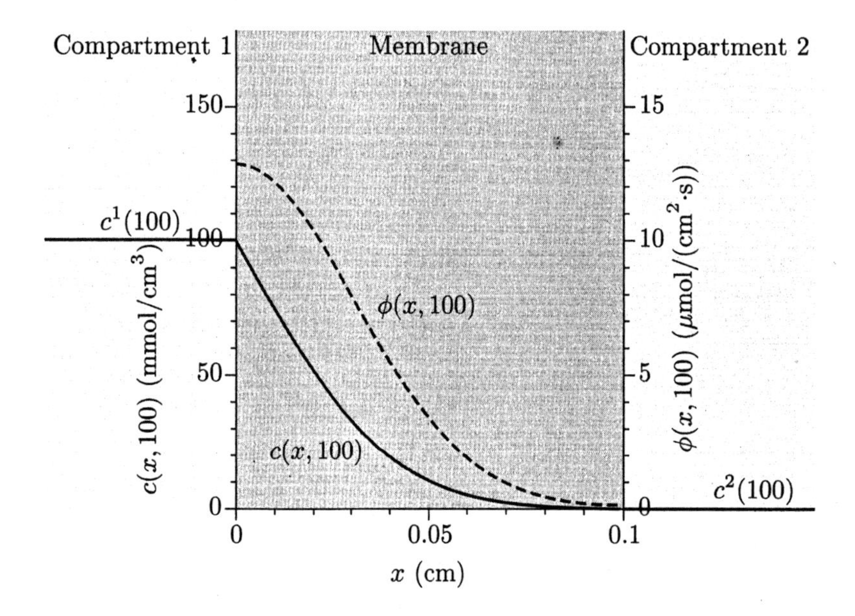

Figure 3.70 The concentrations of solute (solid line) in the two compartments and in the membrane are shown at $t = 100$ seconds (Problem 3.18). The figure illustrates the concentrations in the membrane and in a portion of each compartment only. The concentrations in the compartments are uniform in space. The flux of solute is shown (dashed line) in the membrane only.

3.19 Solute particles contained in water diffuse due to thermal agitation and drift due to the force of gravity so that the total flux is $\phi = \phi_D + \phi_G$, where the flux due to diffusion is ϕ_D and that due to gravity is ϕ_G.

a. If the particles are spherical with radius r, show that at equilibrium

$$\phi_G = -c(x,t)\frac{D}{\lambda} \quad \text{where } \lambda = \frac{kT}{m_{eff}g} \text{ and } m_{eff} = \frac{4}{3}\pi r^3(\rho_p - \rho_w),$$

where the positive sense of x is opposite to the direction of the gravitational force on the particles; $c(x,t)$ is the concentration of particles; D is the diffusion coefficient; m_{eff} is the effective mass of the particles, so that $m_{eff}g$ is the net force on each particle, which equals the gravitational force minus the buoyant force; g is the acceleration due to gravity; k is Boltzmann's constant; T is absolute temperature; and ρ_p and ρ_w are the mass densities of the particles and of water, respectively.

b. Determine the spatial distribution of solute concentration at equilibrium, $c(x)$, in terms of the concentration of particles, $c(0)$, at the reference location $x = 0$.

c. The spatial distribution of particles has a characteristic length λ. Assume that the particles have a density that is twice that of water where the density of water $\rho_w = 1\,\text{g/cm}^3$. At room temperature, 300 K,

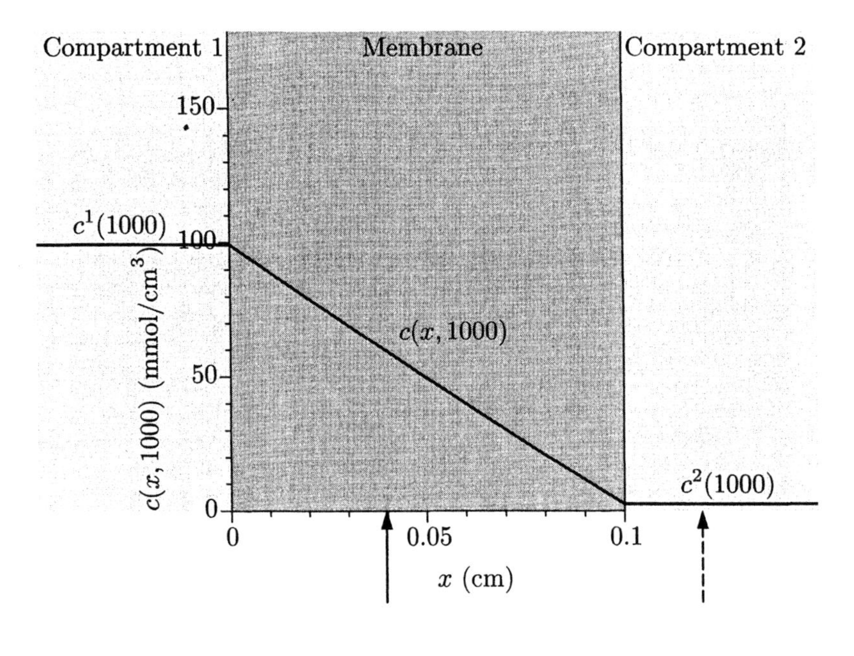

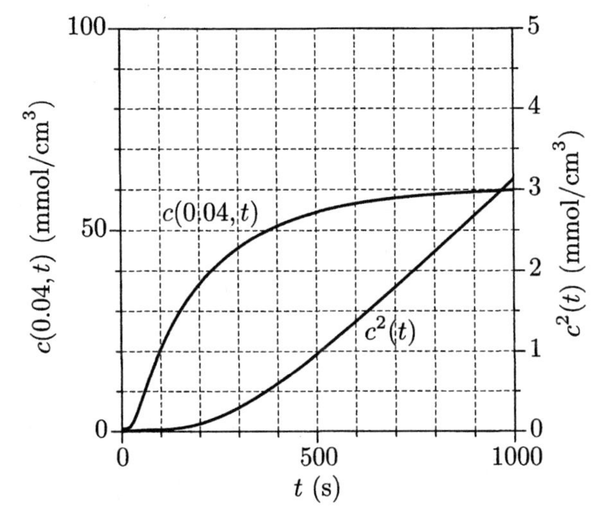

Figure 3.71 The concentrations of solute in the two compartments and in the membrane are shown at $t = 1000$ seconds (upper panel) (Problem 3.18). The figure illustrates the concentration in the membrane and in a portion of each compartment only. The concentrations in the compartments are uniform in space. Also shown (lower panel) are the dependence of concentration on time for the time interval $(0, 1000)$ for a location in the membrane, $c(0.04, t)$ (indicated in the upper panel with a solid arrow), and in compartment 2, $c^2(t)$ (indicated in the upper panel with a dashed arrow).

find the characteristic length for particles of radius 1 mm and 1 nm. Suppose you had two 50 ml beakers; one containing an emulsion of 1 mm particles and the other 1 nm particles. To a first-order approximation, describe the spatial distribution of particles in the two beakers.

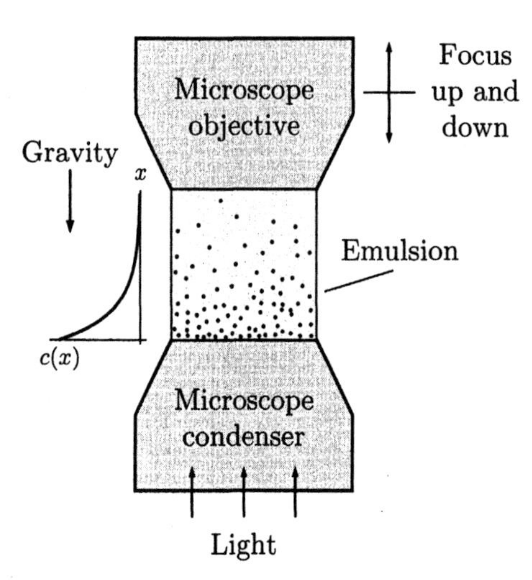

Figure 3.72 Schematic diagram of apparatus used by Perrin to make observations of the concentration of mastic particles in an emulsion as function of depth (Problem 3.19). By focusing at different depths, the number of particles at different depths in the emulsion were counted. The structures are not drawn to scale.

d. With a light microscope (shown schematically in Figure 3.72), Perrin (1909) measured[8] the number of particles as a function of depth in an emulsion consisting of particles of mastic in water. These measurements were an important empirical verification of the theory of Brownian motion. To make these measurements, Perrin chose the radii of his particles very carefully. Which range of particle radii would you choose, and why?

3.20 Collander measured the permeability of the cell membrane of the plant *Nitella mucronata* for a large number of solutes. In this problem you are given characteristics of four solutes (Table 3.8) and asked to estimate their relative permeabilities. Rank order the solutes A–D according to

---

8. Perrin describes some amusing technical problems with these measurements (Perrin, 1909) as follows:

> *Those of the emulsions which were not aseptic were sometimes invaded by elongated and very active protozoa, which, stirring up the emulsion like fish would agitate the mud of a pond, greatly diminished the inequality of distribution between the upper and lower layers. But if one had the patience to wait until, lacking food, these microbes finished by dying and falling inert to the bottom of the preparation, which takes about two or three days, one would again find exactly the first distribution limit, which contains all the characteristics of a distribution of permanent order.*
>
> *Once this state of order occurs, it is easy to see if the concentration decreases in exponential fashion as a function of height.*

Table 3.8 Characteristics of four solutes (Problem 3.20). M is the molecular weight in daltons, and k is the ether:water partition coefficient.

Solute	M	k
A	12	0.1
B	30	1.0
C	50	0.02
D	400	3.0

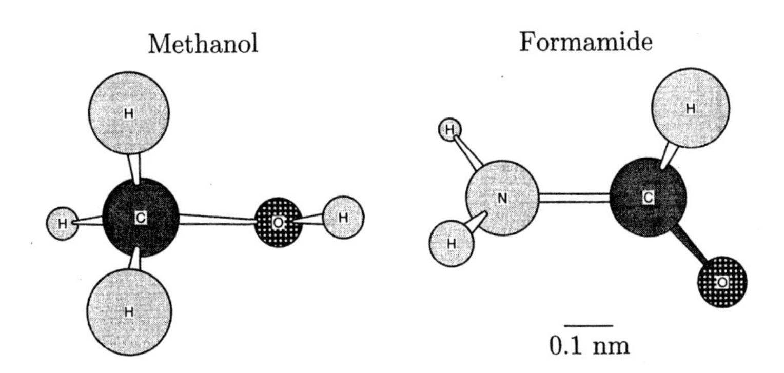

Figure 3.73 Comparison of methanol (CH_4O) and formamide (NH_3O) (Problem 3.21).

your estimate of their permeability through *Nitella* membrane. Explain your method.

3.21 Figure 3.73 shows the molecular structures of methanol and formamide. Which solute is more permeant through cellular membranes? Discuss the factors that affect the permeation of these solutes as quantitatively as possible.

References

Books and Reviews

Baker, R. W. (1987). *Controlled Release of Biologically Active Agents.* John Wiley & Sons, New York.

Baker, R. W. and Lonsdale, H. K. (1974). Controlled release: Mechanisms and rates. In Tanquary, A. C. and Lacey, R. E., eds., *Controlled Release of Biologically Active Agents*, 15–71. Plenum, New York.

Batchelor, G. K. (1967). *An Introduction to Fluid Mechanics.* Cambridge University Press, New York.

Berg, H. C. (1983). *Random Walks in Biology.* Princeton University Press, Princeton, NJ.

Cannon, J. R. (1984). *The One-Dimensional Heat Equation*, vol. 23 of *Encyclopedia of Mathematics.* Addison-Wesley, Reading, MA.

Carslaw, H. S. and Jaeger, J. C. (1959). *The Conduction of Heat in Solids*. Oxford University Press, London.
Cohn, E. J. and Edsall, J. T. (1965). *Proteins, Amino Acids and Peptides as Ions and Dipolar Ions*. Hafner, New York.
Comyn, J., editor (1985). *Polymer Permeability*. Elsevier, New York.
Crank, J. (1964). *The Mathematics of Diffusion*. Oxford University Press, London.
Crank, J. and Park, G. S., eds. (1968). *Diffusion in Polymers*. Academic Press, New York.
Cussler, E. L. (1984). *Diffusion: Mass Transfer In Fluid Systems*. Cambridge University Press, Cambridge, England.
Deen, W. M. (1987). Hindered transport of large molecules in liquid-filled pores. *AIChE J.*, 33:1409–1425.
Dick, D. A. T. (1966). *Cell Water*. Butterworth, Washington, DC.
Harris, E. J. (1972). *Transport and Accumulation in Biological Systems*. University Park Press, Baltimore, MD.
Jacobs, M. H. (1967). *Diffusion Processes*. Springer-Verlag, New York.
Jost, W. (1965). *Diffusion in Solids, Liquids, Gases*. Academic Press, New York.
Lide, D. R. (1990). *CRC Handbook of Chemistry and Physics*. CRC Press, Boston.
Macey, R. I. (1980). Mathematical models of membrane transport processes. In Andreoli, T. E., Hoffman, J. F., and Fanestil, D. D., eds. *Membrane Physiology*, 125–146. Plenum, New York.
Sha'afi, R. I. and Gary-Bobo, C. M. (1973). Water and nonelectrolyte permeability in mammalian red cell membranes. *Prog. Biophys. Mol. Biol.*, 26:103–146.
Shapiro, H. M. (1988). *Practical Flow Cytometry*. Alan R. Liss, New York.
Solomon, A. K. (1968). Characterization of biological membranes by equivalent pores. *J. Gen. Physiol.*, 51:335S–364S.
Solomon, A. K. (1986). On the equivalent pore radius. *J. Membr. Biol.*, 94:227–232.
Stein, W. D. (1967). *The Movement of Molecules across Cell Membranes*. Academic Press, New York.
Stein, W. D. (1986). *Transport and Diffusion across Cell Membranes*. Academic Press, New York.
Stein, W. D. (1990). *Channels, Carriers, and Pumps*. Academic Press, New York.
Sten-Knudsen, O. (1978). Passive transport processes. In Giebisch, G., Tosteson, D. C., and Ussing, H. H., eds., *Membrane Transport in Biology*, vol. 1, *Concepts and Models*, 5–113. Springer-Verlag, New York.
Tanford, C. (1961). *Physical Chemistry of Macromolecules*. John Wiley & Sons, New York.
Villars, F. M. H. and Benedek, G. B. (1974). *Physics with Illustrative Examples from Medicine and Biology*, vol. 2, *Statistical Physics*. Addison-Wesley, Reading, MA.
Widder, D. V. (1975). *The Heat Equation*. Academic Press, New York.

Original Articles

Anderson, J. L. and Quinn, J. A. (1974). Restricted transport in small pores: A model for steric exclusion and hindered particle motion. *Biophys. J.*, 14:130–150.
Batchelor, G. K. (1976). Brownian diffusion of particles with hydrodynamic interaction. *J. Fluid Mech.*, 74:1–29.
Batchelor, G. K. (1977). The effect of brownian motion on the bulk stress in a suspension of spherical particles. *J. Fluid Mech.*, 83:97–117.

Batchelor, G. K. and Green, J. T. (1972). The determination of the bulk stress in a suspension of spherical particles to order c^2. *J. Fluid Mech.*, 56:401–427.

Bean, C. P. (1972). The physics of porous membranes: Neutral pores. In Eisman, G., ed., *Membranes*, vol. 1, *Macroscopic Systems and Models*. Marcel Dekker, New York.

Beck, R. E. and Schultz, J. S. (1972). Hindrance of solute diffusion within membranes as measured with microporous membranes of known pore geometry. *Biochim. Biophys. Acta*, 255:273–303.

Brahm, J. (1982). Diffusional water permeability of human erythrocytes and their ghosts. *J. Gen. Physiol.*, 79:791–819.

Brahm, J. (1983). Permeability of human red cells to a homologous series of aliphatic alcohols. *J. Gen. Physiol.*, 81:283–304.

Brown, R. (1828). A brief account of microscopical observations made in the months of June, July, and August, 1827, on the particles contained in the pollen of plants; and on the general existence of active molecules in organic and inorganic bodies. *Philos. Mag.*, 4:161–173.

Cadée, G. C. (1991). Brownian emotion. *Nature*, 354:180.

Collander, P. R. (1962–63). Ernest Overton (1865–1933): A pioneer to remember. *Leopoldina*, 8–9:242–254.

Collander, R. (1937). The permeability of plant protoplasts to non-electrolytes. *Trans. Far. Soc.*, 33:985–990.

Collander, R. (1949). The permeability of plant protoplasts to small molecules. *Physiol. Plant.*, 2:300–311.

Collander, R. (1950). The permeability of Nitella cells to rapidly penetrating non-electrolytes. *Physiol. Plant.*, 3:45–57.

Collander, R. (1954). The permeability of Nitella cells to non-electrolytes. *Physiol. Plant.*, 7:420–445.

Collander, R. and Bärlund, H. (1933). Permeabilitätsstudien an *Chara ceratophylla*. II. Die permeabilität fur nichtelekrolyte. *Acta Bot. Fenn.*, 11:1–114.

Davidson, M. G. and Deen, W. M. (1988). Hindered diffusion of water-soluble macromolecules in membranes. *Macromolecules*, 21:3474–3481.

Deutsch, D. H. (1991). Did Robert Brown observe Brownian motion: Probably not. *Bull. Am. Phys. Soc.*, 36:1374.

Deutsch, D. H. (1992). Brownian motion. *Nature*, 357:354.

Dix, J. A. and Solomon, A. K. (1984). Role of membrane proteins and lipids in water diffusion across red cell membranes. *Biochim. Biophys. Acta*, 773:219–230.

Einstein, A. (1905). Über die von der molekularkinetischen Theorie der Wärme geforderte Bewegung von in ruhenden Flüssigkeiten suspendierten Teilchen. *Ann. Physik*, 17:549–560 (translation in Einstein, 1956).

Einstein, A. (1906). Sur theorie der brownschen bewegung. *Ann. Physik*, 19:371–381 (translation in Einstein, 1956).

Einstein, A. (1908). Elementare theorie der brownschen bewegung. *Z. Elektrochemie angewandte physik. Chemie*, 14:235–239 (translation in Einstein, 1956).

Einstein, A. (1956). *Investigations on the Theory of the Brownian Movement*. Dover, New York. R. Furthe and A. D. Cowper, eds.

Fick, A. (1855). On liquid diffusion. *Philos. Mag.*, 10:30–39.

Finkelstein, A. (1976). Water and nonelectrolyte permeability of lipid bilayer membranes. *J. Gen. Physiol.*, 68:127–135.

Ford, B. J. (1991). Robert Brown, Brownian movement, and teeth marks on the hatbrim. *Microscope*, 39:161–173.

Ford, B. J. (1992a). Brownian movement in *Clarkia* pollen: A reprise of the first observations. *Microscope*, 40:235–241.

Ford, B. J. (1992b). Brown's observations confirmed. *Nature*, 359:265.

Galey, W. R., Owen, J. D., and Solomon, A. K. (1973). Temperature dependence of nonelectrolyte permeation across red cell membranes. *J. Gen. Physiol.*, 61:727–746.

Goldstein, D. A. and Solomon, A. K. (1960). Determination of equivalent pore radius for human red cells by osmotic pressure measurement. *J. Gen. Physiol.*, 44:1–17.

Graham, T. (1850). On the diffusion of liquids. *Philos. Trans. R. Soc. London, Ser. B*, 140:1–46.

Hodgkin, A. L. and Katz, B. (1949). The effect of sodium ions on the electrical activity of the giant axon of the squid. *J. Physiol.*, 108:37–77.

Jay, A. W. L. (1975). Geometry of the human erythrocyte: I. Effect of albumin on cell geometry. *Biophys. J.*, 15:205–222.

Levitt, D. G. (1973). Kinetics of diffusion and convection in 3.2-Å pores. *Biophys. J.*, 13.

Lieb, W. R. and Stein, W. D. (1986). Non-Stokesian nature of transverse diffusion within human red cell membranes. *J. Membr. Biol.*, 92:111–119.

Malone, D. M. and Anderson, J. L. (1978). Hindered diffusion of particles through small pores. *Chem. Eng. Sci.*, 33:1429–1440.

Mayrand, R. R. and Levitt, D. G. (1983). Urea and ethylene glycol-facilitated transport systems in the human red cell membrane: Saturation, competition, and asymmetry. *J. Gen. Physiol.*, 81:221–237.

Orbach, E. and Finkelstein, A. (1980). The nonelectrolyte permeability of planar lipid bilayer membranes. *J. Gen. Physiol.*, 75:427–436.

Owen, J. D. and Solomon, A. K. (1972). Control of nonelectrolyte permeability in red cells. *Biochim. Biophys. Acta*, 290:414–418.

Paganelli, C. V. and Solomon, A. K. (1957). The rate of exchange of tritiated water across the human red cell membrane. *J. Gen. Physiol.*, 41:259–277.

Pappenheimer, J. R., Renkin, E. M., and Borrero, L. M. (1951). Filtration, diffusion and molecular sieving through peripheral capillary membranes: A contribution to the pore theory of capillary permeability. *Am. J. Physiol.*, 167:13–46.

Park, G. S. (1950). The diffusion of some halo-methanes in polystyrene. *Trans. Faraday Soc.*, 46:684–697.

Park, G. S. (1951). The diffusion of some organic substances in polystyrene. *Trans. Faraday Soc.*, 47:1007–1013.

Perrin, J. (1909). Movement Brownien et réalité moléculaire. *Ann. Chim. Phys.*, 58:5–114.

Renkin, E. M. (1954). Filtration, diffusion, and molecular sieving through porous cellulose membranes. *J. Gen. Physiol.*, 41:225–243.

Ruhland, W. and Hoffman, C. (1925). Die permeabilität von *beggiatoa mirabilis*. Ein beitrag zur ultrafiltertheorie des plasmas. *Planta*, 1:1–83.

Russel, W. B. (1981). Brownian motion of small particles suspended in liquids. *Ann. Rev. Fluid Mech.*, 13:425–455.

Savitz, D. and Solomon, A. K. (1971). Tracer determination of human red cell membrane permeability to small nonelectrolytes. *J. Gen. Physiol.*, 58(3): 259–266.

Sha'afi, R. I., Gary-Bobo, C. M., and Solomon, A. K. (1971). Permeability of red cell membranes to small hydrophilic and lipophilic solutes. *J. Gen. Physiol.*, 58:238–258.

Sha'afi, R. I., Rich, G. T., Sidel, V. W., Bossert, W., and Solomon, A. K. (1967). The effect of the unstirred layer on human red cell water permeability. *J. Gen. Physiol.*, 50:1377–1399.

Solomon, A. K. and Gary-Bobo, C. M. (1972). Aqueous pores in lipid bilayers and red cell membranes. *Biochim. Biophys. Acta*, 255:1019–1021.

Stokes, G. G. (1851). On the effect of the internal friction of fluids on the motion of pendulums. *Trans. Camb. Phil. Soc.*, 9:6–106.

Walter, A. and Gutknecht, J. (1986). Permeability of small nonelectrolytes through lipid bilayer membranes. *J. Membr. Biol.*, 90:207–217.

Weiss, T. F., Trevisan, G., Doering, E. B., Shah, D. M., Huang, D., and Berkenblit, S. I. (1992). Software for teaching physiology and biophysics. *J. Sci. Ed. Tech.*, 1:259–274.

Wolosin, J. M., Ginsburg, H., Lieb, W. R., and Stein, W. D. (1978). Diffusion within egg lecithin bilayers resembles that within soft polymers. *J. Gen. Physiol.*, 71:93–100.

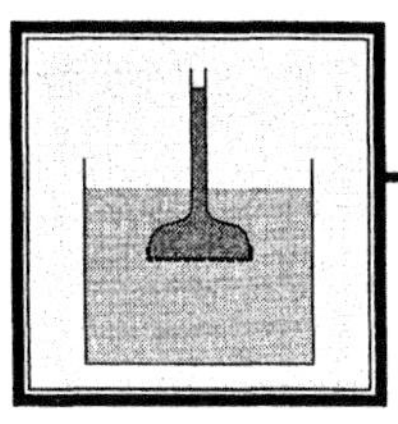

4 Solvent Transport

Again we have the basically pointless question: What exerts osmotic pressure? Really, as already emphasized, I am concerned in the end only with its magnitude; since it has proved to be equal to the gas pressure one tends to think that it comes about by a similar mechanism as with gases. Let he, however, who is led down the false path by this rather quit worrying about the mechanism.
—Van't Hoff, 1892

4.1 Introduction

Life on earth originated in the primordial seas, and this legacy has had a profound effect on all contemporary life forms. Thus, water makes up 60–90% of the mass of plants and animals, and living organisms imbibe and excrete large quantities of water each day. For example, adult humans take in and excrete 2–3 liters of water daily, an amount of water that represents about 3% of body weight. A plant cell may evaporate a quantity of water each hour that is many times its total water content. Yet despite this large flux of water, the constancy of the weight of a living organism implies that its water content is constant. Thus, there is a dynamic steady state in living organisms such that the water content is stable in the face of a large water flux.

Within complex organisms water flows from compartment to compartment. For example, in animal species with circulatory systems, water flows through the circulatory system and exchanges with the interstitial fluids through pores in capillaries. Thus, there is an exchange between compartments separated by the walls of capillaries. Plants have a conductive system so that water that enters through the roots, traverses the plant body, and exits through the leaves. Throughout its course this water exchanges with the interstitial fluids of the plant. Water also exchanges between the intracellular compartment and the extracellular compartment. That is, water readily traverses the membranes of cells. Since more than half of the water in an organism is found inside cells, the stability of an organism's weight suggests that cellular water content is regulated. Precise regulation of cellular water content is required to maintain the concentrations of cytoplasmic constituents that affect cellular processes.

In organisms, water is subjected to two types of forces: hydraulic and osmotic pressure. Hydraulic pressure arises from gravitational and other mechanical forces (e.g., the pumping of the heart), whereas osmotic pressure, which acts like hydraulic pressure, arises from solute concentration gradients. Both of these forces can be enormous. As an example, consider a giant redwood tree that must deliver moisture absorbed in its root structure to its leaves located over 100 meters above the ground. Water (in the form of sap) at the base of the tree must be subject to a pressure that is equivalent to a column of water 100 meters high, which is approximately ten times atmospheric pressure. Since water evaporates continuously from the leaves, there is a significant flow of water (sap) against this gravitational pull, which implies that trees have mechanisms that can overcome this large pressure. Even more impressive are the mangrove trees that grow along the ocean's edge with their roots in contact with seawater (which has a high concentration of salts). The sap of these trees contains a low concentration of salts. Thus, these trees extract water (relatively free of salts) and transport that water against many atmospheres of combined gravitational and osmotic pressure.

This chapter examines the conditions for equilibrium of solvent (water) flow and the basic principles of steady-state solvent flow in porous media. These principles are then applied to an examination of changes in cell volume that result from changes in intracellular or extracellular solute concentration. The present chapter is concerned primarily with solvent transport in the absence of solute transport. Chapter 5 considers concurrent solute and solvent transport. The present chapter ends with a discussion of the molecular mechanisms of water transport through cellular membranes.

4.2 Hydraulic Pressure

Imagine a surface element taken arbitrarily through a material (solid, liquid, or gaseous). The material on one side of this element will, in general, exert some force on the material on the other side of the element. Such a force per unit area is called a *stress*. The direction of the stress vector can be used to categorize different types of stress. If the stress vectors on the two sides of the surface element, point normal to the surface element, then the stress is either a *tension* or a *compression*, depending upon whether the material on one side of the element pulls away from or pushes toward the material on the other side of the surface element. If the stress vectors are parallel to the surface element, the stress is a *shear*. In an arbitrary material, the stress will contain both tension/compression (normal) stress and shear (parallel) stress. Fluids (liquids and gases) do not support a shear stress at equilibrium. The normal stress in a fluid at equilibrium is independent of direction, and therefore is a scalar quantity called the *pressure*. The notion of pressure is readily extended to a moving fluid (Batchelor, 1967). The pressure is the force per unit area, which in SI units is called the *pascal*, defined as a force of 1 newton acting on a square meter of area, i.e., $1\,\text{Pa} = 1\,\text{N/m}^2$.

A variety of units of pressure are widely used; some are referred to standard (at 0°C) atmospheric pressure, which is approximately as follows:

$$
\begin{aligned}
1 \text{ atmosphere} &\approx 14.7\,\text{lbf/inch}^2, \\
&\approx 10^5\,\text{Pa}, \\
&\approx 1\,\text{Bar}, \\
&\approx 760\,\text{mm of Hg}, \\
&\approx 33.5\,\text{ft of } H_2O = 10.2\,\text{m of } H_2O.
\end{aligned}
$$

What are the conditions for mechanical equilibrium in a fluid subjected to some body force? A body force is a force density exerted on the fluid components, e.g., by gravity or by an electric field if the fluid components are charged. Let us define $\mathcal{F}$ as the body force per unit volume, or force density, where $\mathcal{F}$ points in the positive x-direction. Now consider the forces on a volume element as shown in Figure 4.1. The force per unit area of the neighboring fluid on the faces of the volume element is the hydraulic pressure p. At equilibrium the forces on this volume element must sum to zero. Hence,

$$\mathcal{F} A \Delta x + p(x) A - p(x + \Delta x) A = 0. \tag{4.1}$$

If $\Delta x \to 0$, then

$$\mathcal{F} = \frac{dp}{dx}, \tag{4.2}$$

or, in three dimensions,

$$\mathcal{F} = \nabla p. \tag{4.3}$$

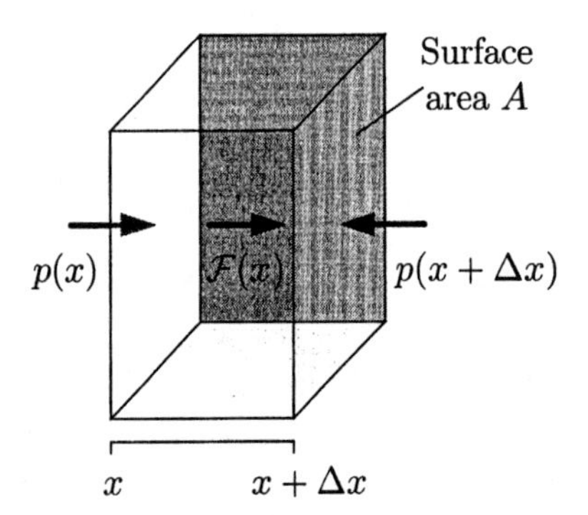

Figure 4.1 An incremental volume element that shows the relation between a body force $\mathcal{F}(x)$ on the element and the hydraulic pressure on the two faces of the volume element.

At equilibrium the body force equals the pressure gradient, which is equivalent to saying that the force density exerted by the pressure gradient on the element is $-\nabla p$ to balance the body force on the element.

4.3 Osmotic Pressure

The transport of solvent from a region of low solute concentration to a region of high solute concentration through a semipermeable medium by convection is called *osmosis*. Osmotic transport of solvent, called osmosis, is in the direction to reduce the solute concentration gradient. As we shall see, the osmotic pressure of a solution is the property that leads to osmosis in the presence of a semipermeable membrane. Though osmotic pressure is less familiar than hydraulic pressure, it is equivalent to a hydraulic pressure in producing a convective solvent transport. Curiously, while the *macroscopic* laws of osmotic pressure and of osmosis have not been controversial, the *microscopic* bases of both osmotic pressure and osmosis have been, and still are, subjects of heated debate.

4.3.1 Historical Perspective

As with diffusion, the basic phenomenon of osmosis is observable every day and must have been observed in antiquity (Hammel and Scholander, 1976). Most people have observed that lettuce leaves in a salad wilt and that the process of wilting is greatly accelerated if the lettuce is salted. Careful observation of the salted lettuce reveals droplets of water on the surfaces. The water droplets come in part from the interiors of lettuce plant cells, which consequently lose their internal rigidity (or turgor pressure); they wilt. The

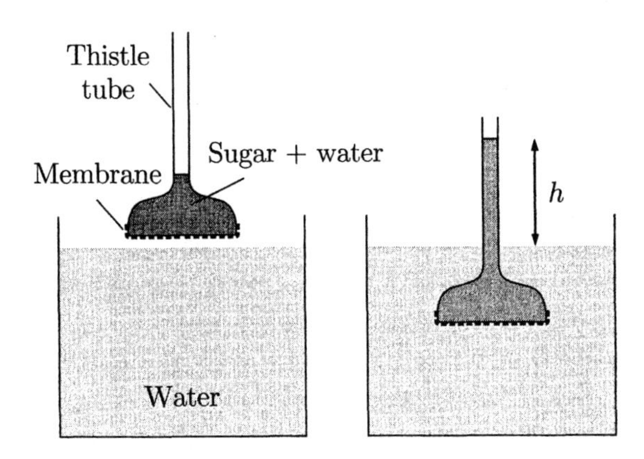

Figure 4.2 Schematic diagram of an osmometer.

process of water transport out of the cells caused by increase in the external salt concentration is an example of osmosis.

Systematic observations of osmotic pressure were made in the 1800's by Dutrochet, who summarized them as follows (Hammel and Scholander, 1976): "One observes that small animal bladders, filled with a dense solution and completely closed and plunged in water swell excessively and become turgid." Dutrochet found that the rate of entry of water into and the hydraulic pressure in such *osmometers* both increased as the concentration of sugar in the bladder was increased. Water flowed into the bladder so as to dilute the solution inside. In the latter part of the nineteenth century, osmometers were constructed from inorganic materials. These gave more reproducible results, and the osmotic pressure of a variety of solutions was measured. The basic phenomena are illustrated in Figure 4.2. A semipermeable membrane is placed on the end of a thistle tube that contains a solution of sugar dissolved in water; the membrane is permeable to water but not to sugar. When the thistle tube is plunged into pure water, the solution in the thistle tube rises to some height and comes to equilibrium. This experiment demonstrates the fundamental properties of osmosis:

- Water flows through the semipermeable membrane in a direction to equalize the sugar concentration.
- As the water flows into the thistle tube, a hydraulic pressure is generated in the thistle tube by the weight of the column of fluid.
- The flow of water eventually ceases, and a hydraulic pressure exists in the thistle tube at equilibrium. At the level of the surface of the water this hydraulic pressure equals $\rho g h$ where ρ is the density of the solution in the

thistle tube, g is the acceleration of gravity, and h is the height of the column of solution in the thistle tube.

The osmotic pressure is defined as the equilibrium hydraulic pressure. That is, the osmotic pressure of a solution is the hydraulic pressure that must be applied to the solution to prevent osmotic flow from a pure solvent through a semipermeable membrane.

Around the end of the nineteenth century, the botanist Pfeffer measured the osmotic pressure of solutions and showed that this pressure was proportional to the concentration of solute and rose slightly when the temperature of the solution was raised. During the nineteenth century, osmotic phenomena were also investigated in cells. In particular, the botanist Hugo de Vries studied osmotic phenomena in plant cells. He made systematic measurements using organic compounds and salts as the solutes. To describe these phenomena, we need to define some nomenclature. As indicated in Figure 4.3, if an animal cell is placed in an aqueous solution with total solute concentration C_{Σ}^{o}, the cell may shrink, swell, or maintain its volume. If the volume remains constant, the external solution is called *isotonic*. If the cell shrinks then the solution is called *hypertonic,* and if it swells then the solution is called *hypotonic*. An animal cell placed in a sufficiently hypotonic solution will swell and may burst. It is found that the solute concentration in hypertonic solutions exceeds that in isotonic solutions, whereas the solute concentration of hypotonic solutions is less than that of isotonic solutions. The changes in cell volume result from the flow of water: in hypertonic solutions water leaves the cell, and in hypotonic solutions water enters the cell.

The swelling and shrinkage of plant cells differs somewhat from that of animal cells. Plant cells contain a relatively rigid, but highly permeable, cell wall that encloses the cell membrane, which, as in animal cells, is relatively deformable and semipermeable. In isotonic solutions, the plasma membrane is adjacent to the cell wall. In hypotonic solutions, the pressure on the cell wall, called the *turgor pressure,* increases. In hypertonic solutions, water leaves the cytoplasm of a plant cell through the cell membrane, the turgor pressure decreases, and the cell membrane shrinks away from the cell wall. The threshold at which such a shrinkage of the cell membrane away from the cell wall of a plant cell is just detectable in a microscope is called *plasmolysis*. De Vries measured the concentration of solute that gave plasmolysis for a variety of solute types. He found that the ratios of concentrations of different solutes required to give plasmolysis tended to be ratios of integers. For example, the concentration of sugar that gave plasmolysis was approximately twice that of a binary salt such as NaCl.

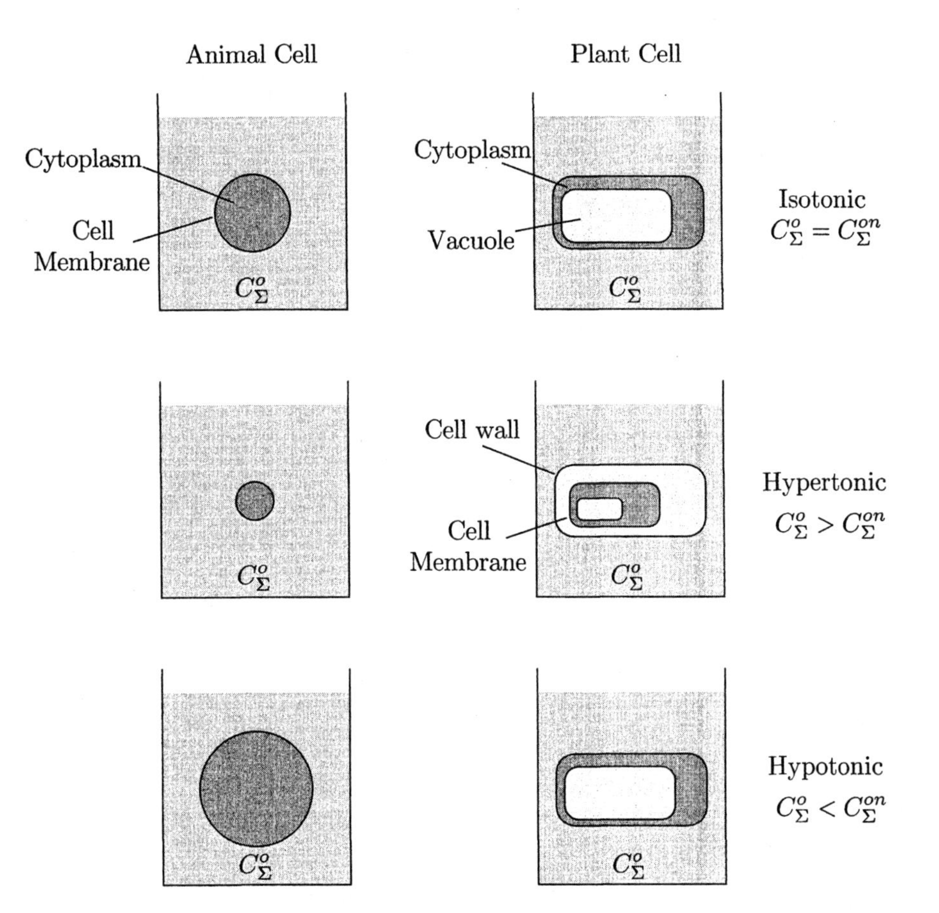

Figure 4.3 Schematic diagrams illustrating osmotic phenomena in isolated plant and animal cells. In isotonic solutions, cell volumes are the same as in situ. In hypertonic solutions, cells shrink; in hypotonic solutions, cells swell. C_Σ^o is the total extracellular concentration of solute, and C_Σ^{on} is its isotonic value. The cytoplasm of both animal and plant cells is surrounded by a cell or plasma membrane; plant cells also have a rigid cell wall that surrounds the cell membrane. In addition, plant cells contain prominent vacuoles, which can exchange their water content with the cytoplasm.

A mathematical expression relating osmotic pressure to solute concentration and to temperature was given first by the physical chemist Jacobus van't Hoff, who noted that the osmotic pressure was proportional to the product of the solute concentration and the absolute temperature with a constant of proportionality that equaled the molar gas constant R. Hence, van't Hoff (1886) made the following analogy: "The pressure which a gas exerts at a given temperature, when a given number of molecules are distributed in a given volume, is equally great as the osmotic pressure, which under the same conditions would be produced by *most solutes,* when they are dissolved in an arbitrary solvent." The qualification *most solutes* was necessary because salts did not appear to follow the *van't Hoff law.* At about the same time the physical chemist Svante Arrhenius was investigating the physical chemistry of salts and proposed, on the basis of measurements of the electrical conductivity of salt solutions, that salts dissociate into ions in water. When the ionization of salts was taken into account, it was found that the van't Hoff law was valid for salts as well. It was simply necessary to compute the concentration of all solutes including both anions and cations.

4.3.2 The Van't Hoff Law of Osmotic Pressure

We shall formulate the van't Hoff law by considering a membrane that separates two compartments (Figure 4.4) filled with aqueous solutions, where p_1 and p_2 are hydraulic pressures and C_Σ^1 and C_Σ^2 are the total solute concentrations on sides 1 and 2, respectively. By *total solute concentration* we mean the sum of all solute species, i.e., $C_\Sigma = \sum_n c_n$ where c_n is the concentration of solute species n. Φ_V is the volume flux from side 1 to side 2 in units of $\text{cm}^3/(\text{cm}^2 \cdot \text{s}) = \text{cm/s}$. The two sides are separated by a semipermeable membrane that is permeable to water but impermeable to the solute. At equilibrium, $\Phi_V = 0$ (there is no net flow of water from side 1 to side 2) and the hydraulic pressure difference equals the osmotic pressure difference

$$p_1 - p_2 = \pi_1 - \pi_2 = RT(C_\Sigma^1 - C_\Sigma^2), \tag{4.4}$$

where

$$\pi = RTC_\Sigma, \tag{4.5}$$

is the osmotic pressure (a derivation based on thermodynamics is given in Appendix 4.1). To see why this is called an osmotic pressure, consider the case when $C_\Sigma^2 = 0$ and $p_2 = 0$. Then the equilibrium condition reduces to $p_1 = \pi_1 = RTC_\Sigma^1$. Thus, a pressure $p_1 = RTC_\Sigma^1$ applied to the solution that contains solute at concentration C_Σ^1 leads to osmotic equilibrium. In this sense, the quantity RTC_Σ^1 is equivalent to a hydraulic pressure; it is, therefore, called the *osmotic pressure.*

4.3.2.1 Osmotic Pressure Is a Colligative Property

According to van't Hoff's law, the osmotic pressure of a solution depends only on the concentration of the solute and not on the chemical properties

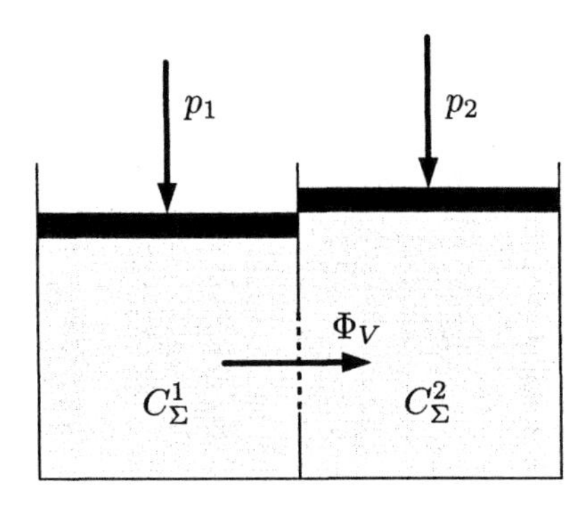

Figure 4.4 Hydraulic and osmotic pressure variables.

of the solute. Properties of solutions that depend only on the concentrations of solutes in solution and not on their chemical nature are called *colligative properties*. Colligative properties include the effect of solute concentration on such properties as freezing point, boiling point, and vapor pressure, as well as the osmotic pressure. In fact, this property is the basis for methods of measuring osmotic pressure. Commercial devices, called osmometers, for measuring osmotic pressure rely primarily on measuring the effect of solute concentration on vapor pressure or freezing point. By calibration of these devices, the effects on vapor pressure or freezing point are used to compute the osmotic pressure.

4.3.2.2 Isosmotic Solutions

For electrolytes, each ion makes a contribution to the osmotic pressure. If salts are fully ionized, then a 0.1 mol/L solution of NaCl is osmotically equivalent or *isosmotic*[1] to a 0.2 mol/L solution of glucose. It is customary to express the osmotic pressure in terms of the solute concentration. That is, the osmotic pressure is expressed as π/RT, and the unit is called the *osmolarity* and abbreviated as osmol.[2] Thus, a 0.1 mol/L solution of glucose has an osmotic pressure of 0.1 osmol/L, a 0.1 mol/L solution of NaCl has an osmotic pressure of about 0.2 osmol/L, and a 0.1 mol/L solution of $CaCl_2$ has an osmotic pressure of 0.3 osmol/L. These relations can be put still another way. The following solutions are all isosmotic: 0.1 mol/L glucose, 0.05 mol/L NaCl,

1. The terms *isotonic* and *isosmotic* are often used interchangeably, although, strictly speaking, two solutions that are *isotonic* have the same effect on cell tonicity, whereas two solutions that are *isosmotic* have the same osmotic pressures.

2. Both the osmolarity, defined in terms of the molarity of the solutes, and the osmolality, defined in terms of the molality of the solutes, are used to express the osmotic pressure.

and 0.033 mol/L $CaCl_2$. These relations are the basis for de Vries's findings on plant cells.

4.3.2.3 The Osmotic Pressure of Physiological Solutions Is Large

The osmotic pressure of a solution whose concentration of solute is comparable to that found in the bodies of plants and animals is a large pressure indeed. A 0.15 mol/L solution of a NaCl, which has approximately the osmolarity of intracellular and extracellular solutions of terrestrial vertebrates, has an osmotic pressure at 27°C (300 K) of

$$\pi = RTC_\Sigma = \left(8.314 \times 10^6 \frac{\text{Pa}}{\text{K} \cdot (\text{mol/cm}^3)}\right)(300\,\text{K})\left(0.3 \times 10^{-3}\,\text{mol/cm}^3\right)$$
$$\approx 7.5 \times 10^5\,\text{Pa} = 7.5 \text{ atmospheres.}$$

Such a pressure is exerted at the base of a column of water 250 feet high. Marine animals may have solutions whose osmolarities are near 1 mol/L. Hence, the osmotic pressure would be about 25 atmospheres, which represents the pressure of a column of water that is about 837 feet high—two-thirds of the height of the Empire State Building in New York City (which is 1,250 feet high without the 204-foot TV tower)!

4.3.2.4 Components of Osmotic Pressure of Physiological Solutions

The osmotic pressure of a solution is determined by the total number of solute particles in solution. Biological solutions, both cytoplasmic and extracellular, contain solutes covering an enormous range of molecular weight from small ions (e.g., hydrogen, sodium, potassium, chloride, calcium, magnesium) to small organic molecules (e.g., amino acids, ATP) to macromolecules (e.g., proteins, lipids, polysaccharides). The osmotic pressures of biological solutions vary a great deal in the plant and animal kingdom. For example, many marine animals have internal solutions with osmolalities like that of seawater, which is approximately 1 osmol/kg of water. Terrestrial mammals typically have solutions with osmolalities of about 300 mosmol/kg. For example, the composition of rat skeletal muscle fibers (Table 4.1) shows that the molality of small solutes accounts for the observed osmotic pressure of about 300 mosmol/kg. Large macromolecules presumably make only a small contribution to the osmotic pressure of cytoplasm. The osmotic pressure due to these macromolecules (with molecular weight greater than 10^4 daltons) is often called the *colloid osmotic pressure*. The colloid osmotic pressure can be

Table 4.1 Composition of cytoplasm of rat skeletal muscle fibers (Burton, 1983). Both the molality in mmol/kg of water and the charge density in mequiv/kg of water are given. The solutes included are those contributing to the osmolality of the cytoplasm. For example, the entry for Mg includes the free magnesium, and entry for amino acids includes only free, protein-forming amino acids.

Solute	Molality (mmol/kg)	Charge density (mequiv/kg)
Na	18	+18
K	165	+165
Mg	3	+6
Cl	6	−6
HCO_3	10	−10
Inorganic phosphate	2	−3
MgATP	9	−18
Phosphorylcreatine	34	−68
Creatine	13	—
Amino acids	24	—
Taurine	18	—
Anserine + carnosine	15	+8
Urea	5	—
Lactate	3	−3
TOTAL	325	+89

measured by an osmometer with a semipermeable membrane that is permeable to the small molecules but not to the macromolecules.

Note that the charge density of the small solutes, which are primarily responsible for the osmotic pressure of cytoplasm, is positive. Therefore, since there must be a balance of charge in cytoplasm (see Chapter 7), the cytoplasmic macromolecules must have a net negative charge. Why is this? In general, many macromolecules in solution are polyelectrolytes because they contain acidic and basic side groups, which, when ionized, are negatively and positively charged, respectively. One macromolecule may contain thousands of charged side groups. The hydrogen ions on the charged groups of the macromolecules are in chemical equilibrium with the hydrogen ions in solution. Therefore, the pH of the solution will determine the extent to which hydrogen ions are associated or dissociated from the side groups, which determines the charge on the side groups. At physiological pH ≈ 7.3, the cytoplasmic macromolecules have a net negative charge. Cytoplasmic ions will be attracted and repelled by these charged groups so that, for example, a negatively charged side group will attract a cloud of positively charged *counterions*. This cloud of ions, or space charge, has dimensions (called the Debye length) in the range of angstroms. Averaged over a distance large compared to the Debye length there is a charge balance in the solution, a property called electroneutrality (Chapter 7). The counterions just balance the charge on the side groups of the macromolecules.

The macromolecular anions are important in the osmotic equilibrium of cells because these solutes are impermeant—they are trapped in the cytoplasm—and they attract mobile counterions. Because of their relatively low concentration, the macromolecules themselves contribute only a small component of the intracellular osmotic pressure. However, the mobile counterions, which electrically neutralize the macromolecule side groups, contribute an appreciable component to the osmotic pressure called the *Donnan osmotic pressure.*

4.3.2.5 Osmotic Pressure of Real Solutions

Real solutions obey van't Hoff's law only at low solute concentration, much as real gases obey the ideal gas law only at low gas concentrations. It is a common practice to express deviations from van't Hoff's law in terms of an osmotic coefficient, χ, a practice similar to the use of such coefficients to express the difference in behavior of real gases from the ideal gas law. Thus, we express the osmotic pressure of a real solution as

$$\pi = \chi RTC_\Sigma. \tag{4.6}$$

The deviations from ideality are all contained in the osmotic coefficient. To get some indication of the deviation from ideality, the osmotic coefficient is shown as a function of concentration for six solutes in Figure 4.5. These data indicate that for concentrated solutions the deviations of the osmotic coefficient from unity can be quite large. However, for physiological concentrations (< 0.5 mol/kg), the osmotic coefficient deviates from 1 by less than 10% for all the solutes except sulfuric acid. The osmotic coefficient of macromolecules (such as hemoglobin) may be significantly greater than 1 and may depend strongly on concentration.

The data shown in Figure 4.5 indicate that at high solute concentration the osmotic pressure is a systematic function of solute concentration and that this function is distinct for different solutions. This dependence on concentration of real solutions is often expressed by the expansion

$$\pi = RT\left(A + BC_\Sigma + CC_\Sigma^2 + DC_\Sigma^3 + \cdots\right),$$

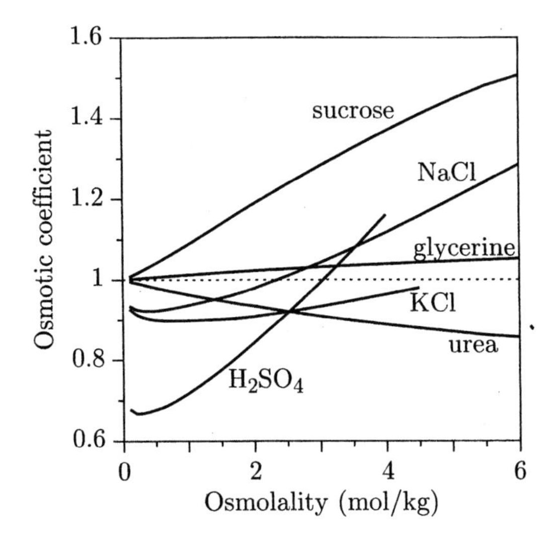

Figure 4.5 Dependence of the osmotic coefficient on concentration for six solutes (Scatchard et al., 1938).

where the coefficients A, B, C, etc., are called *virial coefficients* and express the nonideality of the osmotic pressure of real solutions. Various theories have been developed to predict the values of the virial coefficients for solutions. These virial coefficients take into account such properties of real solutions as the nonzero volume of solute molecules, the interactions between charged solutes, and so forth.

4.4 Osmotic and Hydraulic Flow in Porous Media

4.4.1 Differential Laws of Solvent Transport

4.4.1.1 In the Absence of Concentration Gradients: Darcy's Law

In response to hydraulic pressure, solvent flow in a homogeneous, porous medium has been shown empirically to be analogous to heat flow, diffusive flow, and electrical conduction (see Chapter 3). The convective flux of solvent volume obeys Darcy's law (Scheidegger, 1974):

$$\Phi_V = -\kappa \frac{\partial p}{\partial x}, \tag{4.7}$$

where Φ_V is the flux of solvent volume, i.e., the volume of solvent crossing a unit area per unit time (m/s), p is the hydraulic pressure (Pa), and κ is the *hydraulic permeability* ($m^2/(Pa \cdot s)$).[3] The hydraulic permeability is a material property that reflects the relative ease with which the solvent permeates the porous material. The value of the hydraulic permeability is determined by the microscopic structure of the porous material, such as the ratio of solid to solvent volume and the tortuosity of the channels through which the solvent must flow.

4.4.1.2 In the Presence of Solute Concentration Gradients

If we assume that osmotic and hydraulic pressures are equivalent, then Darcy's law can be generalized to include both hydraulic and osmotic pressures in a porous medium:

$$\Phi_V = -\kappa \frac{\partial (p - \pi)}{\partial x}. \tag{4.8}$$

Thus, volume flow occurs in a porous medium in response to both a hydraulic pressure gradient and an osmotic pressure gradient (which results from a solute concentration gradient).

4.4.2 Conservation of Mass

Conservation of mass relates the solvent volume flux to its density. Consider a volume element whose sides have area A, as shown in Figure 4.6. Let the mass density of the solvent be ρ_m. Then conservation of mass requires that the net mass flowing into the volume element in a time interval of duration Δt must equal the increase in mass in the volume element. This relation can be expressed as follows:

$$\Phi_V(x,t)A\rho_m(x,t)\Delta t - \Phi_V(x+\Delta x,t)A\rho_m(x+\Delta x,t)\Delta t =$$
$$(\rho_m(x,t+\Delta t) - \rho_m(x,t))\, A\Delta x. \tag{4.9}$$

Rearranging the terms in Equation 4.9 and letting $\Delta x \to 0$ and $\Delta t \to 0$, we obtain

$$\frac{\partial(\rho_m \Phi_V)}{\partial x} = -\frac{\partial \rho_m}{\partial t}, \tag{4.10}$$

which is the equation of continuity of mass for the solvent.

3. The *hydraulic permeability* of a solvent is distinct from the *permeability* of a solute (Chapter 3). Unfortunately, this nomenclature is confusing but it is widely used.

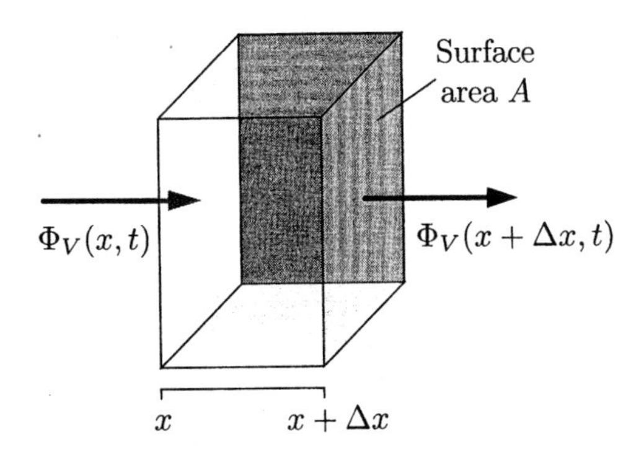

Figure 4.6 Relation of volume flux to mass density. The mass density in the volume element is $\rho_m(x,t)$.

4.4.3 Steady-State Solvent Transport

In the steady state, all solvent transport variables are assumed independent of time. If the solvent density is independent of t, i.e., the solvent is incompressible, then Equation 4.10 implies that

$$\frac{\partial(\rho_m \Phi_V)}{\partial x} = 0. \tag{4.11}$$

If, we assume, furthermore, that the solvent density is also uniform in space, then Equation 4.11 implies that Φ_V is a constant. In the steady state, the volume flux is independent of space and time, and Equation 4.8 becomes

$$\Phi_V = -\kappa \frac{d(p-\pi)}{dx}. \tag{4.12}$$

Since Φ_V is constant, $p - \pi$ must be a linear function of x in the steady state, so that

$$(p(x) - p(x_o)) - (\pi(x) - \pi(x_o)) = -\frac{\Phi_V}{\kappa}(x - x_o), \tag{4.13}$$

where x_o is a reference location at which the hydraulic and osmotic pressures are known.

4.5 Steady-State Solvent Transport Through Thin Membranes

4.5.1 Macroscopic Relations

4.5.1.1 Integration of the Darcy and van't Hoff Laws

Consider a membrane of thickness d that separates two compartments and through which the solvent flows in the steady state (Figure 4.7). We can inte-

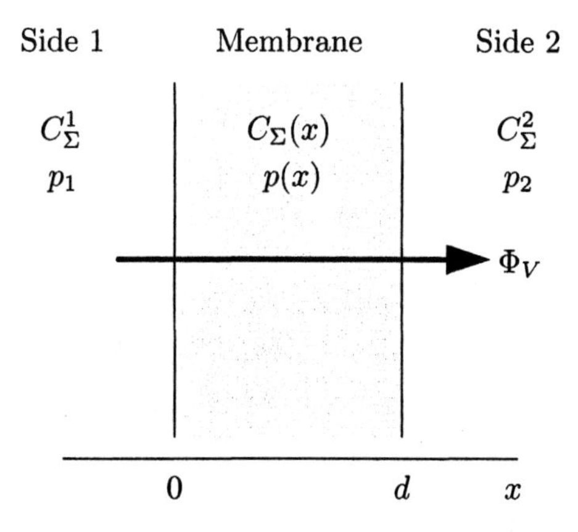

Figure 4.7 Definition of variables for steady-state solvent transport through a membrane. The membrane is shown schematically in this figure and may be thought of as a membrane that allows solvent but not solute to permeate.

grate Equation 4.12 across the membrane as follows:

$$\int_{x=0}^{d} \Phi_V dx = \int_{x=0}^{d} -\kappa \frac{d(p-\pi)}{dx} dx. \tag{4.14}$$

If we assume that the hydraulic permeability is independent of position in the membrane, then integration yields

$$\Phi_V d/\kappa = (p(0) - \pi(0)) - ((p(d) - \pi(d)). \tag{4.15}$$

To relate the hydraulic and osmotic pressures in the membrane at the membrane interface to the pressures in the bulk solution, we need boundary conditions at the interfaces. Equation 4.12 indicates that the quantity $p - \pi$ must be continuous at the interfaces. If it were not continuous then the volume flux would be unbounded at the interface. Since $p - \pi$ must be continuous at each interface, $p(0) - \pi(0) = p_1 - \pi_1$ and $p(d) - \pi(d) = p_2 - \pi_2$. Therefore, Equation 4.15 becomes

$$\Phi_V = \mathcal{L}_V \left((p_1 - \pi_1) - (p_2 - \pi_2)\right), \tag{4.16}$$

or

$$\Phi_V = \mathcal{L}_V \left(\Delta p - \Delta \pi\right), \tag{4.17}$$

where $\Delta\pi = \pi_1 - \pi_2 = RT(C_\Sigma^1 - C_\Sigma^2)$, $\Delta p = p_1 - p_2$, and $\mathcal{L}_V = \kappa/d$. The coefficient $\mathcal{L}_V$, called the *hydraulic conductivity*, has the units m/(Pa·s).[4] The hydraulic conductivity of a membrane is a measure of its capacity to transport

4. The hydraulic conductivity is often denoted as L_p.

volume when the solute is impermeant, i.e., $\mathcal{L}_V$ is the volume flux per unit pressure difference across the membrane.

4.5.1.2 Properties of the Steady-State Relation

Equation 4.17 is a constitutive relation for the flow of volume through a membrane in response to a difference in hydraulic and osmotic pressure across the membrane. We explore some of the properties of this relation.

Equilibrium
Osmotic equilibrium across a membrane occurs when the flow of volume (of water) is zero. Therefore, for equilibrium, we have $\Phi_V = 0$, which implies that either $\mathcal{L}_V = 0$, implying that water cannot flow through the membrane independent of the difference in hydraulic and osmotic pressure across the membrane, or $p_1 - \pi_1 = p_2 - \pi_2$, which can also be written as $p_1 - p_2 = \pi_1 - \pi_2$. We can consider two special cases. First, suppose $\mathcal{L}_V \neq 0$ and that there is no difference in solute concentration on the two sides of the membrane, i.e., $\pi_1 = \pi_2$. Then at equilibrium $p_1 = p_2$, which is a statement of mechanical equilibrium across a membrane. Second, suppose there is no difference of hydraulic pressure across the membrane. Then osmotic equilibrium occurs when $\pi_1 = \pi_2$ or $C_\Sigma^1 = C_\Sigma^2$.

Direction of Osmotic Transport
Equation 4.17 shows that $\Phi_V > 0$ when $p_1 - \pi_1 > p_2 - \pi_2$ and $\Phi_V < 0$ when $p_1 - \pi_1 < p_2 - \pi_2$. For the special case when $p_1 = p_2$, Equation 4.17 can be written as

$$\Phi_V = \mathcal{L}_V RT \left(C_\Sigma^2 - C_\Sigma^1 \right). \tag{4.18}$$

This equation shows that the flux from side 1 to side 2 will be positive if the total concentration of solute on side 2 exceeds that on side 1 so that water will flow in the direction to decrease the concentration of solute.

Relation of Osmotic to Diffusive Transport of Water
When the volume flux is carried entirely by the solvent, water, then the volume flux is proportional to the molar flux of water; i.e., $\Phi_V = \phi_w \overline{V}_w$, where ϕ_w is the number of moles of water flowing through a unit area of membrane per unit time, and $\overline{V}_w$ is the partial molar volume of water, which is the volume of a mole of water. We can express the osmotic molar flux of water as

$$\phi_w = \mathcal{P}_w \left(C_\Sigma^2 - C_\Sigma^1 \right), \tag{4.19}$$

where

$$\mathcal{P}_w = \frac{RT\mathcal{L}_V}{\overline{V}_w}. \tag{4.20}$$

$\mathcal{P}_w$ is called the *osmotic permeability coefficient* or the *filtration permeability coefficient*. Equation 4.19 has a form that resembles (deceptively) Fick's first law for membranes (Equation 3.44), which is

$$\phi_n = P_n\left(c_n^1 - c_n^2\right). \tag{4.21}$$

Both these equations relate a flux to a concentration difference. However, they differ in important ways. Equation 4.19 relates the molar flux of *solvent* to the total molar concentration of *solute*. The constant of proportionality is the *osmotic* permeability, $\mathcal{P}_w$. In contrast, Fick's first law for membranes relates the molar flux of a particular *solute* n to that solute's molar concentration difference across the membrane. However, the diffusion of water through the membrane, in the absence of an osmotic pressure difference, can be measured by means of a radioactive tracer for water. Then the diffusive flux of water can be expressed as

$$\phi_w = P_w\left(c_w^1 - c_w^2\right), \tag{4.22}$$

where P_w is the *diffusive permeability* (called simply the *permeability* in Chapter 3) of the membrane for water. Thus, Equation 4.19 is the osmotic water flux and Equation 4.22 is the diffusive flux of water. Therefore, the two parameters P_w and $\mathcal{P}_w$ measure quite different attributes of water transport through membranes, although both have the same units (cm/s). As we shall see, a direct comparison of P_w to $\mathcal{P}_w$ can be used to assess the mechanisms of water transport across membranes.

4.5.2 Microscopic Mechanisms of Water Transport for Simple Membrane Models

In this section we explore two simple mechanisms for water transport through a membrane (Finkelstein, 1987): (1) the dissolve-diffuse mechanism of water permeation in a homogeneous, hydrophobic membrane, and (2) water transport through a membrane whose matrix is impermeant to water but that contains pores permeant to water. We explore water transport both in the presence of and in the absence of a difference in hydraulic and/or osmotic pressure across the membrane. One purpose of this discussion is to clarify the relation between the diffusion and convection of water. The results will also be important for the interpretation of measurements of water transport through cellular membranes, which will be discussed in a later section.

4.5.2.1 Dissolve-Diffuse Mechanism in a Homogeneous, Hydrophobic Membrane

First we consider a thin membrane consisting of a hydrophobic material, e.g., an organic solvent, such as oil, in which water is poorly soluble. Nevertheless, some water will dissolve in the membrane and diffuse across it.

Water Transport Resulting from an Osmotic Pressure Difference
We assume that the membrane has thickness d and is impermeant to the solutes in the aqueous solutions on the two sides of the membrane. For the sake of simplicity we assume that there is no hydraulic pressure difference across the membrane and that the water molecules will diffuse in the membrane according to Fick's first law for diffusion. Hence, the flux of water is related to the concentration of water at the two interfaces in the interior of the membrane by Fick's first law for membranes:

$$\phi_w = \frac{D_w(h)}{d}\left(c_w^1(h) - c_w^2(h)\right), \tag{4.23}$$

where $D_w(h)$ is the diffusion coefficient of water in the membrane and where $c_w^1(h)$ and $c_w^2(h)$ are the concentrations of water in the hydrophobic membrane at sides 1 and 2, respectively. As shown in Appendix 4.1, the concentration of water in the membrane can be related to the solute concentration in the aqueous solutions on both sides of the membrane as follows:

$$c_w^1(h) = k_w \frac{1 - C_\Sigma^1 \overline{V}_w}{\overline{V}_h} \text{ and } c_w^2(h) = k_w \frac{1 - C_\Sigma^2 \overline{V}_w}{\overline{V}_h}, \tag{4.24}$$

where $\overline{V}_w$ is the partial molar volume of water (which is about 18 mL/mol), and $\overline{V}_h$ is the partial molar volume of the hydrophobic solvent in the membrane. k_w is the partition coefficient of water between two media: the hydrophobic membrane and the aqueous solutions on the two sides of the membrane. We can combine Equations 4.23 with 4.24 to yield

$$\phi_w = \frac{D_w(h)k_w}{d}\left(\frac{1 - C_\Sigma^1 \overline{V}_w}{\overline{V}_h} - \frac{1 - C_\Sigma^2 \overline{V}_w}{\overline{V}_h}\right),$$

which, after simplification, yields

$$\phi_w = \mathcal{P}_w\left(C_\Sigma^2 - C_\Sigma^1\right), \tag{4.25}$$

where the osmotic permeability is

$$\mathcal{P}_w = \frac{D_w(h)k_w\overline{V}_w}{d\overline{V}_h}. \tag{4.26}$$

Equation 4.25 relates the flux of water to the solute concentration. If the solute concentration is lower on side 1 than on side 2 of the membrane, then the water concentration will be higher on side 1 than on side 2 and water will flow through the membrane by diffusion from side 1 to side 2. The volume flux is $\Phi_V = \overline{V}_w\phi_w$, so that

$$\Phi_V = \mathcal{P}_w\overline{V}_w\left(C_\Sigma^2 - C_\Sigma^1\right). \tag{4.27}$$

Therefore, the hydraulic conductivity for this mechanism is

$$\mathcal{L}_V = \frac{\mathcal{P}_w\overline{V}_w}{RT}.$$

Water Diffusion

Water diffusion through the membrane can be measured by means of a radioactive isotope of water (e.g., deuterium, tritium, etc.) whose transport properties are assumed to be identical to those of water molecules. Since the solute concentrations are the same on the two sides of the membrane and the hydraulic pressures are also the same, there are no convective forces on the water molecules. We designate the tracer with an asterisk (*), so that Fick's first law for membranes implies that the tracer flux is

$$\phi_w^* = \frac{D_w(h)}{d}\left(c_w^{1*}(h) - c_w^{2*}(h)\right),$$

where the concentrations of tracer in the hydrophobic membrane are $c_w^{1*}(h)$ and $c_w^{2*}(h)$. The concentration of tracer in the membrane at the interface with the aqueous solution is proportional to the mole fraction of tracer in the aqueous solution so that

$$\phi_w^* = \frac{D_w(h)k_w}{d\overline{V}_h}\left(x_w^{1*} - x_w^{2*}\right).$$

However, the tracer acts as a solute in the aqueous solution so that $x_w^*(w) = c_w^*(w)\overline{V}_w$ and

$$\phi_w^* = \frac{D_w(h)k_w\overline{V}_w}{d\overline{V}_h}\left(c_w^{1*}(w) - c_w^{2*}(w)\right).$$

Therefore, the diffusive permeability is

$$P_w = \frac{D_w(h)k_w\overline{V}_w}{d\overline{V}_h}. \tag{4.28}$$

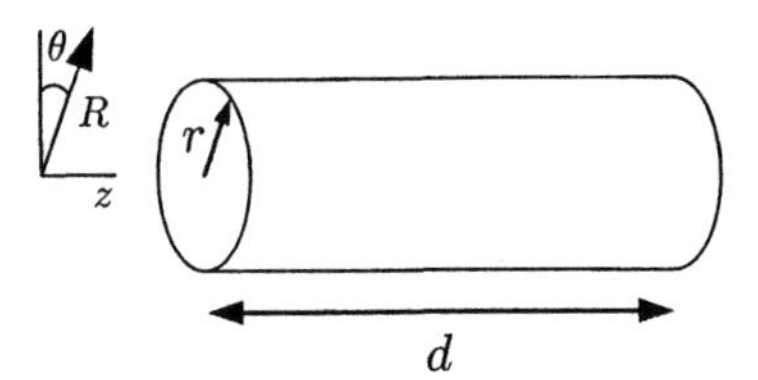

Figure 4.8 Geometry of a cylindrical pore.

Comparison of $\mathcal{P}_w$ and P_w for a Homogeneous, Hydrophobic Membrane
Comparison of Equations 4.26 and 4.28 reveals that for water diffusion in a homogeneous, hydrophobic membrane $\mathcal{P}_w = P_w$. This results because the mechanism of water transport through the membrane is the dissolve-diffuse mechanism whether it results from a difference in osmotic pressure or from a difference in isotope concentration at osmotic equilibrium.

4.5.2.2 Transport Through a Porous Membrane

We next consider a membrane that is made up of a material impermeant to water which contains pores through which water is transported.

Water Convection Through Large Pores: Poiseuille's Law
The simplest model for solvent flow through a porous membrane occurs when the pore area is sufficiently large that continuum hydrodynamics applies and the pore length is sufficiently long that entrance and exit phenomena at the pore have a negligible effect on flow. In this range, Poiseuille's law applies. If there are $\mathcal{N}$ pores per unit area of membrane, and each pore is a cylinder of radius r and length d (Figure 4.8), then Poiseuille's law (Appendix 4.2) relates the volume flux from side 1 to side 2 through a single pore, $(\Phi_V)_p$, to the pore geometry and the viscosity, η, as follows:

$$(\Phi_V)_p = \frac{\pi r^4}{8\eta d}(p_1 - p_2)\,. \tag{4.29}$$

Hence, the volume flux through the membrane is

$$\Phi_V = \mathcal{N}\frac{\pi r^4}{8\eta d}(p_1 - p_2)\,, \tag{4.30}$$

and, therefore,

$$\mathcal{L}_V = \mathcal{N}\frac{\pi r^4}{8\eta d}. \tag{4.31}$$

The osmotic permeability of the porous membrane is

$$\mathcal{P}_w = \mathcal{N}\frac{RT\pi r^4}{8\overline{V}_w\eta d}. \tag{4.32}$$

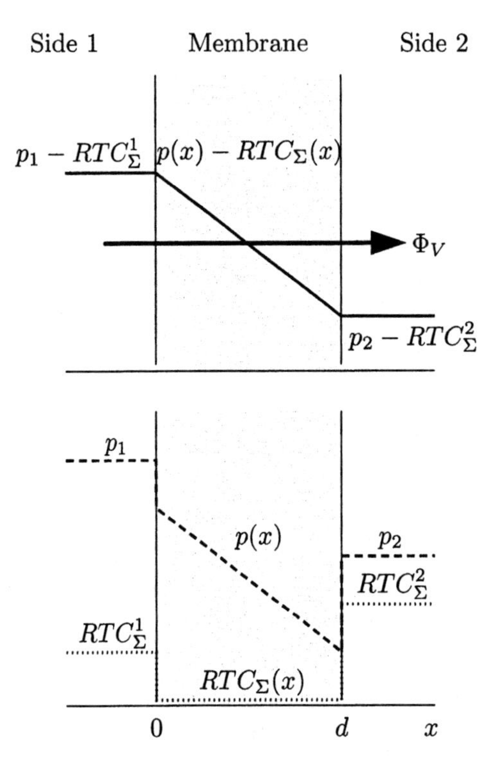

Figure 4.9 Sketch of osmotic and hydraulic pressure through a membrane for steady-state solvent transport. The upper panel shows a plot of the difference of hydraulic and osmotic pressure. This difference is continuous at the interface between the membrane and each bath. The lower panel shows the hydraulic (dashed line) and osmotic pressures (dotted line) individually. Both are discontinuous at each membrane-bath interface.

Because of the equivalence of hydraulic and osmotic pressure in producing a volume flux, the hydraulic conductivity of a porous membrane can be obtained from measurements of volume flux for either a known hydraulic or a known osmotic pressure difference across the membrane.

What Are the Forces That Drive Water Through a Porous Membrane?

Insight into the macroscopic mechanism of osmosis through a porous membrane is obtained by examining the solution for steady-state solvent transport through the membrane as shown in Figure 4.9 (Garby, 1957; Mauro, 1957, 1960, 1965). The quantity $p - \pi$ takes on constant values in each bath solution and, according to Equation 4.13, is a linear function of distance in the membrane for a constant volume flux. The boundary conditions at the interfaces ensure continuity of $p - \pi$ at the interfaces.[5] However, because the solute is

5. Continuity of $p - \pi$ at the interface is equivalent to continuity of the chemical potential at that interface which must occur if the solutions on the two sides of the interface are at equilibrium (Appendix 4.1).

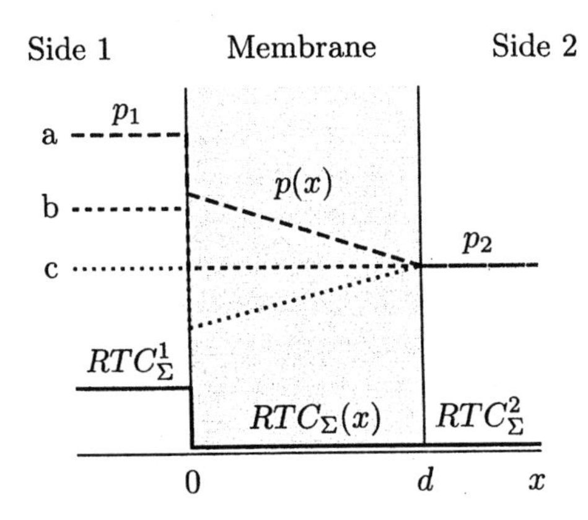

Figure 4.10 Sketch of hydraulic pressure through a membrane for steady-state solvent transport. Profiles of hydraulic pressure are shown (dashed and dotted lines) for three values of p_1. The osmotic pressure (solid line) is the same for all three values of p_1.

impermeant, its concentration in the membrane is zero. Thus, there is a discontinuity in solute concentration, and therefore in the osmotic pressure, at each membrane-solution interface. Since $p - \pi$ is continuous at the interface, there must be a discontinuity in hydraulic pressure at each interface that just equals the discontinuity in osmotic pressure. The solute concentration drops inside the membrane, which results in a drop in the osmotic pressure inside the membrane and an equivalent drop in the hydraulic pressure. Since there is no solute in the membrane, the volume flux through the membrane is proportional to the negative of the pressure gradient.

The relation between osmotic and hydraulic pressure in a membrane is explored further in Figure 4.10, which shows a membrane separating a solution on side 1 from pure solvent on side 2. The pressure on side 2 is maintained constant, as is the concentration of solute on side 1. The figure shows the variation in hydraulic pressure in the membrane as the hydraulic pressure is varied on side 1. For the largest value of p_1 (trace a in Figure 4.10), the pressure gradient in the membrane is negative, so the volume flux is positive, i.e., volume flows in the positive x-direction. As p_1 is decreased, the pressure gradient in the membrane is decreased and the volume flux decreases. If p_1 is reduced so that $p_1 - RTC_\Sigma^1 = p_2$ (trace b in Figure 4.10), then the volume flux is zero. Further reduction in p_1 results in a negative volume flux, i.e., volume flows in the negative x-direction. If $p_1 = p_2$ (trace c in Figure 4.10) then there is no difference in hydraulic pressure between the compartments. However, the discontinuity in osmotic pressure at the interface between side 1 and the membrane causes a discontinuity in the hydraulic pressure at the interface. Hence, there is a positive pressure gradient in the membrane, which drives water in the negative x-direction. Figures 4.9 and 4.10 also imply that if

$RTC_\Sigma > p$ in a bath, then the hydraulic pressure just inside the membrane facing that bath will be negative, i.e., the solvent will be in tension. Such tensile solvent phenomena have been observed (Mauro, 1965).

In summary, the equivalence of hydraulic and osmotic pressure in causing solvent flux through a porous semipermeable membrane implies that in the steady state there is a discontinuity in hydraulic pressure at the interface between the membrane and the solution. The discontinuous reduction in hydraulic pressure equals the osmotic pressure in the bath. Thus, a discontinuity in solute concentration causes a discontinuity in hydraulic pressure at the membrane-solution interface, which drives water through the membrane by convection.

Water Diffusion Through Large Pores

The diffusion of water through a porous membrane can be measured by means of a radioactive tracer of water. We can define the diffusion of tracer water by Fick's law for membranes as follows:

$$\phi_w^* = P_w \left(c_w^{1*} - c_w^{2*} \right), \tag{4.33}$$

where ϕ_w^* is the diffusive flux of tracer water and P_w is the diffusive permeability. In Chapter 3 we found an expression for the diffusive permeability of a porous membrane to solutes that diffuse through the pores if the pores are large (Equation 3.50). We use this expression for the diffusive permeability of such a porous membrane to water to obtain

$$P_w = \frac{D_w \mathcal{N} \pi r^2}{d}, \tag{4.34}$$

where D_w is the diffusion coefficient of water molecules in water.

Comparison of $\mathcal{P}_w$ and P_w for a Membrane with Large Pores

The ratio of osmotic permeability to diffusive permeability for a porous membrane with large pores is

$$\frac{\mathcal{P}_w}{P_w} = \frac{(RT\mathcal{N}\pi r^4)/(8\eta d \overline{V}_w)}{(D_w \mathcal{N} \pi r^2/d)} = \frac{RTr^2}{8\eta D_w \overline{V}_w}. \tag{4.35}$$

$\mathcal{P}_w/P_w$ is plotted versus pore radius r in Figure 4.11 for $R = 8.3144 \times 10^6$ Pa/(K · (mol/cm^3)), $T = 293$ K, $\eta = 1 \times 10^{-3}$ Pa · s, $D_w = 2.24 \times 10^{-5}$ cm^2/s, and $\overline{V}_w = 18$ cm^3/mol. $\mathcal{P}_w/P_w$ increases as r^2 and exceeds 1 for $r > 0.36$ nm. However, a priori it is not clear that this theory for transport in large pores is valid for pores as small as 0.36 nm.

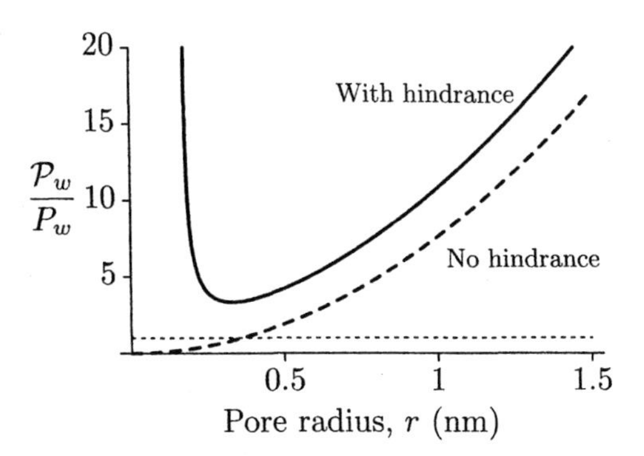

Figure 4.11 Ratio of osmotic to diffusive permeability of a porous membrane as a function of the pore radius. The thin dotted horizontal line is for $\mathcal{P}_w/P_w = 1$. The dashed line is a plot of Equation 4.35 for unhindered diffusion and convection. The solid line is a plot for hindered diffusion and convection (Hill, 1994, from Equation 9).

Convection and Diffusion Through Large Pores

Although we can imagine experiments in pure diffusion of water, it is not possible to design an experiment in which water is convected and does not diffuse. Thus, we need to consider convection and diffusion that occur simultaneously in the presence of a difference of hydraulic and/or osmotic pressure across a porous membrane (Bean, 1972). Suppose some of the water molecules are tagged (e.g., radioactively) so that the concentration of tracer water molecules is c_w^* and the flux of tracer water molecules is ϕ_w^*. Then

$$\phi_w^* = \underbrace{c_w^*(x)\Phi_V}_{\text{convection}} \underbrace{-D_w \mathcal{N}\pi r^2 \frac{dc_w^*(x)}{dx}}_{\text{diffusion}}, \tag{4.36}$$

where the two terms on the right-hand side of Equation 4.36 are water fluxes resulting from convection and diffusion, respectively. The factor $\mathcal{N}\pi r^2$ is the fraction of the membrane surface area that consists of pores. We shall assume that the fluxes change sufficiently slowly that steady-state conditions hold in the membrane. Thus, ϕ_w^* and Φ_V are constant and independent of position, x, in the membrane. We shall find the solution to Equation 4.36 subject to the boundary condition

$$c_w^*(0) = c_w^{1*}, \text{ and } c_w^*(d) = c_w^{2*}. \tag{4.37}$$

To find the total solution we first find the solution to the homogeneous equation, which is

$$\frac{dc_w^*(x)}{dx} - \frac{\Phi_V}{D_w \mathcal{N}\pi r^2} c_w^*(x) = 0, \tag{4.38}$$

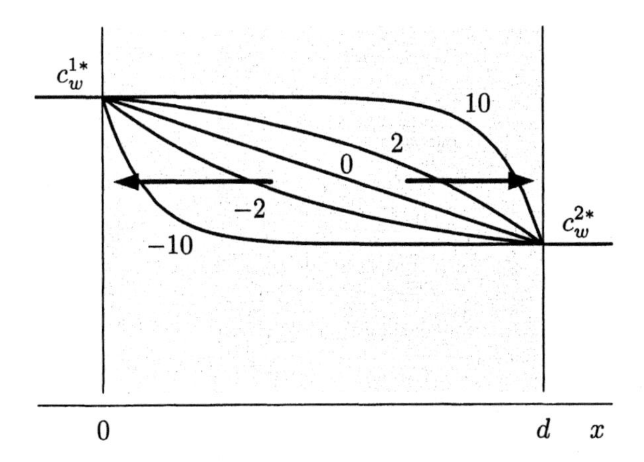

Figure 4.12 Dependence of concentration on position in a pore. The parameter is γd, which is proportional to Φ_V. The sign of γ equals the sign of Φ_V, whose direction is shown by the arrows.

which has the solution

$$c_w^*(x) = Ae^{\gamma x}, \tag{4.39}$$

where $\gamma = \Phi_V/(D_w\mathcal{N}\pi r^2)$. Therefore, the general solution has the form

$$c_w^*(x) = Ae^{\gamma x} + B. \tag{4.40}$$

A and B can be found from the boundary conditions (Equation 4.37) and are

$$A = \frac{c_w^{1*} - c_w^{2*}}{1 - e^{\gamma d}}; \qquad B = \frac{c_w^{2*} - c_w^{1*}e^{\gamma d}}{1 - e^{\gamma d}}. \tag{4.41}$$

With substitution of Equations 4.41 into 4.40, we obtain

$$c_w^*(x) = \frac{(c_w^{1*} - c_w^{2*})e^{\gamma x} + (c_w^{2*} - c_w^{1*}e^{\gamma d})}{1 - e^{\gamma d}}. \tag{4.42}$$

Figure 4.12 shows the form of the concentration of water tracer as a function of position in the pore. Note that γ is proportional to Φ_V. When $\Phi_V = 0$ there is no fluid convection. Hence, the profile of concentration in the membrane is linear with position. This profile is just what occurs with steady-state diffusion through a thin membrane. When there is fluid convection, the profile is exponential and the concentration is continuous at each interface between the membrane and the bath. The direction of Φ_V determines the slope of the concentration near the interface. When $\Phi_V > 0$, the concentration of tracer water remains close to its value in the left bath as the tracer water molecules are swept into the membrane from that bath. Conversely, when $\Phi_V < 0$, the concentration remains near its value in the right bath because water molecules are being swept into the membrane from the right bath.

The water flux can be obtained by substituting Equation 4.42 into 4.36 to yield

$$\phi_w^* = \Phi_V \frac{c_w^{2*} - c_w^{1*} e^{\gamma d}}{1 - e^{\gamma d}}. \tag{4.43}$$

The term γ in the exponent depends upon Φ_V, which can be written in terms of the difference in hydraulic and osmotic pressure, $\Phi_V = \mathcal{L}_V(\Delta p - \Delta\pi)$, so that

$$\phi_w^* = \mathcal{L}_V \frac{c_w^{2*} - c_w^{1*} e^{\beta(\Delta p - \Delta\pi)}}{1 - e^{\beta\Delta p}}(\Delta p - \Delta\pi), \tag{4.44}$$

where

$$\beta = \frac{\mathcal{L}_V}{P_w} \tag{4.45}$$

and

$$P_w = \frac{D_w \mathcal{N}\pi r^2}{d}. \tag{4.46}$$

If Poiseuille's law governs convection through the membrane, the substitution of Equation 4.31 and 4.46 into 4.45 yields

$$\beta = \frac{r^2}{8\eta D_w}. \tag{4.47}$$

Note that β is the ratio of the hydraulic conductivity to the diffusive permeability of the membrane. Therefore, β is a measure of whether the transport is dominated by diffusion or by convection. For large values of β, the transport is dominated by convection; for small values of β the transport is dominated by diffusion. For Poiseuille flow, β is proportional to r^2 (if D_w is independent of the pore radius). Therefore, the value of β depends upon r. We will consider the limiting behavior of transport through such a porous membrane when either convection or diffusion dominates the flow.

Transport Through Small Pores ($\beta(\Delta p - \Delta\pi) \ll 1$) Is Predominantly Due to Diffusion When the pores are sufficiently small that $\beta(\Delta p - \Delta\pi) \ll 1$, we can approximate $e^{\beta(\Delta p - \Delta\pi)} \approx 1 + \beta(\Delta p - \Delta\pi)$ and substitute this into Equation 4.44 to obtain

$$\begin{aligned}\phi_w^* \approx \mathcal{L}_V \frac{c_w^{2*} - c_w^{1*}(1 + \beta(\Delta p - \Delta\pi))}{1 - (1 + \beta(\Delta p - \Delta\pi))}(\Delta p - \Delta\pi) &\approx \frac{\mathcal{L}_V}{\beta}(c_w^{1*} - c_w^{2*}) \\ &= P_w(c_w^{1*} - c_w^{2*}).\end{aligned} \tag{4.48}$$

For small pores, convection becomes negligible ($\mathcal{L}_V$ is very small) and transport is due predominantly to diffusion. As indicated in Equation 4.48, water flux is proportional to concentration difference across the membrane, i.e., the flux satisfies Fick's law for membranes.

Transport Through Large Pores ($\beta(\Delta p - \Delta\pi) \gg 1$) Is Predominantly Due to Convection When the pores are so large that $c_w^{1*}e^{\beta(\Delta p-\Delta\pi)} \gg c_w^{2*}$, and $e^{\beta(\Delta p-\Delta\pi)} \gg 1$, then Equation 4.44 implies that

$$\phi_w^* \approx \mathcal{L}_V c_w^{1*}(\Delta p - \Delta\pi), \tag{4.49}$$

which is the convective water flux from side 1 to side 2. For large pores, water diffusion makes a negligible contribution to water transport, and water flux is proportional to the hydraulic pressure difference.

Transport into an Initially Empty Compartment For the case $c_w^{2*} = 0$ and $\Delta\pi = 0$, Equation 4.44 becomes

$$\phi_w^* = \mathcal{L}_V \frac{c_w^{1*}e^{\beta\Delta p}}{e^{\beta\Delta p} - 1}\Delta p = \frac{\mathcal{L}_V c_w^{1*}}{\beta}\frac{e^{\beta\Delta p}}{e^{\beta\Delta p} - 1}\beta\Delta p. \tag{4.50}$$

This equation can be expressed in normalized variables as follows:

$$\widehat{\phi}_w = \frac{e^{\widehat{p}}}{e^{\widehat{p}} - 1}\widehat{p} \tag{4.51}$$

The dependence of $\widehat{\phi}_w$ on $\widehat{p}$ for this case (Equation 4.51) is plotted in Figure 4.13. As the pressure difference increases, the total flux approaches that obtained for convection alone.

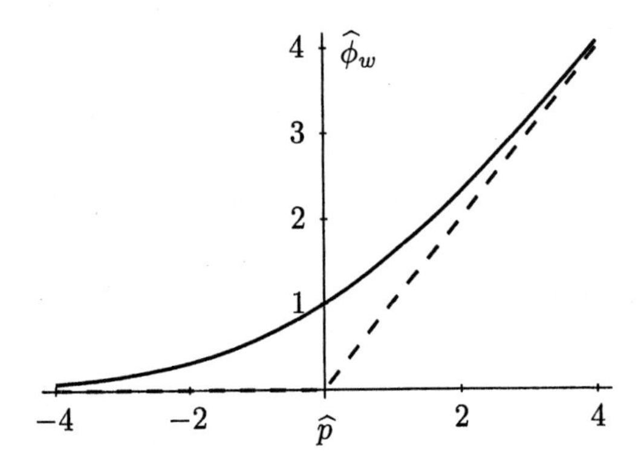

Figure 4.13 Normalized flux of water through a porous membrane as a function of the normalized hydraulic pressure. The solid line shows the flux in the presence of both diffusion and convection; the dashed line shows the flux for convection alone.

Measurements of Transport Through Artificial Porous Membranes The theory of water transport through porous membranes was tested by etching pores into sheets of mica to produce a membrane whose pore geometry was known (Bean, 1972). The range of average pore radii was about 15–150 nm. The etched mica membrane was placed in a measurement cell and separated two compartments filled with water. One compartment contained tritiated water. Diffusion of tritiated water was measured in the absence of a hydraulic pressure difference between the compartments, and total flux of tritiated water was measured in the presence of a hydraulic pressure difference. The flux of tritiated water was computed by sampling the water in the chamber that initially contained no tritiated water and computing the flux from the concentration changes as a function of time and the chamber geometry. The measurements are compared to the predictions of Equation 4.50 in Figure 4.14. The theoretically predicted dependence of water flux on pressure difference (Equation 4.50) fits these measurements to within the scatter of the measurements. Furthermore, measurements of diffusion in the absence of a hydraulic pressure difference and transport in the presence of such a difference each lead to an estimate of the pore radius. These estimates agree with each other to within 20% and agree with direct estimates of the pore radii of the mica membrane (obtained by electron microscopy) to within the same accuracy. Thus, these measurements show that the simple hydrodynamic theory of convection and diffusion in a porous membrane agrees with measurements for pore radii down to 15 nm.

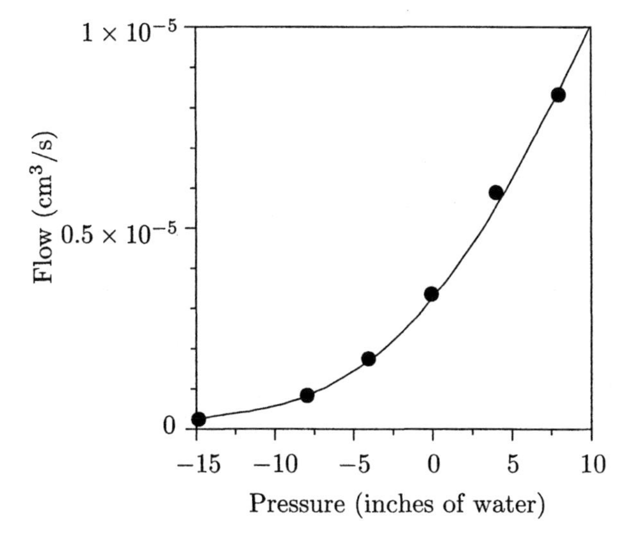

Figure 4.14 Measurements of water flux through porous mica membranes (adapted from Bean, 1972, Figure 9). Measurements of the flow of tritiated water (points) are compared with the predictions of Equation 4.51 (line) for pores etched in mica membranes. The pore diameters were approximately 3 μm in diameter.

Transport Through Pores of Molecular Dimensions

When the dimensions of pores become small, it is no longer valid to assume that the diffusion coefficient in the pore is the same as that in a free solution. The idea that the diffusion coefficient can be expressed as the free solution diffusion coefficient multiplied by a hindrance factor was discussed in Section 3.6.2. Similarly, for small pores Poiseuille's law is no longer valid, but hindrance factors can be used to take into account the effects of small pores on convective transport (Deen, 1987; Finkelstein, 1987). In terms of hindrance factors, the osmotic and diffusive permeabilities can be expressed as

$$\mathcal{P}_w = \mathcal{N}\frac{RT\pi r^4}{8\overline{V}_w \eta d} G(a_w/r) \text{ and } P_w = \frac{D_w \mathcal{N}\pi r^2}{d} F(a_w/r), \quad (4.52)$$

where a_w is the radius of the water molecule and $G(a_w/r)$ and $F(a_w/r)$ are hindrance factors. A number of different hindrance factors have been derived from different models of diffusion and convection in small pores. However, all these hindrance factors approach 1 when $a_w/r \to 0$. For example, for a particular choice of hindrance factors (Hill, 1994) for $a_w/r < 0.01$, the hindrance factors differ from 1 by less than 2%.[6] Thus, $\mathcal{P}_w/P_w$ for the hindered transport approaches that for unhindered transport for pores whose radii are large compared to the radius of a water molecule. However, $\mathcal{P}_w/P_w$ for hindered transport deviates from that for unhindered transport for pore radii that are comparable to the radius of a water molecule (Figure 4.11). Furthermore, for hindered transport the ratio $\mathcal{P}_w/P_w$ no longer increases as the square of pore radius, but has a minimum value at which $\mathcal{P}_w/P_w > 1$. Thus, this theory of hindered transport, which is more accurate for microscopic pores, predicts that $\mathcal{P}_w/P_w > 1$.

The validity of these hindered transport equations—which are derived from continuum hydrodynamics—for pores of molecular dimensions is not entirely clear. However, another theoretical approach that takes the molecular size of pores more explicitly into account has also yielded the conclusion that $\mathcal{P}_w/P_w > 1$. Consider pores that are so small that the diameter of the pore is only slightly greater than that of the water molecules (Figure 4.15). Under these circumstances the water molecules are transported in a single file, i.e., the water molecules cannot pass each other in the pore. As shown elsewhere (Finkelstein, 1987), osmotic and diffusional permeabilities are

---

6. The hindrance factors used in this calculation were $F(a_w/r) = 1 - 2.1019(a_w/r) + 2.829(a_w/r)^3 - 1.6764(a_w/r)^5 + 0.6772(a_w/r)^6$ and $G(a_w/r) = 1 - 0.8341(a_w/r)^2 + 0.8977(a_w/r)^3 - 1.0586(a_w/r)^4$.

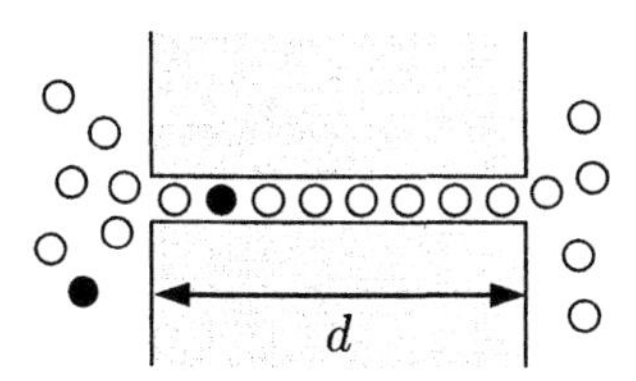

Figure 4.15 Schematic diagram of water transport through a microscopic pore whose diameter is only slightly larger than that of the water molecule. The length of the pore is d. Tracer water molecules are shown as filled circles; other water molecules are shown unfilled. In this schematic diagram $N_p = 8$.

$$\mathcal{P}_w = \frac{\overline{V}_w kT\mathcal{N}}{\gamma d^2} N_p \text{ and } P_w = \frac{\overline{V}_w kT\mathcal{N}}{\gamma d^2}, \tag{4.53}$$

where γ is a frictional coefficient, N_p is the number of water molecules that fit into the pore, and $\mathcal{N}$ is the pore density. The interesting finding is that the ratio of osmotic to diffusive permeability,

$$\frac{\mathcal{P}_w}{P_w} = N_p, \tag{4.54}$$

equals the number of water molecules in the pore. This result can be interpreted qualitatively for a pore with N_p fixed binding sites. Let us assume that the arrival of a water molecule at one end of the pore shifts the water molecules in the pore by one position. Then every time there is an arrival at one end there is a departure at the other end. However, for tracer diffusion of water molecule, the tracer molecule must traverse N_p sites before it will exit at the other end of the pore. Thus, tracer diffusion of molecules differs from convection by a factor of N_p.

In summary, we see that $\mathcal{P}_w/P_w > 1$ both for hydrodynamic theories based on hindered diffusion and convection and for theories of single-file transport through molecular poles.

4.5.2.3 *Comparison of Water Transport in Hydrophobic and Porous Membranes*

Both for a homogeneous, hydrophobic membrane and for a porous membrane, there is net flow of water due to a difference in osmotic pressure across the membrane. However, for the hydrophobic membrane, a careful analysis reveals that the underlying mechanism is diffusive water transport and not convective water transport. Thus, transport of water through the hydrophobic membrane is not by osmosis but by diffusion. This conclusion is quantified by the result that $\mathcal{P}_w/P_w = 1$ for the hydrophobic membrane. In contrast, a porous membrane supports both diffusion and osmosis of water in response to a difference in solute concentration. Although the theory for transport for porous membranes is not entirely satisfactory, several different approaches show that $\mathcal{P}_w/P_w > 1$ for a porous membrane.

4.6 The Physical Basis of Osmotic Pressure and Osmosis

From its inception, the physical basis of van't Hoff's law of osmotic pressure (and the resulting phenomenon of osmosis) was surrounded by controversy. Van't Hoff himself took various positions at various times (Hammel and Scholander, 1976) and apparently became quite exasperated with his critics, as is made clear in the quote at the beginning of this chapter. Many of the great names in nineteenth century physics took part in the discussion of the microscopic basis of osmotic pressure and osmosis (Gibbs, 1897; Kelvin, 1897; Rayleigh, 1897). Curiously enough, the controversy continues to this day. The liveliness of the debate can be appreciated in the tones of recent reviews which can get very contentious (Hildebrand, 1955; Babbitt, 1956; Chinard and Enns, 1956; Andrews, 1976; Hammel, 1976; Soodak and Iberall, 1978; Yates, 1978; Hammel, 1979a, 1979b; Mauro, 1979; Soodak and Iberall, 1979).

Central questions concerning osmotic pressure and osmosis include:

- What is the physical mechanism that gives rise to van't Hoff's law of osmotic pressure?
- Why is osmotic pressure equivalent to hydraulic pressure both in establishing equilibrium between two solutions separated by a semipermeable membrane and in producing volume flux of solvent through the membrane?
- What is the physical mechanism of the discontinuity in hydraulic pressure (see Section 4.5) that just equals the osmotic pressure at the interface between a semipermeable porous membrane and a bath?

4.6.1 Some Proposed Mechanisms

A number of different mechanisms of osmotic pressure and of osmosis have been proposed in the century since van't Hoff proposed his law. In a recent review (Guell, 1991), fourteen different mechanisms were identified, all purporting to explain osmotic pressure and osmosis. We will describe just three of these mechanisms and comment on their strengths and weaknesses.

4.6.1.1 Solute Bombardment

One of the more enticing facets of van't Hoff's law is its analogy to the ideal gas law:

$$p = RTc, \tag{4.55}$$

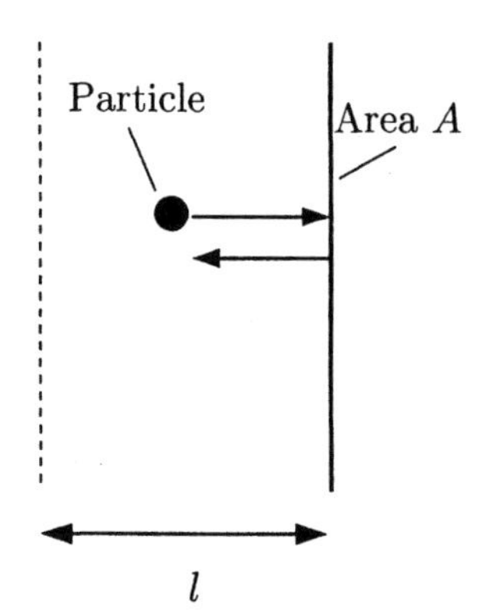

Figure 4.16 Schematic diagram of gas particles bombarding a wall.

where p is the pressure exerted by the gas on the containing wall and c is the concentration of the gas. A comparison between Equations 4.5 and 4.55 reveals that the form of the ideal gas law is the same as van't Hoff's law. To understand whether this analogy implies that the physical basis of osmotic pressure is the same as the physical basis of gas pressure, we need to understand the physical basis of gas pressure. The following argument indicates, in an approximate manner, how such a pressure arises. Consider a wall of area A and the particles contained in the volume Al on one side of the wall (Figure 4.16). We shall compute the change in momentum imparted to the wall by particles in this volume during an observation interval τ where the particles are assumed to undergo a one-dimensional random walk with a step size l and interval τ (see Chapter 3). The change in momentum of the wall in time τ is $pA\tau$, where p is pressure on the wall. The number of particles that hit the wall in the time interval τ equals half the number contained in the volume element and, therefore, equals $\frac{1}{2}cN_AlA$. The change in momentum of a particle that hits the wall is $2mv$, where v is the particle velocity in a direction perpendicular to the wall. The change in momentum of the wall must equal the change in momentum of the particles. Hence,

$$pA\tau = (cN_AlA/2)(2mv). \tag{4.56}$$

If we assume that the average particle velocity is l/τ and substitute this into Equation 4.56, we obtain

$$p = cN_Amv^2. \tag{4.57}$$

By the equipartition theorem,

$$\frac{1}{2}mv^2 = \frac{1}{2}kT. \tag{4.58}$$

Substituting Equation 4.58 into 4.57, we obtain

$$p = cN_A kT = cRT, \tag{4.59}$$

which is the ideal gas law. Thus, the thermal energy of the particles is responsible for their momentum, which is transferred to the wall during a collision between particle and wall.

Beginning with van't Hoff, the analogy between gas pressure and osmotic pressure has prompted a number of explanations that appear in textbooks, such as the following (Fermi, 1936):

> *This simple thermodynamical result* [the analogy of van't Hoff's law and the ideal gas law] *can be easily interpreted from the point of view of the kinetic theory. We consider a container divided into two parts by a semipermeable membrane with pure solvent in each part. Since the solvent can pass freely through the semipermeable membrane, the pressure on both sides of the membrane will be the same. Now let us dissolve some substance in one part and not in the other. Then the pressure on the side of the membrane facing the solution will be increased by the impacts against it of the molecules of the dissolved substances, which cannot pass through the membrane and which move with a velocity that depends on* T. *The larger the number of molecules dissolved and the higher the temperature, the larger will be the number of impacts per unit time, hence, the greater the osmotic pressure.*

This *solute bombardment* theory gives a reasonable explanation of how an additional pressure can develop on the side containing the solute but does not suggest a mechanism by which this additional pressure can cause a flux of solvent into the side that has the larger pressure. Furthermore, this type of explanation ignores the presence of the solvent, to which the momentum incurred by solute bombardment must be transferred.

4.6.1.2 Water Diffusion Theory

If solute is added to a solution then the concentration of water in the solution must be reduced. Consider a semipermeable membrane that separates two solutions having different solute concentrations. From Fick's law, we expect that water will diffuse from the side with higher water concentration (and lower solute concentration) to the side with lower water concentration (and higher solute concentration). We have analyzed just such a case in Section 4.5, where we examined transport through a homogeneous hydrophobic membrane. The argument that osmosis is equivalent to diffusion clearly gives the right direction for osmotic water transport, but it does not in general give the right magnitude. For example, in a porous membrane the difference between P_w and $\mathcal{P}_w$ quantifies the difference between diffusion and osmosis of water. Measurements have been made on porous membranes in which convective flow was measured in response to both hydraulic and osmotic pressure differences and diffusive flow was measured using radioactively labeled water molecules (Mauro, 1957). These experiments showed that the flow of water resulting from a difference in osmotic pressure is equivalent to that resulting from a difference in hydraulic pressure. These results, together with the equivalence of osmotic pressure and hydraulic pressure in establishing osmotic equilibrium, suggest that osmosis is convective rather than diffusive transport.

4.6.1.3 The Water Tension Theory

The "water tension" theory of osmotic pressure is a microscopic theory that, in one form or another, has been proposed at numerous times. This theory, described briefly here, has its proponents (Hammel and Scholander, 1976) but has not been accepted universally.

Solvent molecules are held together by intermolecular forces. This tendency to cohere is the microscopic basis for such macroscopic properties as surface tension and viscosity. The intermolecular forces in water result from the dipole moment of water molecules (Figure 1.5), which results in hydrogen bonds between water molecules that hold water molecules together. At the free surface of water, the molecules are also held together by these cohesive forces. However, the surface molecules experience a net pressure resulting from the bombardment of water molecules from one side of the surface but not from the other. This pressure arises by the same mechanism as that by which an ideal gas produces a pressure on a rigid wall. This type of bombardment of the surface molecules causes a tension on the surface molecules. Since the molecules cohere, this tension is transmitted by intermolecular forces to all the molecules in the bulk of the water. Hence, the water is in tension. The addition of solute molecules merely augments this tension on the surface of the solution and, hence, in the bulk of the solution. A more detailed analysis (Hammel, 1976) indicates that the excess tension caused by the presence of the solute in the solution is $\pi = RTC_{\Sigma}$. Hence, a solution under external compressive, hydraulic pressure p has an internal pressure $p - \pi$.

While there is experimental evidence that water can support a tensile pressure, the theoretical and experimental evidence to support the water tension theory of osmotic pressure is, at best, equivocal. The theory has been dealt a significant blow by a simple experiment (Mauro, 1979) in which a crystal of an impermeant solute was dropped into a solvent on one side of semipermeable membrane. Osmosis occurred only when the solute dissolved near the membrane. This experiment suggests that it is the concentration of solute near the membrane, and not at the free surface of the solution, that causes osmosis.

4.6.2 General Conclusions Concerning the Mechanism of Osmotic Pressure and of Osmosis

The following general conclusions devolve from the above considerations:

- Starting from the equation of state of a solution, the van't Hoff law can be derived from thermodynamics (Appendix 4.1). This derivation gives the empirically observed relation between osmotic pressure and solute concentration (at least for dilute solutions). However, as with all thermodynamic derivations, although the correct macroscopic relations are obtained, the microscopic mechanism remains obscure. This property is at once a strength and a weakness of such derivations. On the one hand, the macroscopic relations are quite general; they are valid for all microscopic models that result in the same equation of state. However, such derivations yield little insight into the physical mechanism of osmotic pressure.
- The notion that osmotic flow results from diffusion of solvent from a region of high solvent concentration to a region of low solvent concentration is intuitively appealing, and gives the right *direction* for osmotic flow. However, this does not predict the magnitude of osmotic flow.
- The ideal gas law is an excellent mnemonic for remembering van't Hoff's law, but the kinetic theory of gases by itself does not reveal the mechanism of osmotic pressure or osmosis.
- The equivalence of hydraulic pressure and osmotic pressure in establishing equilibrium across and in causing flow through a semipermeable membrane suggests that the molecular mechanism is one that equates osmotic pressure and hydraulic pressure. It must follow that the presence of a solute gradient creates an *osmotic* pressure that acts just like a pressure resulting from a body force.
- The presence of a semipermeable membrane is essential for the expression of osmotic pressure. *Osmotic pressure* is an unfortunate term, because it suggests the presence of a pressure in a solution. However, in contrast to hydraulic pressure, osmotic pressure is not observable in the absence of a semipermeable membrane.

4.6.3 An Intuitive Explanation of Osmotic Pressure and Osmosis

The simplest explanation of osmotic pressure and osmosis consistent with the above conclusions is summarized below (Villars and Benedek, 1974); a more precise treatment is available elsewhere (Guell, 1991).

At the interface between the solution and membrane there is a sharp gradient in solute concentration because the solute cannot enter the membrane. Solute particles would diffuse in such a gradient were it not for a force imposed on these particles by the membrane that prevents their transport. We can compute this force if we assume that transport of solute particles at the interface between the solution and membrane can be represented by a sum of diffusive and convective transport (see Equation 3.13) as follows:

$$\phi_n(x) = -D_n\frac{dc_n(x)}{dx} + \frac{D_n}{RT}c_n(x)f(x), \tag{4.60}$$

where ϕ_n, $c_n(x)$, and D_n are the flux, concentration, and diffusion coefficient of solute n, and $f(x)$ is the body force on a mole of particles. Hence, $c(x)f(x)$ is force density, i.e., the force per unit volume. Since the particles do not enter the membrane, the particle flux must be zero and the force imposed on the solute particles by the semipermeable membrane is

$$c(x)f(x) = RT\frac{dc_n(x)}{dx}. \tag{4.61}$$

Since the solute particles are in a viscous solution, the force transmitted to the solutes is coupled to the solution by viscous forces. To compute the forces on the solution at the membrane-solution interface, consider the differential volume element shown in Figure 4.17 at the entrance to a membrane pore. The volume element is subject to forces of fluid origin on both surfaces as well as to a body force due to the effect of the membrane on the solute particles. The total body force per unit area on the volume element is simply the body force density integrated over the length of the volume element as follows:

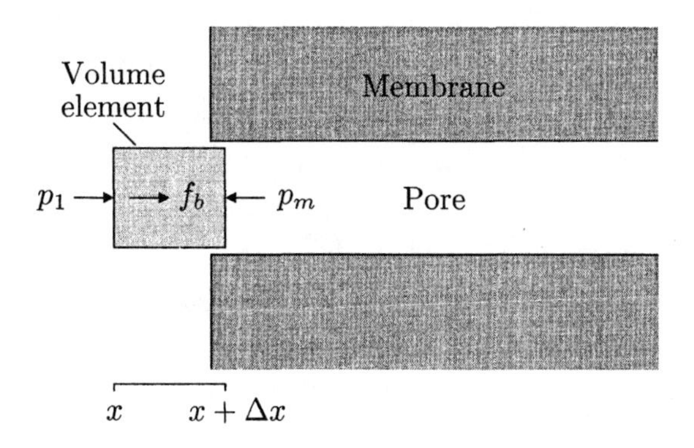

Figure 4.17 Volume element used to compute osmotic pressure. The right boundary of the volume element is just inside a membrane pore at a position where the solute concentration is zero and the pressure in the pore is p_m; the length of the volume element is Δx.

$$f_b = \int_x^{x+\Delta x} c_n(x) f(x)\, dx = \int_x^{x+\Delta x} RT \left(\frac{dc_n(x)}{dx} \right) dx$$
$$= RT \left(c_n(x + \Delta x, t) - c_n(x) \right),$$

which simply equals $-RTc_n(x)$ because the concentration of solute is zero inside the pore. Therefore, if the acceleration of the volume element is zero, then the forces on the volume element must sum to zero so that

$$p_1 - p_m - RTc_n(x) = 0,$$

which implies that the pressure just inside the membrane is $p_m = p_1 - RTc_n(x)$. This argument shows that the discontinuity in hydraulic pressure at the membrane-solution interface results from the force of the membrane on the solute particles, which is transmitted to the solvent.

4.7 Primary Responses of Cells to Changes in Osmotic Pressure

Cellular membranes are highly permeable to water. Therefore, a change in solute concentration either intracellularly or extracellularly causes water to flow through the membrane in response to the change in osmotic pressure across the membrane. If there are no appreciable mechanical restraining forces, as appears to be the case in animal cells but not in plant cells, the volume of the cell changes. The volume response of an animal cell to a change in the concentration of some solute is complex and will require study of other transport mechanisms (in later chapters). However, in general, cells exhibit a *primary response* to a change in solute concentration, and it is that response we investigate first. The primary response is rapid and due to the flow of water in the absence of the flow of any other substance. Thus, in this chapter we focus on the flow of water in response to a change in concentration of an *impermeant* solute.

In this section we investigate the theoretical consequences of the following assumptions:

- The cell membrane is semipermeable, it allows the transport of water, and it is impermeable to all solutes. Therefore, the total intracellular quantity of solutes remains constant as the cell swells.
- The flux of water across the membrane, is proportional to the difference in osmotic pressure across the membrane, which is proportional to the difference in solute concentration across the membrane.

- The cell membrane is freely extensible so that there are no mechanical forces that constrain a change in cell volume. This assumption appears valid for isolated animal cells, but not for animal cells in a tissue and not for plant cells, since the latter are surrounded by a rigid cell wall.
- The cell volume consists of two components: one is due to cell water that is osmotically active, and the other is not due to cell water that is osmotically active.

These assumptions constitute the basis of the theory underlying the primary response of animal cells that is described in this chapter.

4.7.1 Osmotic Equilibrium of Cells

4.7.1.1 *Theory*

We apply the notions of osmotic equilibrium between solutions separated by a semipermeable membrane to cells. At osmotic equilibrium, the net flux of volume out of the cell must be zero ($\Phi_V = 0$). Therefore, at osmotic equilibrium the osmotic pressure on the inside and the outside of the membrane must be the same, i.e.,

$$\pi^o = \pi^i. \tag{4.62}$$

Suppose the cell has a volume $\mathcal{V}_c$ (as shown in Figure 4.18). Let the volume of water in the cell be $\mathcal{V}^i = \mathcal{V}_c - \mathcal{V}_c'$, where $\mathcal{V}_c'$ is the volume of the cell that is not due to the volume of intracellular water. From Equation 4.62 and from the definition of osmotic pressure (Equation 4.5), osmotic equilibrium implies

$$C_\Sigma^o = C_\Sigma^i. \tag{4.63}$$

If N_Σ^i is the number of moles of solute inside the cell, then

$$C_\Sigma^o = \frac{N_\Sigma^i}{\mathcal{V}^i} = \frac{N_\Sigma^i}{\mathcal{V}_c - \mathcal{V}_c'}, \tag{4.64}$$

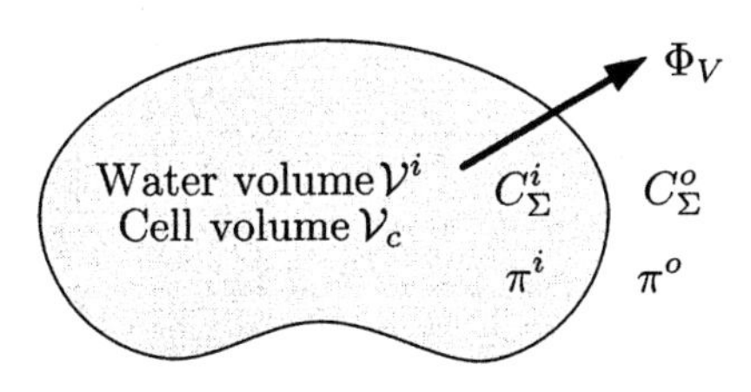

Figure 4.18 Schematic diagram of a cell showing osmotic variables.

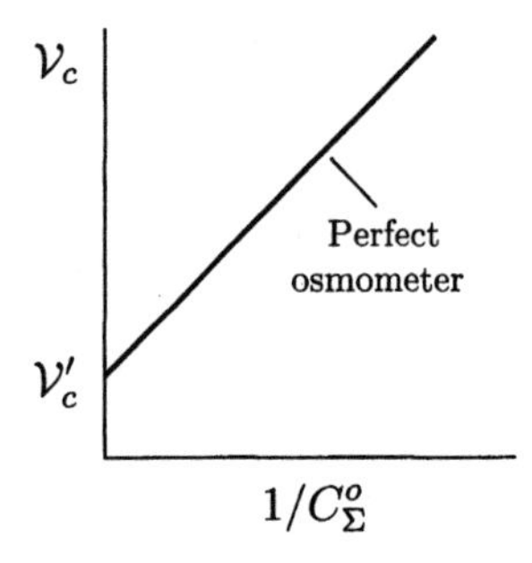

Figure 4.19 The volume of a perfect osmometer in response to a change in solute concentration.

or

$$\mathcal{V}_c = \frac{N^i_\Sigma}{C^o_\Sigma} + \mathcal{V}'_c, \tag{4.65}$$

and N^i_Σ is constant if the membrane is impermeable to the solutes. Thus, according to this simple theory, the volume of a cell is a linear function of the reciprocal of the extracellular solute concentration (Figure 4.19). The slope gives the quantity of intracellular solute, and the intercept on the ordinant gives the cell volume that is not due to osmotically active water. Note that $1/C^o_\Sigma \to 0$ occurs when $C^o_\Sigma \to \infty$, so that all the water flows out of the cell and the remaining volume is the volume not due to water. We shall call an object that obeys such a simple relation a *perfect osmometer*.

4.7.1.2 Measurements

To test the predictions of this simple theory, it is necessary to measure the volume of a cell as a function of the extracellular osmolarity. That is, according to Equation 4.65 the relation between $\mathcal{V}_c$ and $1/C^o_\Sigma$ is a straight line and the intercept where $1/C^o_\Sigma \to 0$ is where $\mathcal{V}_c = \mathcal{V}'_c$. The measurement of volume is clearly done most simply for cells with regular geometries. Egg cells of marine invertebrates are large and almost perfectly spherical (Figure 4.20). Hence, it is comparatively easy to measure the diameters of such cells with a microscope and an ocular micrometer and then to estimate the cell volume. These cells act as almost perfect osmometers, as indicated in Figure 4.21, which shows the effect of various concentrations of seawater on eggs of the sea urchin (*Arbacia punctulata*). The relation of the cell volume to the reciprocal of the concentration is a straight line. The concentration is normalized to the isotonic concentration. Hence, when this normalized concentration is 1, the cell volume is the same as it is in its normal environs in the body. That is the cell volume equals its *isotonic* value. The osmometric behavior of these cells is de-

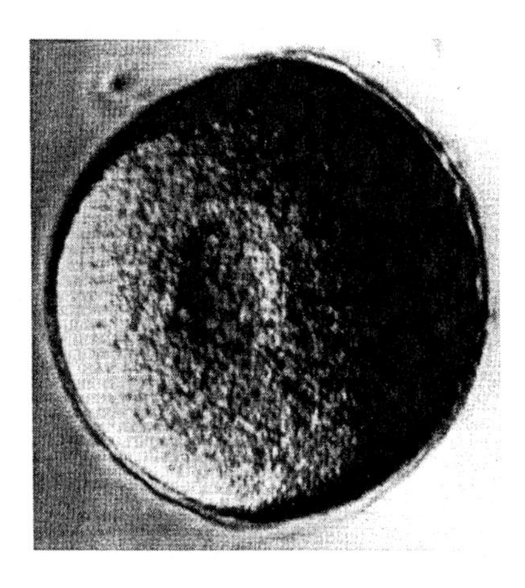

Figure 4.20 Photomicrograph of an egg cell of a sea urchin *Arbacia punctulata*. These egg cells typically have a diameter of 75 μm. The photomicrograph is from © Cabisco/Visuals Unlimited.

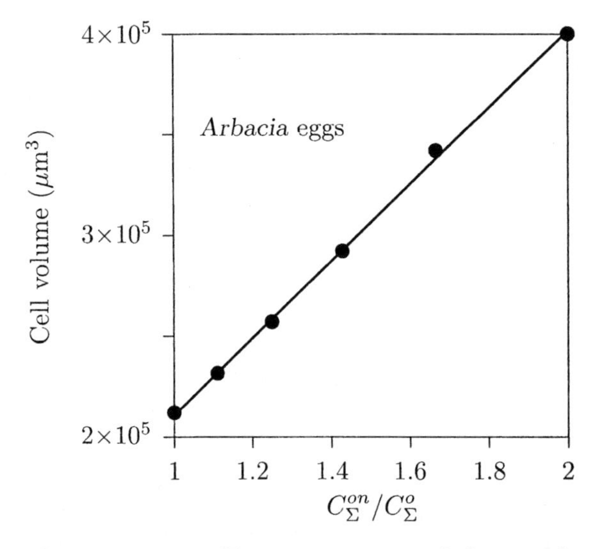

Figure 4.21 Equilibrium osmometric behavior of *Arbacia* eggs (adapted from Lucké and McCutcheon, 1932, p. 90). These cells swell in seawater of reduced osmolarity. The cell volume is plotted versus the reciprocal of the external concentration normalized to the isotonic external concentration. Therefore, when $C_\Sigma^{on}/C_\Sigma^{o} = 1$ the volume is the isotonic volume. The points represent the volume computed from the average diameters obtained from 300 individual cells in each solution (McCutcheon et al., 1931). The line through the points was computed by linear regression and has the equation $\mathcal{V}_c = 1.91 \times 10^5(C_\Sigma^{on}/C_\Sigma^{o}) + 2.03 \times 10^4\,\mu\text{m}^3$. The correlation coefficient of the fit is 0.9996.

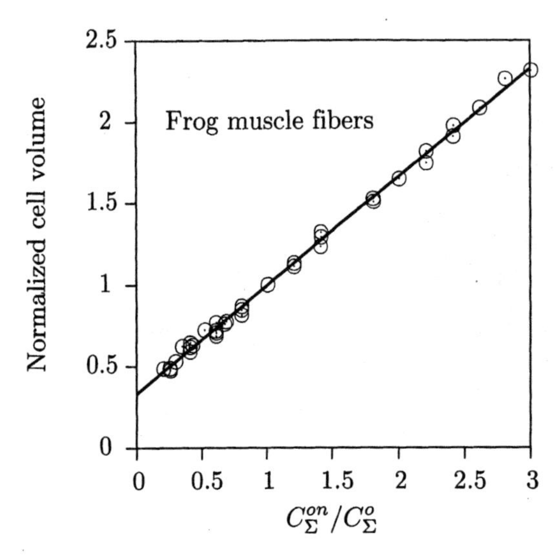

Figure 4.22 Osmometric behavior of frog muscle fibers, anterior tibial muscles of *Rana temporaria* (adapted from Blinks, 1965, Figure 4). The cell volume $\mathcal{V}_c$ for nine different fibers, each normalized to the isotonic volume $\mathcal{V}_c^n$, is plotted versus the reciprocal of the external concentration normalized to the isotonic external concentration as in Figure 4.21. The line through the points was computed by linear regression and has the equation $\mathcal{V}_c/\mathcal{V}_c^n = 0.666(C_\Sigma^{on}/C_\Sigma^o) + 0.333$. The correlation coefficient of the fit is 0.998.

stroyed if the membrane is damaged. Thus, the integrity of the semipermeable cell membrane is crucial to the osmometric behavior.

A muscle fiber is a long cell whose cross section, though not perfectly circular, is nevertheless fairly uniform along its length. An optical method has been used to measure the area of a cross section of a muscle fiber at several positions along its length and to estimate the volume of the fiber. The osmometric behavior of single muscle fibers is shown in Figure 4.22.

It is more difficult to measure the cell volume of a small cell or a cell with an irregular geometry. Human erythrocytes (red blood cells) are small cells shaped like biconcave disks, as shown in Figure 4.23. A variety of methods have been used to measure the volume of a population of erythrocytes. One method is to put the suspension of erythrocytes in their extracellular liquid in a capillary tube and to centrifuge the suspension until the erythrocytes and the liquid are separated. The fraction of the volume that is erythrocytes is measured. This quantity, called the hematocrit reading, is a measure of the relative volume of erythrocytes to the total volume of the suspension. An example of such results is shown in Figure 4.24.

Still other indirect methods rely on the changes in the optical density of a suspension of cells as the cell volume changes. Such methods require that the relation between optical density and cell volume be calibrated in some way. While these indirect methods are more subject to artifacts, they have the advantage that they can also be used to measure rapid changes in cell volume. An example of measurements based on an optical density method

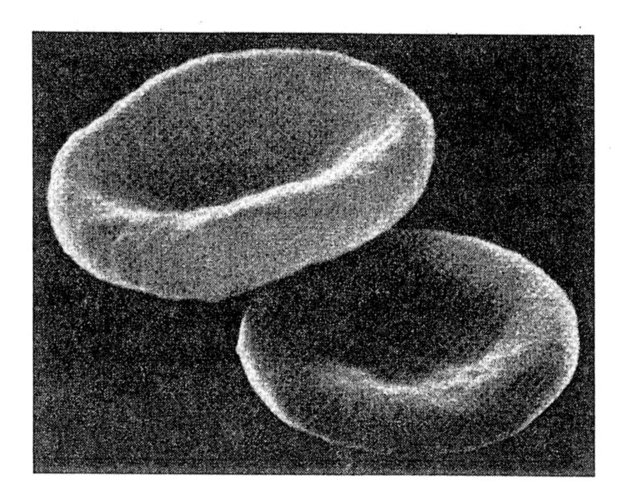

Figure 4.23 Scanning electronmicrograph of two erythrocytes (adapted from Bessis, 1974, Figure 2-1).

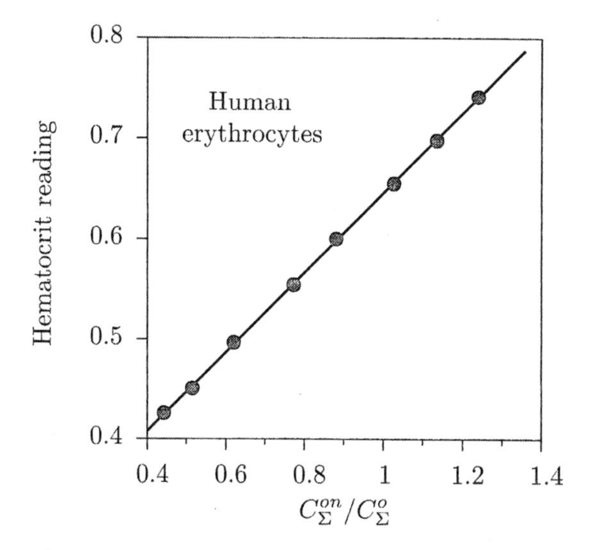

Figure 4.24 Osmometric behavior of red blood cells (adapted from LeFevre, 1964, Figure 3). The hematocrit reading, which is related linearly to the cell volume, is plotted versus the reciprocal of the external concentration normalized to the isotonic external concentration as in Figure 4.21. The osmolarity of the solution was changed by changes in saline concentration. The line through the points was computed by linear regression and has the equation $H = 0.3967(C_{\Sigma}^{on}/C_{\Sigma}^{o}) + 0.2489$ where H is the hematocrit reading. The correlation coefficient of the fit is 0.9999.

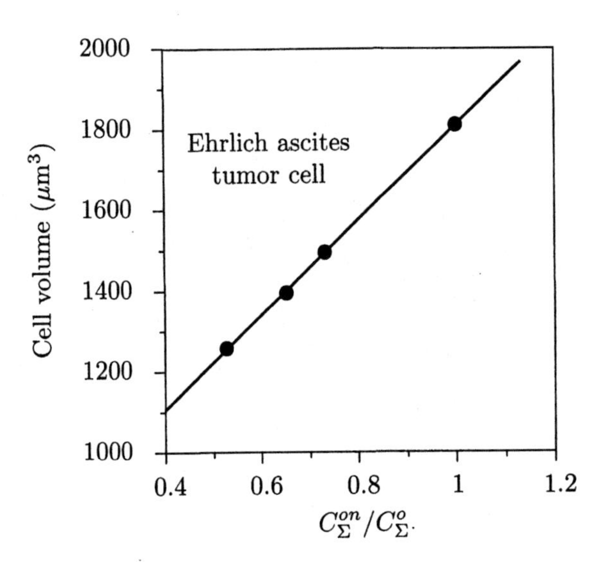

Figure 4.25 Osmometric behavior of Ehrlich ascites tumor cell (adapted from Hempling, 1960, Figure 1). The cell volume is plotted versus the reciprocal of the external concentration normalized to the isotonic external concentration as in Figure 4.21. These cells are shrinking in hyperosmotic saline solutions. The line through the points was computed by linear regression and has the equation $\mathcal{V}_c = 1173(C_\Sigma^{on}/C_\Sigma^o) + 637\,\mu\text{m}^3$. The correlation coefficient of the fit is 0.9999.

is shown in Figures 4.25. In these measurements, the light intensity that has been passed through a suspension of cells is detected with a photoelectric cell. The relation between photoelectric-cell output and cell volume can be obtained by combining the densitometric method with direct observation of the dimensions of single cells or by the centrifuge method combined with measurements of the number of cells centrifuged. The results are similar to those obtained with the more direct methods.

From the data in Figures 4.21 to 4.25, it is apparent that the quantity $\mathcal{V}_c'$, which represents the volume of the cell that does not function osmotically as water, can be a substantial fraction of the cell volume. Therefore, it is important to check the volume of cell water derived from osmotic experiments against the volume of cell water found by direct chemical measurements, e.g., by subtracting the dry weight of cells from the wet weight. A measure of the relation between the osmotically and chemically determined cell water volume is the quantity called *Ponder's R*, or R_P, which is

$$R_P = \frac{\mathcal{V}_c - \mathcal{V}_c'}{\mathcal{V}_w}, \tag{4.66}$$

where $\mathcal{V}_w$ is the volume of cell water determined by chemical means. Values of R_P lie in the range 0.8 to 1.0 (Dick, 1966). A number of factors may be responsible for the deviation of R_P from unity, including (Dick, 1966): (1) Some intracellular water is bound to organelles or macromolecules in the cell and is not, therefore, osmotically active; (2) the osmotic coefficient, χ, deviates from unity; (3) there is some mechanical rigidity to the membrane so that the assumption of complete distensibility is not valid.

To conclude, at equilibrium, cells behave as almost perfect osmometers in the presence of solute molecules to which the membrane is impermeable. The volume of osmotically active cell water is estimated to be within about 20% of that found by chemical means.

4.7.2 Kinetics of Volume Changes of Cells in Response to Osmotic Pressure Changes

We have seen that cells behave as almost perfect osmometers at osmotic equilibrium. But what is the time course of shrinkage or swelling of cells following a change in osmotic pressure? Which membrane and cellular properties affect this time course? To answer these questions, we shall compute the change in volume of a cell as a function of time after the cell has been placed in a hypertonic or hypotonic solution of impermeant solute. Then we will compare these calculations with measurements.

4.7.2.1 *Theory*

If Φ_V is the outward volume flux of water (see Figure 4.18) then, in the absence of a hydraulic pressure,

$$\Phi_V(t) = \mathcal{L}_V\left(\pi^o(t) - \pi^i(t)\right) = \mathcal{L}_V RT\left(C_\Sigma^o(t) - C_\Sigma^i(t)\right). \tag{4.67}$$

The outward flux of water must equal the reduction in intracellular water volume per unit area of membrane.

$$-\frac{1}{A(t)}\frac{d\mathcal{V}^i(t)}{dt} = \mathcal{L}_V RT\left(C_\Sigma^o(t) - C_\Sigma^i(t)\right). \tag{4.68}$$

It is convenient to express this equation in normalized form by first dividing the equation by $\mathcal{V}^{in}$, which is the isotonic water volume, and then factoring C_Σ^{in} out of the right-hand side to yield

$$\frac{d\nu(t)}{dt} = \alpha_\nu \mathcal{A}(t)\left(\frac{1}{\nu(t)} - \tilde{c}(t)\right), \tag{4.69}$$

where $\nu(t) = \mathcal{V}^i(t)/\mathcal{V}^{in}$ is the cell water volume normalized to the isotonic volume; $\mathcal{A}(t) = A(t)/A^n$ is the surface area of the cell normalized to its isotonic value; $\tilde{c}(t) = C^o_\Sigma(t)/C^{in}_\Sigma = C^o_\Sigma(t)/C^{on}_\Sigma$ is the extracellular solute concentration normalized to its isotonic value; and $\alpha_\mathcal{V} = \mathcal{L}_V RTA^n C^{in}_\Sigma/\mathcal{V}^{in}$. Note that $C^i_\Sigma(t)/C^{in}_\Sigma = (N^i_\Sigma(t)/\mathcal{V}^i(t))/(N^{in}_\Sigma/\mathcal{V}^{in}) = 1/v(t)$, since $N^i_\Sigma(t) = N^{in}_\Sigma$ for impermeant solutes.

Equation 4.69 is a kinetic equation that expresses the volume change of a cell of arbitrary geometry as a function of time for some imposed change in extracellular concentration of impermeant solute. The factor $\alpha_\mathcal{V}$ in Equation 4.69 determines the time course of the volume change only. Hence, two solutions of this equation for different values of this parameter differ only in time scale. The larger is $\alpha_\mathcal{V}$, the more rapid the change in volume. Thus, we see that increasing $\mathcal{L}_V$, T, C^{in}_Σ, or $A^n/\mathcal{V}^{in}$ results in an increase in $\alpha_\mathcal{V}$ and hence an increase in the response rate. Furthermore, the definition of $\alpha_\mathcal{V}$ determines how the various factors interact to produce a given rate of response. For example, doubling the hydraulic conductivity and halving the ratio of surface area to volume leaves the response unchanged.

To solve Equation 4.69, $\mathcal{A}(t)$ must be expressed in terms of $v(t)$. Hence, $\mathcal{A}(t)$ characterizes the shape of the cell. Equation 4.69 is readily integrable for several different assumptions about cell geometry, including the following:[7] (1) A cell for which the surface area remains constant as the cell volume changes, $\mathcal{A}(t) = 1$; (2) a spherical cell that swells uniformly, $\mathcal{A}(t) = (v(t))^{2/3}$; (3) a long cylindrical cell that swells radially only, $\mathcal{A}(t) = (v(t))^{1/2}$.

Cell with Constant Surface Area

Cells with complex shapes, such as erythrocytes (Figure 4.23), may not be expected to change their surface areas much as their volumes change, at least for a range of volume changes. As they swell, they become more spherical so that their volumes can increase substantially without a large change in surface area. In this section we analyze the kinetic changes in volume for such cells.

For constant surface area, Equation 4.69 reduces to

$$\frac{dv(t)}{dt} = \alpha_\mathcal{V}\left(\frac{1}{v(t)} - \tilde{c}(t)\right). \tag{4.70}$$

A solution to Equation 4.70 can be obtained by separation of variables, which yields

---

7. For a spherical cell the volume is $(4/3)\pi r^3$ and the surface area is $4\pi r^2$, where r is the radius, so that the area can be expressed in terms of the volume as $A = (4\pi)^{1/3}(3\mathcal{V})^{2/3}$. Similarly, for a long cylindrical cell, neglecting the surface area of the end caps, the volume is $\pi r^2 l$ and the surface area is $2\pi r l$, where l is the length of the cylinder and r is its area, so that the surface area can expressed as $A \approx 2(\pi l \mathcal{V})^{1/2}$.

$$\frac{v(t)\,dv(t)}{\tilde{c}(t)v(t)-1} = -\alpha_v\,dt, \tag{4.71}$$

To obtain a solution, we first divide the denominator into the numerator, and, in order to express the result in terms of t and $v(t)$, we use dummy integration variables and integrate from $t' = 0, v'(t) = 1$ to $t' = t$ and $v'(t) = v(t)$, as follows:

$$\int_1^{v(t)} \left(1 + \frac{1}{\tilde{c}(t)v'(t)-1}\right) dv'(t) = -\int_o^t \alpha_v \tilde{c}(t')\,dt',$$

which, for any interval for which $\tilde{c}(t)$ is constant, yields

$$\left[v'(t) + \frac{1}{\tilde{c}}\ln(\tilde{c}v'(t)-1)\right]_1^{v(t)} = -\left[\alpha_v \tilde{c}t'\right]_0^t$$

and can be evaluated to give

$$H(\tilde{c}, v) = \left(v(t) - 1 + \frac{1}{\tilde{c}}\ln\left(\frac{\tilde{c}v(t)-1}{\tilde{c}-1}\right)\right) = -\alpha_v \tilde{c}t. \tag{4.72}$$

This result also suggests that to test this theory, measurements of volume can be used to calculate the quantity $H(\tilde{c}, v)$, which should be proportional to time.

Solutions to the equation for a cell with constant surface area, which correspond to Equation 4.72 but were obtained by numerical integration of Equation 4.70, are shown in Figure 4.26. The solutions show that, as expected, if the normalized concentration is 1 (i.e., the external concentration is maintained at the isotonic value), then the volume remains at its isotonic value $v(t) = 1$. If the external solution concentration is greater than isotonic, then the cell shrinks, and if the concentration is smaller than the isotonic value, the cell swells. Interestingly, the kinetics of volume changes depend upon the change in concentration. For example, the volume transient time course is more rapid for larger values of $\tilde{c}(t)$.

In general, the solution for $v(t)$ in Equation 4.70 (i.e., Equation 4.72) is not an exponential function of t. However, for small changes in volume, i.e., $v(t) \approx 1$, the volume is an exponential function of time. For $v(t) \approx 1$, Equation 4.72 reduces to

$$\frac{\tilde{c}v(t)-1}{\tilde{c}-1} \approx \exp(-\alpha_v \tilde{c}^2 t), \tag{4.73}$$

which is an exponential function of time with time constant

$$\tau = \frac{1}{\alpha_v \tilde{c}^2} = \frac{V^{in}}{A\mathcal{L}_V RT C_\Sigma^{in} \tilde{c}^2} = \frac{V^{in} C_\Sigma^{in}}{A\mathcal{L}_V RT (C_\Sigma^o)^2}. \tag{4.74}$$

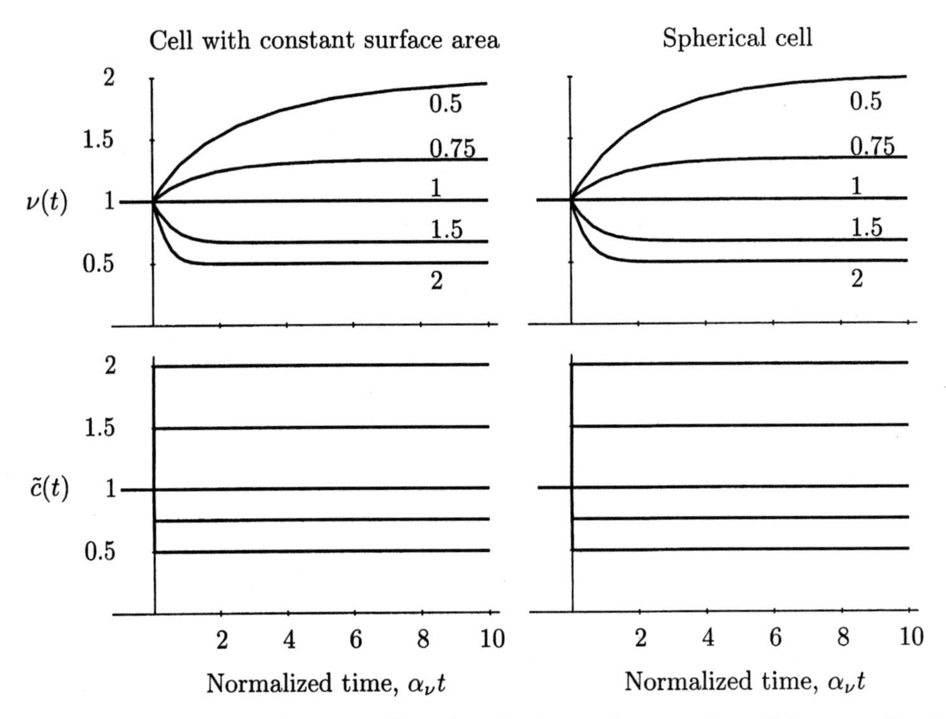

Figure 4.26 Kinetics of osmotically induced volume changes of a cell in normalized coordinates. The normalized volume ($\nu(t) = V^i(t)/V^{in}$) is plotted versus the normalized time ($\alpha_\nu t$) in the upper panels for the five normalized extracellular concentrations ($\tilde{c}(t) = C^o_\Sigma(t)/C^{on}_\Sigma$) shown in the lower panels. The normalized volume was computed according to Equation 4.69 for a cell whose area is constant as its volume changes (left panels) and for a spherical cell (right panels). The normalized concentration is $\tilde{c}(t) = 1$ $t < 0$ and takes on values from 0.5 to 2, for $t > 0$.

Note that the larger the hydraulic conductivity, $\mathcal{L}_V$, the smaller τ will be and the faster the cell will equilibrate. An increase in the ratio of surface area to volume of the cell will also decrease τ and speed up equilibration.

Spherical Cell

For a spherical cell, Equation 4.69 takes on the form

$$\frac{d\nu(t)}{dt} = \alpha_\nu \left(\nu(t)\right)^{2/3} \left(\frac{1}{\nu(t)} - \tilde{c}(t)\right). \tag{4.75}$$

Equation 4.75 can be solved by separation of variables in a manner similar to the method used to find the solution of Equation 4.70. For a constant value of $\tilde{c}(t)$, the solution is

$$F(\tilde{c}) - F(\tilde{c}\nu(t)) = (\tilde{c})^{4/3}\alpha_\nu t, \tag{4.76}$$

where

$$F(y) = \frac{1}{2}\ln\left(\frac{1 + y^{1/3} + y^{2/3}}{(1 - y^{1/3})^2}\right) + \sqrt{3}\tan^{-1}\left(\frac{1 + 2y^{1/3}}{\sqrt{3}}\right) - 3y^{1/3}. \tag{4.77}$$

Despite the fact that the solution for a spherical cell is algebraically messier than that for a cell with constant surface area, solutions for the volume changes do not differ much (Figure 4.26). For a further comparison of the two solutions see Problem 4.18. The form of the solution for a spherical cell (Equation 4.76) also suggests a way to test the ability of this theory to predict volume changes of cells. From measured values of the cell volume, it is possible to compute the value of the left-hand side of Equation 4.76 and plot that result versus time. The result should be a straight line.

4.7.2.2 Measurements

Measurements of the kinetics of volume changes in response to changes in extracellular osmolarity have been conducted on numerous types of cells. A direct test of whether the theory fits the kinetics of volume changes was obtained for egg cells of marine invertebrates. These cells are particularly effective for such measurements because they change their volume rather slowly. Therefore, the diameters of these spherical cells were measured in response to a change in extracellular osmolarity and the volume computed. A simple method to test the theory is suggested by Equation 4.76. Rearrangement of the terms in this equation yields

$$G(\tilde{c}, \nu) = \frac{F(\tilde{c}) - F(\tilde{c}\nu(t))}{RTC_\Sigma^{on}(36\pi)^{1/3}(\mathcal{V}^{in})^{-1/3}(\tilde{c})^{4/3}} = \mathcal{L}_V t. \tag{4.78}$$

$G(\tilde{c}, \nu)$ can be computed directly from measurements of the volume, the known physical constants, the temperature, and the osmolarity of the extracellular medium. $G(\tilde{c}, \nu)$ can be plotted versus time. The theory predicts that the relation should be a straight line whose slope is the hydraulic conductivity. The comparison between the measurements and the prediction is shown in Figure 4.27 and indicates that the swelling of these egg cells in hypotonic seawater can be accounted for by the theory.

Measurements of volumes of individual cells are laborious, slow, and variable. Thus, new methods were developed that, although less direct, were more rapid and averaged the results from many cells simultaneously. Measurements, based on an optical diffraction technique, of volume changes of egg

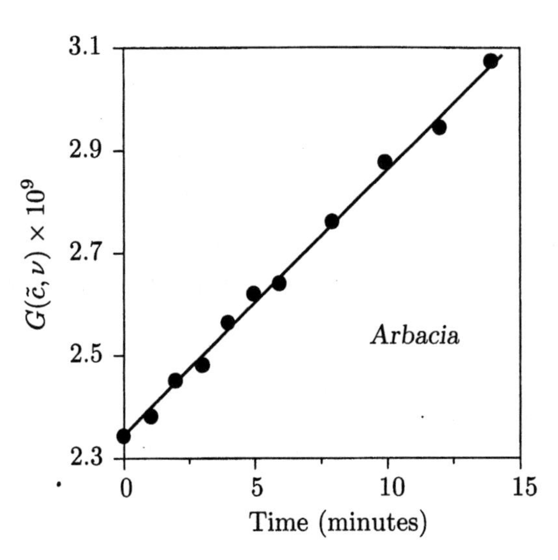

Figure 4.27 The swelling of *Arbacia* egg cells in 60% seawater. The quantity $G(\tilde{c}, \nu)$ is derived from the measured volume of a cell according to Equation 4.78 and is plotted versus time (adapted from Lucké et al., 1931, Figure 1). Each point is based on measurements of the diameters of six cells.

cells from three marine invertebrates are shown in Figure 4.28. A monochromatic light was passed through a solution consisting of a population of egg cells in seawater, and the resultant diffraction pattern was observed. The diameters of the rings in the diffraction pattern are related inversely to the diameters of the cells and yield an estimate of the volume of the cells. The measurements give, at each point in time, the average volume of a large number of cells. Direct measurements of the diameters of a few cells agree with the diffraction measurements. The results suggest that the theory fits the measurements adequately.

Direct measurements by observation with a light microscope are not possible for cells as small as erythrocytes whose changes in volume are very rapid. More indirect methods are necessary. Typically, a type of flow cytometer is used to measure changes in optical properties of the solution of erythrocytes caused by changes in cell volume. Such methods require careful calibration to relate the photic signal to the volume. Figure 4.29 shows an example of such measurements in which the change in light intensity transmitted through a solution containing a population of human erythrocytes was measured in response to an increase in the osmolarity of the bathing solution caused by additional NaCl. The measurements (which are proportional to the cell volume) are transformed to estimate the quantity $H(\tilde{c}, k\nu)$ defined in Equation 4.72, where k is a constant of proportionality between light intensity and volume. The transformed measurements show that the erythrocytes shrink in the higher-osmolarity solution. However, the time course of the volume changes is two orders of magnitude faster than the time course of changes in volume of the

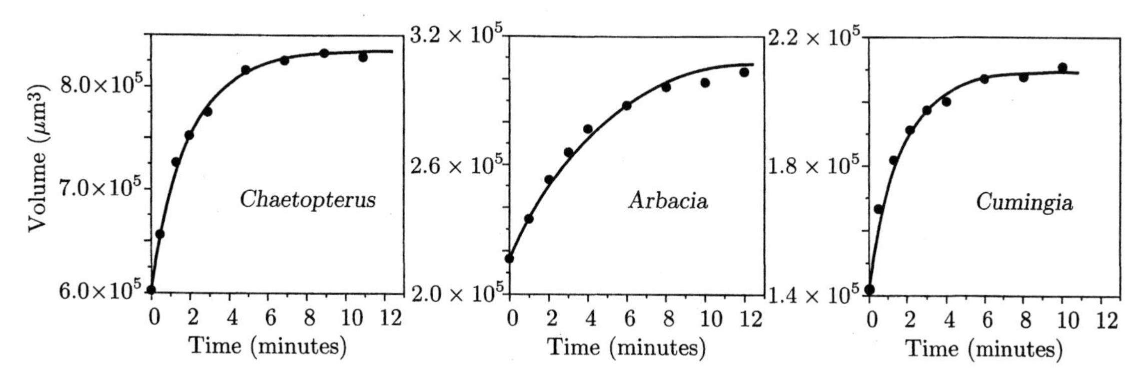

Figure 4.28 Response of marine invertebrate egg cells whose extracellular medium is changed from 100% to 60% seawater at time zero (adapted from Lucké et al., 1939, Figure 3). The results are shown for three species of marine animals: the sandworm *Chaetopterus pergamentaceus,* the sea urchin *Arbacia punctulata,* and the bivalve *Cumingia tellenoides*. The points are the measurements. The line is drawn according to Equation 4.76 with the parameters $\mathcal{V}_c'$ (μm^3) and $\mathcal{L}_V$ (pm · s^{-1}·Pa) as follows: *Chaetopterus*, 2.53×10^5, 7.33×10^{-2}; *Arbacia*, 2.72×10^4, 2×10^{-2}; *Cumingia*, 5.2×10^4, 6.33×10^{-2}.

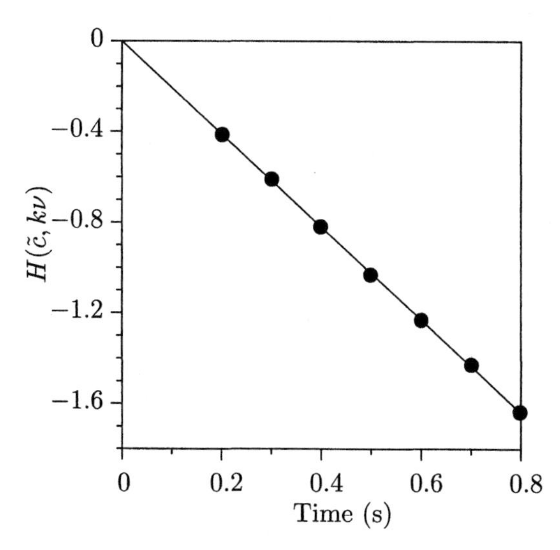

Figure 4.29 Time course of shrinkage of a population of human erythrocytes in hypertonic saline (adapted from Mlekoday et al., 1983, Figure 2). The quantity $H(\tilde{c}, k\nu)$ is plotted versus time (Equation 4.72).

marine eggs shown in Figure 4.28. The fact that the transformed measurements on erythrocytes are proportional to time indicates that Equation 4.72 fits the measurements. The constant of proportionality that relates the transformed measurements to time is proportional to the rate constant α_ν and can be used to estimate $\mathcal{L}_V$ for the erythrocyte membrane.

4.7.3 The Complexity of Cellular Volume Control

Measurements of both the equilibrium and kinetic changes in cell volume in response to changes in concentration clearly show that both the static and the dynamic volume changes are predicted by simple physical principles based on conservation of mass and on van't Hoff's law. These volume changes clearly have implications for the control of cellular volume. However, the simplicity of the primary response of cells to changes in osmotic pressure induced by impermeant solutes, as described thus far, does not reveal the underlying complexity of the control of cellular volume. A number of factors, not yet considered, are important in the control of cellular volume. We introduce some of these here.

4.7.3.1 The Osmotic Response to Permeant Solutes

If the osmolarity of the extracellular solution is increased by the addition of a *permeant* solute, the cells will shrink initially and then swell back to their original volume. This response occurs because the increase in osmolarity initially causes water to flow out of the cell to establish osmotic equilibrium. Concurrently, the permeant solute flows into the cell to reduce the difference in solute concentration, which tends to drive water into the cell to restore the cell volume to its initial value. The concurrent flow of solvent and solute, both when they do not interact directly and when they do, is explored in Chapter 5.

4.7.3.2 The Role of Ions

As shown clearly in Table 4.1, the predominant osmolytes in cytoplasm (and in extracellular media) are ions. These ions play an important role in volume control that cannot be understood until the interrelation of the membrane potential and ion concentrations is understood. This topic will be discussed in Chapter 7.

4.7.3.3 The Effect of Metabolism

It is well known that cells will swell if their metabolism is blocked. Clearly, there is nothing in the theory of primary cellular responses as developed in this chapter that can account for the role of metabolism in the control of cellular volume. We will return to the interpretation of the role of metabolism in volume control in Chapters 7 and 8.

4.7.3.4 Volume Regulation

Most animal cells respond to large changes in osmolarity by a rapid change in volume consistent with the primary response described in this chapter, but then the volume of such a cell slowly returns to near its normal value. The slow response occurs even if the primary volume change is initiated by a change in the osmolarity of an impermeant solute. In order to discuss volume regulation in a meaningful manner, we shall need to understand the additional transport mechanisms discussed in Chapters 6 and 7. Then we will return to volume control in Chapter 8.

4.7.3.5 The Effects of Hormones

The hydraulic conductivity of several types of cells (e.g., epithelial cells in a portion of the kidney) increases appreciably in response to the antidiuretic hormone vasopressin. Thus, in some cells water transport is under hormonal control. The control mechanism has implications for the molecular mechanisms underlying water transport through cellular membranes, which are discussed in the next section.

4.8 Molecular Mechanisms of Water Transport Through Cellular Membranes

4.8.1 Osmotic and Diffusive Permeability of Membranes

Measurements of the kinetic changes in the volume of a cell, as described in Section 4.7, yield a numerical value for the hydraulic conductivity (or, equivalently, for the osmotic permeability). Beginning in the 1920s and continuing to this day, the hydraulic conductivity has been measured in several types of cells as well as for different types of artificial membranes. In a variety of plant and animal cells and in artificial lipid membranes, $\mathcal{L}_V$ generally lies in the range 0.003–3 pm/(Pa·s), which corresponds to an osmotic permeability $\mathcal{P}_w$ in the range 0.4–400 μm/s (Dick, 1966; Stein, 1990). These results show that the ability to transport water in response to an osmotic or hydraulic pressure difference varies by three orders of magnitude in the different types of membranes. Measurements with water tracers have shown that the diffusive permeability P_w of different membranes also has a large range of values. However, it has been found that several characteristics of water transport differ in different types of membranes. These characteristics include the temperature

Table 4.2 Properties of water transport through several types of membranes (Finkelstein, 1987; Stein, 1990; Verkman, 1992). Results are given in the absence of (No Hg) and in the presence of (With Hg) mercury compounds. *Lipid bilayers plus pores* are artificial lipid bilayers, which include pore-forming antibiotics.

	$\mathcal{P}_w$ ($\mu m/s$)		$\mathcal{P}_w/P_w$		Q_{10}	
Membrane type	No Hg	With Hg	No Hg	With Hg	No Hg	With Hg
Unmodified lipid bilayers	1–50	1–50	~ 1	~ 1	1.7–3.1	1.7–3.1
Lipid bilayers plus pores	—	—	3–5.3	—	—	—
Erythrocytes	200	20	3.4	~ 1	1.3	1.7
Proximal tubules, apical	200	80	3.2	—	1.1	1.7
Proximal tubules, basolateral	300	60	9	—	1.2	1.6

dependence of the osmotic permeability, the effect of compounds that contain mercury on the osmotic permeability, the ratio of the osmotic permeability to the diffusive permeability, and the effect of antidiuretic hormones.

4.8.1.1 Lipid Bilayers and Erythrocyte Membranes

In lipid bilayers, the osmotic permeability tends to be comparatively low, is unaffected by mercury compounds, is approximately equal to the diffusive permeability, and has a relatively large dependence on temperature as quantified by Q_{10} (Table 4.2).[8] The temperature dependence is about equal to that for the diffusion of small solutes like water molecules through these lipid bilayers. Thus, the results on lipid bilayers are consistent with the conclusion that water traverses these membranes via the dissolve-diffuse mechanism (see Section 4.5). When the lipid bilayers are modified by the inclusion of antibiotics known to form pores in lipid bilayers (e.g., gramicidin A, nystatin, amphotericin B), $\mathcal{P}_w/P_w$ becomes greater than 1.

In contrast to unmodified lipid bilayers, erythrocyte membranes have an osmotic permeability that is comparatively large, is inhibited substantially by mercury compounds, is appreciably larger than the diffusive permeability, and has a relatively small dependence on temperature. The temperature dependence is about equal to that of the diffusion of water molecules in water. These properties suggest that erythrocyte membranes contain a water trans-

8. Q_{10} is a measure of the dependence of the osmotic permeability on temperature (Section 6.2.2). The osmotic permeability is multiplied by Q_{10} for a 10°C increase in temperature.

port mechanism that differs from that of lipid bilayers. In particular, the fact that $\mathcal{P}_w / P_w > 1$ and the small dependence of the osmotic permeability on temperature suggest that erythrocytes contain aqueous pores or water channels that provide a water-filled pathway that links the cytoplasm to the extracellular fluid. The inhibition of water transport by mercury compounds ($HgCl_2$, p-chloromercuribenzoate, abbreviated PCMB, and p-chloromercuribenzene sulfonate, abbreviated PCMBS) in erythrocytes is largely independent of changes in solute permeation, which suggests that mercury compounds act on these water channels (Macey and Farmer, 1970). As indicated in Table 4.2, in the presence of mercury compounds, $\mathcal{P}_w / P_w = 1$ and the temperature dependence of the osmotic permeability increases to that found for lipid bilayers. These results fit with the idea that water traverses the erythrocyte membrane by at least two mechanisms—via the dissolve-diffuse mechanism through the lipid bilayer and through water channels. The dissolve-diffuse mechanism has a relatively low osmotic permeability and a high temperature coefficient and is unaffected by mercury compounds. The water channels have a high osmotic permeability and a low temperature coefficient and are inhibited by mercury compounds.

4.8.1.2 Epithelial Cells

Water transport through epithelial cells is important for osmotic homeostasis in higher organisms. However, it has been difficult to interpret measurements of water transport in intact epithelia. The complexity of this process is readily appreciated by examination of different pathways for transport, shown schematically in Figure 2.13. First, water can flow through both transcellular (through the epithelial cells) and paracellular (around the epithelial cells but through the tight junctions) pathways. Second, the epithelial cells in a typical epithelium are heterogeneous. Thus, transport through the epithelium reflects the superposition of transport through a population of cells with heterogeneous transport properties. Third, the apical membranes and basolateral membranes of epithelial cells typically have different water transporting properties. It has been difficult to separate all these factors from measurements in whole epithelia.

An important finding is that in certain epithelia (including anuran skin, anuran urinary bladder, and mammalian kidney collecting tubules), water transport is low except in the presence of vasopressin. We shall return to a discussion of the role of vasopressin after we briefly describe the transport of water in the kidney.

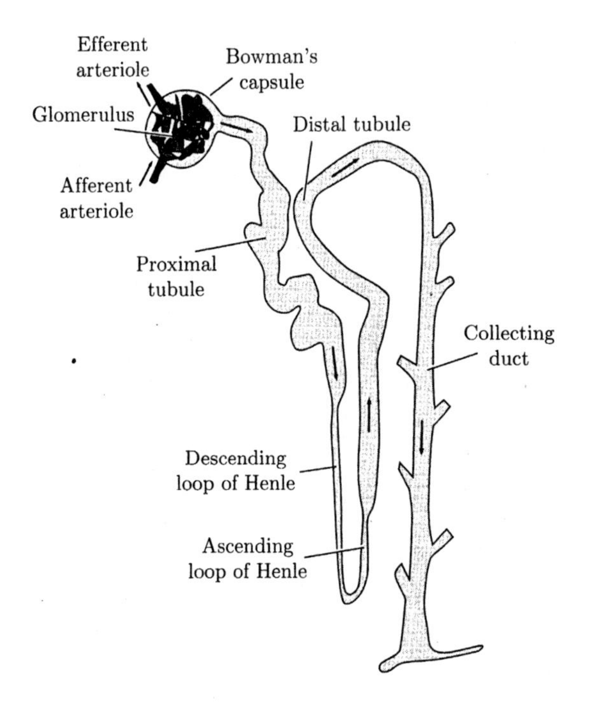

Figure 4.30 Schematic diagram of a nephron. The arrows in the tubule show the direction of flow from the glomerular filtrate that leaves the glomerulus to the urine that leaves the collecting duct.

Functional Anatomy of the Kidney

The main functions of the kidney are to regulate the volume of water in the extracellular space, to regulate the concentrations of ions, to conserve nutrients (e.g., glucose and amino acids), and to eliminate waste products of digestion (e.g., urea, and uric acid). The kidney also secretes several hormones. The principal functional unit of the kidney is the nephron (Figure 4.30). Humans have two kidneys; each kidney contains about one million nephrons. Each nephron consists of several functionally distinct components. The systemic circulation enters the kidney via the renal artery. Each renal artery divides into successively smaller vessels, the smallest of which, the afferent arteriole, enters a plexus of capillaries contained in Bowman's capsule. The capillaries recombine to form the efferent arteriole, by which blood leaves the glomerulus (which is the combination of the capillary bed and capsule). The blood in the capillaries is filtered to produce the *glomerular filtrate,* whose composition is identical to that of blood except that it is free of blood cells and large molecules (those with a molecular weight greater than about 80 kD). The filtrate passes through the tubule in the nephron, which consists of several distinct parts: the proximal tubule, the descending and ascending loops of Henle, and the distal tubule. Several nephrons combine to drain into the collecting duct, which in turn drains into the renal pelvis (the central cavity in the kidney), which connects to the ureter, which carries the urine to the urinary bladder, from which the urine is eliminated via the urethra. The urine that leaves the urethra has a composition that differs radically from that of the glomerular filtrate. The difference in composition results because throughout its course between the glomerulus and the urethra, the filtrate is subjected to both the secretion of substances and the reabsorption of substances from the interstitial fluids in the kidney.

Secretion and reabsorption occur via the epithelium that lines the lumen of the nephron tubule. Both the structure and the function of the epithelial cells that make up the epithelium vary along the tubule. Solute transport varies along the length of the tubule, so that some portions are principally involved in the regulation of sodium, others in the regulation of pH, etc. The properties of these epithelial cells for the transport of water also vary systematically with their location along the tubule.

Water Transport in the Kidney

The transport of water in the kidney has been reviewed elsewhere (Verkman, 1989). The glomerular filtrate is relatively dilute. Because of the active transport of ions out of the lumen of the nephron into the interstitium, the interstitial fluids are slightly hyperosmotic. Because the osmotic permeability of the proximal tubule is very high (Table 4.2), water flows out of the lumen of the nephron into the interstitial fluids, crossing both the mucosal (apical) and serosal (basolateral) membranes of the epithelial cells. Some water also takes a paracellular route. The transport of water results in a concentration of the lumenal fluid. The descending loop of Henle also has a high osmotic permeability, but the ascending loop has a very low osmotic permeability. The osmolarity of the lumenal fluid is reduced by the extrusion of salts.

Water transport across the epithelial cells of the collecting duct is sensitive to the hormone vasopressin. In the absence of vasopressin, the osmotic permeability of the apical membrane of these cells is extremely low whereas the osmotic permeability of the basolateral membrane is high. The net effect is that in the absence of vasopressin the transepithelial osmotic permeability is low. However, in the presence of vasopressin on the basolateral surface of the epithelial cells, the osmotic permeability of the *apical* membrane of these cells increases so as to increase the transepithelial osmotic permeability by as much as two orders of magnitude. The effect of vasopressin is to increase water transport out of the collecting duct. Thus, water is reabsorbed in the kidney and not eliminated.

The response to vasopressin is part of a feedback system that maintains extracellular osmolarity. An increase in osmolarity results in a decrease in the volume of very sensitive osmoreceptors located in the hypothalamus of the brain. The cell bodies of these cells store vasopressin in an inactive form. The osmotic response of these cells results in the release of vasopressin to their terminal endings in the posterior pituitary, where vasopressin is released into the systemic circulation. Vasopressin reaches its target epithelial cells, including those in the collecting ducts of the kidney, which contain vasopressin receptors on their basolateral surfaces. The binding of vasopressin to the vasopressin receptor results in a large increase in the water transport across the collecting duct membrane. Hence, water is reabsorbed to decrease the osmotic pressure. Thus, this feedback system conserves water. In addition, the vasopressin acts on thirst receptors to promote water intake.

Thus, there are stark differences in water transport in the various parts of the nephron. The proximal tubule has a high osmotic permeability that is independent of vasopressin, whereas the collecting duct has a very low osmotic permeability in the absence of vasopressin but a greatly increased osmotic permeability in the presence of vasopressin.

4.8.2 Molecular Biology of Water Channels

A variety of indirect evidence of the type just described pointed to the existence of water channels in erythrocyte membranes as well as in other cell types. However, it was not until the late 1980s that water channel proteins were identified. The pace of research on water channels quickened immediately after identification of the first channel protein, and this work is reviewed in detail elsewhere (Agre et al., 1993; Verkman, 1993).

4.8.2.1 The CHIP Water Channel

The first water channel identified was a 28 kD protein isolated from erythrocyte membranes (Denker et al., 1988; Smith and Agre, 1991) and named CHIP (*ch*annel-forming *i*ntegral *p*rotein).

Structure of CHIP

CHIP is a 28 kD protein found abundantly in erythrocytes, renal proximal tubules, renal descending thin limb, choroid plexus of the brain, as well as in several other epithelia in the eye, gastrointestinal tract, etc. Complementary DNA (cDNA) corresponding to CHIP was isolated and sequenced (Preston and Agre, 1991), and consists of 807 translated base pairs that correspond

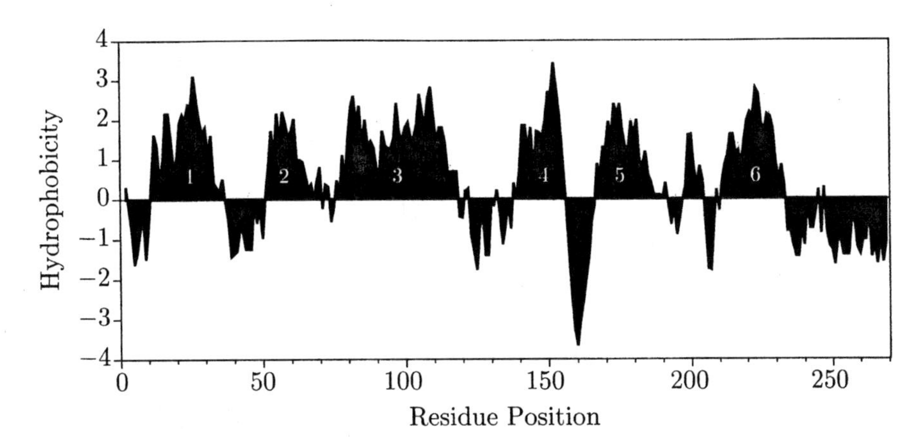

Figure 4.31 Hydrophobicity plot of CHIP. The nucleotide sequence of CHIP (Preston and Agre, 1991) was obtained from the GenBank database (accession no. 77829) and translated to give the amino acid sequence. The hydrophobicity of each residue was assigned (Kyte and Doolittle, 1982), and the resultant sequence of hydrophobicities was averaged with a window that was seven residues long to yield the averaged hydrophobicity shown.

to a protein with 269 amino acids. Analysis of the hydrophobicities of the amino acid sequence (Figure 4.31) shows that the protein contains a number of hydrophobic regions. Hydrophobicity analysis depends upon a numerical scale of hydrophobicity based on a number of physical chemical properties of amino acids. Based on this scale, each amino acid is assigned a hydrophobicity and the sequence of hydrophobicities of the protein can be smoothed to reveal a one-dimensional spatial pattern of hydrophobicity. As shown in Figure 4.31, CHIP exhibits six hydrophobic regions of appropriate length to span the membrane. Based on the hydrophobicity and identification of the site of the two terminal regions of the protein, a two-dimensional structure of CHIP was suggested (Figure 4.32). A variety of evidence (Agre et al., 1993) suggests that each native water channel consists of a cluster of four CHIP subunits, each with a molecular weight of 28 kD. Each CHIP subunit apparently acts as a functional water channel. CHIP is also found in a glycosylated form with a molecular weight of 50–60 kD. The ratio of CHIP to glycosylated CHIP is about 3:1, suggesting that only one CHIP in each tetrameric water channel is glycosylated.

Evidence that CHIP is a Water Channel

To determine whether or not CHIP is a water channel, the mRNA for this protein was injected into toad *Xenopus laevis* oocytes (Preston et al., 1992;

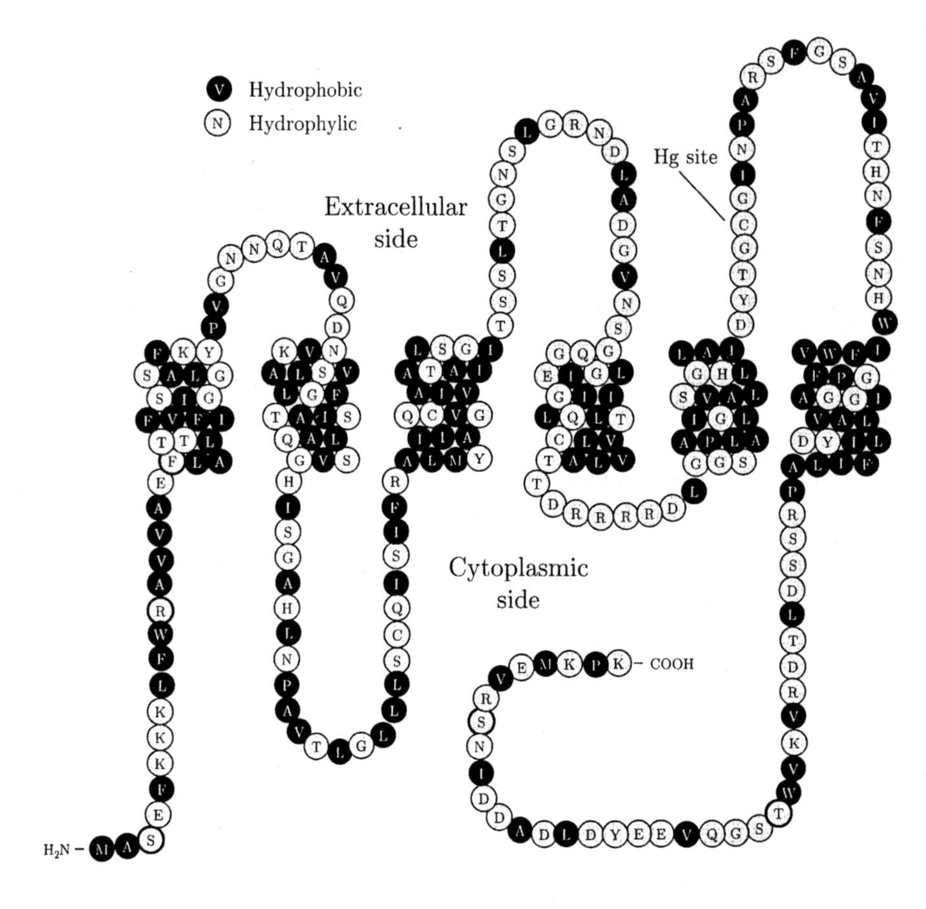

Figure 4.32 Model of the two-dimensional structure of CHIP (adapted from Preston and Agre, 1991, Figure 3). Each amino acid is represented by its single-letter code (see Figure 1.16, this volume). The binding site for mercury compounds is indicated at cysteine 189.

Preston et al., 1993). Water was injected into control oocytes. First it was determined that CHIP was expressed in the oocytes in 72 hours. The osmotic properties of both mRNA-injected and control oocytes were tested 72 hours after injection (Figure 4.33). The oocytes that expressed CHIP showed an eightfold increase in osmotic permeability over the control oocytes; their osmotic permeability was relatively insensitive to temperature changes and was reversibly blocked by $HgCl_2$. The control oocytes had a smaller osmotic permeability that was more temperature dependent and was not affected by incubation in $HgCl_2$. Thus, the oocytes that expressed CHIP behaved as if water transport was mediated by a water channel, whereas the control oocytes behaved as if water were transported by diffusion through the lipid bilayer. These experiments suggest that CHIP is involved somehow in producing water transport, but they do not prove that CHIP is the water channel. However, the exper-

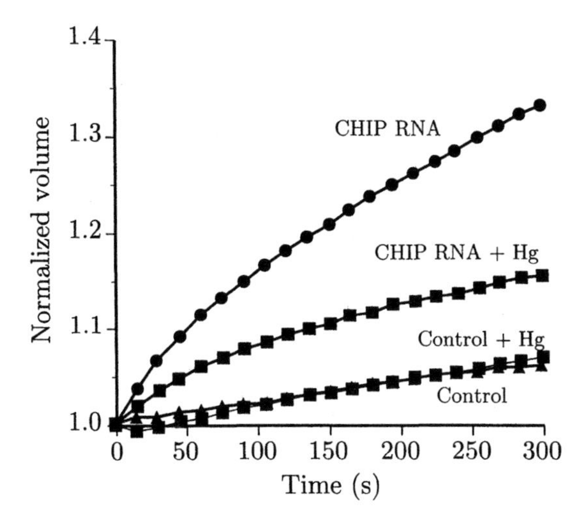

Figure 4.33 Effect on water transport of the expression of CHIP in *Xenopus laevis* oocytes (adapted from Preston et al., 1993, Figure 1). The oocytes were injected with 50 nL of water (control) or with 50 nL of water containing 1 ng CHIP RNA (CHIP RNA). Seventy-two hours later the oocytes were transferred from a 200 mosmol to a 70 mosmol solution and their volumes were measured as a function of time with a video microscopy system. In addition, some oocytes injected with CHIP RNA (CHIP RNA + Hg) and with water (control + Hg) were maintained in a solution containing 0.3 mmol/L of $HgCl_2$.

iments have been repeated on liposomes, which are lipid vesicles, in which CHIP has been incorporated, with similar results (Van Hoek and Verkman, 1992; Zeidel et al., 1992). These liposomes are simpler systems than oocytes and contain no transport proteins other than CHIP. These experiments give further weight to the idea that CHIP is a water-channel-forming protein.

Transport Through CHIP Water Channels

The reconstitution of CHIP in liposomes allows physiological experiments to be performed on a pure population of water channels (Van Hoek and Verkman, 1992; Zeidel et al., 1992, 1994). These experiments reveal that incorporation of CHIP greatly increases water permeability in proportion to the amount of protein expressed, but does not increase the permeability of the membrane to a number of small solutes, including urea, H^+, and NH_3. Hence, CHIP appears to be highly selective for water transport. Transport through these water channels is passive, and the driving force for water transport is the hydraulic-osmotic pressure difference across the membrane. These water channels act physiologically as water pores with a small pore diameter.

Site of Mercury Sensitivity
Site-directed mutagenesis has been used to determine the site of mercury sensitivity on CHIP (Preston et al., 1993). The protein contains four cysteines, which are likely binding sites for the divalent Hg^{++}, at locations 87, 102, 152, and 189. Substitution of a serine for any of these cysteines does not affect the osmotic permeability of water, nor does this substitution affect mercury sensitivity of water transport for three of these locations. However, substitution of serine for cysteine 189 results in a channel whose water transport is no longer blocked by mercury. Thus, cysteine 189 is implicated as the site of mercury sensitivity and may be located near the mouth of the pore (Figure 4.32).

4.8.2.2 The Family of Water Channels

Several water channels, also called *aquaporins,* that are similar to CHIP have been identified. These include WCH-CD, which is thought to be the vasopressin-sensitive water channel of kidney collecting ducts (Fushimi et al., 1993), MIP26 (*m*ajor *i*ntrinsic *p*rotein), a water channel found in lens fiber cells (Reizer et al., 1993), MIWC (Hasegawa et al., 1994) a mercury-insensitive water channel found widely at sites that do not express CHIP, AQP3 (Ishibashi et al., 1994), a water channel found in the basolateral membrane of kidney collecting duct that is also permeant to glycerol and urea, and a number of proteins found in plants that appear to serve as water channels. There are large homologies in the amino sequences of this family of water channel proteins, which are being worked out as new channels are being discovered.

Mechanism of Vasopressin-Dependent Increase in Water Permeability
The mechanism for the increase in water permeability of the apical membranes of epithelial cells in the collecting duct due to the presence of vasopressin on the basolateral surface has been worked out in some detail (Handler, 1988; Brown, 1989; Verkman, 1989). We give a brief description here. Vasopressin binds to vasopressin receptors, located on the basolateral membrane, which activates adenylate cyclase intracellularly to produce the intracellular second messenger 3′, 5′-cyclic monophosphate (cAMP). The action of this messenger, which is not yet entirely understood, results in the insertion of preexisting vesicles containing water channels into the apical membrane of the epithelial cell. Thus, the effect of vasopressin on the basolateral membrane is to recruit water channels in the apical membrane. Reduction of vasopressin concentration on the basolateral membrane reverses this process.

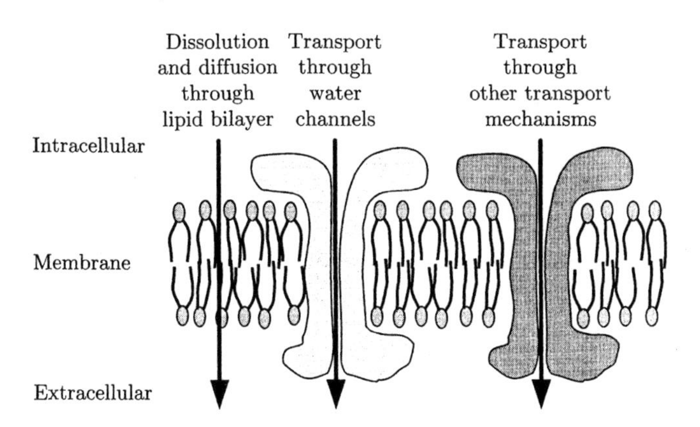

Figure 4.34 Mechanisms of water transport through membranes.

Localization of Water Channels

The availability of the purified water channel proteins has allowed antibodies to be raised against these proteins. When markers are attached to these antibodies, the marked antibodies can be used to localize water channel proteins in tissues. The pattern of localization of CHIP in the kidney is particularly revealing. CHIP is found in abundance in the proximal tubules and in the thin descending limbs of Henle, which have a high water permeability. CHIP is not found abundantly in other parts of the kidney. In contrast, WCH-CD is localized to the apical region of the collecting ducts, suggesting that it is the vasopressin-sensitive water channel.

4.8.3 Summary of Water Transport Mechanisms in Cell Membranes

In summary, three distinctly different mechanisms have been identified in the transport of water across cellular membranes in response to osmotic and hydraulic pressure differences (Figure 4.34). As we have seen, a difference in solute concentration can result in transport of water through a hydrophobic membrane by the dissolve and diffuse mechanism discussed in Chapter 3. This transport mechanism accounts for water transport through lipid bilayers—in both artificial lipid bilayers and in native cellular membranes. A second mechanism is transport through channels that are specific for the transport of water. These are found abundantly in tissues that have special needs to transport water. Finally, there is some evidence that water is transported through transport systems that are used primarily to transport some other substance. Presumably, water transport differs in different cell types because they contain different proportions of these water-transporting mechanisms.

Appendix 4.1 Thermodynamic Relations for an Ideal, Dilute Solution

More detailed descriptions of the thermodynamics of solutions can be found in a number of texts (Moore, 1972; Eisenberg and Crothers, 1979). In this appendix we give a brief introduction to ideal solutions.

A solution is a mixture of components whose composition can be specified by the component concentrations. Let the number of moles of component j be n_j, the mole fraction $x_j = n_j / \sum_j n_j$, the molar concentration $c_j = n_j/\mathcal{V}$, and the partial molar volume $\overline{\mathcal{V}}_j$. The total solution volume is $\mathcal{V} = \sum_j n_j \overline{\mathcal{V}}_j$. An *ideal* solution is defined as one that has a chemical potential

$$\mu = \mu_j^\circ + RT \ln x_j + p\overline{\mathcal{V}}_j. \tag{4.79}$$

Let component j communicate through the interface between solutions a and b. At thermodynamic equilibrium of component j, the chemical potential on the two sides of the interface must be the same. If the two solutions are at the same temperature, then

$$\mu_j^\circ(a) + RT \ln x_j^a + p_a \overline{\mathcal{V}}_j = \mu_j^\circ(b) + RT \ln x_j^b + p_b \overline{\mathcal{V}}_j, \tag{4.80}$$

where $\mu_j^\circ(a)$ and $\mu_j^\circ(b)$ are the standard chemical potentials of the pure component at a standard hydrostatic pressure; x_j^a and x_j^b are the mole fractions of j in the two solutions; and p_a and p_b are the hydraulic pressures in the two solutions.

Now suppose that the two solutions each consist of two types of components, one of which is present at a much higher concentration (the solvent) than the other type (the solutes). Let the solvent be designated by the subscript s, let k be the kth solute, and let the sum of the number of moles of solute be $N_\Sigma = \sum_k n_k$. Then the total volume of dilute solution, for which $n_k \ll n_s$, is

$$\mathcal{V} = \sum_k n_k \overline{\mathcal{V}}_k + n_s \overline{\mathcal{V}}_s \approx n_s \overline{\mathcal{V}}_s. \tag{4.81}$$

The mole fraction for each solute can be approximated as

$$x_k = \frac{n_k}{N_\Sigma + n_s} \approx \frac{n_k}{n_s}, \tag{4.82}$$

because $N_\Sigma \ll n_s$. By combining Equations 4.81 and 4.82 we obtain

$$x_k \approx \frac{n_k}{\mathcal{V}/\overline{\mathcal{V}}_s} = c_k \overline{\mathcal{V}}_s, \tag{4.83}$$

which shows that for an ideal dilute solution the concentration of a solute is proportional to its mole fraction. The mole fraction of solvent can be approximated by

$$x_s = \frac{n_s}{N_\Sigma + n_s} = \frac{1}{1 + (N_\Sigma/n_s)} \approx 1 - (N_\Sigma/n_s),$$

under the assumption that $n_s \gg N_\Sigma$. Finally, if we combine Equations 4.81 and 4.83 we obtain

$$x_s \approx 1 - C_\Sigma \overline{V}_s. \tag{4.84}$$

Osmotic Equilibrium and Van't Hoff's Law

Let solutions a and b be aqueous solutions so that the solvent s is water w. Then, provided water can permeate the interface between the solutions, at thermodynamic equilibrium between the two solutions the water must satisfy

$$\mu_w^\circ(a) + RT \ln x_w^a + p_a \overline{V}_w = \mu_w^\circ(b) + RT \ln x_w^b + p_b \overline{V}_w. \tag{4.85}$$

Note that since both solutions are aqueous the two standard chemical potentials are equal, $\mu_w^\circ(a) = \mu_w^\circ(b)$. By substitution of the expression for the mole fraction of water for a dilute solution, we obtain

$$RT \ln(1 - C_\Sigma^a \overline{V}_w) + p_a \overline{V}_w = RT \ln(1 - C_\Sigma^b \overline{V}_w) + p_b \overline{V}_w.$$

The logarithm of $\ln(1 + x) \approx x$ for $x \ll 1$. Therefore, with this approximation and after factoring out the molar volume of water and rearranging terms, we obtain

$$p_a - RTC_\Sigma^a = p_b - RTC_\Sigma^b,$$

which identifies the osmotic pressure as RTC_Σ (van't Hoff's law) and is a statement of osmotic equilibrium between solutions a and b.

Partitioning of Solute between Two Solutions

Substitution for the mole fraction of solute into Equation 4.80 yields

$$\mu_k^\bullet(a) + RT \ln c_k^a + p_a \overline{V}_k = \mu_k^\bullet(b) + RT \ln c_k^b + p_b \overline{V}_k,$$

where a new standard chemical potential has been defined as $\mu_k^\bullet = \mu_k^\circ + RT \ln \overline{V}_w$. If there is no hydraulic pressure difference between the two solutions, then we can solve this relation for the ratio of solute concentrations, which is a constant, the solute partition coefficient:

$$k_k = \frac{c_k^b}{c_k^a} = \exp\left(\frac{\mu_k^{\bullet}(a) - \mu_k^{\bullet}(b)}{RT}\right).$$

Partitioning of Water between an Aqueous and a Hydrophobic Solution

Suppose the solvent in solution a is water and the one in solution b is a hydrophobic solvent e.g., an organic solvent such as oil. We shall designate the solvent in solution b by the subscript h for *hydrophobic*. We wish to compute the partitioning of water at an interface between a, where water acts as the solvent, and b, where water acts as a solute dissolved in the hydrophobic solvent (Finkelstein, 1987). Thermodynamic equilibrium requires that Equation 4.85 be satisfied. Therefore, we can solve for the ratio of mole fractions to obtain

$$\frac{x_w^b}{x_w^a} = \exp\left(\frac{(\mu_w^a - \mu_w^b) + (p_a - p_b)\overline{V}_w}{RT}\right).$$

For the special case when $p_a = p_b$, we have

$$\frac{x_w^b}{x_w^a} = k_w = \exp\left(\frac{\mu_w^a - \mu_w^b}{RT}\right), \qquad (4.86)$$

where k_w is the partition coefficient of water in the hydrophobic and aqueous solutions. To relate the partitioning to solute concentration in the water, we need only find the mole fraction of water in the two solutions. In solution a, water acts as the solvent and Equation 4.84 shows that

$$x_w^a = 1 - C_\Sigma^a \overline{V}_w.$$

In solution b, water acts as a solute in the solvent h and Equation 4.83 shows that

$$x_w^b = c_w^b \overline{V}_h.$$

Substituting these expressions for the mole fractions into Equation 4.86, we obtain, after some rearrangement of terms,

$$c_w^b = k_w \frac{1 - C_\Sigma^a \overline{V}_w}{\overline{V}_h}. \qquad (4.87)$$

Appendix 4.2 Poiseuille's Law

We shall determine the relation between pressure and total volume flux for the steady flow of a viscous, incompressible fluid in a cylindrical pore of radius r. Our starting points are the linearized equation of motion of a fluid and the continuity equation (equation of conservation of mass). In vector form the equation of motion is (Batchelor, 1967)

$$\rho_m \frac{\partial \mathbf{u}}{\partial t} = -\nabla p + \eta \nabla^2 \mathbf{u}, \tag{4.88}$$

where $\mathbf{u}$ is the fluid velocity, p is the pressure, ρ_m is the fluid mass density, η is the fluid viscosity, and ∇ is the del operator. The term on the left-hand side of Equation 4.88 is the inertial force, the first term on the right-hand side is the force due to a pressure gradient, and the second term on the right is the viscous force. The continuity equation is

$$\nabla \cdot \mathbf{u} = -\frac{\partial \rho_m}{\partial t}. \tag{4.89}$$

For steady flow $\partial \mathbf{u}/\partial t = 0$, and for an incompressible fluid $\partial \rho_m/\partial t = 0$, so that the equations become

$$\nabla p = \eta \nabla^2 \mathbf{u} \tag{4.90}$$

and

$$\nabla \cdot \mathbf{u} = 0. \tag{4.91}$$

We now express these equations in cylindrical coordinates as shown in Figure 4.8. If flow is in the z-direction, then $\mathbf{u} = u_z \epsilon_z$ where ϵ_z is a unit vector in the z-direction. We assume the flow has circular symmetry so that p and u_z are independent of θ. From Equation 4.91 we obtain

$$\frac{\partial u_z}{\partial z} = 0. \tag{4.92}$$

Hence, u_z is a function of R only. Therefore, Equation 4.90 can be written as

$$\frac{\partial p}{\partial z} = \eta \nabla^2 u_z(R). \tag{4.93}$$

If we integrate Equation 4.93 from $z = 0$ to $z = d$, we obtain

$$p(d) - p(0) = -\Delta p = d\eta \nabla^2 u_z(R). \tag{4.94}$$

But

$$\nabla^2 u_z(R) = \frac{1}{R}\frac{d}{dR}\left(R\frac{du_z}{dR}\right). \tag{4.95}$$

If we substitute Equation 4.95 into 4.94 and integrate once, we obtain

$$\int d\left(R\frac{du_z}{dR}\right) = -\frac{\Delta p}{\eta d}\int R\,dR,$$

$$R\frac{du_z}{dR} = -\frac{\Delta p}{\eta d}\frac{R^2}{2} + C_1,$$

$$\frac{du_z}{dR} = -\frac{\Delta p}{\eta d}\frac{R}{2} + \frac{C_1}{R}. \tag{4.96}$$

Integrating Equation 4.96 on R, we obtain

$$u_z = -\frac{\Delta p}{\eta d}\frac{R^2}{4} + C_1 \ln R + C_2. \tag{4.97}$$

The term lnR diverges when $R = 0$, hence we let $C_1 = 0$. The boundary condition at $R = r$ is the no-slip condition, $u_z = 0$. Therefore,

$$0 = -\frac{\Delta p}{\eta d}\frac{r^2}{4} + C_2. \tag{4.98}$$

Substituting expressions for C_1 and C_2 into Equation 4.97, we get

$$u_z = \frac{\Delta p}{\eta d}\left(\frac{r^2 - R^2}{4}\right). \tag{4.99}$$

Therefore, the profile of fluid velocity through the cylinder is parabolic, as shown in Figure 4.35.

The volume flux through the pore, $(\Phi_V)_p$, is

$$(\Phi_V)_p = \int_0^r u_z(R) 2\pi R\,dR = \frac{\pi \Delta p}{2\eta d}\int_0^r (r^2 R - R^3)\,dR, \tag{4.100}$$

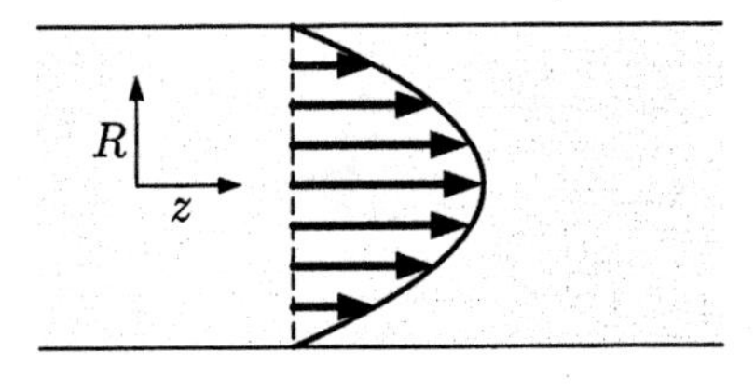

Figure 4.35 Parabolic profile of velocity in a cylindrical pore as given by Poiseuille's law. The thick arrows represent the fluid velocity in the pore, which is in the z-direction but whose magnitude varies parabolically in the R-direction and takes on a value of zero at the cylinder, where $R = r$.

$$(\Phi_V)_p = \frac{\pi \Delta p}{2\eta d}\left[\frac{r^2R^2}{2} - \frac{R^4}{4}\right]_0^r = \frac{\pi r^4}{8\eta d}\Delta p. \tag{4.101}$$

Exercises

4.1 In the schematic diagram of the plant cell shown in Figure 4.3, what is the significance of the light annular region that lies between the cell wall and cell membrane when the cell is in a hypertonic solution? Why is this region not present when the cell is in a hypotonic solution?

4.2 What is a *colligative property* of a solution? List three colligative properties of solutions.

4.3 What does it mean that two solutions are *isosmotic*?

4.4 What is the distinction between *isosmotic* and *isotonic* solutions?

4.5 Discuss the analogy between van't Hoff's law and the ideal gas law.

4.6 Define *osmolarity*.

4.7 Figure 4.5 shows the osmotic coefficient as a function of solute osmolality for several solutes.

a. For which of these solutes is the solution most like that predicted by van't Hoff's law?

b. Is the osmolality of a 1 mol/kg solution of sucrose less than, equal to, or greater than for a 1 mol/kg solution of urea? Explain.

4.8 Figure 4.36 shows two chambers, each containing a semipermeable membrane—permeable to water but not to the solute—separating two compartments. One of the compartments contains a weight W on a piston; the other does not. The only difference between the two arrangements is the location of the solute particles. Which of the two chambers could be in osmotic equilibrium? Explain.

4.9 In a brief paragraph, describe the distinction between P_w and $\mathcal{P}_w$. Your description should be independent of specific mechanisms for water transport, but should indicate methods for measuring both quantities.

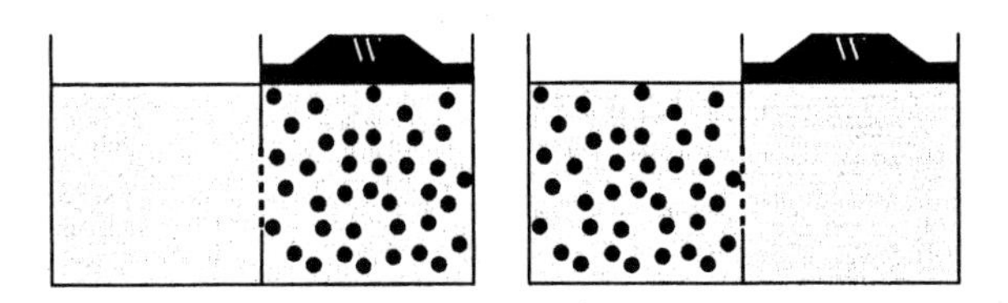

Figure 4.36 Arrangement of weight and solute for two chambers each with a semipermeable membrane separating two compartments (Exercise 4.8).

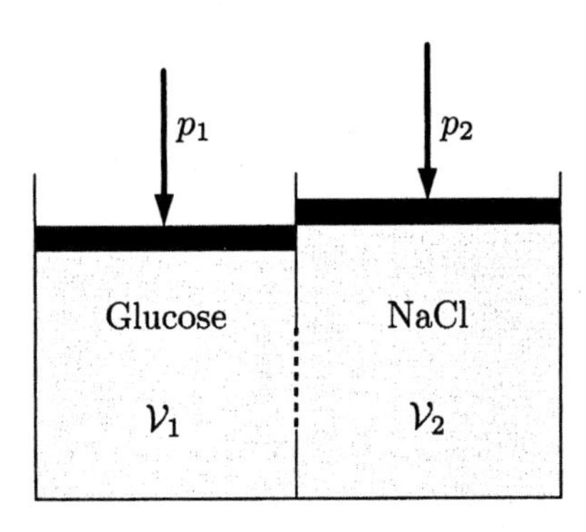

Figure 4.37 Experimental apparatus for testing a membrane (Exercise 4.13). Volumes $\mathcal{V}_1$ and $\mathcal{V}_2$ contain well-stirred aqueous solutions of glucose and NaCl, respectively. These volumes are separated by a membrane that is permeable only to water. Assume that the pistons are ideal, i.e., that they are frictionless and faithfully transmit the pressures p_1 and p_2 to $\mathcal{V}_1$ and $\mathcal{V}_2$, respectively. Also assume that effects of gravity are negligible.

4.10 A membrane separates two solutions, and the steady-state volume flux from side 1 to side 2 can be written as $\Phi_V = \mathcal{L}_V RT(C_\Sigma^2 - C_\Sigma^1)$. This volume flux bears a formal resemblance to solute diffusion according to Fick's law for membranes. Discuss the distinctions between these two flow laws.

4.11 Water transport through a homogeneous, hydrophobic membrane and a porous membrane are contrasted in Section 4.5.

a. What is the key distinction between the *mechanisms* of water transport in response to a difference in osmotic pressure for each of these two simple membrane models?

b. Suppose we have a membrane whose water transport mechanism is unknown. Measurements are made with radioactive tracers of the transport of water both in the absence and in the presence of an osmotic pressure difference across the membrane. Describe how these measurements could be used to determine whether the membrane is a homogeneous, hydrophobic membrane or a porous membrane.

4.12 What is wrong with the solute bombardment theory for explaining the physical basis of osmosis?

4.13 Figure 4.37 shows an experimental apparatus for testing a semipermeable membrane.

a. When the system is in equilibrium, what will be the relation between hydraulic pressures (p_1 and p_2) and solute concentrations in each compartment?

b. Volumes $\mathcal{V}_1$ and $\mathcal{V}_2$ are initially equal, with $\mathcal{V}_1 = \mathcal{V}_2 = 1$ L. At time $t = 0$, the concentration of glucose in compartment 1 is 0.01 mol/L and the concentration of NaCl in compartment 2 is 0.01 mol/L. If $p_1 = p_2$, what is the final volume of compartment 2? Sketch $\mathcal{V}_2$ as a function of time for $t \geq 0$.

4.14 Define the *isotonic volume* of a cell.

4.15 Figure 4.21 shows the volume of *Arbacia* eggs as a function of extracellular osmolarity.

a. What is the isotonic volume of these cells?

b. What is the fraction of the isotonic volume of these cells that is not due to intracellular water?

4.16 Under which circumstances is it reasonable to assume that a cell whose volume is changing as it either shrinks or swells has a constant surface area?

4.17 Using no mathematical formulas or equations, describe the meaning of Equation 4.68 in a few well-chosen English sentences.

4.18 A cell in a bath is subjected to changes in extracellular osmolarity. Assume that the flux of volume is given by $\Phi_V = \mathcal{L}_V RT(C_\Sigma^o - C_\Sigma^i)$. A step change in the osmolarity of the bath produces a change in the normalized volume of the cell as shown in the $v(t)$ waveform in Figure 4.38. Twelve possible normalized concentration waveforms $\tilde{c}(t)$ are shown in the lower panel of the figure. Assume all solutes are impermeant and that the cell acts as an ideal osmometer so that

$$\mathcal{V}_c = \frac{N_\Sigma^i}{C_\Sigma^o} + \mathcal{V}_c',$$

where $\mathcal{V}_c'$ is the osmotically inactive portion of the total cell volume $\mathcal{V}_c$.

a. Which of the twelve $\tilde{c}(t)$ waveforms are possible causes of the $v(t)$ waveform? Explain your choices briefly.

b. The temperature of all the components (cell and bath) is decreased by 10°C. Are the initial slope of the normalized volume, $dv(t)/dt$ at

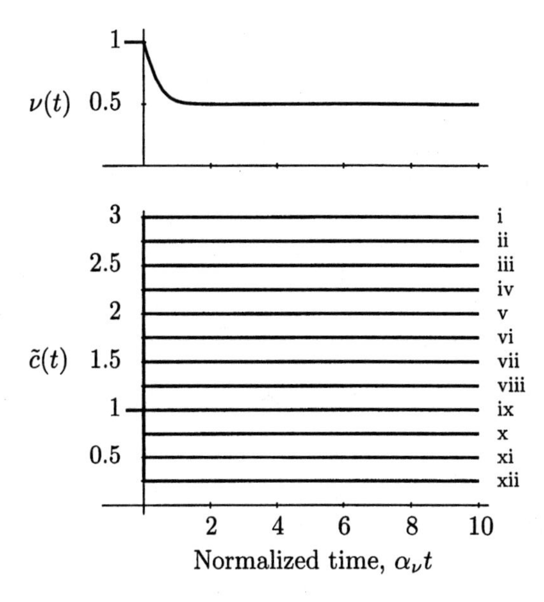

Figure 4.38 Change in normalized volume (upper panel) in response to a change in normalized external solute concentration (lower panel) (Exercise 4.18). The normalized volume is $\nu(t) = \mathcal{V}^i(t)/\mathcal{V}^{in}$, where $\mathcal{V}^i(t)$ and $\mathcal{V}^{in}$ are the cell water volume and its isotonic value. The normalized extracellular concentration is $\tilde{c}(t) = C^o_\Sigma(t)/C^{on}_\Sigma$, where $C^o_\Sigma(t)$ and C^{on}_Σ are the total extracellular solute concentration and its isotonic value.

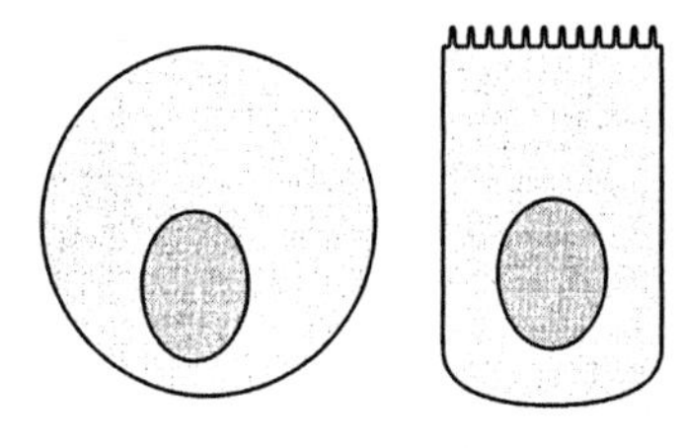

Figure 4.39 Two cells that have the same volume but different shapes (Exercise 4.19). One cell (left panel) is spherical; the other is approximately cylindrical but contains a large number of microvilli.

$t = 0+$, and $\nu(\infty)$ increased, decreased, or unchanged? Explain your reasoning.

4.19 Figure 4.39 shows schematic diagrams of two cells that have the same volume but quite different shapes. The cell membranes have the same hydraulic conductivity to water. If the two cells are subjected to the same decrease in extracellular osmolarity, which cell swells more rapidly? Explain.

4.20 What is the point of transforming volume measurements by the complicated transformation given in Equation 4.78 and plotted in Figure 4.27?

4.21 A cylindrical cell is bathed in solutions of a single impermeant substance that does not ionize. For each solution, the concentration C^o_Σ of the impermeant substance and the equilibrium cell volume are measured; the

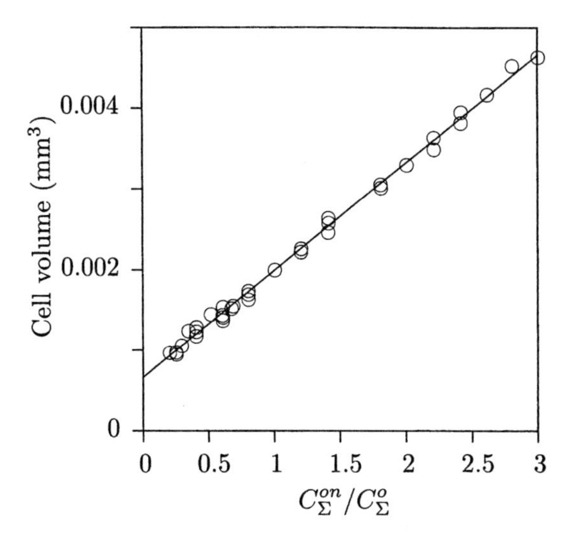

Figure 4.40 Relation between cell volume and extracellular solute concentration (Exercise 4.21).

results are shown in Figure 4.40, where C_Σ^{on} represents the isotonic concentration.

a. Estimate the isotonic volume of the cylindrical cell.

b. Assume that the cell volume $\mathcal{V}_c$ contains a portion $\mathcal{V}^i$ that consists of water that is osmotically active and a portion $\mathcal{V}_c'$ that is osmotically inactive. Estimate $\mathcal{V}_c'$ and explain the basis of your estimate.

c. Assume that $C_\Sigma^{on} = 200$ mosmol/L and that the cell membrane is impermeable to NaCl. Estimate the equilibrium cell volume that would result if the cell were bathed in a solution of NaCl with concentration 200 mmol/L. Explain your reasoning.

d. Assume that the cell bath is changed from an isotonic solution to a 200 mmol/L NaCl solution at time $t = 0$. Sketch the time course for equilibration of the cell volume, indicating numerical values where possible. List the physical parameters of the cell that influence the time course.

4.22 The properties of water transport through lipid bilayers and through cell membranes containing water channels are distinctly different. In a brief paragraph, describe the differences.

4.23 What is the significance of the results shown in Figure 4.33?

4.24 What does vasopressin do?

Problems

4.1 Measurements of the osmotic pressure of a solution of egg albumin at a temperature of 25°C are shown in Figure 4.41. Estimate the molecular weight of egg albumin from these data.

4.2 Experiments show that at 0°C a 0.2 molal sucrose solution has an osmotic pressure (relative to pure water) of 4.76 atmospheres whereas a 0.2 molal NaCl solution has an osmotic pressure of 8.75 atmospheres—not quite twice as great. Estimate the fraction of the NaCl molecules that are dissociated at this temperature and concentration.

4.3 The following three formulas for the sugar (trisaccharide) raffinose (found in sugar beets) were proposed:

$C_{12}H_{22}O_{11} + 3H_2O$,

$C_{18}H_{32}O_{16} + 5H_2O$,

$C_{36}H_{64}O_{32} + 10H_2O$.

In 1988, de Vries used a plant cell to determine that plasmolysis occurred with a solution containing 59.4 grams of raffinose per liter of water whereas plasmolysis occurred in a solution of sucrose at a concentration of 0.1 mol/L. Based on these measurements, de Vries determined

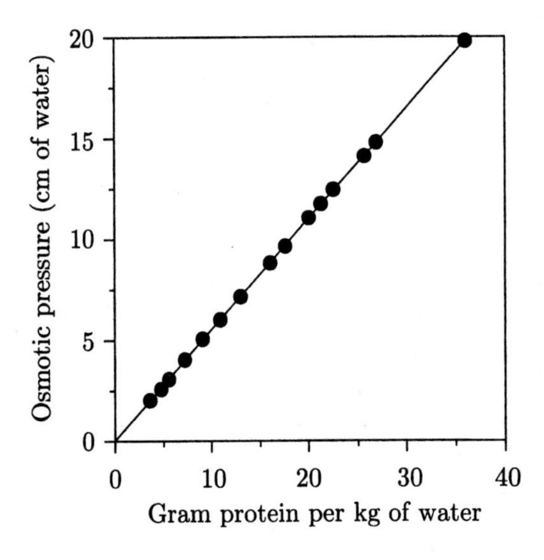

Figure 4.41 Osmotic pressure of a solution of egg albumin (Bull, 1964) (Problem 4.1). The osmotic pressure is measured in cm of water, and the concentration of egg albumin is expressed in grams per kg of water.

Table 4.3 Composition of human tears (Lentner, 1981) (Problem 4.4).

Solute	Concentration (mmol/L)
Inorganic substances	
Bicarbonate	26
Chloride	130
Phosphorus	8
Potassium	16
Sodium	144
Organic substances	
Total nitrogen	113
Nonprotein nitrogen	36
Urea	5
Ammonia	3

the correct formula for raffinose. Which formula would you choose and why would you choose it?

4.4 The principal components of human tears are shown in Table 4.3.

a. Estimate the osmolarity of human tears based on the composition. Be careful to state your assumptions.

b. Measurements show that the osmolarity of human tears is 320 mosmol/L. Discuss the differences (if any) between your estimate and the measured value of the osmolarity.

c. Determine the value of the osmotic pressure of human tears in pascals.

4.5 The *New York Times* on Sunday, November 18, 1990, in its section on technology (Section F, page 8), carried a feature article on desalination entitled "A thirsty California is trying desalination." The article describes a scheme in which salt water is prefiltered to remove large suspended particles and then desalinated by the "reverse osmosis unit" shown schematically in Figure 4.42 to give water fit for drinking.

"In reverse osmosis, the salt water is pumped along a porous cylindrical membrane at high pressure. The water molecules pass through the membrane, while salt molecules, bacteria and pollutants, which are larger, are left behind. It takes two or three gallons of salt water to make one gallon of drinkable water. . . . It takes just one pass through the reverse osmosis

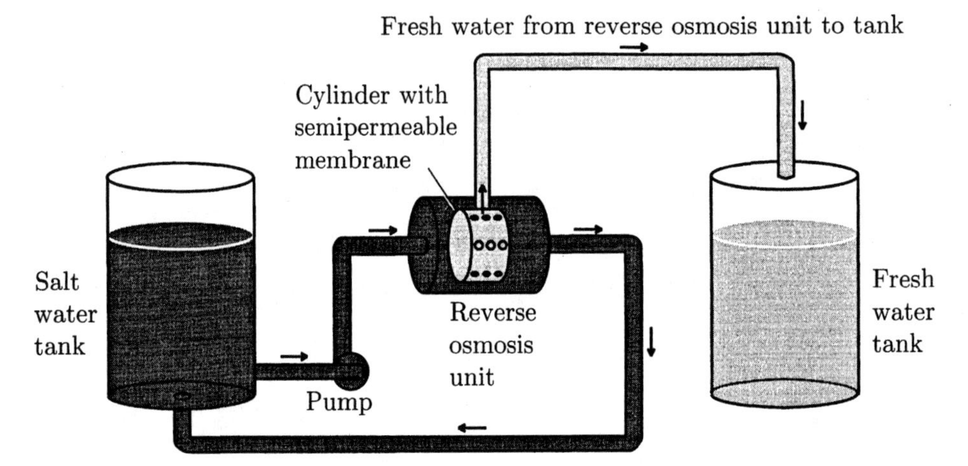

Figure 4.42 Use of reverse osmosis for extracting fresh water from salt water (Problem 4.5). Salt water is pumped from the salt water tank past a cylinder covered with a semipermeable membrane that allows transport of water only. The fresh water is collected in the fresh water tank.

unit to reduce total dissolved solids from 35,000 to 40,000 parts per million[9] typical of seawater to less than 500, the level considered safe to drink."

a. The state of California requires your help in the design of this system. An important technical issue is to determine the size of the pump necessary for the reverse osmosis unit. *Estimate* the minimum hydraulic pressure that the pump must generate in the reverse osmosis unit to produce fresh water from the seawater off the coast of Los Angeles. You may assume that seawater has a density of $\approx 1\,\text{g/cm}^3$ and that the dissolved solids in the seawater are composed primarily of NaCl, which has a molecular weight of 58 daltons.

b. When you present your calculations of the pump requirements at a state hearing on plans for desalination, it is pointed out by a state legislator that the system shown in Figure 4.42 requires a pump to produce fresh water. To cut the cost of the project, he proposes that you use gravity to produce the hydraulic pressure by raising the height of the salt water tank above that of the fresh water tank. Estimate the

9. "Parts per million" is defined as the total number of grams of solute in 10^6 grams of solution.

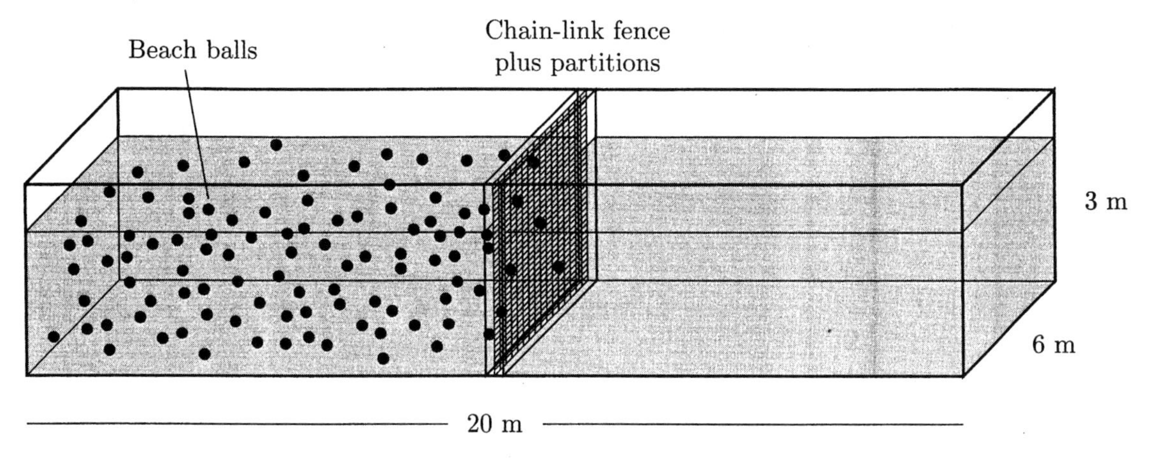

Figure 4.43 A test of osmosis on a macroscopic scale (Problem 4.6).

minimum height of the salt water tank above the fresh water tank so that fresh water will be produced.

c. The state of Utah hears about your plans for the gravity-feed system described in b and is interested in extracting fresh water from the Great Salt Lake using the system outlined in b. On a site inspection trip to Salt Lake City you take a dip in the Great Salt Lake and discover that you float easily with about half your body underwater. This behavior is in stark contrast to your experience at Laguna Beach near Los Angeles, where you find that you are not neutrally buoyant—you sink. You announce to the state of Utah that the Los Angeles system will not work in Salt Lake City. Why?

4.6 It is proposed to test the theory of osmotic pressure on a truly macroscopic scale by using a swimming pool, a chain-link fence, and some beach balls (Figure 4.43). The swimming pool is 20 m long, 6 m wide, and 4 m deep. A chain-link fence, surrounded on both sides by plexiglass partitions, is put in the center of the pool so as to create two compartments each 10 m long. The space between the partitions is negligibly small. The plexiglass partitions are watertight. One hundred neutrally buoyant beach balls, each 25 cm in diameter, are put into one of the compartments of the pool. Now the pool on both sides of the partition is filled with water to a depth of exactly 3 m. Then the plexiglass partitions are removed quickly. The balls are too large to go through the chain link fence.

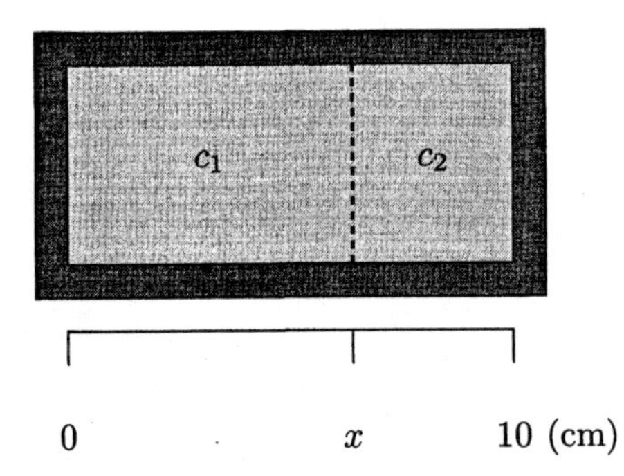

Figure 4.44 Volume element with rigid walls and movable partition (Problem 4.7).

a. Explain what will happen.

b. Estimate the height of the water on the two sides of the pool at equilibrium.

c. Is this a good test of the theory of osmotic equilibrium?

4.7 A volume element with constant cross-sectional area A has rigid walls and is divided into two parts by a rigid, semipermeable membrane that is mounted on frictionless bearings so that the membrane is free to move in the x-direction as shown in Figure 4.44. The semipermeable membrane is permeable to water but not to the solutes (glucose or NaCl). At $t = 0$, solute 1 is added to side 1 to give an initial concentration of $c_1(0)$ and solute 2 is added to side 2 to give an initial concentration of $c_2(0)$. Concentrations are specified as the number of millimoles of glucose or NaCl per liter of solution. The initial position of the membrane is $x(0)$. For each of the initial membrane positions and concentrations given in Table 4.4, find the final (equilibrium) values of the membrane position $x(\infty)$, and the concentrations $c_1(\infty)$ and $c_2(\infty)$. State your assumptions.

4.8 The water content of plants is very high (up to 90% by weight), but this water is in flux; water is absorbed through the roots, rises as sap, and evaporates from the leaves. The total water content of a plant can be replaced many times per hour. The mechanisms that determine water flow include gravity, osmosis, and capillarity. In this problem we will consider the effects of gravity and osmosis. Assume that water flow is steady and is due to gravitational, osmotic, and other forces. Let the pressure due to gravity be p_g and the osmotic pressure be π. Let the pressure due to other sources in the trees be p_o; assume these sources are not appreciable in the medium that is in contact with the roots. Consider two trees:

Table 4.4 Initial solute concentrations and partition position (Problem 4.7).

Solutes		*Initial values*			*Final values*		
1	2	$c_1(0)$ mmol/L	$c_2(0)$ mmol/L	$x(0)$ cm	$c_1(\infty)$ mmol/L	$c_2(\infty)$ mmol/L	$x(\infty)$ cm
Glucose	Glucose	0	10	5			
Glucose	Glucose	10	10	8			
Glucose	Glucose	30	10	5			
Glucose	NaCl	20	10	2			
Glucose	NaCl	80	10	2			
NaCl	NaCl	30	10	5			

Table 4.5 Composition of sap and extracellular solutions for the giant sequoia and the red mangrove trees (Problem 4.8).

	Concentration (mmol/L)			
	Giant sequoia		*Red mangrove*	
Ion	soil	sap	seawater	sap
K^+	1	20	10	20
Na^+	2	0	450	0
Cl^-	30	2	530	2
NO_3^-	3	20	0	20
Ca^{++}	7	1	10	1
Mg^{++}	8	0	50	0

a. The General Sherman giant sequoia *(Sequoia gigantea)*, which was located in Sequoia National Park in California, stood 272 feet high and had a diameter of 30.7 feet at its base; you could drive your car through a tunnel that had been cut through the base. The tree was estimated to be 3,800 years old.

b. The red mangrove *(Rhizophora mangle)* grows in the tropics to a height of up to 80 feet. It is found in tidal creeks and estuaries, and it can grow with its roots in seawater. Yet the sap has the composition of fresh water.

The compositions of the relevant media are given in Table 4.5.

a. Find a numerical bound on p_o (expressed in atmospheres) such that the sap will rise in each tree.

b. Which tree requires the larger value of p_o? Explain.

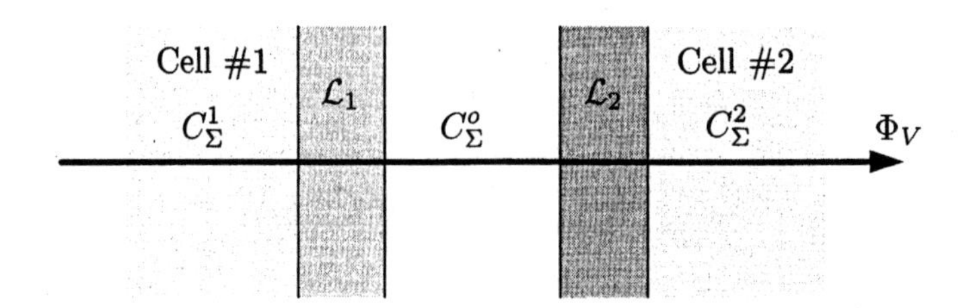

Figure 4.45 Two membranes in series (Problem 4.9).

4.9 Two adjoining cells have closely apposed membranes (Figure 4.45). The total concentrations of impermeant solute are C_Σ^1 and C_Σ^2 inside cells 1 and 2, respectively, and C_Σ^o in the intercellular space. The membrane hydraulic conductivities are $\mathcal{L}_1$ and $\mathcal{L}_2$ for the membranes of cells 1 and 2, respectively. Find the net hydraulic conductivity, $\mathcal{L}_V$, between the inside of cell 1 and the inside of cell 2 in terms of $\mathcal{L}_1$ and $\mathcal{L}_2$, where

$$\Phi_V = \mathcal{L}_V RT \left(C_\Sigma^2 - C_\Sigma^1\right).$$

and Φ_V is the steady-state volume flux in cm/s across both membranes. There is no difference in hydraulic pressure across either membrane.

4.10 A membrane separates two solutions containing total impermeant solute as shown in Figure 4.46. There is no difference in hydraulic pressure across the membrane. The membrane consists of two contiguous patches having surface areas and hydraulic conductivities of A_1 and $\mathcal{L}_1$ for one patch and A_2 and $\mathcal{L}_2$ for the other patch. Assume that steady-state volume flow occurs through the membrane. Determine the hydraulic conductivity of the membrane $\mathcal{L}_V$, defined as

$$\Phi_V = \mathcal{L}_V RT \left(C_\Sigma^2 - C_\Sigma^1\right).$$

4.11 A method for measuring the hydraulic conductivity, $\mathcal{L}_V$, of an artificial membrane is shown in Figure 4.47. An impermeable, cylindrical glass tube whose internal cross-sectional area $A = 0.5\text{cm}^2$ is closed at one end by an artificial membrane and is filled with 2 cm^3 of a sucrose solution of concentration 0.1 mol/L. The membrane is impermeable to sucrose but is permeable to water. At $t = 0$ the glass tube is inserted into a large vessel of pure water. The height, $h(t)$, of the solution in the tube is measured as a function of time. Assume that the effects of hydraulic pressure are negligible. Assume $T = 300$ K.

a. Derive a differential equation for $h(t)$.

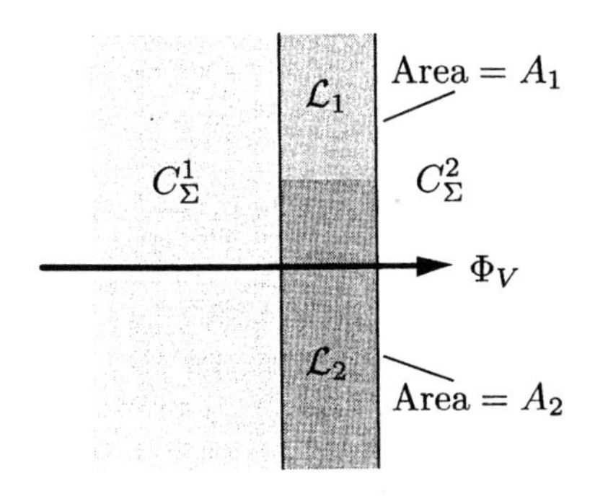

Figure 4.46 Two membranes in parallel (Problem 4.10).

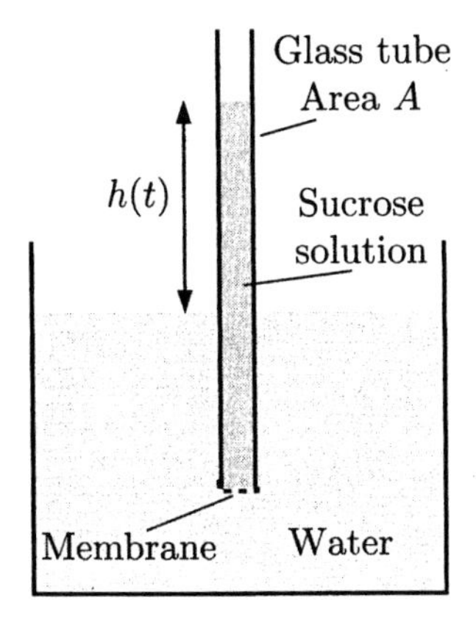

Figure 4.47 Method used to measure osmotic pressure (Problem 4.11).

b. Find and sketch $h(t)$, the solution of the differential equation you derived in part a.

c. Measurements show that for a brief time interval after the tube is placed in the water, the height of the solution in the tube rises linearly with time, i.e., $\lim_{t \to 0} h(t) = h(0) + 0.02t$ cm. Find the numerical value of $\mathcal{L}_V$. *Note:* You do not need to solve the differential equation of part a to do this part of the problem.

4.12 A long, steel pipe filled with fresh water to a height h_o is lowered quickly into the ocean until its bottom end is a distance h_s below the surface of the ocean as shown in Figure 4.48. The bottom end of the pipe is closed with a semipermeable membrane permeable only to water.

a. Derive an expression for the inward flux of water Φ_V, in terms of the variables shown in Figure 4.48, the acceleration of gravity g, and any other necessary constants.

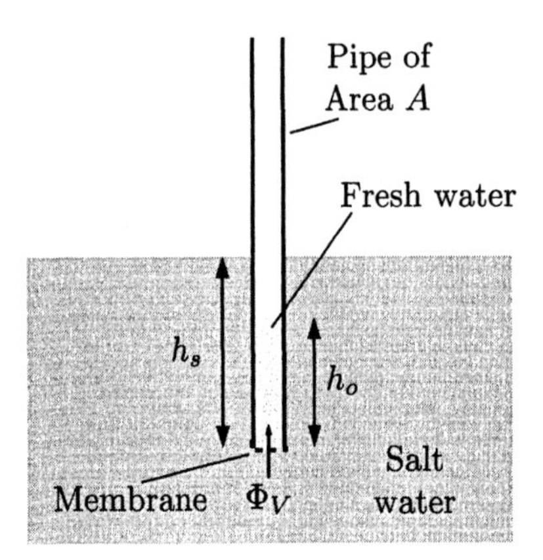

Figure 4.48 Schematic diagram of a pipe terminated in a semipermeable membrane and immersed in the ocean (Problem 4.12). The fresh water in the pipe has mass density ρ_o and osmolarity 0; the salt water has mass density ρ_s and osmolarity C_Σ; the semipermeable membrane has hydraulic conductivity $\mathcal{L}_V$.

b. Determine the magnitude and sign of the initial derivative of $h_o(t)$, i.e., dh_o/dt evaluated at $t = 0$, if

$$
\begin{aligned}
h_s &= 100 \text{ m} & \rho_s &= 1.03 \text{ g/cm}^3, \\
h_o(0) &= 1 \text{ m} & \rho_o &= 1.00 \text{ g/cm}^3, \\
A &= 10 \text{ cm}^2 & \mathcal{L}_V &= 3 \times 10^{-12} \text{ m/(Pa} \cdot \text{s)}, \\
C_\Sigma &= 1 \text{ osmol/L} & g &= 980 \text{ cm/s}^2, \\
T &= 300\text{K}. & &
\end{aligned}
$$

c. Show that, provided h_s is greater than some critical depth, h_c, the final equilibrium value of h_o is greater than h_s. Find the value of h_c.

d. If $h_s > h_c$, it appears that both power and fresh water could be obtained, free, from the ocean. Is this reasonable? If not, which assumption(s) in the problem statement would you consider to be invalid?

4.13 Figure 4.12 shows the concentration of tracer water in a porous membrane in the presence of both tracer diffusion and convection. For this membrane, assume that total concentrations of solute on the two sides of the membrane are related by $C_\Sigma^2 = 3C_\Sigma^1$, and the hydraulic pressures are $p_1 = (3/2)p_2$ and $p_2 = RTC_\Sigma^2$.

a. Sketch the hydraulic pressure in the solutions and in the membrane as a function of position.

b. Indicate the direction of volume flux of water on your sketch.

c. Which of the traces shown in Figure 4.12 could represent the concentration of tracer water in the membrane for the parameters given in this problem? Explain.

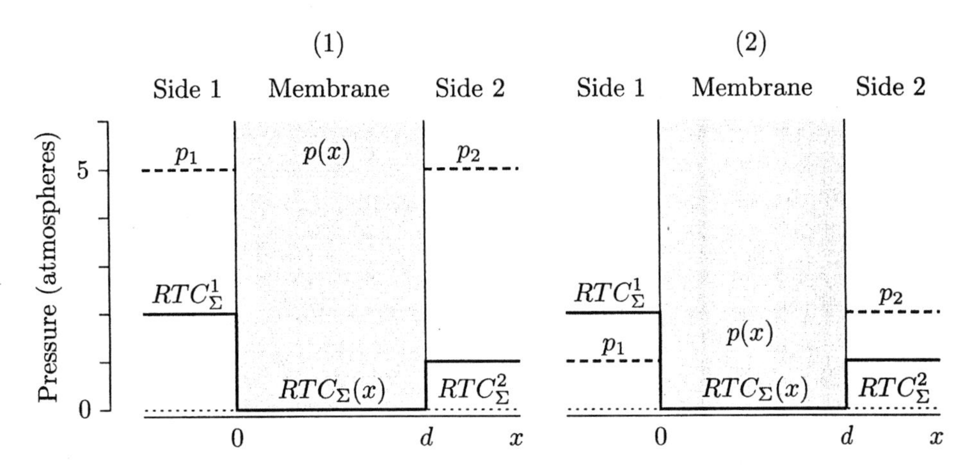

Figure 4.49 Membrane hydraulic and osmotic pressure (Problem 4.14).

4.14 A semipermeable membrane separates two solutions as shown in Figure 4.49 and is subject to two sets of solute concentrations and hydraulic pressures: (1) and (2). For both cases (1) and (2), the spatial profiles of osmotic pressure are given, and the hydraulic pressure in the two baths is also given.

a. Determine the complete profile of hydraulic pressure in the membrane for both cases (1) and (2).

b. Determine the profile of $p(x) - RTC_\Sigma(x)$ in the membrane for both cases (1) and (2).

c. The concentration of solute on side 1 is the same for both cases (1) and (2). Find the total concentration of solute on side 1.

d. Is the volume flux through the membrane in case (1) smaller, larger, or equal to that in case (2)? Explain.

4.15 It is known that the membrane of a certain type of cell is highly permeable to water but relatively impermeable to L-glucose, sodium ions, and chloride ions. When the cell is removed from interstitial fluids and placed in a 150 mmol/L NaCl solution, the cell neither shrinks nor swells.

a. Would the cell shrink, swell, or remain at constant volume if placed in a 150 mmol/L solution of L-glucose? Explain.

b. Would the cell shrink, swell, or remain at constant volume if placed in a 300 mmol/L solution of L-glucose? Explain.

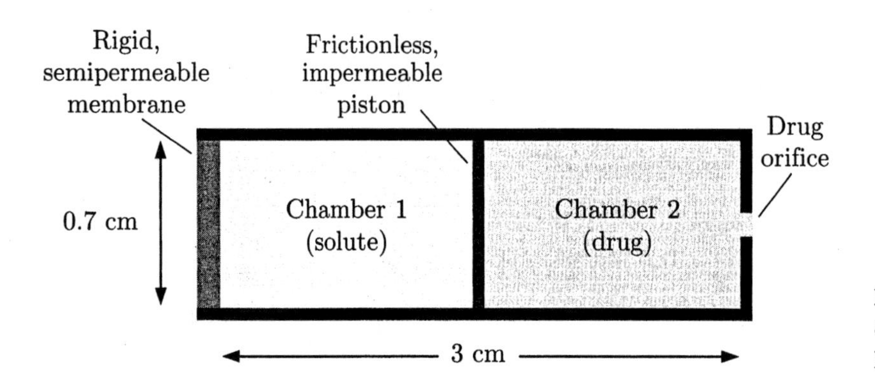

Figure 4.50 Design of a miniature implantable pump (Problem 4.16).

4.16 Figure 4.50 shows the design of a miniature pump that can be implanted in the body to deliver a drug. No batteries are required to run this pump! The pump contains two cylindrical chambers filled with incompressible fluids; the two chambers together have a length of 3 cm and a diameter of 0.7 cm. Chamber 1 is filled with a solution whose concentration is 10 mol/L; the osmolarity of this solution greatly exceeds that of body fluids. Chamber 2 is filled with the drug solution. The two chambers are separated by a frictionless, massless, and impermeable piston. The piston moves freely and supports no difference in hydraulic pressure between the chambers; the piston allows no transport of water, solute, or drug between chambers. The pump walls are rigid, impermeable, and cylindrical with an orifice at one end for delivering the drug and a rigid, semipermeable membrane at the other end. The orifice diameter is sufficiently large that the hydraulic pressure drop across the orifice is negligible and sufficiently small so that the diffusion of drug through the orifice is also negligible. The semipermeable membrane is permeable to water only; it is not permeable to the solute. Assume that $T = 300$ K.

a. Provide a discussion of fifty words or less for each of the following:
   i. What is the physical mechanism of drug delivery implied by the pump design?
   ii. What is (are) the source(s) of energy for pumping the drug?
   iii. Assume there is an adequate supply of drug in the pump for the lifetime of the implanted subject and that it is necessary to provide a constant rate of drug delivery. Which fundamental factors limit the useful lifetime of this pump in the body?

b. When implanted in the body, the pump delivers the drug at a rate of 1 μL/h. Find the value of the hydraulic conductivity, $\mathcal{L}_V$, of the semipermeable membrane.

4.17 A long and thin cylindrical cell of radius r and length l contains N_Σ^i moles of the impermeant solute S and is immersed in a bath whose concentration of S is C_Σ^o. The membrane has a hydraulic conductivity $\mathcal{L}_V$. Assume that the membrane of the cell is permeable only to water and that the volume of water in the cell equals the volume of the cell. Also assume that flow of water through the ends of the cylindrical cell is negligible and that, as the cell changes its volume, only the radius changes while the length of the cell is constant. Assume that the hydraulic pressure difference across the membrane is zero.

a. Show that the cell radius r satisfies the differential equation

$$\frac{dr(t)}{dt} + \frac{A}{r^2(t)} = B.$$

b. Find A and B in terms of l, $\mathcal{L}_V$, N_Σ^i, C_Σ^o, R, and T, where R is the molar gas constant and T is absolute temperature.

4.18 Figure 4.26 indicates that according to the theory developed in Section 4.7.2, the kinetics of swelling of a cell with constant surface area and of a spherical cell are quite similar. However, a careful comparison indicates that some differences do occur (Figure 4.51).

a. By examining Equation 4.69, explain the difference in kinetics for the two types of cell shapes.

b. On the basis of your explanation, compare the swelling kinetics of a cylindrical cell that swells only in its radial direction to that of the spherical and constant-area cell types shown in Figure 4.51.

4.19 A spherical cell is subjected to four different aqueous solutions of impermeant solutes, and its equilibrium radius is measured as shown in Figure 4.52. The isotonic radius is 80 μm. The four solutions are 150-mmol/L NaCl, 200-mmol/L $CaCl_2$, 800-mmol/L sucrose, and 150-mmol/L xylose. The latter two solutes are sugars. You may assume that the intracellular quantity of solute does not change during these measurements.

a. Determine the compositions of solutions 1–4.

b. Find the total quantity of intracellular solute.

c. What fraction of the isotonic volume of the cell is due to osmotically active water?

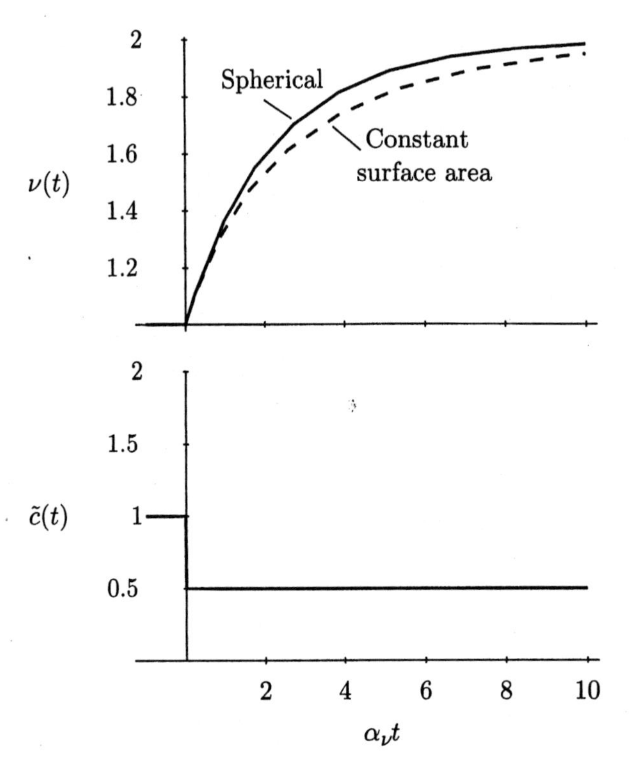

Figure 4.51 Comparison of swelling kinetics of a cell with constant surface area with that of a spherical cell (Problem 4.18). The normalized volume is shown as a function of time in response to a step change in extracellular concentration of impermeant solute. The solutions shown are taken directly from Figure 4.26.

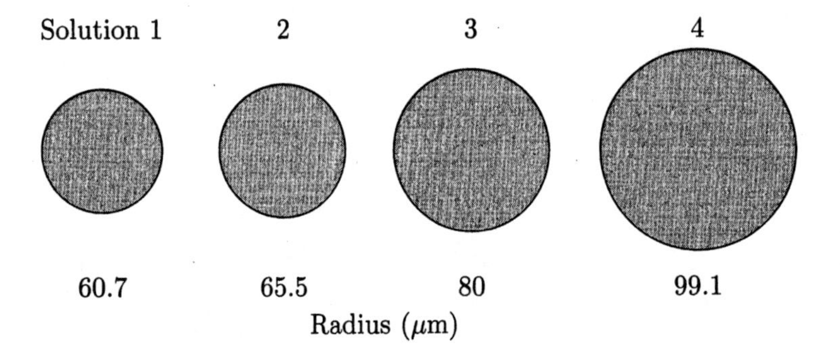

Figure 4.52 A spherical cell placed in four different solutions of impermeant solutes (Problem 4.19). A schematic diagram of the cell is shown above; the measured radius of the cell is shown below.

4.20 This problem involves interpretation of the measurements of the swelling of sea urchin (*Arbacia*) eggs shown in Figure 4.28. The measurements show the volume as a function of time after the bathing solution was changed from isotonic $C^o_\Sigma(t) = C^i_\Sigma(t)$ for $t < 0$ to hypotonic $C^o_\Sigma(t) < C^o_\Sigma(t)$ for $t > 0$.

a. Derive an expression for $C^i_\Sigma(t)/C^i_\Sigma(0)$ in terms of the volume $\mathcal{V}_c(t)$ of the cell and its initial volume $\mathcal{V}_c(0)$. Assume only part of this volume, $\mathcal{V}_c(t) - \mathcal{V}_c'$, is osmotically active, where $\mathcal{V}_c'$ is a constant. Assume that the number of moles, N_Σ^i, of solutes inside the cell is fixed.

b. Derive a first-order differential equation for $\mathcal{V}_c(t)$. Assume the eggs are initially spherical and remain so as they expand. Assume that the hydraulic conductivity, $\mathcal{L}_V$, is a constant independent of radius.

c. Explain with a sketch how you would determine values for $\mathcal{L}_V$ and $\mathcal{V}_c'$ from measurements such as those in Figure 4.28. Is there any information about this experiment other than that given in the figure that you will need? (Solution of the differential equation is not necessary. Actual numerical values for $\mathcal{L}_V$ and $\mathcal{V}_c'$ are not required.)

4.21 The pore radius of water channels in erythrocytes have been estimated repeatedly from different types of measurements. The point of this problem is to use a particular set of measurements to estimate the pore radius based on the particular theories of convection and diffusion through porous membranes given in Section 4.5. Measurements of water convection and diffusion in human erythrocytes have given values for the hydraulic conductivity of $\mathcal{L}_V = 1.4$ pm/(Pa · s) (Terwilliger and Solomon, 1981; Mlekoday et al., 1983) and of the diffusive permeability of $P_w = 24$ μm/s (Brahm, 1982).

a. Assume that both convection and diffusion of water through the erythrocyte membrane is via aqueous pores (water channels). Estimate the pore radius of the water channels in erythrocytes based on a model for water transport in unhindered large pores.

b. Assume that both convection and diffusion of water through the erythrocyte membrane is via water channels. Estimate the pore radius of the water channels in erythrocytes based on a model for water transport in microscopic pores with hindrance.

c. Water transport through erythrocytes occurs both via the dissolve-diffuse mechanism and through water channels. Suppose that the addition of a mercury compound reduces the value of P_w by 50%. Taking this into account, estimate the pore radius of the water channels in erythrocytes based on a model for water transport in microscopic pores with hindrance.

d. Briefly discuss the implications of the results of parts a–c for the transport of water through erythrocyte membranes.

References

Books and Reviews

Agre, P., Preston, G. M., Smith, B. L., Jung, J. S., Raina, S., Moon, C., Guggino, W. B., and Nielsen, S. (1993). Aquaporin CHIP: The archetypal molecular water channel. *Am. J. Physiol.*, 265:F463–F476.

Batchelor, G. K. (1967). *An Introduction to Fluid Mechanics*. Cambridge University Press, New York.

Bessis, M. (1974). *Corpuscles*. Springer-Verlag, New York.

Bull, H. B. (1964). *An Introduction to Physical Biochemistry*. F. A. Davis, Philadelphia.

Deen, W. M. (1987). Hindered transport of large molecules in liquid-filled pores. *AIChE J.*, 33:1409–1425.

Dempster, J. A., Van Hoek, A. N., and Van Os, C. H. (1992). The quest for water channels. *News Physiol. Sci.*, 7:172–176.

Dick, D. A. T. (1966). *Cell Water*. Butterworths, Washington, DC.

Dittmer, D. S., ed. (1961). *Blood and Other Body Fluids*. Federation of American Societies for Experimental Biology, Bethesda, MD.

Dowben, R. M. (1969). *General Physiology: A Molecular Approach*. Harper & Row, New York.

Eisenberg, D. and Crothers, D. (1979). *Physical Chemistry with Applications to the Life Sciences*. Benjamin-Cummings, Reading, MA.

Fermi, E. (1936). *Thermodynamics*. Dover, New York.

Fettiplace, R. and Haydon, D. A. (1980). Water permeability of lipid membranes. *Physiol. Rev.*, 60:510–550.

Finkelstein, A. (1987). *Water Movement Through Lipid Bilayers, Pores, and Plasma Membranes*. John Wiley & Sons, New York.

Hammel, H. T. and Scholander, P. F. (1976). *Osmosis and Tensile Solvent*. Springer-Verlag, New York.

House, C. R. (1974). *Water Transport in Cells and Tissues*. Williams & Wilkins, Baltimore, MD.

Kepner, G. R. (1979). *Cell Membrane Permeability and Transport*. Dowden, Hutchinson and Ross, Stroudsburg, PA.

Lentner, C. (1981). *Geigy Scientific Tables*, vol. 1. Ciba-Geigy Corp., West Caldwell, NJ.

Lucké, B. and McCutcheon, M. (1932). The living cell as an osmotic system and its permeability to water. *Physiol. Rev.*, 12:68–139.

Milburn, J. A. (1979). *Water Flow in Plants*. Longman, New York.

Moore, W. J. (1972). *Physical Chemistry*. Prentice-Hall, Englewood Cliffs, NJ.

Murrell, J. N. and Boucher, E. A. (1982). *Properties of Liquids and Solutions*. John Wiley & Sons, New York.

Olmstead, E. G. (1966). *Mammallian Cell Water: Physiologic and Clinical Aspects*. Lea and Febiger, Philadelphia.

Passioura, J. B. (1988). Water transport in and to roots. *Ann. Rev. Plant Physiol. Plant Mol. Biol.*, 39:245–265.

Pfeffer, W. (1877). *Osmotic Investigations: Studies on Cell Mechanics*. Trans. G. R. Kepner and E. J. Stadelmann (1985). Van Nostrand Reinholt, New York.

Quan, A. H. and Cogan, M. G. (1993). Body fluid compartment and water balance. In Seldin, D. W. and Giebisch, G., eds., *Clinical Disturbances of Water Metabolism*. Raven, New York.

Richards, E. G. (1980). *An Introduction to the Physical Properties of Large Molecules in Solution*. Cambridge University Press, New York.

Scheidegger, A. E. (1974). *The Physics of Flow through Porous Media*. University of Toronto Press, Toronto, Canada.

Solomon, A. K. (1968). Characterization of biological membranes by equivalent pores. *J. Gen. Physiol.*, 51:335S–364S.

Solomon, A. K. (1986). On the equivalent pore radius. *J. Membr. Biol.*, 94:227–232.

Stein, W. D. (1990). *Channels, Carriers, and Pumps*. Academic Press, New York.

Tombs, M. P. and Peacocke, A. R. (1974). *The Osmotic Pressure of Biological Macromolecules*. Oxford University Press, New York.

Verkman, A. S. (1992). Water channels in cell membranes. *Ann. Rev. Physiol.*, 54:97–108.

Verkman, A. S. (1993). *Water Channels*. R. G. Landes, Austin, TX.

Villars, F. M. H. and Benedek, G. B. (1974). *Physics with Illustrative Examples from Medicine and Biology*, vol. 2, *Statistical Physics*. Addison-Wesley, Reading, MA.

Wolf, A. V. (1966). *Aqueous Solutions and Body Fluids: Their Concentrative Properties and Conversion Tables*. Harper & Row, New York.

Original Articles

Abrami, L., Capurro, C., Ibarra, C., Buhler, J. M., and Ripoche, P. (1995). Distribution of mRNA encoding the FA-CHIP water channel in amphibian tissues: Effects of salt adaptation. *J. Membr. Biol.*, 143:199–205.

Adair, G. S. (1928). A theory of partial osmotic pressures and membrane equilibria, with special reference to the application of Dalton's law to hæmoglobin solutions in the presence of salts. *Proc. R. Soc. London, Ser. A*, 120:573–603.

Adair, G. S. and Robinson, M. E. (1930). The analysis of the osmotic pressures of the serum proteins, and the molecular weights of albumins and globulins. *Biochem. J.*, 24:1864–1889.

Agre, P., Smith, B. L., Baumgarten, R., Preston, G. M., Pressman, E., Wilson, P., Illum, N., Anstee, D. J., Lande, M. B., and Zeidel, M. L. (1994). Human red cell aquaporin CHIP: II. Expression during normal fetal development and in a novel form in congenital dyserythropoietic anemia. *J. Clin. Invest.*, 94:1050–1058.

Andersen, B. and Ussing, H. H. (1957). Solvent drage on non-electrolytes during osmotic flow through isolated toad skin and its response to antidiuretic hormone. *Acta Physiol. Scand.*, 39:228–239.

Andrews, F. C. (1976). Colligative properties of simple solutions. *Science*, 194:567–571.

Babbitt, J. D. (1956). Osmotic pressure. *Science*, 122:285–287.

Barry, P. H. and Diamond, J. M. (1984). Effects of unstirred layers on membrane phenomena. *Physiol. Rev.*, 64:763–872.

Bean, C. P. (1972). The physics of porous membranes: Neutral pores. In Eisenman, G., ed., *Membranes*, vol. 1, *Macroscopic Systems and Models*. Marcel Dekker, New York.

Benga, G., Pop, V. I., Popescu, O., and Borza, V. (1990). On measuring the diffusional water permeability of human red blood cells and ghosts by nuclear magnetic resonance. *J. Biochem. Biophys. Meth.*, 21:87–102.

Bentley, P. J. (1958). The effects of neurohypophysial extracts on water transfer across the wall of the isolated urinary bladder of the toad *Bufo marinus*. *J. Endocrin.*, 17:201–209.

Blinks, J. R. (1965). Influence of osmotic strength on cross-section and volume of isolated single muscle fibres. *J. Physiol.*, 177:42–57.

Bondy, C., Chin, E., Smith, B. L., Preston, G. M., and Agre, P. (1993). Developmental gene expression and tissue distribution of the CHIP28 water-channel protein. *Proc. Natl. Acad. Sci. U.S.A.*, 90:4500–4504.

Brahm, J. (1982). Diffusional water permeability of human erythrocytes and their ghosts. *J. Gen. Physiol.*, 79:791–819.

Brown, D. (1989). Membrane recycling and epithelial cell function. *Am. J. Physiol.*, 256:F1–F12.

Burton, R. F. (1983). The composition of animal cells: Solutes contributing to osmotic pressure and charge balance. *Comp. Biochem. Physiol.*, 76B:663–671.

Cass, A. and Finkelstein, A. (1967). Water permeability of thin lipid membranes. *J. Gen. Physiol.*, 50:1765–1784.

Chinard, F. P. and Enns, T. (1956). Osmotic pressure. *Science*, 124:472–474.

Cohen, F. S., Niles, W. D., and Akabas, M. H. (1989). Fusion of phospholipid vesicles with a planar membrane depends on the membrane permeability of the solute used to create the osmotic pressure. *J. Gen. Physiol.*, 93:201–210.

Dainty, J. (1963). Water relations of plant cells. *Advan. Botan. Res.*, 1:279–326.

Dempster, J. A., Van Hoek, A. N., de Jong, M. D., and Van Os, C. H. (1991). Glucose transporters do not serve as water channels in renal and intestinal epithelia. *Pflügers Arch.*, 419:249–255.

Denker, B. M., Smith, B. L., Kuhajda, F. P., and Agre, P. (1988). Identification, purification, and partial characterization of a novel M_r 28,000 integral membrane protein from erythrocytes and renal tubules. *J. Biol. Chem.*, 263:15634–15642.

Durbin, R. P. (1960). Osmotic flow of water across permeable cellulose membranes. *J. Gen. Physiol.*, 44:315–326.

Echevarria, M. and Verkman, A. S. (1992). Optical measurement of osmotic water transport in cultured cells. *J. Gen. Physiol.*, 99:573–589.

Essig, A. and Caplan, S. R. (1989). Water movement: Does thermodynamic interpretation distort reality? *Am. J. Physiol.*, 256:C694–C698.

Farinas, J., Van Hoek, A. N., Shi, L. B., Erickson, C., and Verkman, A. S. (1993). Nonpolar environment of tryptophans in erythrocyte water channel CHIP28 determined by fluorescence quenching. *Biochem.*, 32:11857–11864.

Ferrier, J. (1984). Osmosis and intermolecular force. *J. Theor. Biol.*, 106:449–453.

Fischbarg, J., Kuang, K., Vera, J. C., Arant, S., Silverstein, S. C., Loike, J., and Rosen, O. M. (1990). Glucose transporters serve as water channels. *Proc. Natl. Acad. Sci. U.S.A.*, 87:3244–3247.

Fushimi, K., Uchida, S., Hara, Y., Hirata, Y., Marumo, F., and Sasaki, S. (1993). Cloning and expression of apical membrane water channel of rat kidney collecting tubule. *Nature*, 361:549–552.

Galey, W. R. and Brahm, J. (1985). The failure of hydrodynamic analysis to define pore size in cell membranes. *Biochim. Biophys. Acta*, 818:425–428.

Garby, L. (1957). Studies on transfer of matter across membranes with special reference to the isolated human amniotic membrane and the exchange of amniotic fluid. *Acta Physiol. Scand*, 40(Suppl. 137): 1–84.

Gibbs, J. W. (1897). Semi-permeable films and osmotic pressure. *Nature*, 55:461–462.
Guell, D. C. (1991). The physical mechanism of osmosis and osmotic pressure: A hydrodynamic theory for calculating the osmotic reflection coefficient. PhD thesis, Department of Chemical Engineering, Massachusetts Institute of Technology.
Hammel, H. T. (1976). Colligative properties of a solution. *Science*, 192:748–756.
Hammel, H. T. (1979a). Forum on osmosis: I. Osmosis: Diminished solvent activity or enhanced solvent tension? *Am. J. Physiol.*, 237:R95–R107.
Hammel, H. T. (1979b). Forum on osmosis: V. Epilogue. *Am. J. Physiol.*, 237:R123–R125.
Handler, J. S. (1988). Antidiuretic hormone moves membranes. *Am. J. Physiol.*, 255:F375–F382.
Hasegawa, H., Ma, T., Skach, W., Matthay, M. A., and Verkman, A. S. (1994). Molecular cloning of a mercurial-insensitive water channel expressed in selected water-transporting tissues. *J. Biol. Chem.*, 269:5497–5500.
Hays, R. M. and Leaf, A. (1962). Studies on the movement of water through the isolated toad bladder and its modification by vasopressin. *J. Gen. Physiol.*, 45:905–919.
Hempling (1960). Permeability of the Ehrlich ascites tumor cell to water. *J. Gen. Physiol.*, 44:365–379.
Heyer, E., Cass, A., and Mauro, A. (1970). A demonstration of the effect of permeant and impermeant solutes, and unstirred boundary layers on osmotic flow. *Yale J. Biol. Med.*, 42:139–153.
Hildebrand, J. H. (1955). Osmotic pressure. *Science*, 12:116–119.
Hildebrand, J. H. (1979). Forum on osmosis: II. A criticism of solvent "tension" in osmosis. *Am. J. Physiol.*, 237:R108–R109.
Hill, A. E. (1994). Osmotic flow in membrane pores of molecular size. *J. Membr. Biol.*, 137:197–203.
Ishibashi, K., Sasaki, S., Fushimi, K., Uchida, S., Kuwahara, M., Saito, H., Furukawa, T., Nakajima, K., Yamaguchi, Y., Gojobori, T., and Marumo, F. (1994). Molecular cloning and expression of a member of the aquaporin family with permeability to glycerol and urea in addition to water expressed at the basolateral membrane of kidney collecting duct cells. *Proc. Natl. Acad. Sci. U.S.A.*, 91:6269–6273.
Jacobs, M. H. and Stewart, D. R. (1947). Osmotic properties of the erythrocyte. *J. Cell. Comp. Physiol.*, 30:79–103.
Jung, J. S., Bhat, R. V., Preston, G. M., Guggino, W. B., Baraban, J. M., and Agre, P. (1994). Molecular characterization of an aquaporin cDNA from brain: Candidate osmoreceptor and regulator of water balance. *Proc. Natl. Acad. Sci. U.S.A.*, 91:13052–13056.
Kelvin, L. (1897). On osmotic pressure against an ideal semi-permeable membrane. *Nature*, 55:272–273.
Knepper, M. A. (1994). The aquaporin family of molecular water channels. *Proc. Natl. Acad. Sci. U.S.A.*, 91:6255–6258.
Koefoed-Johnsen, V. and Ussing, H. H. (1953). The contributions of diffusion and flow to the passage of D_2O through living membranes: Effect of neurohypophyseal hormone on isolated anuran skin. *Acta Physiol. Scand.*, 28:60–76.
Kyte, J. and Doolittle, R. F. (1982). A simple method for displaying the hydropathic character of a protein. *J. Mol. Biol.*, 157:105–132.
Leaf, A. and Hays, R. M. (1962). Permeability of the isolated toad bladder to solutes and its modification by vasopressin. *J. Gen. Physiol.*, 45:921–932.

LeFevre, P. G. (1964). The osmotically functional water content of the human erythrocyte. *J. Gen. Physiol.*, 47:585–603.

Levine, S. D., Jacoby, M., and Finkelstein, A. (1984a). The water permeability of the toad urinary bladder: I. Permeability of barriers in series with the luminal membrane. *J. Gen. Physiol.*, 83:529–541.

Levine, S. D., Jacoby, M., and Finkelstein, A. (1984b). The water permeability of the toad urinary bladder: II. The value of $P_f/P_d(w)$ for the antidiuretic hormone-induced water permeation pathway. *J. Gen. Physiol.*, 83:543–561.

Levitt, D. G. (1984). Kinetics of movement in narrow channels. In Stein, W. D., ed., *Ion Channels: Molecular and Physiological Aspects*, 181–197. Academic Press, New York.

Lightfoot, E. N., Bassingthwaighte, J. B., and Grabowski, E. F. (1976). Hydrodynamic models for diffusion in microporous membranes. *Ann. Biomed. Eng.*, 4:78–90.

Loike, J. D., Cao, L., Kuang, K., Vera, J. C., Silverstein, S. C., and Fischbarg, J. (1993). Role of facilitative glucose transporters in diffusional water permeability through J774 cells. *J. Gen. Physiol.*, 102:897–906.

Longuet-Higgins, H. C. and Austin, G. (1966). The kinetics of osmotic transport through pores of molecular dimensions. *Biophys. J.*, 6:217–224.

Lucké, B., Hartline, H. K., and McCutcheon, M. (1931). Further studies on the kinetics of osmosis in living cells. *J. Gen. Physiol.*, 14:405–419.

Lucké, B., Hartline, H. K., and Ricca, R. A. (1939). Comparative permeability to water and to certain solutes of the egg cells of three marine invertebrates, *Arbacia, Cumingia* and *Chaetopterus*. *J. Cell. Comp. Physiol.*, 14:237–252.

Lucké, B., Larrabee, M. G., and Hartline, H. K. (1935). Studies on osmotic equilibrium and on the kinetics of osmosis in living cells by a diffraction method. *J. Gen. Physiol.*, 19:1–17.

Ma, T., Frigeri, A., Skach, W., and Verkman, A. S. (1993). Cloning of a novel rat kidney cDNA homologous to CHIP28 and WCH-CD water channels. *Biochem. Biophys. Res. Comm.*, 197:654–659.

Macey, R. I. (1984). Transport of water and urea in red blood cells. *Am. J. Physiol.*, 246:C195–C203.

Macey, R. I. and Farmer, R. E. L. (1970). Inhibition of water and solute permeability in human red cells. *Biochim. Biophys. Acta*, 211:104–106.

Manning, G. S. (1975). The relation between osmotic flow and tracer solvent diffusion for single-file transport. *Biophys. Chem.*, 3:147–152.

Mauro, A. (1957). Nature of solvent transfer in osmosis. *Science*, 126:252–253.

Mauro, A. (1960). Some properties of ionic and nonionic semipermeable membranes. *Circulation*, 21:845–854.

Mauro, A. (1965). Osmotic flow in a rigid porous membrane. *Science*, 149:867–869.

Mauro, A. (1979). Forum on osmosis: III. Comments on Hammel and Scholander's solvent tension theory and its application to phenomenon of osmotic flow. *Am. J. Physiol.*, 237:R110–R113.

Mauro, A. (1981). The role of negative pressure in osmotic equilibrium and osmotic flow. In *Water Transport across Epithelia*, 107–119. Alfred Benzon Symposium 15. Munksgaard, Copenhagen.

McCutcheon, M., Lucké, B., and Hartline, H. K. (1931). The osmotic properties of living cells (eggs of *arbacia punctulata*). *J. Gen. Physiol.*, 14:393–403.

Mlekoday, H. J., Moore, R., and Levitt, D. G. (1983). Osmotic water permeability of the human red cell. *J. Gen. Physiol.*, 81:213–220.

Nielsen, S., Smith, B. L., Christensen, E. I., Knepper, M., and Agre, P. (1993). CHIP28 water channels are localized in constitutively water-permeable segments of the nephron. *J. Cell Biol.*, 120:371–383.

Niles, W. D., Cohen, F. S., and Finkelstein, A. (1989). Hydrostatic pressures developed by osmotically swelling vesicles bound to planar membranes. *J. Gen. Physiol.*, 93:211–244.

Paganelli, C. V. and Solomon, A. K. (1957). The rate of exchange of tritiated water across the human red cell membrane. *J. Gen. Physiol.*, 41:259–277.

Parker, J. C. (1971). Ouabain-insensitive effects of metabolism on ion and water content of red blood cells. *Am. J. Physiol.*, 221:338–342.

Pedley, T. J. (1978). The development of osmotic flow through an unstirred layer. *J. Theor. Biol.*, 70:427–447.

Pedley, T. J. (1980). The interaction between stirring and osmosis. Part 1. *J. Fluid Mech.*, 101:843–861.

Pedley, T. J. (1981). The interaction between stirring and osmosis. Part 2. *J. Fluid Mech.*, 107:281–296.

Preston, G. M. and Agre, P. (1991). Isolation of the cDNA for the erythrocyte integral membrane protein of 28 kilodaltons: Member of an ancient channel family. *Proc. Natl. Acad. Sci. U.S.A.*, 88:11110–11114.

Preston, G. M., Carroll, T. P., Guggino, W. B., and Agre, P. (1992). Appearance of water channels in *Xenopus* oocytes expressing red cell CHIP28 protein. *Science*, 256:385–387.

Preston, G. M., Jung, J. S., Guggino, W. B., and Agre, P. (1993). The mercury-sensitive residue at cysteine 189 in the CHIP28 water channel. *J. Biol. Chem.*, 268:17–20.

Preston, G. M., Jung, J. S., Guggino, W. B., and Agre, P. (1994a). Membrane topology of aquaporin CHIP. *J. Biol. Chem.*, 269:1668–1673.

Preston, G. M., Smith, B. L., Zeidel, M. L., Moulds, J. J., and Agre, P. (1994b). Mutations in *aquaporin-1* in phenotypically normal humans without functional CHIP water channels. *Science*, 265:1585–1587.

Raina, S., Preston, G. M., Guggino, W. B., and Agre, P. (1995). Molecular cloning and characterization of an aquaporin cDNA from salivary, lacrimal, and respiratory tissues. *J. Biol. Chem.*, 270:1908–1912.

Ray, P. M. (1960). On the theory of osmotic water movement. *Plant Physiol.*, 35:783–795.

Rayleigh, L. (1897). The theory of solutions. *Nature*, 55:253–254.

Reizer, J., Reizer, A., and Saier, M. H. J. (1993). The MIP family of integral membrane channel proteins: Sequence comparisons, evolutionary relationships, reconstructed pathway of evolution, and proposed functional differentiation of the two repeated halves of the protein. *Crit. Rev. Biochem. Mol. Biol.*, 28:235–257.

Robbins, E. and Mauro, A. (1960). Experimental study of the independence of diffusion and hydrodynamic permeability coefficients in collodion membranes. *J. Gen. Physiol.*, 43:523–532.

Savitz, D., Sidel, V. W., and Solomon, A. K. (1964). Osmotic properties of human red cells. *J. Gen. Physiol.*, 48:79–94.

Scatchard, G., Hamer, W. J., and Wood, S. E. (1938). Isotonic solutions. I. The chemical potential of water in aqueous solutions of sodium chloride, potassium chloride, sulfuric acid, sucrose, urea and glycerol at 25°. *J. Am Chem. Soc.*, 60:3061–3070.

Schick, M. J., Doty, P., and Zimm, B. H. (1950). Thermodynamic properties of concentrated polystyrene solutions. *J. Am. Chem. Soc.*, 72:530–534.

Serve, G., Endres, W., and Grafe, P. (1988). Continuous electrophysiological measurements of changes in cell volume of motoneurons in the isolated frog spinal cord. *Pflügers Arch.*, 411:410–415.

Sha'afi, R. I. and Gary-Bobo, C. M. (1973). Water and nonelectrolyte permeability in mammalian red cell membranes. *Prog. Biophys. Mol. Biol.*, 26:103–146.

Sha'afi, R. I., Rich, G. T., Sidel, V. W., Bossert, W., and Solomon, A. K. (1967). The effect of the unstirred layer on human red cell water permeability. *J. Gen. Physiol.*, 50:1377–1399.

Sidel, V. W. and Solomon, A. K. (1957). Entrance of water into human red cells under an osmotic pressure gradient. *J. Gen. Physiol.*, 41:243–257.

Smith, B. L. and Agre, P. (1991). Erythrocyte M_r 28,000 transmembrane protein exists as a multisubunit oligomer similar to channel proteins. *J. Biol. Chem.*, 266:6407–6415.

Soodak, H. and Iberall, A. (1978). Osmosis, diffusion, convection. *Am. J. Physiol.*, 235:R3–R17.

Soodak, H. and Iberall, A. (1979). Forum on osmosis: IV. More on osmosis and diffusion. *Am. J. Physiol.*, 237:R114–R122.

Starling, E. H. (1896). On the absorption of fluids from the connective tissue spaces. *J. Physiol.*, 19:312–326.

Terwilliger, T. C. and Solomon, A. K. (1981). Osmotic water permeability of human red cells. *J. Gen. Physiol.*, 77:549–570.

Theeuwes, F. and Yum, S. I. (1976). Principles of the design and operation of generic osmotic pumps for the delivery of semisolid or liquid drug formulations. *Ann. Biomed. Eng.*, 4:343–353.

Tsai, S. T., Zhang, R., and Verkman, A. S. (1991). High channel-mediated water permeability in rabbit erythrocytes: Characterization in native cells and expressions in *Xenopus* oocytes. *Biochem.*, 30:2087–2092.

Ussing, H. H., Fischbarg, J., Sten-Knudsen, O., Larsen, E. H., and Willumsen, N. J., eds. (1993). *Isotonic Transport in Leaky Epithelia*. Alfred Benzon Symposium 34. Munksgaard, Copenhagen, Denmark.

Van Hoek, A. N. and Verkman, A. S. (1992). Functional reconstitution of the isolated erythrocyte water channel CHIP28. *J. Biol. Chem.*, 267:18267–18269.

Van Hoek, A. N., Wiener, M., Bicknese, S., Miercke, L., Biwersi, J., and Verkman, A. S. (1993). Secondary structure analysis of purified functional CHIP28 water channels by CD and FTIR spectroscopy. *Biochem.*, 32:11847–11856.

Van't Hoff, J. H. (1886). Une propriété général de la matière diluée. *Svenska Vet. Akad. Handl.*, 21:42–49 (translation in Hammel and Scholander, 1976).

Van't Hoff, J. H. (1887). Die Rolle des osmotischen Druckes in der Analogie zwischen Lösungen und Gasen. *Z. Phys. Chem.*, 1:481–508 (translation in Keppner, 1979).

Van't Hoff, J. H. (1892). Zur Theorie der Lösungen. *Z. Phys. Chem.*, 9:477 (translation in Hammel and Scholander, 1976).

Vegard, L. (1910). On the free pressure in osmosis. *Proc. Camb. Phil. Soc.*, 15:13–23.

Verbavatz, J. M., Brown, D., Sabolić, I., Valenti, G., Ausiello, D. A., Van Hoek, A. N., Ma,

T., and Verkman, A. S. (1993). Tetrameric assembly of CHIP28 water channels in liposomes and cell membranes: A freeze-fracture study. *J. Cell Biol.*, 123:605–618.

Verkman, A. S. (1989). Mechanisms and regulation of water permeability in renal epithelia. *Am. J. Physiol.*, 257:C837–C850.

Verkman, A. S. and Wong, K. R. (1987). Proton nuclear magnetic resonance measurement of diffusional water permeability in suspended renal proximal tubules. *Biophys. J.*, 51:717–723.

Walz, T., Smith, B. L., Zeidel, M. L., Engel, A., and Agre, P. (1994). Biologically active two-dimensional crystals of aquaporin CHIP. *J. Biol. Chem.*, 269:1583–1586.

Wong, K. R. and Verkman, A. S. (1987). Human platelet diffusional water permeability measured by nuclear magnetic resonance. *Am. J. Physiol.*, 252:C618–C622.

Yates, F. E. (1978). Osmosis: A transport of confusion. *Am. J. Physiol.*, 235:R1–R2.

Ye, R. and Verkman, A. S. (1989). Simultaneous optical measurement of osmotic and diffusional water permeability in cells and liposomes. *Biochem.*, 28:824–829.

Zeidel, M. L., Ambudkar, S. V., Smith, B. L., and Agre, P. (1992). Reconstitution of functional water channels in liposomes containing purified red cell CHIP28 protein. *Biochem.*, 31:7436–7440.

Zeidel, M. L., Nielsen, S., Smith, B. L., Ambudkar, S. V., Maunsbach, A. B., and Agre, P. (1994). Ultrastructure, pharmacologic inhibition, and transport selectivity of aquaporin channel-forming integral protein in proteoliposomes. *Biochemistry*, 33:1606–1615.

Zhang, R., Skach, W., Hasegawa, H., Van Hoek, A. N., and Verkman, A. S. (1993). Cloning, functional analysis and cell localization of a kidney proximal tubule water transporter homologous to CHIP28. *J. Cell Biol.*, 120:359–369.

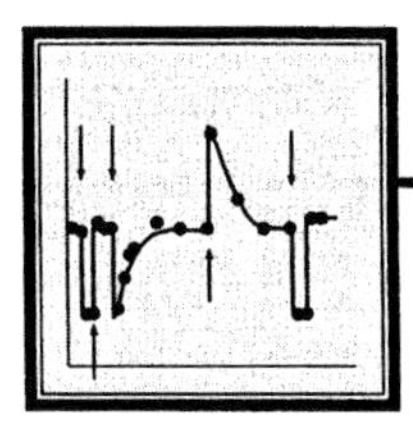

5 Concurrent Solute and Solvent Transport

Although irreversible thermodynamics has made some positive contributions to the subject of osmotic flow across membranes, particularly in lending its imprimatur to certain general results and conclusions, it has also, in my opinion, been a source of some serious confusions, and has acted as a fig leaf behind which underlying physical mechanisms are hidden.

—Finkelstein, 1987

5.1 Introduction

Chapters 3 and 4 dealt with solute and solvent transport, respectively, in isolation. Chapter 3 was concerned with *solute* transport that results from a *solute* concentration gradient (or diffusion) under conditions of osmotic equilibrium. Transmembrane transport by this mechanism is summarized by the equations

$$\Phi_V = 0 \quad \text{and} \quad \phi_n = P_n(c_n^i - c_n^o), \tag{5.1}$$

where Φ_V is the solvent volume flux, ϕ_n is the solute molar flux, P_n is the permeability of the membrane to solute n, and $c_n^i - c_n^o$ is the difference of concentration of solute n across the membrane (Figure 5.1). Chapter 4 dealt with *solvent* transport resulting from a hydraulic pressure and a *solute* concentration gradient across a membrane (or hydraulic and osmotic flow) under the assumption that the membrane is impermeable to the solute. Transmembrane transport by this mechanism is summarized by the equations

$$\Phi_V = \mathcal{L}_V\left((p_i - p_o) - RT(C_\Sigma^i - C_\Sigma^o)\right) \quad \text{and} \quad \phi_n = 0, \tag{5.2}$$

where $\mathcal{L}_V$ is the hydraulic conductivity of the membrane, $p_i - p_o$ is the difference in hydraulic pressure across the membrane, $C_\Sigma^o = \sum_n c_n^o$ and $C_\Sigma^i = \sum_n c_n^i$ are the total extracellular and intracellular solute concentrations, R is the mo-

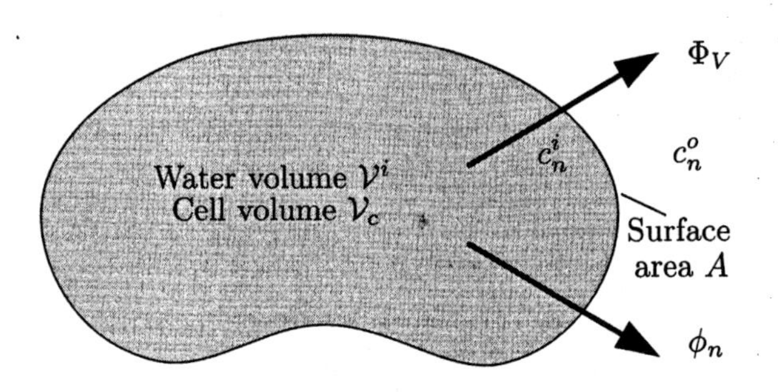

Figure 5.1 Schematic diagram of a cell showing solute and solvent transport variables.

lar gas constant, and T is the absolute temperature. Since all the solutes are assumed impermeant, the flux of each solute is zero.

In this chapter we consider the question, What happens if there is a solute gradient across a membrane that is permeable to both the solute and the solvent? We shall find that new phenomena occur under these circumstances; some of these are illustrated in Figure 5.2. In each panel, the external osmolarity of the Ringer's solution (primarily NaCl) bathing a muscle fiber is doubled at the time indicated by 2R. Doubling the osmolarity with Ringer's solution results in a reduction in cell volume by about 30–35%, presumably because water flows out of the cell. This change in volume is sustained as long as the osmolarity is doubled. When the normal Ringer's solution is replaced at R, the cell volume returns to its original value. At a later time the osmolarity of normal Ringer's is doubled using a test solute that is added to the Ringer's solution; the results are shown for four different nonelectrolyte test solutes in panels (a)–(d). In (a) the test solute is the five-carbon sugar xylose (Xy). The volume change produced by xylose is about equal to that produced by doubling the osmolarity with just the Ringer's solution. The change in volume caused by xylose is maintained for over an hour. When the normal Ringer's is restored, the volume returns to its isotonic value. Thus, the effect of xylose on the volume of the cell is similar to the effect of the Ringer's solution. Both responses resemble, at least qualitatively, the responses of cells to changes in osmolarity induced by impermeant solutes as described in Chapter 4. In contrast, in panel (b) the test solute is ethyl alcohol (Eth). Recall that ethyl alcohol (ethanol) has a relatively high oil:water partition coefficient; hence, it is a highly permeant solute (see Table 3.6). Despite the apparent initial difference in osmolarity caused by the ethyl alcohol, there is no change in the volume of the muscle cell. In panels (c) and (d) the test solutes glycerol and urea are used to double the osmolarity. The muscle membrane is poorly permeable to both of these solutes. The initial effect of these solutes is to produce a volume change that is about equivalent to that produced by the relatively imperme-

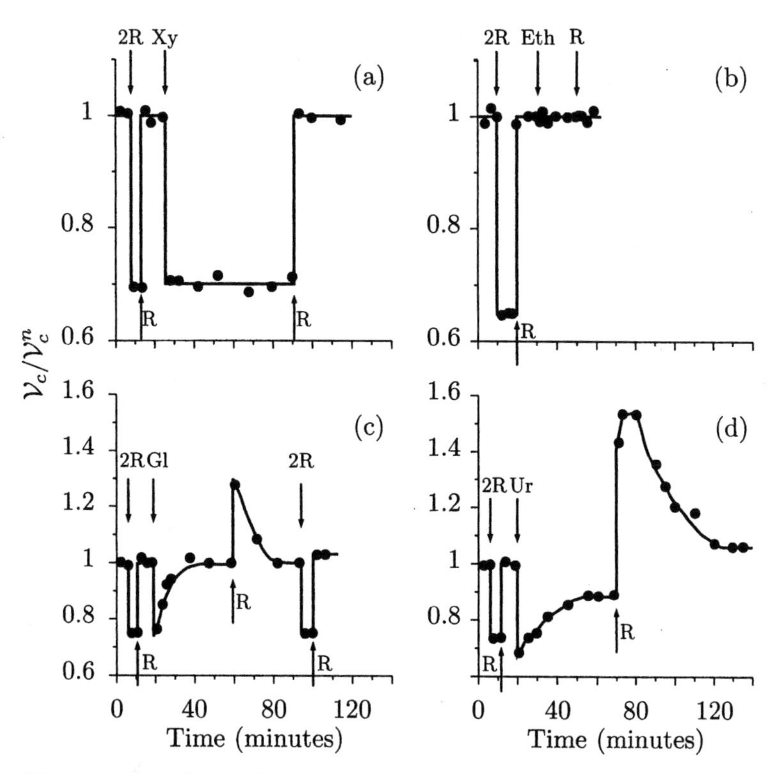

Figure 5.2 Volume changes of a muscle cell caused by changing the extracellular composition (adapted from Davson, 1970, Figure 215). The measured ratio of the volume of a muscle fiber to its isotonic volume (points) is shown as a function of time. In each panel, the muscle is in a normal Ringer's solution initially and is then in a solution of Ringer's with twice the normal osmolarity (2R). Following this sequence, in each panel a different test solute is used to double the osmolarity of normal Ringer's. The test solutes are (a) xylose (Xy); (b) ethyl alcohol (Eth); (c) glycerol (Gl); and (d) urea (Ur). In (c) the response to doubling the osmolarity with Ringer's is repeated at the end of the test sequence. The test were performed in the time sequence (b), (a), (c), and then (d).

ant solutes (the electrolytes in Ringer's and the xylose). However, the volume slowly returns to its isotonic value as the solute presumably penetrates the cell. If the external solution is now replaced with normal Ringer's solution, the volume of the cell increases 30–50% above its normal isotonic value. The cell swells because its cytoplasm is now hypertonic because of the added solute that has diffused inside. As this solute leaks back out of the cell, the cell volume returns toward its normal isotonic value. These measurements show that for this muscle fiber some solutes (e.g., the salts in Ringer's solutions as well as xylose) produce a maintained change in volume, others (e.g., ethyl alcohol) produce no volume change, and still others (e.g., glycerol and urea) produce transient changes in volume. In this chapter we shall develop theories to account for such phenomena.

5.2 Concurrent, Uncoupled Transport of Solute and Solvent

5.2.1 Derivation of Equations

In this section, we shall examine the equations that govern concurrent transport of solute and water across the membrane of a cell when the two transports are not directly coupled but occur simultaneously, i.e., concurrently. In the following section, we examine circumstances under which these uncoupled equations lead to paradoxical results, which will provide the impetus for examining the equations of coupled transport.

To focus on phenomena that occur with concurrent transport of solute and water in a relatively simple context, we examine the changes in cell volume and concentration when the membrane is permeant to no more than one of the solute species. Let species k denote a generic species, i.e., either permeant or impermeant, and species j denote the permeant species. Then the total external solute concentration can be written as follows:

$$C_{\Sigma}^{o}(t) = c_j^{o}(t) + \sum_{k \neq j} c_k^{o}(t),$$

where the impermeant and permeant species have been isolated. Similarly, the total isotonic external solute concentration is

$$C_{\Sigma}^{on} = c_j^{on} + \sum_{k \neq j} c_k^{on}.$$

Because the internal concentration of solute changes either if there is a change in the number of solute molecules in the cell or if the volume changes as a result of water transport, it is simpler to develop the equations for solute and volume changes in terms of the total number of moles of solute in the cell, which we designate as $N_{\Sigma}^{i}(t)$ and refer to as the total *quantity* of solute. Again, we isolate the permeant species as follows:

$$N_{\Sigma}^{i}(t) = n_j^{i}(t) + \sum_{k \neq j} n_k^{i}(t),$$

where $n_k^i(t)$ is the intracellular quantity of solute k at time t. We allow the quantity of impermeant intracellular solute of species k to be a function of time so that we can represent the injection of an impermeant species into a cell. The total isotonic internal solute quantity is

$$N_\Sigma^{in} = n_j^{in} + \sum_{k \neq j} n_k^{in}.$$

The total volume of water in the cell is $\mathcal{V}^i(t)$, and the isotonic volume is $\mathcal{V}^{in}$.[1] Finally, we know that at osmotic equilibrium $C_\Sigma^o = C_\Sigma^i$, and under isotonic conditions $C_\Sigma^{on} = C_\Sigma^{in}$ and $\mathcal{V}^i(t) = \mathcal{V}^{in}$. This derivation defines all the variables, and we now proceed to write the equations in normalized variables to expose the essential behavior of the cell in response to a change in permeant solute concentration.

5.2.1.1 Solute Diffusion

From Chapter 3, the flux of impermeant solute is zero and the flux of permeant solute satisfies both Fick's law for membranes and a conservation law that yields the following:

$$\phi_k(t) = \begin{cases} -\frac{1}{A(t)} \frac{dn_k^i(t)}{dt} = P_k \left(c_k^i(t) - c_k^o(t) \right) & \text{for } k = j \\ 0 & \text{for } k \neq j. \end{cases}$$

If we divide the equation for the permeant solute by the total quantity of solute under isotonic conditions $C_\Sigma^{in}\mathcal{V}^{in} = C_\Sigma^{on}\mathcal{V}^{in}$ and rearrange the terms, we obtain

$$\frac{dn(t)}{dt} = \alpha_n \mathcal{A}(t) \left(c(t) - \frac{n(t)}{v(t)} \right), \tag{5.3}$$

where $n(t) = n_j^i(t)/(C_\Sigma^{in}\mathcal{V}^{in})$, which is the ratio of the intracellular quantity of permeant solute to the total isotonic solute quantity; $\mathcal{A}(t) = A(t)/A^n$ is the surface area of the cell normalized to its isotonic value; $c(t) = c_j^o(t)/C_\Sigma^{on}$ is the ratio of the extracellular concentration of permeant solute to the total isotonic extracellular concentration; $v(t) = \mathcal{V}^i(t)/\mathcal{V}^{in}$ is the ratio of the volume to the

1. In this section, we ignore the distinction between the volume of the cell and the volume of the osmotically active water in the cell. Inclusion of this effect does not materially affect the main issues and can easily be accounted for at the end of the derivations, but it does add to the algebraic burden of the expressions.

isotonic volume; and $\alpha_n = P_j A^n / \mathcal{V}^{in}$ is a rate constant for n. Information about the shape of the cell is expressed by the *shape factor* $\mathcal{A}(t)$, which is expressed in terms of $v(t)$.

5.2.1.2 Solvent Transport

From Chapter 4, the flux of water satisfies a constitutive equation for volume transport through membranes and a conservation law that yields the following differential equation:

$$\Phi_V(t) = -\frac{1}{A(t)}\frac{d\mathcal{V}^i(t)}{dt} = \mathcal{L}_V RT \sum_k \left(c_k^o(t) - c_k^i(t)\right).$$

If we divide this equation by $\mathcal{V}^{in}$ and rearrange the terms, we obtain

$$\frac{dv(t)}{dt} = \alpha_v \mathcal{A}(t)\left(\frac{n(t) + \tilde{n}(t)}{v(t)} - (c(t) + \tilde{c}(t))\right), \tag{5.4}$$

where $\tilde{n}(t) = \sum_{k \neq j} n_k^i(t)/(C_\Sigma^{in}\mathcal{V}^{in})$ is the ratio of the intracellular quantity of solute that is impermeant to the total isotonic solute; $\tilde{c}(t) = \sum_{k \neq j} c_k^o(t)/C_\Sigma^{on}$ is the ratio of the extracellular concentration of solute that is impermeant to the total isotonic concentration; and $\alpha_v = \mathcal{L}_V RT C_\Sigma^{on} A^n / \mathcal{V}^{in}$ is a rate constant for v. The normalized equations of solute quantity $n(t)$ and cell water volume $v(t)$ (Equations 5.3 and 5.4) are a pair of differential equations that describe the time variation of the dependent variables $n(t)$ and $v(t)$ as a function of the independent variables $c(t)$, $\tilde{c}(t)$, and $\tilde{n}(t)$; the constants α_n and α_v; and the shape factor $\mathcal{A}(t)$. These equations show that, for example, a change in the extracellular concentration, $c(t)$, of the permeant solute species results in both a transport of solute through a change in $n(t)$ and a transport of water that results in a change in $v(t)$. The derivatives of $n(t)$ and $v(t)$ depend upon both $n(t)$ and $v(t)$. Furthermore, both differential equations are nonlinear. Finally, note also that both rate constants, α_n and α_v, depend on physical constants, membrane characteristics, and cell dimensions. The shape of the cell enters through the quantity $\mathcal{A}(t)$.

5.2.2 Solutions for a Cell with Constant Surface Area

If the surface area is constant as the cell swells, then $\mathcal{A}(t) = 1$. We consider this case in detail. Other cell geometries give qualitatively similar results and show only quantitative differences.

5.2.2.1 Impermeant Solutes

We begin by reviewing what happens when the external concentration of impermeant solute is changed. This situation was explored in Chapter 4. When there is no permeant solute present, the equations for concurrent transport of solute and water are much simplified. For impermeant solutes, $P_k = 0$ for all k and hence $\alpha_n = 0$. Since all the intracellular solute quantity is due to impermeant solute, $n(t) = 0$ and $\tilde{n}(t) = 1$. Since the extracellular composition is due to impermeant solute, $c(t) = 0$, so that

$$\frac{dn(t)}{dt} = 0,$$

$$\frac{dv(t)}{dt} = \alpha_v \left(\frac{1}{v(t)} - \tilde{c}(t) \right), \tag{5.5}$$

which are identical to equations described in Chapter 4. The solutions to these equations are shown in Figure 4.26 for a step of concentration and in Figure 5.3 for a pulse of concentration. An increase in the extracellular concentration of impermeant solute results in a decrease in cell volume, which reaches a steady level such that the intracellular solute concentration equals the extracellular concentration. However, no solute crosses the membrane. The change in intracellular concentration is entirely due to the transport of water. Conversely, if the extracellular concentration of impermeant solute is decreased, then the water flows into the cell and its volume increases so that the intracellular concentration decreases to equal the extracellular concentration.

5.2.2.2 One Permeant Solute

For the next, more complicated, case we consider that there is a single permeant species that is the only solute present. For this case, all the impermeant solutes have zero concentrations, and the equations reduce to

$$\frac{dn(t)}{dt} = \alpha_n \left(c(t) - \frac{n(t)}{v(t)} \right),$$

$$\frac{dv(t)}{dt} = \alpha_v \left(\frac{n(t)}{v(t)} - c(t) \right). \tag{5.6}$$

Numerical solutions to these differential equations are shown in Figure 5.3 for a cell at its isotonic volume for $t < 0$, i.e., $v(t) = 1$, $n(t) = 1$, and $c(t) = 1$. At $t = 0$, the extracellular concentration of solute is doubled, i.e., $c(t) = 2$. At $\alpha_v t = 3$, $c(t)$ is reduced back to 1. The parameter is $\alpha = \alpha_n / \alpha_v$. For all

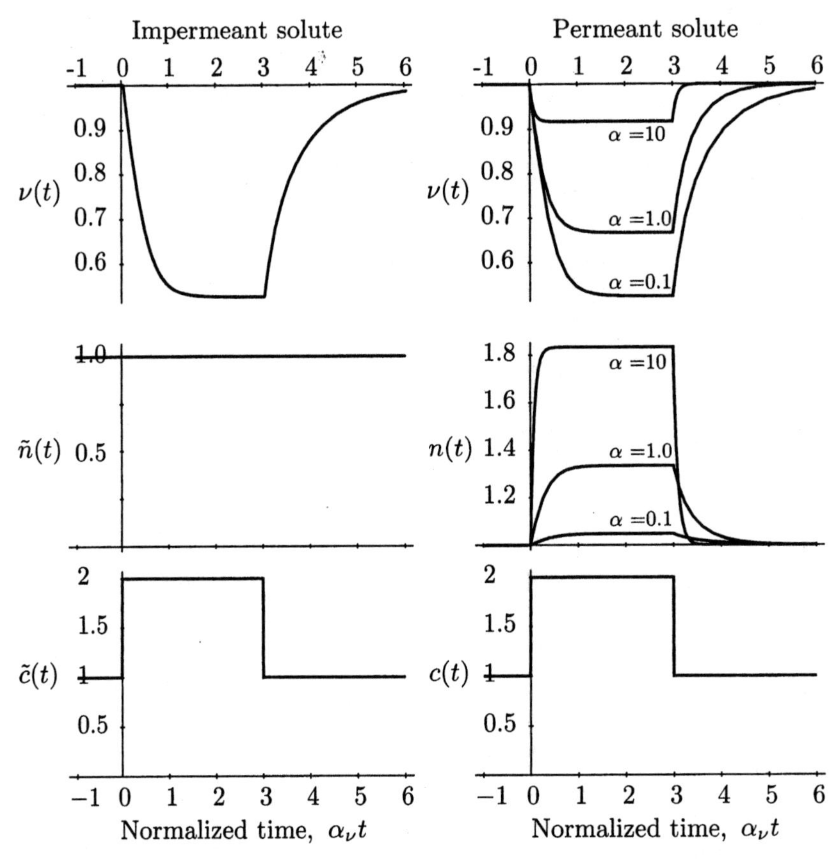

Figure 5.3 The cell volume and the quantity of intracellular solute as a function of time in response to a pulse change in concentration of a single impermeant (left panel) and a single permeant (right panel) solute. For both solutes, the volume is the normalized volume $v(t)$ and the time axis is normalized to $\alpha_v t$. For the impermeant solute, the quantity of intracellular solute is the normalized quantity $\tilde{n}(t)$. For the permeant solute, the solutions are shown for three values of the ratio α. The quantity of solute is the normalized quantity $n(t)$. The external concentration ($\tilde{c}(t)$ for the impermeant solute and $c(t)$ for the permeant solute) is increased to twice its isotonic value at $t = 0$ and returned to the isotonic value at $\alpha_v t = 3$.

values of this parameter, the increase in extracellular concentration of solute results in an increase in the quantity of solute in the cell and in a decrease of cell volume. The parameter α controls the relative percentage change in volume and solute quantity. Note that $\alpha = P_j/(\mathcal{L}_V RTC_\Sigma^{on})$ is a ratio of the rate constants for diffusion and osmosis. For a large value of α, the diffusion rate constant is large compared to that for osmosis. This range of α results in a relatively small decrease in volume but a large increase in intracellular solute quantity. For a small value of α, the diffusion rate constant is small compared to that for osmosis. This range of α results in a large change in volume and a small change in intracellular solute quantity.

If we take the quotient of Equations 5.6, then

$$\frac{dn(t)}{dt} = -\alpha\frac{dv(t)}{dt},$$

which shows that the time course of the change in volume is proportional to that of the change in solute quantity, i.e., the time dependence of $n(t)$ and $v(t)$ are the same. However, the relative magnitude of the rate of change is governed by α. To put it another way, diffusive and osmotic equilibrium occurs with transport of either water or solute or both. The degree of water versus solute transport to achieve equilibrium is governed by the quantity α. In the race between the flow of solute and the flow of water to achieve equilibrium, if the solute is slow, water flow will determine equilibrium before much solute has been transferred. If, on the other hand, the solute flow is fast, then solute flow will equilibrate the two solutions and not much water will be transferred.

5.2.2.3 One Permeant Solute in the Presence of Impermeant Solutes

With the introduction of multiple solutes, one of which is permeant, new phenomena occur. To see this, we explore the equations for concurrent transport:

$$\frac{dn(t)}{dt} = \alpha_n\left(c(t) - \frac{n(t)}{v(t)}\right),$$

$$\frac{dv(t)}{dt} = \alpha_v\left(\frac{n(t)+\tilde{n}(t)}{v(t)} - (c(t)+\tilde{c}(t))\right). \tag{5.7}$$

Numerical solutions to these equations are shown in Figures 5.4 and 5.5. In all of these computations, $\tilde{c}(t) = 1$ and $\tilde{n}(t) = 1$. The only change is the addition of a pulse of a permeant solute, which was not previously present in the system, to the extracellular solution beginning at $t = 0$ (lowest panel). Because the pulse amplitude of the permeant solute is 1, during the pulse interval the extracellular osmolarity is doubled. First, we note that doubling the extracellular osmolarity with a permeant solute results in a change in both the volume of the cell and the quantity of intracellular permeant solute (Figure 5.4). However, the volume change is transient. Doubling the osmolarity results in an initial efflux of water to shrink the cell. But the permeant solute diffuses into

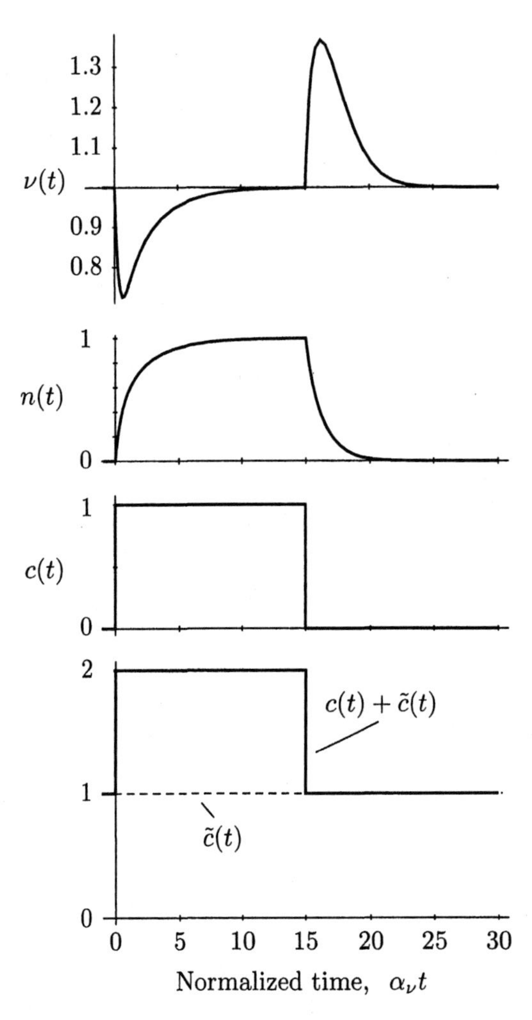

Figure 5.4 The cell volume and the quantity of intracellular solute as a function of time in response to a pulse change in concentration of a single permeant solute in the presence of impermeant solutes for $\alpha = 1$. The volume is the normalized volume $\nu(t)$; the quantity of solute is the normalized quantity $n(t)$; and the time axis is normalized to $\alpha_\nu t$. The external concentration of impermeant solute is constant, $\tilde{c}(t) = 1$. The external concentration of permeant solute $c(t)$ is zero for $t < 0$, is increased at $t = 0$ to double the osmolarity of the extracellular solution, and is returned to zero at $\alpha_\nu t = 15$.

the cell, as indicated by the increase in $n(t)$. Therefore, the intracellular osmolarity rises and water flows into the cell to restore the volume to its isotonic value. When, at $\alpha_\nu t = 15$, the extracellular concentration of permeant solute is reduced to zero, then the intracellular osmotic pressure is twice the extracellular osmotic pressure and water flows into the cell to equalize the osmotic pressure on the two sides of the membrane. However, the permeant solute now diffuses out of the cell and is followed by water to reduce the volume of the cell. Further insight is obtained from examination of the responses for both a poorly permeant and a highly permeant solute (Figure 5.5). For a poorly permeant solute ($\alpha = 0.01$), doubling the extracellular osmolarity results in a

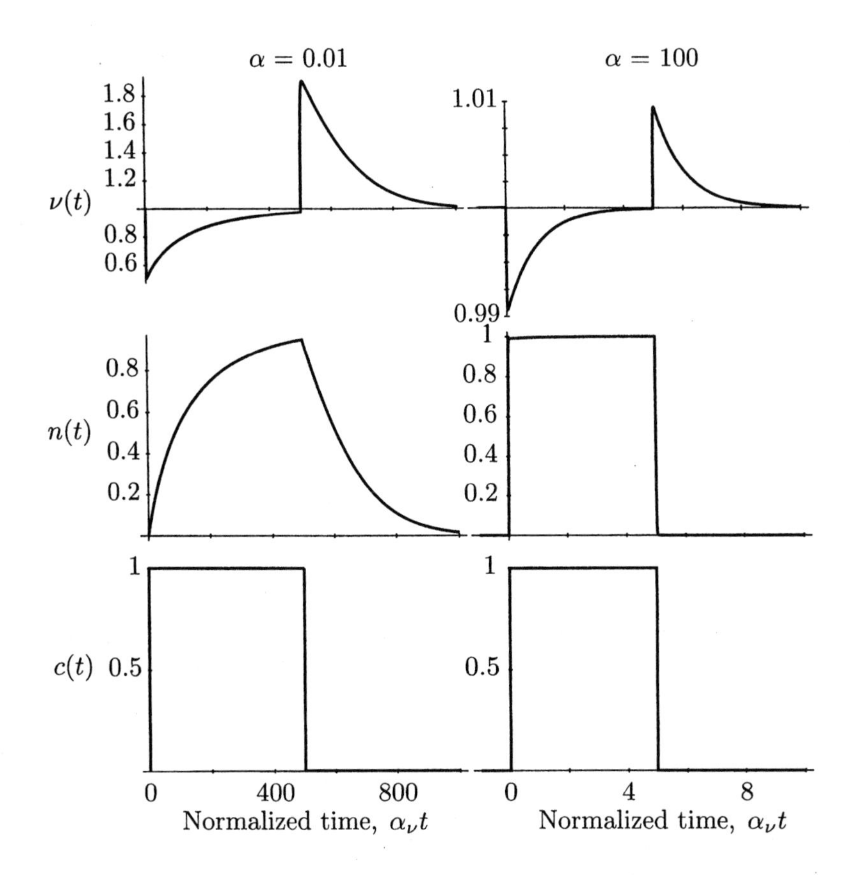

Figure 5.5 The normalized cell volume $\nu(t)$ and the normalized quantity of intracellular solute $n(t)$ as a function of time in response to a pulse change in normalized extracellular concentration $c(t)$ of a permeant solute in the presence of impermeant solutes with constant concentration $\tilde{c}(t) = 1$. Results are shown for two values of the relative permeability of the permeant solute: a relatively impermeant solute ($\alpha = 0.01$) and a relatively permeant solute ($\alpha = 100$). At $t = 0$, $c(t)$ increases from 0 to 1 and thereby doubles the osmolarity of the extracellular solution as shown in Figure 5.4.

rapid, large decrease in volume of about 50%, as we would expect for an impermeant solute. Solute starts to enter the cell following the initial volume transient, but at a rate that is much slower than the rate of initial volume decrease. The increase in intracellular solute increases the intracellular osmolarity, and hence the volume returns to its isotonic value ($\nu(t) = 1$) at a rate that is the same as that of the increase in quantity of solute. Thus, for a highly impermeant solute, water follows the diffusion of the solute at the rate of permeation of the solute. When the extracellular concentration of permeant solute is returned to zero, the intracellular osmolarity is now about double the extracellular osmolarity and there is a rapid, large increase in cell volume of about a factor of 2 as water flows into the cell to reduce the difference in osmotic pressure across the membrane. However, the difference in permeant solute concentration across the membrane now results in the diffusion of this solute out of the cell, and the volume once again returns to its isotonic value. For a highly permeant solute ($\alpha = 100$), the rapid influx of solute in response to the increase in extracellular osmolarity results in an equilibration of the osmotic pressure difference across the membrane, caused by solute influx. Thus, there is relatively little volume change. Note that doubling the osmolarity for the highly permeant solute leads to a change in volume of less than 1%. The time course of this small volume change is now much slower than that of the solute diffusion into the cell. Similarly, the response to removing the permeant solute results in a small, slow volume transient.

5.2.3 Measurements

The theory of concurrent solute and solvent transport might account for the measurements shown in Figure 5.2, at least qualitatively. For example, on the basis of those measurements we could conclude that both the solutes in Ringer's solution and xylose act as impermeant solutes for the time scale shown. Thus, these results can, in principle, be accounted for by Equation 5.5, whose solutions were examined in Chapter 4. Ethyl alcohol, which produced no discernible change in volume and is a highly permeant solute, might simply equilibrate before any water is transported, i.e., $\alpha \gg 1$. This type of response is shown in Figure 5.5 (right panel). Glycerol and urea have intermediate permeabilities, and hence they produce transient volume changes such as those shown in Figure 5.4.

A more quantitative comparison of measurements with the predictions of the theory of concurrent solute and solvent transport is shown in Figure 5.6 for egg cells of three species of marine invertebrates. The figure shows volume changes as a function of time in response to a hypertonic solution that contains the permeant solute ethylene glycol. The methods used were the same as the light diffraction methods described in Section 4.7.2. As indicated in that section, these egg cells are almost perfect spheres. Addition of the ethylene glycol solution results in an initial reduction in cell volume because the hypertonic medium causes an efflux of water from the cells. However, ethylene glycol is a permeant solute and enters these cells. The influx of ethylene glycol raises the intracellular osmotic pressure, which causes water to flow into these cells, which causes the cells to swell. As indicated in Figure 5.6, the equations of concurrent, uncoupled flow appear to fit these measurements adequately. Furthermore, these measurements yield values for the hydraulic conductivity of the membrane to water, $\mathcal{L}_V$. It is interesting to note that the hydraulic conductivity $\mathcal{L}_V$ of the membrane to water can be estimated both from measurements of volume changes caused by impermeant solutes (Figure 4.28) and from measurements with permeant solutes (Figure 5.6). For these invertebrate egg cells, the values of $\mathcal{L}_V$ obtained in both of these types of measurements are similar, as can be seen from the estimated values given in the captions of

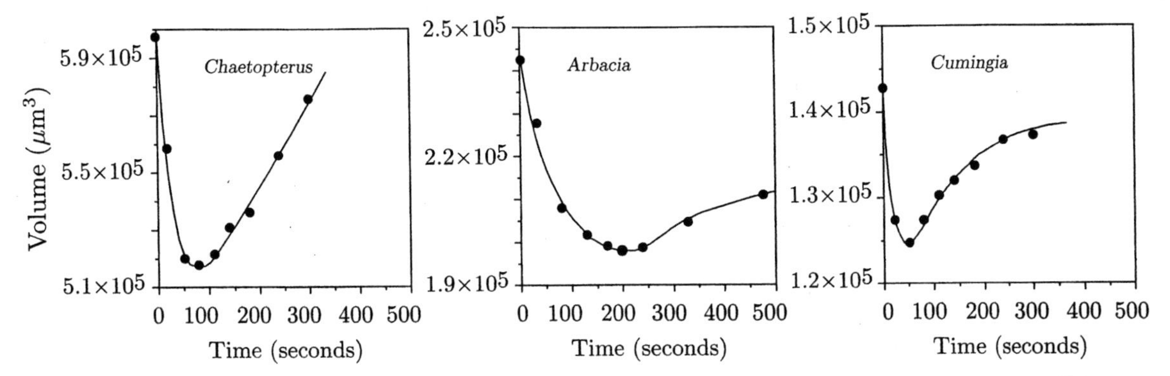

Figure 5.6 Change in volume of egg cells in response to a change in extracellular solution from seawater to seawater containing 0.5 mol/L ethylene glycol (adapted from Lucké et al., 1939, Figure 4). The results are shown for three species of marine animals: the sandworm *Chaetopterus pergamentaceus*, the sea urchin *Arbacia punctulata*, and the bivalve *Cumingia tellenoides*. The points are the measurements representing averages over a population of cells. The line is drawn according to the solutions to the equations of uncoupled concurrent transport for a spherical cell (Equations 5.3 and 5.4) with the parameters P (cm/s) and $\mathcal{L}_V$ (pm · s^{-1}·Pa) as follows: *Chaetopterus*, 2.6×10^{-5}, 7.7×10^{-2}; *Arbacia*, 5.3×10^{-6}, 2.8×10^{-2}; *Cumingia*, 2.5×10^{-5}, 6.7×10^{-2}.

those two figures. Thus, the theory of concurrent solute and solvent transport is certainly capable of accounting for the volume changes measured in some cells in response to changes in solute concentrations both for some impermeant and for some permeant solutes.

5.3 Inadequacies of Uncoupled Flow Equations

The equations of concurrent uncoupled transport of solute and solvent explain some, but not all, measurements of cellular volume changes. In the 1950s it became clear that these equations were incomplete and could not, in principle, account for certain phenomena (Kedem and Katchalsky, 1958).

5.3.1 Conceptual Problems

5.3.1.1 Tracers

In order to understand membrane transport mechanisms, there was great interest in measuring the diffusion of water through cellular membranes. In this type of experiment, cells are placed in radioactively labeled water, and the time course of influx of this radioactively labeled water is measured. The aim of using tracers is to produce a solute that is, for the purposes of interest, indistinguishable from the solute under study, except that the tracer can be detected. The permeability of the membrane to this tracer is estimated from the time constant for its equilibration across the membrane. In this process, the tracer is treated as a solute, and the equations of solute transport described in Chapter 3 are found to adequately fit measurements. These methods were used to estimate the permeability of the membrane to water. But it was found that the radioactively labeled water produced no change in volume over the time course during which there was a clear change in the intracellular quantity of radioactive water. This result seems reasonable, since the substitution of essentially identical solutions, one with tagged water and one without, ought not to produce an osmotic effect. This result is similar to that seen with ethyl alcohol, as shown in Figure 5.2, except that it appears unreasonable to explain the effect of the tracer by postulating that a transport of heavy water equilibrates the intracellular and extracellular solutions before any appreciable volume change occurs through the transport of untagged water. It appears that the heavy water produced *no osmotic effect*. The uncoupled equations of solute and solvent transport,

$$\phi_j(t) = \underbrace{P_j\left(c_j^i(t) - c_j^o(t)\right)}_{\text{diffusion}},$$

$$\Phi_V(t) = \underbrace{\mathcal{L}_V\left(p_i(t) - p_o(t)\right)}_{\text{hydraulic flow}} - \underbrace{\mathcal{L}_V RT\left(c_\Sigma^i(t) - c_\Sigma^o(t)\right)}_{\text{osmotic flow}}, \tag{5.8}$$

cannot account for a solute that produces no osmotic effect. Such solutes are termed *indistinguishable* because the membrane does not distinguish between them and the solvent.

5.3.1.2 The Uncoupled Equations Do Not Apply to Simple Porous Membranes

Examination of Equations 5.8, the equations of uncoupled transport, reveals a problem. The flux of solute contains only a diffusive term independent of the flux of volume. Thus, in the absence of a solute concentration across the membrane, a hydraulic pressure that results in volume flow through the membrane results in no transport of solute. That is, in this model, the membrane is a perfect sieve that allows solvent and not solute to permeate. Although this might be a reasonable model for some membranes, it clearly does not apply to simple porous membranes of arbitrary pore dimensions. Such a membrane would allow some transport of solute by convection. Thus, the equations appear to be missing a term that can account for convection of solute.

5.3.1.3 Different Estimates of the Permeability

Estimates of the permeability of the membrane to a solute gave somewhat different results in different types of experiments. For example, imagine the experiment with a moderately permeant solute such as glycerol (Figure 5.2c). The initial effect of the addition of this solute is to rapidly reduce the volume of the cell. Then the solute begins to penetrate slowly, and water enters essentially simultaneously with the solute because the membrane is so much more permeable to the water. Hence, the relatively slow time course of swelling is due to the slow permeation of this solute through the membrane. From this time course, one can estimate the permeability of the membrane to this solute (in the presence of water transport) as outlined in Section 5.2. One can also estimate the permeability of the membrane to the same solute in a pure diffusion experiment in which no water is transported—for example, by labeling some proportion of the solutes in some manner. By the 1950s, evidence was found that indicated that these two permeabilities, one estimated in the presence of water transport and the other in the absence of this transport, could have different values for certain solutes. Thus, the flux of solute through the membrane might depend not just on the concentration gradient of solute but also on the flux of water—apparently, for some solutes, the transport of solute and water were coupled.

5.3.2 The Distinction Between Uncoupled and Coupled Transport

We have drawn a distinction between concurrent, *uncoupled* transport of solvent and solute and concurrent, *coupled* transport. It is this distinction with which the remainder of this chapter is concerned. As we have seen in Section 5.2, both solute and solvent fluxes depend upon the concentration differences of solutes across the membrane. Thus, these fluxes are coupled *implicitly* through their dependence on common variables—the concentration differences. However, in the theories of uncoupled, concurrent flow, these fluxes are not coupled *explicitly*. By this we mean that a change in the volume flux (for example, as caused by a difference in hydraulic pressure across the membrane) does not directly result in a change in solute flux. In coupled flow, the fluxes are coupled explicitly.

To illustrate the distinction between coupled and uncoupled flow in a particularly simple case, consider the schematic diagram of a porous membrane shown in Figure 5.7. In uncoupled flow, we imagine the flux of solute

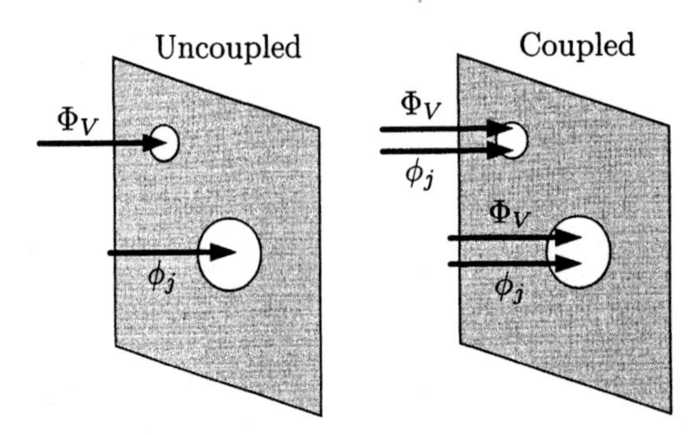

Figure 5.7 Illustration of the difference between uncoupled and coupled concurrent solute and solvent transport. ϕ_j is the flux of solute j, and Φ_V is the volume flux of solvent. In this schematic diagram the solute and solvent traverse different pathways, shown as holes, through the membrane when the fluxes are uncoupled but traverse the same pathways when the fluxes are coupled.

and solvent traversing different pathways, here shown as pores, through the membrane. In coupled flow both fluxes traverse the same pathway through the membrane, and hence they interact directly. Thus, in uncoupled flow the convection of solvent carries no solute through the membrane—the solute is sieved out perfectly. The solute traverses the membrane only by diffusion and not by convection. In the diagram this is illustrated by solute and solvent flow through separate pathways. In coupled flow, the convection of solution allows both the solute and the solvent to flow through the membrane by convection, and the two flows are coupled directly. In addition, the solute can also diffuse through the membrane.

5.3.3 Indistinguishable and Impermeant Solutes

When solute is transported both by convection and by diffusion, it is useful to consider three types of solutes as illustrated by the simple porous membrane shown in Figure 5.8. In this simple model, *impermeant* solutes have radii that exceed the pore radii. These solutes are not transported through the membrane by either convection or diffusion. *Indistinguishable* solutes are indistinguishable from water molecules. Hence, they diffuse through the membrane and are convected through the membrane as if they were water molecules. The membrane cannot distinguish these solutes from water molecules. *Intermediate* solutes are transported through the membrane by both convection and diffusion, but the solute transport is distinguishable from that of water transport.

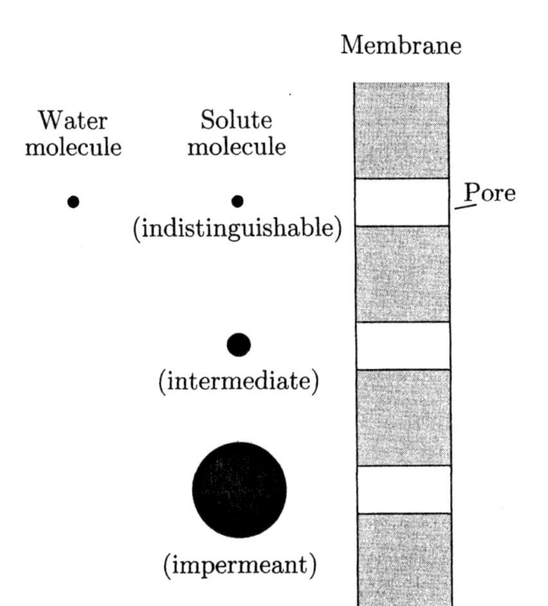

Figure 5.8 Schematic diagram of a porous membrane, water molecules, and solutes of different dimensions.

5.4 Diffusion and Convection Through a Porous Membrane: Indistinguishable Solute

In this section we consider solute and volume flux through a porous membrane when the solvent is convected through the membrane and the solute is both convected and diffuses for the condition that the solute and solvent are indistinguishable. We can imagine that the solute is a radioactive tracer of water molecules, e.g., tritium. Thus, water is the solvent and tritium is the solute. Since these molecules are so similar, it is reasonable to suppose that the membrane cannot tell them apart. This simple problem will indicate the role of convection in solute transport and will be a springboard for a more general formulation of the equations of coupled solute and solvent transport.

5.4.1 Derivation of Flux Equations

For solute molecules that are indistinguishable from the solvent molecules, the volume flow is due to flow of the *solution*, and hence is due only to the hydraulic pressure. Thus, the volume flow is simply

$$\Phi_V = \mathcal{L}_V \Delta p, \tag{5.9}$$

where $\Delta p = p_1 - p_2$. If the pores are sufficiently large so that flow through the pores can be computed from continuum hydrodynamics, $\mathcal{L}_V$ can be computed directly from knowledge of the pore geometry and density.

If there are $\mathcal{N}$ pores per unit area of membrane, and each pore is a cylinder of radius r and length d (Figure 4.8), then Poiseuille's law (Section 4.5) can be used to show that

$$\mathcal{L}_V = \mathcal{N}\frac{\pi r^4}{8\eta d}. \tag{5.10}$$

There are two mechanisms of transport of indistinguishable solutes across porous membranes. Solute is transported with the solution by bulk flow or convection, and by diffusion due to a solute concentration gradient. The total flux of solute j is a sum of these fluxes:

$$\phi_j = \underbrace{c_j(x)\Phi_V}_{\text{convection}} \underbrace{- D_j\mathcal{N}\pi r^2\frac{dc_j(x)}{dx}}_{\text{diffusion}}, \tag{5.11}$$

where the first term on the right-hand side of Equation 5.11 is solute flux resulting from convection of the solution, and the second term is the solute flux due to diffusion in the pores. The factor $\mathcal{N}\pi r^2$ is the fraction of the membrane surface area that consists of pores. D_j is the diffusion coefficient of the solute in the pore. We shall assume that the fluxes change sufficiently slowly that steady-state conditions hold in the membrane. Thus, ϕ_j and Φ_V are constant and independent of position, x, in the membrane. We shall find the solution to Equation 5.11 subject to the boundary condition

$$c_j(0) = c_j^1, \text{ and } c_j(d) = c_j^2. \tag{5.12}$$

This problem is identical to the convection and diffusion of tracer water, a problem whose solution was obtained in Section 4.5. The concentration as a function of position in the pore is

$$c_j(x) = \frac{(c_j^1 - c_j^2)e^{\gamma x} + (c_j^2 - c_j^1 e^{\gamma d})}{1 - e^{\gamma d}}, \tag{5.13}$$

where $\gamma = \Phi_V/(D_j\mathcal{N}\pi r^2) = (\mathcal{L}_V\Delta p)/(D_j\mathcal{N}\pi r^2)$. The water flux is

$$\phi_j = \Phi_V\frac{c_j^2 - c_j^1 e^{\gamma d}}{1 - e^{\gamma d}}, \tag{5.14}$$

which can also be written as

$$\phi_j = \mathcal{L}_V\frac{c_j^2 - c_j^1 e^{\beta\Delta p}}{1 - e^{\beta\Delta p}}\Delta p, \tag{5.15}$$

where

$$\beta = \frac{\mathcal{L}_V}{P_j}, \tag{5.16}$$

and

$$P_j = \frac{D_j \mathcal{N} \pi r^2}{d}. \tag{5.17}$$

5.4.2 The Linearized Equation of Coupled Flow for an "Indistinguishable" Solute

Even for the relatively simple case of flow through a porous membrane, we have found that the flux of an indistinguishable solute (Equation 5.15) does not obey Fick's law for membranes (Equation 3.44). Furthermore, the flux depends nonlinearly on the hydraulic pressure difference. This result shows that even relatively simple membranes can yield fairly complex relations between solute flux and solute concentration. In this section we consider a special and simple case of transport of indistinguishable solutes through porous membranes in which the solute flux is the sum of two fluxes. One flux is proportional to concentration difference and represents diffusion through the porous membrane; the other flux is proportional to the hydraulic pressure difference and represents convection through the porous membrane. That is, we shall linearize the flux equations.

Let us define

$$\overline{C_j} = \frac{c_j^1 + c_j^2}{2}, \tag{5.18}$$

$$\frac{\Delta c_j}{2} = \frac{c_j^1 - c_j^2}{2}. \tag{5.19}$$

Then

$$c_j^1 = \overline{C_j} + \Delta c_j/2 \quad \text{and} \quad c_j^2 = \overline{C_j} - \Delta c_j/2. \tag{5.20}$$

Substitution of Equation 5.20 into 5.15 yields

$$\phi_j = \mathcal{L}_V \frac{(\overline{C_j} - \Delta c_j/2) - (\overline{C_j} + \Delta c_j/2)e^{\beta \Delta p}}{1 - e^{\beta \Delta p}} \Delta p, \tag{5.21}$$

$$\phi_j = \mathcal{L}_V \overline{C_j} \Delta p + \mathcal{L}_V \frac{\Delta c_j}{2} \left(\frac{1 + e^{\beta \Delta p}}{1 - e^{\beta \Delta p}} \right) \Delta p. \tag{5.22}$$

Now suppose the transport is near equilibrium, so that the flux is small. For this case, let

$$c_j^1 = C_j + \Delta^1 \quad \text{and} \quad c_j^2 = C_j + \Delta^2,$$

where we assume $\Delta^1 \ll C_j$ and $\Delta^2 \ll C_j$. We shall also assume that $\beta \Delta p \ll 1$, so that Equation 5.22 can be written as

$$\phi_j \approx \mathcal{L}_V \overline{C}_j \Delta p - \mathcal{L}_V \frac{\Delta c_j}{2} \left(\frac{1 + (1 + \beta \Delta p)}{1 - (1 + \beta \Delta p)} \right) \Delta p,$$

$$\phi_j \approx \mathcal{L}_V \overline{C}_j \Delta p + \frac{\mathcal{L}_V}{\beta} \Delta c_j = \mathcal{L}_V \overline{C}_j \Delta p + P_j \Delta c_j. \tag{5.23}$$

Thus, two macroscopic parameters, $\mathcal{L}_V$ and P_j, characterize the flux of volume and solute through the membrane.

Equations 5.9 and 5.23 describe transport of solvent and solute near equilibrium for indistinguishable solutes. The flux through the porous membrane has a relatively simple characterization for these solutes. The volume flux is proportional to the hydraulic pressure difference. The solute flux is a superposition of linear diffusion and linear convection. The solute flux equation can also be written in another form. Substitution of Equation 5.9 into 5.23 yields

$$\phi_j = \overline{C}_j \Phi_V + P_j \Delta c_j. \tag{5.24}$$

This equation is the starting point for a discussion of more generalized equations of coupled flow.

5.5 The Kedem-Katchalsky Equations for Linear, Coupled Flow Through a Membrane

5.5.1 Macroscopic Laws of Transport

The equations for flux through a porous membrane for the special cases corresponding to impermeant solutes (Equation 5.2) and indistinguishable solutes under linear conditions (Equations 5.9 and 5.23) describe a linear relation between two *flow* variables, Φ_V and ϕ_j, and two *force* variables, Δp and Δc_j, in terms of two constitutive constants, $\mathcal{L}_V$ and P_j. However, they do not describe the most general possible linear relation between two flow and two force variables. The most general description of linear, coupled flow having two flux variables and two forces is

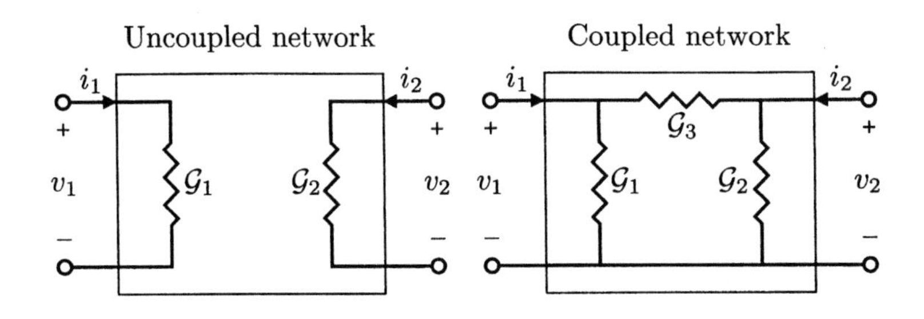

Figure 5.9 Illustration of uncoupled and coupled electrical networks. v_1 and i_1 are the voltage and current at port 1; v_2 and i_2 are the voltage and current at port 2; the $\mathcal{G}$s are conductances.

$$\Phi_V = L_{11}\Delta p + L_{12}\Delta c_j,$$
$$\phi_j = L_{21}\Delta p + L_{22}\Delta c_j, \tag{5.25}$$

and there are four constants that specify the relation of flux to force. One can specify general theoretical conditions for which the flow system is reciprocal (i.e., for which the Onsager relations hold) in the appropriate force and flow variables (Katchalsky and Curran, 1965). Under these conditions, only three constants are required to specify the flow. Pragmatically, linear coupled flow equations with three constitutive constants are the minimum necessary to eliminate internal inconsistencies of coupled flow measurements described in Section 5.3. We explore these equations in the remainder of this section.

The description of linear, coupled reciprocal systems with three constitutive constants is exactly analogous to the description of electrical networks having two ports (Figure 5.9). Here the distinction between systems that can be described by two rather than three constitutive constants is quite clear. Consider the electrical networks in Figure 5.9. The relation between the two currents and the two voltages can be expressed in terms of the conductances. For the uncoupled network,

$$i_1 = \mathcal{G}_1 v_1,$$
$$i_2 = \mathcal{G}_2 v_2,$$

and therefore, the current at port 1 depends only on the voltage at port 1, and the current at port 2 depends only on the voltage at port 2. The addition of a conductance that bridges the two ports produces a coupled network for which

$$i_1 = (\mathcal{G}_1 + \mathcal{G}_3)v_1 - \mathcal{G}_3 v_2,$$
$$i_2 = -\mathcal{G}_3 v_1 + (\mathcal{G}_2 + \mathcal{G}_3)v_2,$$

so that both currents depend upon both voltages. In general, three conductances are required to specify the relations of currents to voltages in an arbi-

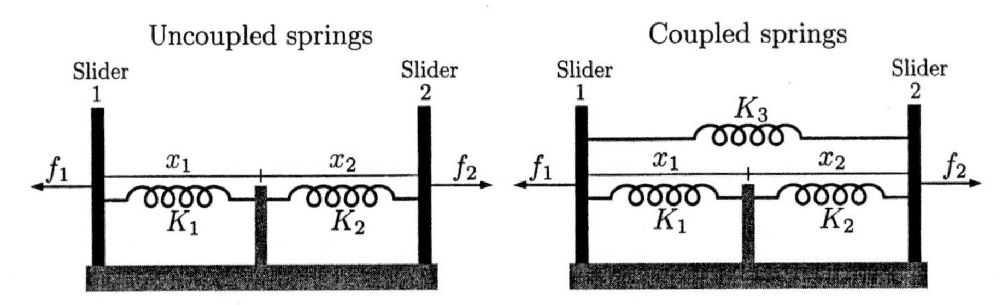

Figure 5.10 Illustration of uncoupled and coupled mechanical systems. f_1 and x_1 are the force on and position of slider number 1; f_2 and x_2 are the force on and position of slider number 2; the Ks are spring stiffnesses.

trary reciprocal, resistive two-port electrical network such as that in Figure 5.9 (right panel).

A similar analysis applies to a mechanical system with two forces resulting in the displacement of two structures coupled by springs (Figure 5.10). Let the force on the springs be zero when $x_1 = l_1$ and $x_2 = l_2$ and the increment in spring lengths be $y_1 = x_1 - l_1$ and $y_2 = x_2 - l_2$. Therefore, for the uncoupled system the relation between the two forces and the incremental displacements is

$$f_1 = K_1 y_1,$$
$$f_2 = K_2 y_2,$$

and the force on spring 1 depends only on the displacement of slider 1 and not on the displacement of slider 2. The addition of another spring couples the slider motions so that

$$f_1 = (K_1 + K_3) y_1 + K_3 y_2,$$
$$f_2 = K_3 y_1 + (K_2 + K_3) y_2,$$

where the force on spring 3 is zero when $x_1 + x_2 = l_3$ and for the condition that $l_3 = l_1 + l_2$. In the coupled system, the force on each slider depends on the displacements of both sliders. In general, three stiffnesses are required to specify the relations between the forces on and displacements of two sliders that are spring-loaded such as that in Figure 5.10 (right panel).

5.5.1.1 *Single Solute*

The linearized, reciprocal equations for coupled flow of solvent and for a single solute species, known as the Kedem-Katchalsky equations, are usually written as follows:

$$\Phi_V = \underbrace{\mathcal{L}_V \Delta p}_{\text{hydraulic flow}} \underbrace{- \sigma_j \mathcal{L}_V RT \Delta c_j}_{\text{osmotic flow}}, \tag{5.26}$$

$$\phi_j = \underbrace{\overline{C}_j (1 - \sigma_j) \Phi_V}_{\text{convection}} + \underbrace{P_j \Delta c_j}_{\text{diffusion}} . \tag{5.27}$$

Three constants define the relation between fluxes and forces: $\mathcal{L}_V$ and P_j plus the new transport parameter σ_j, which is called the *reflection coefficient*. As we shall show, σ_j is a measure of the ability of the membrane to discriminate the solute from the solvent. The equation for volume flux consists of two terms: a hydraulic flow that results from a hydraulic pressure difference and an osmotic flow that results from a difference in osmotic pressure. Similarly, the solute flux consists of two terms: a term that represents solute convection resulting from solution transport and a term that represents solute diffusion.

Special Cases

Next we show that the Kedem-Katchalsky equations *do* include the linearized flux equations for impermeant and indistinguishable solutes, as special cases.

Impermeant Solutes; $\sigma_j = 1$ and $P_j = 0$ By an impermeant solute we mean one to which the membrane is impermeable. Since the Kedem-Katchalsky equations contain two mechanisms that transport the solute, both must yield no transport for an impermeant solute. The condition $P_j = 0$ guarantees that no solute will be transported by diffusion, and the condition $\sigma_j = 1$ (i.e., all the solutes are reflected from the membrane) guarantees that no solute will be convected across the membrane. Under these conditions, even if there is flow of volume, the solute is filtered out. Substituting these values into Equations 5.26 and 5.27 makes those equations identical to Equation 5.2, as expected. Note that impermeant solutes exert their *full osmotic effect,* since the term $\sigma_j RT \Delta c_j$ equals the osmotic pressure difference when $\sigma_j = 1$.

Indistinguishable Solutes; $\sigma_j = 0$ Indistinguishable solutes are not reflected at all by the membrane, hence $\sigma_j = 0$. The molecules flow through the membrane with the solvent and are not filtered or sieved out of the solvent by the membrane. Substitution of this condition into Equations 5.26 and 5.27 reduces them to Equations 5.9 and 5.24.

Interpretation of the Reflection Coefficient

The significance of $\mathcal{L}_V$ and P_j has already been discussed. The reflection coefficient, σ_j, which was first introduced by Staverman (1951), can be interpreted phenomenologically by examining each of the Kedem-Katchalsky equations. From Equation 5.26, we obtain

$$\sigma_j = \left(\frac{\Delta p}{RT\Delta c_j}\right)_{\Phi_V=0}. \tag{5.28}$$

Therefore, σ_j is a measure of the effectiveness of a solute in producing an osmotic pressure at osmotic equilibrium. If $\sigma_j = 1$, then the solute produces its full osmotic effect in the following sense: to prevent volume flow due to some hydraulic pressure difference Δp we require a concentration difference such that $RT\Delta c_j = \Delta p$. However, if $\sigma_j < 1$, then we require a larger concentration difference such that $RT\Delta c_j = \Delta p/\sigma_j$, i.e., the solute does not exert its full osmotic effect. Therefore, we can regard the term $\sigma_j RT\Delta c_j$ as the *effective osmotic pressure.*

The reflection coefficient can also be given an interpretation by examining Equation 5.27. From the convective component of the solute flux, note that if $\sigma_j = 1$ this term is zero and no solute is convected as the solution flows through the membrane, i.e., the solute is filtered or sieved out. Conversely, if $\sigma_j = 0$ then no solute is filtered out as the solution flows through the membrane. Thus, σ_j is a measure of the filtration of the membrane when the transport is dominated by convection, i.e., when diffusion is negligible.

Summary of Main Conclusions

Therefore, we see that the Kedem-Katchalsky equations include the impermeant and indistinguishable solutes as special cases of linear, coupled solute and solvent transport. This formulation of volume and solute flux through a membrane requires that the membrane transport be characterized by three quantities, $\mathcal{L}_V$, P_j, and σ_j.

5.5.1.2 Multiple Solutes

If a number of different types of solutes are simultaneously transported along with the solvent, the linearized, coupled equations of Kedem and Katchalsky take the form

$$\Phi_V = \mathcal{L}_V(\Delta p - RT\sum_k \sigma_k \Delta c_k), \tag{5.29}$$

$$\phi_k = \overline{C}_k(1-\sigma_k)\Phi_V + \sum_i P_{ki}\Delta c_i, \tag{5.30}$$

where the subscript identifies the solute species. Equation 5.30 allows not only for interactions between the transport of solute and solvent but also for interactions between different species of solute.

5.5.1.3 *One Permeant and Multiple Impermeant Solutes*

Equations 5.29 and 5.30 can be put into a somewhat simpler form for the situation where all but one solute are impermeant. Let the generic solute be indicated by the index k where the solute j is the only permeant species. Therefore, $\sigma_k = 1$ for $k \neq j$ so that

$$\Phi_V = \mathcal{L}_V(\Delta p - RT \sum_{k \neq j} \Delta c_k - \sigma_j RT \Delta c_j),$$

$$\phi_j = \overline{C}_j(1 - \sigma_j)\Phi_V + P_j \Delta c_j, \quad (5.31)$$

$$\phi_k = 0 \quad \text{for } k \neq j. \quad (5.32)$$

5.5.2 Microscopic Mechanisms of Transport of Water and a Permeant Solute in Simple Membrane Models

In this section we explore two simple mechanisms for concurrent water and solute transport through a membrane: (1) the dissolve and diffuse mechanism of permeation in a homogeneous, hydrophobic membrane, and (2) water and solute transport through a membrane whose matrix is impermeant to both the water and the solute but which contains pores that are permeant to both. The discussion in this section parallels the discussion in Section 4.5 but considers the case when both the solute and water permeate the membrane (Finkelstein, 1987). We consider a simple membrane as shown in Figure 5.11. A single solute and water permeate this membrane. The purpose of this section is to provide some further physical intuition for the reflection coefficient σ in terms of two simple microscopic mechanisms of transport.

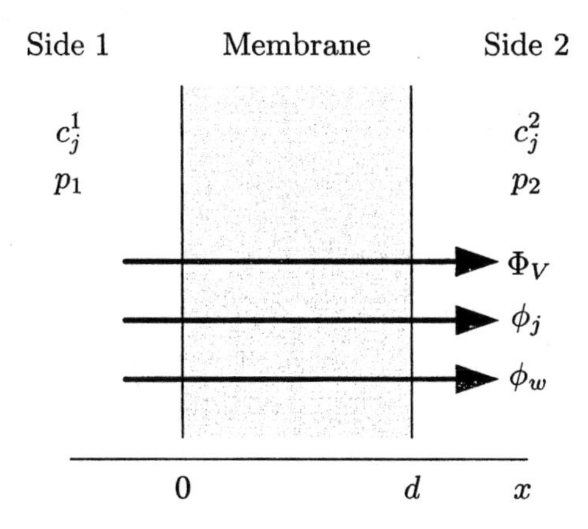

Figure 5.11 Definition of variables for steady-state solute and water transport through a membrane. The membrane allows a single solute (solute j) and water to permeate.

5.5.2.1 *Transport of Water and a Permeant Solute in a Homogeneous, Hydrophobic Membrane*

We consider a homogeneous, hydrophobic membrane through which both water and the solute diffuse independently. Therefore, the flux of water and solute j from side 1 to side 2 inside the membrane is given by Fick's law for membranes:

$$\phi_W = \frac{D_W(h)}{d}\left(c_W^1(h) - c_W^2(h)\right) \text{ and } \phi_j = \frac{D_j(h)}{d}\left(c_j^1(h) - c_j^2(h)\right),$$

where $D_W(h)$ and $D_j(h)$ are the diffusion coefficients for water and solute j in the membrane; and $c_W^1(h)$, $c_W^2(h)$, $c_j^1(h)$, and $c_j^2(h)$ are the concentrations of water and solute j in the membrane at the interface with sides 1 and 2. Using the results shown in Section 4.5 and Appendix 4.1, these equations can be expressed in terms of the solute concentration on the two sides of the membrane as follows:

$$\phi_W = -\mathcal{P}_W\left(c_j^1 - c_j^2\right) \text{ and } \phi_j = P_j\left(c_j^1 - c_j^2\right), \tag{5.33}$$

where $\mathcal{P}_W = D_W(h)k_W\overline{V}_W/(d\overline{V}_h)$ and $P_j = D_j(h)k_j/d$; k_W is the partition coefficient of water in the membrane to that in the solutions; k_j is the partition coefficient of solute j in the membrane to that in the solutions; and $\overline{V}_W$ and $\overline{V}_h$ are the partial molar volumes of water and of the hydrophobic solvent that makes up the membrane. The total volume flux is, in general, carried by both water and solute and is

$$\Phi_V = \overline{V}_W\phi_W + \overline{V}_j\phi_j. \tag{5.34}$$

Substitution of Equation 5.33 into 5.34, rearrangement of terms, and use of the property that for a homogeneous, hydrophobic membrane $\mathcal{P}_W = P_W$, yields

$$\Phi_V = -\mathcal{P}_W\overline{V}_W\left(1 - \frac{P_j\overline{V}_j}{P_W\overline{V}_W}\right)\left(c_j^1 - c_j^2\right).$$

Multiplication and division by RT and simplifying terms yields the relation

$$\Phi_V = -\mathcal{L}_V\sigma_j\left(\pi^1 - \pi^2\right),$$

where

$$\sigma_j = 1 - \frac{P_j\overline{V}_j}{P_W\overline{V}_W}. \tag{5.35}$$

Thus, for a homogeneous, hydrophobic membrane in which both water and solute j diffuse independently, they diffuse in opposite directions. Therefore, the volume flux is less than the volume flux due to water alone because some volume is transported in the opposite direction by the solute. This effect is summarized by $\sigma_j < 1$, and the deviation is quantified by the ratio of the product of permeability and the molar volume for the solute and water. Equation 5.35 gives an intuitive interpretation of the reflection coefficient. If the solute permeability is arbitrarily smaller than the permeability of water, then σ_j approaches arbitrarily close to 1. If the solute is indistinguishable from water, as in the case of a tracer for the water, so that $P_j = P_w$ and $\overline{V}_j = \overline{V}_w$, then $\sigma_j = 0$. Furthermore, if the solute is more permeant than water—more precisely, if $P_j\overline{V}_j > P_w\overline{V}_w$—then volume flows in a direction opposite to the direction of water flow, and $\sigma_j < 0$.

5.5.2.2 Transport of Water and a Permeant Solute in a Porous Membrane

The transport of water and permeant solute in a porous membrane is reviewed elsewhere (Finkelstein, 1987) and is discussed in detail in a number of publications (Anderson and Malone, 1974; Anderson, 1981; Anderson and Brannon, 1981; Adamski and Anderson, 1983; Anderson and Adamski, 1983; Yan et al., 1986; Adamski and Anderson, 1987; Guell, 1991). These and earlier studies (Vegard, 1910; Garby, 1957) propose that water flow caused by a permeant solute—as for that caused by an impermeant solute—results because a solute concentration gradient at the pore entrance produces a hydraulic pressure gradient that drives water through the pore. However, a permeant solute produces a smaller pressure gradient than does an impermeant solute of the same osmolarity. Thus, the effective osmotic pressure across the membrane is less than the difference of osmotic pressure across the membrane. A simple version of this argument is presented in further detail below. However, this picture of osmosis resulting from an impermeant solute is not without controversy (Hill, 1982; 1989a; 1989b; 1994).

What Are the Forces That Drive Water Through a Porous Membrane When the Osmotic Pressure is Due to a Permeant Solute?

The discussion of water flow due to a permeant solute follows closely that for an impermeant solute, which is found in Section 4.5. We consider a porous membrane made of an impermeant matrix and pores through which both water and the solute j can permeate, as shown in Figure 5.11. The solutions on the two sides of the membrane have concentrations c_j^1 and c_j^2. Because the pore is permeable to the solute, the concentration of solute in the pore is not zero as it is for an impermeant solute (see Figure 4.9). However, for steric or other reasons, the concentration of solute in the pore is less than that in the bath. We define that relation between these concentrations by a partition coefficient, so that $k_j = c_j(0)/c_j^1 = c_j(d)/c_j^2$. Once again, we assume that $p - RTc_j$ is continuous at each interface between the porous membrane and the baths. Therefore,

$$p_1 - RTc_j^1 = p(0) - RTc_j(0) \text{ and } p_2 - RTc_j^2 = p(d) - RTc_j(d).$$

Rearranging terms, we obtain

$$p_1 - p(0) = RT\left(c_j^1 - c_j(0)\right) \text{ and } p_2 - p(d) = RT\left(c_j^2 - c_j(d)\right).$$

Using the boundary conditions, we obtain

$$p_1 - p(0) = RTc_j^1(1 - k_j) \text{ and } p_2 - p(d) = RTc_j^2(1 - k_j).$$

Therefore, the difference in pressure between the two ends of the pore is

$$p(0) - p(d) = (p_1 - p_2) - \sigma_j RT\left(c_j^1 - c_j^2\right), \tag{5.36}$$

where $\sigma_j = 1 - k_j$. Equation 5.36 shows that the hydraulic pressure in the pore, which drives water through the pore, has a component that equals the osmotic pressure difference across the membrane multiplied by the reflection coefficient.

5.6 Coupled Solute and Solvent Transport for a Cell

We are now in a position to discuss the changes in volume and in concentration of permeant solutes for a cell, given that the transport is coupled. We shall develop the relevant equations first and then examine measurements.

5.6.1 Theory

We derive the transport equations for a single permeant solute in the presence of impermeant solutes using the equations of coupled membrane transport of Kedem and Katchalsky. We designate a generic solute species as k and the permeant solute as j. Therefore, $\sigma_k = 1$ for $k \neq j$, and we call the reflection coefficient of the permeant species σ_j. We use the same notation as in Section 5.2 and will derive equations in normalized variables. We also assume that the hydraulic pressure difference across the membrane is zero ($\Delta p = 0$).

5.6.1.1 Solvent Transport

The volume flux satisfies conservation of volume and a constitutive relation now given by the Kedem-Katchalsky equation to yield

$$\Phi_V = -\frac{1}{A(t)}\frac{d\mathcal{V}^i(t)}{dt} = \mathcal{L}_V RT \sum_{k \neq j} \left(c_k^o(t) - c_k^i(t)\right) + \sigma_j \mathcal{L}_V RT (c_j^o - c_j^i). \tag{5.37}$$

If we divide Equation 5.37 by $\mathcal{V}^{in}$ and rearrange the terms, we obtain

$$\frac{dv(t)}{dt} = \alpha_v \mathcal{A}(t) \left(\frac{\sigma_j n(t) + \tilde{n}(t)}{v(t)} - \left(\sigma_j c(t) + \tilde{c}(t)\right)\right), \tag{5.38}$$

where $n(t) = n_j^i(t)/(C_\Sigma^{in}\mathcal{V}^{in})$, which is the ratio of the intracellular quantity of permeant solute to the total isotonic solute quantity; $\tilde{n}(t) = \sum_{k \neq j} n_k^i(t)/(C_\Sigma^{in}\mathcal{V}^{in})$ is the ratio of the intracellular quantity of solute that is impermeant to the total isotonic solute; $c(t) = c_j^o(t)/C_\Sigma^{on}$ is the ratio of the extracellular concentration of permeant solute to the total isotonic extracellular concentration; $\tilde{c}(t) = \sum_{k \neq j} c_j^o(t)/C_\Sigma^{on}$ is the ratio of the extracellular concentration of solute that is impermeant to the total isotonic concentration; $\mathcal{A}(t) = A(t)/A^n$ is the surface area of the cell normalized to its isotonic value; $v(t) = \mathcal{V}^i(t)/\mathcal{V}^{in}$ is the ratio of the volume to the isotonic volume; and $\alpha_v = \mathcal{L}_V RT C_\Sigma^{on} A^n/\mathcal{V}^{in}$. This equation is similar to the comparable equation for concurrent, uncoupled transport (Equation 5.4) except for the factor of σ_j that multiplies two terms.

5.6.1.2 Solute Transport

The solute flux satisfies conservation of volume and a constitutive relation now given by the Kedem-Katchalsky equation to yield

$$\phi_k(t) = \begin{cases} -\frac{1}{A}\frac{dn_k^i(t)}{dt} = \overline{C}_k(1 - \sigma_k)\Phi_V + P_k\left(c_k^i(t) - c_k^o(t)\right) & \text{for } k = j \\ 0 & \text{for } k \neq j\,. \end{cases} \tag{5.39}$$

If we substitute the average concentration for $\overline{C}_k$, substitute the expression for Φ_V from the equation of volume transport, divide the equation for the permeant solute by the total quantity of solute under isotonic conditions $C_{\Sigma}^{in}\mathcal{V}^{in} = C_{\Sigma}^{on}\mathcal{V}^{in}$, and rearrange the terms, we obtain

$$\frac{dn(t)}{dt} = \alpha_n \mathcal{A}(t)\left(c(t) - \frac{n(t)}{v(t)}\right)$$
$$+ \alpha_v \mathcal{A}(t)\left(\frac{1-\sigma_j}{2}\right)\left(c(t) + \frac{n(t)}{v(t)}\right)\left(\frac{\sigma_j n(t) + \tilde{n}(t)}{v(t)}\right.$$
$$\left. - \left(\sigma_j c(t) + \tilde{c}(t)\right)\right), \qquad (5.40)$$

where $\alpha_n = P_j A^n/\mathcal{V}^{in}$. Note that the first term on the right-hand side of Equation 5.40 is due to solute diffusion, and the second term is due to solute convection.

5.6.1.3 Properties of the General Equations

It is now important to check that these equations can, in principle, eliminate the ambiguities that can arise with the uncoupled equations and that these equations are truly a generalization of our previous equations for transport across cellular membranes.

Impermeant Solutes

If the solute j is impermeant, so that all solutes are impermeant, then $\sigma_j = 1$, $P_j = 0$, $n(t) = 0$, $\tilde{n}(t) = 1$, and $c(t) = 0$, so that Equations 5.38 and 5.40 become

$$\frac{dv(t)}{dt} = \alpha_v \mathcal{A}(t)\left(\frac{1}{v(t)} - \tilde{c}(t)\right),$$
$$\frac{dn(t)}{dt} = 0, \qquad (5.41)$$

which reduces to the equations for an impermeant solute developed in Sections 4.7.2 and 5.2. Solutions to these equations are shown in Figure 4.26. Solutions for the volume resemble the measurements shown in Figure 5.2 for doubling the osmolarity using either Ringer's solution or xylose. Both solutes give a persistent decrease in volume in response to doubling the osmolarity.

Indistinguishable Solute

For a permeant solute that is indistinguishable from water ($\sigma_j = 0$), the equation for volume transport becomes

$$\frac{dv(t)}{dt} = \alpha_v A(t) \left(\frac{\tilde{n}(t)}{v(t)} - \tilde{c}(t) \right). \tag{5.42}$$

These equations can in principle explain the effect of radioactive tracers that diffuse through the membrane but produce no change in volume. This type of explanation could also account for the measurements with ethyl alcohol shown in Figure 5.2. In that experiment, the muscle fiber was in an isotonic medium and then the extracellular osmolarity was doubled with the addition of ethyl alcohol. Let us assume that ethyl alcohol can be regarded as an indistinguishable solute with a value of $\sigma_j = 0$. Assume that the other solutes, both intracellular and extracellular, are impermeant: then $\tilde{n}(0)/v(0) = \tilde{c}(0)$. Then Equation 5.42 indicates that $dv(t)/dt = 0$ at $t = 0$, so that the cell volume remains constant. Thus, if a cell is in osmotic equilibrium, the addition of a solute with a $\sigma_j = 0$ will not result in a volume change.[2]

The equation for solute diffusion for an indistinguishable solute simplifies greatly. Because $\tilde{n}(t)/v(t) = \tilde{c}(t)$, only one term arises:

$$\frac{dn(t)}{dt} = \alpha_n A(t) \left(c(t) - \frac{n(t)}{v(t)} \right),$$

which is the equation for solute transport in the absence of water transport.

5.6.1.4 Solutions for a Cell with Constant Surface Area

The equations for coupled transport of solute and solvent across the membrane of a cell are generalizations of the equations for uncoupled but concurrent transport, which were examined in Section 5.2. The new notion in the coupled equations is the inclusion of the parameter σ, which simultaneously allows for filtration of solute convection and for the possibility that certain solutes do not exert their full osmotic effect. The effect of σ on the transport of solute and solvent is examined in Figure 5.12. The left-hand panels show responses for a solute whose transport is uncoupled to that of solvent ($\sigma = 1$), a solute that is indistinguishable from the solvent ($\sigma = 0$), and an intermediate solute ($\sigma = 0.5$). As expected, the indistinguishable solute produces no change in volume. The other solutes produce a decrease in volume as the extracellular osmolarity is raised by the permeant solute. The intermediate solute produces

2. There is one caveat. In our development of the equations of solute and solvent transport, it has been assumed that the transport of solute does not contribute to a change in cell volume; i.e., only a transport of water changes the cell volume.

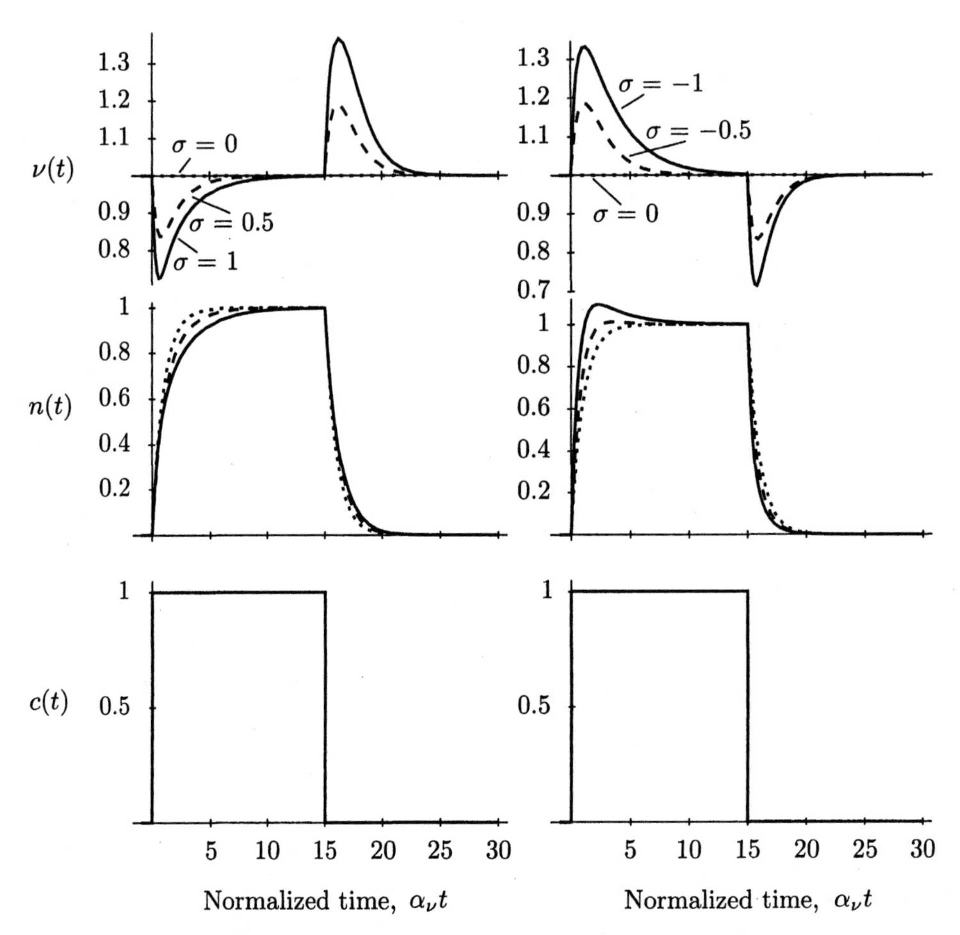

Figure 5.12 Solutions to the equations of coupled transport of solute and solvent (Equations 5.38 and 5.40). The cell volume and the quantity of intracellular solute are plotted as a function of time in response to a pulse change in concentration of a single permeant solute in the presence of impermeant solutes for $\alpha = 1$. The volume is the normalized volume $\nu(t)$; the quantity of solute is the normalized quantity $n(t)$; and the time axis is normalized to $\alpha_\nu t$. The external concentration $c(t)$ is zero for $t < 0$, is increased at $t = 0$ to double the osmolarity of the extracellular solution, and then is returned to zero. Solutions are shown for positive (left panels) and negative (right panels) values of σ.

a smaller osmotic response than does the solute with $\sigma = 1$. The increase in quantity of solute in the cell is greater for smaller, positive values of σ because for these values the contribution of convection to solute flow is greater.

The equations also have well-defined solutions for values of $\sigma < 0$; these are shown in the right-hand panels. The result is that an increase in the extracellular osmolarity of a permeant solute with a negative value of σ results in a simultaneous increase in intracellular concentration of the solute and an *increase* in cell volume. This response can occur if the solute is transported across the membrane more rapidly than is the solvent. Under these conditions, the solute enters rapidly to increase the intracellular osmolarity. Therefore, water flows into the cell to restore osmotic equilibrium. Thus, such solutes have negative reflection coefficients and give rise to this phenomenon, which is called *negative anomalous osmosis.*

5.6.2 Measurements

Volume changes of cells in response to a change in extracellular osmolarity caused by a permeant solute have been measured for a number of years. Many were obtained (e.g., see Figure 5.6) prior to the development of the equations for coupled flow. Since the 1960s the equations of coupled flow have been used to estimate transport parameters from such measurements, with a particular emphasis on the reflection coefficient σ. For example, Figure 5.13 shows measurements of the volume of a population of erythrocytes exposed to an increase in osmolarity by the addition of the permeant solute ethylene glycol. The measurements were performed using a light-scattering method in a stop-flow apparatus. In this apparatus, a solution of cells in an isotonic salt solution is rapidly mixed with a solution that is hyperosmotic by the addition of ethylene glycol. The mixture flows into a chamber where flow stops, and the solution is illuminated. The light scattered by the solution is measured with a photodiode detector. Ideally, the photodiode output is related linearly to the change in volume of the cells. In response to the increase in osmolarity, the cells initially shrink and then swell back to their original volume as the solute permeates the cell membrane. Solutions to the equations of coupled transport (Equations 5.38 and 5.39) were fit to the measurements.[3] The method of fit was to pick different values of σ and to choose other parameters so as to minimize the mean-squared error between the measurements and the calcu-

3. The flux of ethylene glycol saturates at high ethylene glycol concentration. This factor was incorporated by a simple modification of the equations.

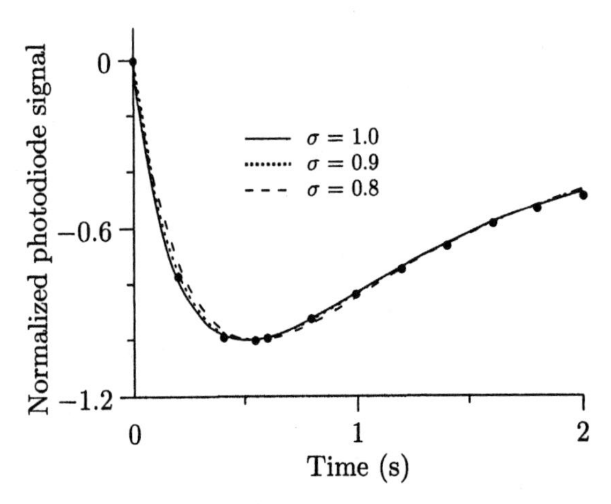

Figure 5.13 Volume change of erythrocytes as a function of time in response to a change in osmolarity: measurements (points) and fit of theory to measurements (lines) for different values of σ (adapted from Levitt and Mlekoday, 1983, Figure 4). At $t = 0$ the solution containing erythrocytes immersed in saline with an osmolarity of 220 mosm/L is mixed with an equal volume of solution containing 220 mosm/L saline plus 206 mosm/L ethylene glycol. The normalized photodiode signal is related linearly to the change in cell volume.

lations. All three values of σ shown fit the data approximately, although the value of $\sigma = 1$ fits these data best (with the minimum value of the minimum mean-squared error). Thus, these results suggest that $\sigma \approx 1$ for ethylene glycol, although the sensitivity of the measurements to changes in σ is not very great. A value of $\sigma \approx 1$ suggests that the coupling of water and ethylene glycol transport is minimal.

A principal aim of these types of studies has been to determine if there are solutes whose transport through the membrane is coupled to that of water. Measurements of σ should, in principle, provide direct evidence to help decide this question. For example, if σ is appreciably less than unity for a solute, that constitutes evidence of the coupling of the transports of that solute and water. If it could also be shown that the transport of water is through water channels, that would constitute evidence that some solutes can permeate water channels. However, methodological problems have plagued resolution of this question, and the measurements of σ for certain solutes remain in dispute. This line of evidence of the coupling of water and solute transport remains controversial (Solomon, 1968; Brahm and Wieth, 1977; Levitt and Mlekoday, 1983; Mayrand and Levitt, 1983; Chasan and Solomon, 1985; Solomon, 1986; Toon and Solomon, 1987, 1990; Macey and Karan, 1993). Despite over three decades of research on the question, we do not know unambiguously whether there exist solutes for which $\sigma < 1$. Thus, the question of the coupling of water and solute transport through cellular membranes is still open. Recent measurements of transport through vesicles containing only one type of water channel may ultimately resolve the question about the coupling of water and solute transport. Results of these studies suggest that some types of water channels are highly selective for water, i.e., they show no appreciable permeability to small hydrophilic solutes. However, at least one type of water channel has been described that shows some permeability to small hydrophilic solutes (see Chapter 4).

5.7 Conclusions

The description of concurrent transport of solute and solvent is more complex than that of the transport of either solute alone or solvent alone. However, under normal physiological conditions, these two transports *do* occur concurrently. In many instances, the equations of uncoupled, concurrent transport are adequate to describe the concurrent transport. These equations depend upon two properties of the membrane: the hydraulic conductivity ($\mathcal{L}_V$), which describes the transport of water through the membrane; and the solute permeability (P_n), which describes the transport of the solute by diffusion. However, these equations may not describe the transport in every instance. The equations of coupled, concurrent transport—which are generalizations of the equations of uncoupled, concurrent transport—allow for the possibility that the transport of water and solute are coupled. This generalization requires the specification of one more parameter, the reflection coefficient (σ), which is a measure of the coupling of water and solute transports. The equations of coupled, concurrent transport also allow for a self-consistent description of the transport of a solute that is indistinguishable from the solvent molecules, such as a radioactive tracer for water.

The equations of concurrent transport of solute and solvent—either the uncoupled or the coupled equations—indicate that the effect of a change in osmolarity produced by a solute can have three different kinds of effects on cellular volume (Figure 5.14). Solutes that produce a maintained change in volume are impermeant. Those that produce a transient change in volume are permeant. Solutes that produce no change in volume are either highly permeant or are indistinguishable from the solvent molecules.

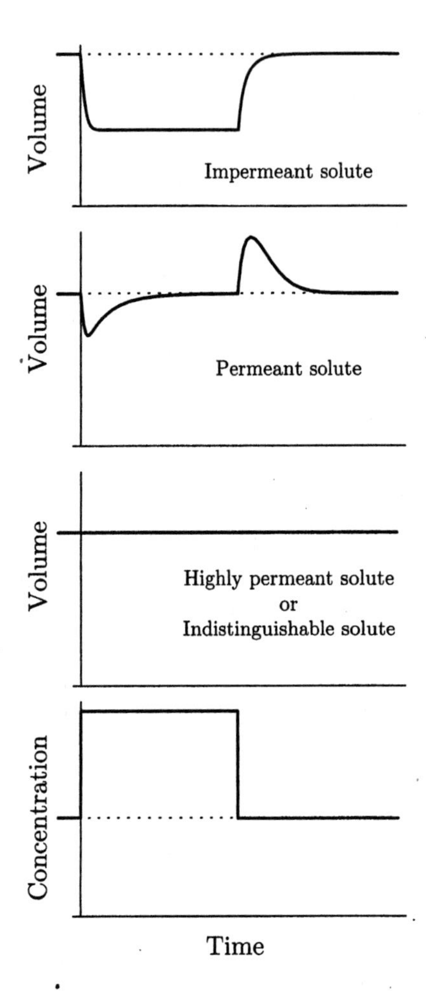

Figure 5.14 Characteristic volume responses of a cell to different types of solutes. A pulse of extracellular test solute is superimposed on the isotonic solute concentration. Each of the upper three traces is the response to a different solute test pulse. The traces show the volume response for an impermeant test solute (upper trace); a permeant test solute (middle trace); and a highly permeant or indistinguishable solute (lower trace).

Exercises

5.1 As shown in Figure 5.15, a membrane separates two solutions subjected to hydraulic pressures p_1 and p_2. The membrane is permeable to water and to solute j, which is the only solute in the solutions. Is thermodynamic equilibrium possible for $c_j^1 \neq c_j^2$ for some choice of $p_1 - p_2$? Explain.

5.2 In Figure 5.3, for the impermeant solute (left-hand panels), why is $\tilde{n}(t) = 1$ and why is the steady-state value of $\nu(t) = 0.5$?

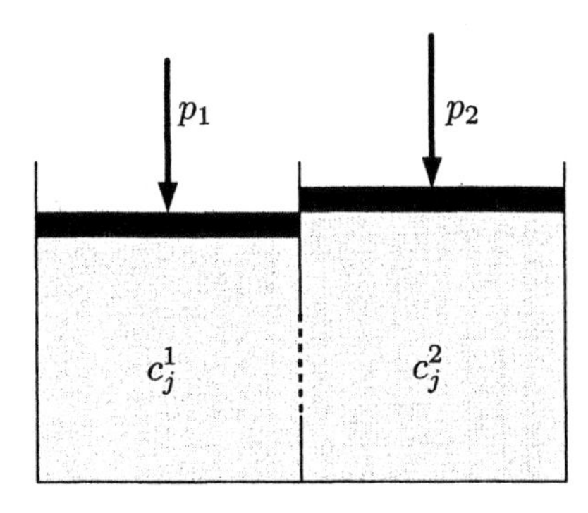

Figure 5.15 Arrangement of compartments (Exercises 5.1 and 5.9).

5.3 In Figure 5.3, for the permeant solute (right-hand panels), why is the steady-state value of $\nu(t) \approx 0.5$ for $\alpha = 0.1$?

5.4 The differential equations of uncoupled transport of solute and solvent for a cell are nonlinear. Therefore, the relations between extracellular concentration and both the intracellular quantity of permeant solute and the cell water volume are nonlinear. Briefly describe the indications of this nonlinearity that are evident in the solutions shown in Figure 5.4.

5.5 An egg cell is placed in an isotonic solution until it equilibrates. At $t = 0$, the concentration of an *impermeant* solute in the extracellular solution, $\tilde{c}^o(t)$, is doubled. The cell water volume, $\nu(t)$, is plotted versus normalized time in Figure 5.16. All the variables in this problem are normalized to their isotonic values. The initial quantity of impermeant solute in the cell $\tilde{n} = 1$ for $t < 0$, and the initial intracellular concentration of impermeant solute $\tilde{c}^i(t) = 1$ for $t < 0$.

a. Sketch $\tilde{n}^i(t)$ for all t.

b. Sketch $\tilde{c}^i(t)$ for all t.

5.6 Two cells have identical membranes. One cell contains *permeant* solute A only and is placed in an isotonic solution that contains only solute A at a concentration $c_A(t)$. The second cell contains *permeant* solute B only and is placed in an isotonic solution that contains only solute B at a concentration $c_B(t)$. The membrane is more permeable to solute A than to solute B.

At time $t = 0$, the concentrations of A and B are doubled; the cell water volume $\nu(t)$ and the quantity of intracellular solute $n(t)$ are shown as a function of normalized time in Figure 5.17. All the variables are normalized to their isotonic values. Determine which volume change

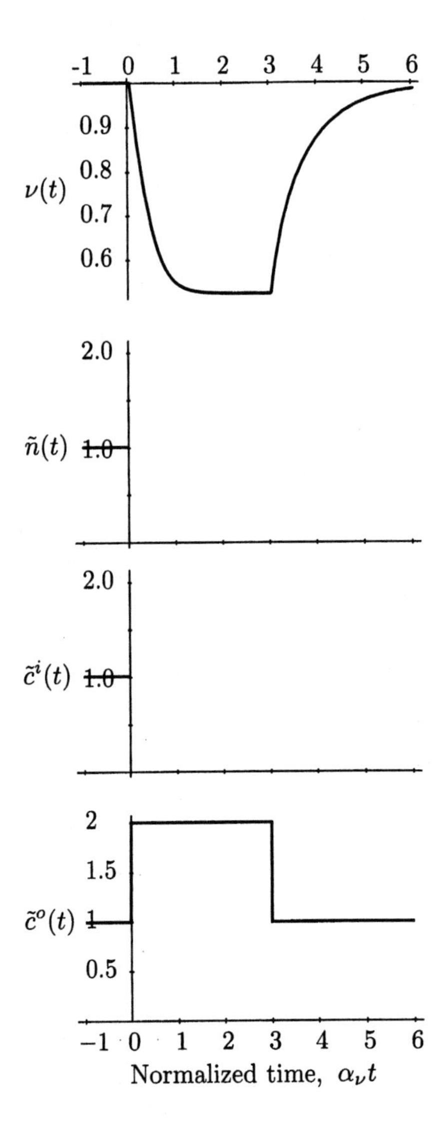

Figure 5.16 Osmotic responses to an increase in concentration of an impermeant extracellular solute (Exercise 5.5).

and which change in intracellular quantity correspond to solute A and B, respectively. Explain briefly.

5.7 In the measurements shown in Figure 5.6, for $t > 0$ the extracellular solution consists of seawater and ethylene glycol. The theoretical solutions are based on Equations 5.3 and 5.4 for the transport of a single permeant solute in the presence of impermeant solutes. Identify the permeant and impermeant solutes in this experiment.

5.8 Define the distinction between solute convection and solute diffusion.

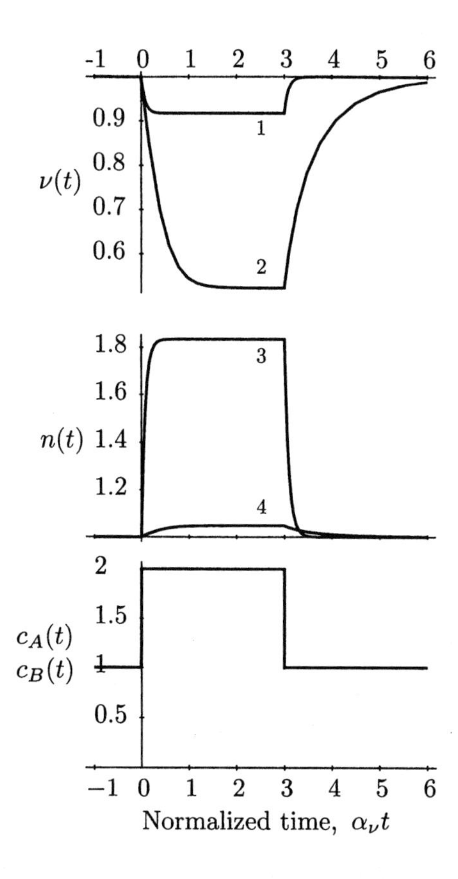

Figure 5.17 Osmotic responses to increases in the concentrations of permeant extracellular solutes (Exercise 5.6).

5.9 The Kedem-Katchalsky equations describe coupled flow of solute and solvent through membranes. For the case of a single solute, the equations are

$$\Phi_V = \mathcal{L}_V(\Delta p - \sigma_j RT \Delta c_j),$$

$$\phi_j = \overline{C}_j(1 - \sigma_j)\Phi_V + P_j \Delta c_j,$$

where $\mathcal{L}_V \geq 0$, and $P_j \geq 0$.

a. Figure 5.15 shows a membrane separating two compartments with $\Delta p = p_1 - p_2$ and $\Delta c_j = c_j^1 - c_j^2$. Indicate on a sketch the reference directions for Φ_V, the volume flux of solvent, and ϕ_j, the molar flux of solute.

b. In these equations the membrane is characterized by three constants. If $\Delta p = 0$ and $\Delta c_j \neq 0$, for what values of the constants is the flow through the membrane only *osmotic*?

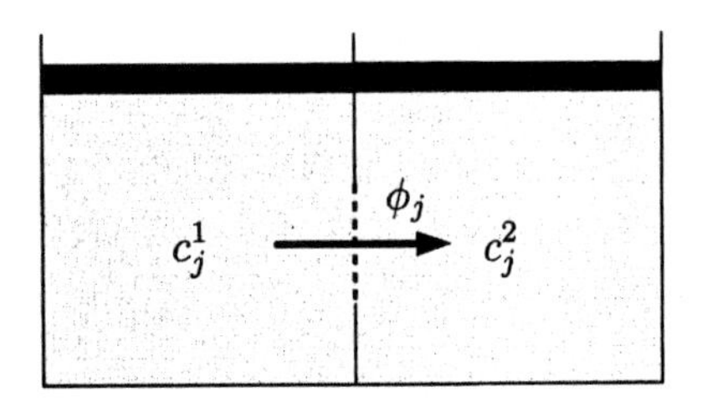

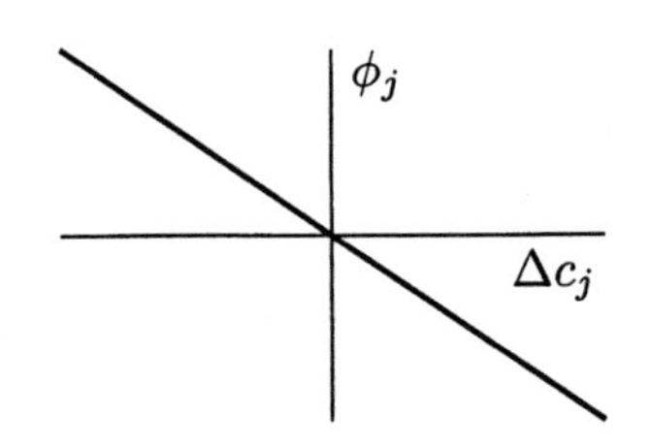

Figure 5.18 Solute flux (Exercise 5.9).

c. It has been reported for a particular membrane, with $\Delta p = 0$, that the solute flux is proportional to Δc_j, as indicated in Figure 5.18. Is this result consistent with the Kedem-Katchalsky equations? Explain your reasoning briefly.

Problems

5.1 A cell is in an isotonic solution of impermeant solute A. At $t = 0$, the extracellular solution is changed so that half of the concentration of solute A is replaced with an equal concentration of solute B, to which the membrane of the cell is permeable. The solution osmolarities before and after solution change are the same. The transport of solute B and of water are uncoupled. The normalized volume of the cell and the normalized quantity of solute B in the cell are shown in Figure 5.19.

a. Why does the cell swell for $0 < \alpha_v t < 25$ despite the fact that the osmolarity of the extracellular solution has not changed?

b. Why does the normalized quantity of solute B approach 1 during the interval $0 < \alpha_v t < 25$?

c. Why does the volume of the cell double during the interval $0 < \alpha_v t < 25$?

5.2 The equations for uncoupled solute and solvent transport (Equations 5.3 and 5.4) can be simplified for a poorly permeant solute. Under these conditions, water transport is so rapid, compared to solute transport, that it can be assumed that osmotic equilibrium holds at each instant in time. Use the normalized equations of uncoupled solute and solvent transport in the following parts.

a. Show that

$$v(t) = \frac{n(t) + \tilde{n}(t)}{c(t) + \tilde{c}(t)}.$$

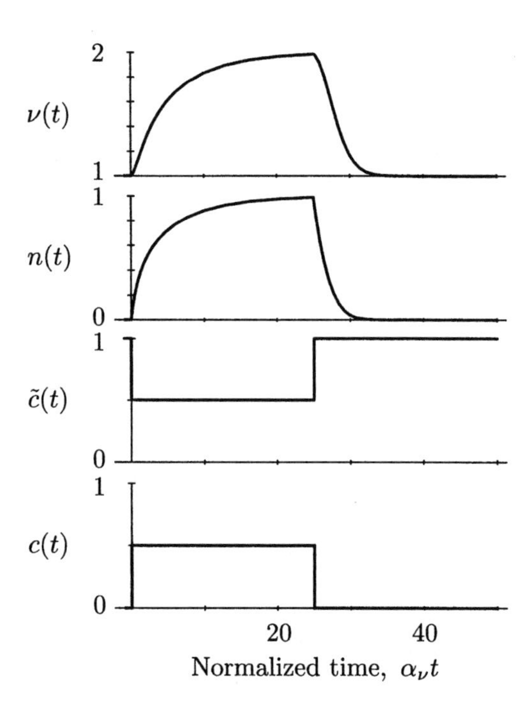

Figure 5.19 Isosmotic swelling of a cell (Problem 5.1). The waveforms show computations of $\nu(t)$ and $n(t)$ according to Equation 5.7 for the values of $c(t)$ and $\tilde{c}(t)$ shown.

b. Show that

$$\frac{dn(t)}{dt} = \alpha_n A(t) \left(\frac{c(t)\tilde{n}(t) - \tilde{c}(t)n(t)}{n(t) + \tilde{n}(t)} \right).$$

c. Consider the case of a cell with constant surface area, with a constant number of impermeant solutes $\tilde{n}(t) = \tilde{N}$, a constant concentration of impermeant extracellular solute $\tilde{c}(t) = \tilde{C}$, and a concentration of permeant extracellular solute

$$c(t) = \begin{cases} 0 & \text{for } t < 0 \\ \tilde{C} & \text{for } t \geq 0 \end{cases}.$$

Assume that the isotonic, total, extracellular concentration is $\tilde{C}$.

i. Determine $dn(t)/dt$ for $t = 0+$.

ii. Determine $n(\infty)$.

iii. Sketch both $n(t)$ and $\nu(t)$ for all t.

d. Explain how the permeability of the membrane for a poorly permeant solute can be obtained from measurements of the volume response that results from the addition of the permeant solute to the extracellular solution.

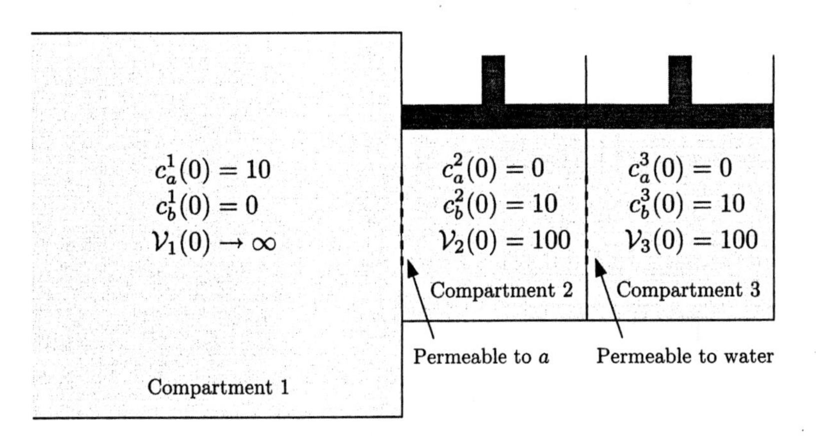

Figure 5.20 Three compartments separated by semipermeable membranes (Problem 5.3). Initial concentrations of each solute are given in mmol/L, and initial volumes are given in cm^3.

5.3 Three compartments filled with aqueous solutions are separated by semipermeable membranes as shown in Figure 5.20. Two nonelectrolyte solutes a and b are contained in the solutions. The membrane separating compartments 1 and 2 is permeable to solute a only and not permeable either to solute b or to water. The membrane separating compartments 2 and 3 is permeable to water only and not permeable to either solute a or b. The volume of compartment 1, V_1, is much larger than that of the other compartments. You may assume that it has an infinite volume. Compartments 2 and 3 contain pistons with frictionless bearing surfaces. Therefore, the hydraulic pressure in these compartments is zero.

a. Is the system in diffusive equilibrium at $t = 0$? Explain.

b. Is the system in osmotic equilibrium at $t = 0$? Explain.

c. Determine the equilibrium concentrations ($c_a^1(\infty)$, $c_b^1(\infty)$, $c_a^2(\infty)$, $c_b^2(\infty)$, $c_a^3(\infty)$, and $c_b^3(\infty)$) and volumes ($V_2(\infty)$ and $V_3(\infty)$).

5.4 A cell is placed in an isotonic solution of impermeant uncharged solute of concentration C, to which is added a pulse of an unknown nonelectrolyte test solute x with concentration $c_x(t)$ so that the concentration of x has a peak amplitude C_x. For the different test solutes A–F, the cell's volume response shows characteristic differences, as shown in Figure 5.21. You may assume that the transport of water and of solute are uncoupled.

a. Which one of the volume responses A–F corresponds best to the results that would be obtained with an impermeant test solute? Explain.

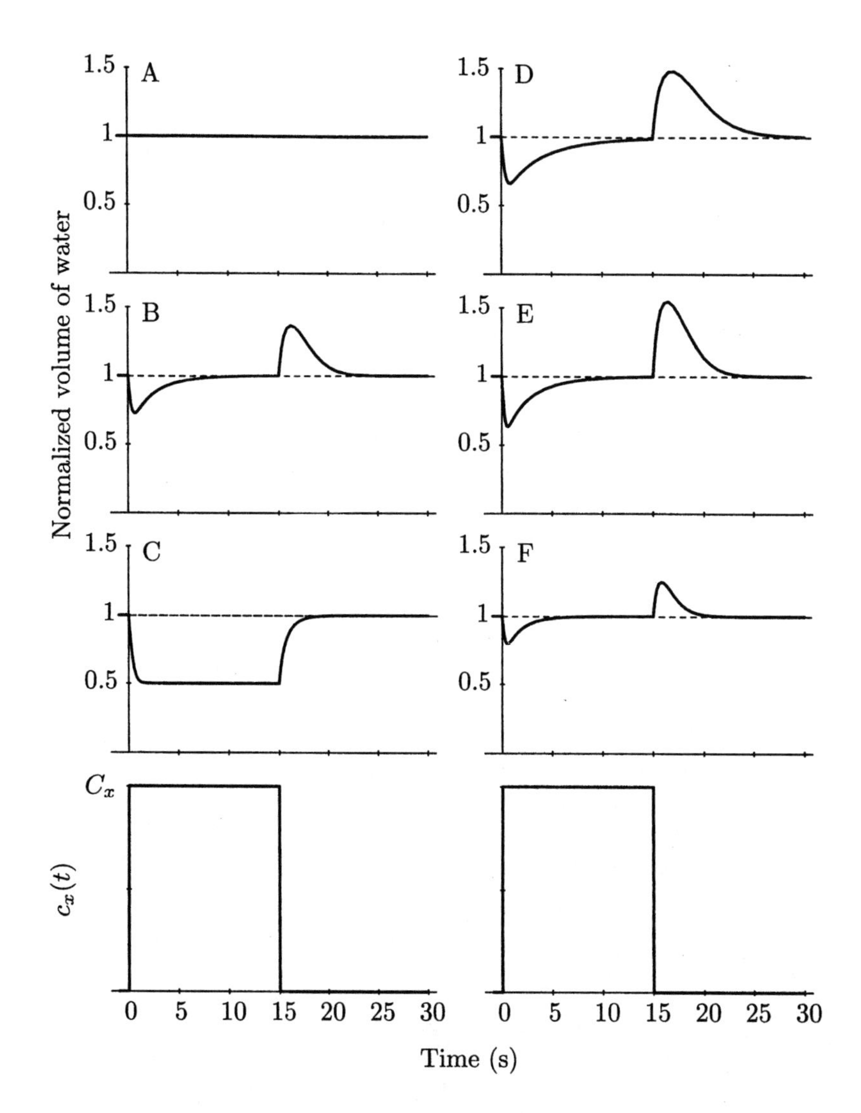

Figure 5.21 Response of a cell to six nonelectrolyte solutes (Problem 5.4). The normalized volume represents the volume of cell water normalized to its isotonic value.

b. Assume that the concentration C_x is the same for A, B, and C. Which of these test solutes has the highest membrane permeability? Explain.

c. Assume that the concentration C_x is the same for A, B, and C. Find the numerical value of C_x/C. Explain.

d. Assume that C_x is the same for B and F. Is the permeability of test solute B greater than or less than that of test solute F? Explain.

e. Assume that the same species of test solutes are used to obtain responses B and E. For which of these responses is C_x largest? Explain.

f. The time course of the onset of the response at $t = 0$ appears to be about the same in B–F. Why?

5.5 A solution of uncharged solute molecules, j, in water is in the steady state, with volume flux Φ_V and solute flux ϕ_j both in the $+x$ direction. Both Φ_V and ϕ_j are constant. The solute diffusion coefficient is D. The solute concentration, $c_j(x)$, is a function of x where $\lim_{x \to -\infty} c_j(x) = C_1$ and $c_j(0) = C_o$.

a. Determine a differential equation for $c_j(x)$.

b. Determine and sketch $c_j(x)$ for $-\infty < x \le 0$.

5.6 This problem deals with conceptual difficulties that can arise when the uncoupled equations of solute and volume flow are used for permeable solutes. Consider the uncoupled equations for solute and volume flow:

$$-\frac{1}{A}\frac{d\left(c_j^i \mathcal{V}^i\right)}{dt} = \phi_j = P_j\left(c_j^i - c_j^o\right); \tag{5.43}$$

$$-\frac{1}{A}\frac{d\mathcal{V}^i}{dt} = \Phi_V = \mathcal{L}_V RT\left(c_j^o - c_j^i\right). \tag{5.44}$$

In experiments by Prescott and Zeuthen (1953), fish and amphibian eggs were equilibrated in an isotonic solution and then transferred to an identical solution, except that a fraction of the water molecules was replaced by radioactive water (D_2O, where D is deuterium, which, in contrast to hydrogen, has one neutron in its nucleus). It was found that the D_2O molecules were transported into the egg cells with a time course predicted by Equation 5.43. It was also observed that the volume of the cells remained constant at all times.

a. Show that the outcome of this experiment is not simultaneously consistent with both Equations 5.43 and 5.44.

b. Show that the Kedem-Katchalsky equations for coupled flow remove the inconsistency implicit in Equations 5.43 and 5.44.

c. What is the value of σ for D_2O if $d\mathcal{V}^i/dt = 0$?

5.7 Assume that simultaneous solute and solvent transport through a membrane from side 1 to side 2 can be represented by the Kedem-Katchalsky equations:

$$\Phi_V = \mathcal{L}_V(\Delta p - \sigma_j \Delta\pi),$$

$$\phi_j = P_j \Delta c_j + \overline{C}_j(1 - \sigma_j)\Phi_V,$$

where $\Delta\pi = \pi_1 - \pi_2, \Delta c_j = c_j^1 - c_j^2$, and Φ_V and ϕ_j are positive for flow from side 1 to side 2. If the pressure difference, Δp, is zero, the solute flux, ϕ_j, can be shown to be proportional to the concentration difference, Δc_j, and an effective permeability can be defined as

$$P_{eff} = \frac{\phi_j}{\Delta c_j}.$$

a. Determine P_{eff} in terms of the parameters ($\mathcal{L}_V$, σ_j, P_j, $\overline{C}_j$) of the Kedem-Katchalsky equations.

b. Assuming $0 \le \sigma_j \le 1$ and $\mathcal{L}_V \ge 0$, is P_{eff} greater than or less than P_j? In a short paragraph explain qualitatively the mechanism(s) that make(s) this difference between P_{eff} and P_j.

5.8 A cell is in 10 mL of isotonic solution with a total concentration of solute of 200 mmol/L. The volume of the cell is negligible compared with the volume of external solution. Solute k, which is distinct from any solute present in the cell cytoplasm, is added to the external solution at the time indicated by the arrow in Figure 5.22, and the change in the cell's volume, as a fraction of its isotonic volume, is measured. You may assume that the solute does not ionize. Also assume that the volume of the cell equals the volume of water in the cell, i.e., the volume that is *not* water is negligible. The experiment is repeated with 6 different solutes with the results shown in Figure 5.22.

Let n_k be the number of moles of solute k added. P_k is the permeability of the membrane for solute k and σ_k is the reflection coefficient for solute k. Answer the following questions by filling in the blank or answering T for true or F for false. Explain your reasons briefly.

a. ______ moles of solute 2 were added to the external solution.

b. (T or F) The membrane is impermeable to solute 3.

c. (T or F) The membrane is more permeable to solute 4 than to solute 6.

d. (T or F) The membrane is more permeable to solute 5 than to solute 4.

e. (T or F) $\sigma_3 = 0$.

f. (T or F) The amount of solute 2 added is equal to the amount of solute 5 added.

g. (T or F) $\sigma_1 \neq 0$.

h. (T or F) $P_4 \neq 0$.

5.9 A mosaic membrane is made of two types of membrane. Membrane type X is semipermeable and allows water to permeate but does not allow

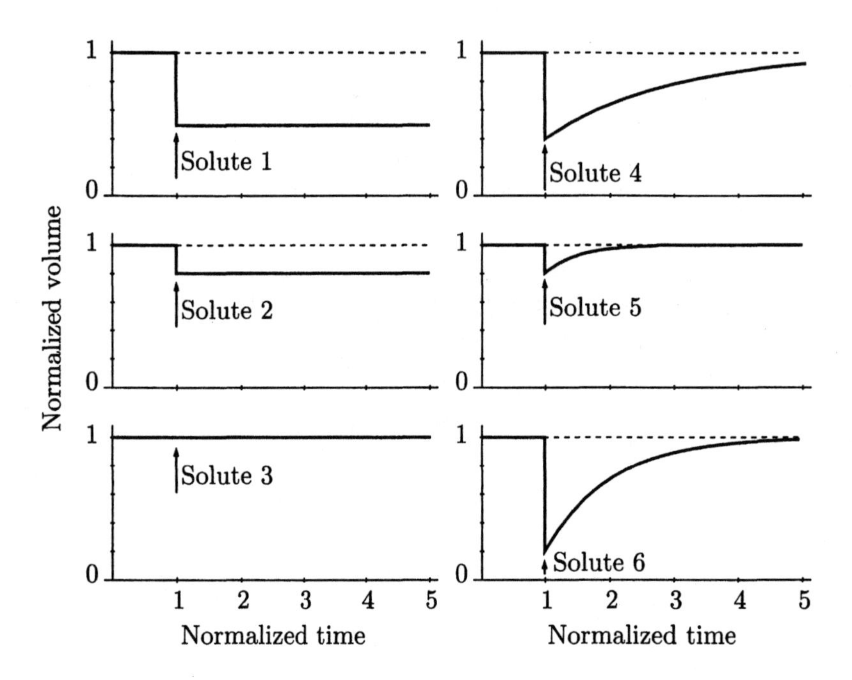

Figure 5.22 Change in cell volume caused by six different solutes (Problem 5.8). The cell volume has been normalized to its isotonic value, and the time scale is normalized but the same for all six panels.

solute j to permeate. The hydraulic conductivity of membrane type X is $\mathcal{L}_X$. Membrane type Y cannot distinguish water from solute j. Membrane type Y has hydraulic conductivity $\mathcal{L}_Y$ and permeability P_Y. These two types of membranes are used to make the mosaic membrane, whose surface area consists of 50% membrane type X and 50% membrane type Y.

a. Show that the flux of volume through the mosaic membrane, Φ_V, can be written as

$$\Phi_V = \mathcal{L}_m(\Delta p - \sigma_m RT\Delta c_j),$$

where $\Delta p = p_1 - p_2$ and $\Delta c_j = c_j^1 - c_j^2$. p_1 and p_2 are the hydraulic pressures on side 1 and 2 of the membrane, and c_j^1 and c_j^2 are the solute concentrations on side 1 and side 2. Φ_V is the volume flux from side 1 to side 2 of the membrane. Find the parameters $\mathcal{L}_m$ and σ_m in terms of $\mathcal{L}_X$, $\mathcal{L}_Y$, and P_Y.

b. Show that the flux of solute ϕ_j can be written as

$$\phi_j = \overline{C}_j(1 - \sigma_m)\Phi_V + P_m\Delta c_j,$$

where $\overline{C}_j = (c_j^1 + c_j^2)/2$. Find P_m in terms of $\mathcal{L}_X$, $\mathcal{L}_Y$, and P_Y.

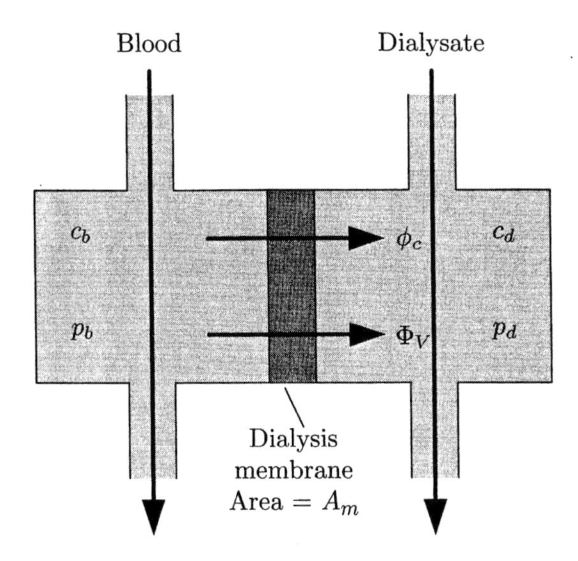

Figure 5.23 Schematic diagram of a dialysis machine (Problem 5.10).

5.10 Creatinine (molecular weight, 100 g/mol) is a metabolic waste product of the metabolism of nitrogen-containing substances (e.g., amino acids and nucleotides) and is normally eliminated by the kidneys. Each day, approximately 1 gram of creatinine is generated in the body and eliminated with at least 1 liter of water. You are asked to decide on the adequacy of a new dialysis membrane that is designed to remove creatinine from the blood of patients with kidney failure (normal body temperature of 37°C). The membrane is designed to work in the dialysis machine shown schematically in Figure 5.23.

Figure 5.24 shows performance test data on the dialysis membrane, which separates two solutions, blood and dialysate (the solution in the dialysis machine). Under the test conditions, creatinine is the only solute that crosses the membrane. The contribution of creatinine to the total osmotic pressure of the blood and dialysate is small, and the osmotic pressures of the dialysate and of the blood are equal. Concentrations of creatinine in the blood ($c_b = 0.1$ mg/cm^3) and in the dialysate ($c_d = 0.4$ mg/cm^3) are maintained constant by recirculation. The measured volume flux, Φ_V, and the solute flux, ϕ_c, are plotted as a function of imposed hydrostatic pressure drop, Δp, in Figure 5.24.

a. Determine the following membrane parameters: the hydraulic conductivity, $\mathcal{L}_V$; the reflection coefficient, σ; and the membrane permeability to creatinine, P_c.

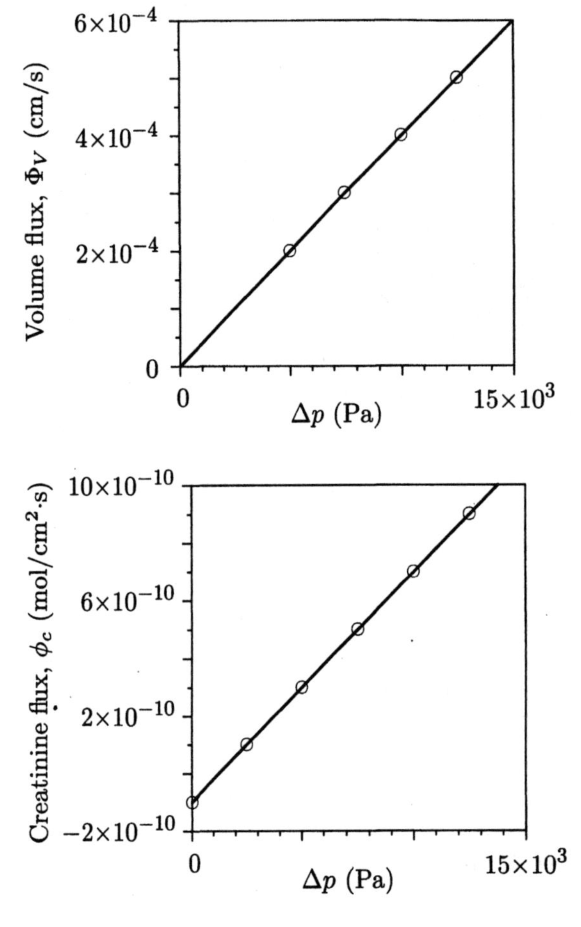

Figure 5.24 Volume flux (top) and solute flux (bottom) across dialysis membrane (Problem 5.10). $\Delta p = p_b - p_d$.

b. Assume that the normal time average blood pressure is 1.36×10^4 Pa, the dialysate pressure is held at 1.2×10^3 Pa, and the membrane area $A_m = 100$ cm^2. Determine how much creatinine and water are removed in six hours.

c. It is proposed that a patient be connected to this dialysis machine for a six-hour period each day. Discuss the effectiveness of the machine and indicate design improvements, if any, that you would recommend.

References

Books and Reviews

Davson, H. (1970). *A Textbook of General Physiology*, 4th ed., 2 vols. J. and A. Churchill, London.

Finkelstein, A. (1987). *Water Movement Through Lipid Bilayers, Pores, and Plasma Membranes*. John Wiley & Sons, New York.

Jacobs, M. H. (1952). The measurement of cell permeability with particular reference to the erythroctye. In Barron, E. S. G., ed., *Modern Trends in Physiology and Biochemistry*, 149–171. Academic Press, New York.

Katchalsky, A. and Curran, P. F. (1965). *Nonequilibrium Thermodynamics in Biophysics*. Harvard University Press, Cambridge, MA.

Scheidegger, A. E. (1960). *The Physics of Flow through Porous Media*. University of Toronto Press, Toronto, Canada.

Sha'afi, R. I. and Gary-Bobo, C. M. (1973). Water and nonelectrolyte permeability in mammalian red cell membranes. *Prog. Biophys. Mol. Biol.*, 26:103–146.

Solomon, A. K. (1968). Characterization of biological membranes by equivalent pores. *J. Gen. Physiol.*, 51:335–364.

Solomon, A. K. (1986). On the equivalent pore radius. *J. Membr. Biol.*, 94:227–232.

Villars, F. M. H. and Benedek, G. B. (1974). *Physics with Illustrative Examples from Medicine and Biology*, vol. 2, *Statistical Physics*. Addison-Wesley, Reading, MA.

Original Articles

Adamski, R. P. and Anderson, J. L. (1983). Solute concentration effect on osmotic reflection coefficient. *Biophys. J.*, 44:79–90.

Adamski, R. P. and Anderson, J. L. (1987). Configurational effects on polystyrene rejection from microporous membranes. *J. Polymer Sci. Polymer Physics Ed.*, 25:765–775.

Anderson, J. L. (1981). Configurational effect on the reflection coefficient for rigid solutes in capillary pores. *J. Theor. Biol.*, 90:405–426.

Anderson, J. L. and Adamski, R. P. (1983). Solute concentration effects on membrane reflection coefficients. *AICHE Symp. Series*, 79:84–92.

Anderson, J. L. and Brannon, J. H. (1981). Concentration dependence of the distribution coefficient for macromolecules in porous media. *J. Polymer Sci. Polymer Physics Ed.*, 19:405–421.

Anderson, J. L. and Malone, D. M. (1974). Mechanism of osmotic flow in porous membranes. *Biophys. J.*, 14:957–982.

Anderson, J. L. and Quinn, J. A. (1972). Ionic mobility in microcapillaries. A test for anomalous water structures. *J. Chem. Soc. Faraday Trans. I*, 68:744–748.

Bean, C. P. (1972). The physics of porous membranes: Neutral pores. In Eisenman, G., ed., *Membranes*, vol. 1, *Macroscopic Systems and Models*. Marcel Dekker, New York.

Beck, R. E. and Schultz, J. S. (1970). Hindered diffusion in microporous membranes with known pore geometry. *Science*, 170:1302–1305.

Brahm, J. and Wieth, J. O. (1977). Separate pathways for urea and water, and for chloride in chicken erythrocytes. *J. Physiol.*, 266:727–749.

Brenner, H. and Gaydos, L. J. (1977). The constrained Brownian movement of spherical particles in cylindrical pores of comparable radius. *J. Colloid Interface Sci.*, 58:312–356.

Bungay, P. M. and Brenner, H. (1973). The motion of a closely-fitting sphere in a fluid-filled tube. *Int. J. Multiphase Flow*, 1:25–56.

Chasan, B. and Solomon, A. K. (1985). Urea reflection coefficient for the human red cell membrane. *Biochim. Biophys. Acta*, 821:56–62.

Ferry, J. D. (1936). Statistical evaluation of sieve constants in ultrafiltration. *J. Gen. Physiol.*, 20:95–104.

Garby, L. (1957). Studies on transfer of matter across membranes with special reference to the isolated human amniotic membrane and the exchange of amniotic fluid. *Acta Physiol. Scand.*, 40 Suppl. 137:1–84.

Goldstein, D. A. and Solomon, A. K. (1960). Determination of equivalent pore radius for human red cells by osmotic pressure measurement. *J. Gen. Physiol.*, 44:1–17.

Guell, D. C. (1991). The physical mechanism of osmosis and osmotic pressure: A hydrodynamic theory for calculating the osmotic reflection coefficient. PhD thesis, Department of Chemical Engineering, Massachusetts Institute of Technology.

Hill, A. (1982). Osmosis: A bimodal theory with implications for symmetry. *Proc. R. Soc. London, Ser. B*, 215:155–174.

Hill, A. E. (1989a). Osmosis in leaky pores: The role of pressure. *Proc. R. Soc. London, Ser. B*, 237:363–367.

Hill, A. E. (1989b). Osmotic flow equations for leaky porous membranes. *Proc. R. Soc. London, Ser. B*, 237:369–377.

Hill, A. E. (1994). Osmotic flow in membrane pores of molecular size. *J. Membr. Biol.*, 137:197–203.

Jacobs, M. H. (1933a). The relation between cell volume and penetration of a solute from an isosmotic solution. *J. Cell. Comp. Physiol.*, 3:29–43.

Jacobs, M. H. (1933b). The simultaneous measurement of cell permeability to water and to dissolved substances. *J. Cell. Comp. Physiol.*, 2:427–444.

Jacobs, M. H. (1934). The quantitative measurement of the permeability of the erythrocyte to water and to solutes by the hemolysis method. *J. Cell. Comp. Physiol.*, 4:161–183.

Jacobs, M. H. and Stewart, D. R. (1932). A simple method for the quantitative measurement of cell permeability. *J. Cell. Comp. Physiol.*, 1:71–82.

Kedem, O. and Katchalsky, A. (1958). Thermodynamic analysis of the permeability of biological membranes to non-electrolytes. *Biochim. Biophys. Acta*, 27:229–246.

Kedem, O. and Katchalsky, A. (1961). A physical interpretation of the phenomenological coefficients of membrane permeability. *J. Gen. Physiol.*, 45:143–179.

Levin, S. W., Levin, R. L., and Solomon, A. K. (1980). Improved stop-flow apparatus to measure permeability of human red cells and ghosts. *J. Biochem. and Biophys. Meth.*, 3:255–272.

Levitt, D. G. (1974). A new theory of transport for cell membrane pores: I. General theory and application to red cell. *Biochim. Biophys. Acta*, 373:115–131.

Levitt, D. G. (1975). General continuum analysis of transport through pores: I. Proof of Onsager's reciprocity postulate for uniform pore. *Biophys. J.*, 15:533–551.

Levitt, D. G. and Mlekoday, H. J. (1983). Reflection coefficient and permeability of urea and ethylene gycol in the human red cell membrane. *J. Gen. Physiol.*, 81:239–253.
Levitt, D. G. and Subramanian, G. (1974). A new theory of transport for cell membrane pores: II. Exact results and computer simulation (molecular dynamics). *Biochim. Biophys. Acta*, 373:132–140.
Long, T. D. and Anderson, J. L. (1984). Flow-dependent rejection of polystyrene from microporous membranes. *J. Polymer Sci. Polymer Physics Ed.*, 22:1261–1281.
Long, T. D. and Anderson, J. L. (1985). Effects of solvent goodness and polymer concentration of polystyrene from small pores. *J. Polymer Sci. Polymer Physics Ed.*, 23:191–197.
Long, T. D., Jacobs, D. L., and Anderson, J. L. (1981). Configurational effects on membrane rejection. *J. Memb. Sci.*, 9:13–27.
Lucké, B., Hartline, H. K., and Ricca, R. A. (1939). Comparative permeability to water and to certain solutes of the egg cells of three marine invertebrates, *Arbacia*, *Cumingia* and *Chaetopterus*. *J. Cell. Comp. Physiol.*, 14:237–252.
Macey, R. I. and Karan, D. M. (1993). Independence of water and solute pathways in human RBCs. *J. Membr. Biol.*, 134:241–250.
Mayrand, R. R. and Levitt, D. G. (1983). Urea and ethylene glycol-facilitated transport systems in the human red cell membrane. *J. Gen. Physiol.*, 81:221–237.
Onsager, L. (1945). Theories and problems of liquid diffusion. *Ann. N.Y. Acad. Sci.*, 46:241–265.
Owen, J. D. (1978). Reflection coefficients in red cells. *J. Membr. Biol.*, 40:191.
Owen, J. D. and Eyring, E. M. (1975). Reflection coefficients of permeant molecules in human red cell suspensions. *J. Gen. Physiol.*, 66:251–265.
Pappenheimer, J. R., Renkin, E. M., and Borrero, L. M. (1951). Filtration diffusion and molecular sieving through peripheral capillary membranes: A contribution to the pore theory of capillary permeability. *Am. J. Physiol.*, 167:13–46.
Prescott, D. M. and Zeuthen, E. (1953). Comparison of water diffusion and water filtration across cell surfaces. *Acta Physiol. Scand.*, 28:77–94.
Sha'afi, R. I., Gary-Bobo, C. M., and Solomon, A. K. (1971). Permeability of red cell membranes to small hydrophilic and lipophilic solutes. *J. Gen. Physiol.*, 58:238–258.
Sha'afi, R. I., Rich, G. T., Mikulecky, D. C., and Solomon, A. K. (1970). Determination of urea permeability in red cells by minimum method. *J. Gen. Physiol.*, 55:427–450.
Staverman, A. J. (1951). The theory of measurement of osmotic pressure. *Recueil*, 70:344–352.
Toon, M. R. and Solomon, A. K. (1987). Interrelation of ethylene glycol, urea and water transport in the red cell. *Biochim. Biophys. Acta*, 898:275–282.
Toon, M. R. and Solomon, A. K. (1990). Transport parameters in the human red cell membrane: Solute-membrane interactions of hydrophilic alcohols and their effect on permeation. *Biochim. Biophys. Acta*, 1022:57–71.
Vegard, L. (1910). On the free pressure in osmosis. *Proc. Camb. Phil. Soc.*, 15:13–23.
Weinbaum, S. (1981). Strong interaction theory for particle motion through pores and near boundaries in biological flows at low Reynolds number. *Lect. Math. Life Sci.*, 14:119–192.
Yan, Z., Weinbaum, S., and Pfeffer, R. (1986). On the fine structure of osmosis including three-dimensional pore entrance and exit behaviour. *J. Fluid Mech.*, 162:415–438.

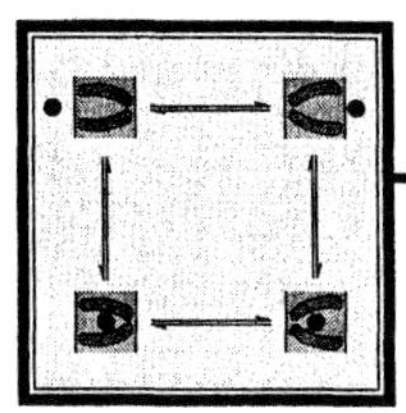

6 Carrier-Mediated Transport

The Channel
A channel's a hole
With a physiological role;
A peptide chain
In a lipid domain,
With its helical coil
Twisted through oil,
And its aqueous core
An open pore,
A path for flow
Where ions may go
If not too large
With electrostatic charge.
Examined kinetically,
Determined frenetically,
With equipment deluxe
Measuring current and flux.
Just a spike on a screen
In fluorescent green,
Does it really exist?
Does the channel persist?
Or is it invention
To bemuse this convention?

The Carrier
A carrier's a gate
For a special substrate;
In the past, hypothetical
or even heretical,
Its now accepted in fact
That a peptide can act
As a transporting entity
With a carrier identity.
In the lipid submerged,
Hydrophobically urged,
It pulls and it tugs
While its substrate it hugs.
It twists and it turns
and thermodynamically yearns.
Until seriously contorted
Unless too soon aborted
The site that was *out*
Becomes *in* without doubt.
Conformations are changed
and sites rearranged,
Done exceedingly quick
Like a magical trick.
The kinetics are right
For defining the site,
And the odds are all long
That the mode is ping-pong.
It's a model supreme
That wise heads did dream;
It stands to the test
And so far it's best.

—Rothstein, 1980

6.1 Introduction

Certain molecules are transported through cellular membranes in a manner that differs substantially from simple diffusion or even solute diffusion coupled to osmosis, which we have discussed in Chapters 3-5. In fact, many biologically important molecules (such as monosaccharides, amino acids, mononucleotides, phosphates, uric acid, choline, etc.) that cells must either acquire or eliminate are transported not by diffusion but by specialized transport mechanisms mediated by macromolecules that reside in the membrane and are called *carriers* or *transporters*. Each type of carrier transports a restricted family of molecules. Of all the carrier-mediated transport mechanisms, the ones responsible for sugar transport in cells are probably the best understood (LeFevre, 1961; Carruthers, 1984; Stein, 1986, 1990). Because these mechanisms transport primarily 6-carbon sugars, they are called *hexose transport* systems or *hexose carriers*. Since glucose is the most important hexose for cells, these mechanisms are often called *glucose transport* systems or *glucose carriers*. As we shall see in this and in later chapters, there are at least two broad categories of hexose transport systems: passive and active transport systems. In this chapter we deal primarily with the passive hexose transport systems.

6.1.1 Distinguishing Characteristics

The hexose transport mechanism has a number of characteristics, some of which indicate that the mechanism of transport differs from simple diffusion.

6.1.1.1 Flux Is Facilitated

The transmembrane flux of sugars exceeds by several orders of magnitude that which would be expected on the basis of diffusion. For example, D-glucose with its polar OH groups (Figure 1.6) is readily soluble in water but is poorly lipid soluble. Consistent with the poor solubility of glucose in lipids, estimates based on diffusion in artificial lipid bilayers indicate that the permeability of glucose is $2\text{–}4 \times 10^{-10}$ cm/s, whereas the permeability of glucose through membranes of animal cells is two to five orders of magnitude higher. Furthermore, glucose has a molecular radius that exceeds that of the small polar molecules that may traverse membranes via water channels. Despite its poor lipid solubility and its comparatively large dimensions, D-glucose is a highly permeant solute.

6.1.1.2 Flux Is Passive

The net flux of the passive hexose transport system is down the concentration gradient of the hexose. If the hexose transport system is allowed to come to equilibrium, the flux goes to zero when the concentration of hexose on both sides of the membrane is the same. This behavior is distinct from that of the active hexose transport systems, in which the flux of hexose is coupled to that of other solutes so that there can be a net flux of hexose up the hexose concentration gradient (Section 2.3).

6.1.1.3 Flux of Structurally Similar Molecules Can Differ Widely

The hexose mechanism transports sugar molecules exclusively; other solutes are not transported by this system. Even among sugar molecules, only the monosaccharides are transported; e.g., molecules that consist of polymers of simple sugars, for example—even those as small as the disaccharides—are excluded. Of the monosaccharides, the sugar-transporting mechanism can transport many hexoses, some pentoses, and some tetroses. Even among the hexoses, the flux of structurally similar molecules can differ widely. For example, the hexoses all have the same molecular formula, i.e., $C_6(H_2O)_6$. Different hexoses vary only in the molecular arrangement of the atoms. The examples of hexoses shown in Figure 1.7 (which are all aldoses) differ only in the orientation of the H and OH groups on the carbon atoms in the ring. All the hexoses shown in Figure 1.7 are in the "D" configuration. "D-" and "L-" identify stereoisomers that differ only in having mirror-image symmetry. For example, L-glucose is the mirror image of D-glucose, as shown in Figure 1.8.

Table 6.1 gives erythrocyte permeabilities (K_{eq} is a measure of *impermeability,* which we shall define later) for a number of different sugar molecules, some of whose structural formulas are shown in Figure 1.7. Note the large variation of the permeabilities of the erythrocyte membrane to these different sugars. Note especially that the hexoses, which all have the same molecular formula, vary in permeability by about 10^3. Although not shown in the table, the permeability to D-glucose greatly exceeds that to L-glucose.

Thus, the hexose transport mechanism transports simple sugars, primarily the hexoses in the D configuration, of which the most biologically

Table 6.1 The permeability of the membrane of human erythrocytes to different sugars (adapted from Stein, 1990, Table 4.1). K_{eq} is a measure of impermeability. The sugars marked with "6" are 6-carbon sugars (hexoses); those marked with "5" are 5-carbon sugars (pentoses). These estimates were obtained with similar measurement methods—using outdated transfusion blood at a temperature of 37°C—for all the sugars except for L-sorbose and D-fructose, for which measurements were obtained at 25°C.

Sugar	K_{eq} (mmol/L)
D-glucose(6)	4–10
D-mannose(6)	14
D-galactose(6)	40–60
D-xylose(5)	60
L-arabinose(5)	250
D-ribose(5)	2,000
L-sorbose(6)	3,100
D-fructose(6)	9,300

important is D-glucose. The mechanism can distinguish between even these structurally similar hexoses. Such sensitivity of the transport of a group of molecules to the detailed arrangement of the atoms in the molecules is incompatible with a diffusion mechanism.

6.1.1.4 Relation of Flux to Concentration Saturates

The flux of solute is not a linear function of concentration difference, but the flux saturates at high solute concentration. This behavior is demonstrated in Figures 6.1 and 6.2, which show measurements of galactose transport in human erythrocytes. In these measurements, a population of erythrocytes was loaded with radioactive galactose and rapidly mixed with solutions that initially contained no galactose. The extracellular composition was sampled and counted. Figure 6.1 shows the accumulation of galactose extracellularly as a function of time for different initial concentrations of galactose in the erythrocytes. The increase in extracellular concentration indicates that there is galactose efflux from the cells. These results clearly show that the rate at which extracellular galactose increases with time increases as the intracellular concentration of galactose increases. However, whereas the slope is approximately proportional to intracellular galactose concentration at the lower concentrations, the slope clearly saturates at the highest concentrations. For

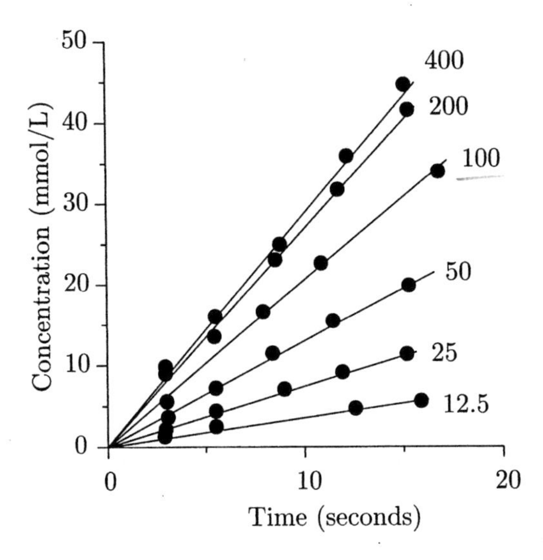

Figure 6.1 Efflux of galactose from human erythrocytes (adapted from Ginsburg, 1978, Figure 1). The extracellular concentration of galactose is plotted versus time. The concentration is expressed in mmol of galactose per liter of cell water, i.e., the concentration of galactose in a solution containing the erythrocytes. The rate of increase of extracellular galactose concentration is proportional to galactose efflux. The parameter is the initial intracellular concentration of galactose (in mmol/L). Straight lines were fit to the measurements (points) by linear regression analysis.

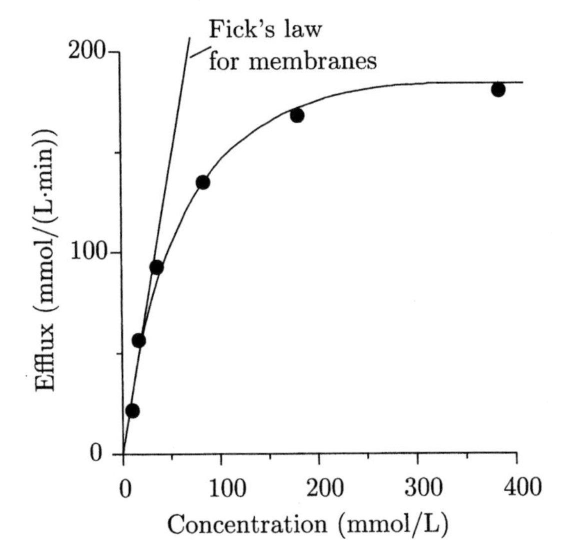

Figure 6.2 The efflux of galactose from erythrocytes as a function of intracellular galactose concentration (adapted from Ginsburg, 1978, Figure 2). The efflux was computed from the slopes of the curves shown in Figure 6.1. The ordinate scale is the rate of change of galactose concentration, which is proportional to galactose efflux. The curved line is a rectangular hyperbola fit to the measurements (points), and the straight line has been fit to the onset of this curved line.

example, note that while the slope for 25 mmol/L is about twice that for 12.5 mmol/L, the slope for 400 mmol/L exceeds that for 200 mmol/L by very little. The rate of increase in extracellular concentration is proportional to the efflux of galactose. Thus, the slope of the extracellular concentration is proportional to galactose efflux. The dependence of the efflux on concentration is shown in Figure 6.2. The flux *is* proportional to the concentration difference across the membrane, consistent with Fick's first law for membranes (Equation 3.44), but only at very low concentrations of galactose. At higher concentrations the flux is a saturating function of concentration. Thus, the transport of galactose in this experiment is incompatible with a diffusion mechanism. This type of saturation of flux with concentration has been widely found for sugar transport in a large variety of cell types.

6.1.1.5 Inhibition

The solute flux of one sugar can be inhibited by the addition of another sugar. For example, curve 1 in Figure 6.3 shows the intracellular accumulation of sorbose into yeast cells as a function of time in the presence of extracellular sorbose but in the absence of glucose. Curve 3 shows the accumulation of sorbose in the presence of both sorbose and glucose. Clearly the accumulation of sorbose is much reduced by the presence of glucose. Curve 2 results if the accumulation of sorbose is measured in the initial absence of glucose, which is added at about 60 minutes after the beginning of the measurement. The addition of glucose decreases the accumulation of sorbose; i.e., sorbose is transported out of the cells in the presence of extracellular glucose. Since glucose is also transported across the membranes of yeast cells, these results suggest that the two sugars compete for the transport mechanism.

The transport of a solute can also be inhibited by the addition of a compound that is not itself transported across the membrane. For instance, certain sugar molecules that are not themselves transported, such as the disaccharide maltose, can inhibit the transport of monosaccharides. In addition, the substances cytochalasin B and phloretin are highly specific inhibitors of sugar transport in cells.

6.1.1.6 Regulation of Transport

Sugar transport in cells is highly regulated, and the regulation is specific to different cell types. For example, sugar accumulation in skeletal muscle cells is increased during exercise; in response to metabolic poisons; during anoxia; and, most importantly, upon an increase in the concentration of the hormone insulin in the extracellular solution of the muscle. If secretion of insulin by the pancreas is reduced (as occurs in certain forms of diabetes mellitus), the concentration of insulin in the blood is reduced. Thus, the insulin concentration at the muscle surface decreases, and the absorption of glucose into the muscle cells is reduced. The muscle cells are literally starved. The excess glucose in the body that results from the reduced transport into the muscle cells appears in the blood and urine, which signals the occurrence of the disease. Thus, the presence of insulin facilitates glucose transport into muscle cells.

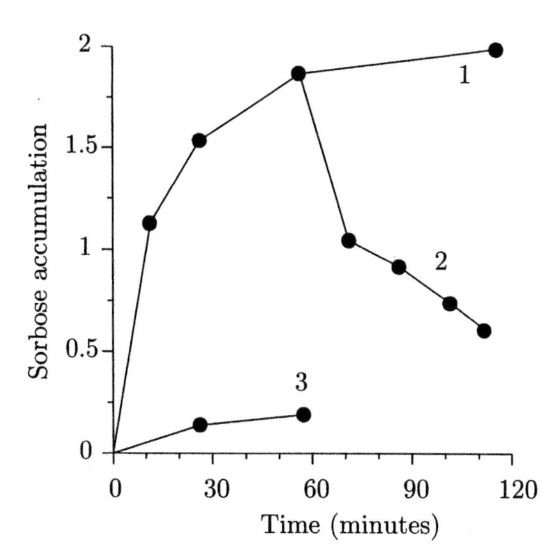

Figure 6.3 Inhibition of sorbose accumulation by yeast cells resulting from the addition of glucose to the extracellular solution (adapted from Cirillo, 1961, Figure 2). The accumulation is approximately proportional to the intracellular concentration and was defined as follows: the intracellular quantity of sorbose (mg) was measured in yeast by chemical means and divided by the volume of yeast (mL); this quotient, multiplied by 100, was divided by the external sorbose concentration (which was constant) to give the accumulation.

Although glucose transport in adipose (fat) tissue is similarly regulated by insulin, the response of other tissues (e.g., liver) to insulin is quite different. The response of a given cell to a hormone depends not only upon the type of hormone but upon the properties of the hormone receptors present on the cell surface.

6.1.2 The Notion of a Carrier

Transport with the above characteristics apparently involves a chemical reaction in which the solute molecule binds to (combines with) some membrane macromolecule, called a *carrier*, at one face of the membrane and is released at the opposite face. The molecular mechanisms of the binding and release steps have not been worked out in detail yet, but we can visualize the steps schematically as shown in Figure 6.4. The solute binds to a membrane carrier macromolecule. The complex translocates via a conformational change. After translocation, the solute is released and the carrier macromolecule resets to its initial conformation and location. In these mechanisms, the binding/unbinding sites are accessible to only one side of the membrane at each instant of time. The carrier bound to solute translocates so that the site is accessible to the opposite side of the membrane, from which the unbound carrier must return in the reset part of the cycle. Such a mechanism has been described as a *ping-pong* mechanism.

The characteristics of the carrier account qualitatively for the distinguishing properties listed in Section 6.1. Clearly the permeability of the membrane

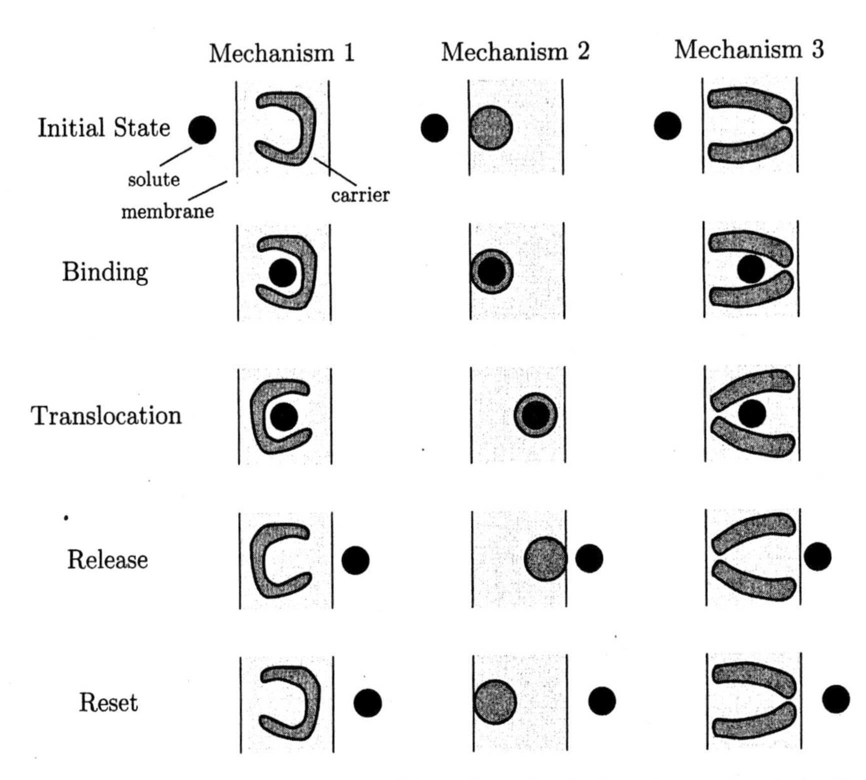

Figure 6.4 Schematic diagrams of three hypothetical carrier mechanisms. Each panel shows a carrier molecule in a membrane. The rows show different carrier states for each of three different mechanisms. The mechanisms differ in their translocation and reset methods. In mechanism 1, the carrier translocates and resets by rotating; in mechanism 2 by translating; and in mechanism 3 by changing its conformation. Evidence based on the molecular structure of carriers suggests that of the three mechanisms shown, mechanism 3 is the most likely.

need not be related either to the molecular radius or to the lipid solubility of a solute, but rather is related to the binding affinity of the solute for the membrane macromolecule. Hence, the permeability can be highly structure-specific. The saturable flux results because transport is through a discrete number of carriers. If a membrane contains $\mathcal{N}$ carriers/cm^2, each of which transports solute at some maximum rate, then the net transport through the membrane is limited. Competition arises because the carriers are limited in number and can bind different solutes (ligands). Hence, different ligands compete for the same site. Transport inhibitors are substances that bind to the carrier; but, because of the inhibitors' structure, the binding does not result in translocation or release. In general, transport regulators, such as hormones, act in one of the following two manners: (1) The hormone binds to a receptor site on the carrier (not the site at which the solute binds), and this binding results in a conformational change of the carrier that either facilitates or inhibits the transport of the solute. (2) The hormone binds to a receptor molecule on the membrane that is distinct from the carrier. The binding of the hormone to the receptor on the outer surface of the membrane results in a conformational change of the receptor, which releases a substance, the *second messenger*, into the cytoplasm, where it can act on the internal surface of the carrier to affect transport of solute or on intracellular stores to recruit more carriers into the membrane.

6.2 Chemical Reactions: A Macroscopic Description

In order to understand transport mechanisms that involve chemical reactions, it will be necessary to first review elementary properties of chemical reactions. In a chemical reaction, a number of chemical compounds, called reactants (or R_ks), combine to form new compounds, called products (or P_ks). The reaction is depicted in a mass balance relation (Equation 6.1),

$$\nu_1 R_1 + \nu_2 R_2 + \nu_3 R_3 \rightleftharpoons \nu_1' P_1 + \nu_2' P_2 + \nu_3' P_3, \tag{6.1}$$

where, arbitrarily but conventionally, the reactants are shown at the left and the products at the right, and where the $\nu's$ are the stoichiometric coefficients which are usually small integers and indicate the proportions by which the molecules combine. The symbol $\rightleftharpoons$ implies that the chemical reaction can proceed in either direction, in general. For example, the oxidation of glucose has the following mass balance

$$C_6(H_2O)_6 + 6O_2 \rightleftharpoons 6CO_2 + 6H_2O$$

which indicates that one molecule of glucose, $C_6(H_2O)_6$, combines with 6 molecules of oxygen to yield 6 molecules of carbon dioxide plus 6 molecules of water.

We shall be concerned with two important properties of chemical reactions:

- What is the equilibrium state of the reaction? That is, if we let the reaction proceed and if it comes to equilibrium, what are the equilibrium concentrations of the reactants and products?
- What is the rate of the reaction? How fast does the reaction proceed to its equilibrium state?

We shall present a macroscopic description of simple chemical reactions in hierarchical order. The rate of advancement of a chemical reaction, ν, is defined as the rate of formation of product:

$$\nu = \frac{1}{\nu_1'}\frac{dc_{P_1}}{dt} = \frac{1}{\nu_2'}\frac{dc_{P_2}}{dt} = \frac{1}{\nu_3'}\frac{dc_{P_3}}{dt} = -\frac{1}{\nu_1}\frac{dc_{R_1}}{dt} = -\frac{1}{\nu_2}\frac{dc_{R_2}}{dt} = -\frac{1}{\nu_3}\frac{dc_{R_3}}{dt}. \quad (6.2)$$

Positive signs are used to indicate the rate of production of products, and negative signs are used to indicate the rate of production of reactants. In a general chemical reaction, the rate ν may depend on the concentration of all the reactants and all the products. The dependence of reaction rate on these concentrations is used to define the order of a chemical reaction. Consider the concentration of A (either a reactant or a product). In general,

$$\frac{dc_A}{dt} = f(c_{P_1}, c_{P_2}, c_{P_3}, \ldots, c_{R_2}, c_{R_2}, c_{R_3}, \ldots). \quad (6.3)$$

Often this function has the form

$$\frac{dc_A}{dt} = k' c_{P_1}^a c_{P_2}^b c_{P_3}^c \cdots c_{R_1}^A c_{R_2}^B c_{R_3}^C \quad (6.4)$$

(or perhaps a sum of such terms). The order of a chemical reaction is defined as the sum of all the exponents, i.e., order $= a + b + c + \ldots + A + B + C + \ldots$. For some simple chemical reactions these exponents equal the stoichiometric coefficients.[1]

6.2.1 Chemical Reactions of Low Order

6.2.1.1 Zeroth-Order Irreversible Reaction

In a zeroth-order irreversible reaction,

$$R \rightarrow P, \quad (6.5)$$

and reactant R produces product P with a rate that is independent of concentration:

1. In general, knowledge of the equation of mass balance of a chemical reaction (Equation 6.1) is not sufficient to determine the order of the reaction. The mass balance may represent the relation between reactants and products that have undergone a complex sequence of reactions with multiple intermediate stages. For example, the oxidation of glucose takes place in cells via an elaborate sequence of enzyme-catalyzed reactions.

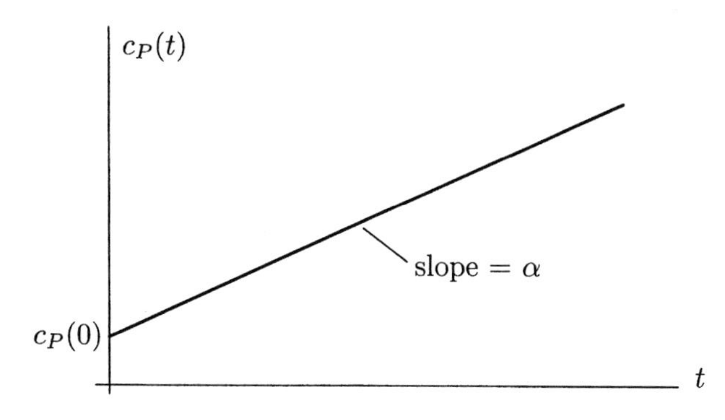

Figure 6.5 Solution for a zeroth-order chemical reaction.

$$\frac{dc_P}{dt} = -\frac{dc_R}{dt} = \alpha. \tag{6.6}$$

This equation is easily integrated to yield

$$c_P(t) = c_P(0) + \alpha t, \tag{6.7}$$

and the result is sketched in Figure 6.5.

6.2.1.2 *First-Order Irreversible Reaction*

In a first-order reaction,

$$R \to P, \tag{6.8}$$

but the rate of the reaction depends on the concentration of R:

$$-\frac{dc_R}{dt} = \alpha c_R. \tag{6.9}$$

We can integrate Equation 6.9 to yield

$$\int_{c_R(0)}^{c_R(t)} \frac{dc_R'}{c_R'} = -\int_o^t \alpha dt',$$

$$\ln \frac{c_R(t)}{c_R(0)} = -\alpha t,$$

$$c_R(t) = c_R(0)e^{-\alpha t}. \tag{6.10}$$

The concentration is sketched in Figure 6.6. Thus, in a first-order irreversible reaction, *all* the reactant is converted to product, and the time course is exponential.

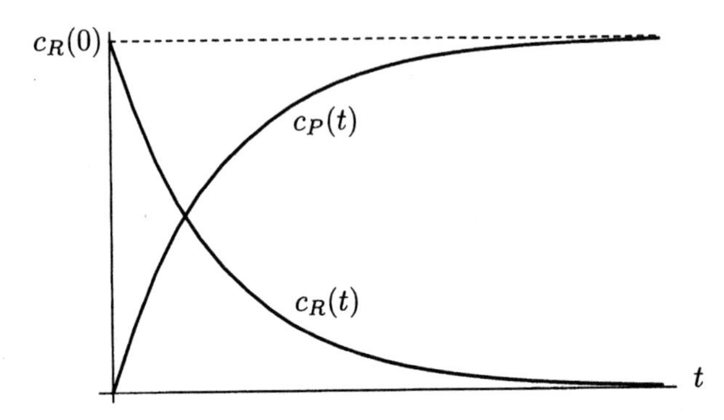

Figure 6.6 Solution for a first-order irreversible chemical reaction.

6.2.1.3 First-Order Reversible Reaction

A first-order reversible reaction can be represented as

$$R \underset{\beta}{\overset{\alpha}{\rightleftharpoons}} P, \tag{6.11}$$

where α and β are the forward and reverse rate constants. In such a reaction, R is consumed by the forward reaction and produced by the reverse reaction. Hence,

$$-\frac{dc_R}{dt} = \alpha c_R - \beta c_P, \tag{6.12}$$

$$-\frac{dc_P}{dt} = \beta c_P - \alpha c_R.$$

Equilibrium
At equilibrium, $dc_R/dt = dc_P/dt = 0$. Hence, the equilibrium values of concentrations, $c_R(\infty)$ and $c_P(\infty)$, satisfy the relation

$$\frac{c_P(\infty)}{c_R(\infty)} = \frac{\alpha}{\beta} = K_a, \tag{6.13}$$

where K_a has a variety of names including the *association, equilibrium, affinity, stability, binding,* and *formation* constant of the reaction. This result is a special case of the *law of mass action.* Note that K_a determines the equilibrium state of the reaction. If $K_a > 1$, then $c_P(\infty) > c_R(\infty)$ and there is more product than reactant at equilibrium. If $K_a < 1$, then $c_P(\infty) < c_R(\infty)$ and more reactant than product occurs at equilibrium.

Kinetics

To solve the kinetic equations for $c_R(t)$ and $c_P(t)$, we shall assume that the combination of reactant and product is conserved so that $c_R(t) + c_P(t) = C$ (constant). Therefore,

$$-\frac{dc_R(t)}{dt} = \alpha c_R(t) - \beta\,(C - c_R(t))\,, \tag{6.14}$$

or

$$\frac{dc_R(t)}{dt} + (\alpha + \beta)c_R(t) = \beta C.$$

Hence, the solution is

$$c_R(t) = c_R(\infty) - (c_R(\infty) - c_R(0))\,e^{-t/\tau}, \text{ for } t > 0, \tag{6.15}$$

where

$$c_R(\infty) = \frac{\beta}{\alpha + \beta}C = \frac{1}{1 + K_a}C, \tag{6.16}$$

and

$$\tau = \frac{1}{\alpha + \beta}. \tag{6.17}$$

Note that the equilibrium value of c_R depends only on the association constant K_a and the total concentration of R and P, and not explicitly on the rate constants. However, the rate constant of the reaction, $\alpha + \beta = 1/\tau$, equals the sum of the forward and reverse rate constants. The concentration of P is $c_P(t) = C - c_R(t)$. As shown in Figure 6.7, both $c_R(t)$ and $c_P(t)$ change exponentially from their initial to their equilibrium values.

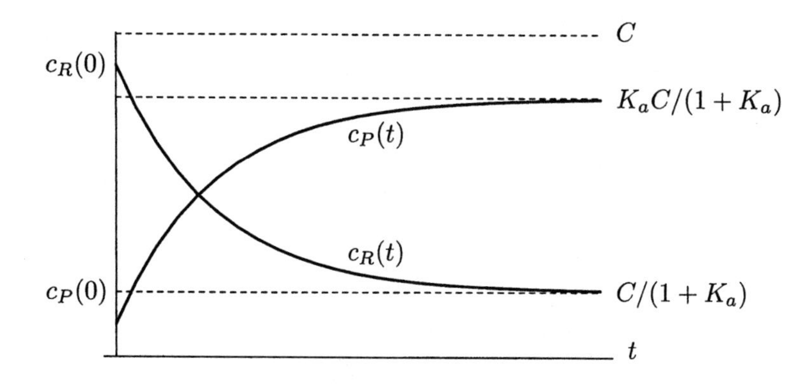

Figure 6.7 Solution for a first-order reversible chemical reaction.

6.2.1.4 Second-Order Reversible Reaction; Binding of an Enzyme with Its Substrate

In the remaining chapters, we will apply models of second-order reactions to many phenomena. Here, we will assume that the second-order reaction corresponds to the binding of an enzyme (E) with its substrate (S) to form the bound complex (ES), which is depicted as

$$S + E \underset{\beta}{\overset{\alpha}{\rightleftharpoons}} ES. \tag{6.18}$$

The kinetic equations are as follows:

$$\begin{aligned} \frac{dc_{ES}(t)}{dt} &= \alpha c_S(t)c_E(t) - \beta c_{ES}(t), \\ \frac{dc_S(t)}{dt} &= \beta c_{ES}(t) - \alpha c_S(t)c_E(t), \\ \frac{dc_E(t)}{dt} &= \beta c_{ES}(t) - \alpha c_S(t)c_E(t). \end{aligned} \tag{6.19}$$

Since each equation contains a term that is the product of two concentrations each raised to the first power, this is a second-order reaction.

If we make the additional assumption that the enzyme is conserved, then

$$C_{ET} = c_E(t) + c_{ES}(t), \tag{6.20}$$

where C_{ET} is the total quantity of enzyme.

Equilibrium
At equilibrium,

$$\frac{dc_{ES}}{dt} = \frac{dc_S}{dt} = \frac{dc_E}{dt} = 0,$$

and hence,

$$\alpha c_S(\infty)c_E(\infty) - \beta c_{ES}(\infty) = 0,$$

or

$$\frac{c_{ES}(\infty)}{c_S(\infty)c_E(\infty)} = \frac{\alpha}{\beta} = K_a = \frac{1}{K}. \tag{6.21}$$

K_a is the association constant, traditionally defined as the ratio of product concentrations to reactant concentration. K is the ratio of reactant concentrations to product concentrations and also has several names, including the *dissociation* and *unbinding* constant. For the second-order reversible reaction,

$$K = \frac{c_S(\infty)c_E(\infty)}{c_{ES}(\infty)}, \tag{6.22}$$

and K has the units of concentration mol/cm^3.

Examination of Equations 6.18 and 6.22 reveals that we can also represent this reaction as a first-order reversible reaction with a concentration-dependent rate constant, as follows:

$$E \underset{\beta}{\overset{\alpha c_S(\infty)}{\rightleftharpoons}} ES. \tag{6.23}$$

The value of K controls the ratio of the concentrations of reactants to the concentrations of products at equilibrium. If K is large, then the substrate and enzyme are found largely in their dissociated forms, S and E. Since we assume the enzyme is conserved, Equation 6.20 implies that

$$C_{ET} = c_E(\infty) + c_{ES}(\infty),$$

$$C_{ET} = \frac{Kc_{ES}(\infty)}{c_S(\infty)} + c_{ES}(\infty) = \left(\frac{K}{c_S(\infty)} + 1\right) c_{ES}(\infty),$$

or, at equilibrium,

$$c_{ES}(\infty) = \left(\frac{c_S(\infty)}{K + c_S(\infty)}\right) C_{ET}. \tag{6.24}$$

If $c_S(\infty)$ is large ($c_S(\infty) \gg K$), then $c_{ES}(\infty) \approx C_{ET}$, i.e., all the enzyme is bound to substrate. If $c_S(\infty)$ is small ($c_S(\infty) \ll K$), then $c_{ES}(\infty) \approx (C_{ET}/K)c_S(\infty)$, i.e., the amount of enzyme bound to substrate is proportional to the amount of substrate. For $c_S(\infty) = K$, half are bound to substrate and the other half are unbound. The relation of $c_{ES}(\infty)$ to $c_S(\infty)$ is that of a rectangular hyperbola and is shown in Figure 6.8. This relation is known as the Michaelis-Menton relation. Alternatively, the relation between $c_{ES}(\infty)$ and $c_S(\infty)$ can be represented by plotting $c_{ES}^{-1}(\infty)$ versus $c_S^{-1}(\infty)$, as follows:

$$\frac{1}{c_{ES}(\infty)} = \left(1 + \frac{K}{c_S(\infty)}\right)\frac{1}{C_{ET}} = \left(\frac{K}{C_{ET}}\right)\frac{1}{c_S(\infty)} + \frac{1}{C_{ET}}. \tag{6.25}$$

In these doubly reciprocal coordinates, the relation is a straight line (Figure 6.9), and the plot is often called a *Lineweaver-Burk* plot. The Lineweaver-Burk plot is useful for testing whether measurements obey the the Michaelis-Menton relation. In addition, linear regression analysis of data plotted in reciprocal coordinates is useful for determining the parameters K and C_{ET}.

Binding and dissociation reactions are often examined over a large range of concentrations, so that logarithmic scales of concentration are useful. For

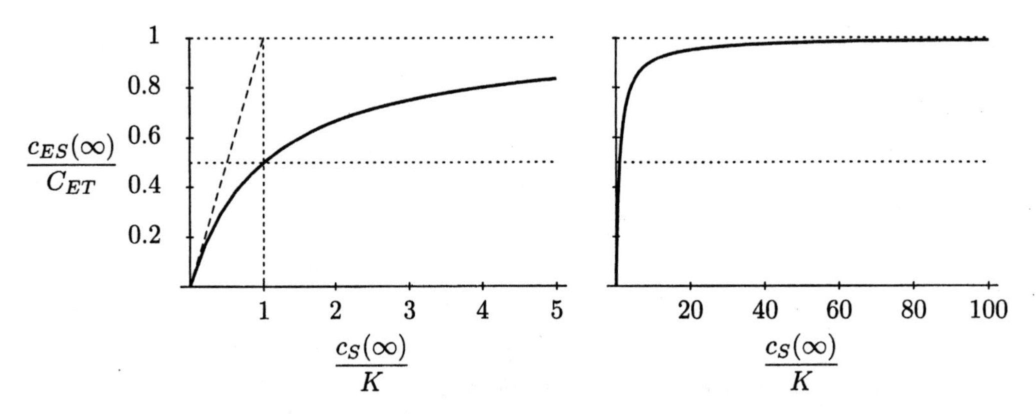

Figure 6.8 The Michaelis-Menton relation of the concentration of bound enzyme to the concentration of substrate shown on two abscissa scales.

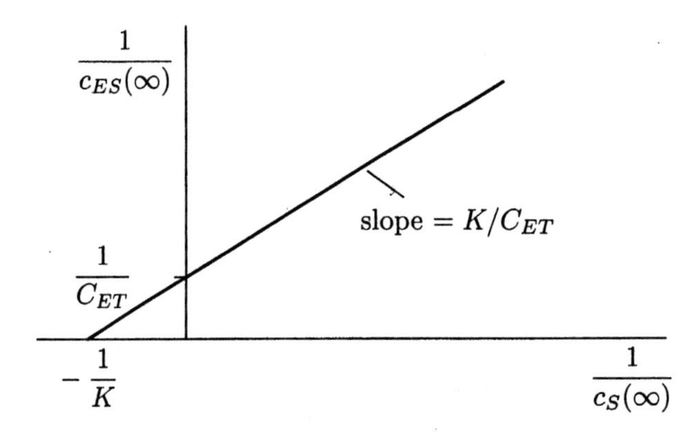

Figure 6.9 The Lineweaver-Burk plot shows the reciprocal of the bound enzyme concentration plotted versus the reciprocal of the substrate concentration.

example, consider the dissociation of an acid in water as an example of a second-order reaction:

$$HA \rightleftharpoons H^+ + A^-, \tag{6.26}$$

where HA is the acid and A^- is the conjugate base. The dissociation constant for this reaction is

$$K = \frac{c_{H^+}c_{A^-}}{c_{HA}}. \tag{6.27}$$

Therefore, for this reaction, Equation 6.24 becomes

$$\frac{c_{HA}}{C_{AT}} = \frac{c_{H^+}}{K + c_{H^+}} = \left(1 + \frac{K}{c_{H^+}}\right)^{-1}, \tag{6.28}$$

where C_{AT} is the total concentration of A in bound and dissociated form.

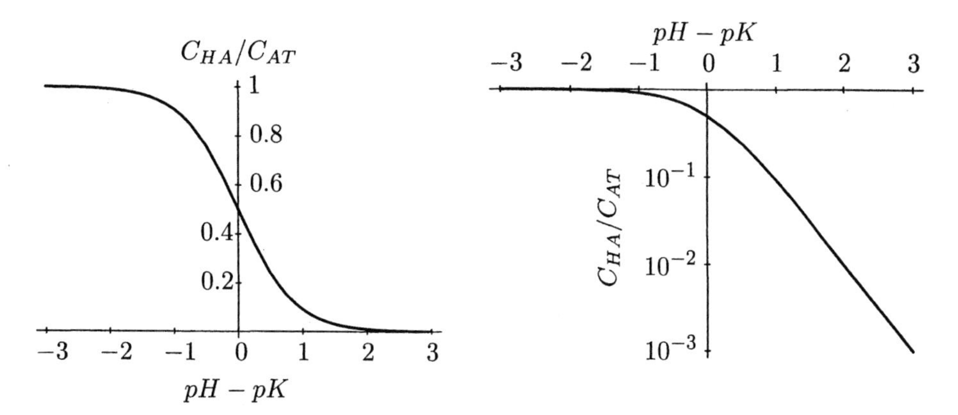

Figure 6.10 Dependence of an acid-base reaction on pH. Plotted versus $pH - pK$ are C_{HA}/C_{AT} (left panel) and the logarithm of this quantity (right panel).

We define

$$pH = -\log_{10}\left(\frac{c_{H^+}}{C_{ref}}\right) \quad \text{and} \quad pK = -\log_{10}\left(\frac{K}{C_{ref}}\right), \tag{6.29}$$

where the reference concentration C_{ref} is usually taken as 1 mol/L. If we combine Equations 6.28 and 6.29, we obtain

$$\frac{c_{HA}}{C_{AT}} = \frac{1}{1 + 10^{pH-pK}}. \tag{6.30}$$

Equation 6.30 and Figure 6.10 show that for $pH \ll pK$ (a high hydrogen ion concentration), the solution tends to be acid and all the the conjugate base is complexed with hydrogen ions. For $pH \gg pK$ (a low hydrogen ion concentration), the solution tends to be basic and all the conjugate base is uncomplexed with hydrogen ions. When $pH = pK$, half the conjugate base is complexed with hydrogen ions. The value of pK determines the range of pH for which the solution is acid.

Kinetics

We examine the kinetics of binding of an enzyme to its substrate when the substrate concentration changes as a step function of time, i.e., $c_S(t) = C_S u(t)$ where $u(t)$ is the unit step function. From Equations 6.19 and 6.20 we can obtain the following differential equation for $c_{ES}(t)$:

$$\frac{dc_{ES}(t)}{dt} = \alpha c_S(t) C_{ET} - (\alpha c_S(t) + \beta)\, c_{ES}(t).$$

From the differential equation we can see that if $c_S(t) = 0$ for $t < 0$, then $c_{ES}(t) = 0$ for $t < 0$. For $t > 0$, the differential equation is

$$\frac{dc_{ES}(t)}{dt} + (\alpha C_S + \beta)\, c_{ES}(t) = \alpha C_S C_{ET}.$$

Thus, the solution has the form

$$c_{ES}(t) = c_{ES}(\infty)\left(1 - e^{-t/\tau}\right).$$

The equilibrium value of the concentration of bound enzyme is

$$c_{ES}(\infty) = \frac{\alpha C_S}{\alpha C_S + \beta} C_{ET} = \frac{C_S}{C_S + K} C_{ET},$$

which is consistent with the discussion of equilibrium of the binding reaction. The equilibrium value of the concentration of bound enzyme as a function of the concentration of the substrate is a rectangular hyperbola. The time constant of the reaction is

$$\tau = \frac{1}{\alpha C_S + \beta}.$$

Thus, the time constant to reach equilibrium is inversely related to the substrate concentration. The larger the substrate concentration, the faster the reaction goes to equilibrium.

6.2.2 Reaction Rates

Near the end of the nineteenth century, Svante Arrhenius found that the rates of many chemical reactions are related to the temperature by an equation of the form

$$\ln \alpha = -\frac{C}{T} + D, \tag{6.31}$$

now called the *Arrhenius relation,* where α is the rate constant, T is the absolute temperature, and C and D are constants that depend upon the specific chemical reaction. An example of measurements of the dependence of the rate constant on temperature is shown in Figure 6.11. The Arrhenius relation can be recast in the form

$$\alpha = Be^{-E_a/kT}, \tag{6.32}$$

in terms of the new constant B and E_a. Arrhenius interpreted the quantity E_a, called the *activation energy,* as the energy the reactants had to acquire to

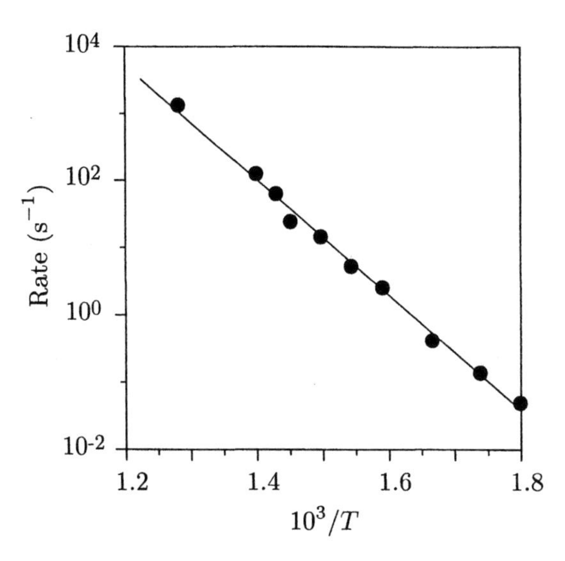

Figure 6.11 Temperature dependence of rate constant for the chemical reaction $H_2 + I_2 \rightarrow 2HI$ (Moore, 1972). The logarithm of the rate constant is plotted on the ordinate. T is the absolute temperature.

make the reaction proceed. The rationale was that the reactants were assumed to combine to form an unstable intermediate *activated complex* that had an energy E_a above the energy of the reactants. If the reactants are at equilibrium with the activated complex, then the concentration of the activated complex is proportional to a Boltzmann factor, $\exp(-E_a/kT)$. Finally, if the rate of the reaction is assumed to be proportional to the concentration of the activated complex, then the rate satisfies Equation 6.32. This formulation implied that as the activation energy increased, the rate of the chemical reaction decreased. The general notion embodied in the Arrhenius relation is essentially correct, although it has been refined in the twentieth century.

6.2.2.1 Potential Energy Surfaces

The reactants and products engaged in a chemical reaction can be viewed as a population of atoms whose potential energy depends on all the interatomic distances. Thus, we can imagine a multidimensional space, called a *potential energy hyperspace* or a *configuration space,* that relates the potential energy of the collection of these atoms to all the interatomic distances. The potential energy hyperspace contains critical or stationary points at which the gradient of the potential energy is zero. Minima in this space occur at points for which an incremental change in position in any direction results in an increase in the potential energy. Maxima occur at points for which an incremental change in position in any direction results in a decrease in the potential energy. Saddle

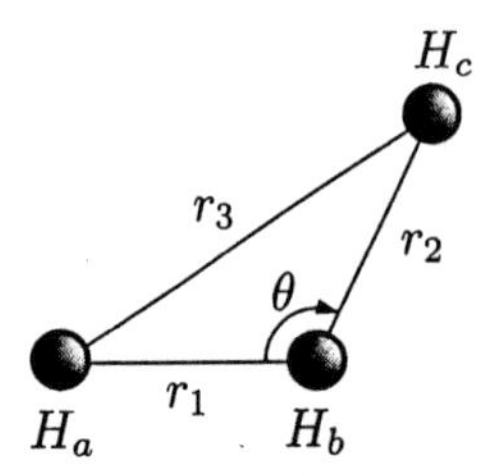

Figure 6.12 Geometric relations among three hydrogen atoms, H_a, H_b, and H_c. The interatomic distances are r_1, r_2, and r_3, where $r_3^2 = r_1^2 + r_2^2 - r_1 r_2 \cos\theta$.

points occur when the second derivative of the potential is negative in one direction and positive in all other directions. Minima correspond to the stable configurations of the atoms, which include the reactants and the products. Maxima or saddle points correspond to unstable intermediate compounds. The *reaction path* is defined as the path traversed in configuration space between the reactants and the products that requires the least energy. This path is the most probable path taken by the chemical reaction in converting reactants to products and vice-versa.

The potential energy hyperspace has been computed from quantum mechanical principles for only a few simple reactions (Murrell et al., 1984). Among the best understood of these is the potential energy hyperspace of H_3, corresponding to the reaction $H + H_2 \rightleftharpoons H_2 + H$. To make the more general notion of the potential energy hyperspace of a chemical reaction more tangible, we shall examine the hyperspace for this reaction. The spatial relation of the three hydrogen atoms is shown in Figure 6.12. The potential energy of the three atoms can be expressed as a function of the two interatomic distances, r_1 and r_2, plus the angle θ. The potential energy hyperspace was computed by numerical solution of the quantum mechanical equations (Liu, 1973; Siegbahn and Liu, 1978), and the resulting numerical solutions have been fit by analytic functions by several workers (Truhlar and Horowitz, 1978; Varandas, 1979; Varandas et al., 1987). The potential energy hyperspace for the electronic ground state is shown as a three-dimensional potential energy surface in Figure 6.13 and as a contour plot in Figure 6.14 as a function of the two interatomic distances for $\theta = 180°$. This particular arrangement of the hydrogen atoms, with three colinear nuclei, is the most energetically favorable for the chemical reaction.

The potential energy surface for this reaction has two stable configurations where the potential energy is a minimum. One stable configuration corresponds to the valley parallel to the r_1-axis shown in Figure 6.13 and the dark region parallel to the r_1-axis in Figure 6.14. This region of minimal potential energy occurs for a value of $r_2 \approx 0.74$ Å. This value of r_2 occurs when H_b and H_c form molecular hydrogen H_2, which has a bond length of 0.74 Å,

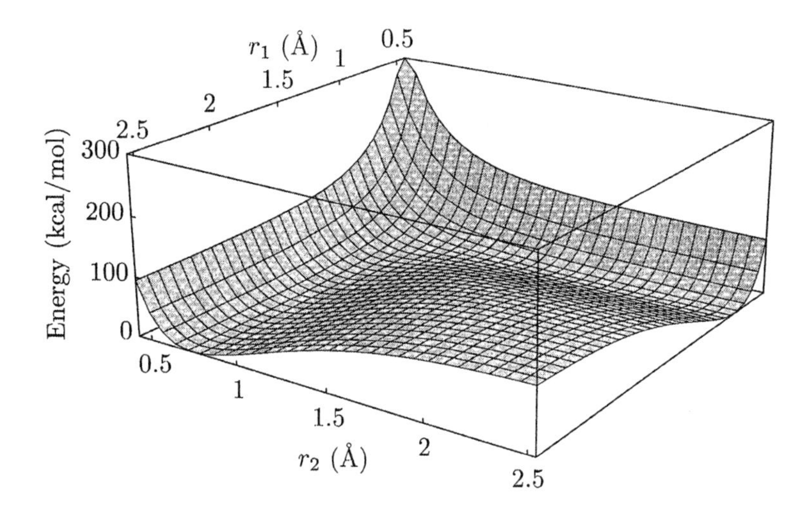

Figure 6.13 Potential energy surface for the reaction $H + H_2 \rightleftharpoons H_2 + H$. The potential energy is shown as a function of the interatomic distances r_1 and r_2 when all three nuclei are aligned linearly ($\theta = 180°$). The potential energy surface was computed from an analytical fit to numerical computations of the potential energy (Varandas, 1979).

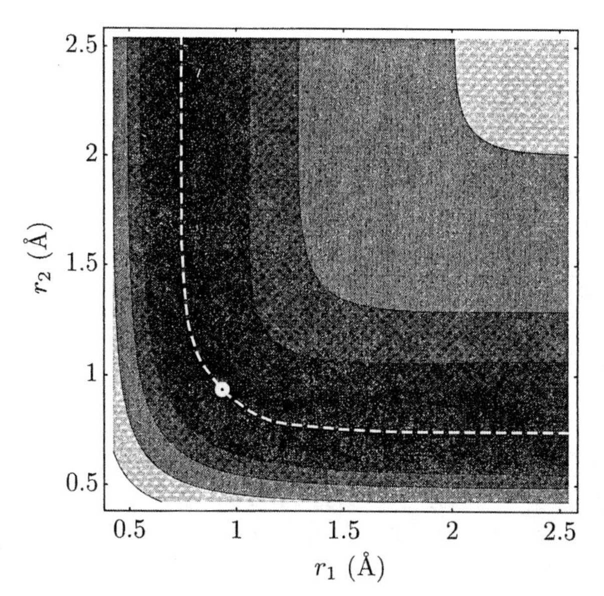

Figure 6.14 Contours of constant potential energy for the reaction $H + H_2 \rightleftharpoons H_2 + H$ for the same conditions as shown in Figure 6.13. The contours were computed at the energies 3, 6, 12, 24, 48, 96, and 192 kcal/mol. The lowest potential energies are shown with the darkest shading. The reaction path is indicated by a dashed white line. The location of the saddle point is marked with a white dot.

and where H_a is atomic hydrogen. The other stable configuration corresponds to the valley parallel to the r_2-axis, for which $r_1 \approx 0.74$ Å, so that H_a and H_b form molecular hydrogen and H_c is atomic hydrogen. The two minimal potential energy regions are connected at a saddle point in the three-dimensional surface, which occurs at $r_1 = r_2 = 0.93$ Å.

The two regions of minimal potential energy are linked by a path of minimal potential energy, the *reaction path*, which is indicated in Figure 6.14.

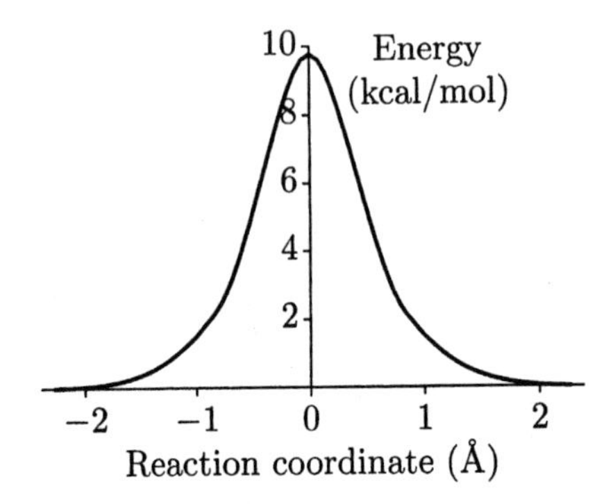

Figure 6.15 Energy profile along the reaction path for the reaction $H + H_2 \rightleftharpoons H_2 + H$ for the same conditions as shown in Figure 6.13. Distance along the reaction path is defined as zero for $r_1 = r_2 = 0.93$ Å, which corresponds to the location of the saddle point in the potential energy surface.

Because this path has minimal potential energy, it is the most likely path taken by the chemical reaction. A plot of the potential energy versus distance along the reaction path, called the *reaction coordinate,* is shown in Figure 6.15. The potential energy approaches zero when the distance along the reaction coordinate is either large and positive or large and negative. Both of these extrema occur when two of the hydrogen atoms form molecular hydrogen. Since the products and reactants in this simple chemical reaction are the same, the potential energy is symmetrical about the maximum potential energy. The maximum of the potential energy is 9.77 kcal/mol and corresponds to an unstable configuration of the three hydrogen atoms when all three atoms are colinear and separated by distances of 0.93 Å. The potential energy of this configuration is larger than that of the two stable configurations, and the difference of energy between the maximum energy and the minimum energy is called the *activation energy*. Thus, in order for the reaction to occur with high probability, the reactants must acquire an energy in excess of the activation energy.

In principle, if the initial positions and momenta of the reactants are known at any time, their dynamics can be computed with knowledge of the potential energy surface. Using a classical (Hamiltonian) or quantum mechanical approach, the trajectory of these atoms can be computed. They will, in general, take a variety of paths from the stable reactants to the stable products. The reaction rate can be obtained by computing the rate of transition of reactants to products taken over all possible paths. While such computations are possible in principle, they are not practically feasible for many reactions.

6.2.2.2 Theory of Absolute Reaction Rates

The *theory of absolute reaction rates*, also called *transition state theory*, makes some simplifying assumptions to allow an approximate computation of the

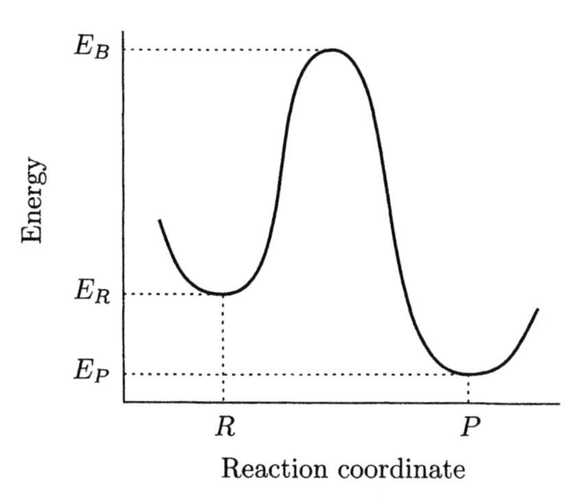

Figure 6.16 Energy profile along the reaction path with a single energy barrier.

reaction rate of a chemical reaction (Moore, 1972; Eyring et al., 1980). The idea is that since the reaction path is the minimum potential energy path from the reactants to the product, the reaction is most likely to occur along this path. Therefore, the rate of the reaction can be estimated by determining the rate at which the reaction passes the peak along the reaction path.

In contrast to the $H + H_2 \rightleftharpoons H_2 + H$ reaction, for an arbitrary chemical reaction that has a single maximum of energy along the reaction path, the minima in potential energy need not be the same. A plot of the energy profile traversed along the reaction path for a generic reaction is shown in Figure 6.16. E_R is the energy of the reactants, and E_P is the energy of the products. The potential energy at the peak of the reaction path, E_B, is called the *barrier energy*. The difference in the barrier energy and the energy of the reactants, $E_B - E_R$, is the activation energy. For the reaction path shown in Figure 6.16, the energy of the reactants is shown as exceeding that of the products. Therefore, the reaction results in a release of energy, and it is called *exergonic*. If energy is required for the reaction to proceed, the reaction is termed *endergonic*.

We have the following qualitative picture of the relation of reaction rate to barrier energy. While the population of reactant molecules will have average energy E_R, the energies of individual molecules may exceed E_B. Hence, some molecules can traverse the barrier and form product. The larger E_B, the smaller will be the number of molecules that have an energy in excess of E_B. Hence, we expect the rate constant for transition from R to P to decrease as E_B increases.

The theory of absolute reaction rates of chemical reactions provides a quantitative relation between reaction rate and barrier energy. Consistent with the Arrhenius relation, the rate constant of a chemical reaction depends exponentially on the ratio of the energy difference across a barrier to the thermal energy. If α is the rate constant at which reactants turn into products, and if β is the rate constant at which products turn into to reactants, then, according to this theory,

$$\alpha = Ae^{(E_R - E_B)/kT}, \tag{6.33}$$

$$\beta = Ae^{(E_P - E_B)/kT}, \tag{6.34}$$

where k is Boltzmann's constant, T is absolute temperature, and A is a rate factor of the form

$$A = \kappa \frac{kT}{h}, \tag{6.35}$$

where h is Planck's constant, and the factor κ depends on properties of the reactants and of the activated complex.

6.2.2.3 Temperature Dependence of Rate Constants

For most organisms, biological reactions take place over a narrow range of *absolute temperature*. To determine the dependence of biochemical reactions on temperature for the normal biological range of temperature, consider the ratio of rate constants at the two absolute temperatures T_1 and T_2:

$$\kappa_t = \frac{\alpha_1}{\alpha_2} = \frac{A(T_1)e^{(E-E_B)/kT_1}}{A(T_2)e^{(E-E_B)/kT_2}} \approx e^{(E-E_B)(1/kT_1 - 1/kT_2)} = e^{(E-E_B)(T_2-T_1)/(kT_1T_2)}, \tag{6.36}$$

where it is assumed that $A(T_1)/A(T_2)$ does not differ appreciably from 1. Equation 6.36 can be expressed in terms of temperature as follows:

$$\kappa_t = (Q_{10})^{(t_1-t_2)/10}, \tag{6.37}$$

where

$$Q_{10} = e^{10(E_B-E)/(kT_1T_2)}, \tag{6.38}$$

and t_1 and t_2 are the two temperatures in degrees Celsius: $T_1 = 273 + t_1$ and $T_2 = 273 + t_2$. Thus, the rate constant of a chemical reaction increases by a factor of Q_{10} for a change in temperature of 10°C. Of course, Q_{10} itself depends upon temperature (Equation 6.38), but only upon the *absolute temperature*. Hence, Q_{10} itself changes rather little with temperature for the range of temperatures at which biochemical reactions normally occur. Q_{10} is often determined empirically by measuring the reaction rates at two temperatures and computing Q_{10} from Equation 6.37.

6.2.2.4 Dependence of Equilibrium and Rate Constants on Activation Energy

The theory of absolute reaction rates is widely used to represent the rates of chemical reactions. We shall adopt this theory in our considerations. Note that, using this theory, we can interpret the association (equilibrium) constant in energetic terms. As an example, let us consider the equilibrium constant for a reversible first-order reaction. Substituting Equation 6.33 and 6.34 into Equation 6.13, we obtain

$$\frac{c_P(\infty)}{c_R(\infty)} = K_a = \frac{\alpha}{\beta} = \frac{Ae^{(E_R - E_B)/kT}}{Ae^{(E_P - E_B)/kT}} = e^{(E_R - E_P)/kT}. \tag{6.39}$$

Equation 6.39 shows that the equilibrium constant, which determines the equilibrium concentration ratio of product to reactant, depends only on the difference in energy between the product and the reactant, and not on the activation energy. Note that if $E_R > E_P$, then $K_a > 1$ and the equilibrium concentration of product exceeds that of the reactant. The reaction proceeds toward the product. If $E_R < E_P$, then $K_a < 1$ and the equilibrium concentration of reactant exceeds that of the product. The reaction proceeds toward the reactant. Equation 6.39 also indicates that at equilibrium the reactants and products distribute themselves according to a factor of the form

$$\frac{c_P(\infty)}{c_R(\infty)} = e^{(E_R - E_P)/kT} = \frac{e^{-E_P/kT}}{e^{-E_R/kT}}.$$

Thus,

$$c_P(\infty) \propto e^{-E_P/kT}, \tag{6.40}$$

and a similar expression can be found for $c_R(\infty)$. Thus, at equilibrium, the product and reactant distribute themselves according to their *Boltzmann factors*.

In contrast to the lack of dependence of the equilibrium constant on the activation energy, the rate constant of the reaction must depend on the activation energy, because that is the basic implication of the theory of absolute reaction rates. This property is seen by substituting Equations 6.33 and 6.34 into Equation 6.17, which yields

$$\frac{1}{\tau} = \alpha + \beta = Ae^{(E_R - E_B)/kT} + Ae^{(E_P - E_B)/kT}.$$

To summarize, the equilibrium state of a chemical reaction depends only on the energies of the reactants and products. However, the time it takes to reach equilibrium depends on the heights of energy barriers that separate the equilibrium states.

6.3 Discrete Diffusion Through Membranes

As indicated in Section 1.4, cellular membranes are of molecular thickness (≈ 75 Å). Thus, it might appear that models of diffusion through such membranes that treat the membrane as a continuous medium are unrealistic. An alternative view of permeation of solutes through such membranes is based on the theory of absolute reaction rates. In this type of model, the membrane is viewed as a series of potential energy barriers and wells that must be traversed by a solute. For example, the potential energy wells might represent binding sites for the solutes. A potential energy profile across such a hypothetical membrane is shown in Figure 6.17. There are M internal potential energy wells. The zeroth well is in contact with the solution on the inside of the membrane, and well $M + 1$ is in contact with the solution on the outside of the membrane. The density of the solute in the well m is $\mathfrak{N}_m$ mol/cm^2. The forward rate constants for transition from well m to well $m + 1$ is α_m and the backward rate constant for transition from well $m + 1$ to well m is β_{m+1}. For steady-state conditions, the solute cannot accumulate in the wells. Therefore, the flux across each barrier must be the same. This conclusion leads to the following set of equations for the flux in terms of the solute densities:

$$\begin{aligned}\phi_s &= \alpha_0\mathfrak{N}_0 - \beta_1\mathfrak{N}_1,\\ \phi_s &= \alpha_1\mathfrak{N}_1 - \beta_2\mathfrak{N}_2,\\ \phi_s &= \alpha_2\mathfrak{N}_2 - \beta_3\mathfrak{N}_3,\end{aligned} \tag{6.41}$$

and, in general,

$$\phi_s = \alpha_m\mathfrak{N}_m - \beta_{m+1}\mathfrak{N}_{m+1}, \quad \text{for } 0 \le m \le M. \tag{6.42}$$

Equations 6.41 and 6.42 can be solved recursively to yield

$$\phi_s = \alpha_0\left(\mathfrak{N}_0 - (\beta_1/\alpha_0)\mathfrak{N}_1\right),$$

$$\phi_s = \frac{\alpha_0\left(\mathfrak{N}_0 - (\beta_1\beta_2/\alpha_0\alpha_1)\mathfrak{N}_2\right)}{1 + (\beta_1/\alpha_1)},$$

$$\phi_s = \frac{\alpha_0\left(\mathfrak{N}_0 - (\beta_1\beta_2\beta_3/\alpha_0\alpha_1\alpha_2)\mathfrak{N}_3\right)}{1 + (\beta_1/\alpha_1) + (\beta_1\beta_2/\alpha_1\alpha_2)},$$

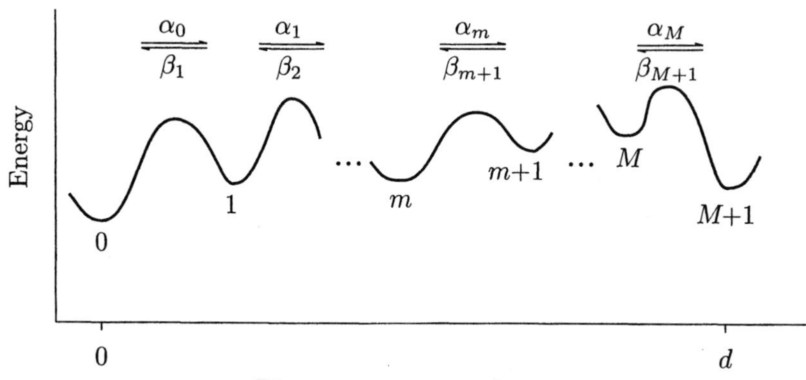

Figure 6.17 Membrane with multiple energy barriers.

or, in general,

$$\phi_s = \alpha_0 \frac{\mathfrak{N}_0 - \left(\prod_{m=1}^{M+1} (\beta_m/\alpha_{m-1})\right) \mathfrak{N}_{M+1}}{1 + \sum_{m=1}^{M} \prod_{k=1}^{m} (\beta_k/\alpha_k)}. \tag{6.43}$$

We now assume that the density of solute in the wells at 0 and at $M+1$ are proportional to the concentrations of solute in the solutions, i.e., $\mathfrak{N}_0 = \lambda c_s^i$ and $\mathfrak{N}_{M+1} = \lambda c_s^o$. Furthermore, we assume that when $c_s^i = c_s^o$, then $\phi_s = 0$. Substituting this condition into Equation 6.43 constrains

$$\prod_{m=1}^{M+1} \left(\frac{\beta_m}{\alpha_{m-1}}\right) = 1. \tag{6.44}$$

The significance of Equation 6.44 can be appreciated if we note, from Equation 6.39, that

$$\frac{\beta_m}{\alpha_{m-1}} = e^{-(E_m - E_{m-1})/kT}, \tag{6.45}$$

where E_m is the energy of potential well m. Substitution of Equation 6.45 into 6.44 and summation of terms in the exponent yields

$$e^{-(E_{M+1} - E_0)/kT} = 1.$$

Thus, the equilibrium condition $\phi_s = 0$ is equivalent to the condition that the two energy wells in contact with the solutions are at the same energy, $E_{M+1} = E_0$. Substitution of the condition that guarantees that the flux will be zero when the concentration difference is zero, Equation 6.44, into Equation 6.43 gives

$$\phi_s = \frac{\lambda \alpha_0}{1 + \sum_{m=1}^{M} \prod_{k=1}^{m} (\beta_k / \alpha_k)} (c_s^i - c_s^o). \tag{6.46}$$

Therefore, this mechanism of discrete diffusion of the solute through the membrane satisfies Fick's law for membranes in that the solute flux is proportional to concentration difference, although the microscopic mechanism of permeation is quite different from diffusion through a continuum. From Equation 6.46 we can identify the permeability of the membrane for solute s as

$$P_s = \frac{\lambda \alpha_0}{1 + \sum_{m=1}^{M} \prod_{k=1}^{m} (\beta_k / \alpha_k)}. \tag{6.47}$$

Hence, in the discrete diffusion mechanism, the permeability of the membrane depends on the rate constants of transitions between wells and, therefore, on the energy profile in the membrane. This profile will depend on the microscopic composition of the permeation pathway and the interaction of that pathway with the solute.

6.4 Carrier Models

Many of the phenomena exhibited by carrier-mediated transport systems described in Section 6.1 can be explained by carrier models, which postulate the existence of a chemical intermediate, a carrier or membrane-bound enzyme, that binds the solute. The bound complex translocates across the membrane. The mechanism of transport of these models is often called *facilitated diffusion* because transport is facilitated above that expected from diffusion alone. First, we shall examine in detail one of the simplest of these models, the *simple, symmetric, four-state carrier model*. This model illustrates many of the important features of carrier models without undue algebraic complexity. With a clear understanding of this model as background, we then investigate a more general four-state carrier model. The simple model is adequate to account for hexose transport in some cells but is inadequate to account for measurements of hexose transport in many cell types. The more general model is much more successful in this respect.

6.4.1 Simple, Symmetric, Four-State Carrier Model with One Solute

6.4.1.1 *Formulation of the Model*

To formulate a model for carrier-mediated transport, we assign specific chemical kinetics to each of the steps in the schematic diagram of carrier mecha-

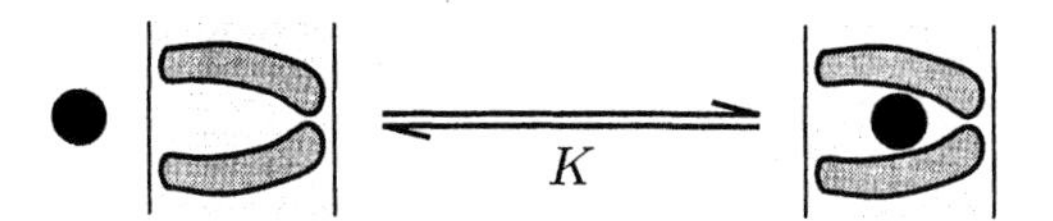

Figure 6.18 Kinetic diagram of the binding reaction of the simple, symmetric, four-state carrier.

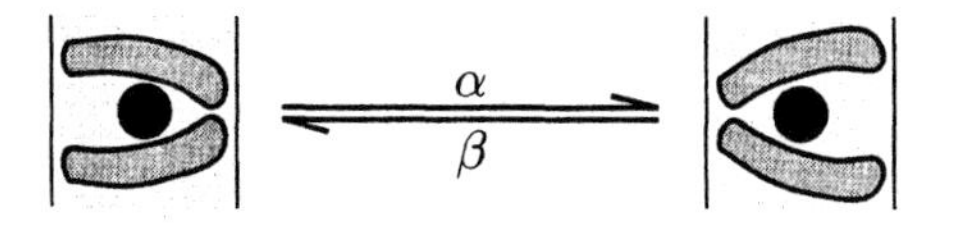

Figure 6.19 Kinetic diagram of the translocation step of the simple, symmetric, four-state carrier.

nisms shown in Figure 6.4. In the simple model, we assume that the binding of solute to carrier is so rapid that the reaction is at equilibrium at each instant in time. Therefore, the binding reaction can be fully characterized by a single dissociation constant K, as shown in Figure 6.18. Suppose there are a number of such carriers in the membrane. Let the concentrations of bound and unbound carriers in the membrane be c_{ES} and c_E, respectively. Let the concentration of solute be c_S. Then, at equilibrium,

$$K = \frac{c_S c_E}{c_{ES}}.$$

If the thickness of the membrane is d, then the densities of bound and unbound carriers in the membrane, i.e., the number of moles of carrier per unit area of membrane, are $\mathfrak{N}_{ES} = c_{ES}d$ and $\mathfrak{N}_E = c_E d$. Thus, the binding reaction can be expressed in terms of the carrier densities as

$$K = \frac{c_S \mathfrak{N}_E}{\mathfrak{N}_{ES}}.$$

We will express the model in terms of the densities of these carriers.

We shall assume that the translocation step for the scheme shown in Figure 6.4 is governed by first-order kinetics as shown in Figure 6.19. If the density of bound carrier that is open to the left side in Figure 6.19 is designated as $\mathfrak{N}_{ES}^i$ and that to the right side is $\mathfrak{N}_{ES}^o$, then the kinetic relation given by this model is

$$\frac{d\mathfrak{N}_{ES}^o}{dt} = \alpha \mathfrak{N}_{ES}^i - \beta \mathfrak{N}_{ES}^o.$$

The binding and translocation steps can be combined into a kinetic scheme by which the carrier transports the solute across a membrane, as shown in Figure 6.20. The solute binds with the carrier rapidly with dissociation constant K. The bound complex translocates across the membrane according to a first-order reversible reaction with forward and reverse rate constants of α and β, respectively. The bound complex dissociates at the

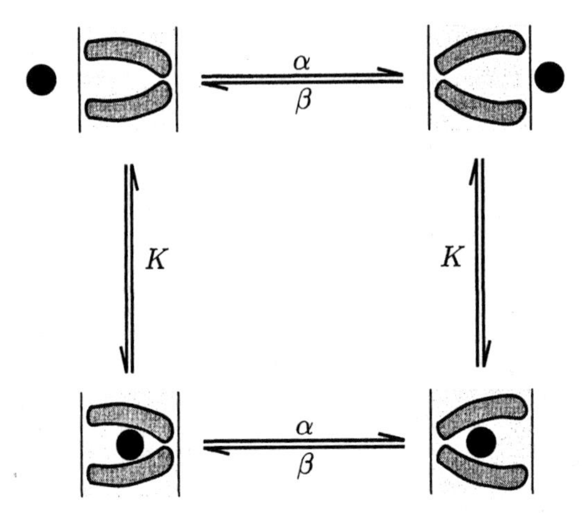

Figure 6.20 Schematic diagram of the simple, symmetric, four-state carrier. The four panels show schematic diagrams of a single carrier in its four states. The panels are connected by chemical kinetics with dissociation constant K and rate constants α and β. The left panels show the carrier in states accessible to the left face of the membrane both unbound (upper left) and bound (lower left) to the solute. The right panels show the carrier in states accessible to the right face of the membrane both unbound (upper right) and bound (lower right) to the solute.

opposite face of the membrane with the same dissociation constant. The unbound carrier translocates across the membrane according to a first-order reversible reaction with rate constants that are the same as those for the translocation of the bound carrier.

We shall analyze this model in the steady state, which implies that the density of carrier in each of its four states is independent of time. Thus, in the steady state, solute molecules that are translocated from the inside of the membrane to the outside are replaced by an equal number that bind on the inside surface, so that the density of solute bound to carrier is constant. Similarly, the density of bound solute molecules that arrive on the outside of the membrane is balanced by an instantaneous unbinding reaction that removes these solutes, so that the density of solute bound to carrier on the outside surface of the membrane remains constant.

This model is portrayed, somewhat more abstractly, superimposed on the membrane in the kinetic diagram shown in Figure 6.21.[2] This kinetic model represents a population of carriers that transport solute molecules.

2. The macroscopic transport equations we shall derive from this kinetic mechanism can also be obtained from other kinetic mechanisms (e.g., see Problem 6.5).

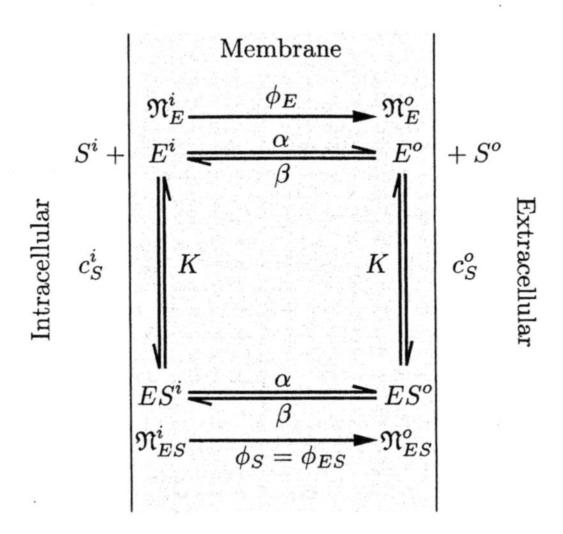

Figure 6.21 Kinetic diagram of the simple, symmetric, four-state carrier. This kinetic diagram describes the kinetics of a population of carriers.

6.4.1.2 Derivation

We now derive the relation between the flux of solute and the concentration of solute implied by Figure 6.21. We assume that the density of carrier molecules, or enzymes, in the membrane is $\mathfrak{N}_{ET}$ moles of carrier per unit area and that each of these carriers exists in one of four states, which we label ES^i, ES^o, E^i, and E^o. In the ES state, the solute S is bound to the carrier E; in the E state, the solute is not bound to the carrier. In the ES^i and E^i states, the carrier, bound and unbound, communicates with the solution on the inner side of the membrane. In the ES^o and E^o states, the carrier, bound and unbound, communicates with the solution on the outer side of the membrane. The densities of carrier in the four states are $\mathfrak{N}_{ES}^i$, $\mathfrak{N}_{ES}^o$, $\mathfrak{N}_E^i$, and $\mathfrak{N}_E^o$. The fluxes of bound and unbound carrier are ϕ_{ES} and ϕ_E, and the flux of solute is ϕ_S; all defined as positive when the flux is in the outward direction. The model is defined by the following assumptions:

- The total carrier density, bound and unbound, is constant, i.e., the sum of the density of carrier over all of its states equals the total density of carrier

$$\mathfrak{N}_{ET} = \mathfrak{N}_{ES}^i + \mathfrak{N}_{ES}^o + \mathfrak{N}_E^i + \mathfrak{N}_E^o. \tag{6.48}$$

- Since the carrier resides permanently in the membrane, the total flux of carrier must be zero, i.e.,

$$\phi_{ES} + \phi_E = 0. \tag{6.49}$$

- The only time the solute can cross the membrane is when it is bound to the carrier; ES^o is assumed to undergo a reversible change in conformation to the form ES^i. We assume that the unbound carrier undergoes a similar reaction between the two conformations E^i and E^o. These two reactions are assumed to be first-order reactions with the same forward and reverse rate constants, α and β, so that the fluxes are defined as follows:

$$\phi_{ES} = \phi_S = \alpha \mathfrak{N}_{ES}^i - \beta \mathfrak{N}_{ES}^o; \tag{6.50}$$

similarly,

$$\phi_E = \alpha \mathfrak{N}_E^i - \beta \mathfrak{N}_E^o. \tag{6.51}$$

- The binding reactions at the membrane interfaces are assumed to be rapid compared to the rate of transport of solute across the membrane, so that the membrane interface reactions are assumed to be at equilibrium; i.e.,

$$\frac{c_S^o \mathfrak{N}_E^o}{\mathfrak{N}_{ES}^o} = \frac{c_S^i \mathfrak{N}_E^i}{\mathfrak{N}_{ES}^i} = K, \tag{6.52}$$

where K is the dissociation constant and is assumed to be the same at both membrane interfaces.

This model is called "simple" because it has been assumed that the interfacial binding reactions are fast. It is called "symmetric" because it has been assumed that the binding reactions at the interfaces have the same dissociation constants and that the forward rate constants are the same for the bound and unbound carrier, as are the reverse rate constants.

We wish to obtain a relation between flux of solute and solute concentration from Equations 6.48–6.52. This result can be accomplished if we use Equation 6.50 and express $\mathfrak{N}_{ES}^i$ and $\mathfrak{N}_{ES}^o$ in terms of the solute concentration plus other constant factors.

First, we can combine Equations 6.50, 6.51, and 6.49 to give

$$\alpha \mathfrak{N}_{ES}^i - \beta \mathfrak{N}_{ES}^o = \beta \mathfrak{N}_E^o - \alpha \mathfrak{N}_E^i,$$

or

$$\frac{\alpha}{\beta}(\mathfrak{N}_{ES}^i + \mathfrak{N}_E^i) = \mathfrak{N}_{ES}^o + \mathfrak{N}_E^o. \tag{6.53}$$

Now substitute Equation 6.48 into Equation 6.53 to eliminate the factor $\mathfrak{N}_{ES}^o + \mathfrak{N}_E^o$:

$$\frac{\alpha}{\beta}(\mathfrak{N}_{ES}^i + \mathfrak{N}_E^i) = \mathfrak{N}_{ET} - (\mathfrak{N}_{ES}^i + \mathfrak{N}_E^i),$$

which gives the following equation involving two unknowns, $\mathfrak{N}^i_{ES}$ and $\mathfrak{N}^i_E$:

$$\left(\frac{\alpha}{\beta}+1\right)(\mathfrak{N}^i_{ES}+\mathfrak{N}^i_E)=\mathfrak{N}_{ET}.$$

This equation is interesting in its own right, since, after shuffling the terms, it yields

$$\frac{\mathfrak{N}^i_{ES}+\mathfrak{N}^i_E}{\mathfrak{N}_{ET}}=\frac{\beta}{\alpha+\beta}, \tag{6.54}$$

which indicates that the fraction of carriers on the inside face of the membrane is given by the quantity $\beta/(\alpha+\beta)$. We can eliminate $\mathfrak{N}^i_E$, using Equation 6.52, to yield

$$\left(\frac{\alpha}{\beta}+1\right)\left(\mathfrak{N}^i_{ES}+\frac{K\mathfrak{N}^i_{ES}}{c^i_S}\right)=\mathfrak{N}_{ET}.$$

We can solve for $\mathfrak{N}^i_{ES}$ to obtain

$$\mathfrak{N}^i_{ES}=\left(\frac{\beta}{\alpha+\beta}\right)\left(\frac{c^i_S}{c^i_S+K}\right)\mathfrak{N}_{ET}. \tag{6.55}$$

Thus, the fraction of bound carriers on the inside face of the membrane is a product of two ratios. One ratio, $\beta/(\alpha+\beta)$, gives the fraction of the carriers on the inside face of the membrane, and the other ratio, $c^i_S/(c^i_S+K)$, gives the fraction of the carriers on the inside face of the membrane that are bound to solute. By substituting Equation 6.55 into Equation 6.54, we obtain

$$\mathfrak{N}^i_E=\left(\frac{\beta}{\alpha+\beta}\right)\left(\frac{K}{c^i_S+K}\right)\mathfrak{N}_{ET}. \tag{6.56}$$

Therefore, as c^i_S increases, $\mathfrak{N}^i_{ES}$ increases and $\mathfrak{N}^i_E$ decreases hyperbolically. For the simple, symmetric four-state carrier model, the state of the carrier on the inside face of the membrane depends only on the solute concentration on the inside and is independent of the solute concentration on the outside face of the membrane.

By an analogous method we can solve for $\mathfrak{N}^o_{ES}$:

$$\mathfrak{N}^o_{ES}=\left(\frac{\alpha}{\alpha+\beta}\right)\left(\frac{c^o_S}{c^o_S+K}\right)\mathfrak{N}_{ET}, \tag{6.57}$$

and $\mathfrak{N}^o_E$:

$$\mathfrak{N}^o_E=\left(\frac{\alpha}{\alpha+\beta}\right)\left(\frac{K}{c^o_S+K}\right)\mathfrak{N}_{ET}. \tag{6.58}$$

Therefore, as c_S^o increases, $\mathfrak{N}_{ES}^o$ increases and $\mathfrak{N}_E^o$ decreases hyperbolically. The state of the carrier on the outside face of the membrane depends only on the solute concentration on the outside and is independent of the solute concentration on the inside face of the membrane.

The solute flux can be obtained by substitution of Equations 6.55 and 6.57 into Equation 6.50 to obtain

$$\phi_S = (\phi_S)_{max}\left(\frac{c_S^i}{c_S^i + K} - \frac{c_S^o}{c_S^o + K}\right), \tag{6.59}$$

where

$$(\phi_S)_{max} = \frac{\alpha\beta}{\alpha + \beta}\mathfrak{N}_{ET}. \tag{6.60}$$

6.4.1.3 The State of the Carrier

We have seen (Equations 6.50 and 6.51) that the fluxes depend on the number of carriers in each of the four possible states and that the number of carriers in each of the states depends on the concentration of solute on the two sides of the membrane (Equations 6.55–6.58). These equations can be understood qualitatively by means of the schematic diagrams shown in Figure 6.22. All seven panels show a membrane with 20 carriers: 10 open to the inside of the membrane and 10 open to the outside of the membrane. In panel 1, $c_S^i = c_S^o = 0$, so that none of the carriers are bound to solutes. In panel 2, the concentration of solute has been increased on the inside of the membrane, but $c_S^i < K$. Hence, less than half of the carriers open to the inside are bound to carrier. None of those open to the outside are bound to carrier, since $c_S^o = 0$. In panel 3, $c_S^i = K$, so that half of the carriers open to the inside are bound to solute and half are not. In panel 4, $c_S^i \gg K$ and all the carriers open to the inside are bound to solutes. In panels 5–7, $c_S^i \gg K$ and c_S^o is progressively increased, so that the number of carriers open to the outside of the membrane is progressively increased. In panel 7, the concentrations on both sides of the membrane are so high that all the carriers are bound to solutes.

The carrier mechanism can be understood more quantitatively with the aid of the more abstract diagram of carrier states shown in Figure 6.23 and the results shown in Figure 6.24. Let us examine how the state diagram changes as the concentration of solute is changed on the two sides of the membrane. Suppose we start with no solute on either side of the membrane, $c_S^i = c_S^o = 0$. In that case the interfacial binding reactions at the two membrane surfaces

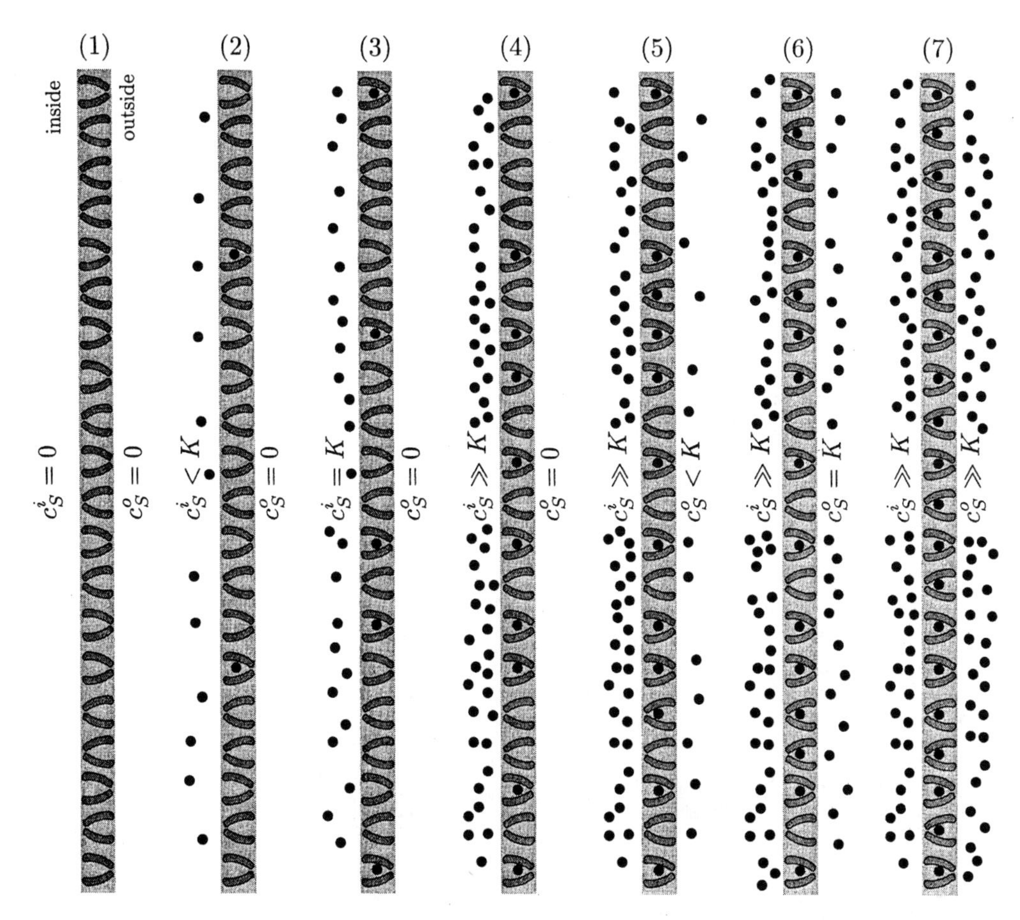

Figure 6.22 Schematic diagram of a membrane with twenty simple, symmetric, four-state carriers. The seven panels show the states of the carriers for different concentrations of solute on the two sides of the membrane.

are driven to the dissociated states, and no solute is bound to carrier. Thus, all the carrier is found in the uncomplexed state on both sides of the membrane; i.e., $\mathfrak{N}_{ES}^i = \mathfrak{N}_{ES}^o = 0$. Therefore, $\mathfrak{N}_E^i/\mathfrak{N}_{ET} = \beta/(\alpha+\beta)$ and $\mathfrak{N}_E^o/\mathfrak{N}_{ET} = \alpha/(\alpha+\beta)$. For the case $\alpha = \beta$, which is assumed in the schematic diagram of Figure 6.23, this implies that $\mathfrak{N}_E^i = \mathfrak{N}_E^o = \mathfrak{N}_{ET}/2$. Only two of the four possible carrier states are occupied, and each is occupied with half the total carrier density. Since $\mathfrak{N}_{ES}^i = \mathfrak{N}_{ES}^o = 0$, the solute flux through the membrane is zero.

If c_S^i is increased, then $\mathfrak{N}_{ES}^i$ increases and $\mathfrak{N}_E^i$ decreases. The carriers on the outer surface of the membrane do not change state, so that $\mathfrak{N}_{ES}^o$ remains

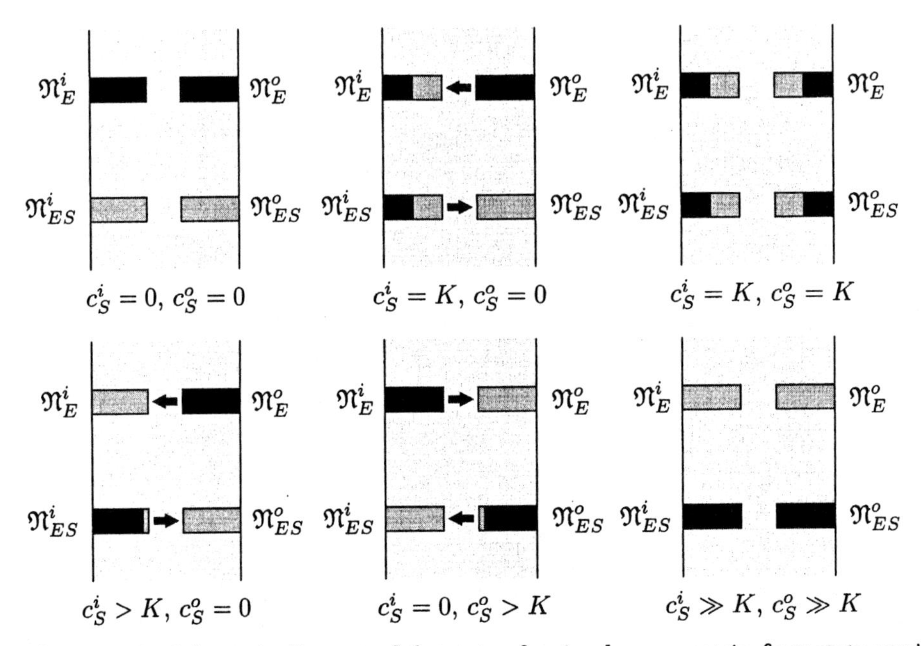

Figure 6.23 Schematic diagram of the state of a simple, symmetric, four-state carrier. The four possible states of the carrier are indicated by bar graphs superimposed on a schematic diagram of the membrane (see Figure 6.21). The length of each dark bar is proportional to the fraction of the carrier density that is in that state; the thin open rectangle indicates the maximum carrier density in each of the states. In this diagram we have assumed that $\alpha = \beta$ so that the thin open rectangles have the same widths. The arrows indicate the direction of the fluxes of E and ES. If the flux is zero the arrow is omitted.

at zero. Therefore, $\mathfrak{N}_{ES}^i > \mathfrak{N}_{ES}^o$ and $\mathfrak{N}_E^i < \mathfrak{N}_E^o$. This state leads to an efflux of bound carrier, and hence to an outward flux of solute. Since the influx of solute remains at zero, the net flux of bound carrier and solute increase as c_S^i is increased. That is, $\phi_{ES} = \phi_S > 0$. Also, there is an inward flux of unbound carrier, i.e., $\phi_E < 0$. If c_S^i is increased so that $c_S^i = K$, then $\mathfrak{N}_E^i/\mathfrak{N}_{ET} = \mathfrak{N}_{ES}^i/\mathfrak{N}_{ET} = \beta/(2(\alpha+\beta))$. Thus, half the carriers on the inside surface of the membrane are bound and half are unbound to solute. If, with $c_S^i = K$, c_S^o is increased from zero, then $\mathfrak{N}_{ES}^o$ will increase and $\mathfrak{N}_E^o$ will decrease. The increase in $\mathfrak{N}_{ES}^o$ will reduce the outward flux of solute until $c_S^o = K$, when half the carrier on the outer surface of the membrane is in the bound and half is in the unbound form. For the case shown, this implies that all four carrier states are occupied equally with one-quarter of the carriers. Under these circumstance the flux of both bound and unbound carrier is zero.

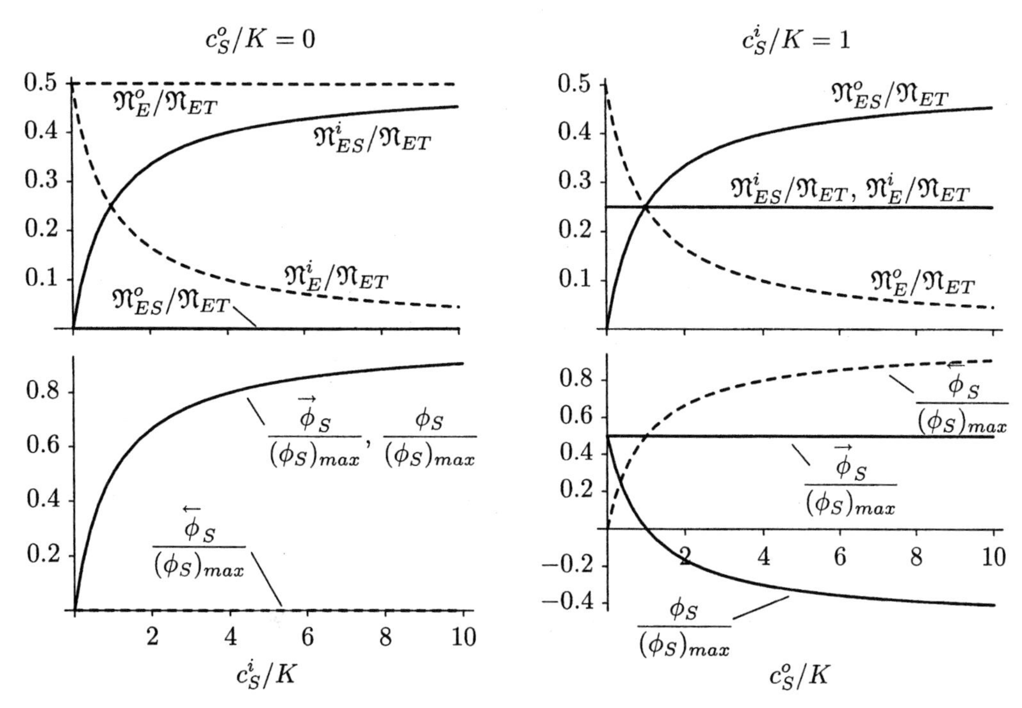

Figure 6.24 Dependence of the four carrier state densities and the solute fluxes on solute concentration. The left panel shows the dependence of these variables on c_S^i for $c_S^o/K = 0$, and the right panel shows the dependence on c_S^o for $c_S^i/K = 1$. $\vec{\phi}_S$ and $\overleftarrow{\phi}_S$ are the unidirectional efflux and influx of S defined in Equations 6.63 and 6.64.

Further increases in solute concentration on either or both sides of the membrane drive the carrier into its bound states. The flux depends upon the relative number of carriers in the two bound states.

6.4.1.4 Properties of the Relation of Flux to Concentration

The kinetic scheme shown in Figure 6.21 leads to the macroscopic transport Equation 6.59.[3] In this section we examine properties of the relation of solute flux to solute concentration for the simple, symmetric, four-state carrier that is implied by Equation 6.59.

3. A number of different kinetic schemes will lead to the same macroscopic relation between flux and concentration.

Passive Transport
Equation 6.59 can be rewritten in the form

$$\phi_S = K(\phi_S)_{max} \frac{c_S^i - c_S^o}{(c_S^i + K)(c_S^o + K)}.$$

Note that, for this model, $\phi_S > 0$ if $c_S^i > c_S^o$; i.e., flow of solute is down the concentration gradient as for diffusion of solute. Therefore, this model for carrier-mediated transport exhibits passive transport.

Relation to Fick's Law for Membranes
At low solute concentrations, $c_S^i < K$, $c_S^o < K$, Equation 6.59 can be approximated as

$$\phi_S \approx \frac{(\phi_S)_{max}}{K} \left(c_S^i - c_S^o \right), \tag{6.61}$$

so that the transport of solute cannot be distinguished from diffusion, and the effective permeability of the membrane for the solute, P_S, is

$$P_S = \frac{(\phi_S)_{max}}{K}. \tag{6.62}$$

However, even under these conditions, for which the relation of flux to concentration obeys Fick's law for membranes, the transport can show structural specificity not seen for diffusion processes. The structural specificity of P_S results from its dependence on the dissociation constant K. Note that P_S depends inversely on K so that as K increases, P_S decreases. Thus, K is a measure of *im*permeability. Since K can show strong structural specificity (Table 6.1), so can P_S.

Unidirectional Flux
Measurements of fluxes are often obtained with radioactive tracers that are used to tag, for instance, the intracellular solute. The efflux of radioactive solute is measured as a function of experimental variables, and the radioactive solute on the extracellular face of the membrane is rapidly removed so that it is not transported intracellularly. Thus, the measured flux is an efflux with no appreciable influx. The total unidirectional flux can be estimated by determining the specific activity of the radioactive tracer, i.e., the fraction of the total solute that is tracer. Thus, we define the outward unidirectional flux, or efflux, of tracer as $\vec{\phi}_S = \alpha \mathfrak{N}_{ES*}^i$ and the total efflux as $\vec{\phi}_S = \alpha \mathfrak{N}_{ES*}^i (c_S^i / c_{S*}^i)$, where $S*$ is the tracer solute. Since $S*$ obeys the same binding reaction as S, $\mathfrak{N}_{ES*}^i = c_{S*}^i \mathfrak{N}_E^i / K$. Combining these relations, we obtain $\vec{\phi}_S = \alpha \mathfrak{N}_E^i c_S^i / K$, which leads to the relation

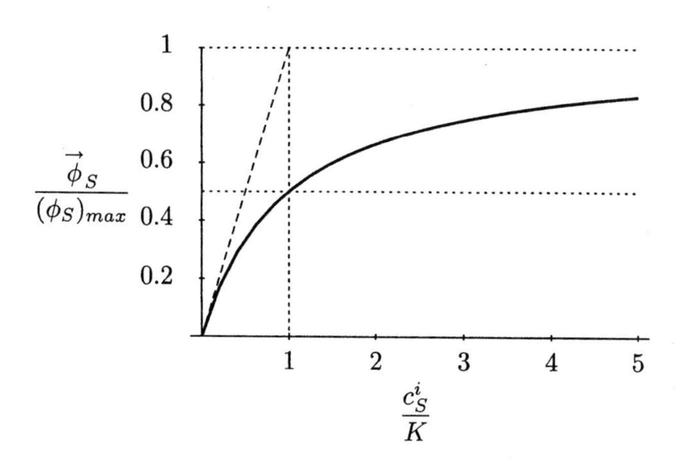

Figure 6.25 Unidirectional flux versus concentration according to Equation 6.63. In these normalized coordinates, the normalized efflux has a slope of 1 (indicated by the dashed line) at $c_S^i/K = 0$ and increases to 1 for large values of c_S^i/K. Therefore, the line tangent to the normalized flux at $c_S^i/K = 0$ has the value 1 when $c_S^i/K = 1$.

$$\overrightarrow{\phi}_S = (\phi_S)_{max} \frac{c_S^i}{c_S^i + K}. \tag{6.63}$$

Similarly, the inward unidirectional flux or influx, $\overleftarrow{\phi}_S$, is

$$\overleftarrow{\phi}_S = (\phi_S)_{max} \frac{c_S^o}{c_S^o + K}. \tag{6.64}$$

Therefore, the net outward flux is the difference of the unidirectional fluxes:

$$\phi_S = \overrightarrow{\phi}_S - \overleftarrow{\phi}_S. \tag{6.65}$$

Note that for the simple, symmetric, four-state model we have analyzed, $\overrightarrow{\phi}_S = \phi_S$ for $c_S^o = 0$, and $\overleftarrow{\phi}_S = \phi_S$ for $c_S^i = 0$. According to the above equations, the relation of unidirectional flux to concentration is a rectangular hyperbola, as shown in Figure 6.25. To test whether this relation fits measurements, it is useful to plot the results of measurements in reciprocal coordinates. Equation 6.63 can be transformed into

$$\frac{1}{\overrightarrow{\phi}_S} = \frac{1}{(\phi_S)_{max}} \left(1 + \frac{K}{c_S^i}\right). \tag{6.66}$$

Hence, by plotting $1/\overrightarrow{\phi}_S$ versus $1/c_S^i$ and calculating the slope and the intercept, one can obtain $(\phi_S)_{max}$ and K (Figure 6.26).

Macroscopic and Microscopic Transport Parameters

The simple, symmetric, four-state carrier model is characterized by four parameters—α, β, K, and $\mathfrak{N}_{ET}$. We refer to these as the *microscopic* parameters because they characterize the kinetic mechanism for a single carrier.

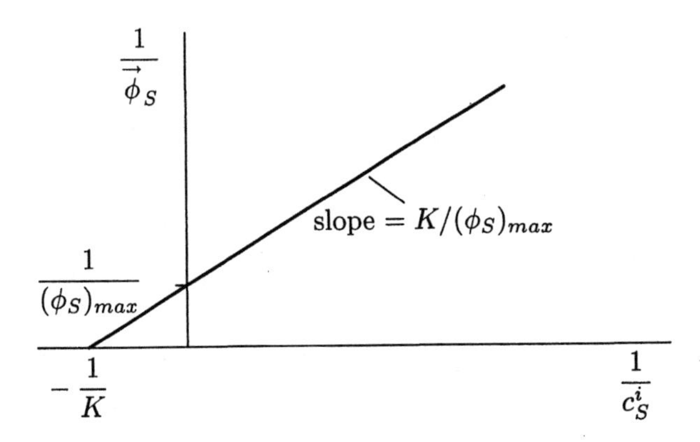

Figure 6.26 Unidirectional flux versus concentration in reciprocal coordinates according to Equation 6.66. The intercepts of the line are at $-1/K$ and $1/(\phi_S)_{max}$.

However, from measurements of the relation of flux to concentration for a population of carriers (as in Equation 6.59), only two parameters can be determined: K and $(\phi_S)_{max}$. We call these the *macroscopic* parameters, because they result from measurements of the flux of many carriers simultaneously. Thus, the parameters α, β, and $\mathfrak{N}_{ET}$ cannot be determined uniquely from just macroscopic measurements of transport.

6.4.2 Simple, Symmetric, Six-State Carrier Model with Two Ligands

We next consider that the carrier can bind more than one substance or *ligand*. First, we consider the case in which both ligands are transported through the membrane by the carrier, i.e., where both ligands are solutes. Then we consider two cases for which one ligand is transported and the other is not.

6.4.2.1 Both Ligands Are Transported

Derivation

The simple, symmetric carrier shown in Figure 6.21 can be extended to deal with two solutes that are both transported by the carrier (Figure 6.27). In this scheme, solutes S and R can combine with carrier E, but with different affinities. The binding to solute S has dissociation constant K_S, and the binding to R has dissociation constant K_R. All other kinetic parameters are symmetric, as in the one-solute case considered previously.

The kinetic equations are analogous to those derived for the single-solute case, except that the carrier now has six states:

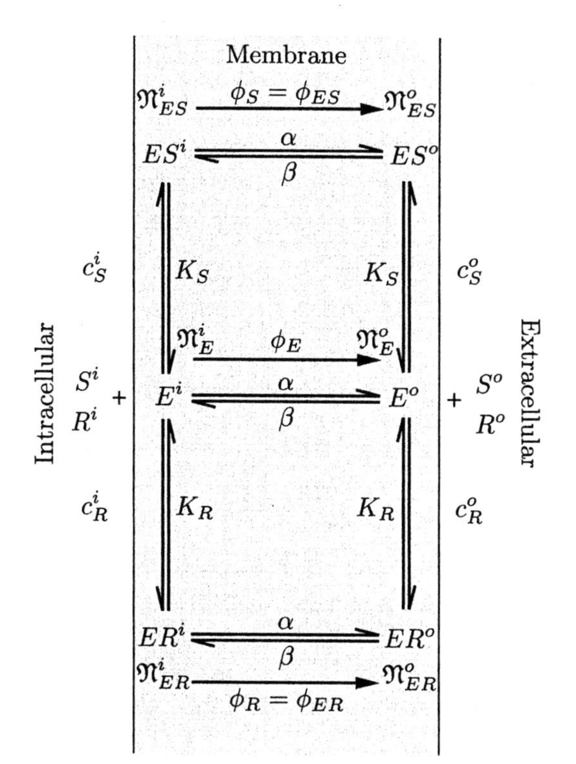

Figure 6.27 Kinetic diagram for a simple, symmetric carrier that can bind two ligands (S and R), both of which are transported.

- We assume that the total amount of carrier, bound and unbound to solute, is constant:

$$\mathfrak{N}_{ET} = \mathfrak{N}^i_{ES} + \mathfrak{N}^o_{ES} + \mathfrak{N}^i_{ER} + \mathfrak{N}^o_{ER} + \mathfrak{N}^i_E + \mathfrak{N}^o_E, \tag{6.67}$$

where $\mathfrak{N}_{ET}$ is the total density of carrier in the membrane.

- Since the carrier remains in the membrane, the net flux of carrier must be zero:

$$\phi_{ES} + \phi_{ER} + \phi_E = 0. \tag{6.68}$$

- The fluxes are governed by first-order kinetics, so that for solute S,

$$\phi_{ES} = \phi_S = \alpha \mathfrak{N}^i_{ES} - \beta \mathfrak{N}^o_{ES}. \tag{6.69}$$

For solute R,

$$\phi_{ER} = \phi_R = \alpha \mathfrak{N}^i_{ER} - \beta \mathfrak{N}^o_{ER}. \tag{6.70}$$

Also,

$$\phi_E = \alpha \mathcal{N}_E^i - \beta \mathcal{N}_E^o. \tag{6.71}$$

- The reactions at the membrane interfaces are assumed to take place so rapidly compared to the rate of transport of solute across the membrane that the membrane interface reactions are assumed to be at equilibrium; i.e.,

$$\frac{c_S^o \mathcal{N}_E^o}{\mathcal{N}_{ES}^o} = \frac{c_S^i \mathcal{N}_E^i}{\mathcal{N}_{ES}^i} = K_S \quad \text{and} \quad \frac{c_R^o \mathcal{N}_E^o}{\mathcal{N}_{ER}^o} = \frac{c_R^i \mathcal{N}_E^i}{\mathcal{N}_{ER}^i} = K_R, \tag{6.72}$$

where K_S and K_R are assumed to be the same at both membrane interfaces.

By combining Equations 6.68, 6.69, 6.70, and 6.71, we obtain

$$(\alpha \mathcal{N}_{ES}^i - \beta \mathcal{N}_{ES}^o) + (\alpha \mathcal{N}_{ER}^i - \beta \mathcal{N}_{ER}^o) + (\alpha \mathcal{N}_E^i - \beta \mathcal{N}_E^o) = 0. \tag{6.73}$$

These equations have a large number of variables, and they are a chore to solve unless a systematic approach is taken. Recall that we wish to obtain the transport of solutes S and R as a function of their concentrations and as a function of the transport parameters K_S, K_R, α, β, and $\mathcal{N}_{ET}$. Therefore, it is useful to regard the system of algebraic equations given by Equations 6.68 through 6.73 as a set of six equations in the six unknowns $\mathcal{N}_{ES}^i$, $\mathcal{N}_{ES}^o$, $\mathcal{N}_{ER}^i$, $\mathcal{N}_{ER}^o$, $\mathcal{N}_E^i$, and $\mathcal{N}_E^o$. With this in mind we can rewrite the equations in matrix form as follows:

$$\begin{bmatrix} K_S & 0 & 0 & 0 & -c_S^i & 0 \\ 0 & K_S & 0 & 0 & 0 & -c_S^o \\ 0 & 0 & K_R & 0 & -c_R^i & 0 \\ 0 & 0 & 0 & K_R & 0 & -c_R^o \\ \alpha & -\beta & \alpha & -\beta & \alpha & -\beta \\ 1 & 1 & 1 & 1 & 1 & 1 \end{bmatrix} \begin{bmatrix} \mathcal{N}_{ES}^i \\ \mathcal{N}_{ES}^o \\ \mathcal{N}_{ER}^i \\ \mathcal{N}_{ER}^o \\ \mathcal{N}_E^i \\ \mathcal{N}_E^o \end{bmatrix} = \begin{bmatrix} 0 \\ 0 \\ 0 \\ 0 \\ 0 \\ \mathcal{N}_{ET} \end{bmatrix}. \tag{6.74}$$

The first four rows correspond to the four relations in Equation 6.72. The fifth row results from Equation 6.73, and the sixth row corresponds to Equation 6.67. This set of simultaneous equations is tedious to solve by hand.[4] However, the resulting solutions are simple to interpret. The solutions are as follows:

4. Solutions were obtained using the symbolic mathematics software packages Macsyma and Mathematica.

$$\mathfrak{N}_{ES}^{i} = \mathfrak{N}_{ET}\left(\frac{\beta}{\alpha+\beta}\right)\left(\frac{c_S^i}{c_S^i + K_S(1+\frac{c_R^i}{K_R})}\right), \tag{6.75}$$

$$\mathfrak{N}_{ES}^{o} = \mathfrak{N}_{ET}\left(\frac{\alpha}{\alpha+\beta}\right)\left(\frac{c_S^o}{c_S^o + K_S(1+\frac{c_R^o}{K_R})}\right), \tag{6.76}$$

$$\mathfrak{N}_{ER}^{i} = \mathfrak{N}_{ET}\left(\frac{\beta}{\alpha+\beta}\right)\left(\frac{c_R^i}{c_R^i + K_R(1+\frac{c_S^i}{K_S})}\right), \tag{6.77}$$

$$\mathfrak{N}_{ER}^{o} = \mathfrak{N}_{ET}\left(\frac{\alpha}{\alpha+\beta}\right)\left(\frac{c_R^o}{c_R^o + K_R(1+\frac{c_S^o}{K_S})}\right), \tag{6.78}$$

$$\mathfrak{N}_{E}^{i} = \mathfrak{N}_{ET}\left(\frac{\beta}{\alpha+\beta}\right)\left(\frac{K_S}{c_S^i + K_S(1+\frac{c_R^i}{K_R})}\right), \tag{6.79}$$

$$\mathfrak{N}_{E}^{o} = \mathfrak{N}_{ET}\left(\frac{\alpha}{\alpha+\beta}\right)\left(\frac{K_S}{c_S^o + K_S(1+\frac{c_R^o}{K_R})}\right). \tag{6.80}$$

By substitution of Equations 6.75 through 6.78 into Equations 6.69 and 6.70, we can compute the flux of solutes S and R as follows:

$$\phi_S = (\phi_S)_{max}\left(\frac{c_S^i}{c_S^i + K_S(1+\frac{c_R^i}{K_R})} - \frac{c_S^o}{c_S^o + K_S(1+\frac{c_R^o}{K_R})}\right), \tag{6.81}$$

and

$$\phi_R = (\phi_S)_{max}\left(\frac{c_R^i}{c_R^i + K_R(1+\frac{c_S^i}{K_S})} - \frac{c_R^o}{c_R^o + K_R(1+\frac{c_S^o}{K_S})}\right), \tag{6.82}$$

where $(\phi_S)_{max}$ is defined in Equation 6.60 and represents the maximum rate of solute transport by the carrier. Note that all these equations are symmetric in the variables S and R. That is, substitution of S for R and R for S leaves Equations 6.75–6.82 unchanged.

It is simplest to understand the interactions between the two solutes by examining the unidirectional fluxes, which are

$$\vec{\phi}_S = (\phi_S)_{max}\frac{c_S^i}{c_S^i + K_S(1+\frac{c_R^i}{K_R})} \quad \text{and} \quad \overleftarrow{\phi}_S = (\phi_S)_{max}\frac{c_S^o}{c_S^o + K_S(1+\frac{c_R^o}{K_R})}, \tag{6.83}$$

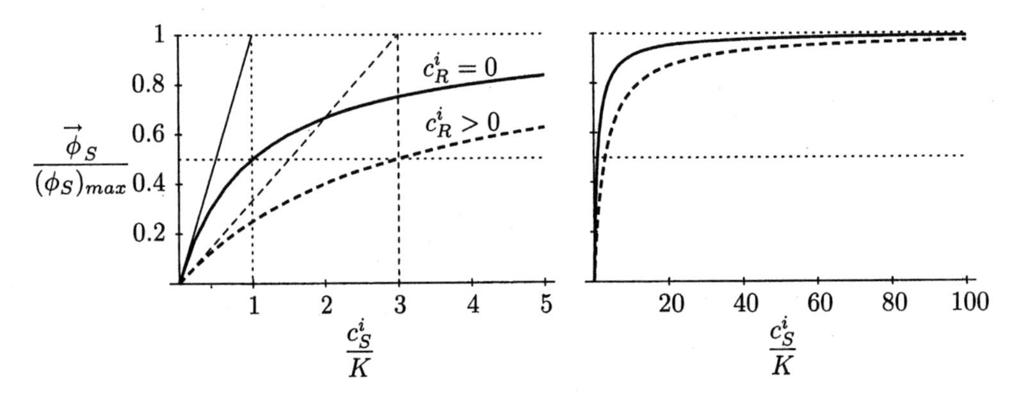

Figure 6.28 Effect of solute R on the relation of the flux of solute S to the concentration of solute S on two abscissa scales. The graph is drawn for a value of $c_R^i/K_R = 2$.

$$\vec{\phi}_R = (\phi_S)_{max}\frac{c_R^i}{c_R^i + K_R(1+\frac{c_S^i}{K_S})} \quad \text{and} \quad \overleftarrow{\phi}_R = (\phi_S)_{max}\frac{c_R^o}{c_R^o + K_R(1+\frac{c_S^o}{K_S})}. \tag{6.84}$$

Consider $\vec{\phi}_S$ to illustrate the effect of solutes S and R on the outward flux of S (Figure 6.28). Note that increasing c_R^i reduces $\vec{\phi}_S$ because R competes with S for the limited number of carrier molecules. Note also that c_R^i has an appreciable effect on $\vec{\phi}_S$ only for values of c_R^i that are appreciable compared with K_R. K_R is inversely proportional to the affinity of carrier for solute R. If this affinity is low, then K_R will be large and a large concentration of R will be required to have an appreciable effect on the efflux of S. Note also that in this model, the efflux of S is affected by the concentration of S and R on the inside of the membrane only. Finally, the effect of the concentration of R on the relation between $\vec{\phi}_S$ and c_S^i is effectively to change the dissociation constant of S from K_S to $K_S(1 + (c_R^i/K_R))$. This type of inhibition of the flux of S by the concentration of R, which acts to change the dissociation constant of S, is called *competitive inhibition.*

The flux of S in the presence of R can also be plotted as a straight line in appropriate coordinates by inverting the unidirectional flux, as follows:

$$\frac{1}{\vec{\phi}_S} = \frac{1}{(\phi_S)_{max}}\left(1 + \frac{K_S}{c_S^i}\left(1 + \frac{c_R^i}{K_R}\right)\right). \tag{6.85}$$

Thus, a plot of $1/\vec{\phi}_S$ versus $1/c_S^i$ is a straight line whose slope depends on c_R^i but whose intercept on the ordinate axis does not (Figure 6.29). In addition,

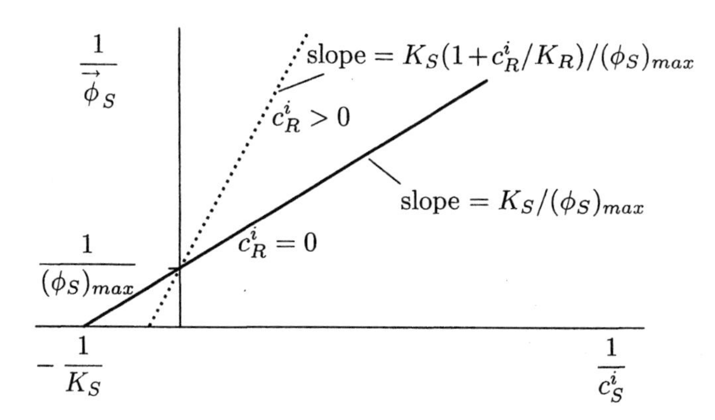

Figure 6.29 Effect of solute R on the relation of the flux of solute S to the concentration of solute S plotted in reciprocal coordinates.

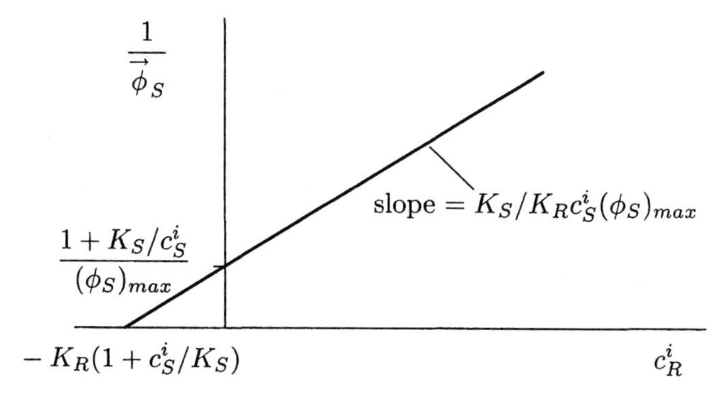

Figure 6.30 Flux of S versus the concentration of R.

a plot of $1/\vec{\phi}_S$ versus c_R^i is also a straight line whose slope depends on c_S^i (Figure 6.30). This type of plot is commonly used to test carrier models.

Countertransport: A Secondary Active-Transport Mechanism

The relation of ϕ_S to the concentrations of S on the two sides of the membrane is given by Equation 6.81. Can S be transported *up* its concentration gradient? To answer this question, we express Equation 6.81 with a common denominator so that

$$\phi_S = (\phi_S)_{max}\left(\frac{c_S^i K_S(1+\frac{c_R^o}{K_R}) - c_S^o K_S(1+\frac{c_R^i}{K_R})}{\left(c_S^i + K_S(1+\frac{c_R^i}{K_R})\right)\left(c_S^o + K_S(1+\frac{c_R^o}{K_R})\right)}\right).$$

The condition for which $\phi_S < 0$ when $c_S^i > c_S^o$ occurs when the numerator is negative, which is the case under the following condition:

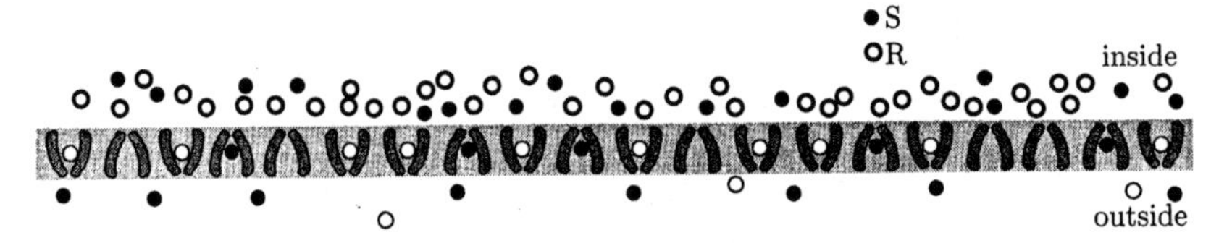

Figure 6.31 Schematic diagram showing a membrane with twenty carriers that can bind two solutes S and R—ten carriers are open to the inside and ten are open to the outside. The concentrations have been arranged so that the flux of S is inward even though the concentration of S is larger on the inside than on the outside.

$$1 < \frac{c_S^i}{c_S^o} < \frac{1 + \frac{c_R^i}{K_R}}{1 + \frac{c_R^o}{K_R}}.$$

Clearly, if c_R^i is sufficiently large and c_R^o is sufficiently small, the flux of S will be negative (or inward) even though the concentration of S inside exceeds that on the outside.

The basis of this transport up a concentration gradient can be appreciated qualitatively by examining Figure 6.31. The concentration of solute R on the inside of the membrane is much larger (compared to its dissociation constant) than that of S. Hence, all the carriers open to the inside are bound to R. Since no carriers are bound to S, the unidirectional efflux of S is zero. On the outside of the membrane, the concentrations of both S and R are less than their concentrations on the inside of the membrane. However, the concentration of S exceeds that of R. Since both concentrations are low, about half of the carriers open to the outside are unbound to solute. Since the concentration of S exceeds that of R, the other half of the carriers open to the outside are bound to S. Therefore, there is an appreciable influx of S up the concentration gradient across the membrane.

The state of the carrier is illustrated somewhat more quantitatively in Figure 6.32. With $c_R^i/K_R > c_S^i/K_S$, more carrier is bound to R than to S on the inside of the membrane. Therefore,

$$\vec{\phi}_R/(\phi_S)_{max} > \vec{\phi}_S/(\phi_S)_{max}.$$

With $c_S^o/K_S > c_R^o/K_R$, more carrier is bound to S than to R on the outside of the membrane. Therefore,

$$\overleftarrow{\phi}_S/(\phi_S)_{max} > \overleftarrow{\phi}_R/(\phi_S)_{max}.$$

Therefore, the efflux of S is small and the influx of S is large, despite the fact that the concentration of S on the inside exceeds that on the outside. Thus,

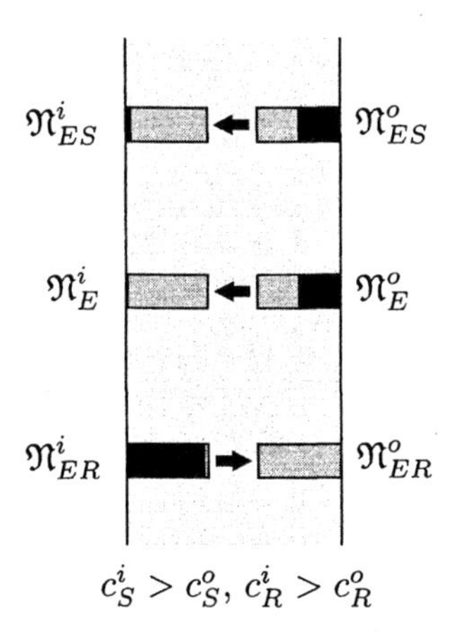

Figure 6.32 Schematic diagram for the state of the carrier for secondary active transport of solute. For this diagram $c^i_S/K_S = 10$, $c^o_S/K_S = 1$, $c^i_R/K_R = 200$, and $c^o_R/K_R = 0.01$. This schematic diagram is similar to that described in Figure 6.23.

S is transported up its concentration gradient. The energy for driving S up its concentration gradient comes from the stored chemical energy of R. As the transport continues, R equilibrates, and S is no longer transported up its concentration gradient.

This type of transport, in which one solute is transported up its concentration gradient because it shares a common transport mechanism with a second solute and for which the energy comes from the stored energy of the second solute, is called *secondary active transport* to distinguish it from *primary active transport,* which is discussed further in a later section and in Chapter 7.

6.4.2.2 One Ligand Is Transported, the Other Is Not

Competitive Inhibition: Inhibitor Binds to Carrier when Carrier Is Not Bound to Solute

Consider the kinetic scheme shown in Figure 6.33. One of the ligands is transported across the membrane. We call it the solute, S. The other ligand is not transported; we call it the inhibitor, I. In this scheme, S and I can combine with carrier E on the outside face of the membrane, but only S can combine with the carrier on the inside face of the membrane. In addition, I can combine with the carrier only when the carrier is not bound to solute S. The binding to solute S has dissociation constant K_S, and the binding to I has dissociation constant K_I. All other kinetic parameters are symmetric in order to minimize algebraic complexity and to focus on this mechanism of competition.

The kinetic equations are analogous to those derived for the two-ligand case where both ligands are transported. The carrier has five states, and the

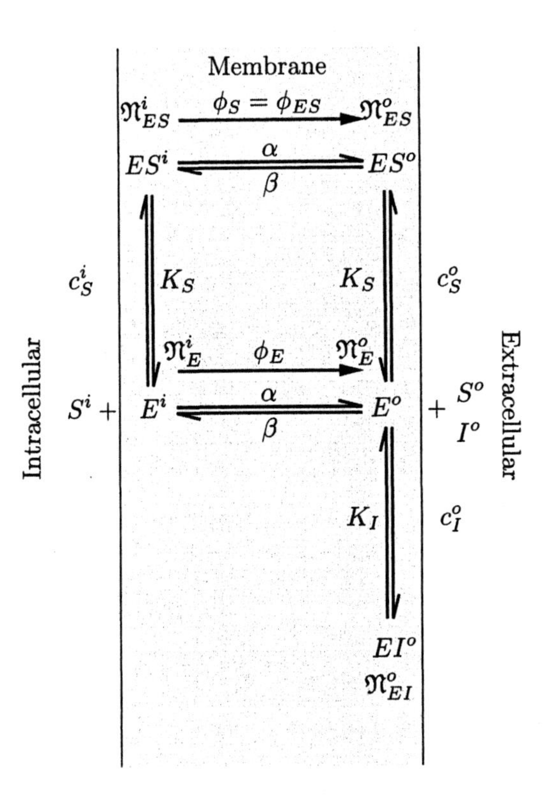

Figure 6.33 Kinetic diagram for a simple, symmetric carrier that can bind two ligands, one of which is transported: competitive inhibition. Ligand S is bound and transported, while ligand I is bound on the outside of the membrane but not transported.

matrix that defines the density of carrier in each state is found by methods analogous to those used in the previous section. The matrix equation is

$$\begin{bmatrix} K_S & 0 & 0 & -c^i_S & 0 \\ 0 & K_S & 0 & 0 & -c^o_S \\ 0 & 0 & K_I & 0 & -c^o_I \\ \alpha & -\beta & 0 & \alpha & -\beta \\ 1 & 1 & 1 & 1 & 1 \end{bmatrix} \begin{bmatrix} \mathfrak{N}^i_{ES} \\ \mathfrak{N}^o_{ES} \\ \mathfrak{N}^o_{EI} \\ \mathfrak{N}^i_E \\ \mathfrak{N}^o_E \end{bmatrix} = \begin{bmatrix} 0 \\ 0 \\ 0 \\ 0 \\ \mathfrak{N}_{ET} \end{bmatrix}. \tag{6.86}$$

The top three rows represent the three binding reactions, the next row guarantees that the net flux of carrier is zero, and the final row guarantees that the sum of the densities of carrier in the five states equals the total carrier density. The solutions to these equations are

$$\mathfrak{N}^i_{ES} = \mathfrak{N}_{ET} \left(\frac{\beta}{\alpha+\beta}\right) \left(\frac{c^o_S + K_S}{c^o_S + K_S(1 + \frac{\alpha}{\alpha+\beta}\frac{c^o_I}{K_I})}\right) \left(\frac{c^i_S}{c^i_S + K_S}\right), \tag{6.87}$$

$$\mathfrak{N}^o_{ES} = \mathfrak{N}_{ET} \left(\frac{\alpha}{\alpha+\beta}\right) \left(\frac{c^o_S}{c^o_S + K_S(1 + \frac{\alpha}{\alpha+\beta}\frac{c^o_I}{K_I})}\right), \tag{6.88}$$

$$\mathfrak{N}_{EI}^{o} = \mathfrak{N}_{ET}\left(\frac{\alpha}{\alpha+\beta}\right)\left(\frac{K_S}{K_I}\right)\left(\frac{c_I^o}{c_S^o + K_S(1 + \frac{\alpha}{\alpha+\beta}\frac{c_I^o}{K_I})}\right), \tag{6.89}$$

$$\mathfrak{N}_{E}^{i} = \mathfrak{N}_{ET}\left(\frac{\beta}{\alpha+\beta}\right)\left(\frac{c_S^o + K_S}{c_S^o + K_S(1 + \frac{\alpha}{\alpha+\beta}\frac{c_I^o}{K_I})}\right)\left(\frac{K_S}{c_S^i + K_S}\right), \tag{6.90}$$

$$\mathfrak{N}_{E}^{o} = \mathfrak{N}_{ET}\left(\frac{\alpha}{\alpha+\beta}\right)\left(\frac{K_S}{c_S^o + K_S(1 + \frac{\alpha}{\alpha+\beta}\frac{c_I^o}{K_I})}\right). \tag{6.91}$$

As expected, for $c_I^o = 0$ these equations reduce to those derived for a single ligand (Equations 6.55-6.58). In general, the effect of the inhibitor I is to reduce the density of the carrier in all states that are not bound by I and to increase the density of the carrier in the state bound by I. The effects of the inhibitor are generally complex, but the effect of the inhibitor on the influx of solute is simple. The unidirectional influx of solute S is

$$\overleftarrow{\phi}_S = \beta\mathfrak{N}_{ES}^{o} = (\phi_S)_{max}\left(\frac{c_S^o}{c_S^o + K_S(1 + \frac{\alpha}{\alpha+\beta}\frac{c_I^o}{K_I})}\right), \tag{6.92}$$

where

$$(\phi_S)_{max} = \mathfrak{N}_{ET}\left(\frac{\alpha\beta}{\alpha+\beta}\right). \tag{6.93}$$

Therefore, this result is identical to that for a single solute (Equation 6.59), except that the dissociation constant now depends upon the concentration of inhibitor. Thus, the net effect of increasing the inhibitor concentration is to reduce the flux by increasing the dissociation constant. This effect results because the inhibitor competes directly with the solute for the carrier. This process is an example of *competitive inhibition* by a ligand that is not a solute.

Noncompetitive Inhibition: Inhibitor Binds to Carrier Whether or Not Carrier Is Bound to Solute

Next, consider the kinetic scheme shown in Figure 6.34. Once again there are two ligands. One ligand, the solute S, is transported across the membrane; the other ligand, the inhibitor I, is not. As in the previous scheme, S and I can combine with carrier E on the outside face of the membrane, but only S can combine with the carrier on the inside face of the membrane. However, in the present scheme, I can combine with the carrier whether or not the carrier is bound to S.

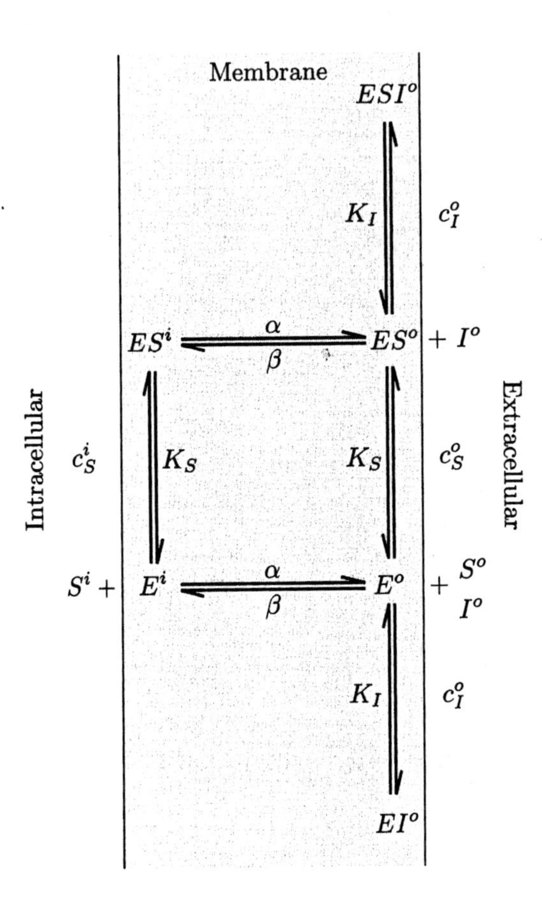

Figure 6.34 Kinetic diagram for a simple, symmetric carrier that can bind two ligands, one of which is transported: noncompetitive inhibition. Ligand S is bound and transported, while ligand I is bound on the outside of the membrane but not transported.

The kinetic equations are analogous to those derived for the previously discussed two-ligand cases. The carrier has six states, and the matrix that defines the density of carrier in each state is found by methods analogous to those used in the previous two sections. The matrix equation is

$$\begin{bmatrix} K_S & 0 & 0 & -c_S^i & 0 & 0 \\ 0 & K_S & 0 & 0 & -c_S^o & 0 \\ 0 & 0 & K_I & 0 & -c_I^o & 0 \\ 0 & -c_I^o & 0 & 0 & 0 & K_I \\ \alpha & -\beta & 0 & \alpha & -\beta & 0 \\ 1 & 1 & 1 & 1 & 1 & 1 \end{bmatrix} \begin{bmatrix} \mathfrak{N}_{ES}^i \\ \mathfrak{N}_{ES}^o \\ \mathfrak{N}_{EI}^o \\ \mathfrak{N}_E^i \\ \mathfrak{N}_E^o \\ \mathfrak{N}_{ESI}^o \end{bmatrix} = \begin{bmatrix} 0 \\ 0 \\ 0 \\ 0 \\ 0 \\ \mathfrak{N}_{ET} \end{bmatrix}. \tag{6.94}$$

The top four rows represent the four binding reactions, the next row guarantees that the net flux of carrier is zero, and the final row guarantees that the sum of the densities of carrier in the six states equals the total carrier density. The solutions to these equations are as follows:

$$\mathfrak{N}_{ES}^{i} = \mathfrak{N}_{ET}\left(\frac{\beta K_I}{\alpha(c_I^o + K_I) + \beta K_I}\right)\left(\frac{c_S^i}{c_S^i + K_S}\right), \tag{6.95}$$

$$\mathfrak{N}_{ES}^{o} = \mathfrak{N}_{ET}\left(\frac{\alpha K_I}{\alpha(c_I^o + K_I) + \beta K_I}\right)\left(\frac{c_S^o}{c_S^o + K_S}\right), \tag{6.96}$$

$$\mathfrak{N}_{EI}^{o} = \mathfrak{N}_{ET}\left(\frac{\alpha c_I^o}{\alpha(c_I^o + K_I) + \beta K_I}\right)\left(\frac{K_S}{c_S^o + K_S}\right), \tag{6.97}$$

$$\mathfrak{N}_{E}^{i} = \mathfrak{N}_{ET}\left(\frac{\beta K_I}{\alpha(c_I^o + K_I) + \beta K_I}\right)\left(\frac{K_S}{c_S^i + K_S}\right), \tag{6.98}$$

$$\mathfrak{N}_{E}^{o} = \mathfrak{N}_{ET}\left(\frac{\alpha K_I}{\alpha(c_I^o + K_I) + \beta K_I}\right)\left(\frac{K_S}{c_S^o + K_S}\right), \tag{6.99}$$

$$\mathfrak{N}_{ESI}^{o} = \mathfrak{N}_{ET}\left(\frac{\alpha c_I^o}{\alpha(c_I^o + K_I) + \beta K_I}\right)\left(\frac{c_S^o}{c_S^o + K_S}\right). \tag{6.100}$$

It is easy to check that with $c_I^o = 0$ these equations reduce to those derived for a single ligand (Equations 6.55–6.58). Note that the effect of the inhibitor I is to reduce the density of the carrier in all states that are not bound by I and to increase the density of carrier in states bound by I. The flux of solute S is

$$\phi_S = (\phi_S)_{max}\left(\frac{c_S^i}{c_S^i + K_S} - \frac{c_S^o}{c_S^o + K_S}\right), \tag{6.101}$$

where

$$(\phi_S)_{max} = \mathfrak{N}_{ET}\left(\frac{\alpha\beta K_I}{\alpha(c_I^o + K_I) + \beta K_I}\right). \tag{6.102}$$

Therefore, this result is identical to that for a single solute (Equation 6.59), except that the maximum flux now depends upon the concentration of inhibitor. Thus, the net effect of increasing the inhibitor concentration is to reduce the maximum flux. This effect results because addition of the inhibitor ties up some of the carrier in states for which there is no transport. This type of inhibition is called *noncompetitive inhibition,* and it differs importantly from competitive inhibition.

6.4.2.3 Summary of Results on Inhibition

In the three two-ligand schemes that have been discussed (Figures 6.27, 6.33, and 6.34), the flux of solute S is inhibited as the concentration of the other ligand is increased. The results of these three schemes are compared with each other and with the results obtained for the transport of S alone in Table 6.2.

Table 6.2 Comparison of unidirectional influx of solute for no inhibition, competitive inhibition (for a transported ligand and for one that is not transported), and noncompetitive inhibition for the simple, symmetric carrier. Each flux equals a product of two terms. The term before the × is the maximum flux.

Ligands	Influx of S
Solute S	$\overleftarrow{\phi}_S = \mathfrak{N}_{ET}\left(\frac{\alpha\beta}{\alpha+\beta}\right) \times \left(\frac{c_S^o}{c_S^o+K}\right)$
Solutes S and R	$\overleftarrow{\phi}_S = \mathfrak{N}_{ET}\left(\frac{\alpha\beta}{\alpha+\beta}\right) \times \left(\frac{c_S^o}{c_S^o+K_S\left(1+\frac{c_R^o}{K_R}\right)}\right)$
Solute S, competitive inhibitor I	$\overleftarrow{\phi}_S = \mathfrak{N}_{ET}\left(\frac{\alpha\beta}{\alpha+\beta}\right) \times \left(\frac{c_S^o}{c_S^o+K_S\left(1+\frac{\alpha}{\alpha+\beta}\frac{c_I^o}{K_I}\right)}\right)$
Solute S, noncompetitive inhibitor I	$\overleftarrow{\phi}_S = \mathfrak{N}_{ET}\left(\frac{\alpha\beta}{\alpha\left(\frac{c_I^o}{K_I}+1\right)+\beta}\right) \times \left(\frac{c_S^o}{c_S^o+K_S}\right)$

To simplify matters, only the influxes are compared in the table. The table shows that the second solute acts essentially to change either the effective dissociation constant of S or the maximum flux of S. For the case when the second ligand is itself a solute, and the case when it is not but binds to the unbound carrier only, the effect of the second ligand is to increase the effective dissociation constant of S. This process is competitive inhibition. However, when the second ligand is not a solute and can bind to carrier that is bound *or* unbound to S, the effect of the second ligand is to reduce the maximum flux of S. This process is noncompetitive inhibition.

The relation between influx of S and its concentration, in both linear and reciprocal coordinates, is shown in Figure 6.35. Both inhibitors reduce the flux so that it is below that obtained in the absence of the inhibitor. However, the competitive and noncompetitive inhibitors reduce the flux in different ways. The competitive inhibitor changes the effective dissociation constant, whereas the noncompetitive inhibitor changes the maximum flux. Therefore, at high solute concentration, the competitive inhibitor results in a vanishingly small inhibition, whereas the noncompetitive inhibitor causes a constant inhibition. In reciprocal coordinates, the relation of flux to solute concentration is a straight line for both types of inhibitors. However, the competitive inhibitor results in a change in the abscissa intercept (which equals the negative reciprocal of the effective dissociation constant of S) without changing the ordinate intercept (which is the reciprocal of the maximum flux of S). In contrast, the

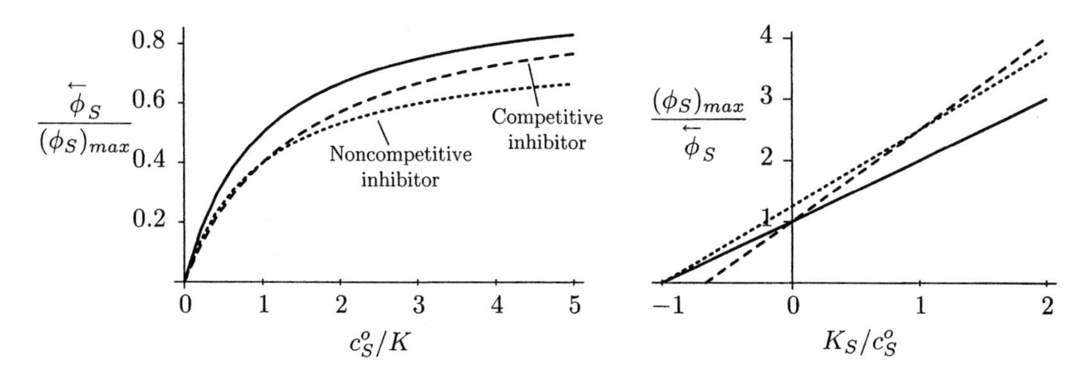

Figure 6.35 The effects of competitive and noncompetitive inhibitors on the relation of influx to concentration of solute in linear (left panel) and reciprocal coordinates (right panel). The solid line shows the relation between influx and concentration in the absence of any inhibitor. For large values of c_S^o/K_S, $\overleftarrow{\phi}_S/(\phi_S)_{max}$ approaches 1 for the competitive inhibitor and 0.8 for the noncompetitive inhibitor. The equivalent dissociation constant for the competitive inhibitor is $1.5K_S$.

noncompetitive inhibitor results in a change in the ordinate intercept without a change in the abscissa intercept.

6.4.3 Introduction to Active Transport

In the simple, symmetric, four-state carrier model for transport of a single solute, transport is *down* the concentration gradient of that solute. However, a simple modification of this model results in transport *up* the solute concentration gradient. Note that Equation 6.59 contains two terms that result from the binding reactions at the two membrane interfaces, which were assumed to have the same dissociation constant, K. Suppose the two dissociation constants differ, so that

$$\mathfrak{N}_{ES}^i = \frac{c_S^i}{c_S^i + K_i}\mathfrak{N}_{ET}, \text{ and } \mathfrak{N}_{ES}^o = \frac{c_S^o}{c_S^o + K_o}\mathfrak{N}_{ET}.$$

Then it is easy to show that

$$\phi_S = (\phi_S)_{max}\left(\frac{c_S^i}{c_S^i + K_i} - \frac{c_S^o}{c_S^o + K_o}\right).$$

Rearrangement of terms yields

$$\phi_S = (\phi_S)_{max}\frac{K_o c_S^i - K_i c_S^o}{(c_S^i + K_i)(c_S^o + K_o)}.$$

Now suppose $c_S^i > c_S^o$, but $K_o c_S^i < K_i c_S^o$: then $\phi_S < 0$. So the simple carrier model can result in transport *up* a concentration gradient. Such transport requires energy and is called *active transport*. If that energy is coupled directly to metabolism, then this form of active transport is called *primary active transport*. We shall return to active transport in Chapter 7.

6.4.4 General, Four-State Carrier Model

Suppose that in the simple, symmetric, four-state carrier model shown in Figure 6.21, we drop the assumptions that the interfacial binding reactions are fast and that the reactions are symmetric. The model that results we call the *general, four-state carrier model,* and the kinetic diagram for this model is shown in Figure 6.36. This general carrier model shows properties not exhibited by the simple, symmetric, four-state carrier model. In this section we briefly discuss this general model; more detailed descriptions are found elsewhere (Lieb and Stein, 1974a; Lieb, 1982; Stein, 1986). This general model still has four carrier states, as did the simple, symmetric, four-state carrier model. But the general model differs from the simple carrier model because the binding reactions are not assumed to be arbitrarily fast. The general model differs from the symmetric carrier model in that the rate constants are not symmetric. The kinetic equations for the general, four-state carrier model can be written for the rate of change of carrier in each of the four states by a matrix differential equation, as follows:

$$\frac{d}{dt}\begin{bmatrix} \mathfrak{N}_E^i \\ \mathfrak{N}_E^o \\ \mathfrak{N}_{ES}^i \\ \mathfrak{N}_{ES}^o \end{bmatrix} = \begin{bmatrix} -(a_i + h_i c_S^i) & a_o & g_i & 0 \\ a_i & -(a_o + h_o c_S^o) & 0 & g_o \\ h_i c_S^i & 0 & -(b_i + g_i) & b_o \\ 0 & h_o c_S^o & b_i & -(b_o + g_o) \end{bmatrix} \begin{bmatrix} \mathfrak{N}_E^i \\ \mathfrak{N}_E^o \\ \mathfrak{N}_{ES}^i \\ \mathfrak{N}_{ES}^o \end{bmatrix}. \tag{6.103}$$

The solute concentrations enter the binding reactions, which are second-order kinetic equations.

6.4.4.1 *Steady-State Distribution of State Density*

In the steady state, the density of carrier in each state is constant, and the derivative terms on the left-hand side of Equation 6.103 are zero. In addition, the four equations shown are not independent; e.g., the fourth can be derived from the first three equations. Thus, we can eliminate one of the equations. However, conservation of carrier leads to an independent constraint that can

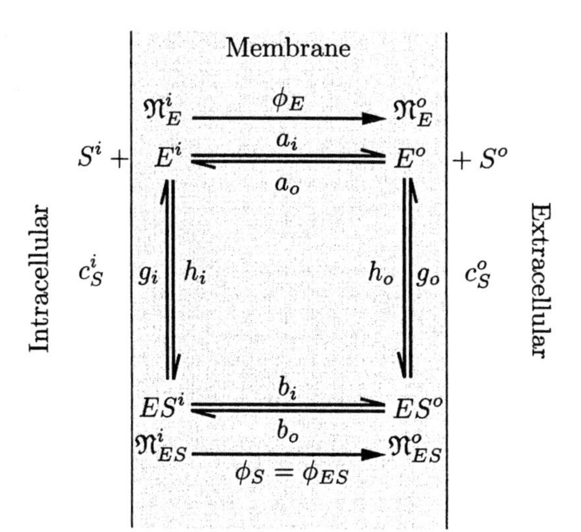

Figure 6.36 Kinetic diagram for a general, four-state carrier model.

be added to the three independent steady-state equations to yield the following four steady-state equations:

$$\begin{bmatrix} -(a_i + h_i c^i_S) & a_o & g_i & 0 \\ a_i & -(a_o + h_o c^o_S) & 0 & g_o \\ h_i c^i_S & 0 & -(b_i + g_i) & b_o \\ 1 & 1 & 1 & 1 \end{bmatrix} \begin{bmatrix} \mathfrak{N}^i_E \\ \mathfrak{N}^o_E \\ \mathfrak{N}^i_{ES} \\ \mathfrak{N}^o_{ES} \end{bmatrix} = \begin{bmatrix} 0 \\ 0 \\ 0 \\ \mathfrak{N}_{ET} \end{bmatrix}. \tag{6.104}$$

The solutions are ratios of sums of products of rate constants:

$$\frac{\mathfrak{N}^i_E}{\mathfrak{N}_{ET}} = \frac{\mathfrak{M}^i_E}{D}, \; \frac{\mathfrak{N}^o_E}{\mathfrak{N}_{ET}} = \frac{\mathfrak{M}^o_E}{D}, \; \frac{\mathfrak{N}^i_{ES}}{\mathfrak{N}_{ET}} = \frac{\mathfrak{M}^i_{ES}}{D}, \; \frac{\mathfrak{N}^o_{ES}}{N_{ET}} = \frac{\mathfrak{M}^o_{ES}}{D}, \tag{6.105}$$

where

$$\begin{aligned} \mathfrak{M}^i_E &= a_o b_i g_o + a_o g_i g_o + a_o b_o g_i + b_o g_i h_o c^o_S, \\ \mathfrak{M}^o_E &= a_i b_i g_o + a_i g_i g_o + a_i b_o g_i + b_i g_o h_i c^i_S, \\ \mathfrak{M}^i_{ES} &= a_i b_o h_o c^o_S + a_o g_o h_i c^i_S + a_o b_o h_i c^i_S + b_o h_i h_o c^i_S c^o_S, \\ \mathfrak{M}^o_{ES} &= a_i b_i h_o c^o_S + a_i g_i h_o c^o_S + a_o b_i h_i c^i_S + b_i h_i h_o c^i_S c^o_S, \\ D &= \mathfrak{M}^i_E + \mathfrak{M}^o_E + \mathfrak{M}^i_{ES} + \mathfrak{M}^o_{ES}. \end{aligned} \tag{6.106}$$

6.4.4.2 Steady-State Unidirectional Flux

The unidirectional efflux of solute can be derived by explicitly defining a tracer, as was done in the method that led to Equation 6.63. The main assumption in that method is that the unidirectional efflux is computed on the assumption that no tracer binds to E^o. Thus, the unidirectional efflux is

$$\vec{\phi}_S = g_o \mathfrak{N}^o_{ES*} \frac{c^i_S}{c^i_{S*}}.$$

In the steady state,

$$\frac{d\mathfrak{N}^o_{ES*}}{dt} = 0 = b_i \mathfrak{N}^i_{ES*} - (b_o + g_o)\mathfrak{N}^o_{ES*},$$

which has included the explicit assumption that no S^* binds to E^o. This leads to the relation

$$\mathfrak{N}^o_{ES*} = \frac{b_i}{b_o + g_o} \mathfrak{N}^i_{ES*}.$$

Steady state also implies that

$$\frac{d\mathfrak{N}^i_{ES*}}{dt} = 0 = h_i c^i_{S*} \mathfrak{N}^i_E + b_o \mathfrak{N}^o_{ES*} - (b_i + g_i)\mathfrak{N}^i_{ES*},$$

which, after elimination of $\mathfrak{N}^o_{ES*}$, can be solved to yield

$$\mathfrak{N}^i_{ES*} = \frac{h_i(b_o + g_o)}{b_i g_o + b_o g_i + g_i g_o} c^i_{S*} \mathfrak{N}^i_E.$$

The efflux of solute is obtained by combining these expressions to give

$$\vec{\phi}_S = \frac{b_i g_o h_i}{b_i g_o + b_o g_i + g_i g_o} c^i_S \mathfrak{N}^i_E. \tag{6.107}$$

By combining Equations 6.105 and 6.107, the unidirectional efflux can be expressed in terms of the the concentrations of solute and the rate constants. Similarly, the unidirectional influx of solute can be obtained directly from Equation 6.107 by making the replacement $i \leftrightarrow o$ to yield

$$\overleftarrow{\phi}_S = \frac{b_o g_i h_o}{b_i g_o + b_o g_i + g_i g_o} c^o_S \mathfrak{N}^o_E. \tag{6.108}$$

By combining Equations 6.105 and 6.108, the unidirectional influx can also be expressed in terms of the concentrations of solute and the rate constants.

6.4.4.3 *Properties of the Steady-State Solutions*

Passivity

We now assume that the net flux of solute is zero when the concentrations of solute on the two sides of the membrane are the same. The consequence of this assumption is found by equating the unidirectional fluxes when $c_S^i = c_S^o = C$:

$$\frac{b_i g_o h_i}{b_i g_o + b_o g_i + g_i g_o} C\mathfrak{N}_E^i = \frac{b_o g_i h_o}{b_i g_o + b_o g_i + g_i g_o} C\mathfrak{N}_E^o.$$

After substitution of Equations 6.105 for both $\mathfrak{N}_E^i$ and $\mathfrak{N}_E^o$ and cancellation of common factors, we obtain

$$a_o b_i g_o h_i = a_i b_o g_i h_o. \tag{6.109}$$

Note that the left-hand side is the product of rate constants going counterclockwise around the reaction cycle shown in Figure 6.36, whereas the right-hand term is the product going clockwise. Therefore, this constraint is equivalent to stating that, at equilibrium, the rate for traversing the reaction cycle clockwise is the same as that for traversing it counterclockwise. This result is a special case of the *principle of detailed balance.*

The net flux is $\phi_S = \vec{\phi}_S - \overleftarrow{\phi}_S$, which, after some simplification, can be written as

$$\phi_S = \frac{\mathfrak{N}_{ET}}{D}(a_o b_i g_o h_i c_S^i - a_i b_o g_i h_o c_S^o).$$

Making use of Equation 6.109, we obtain

$$\phi_S = \frac{a_o b_i g_o h_i \mathfrak{N}_{ET}}{D}(c_S^i - c_S^o), \tag{6.110}$$

which demonstrates that the flux of S is down the concentration gradient of S. To summarize, the general, four-state carrier model, with the assumption that the net flux is zero when the concentrations on the two sides of the membrane are the same, is a *passive transport mechanism.*

Dependence of Unidirectional Flux on Solute Concentrations

The unidirectional fluxes are expressed as relatively messy algebraic functions of the concentrations and rate constants by combining Equations 6.107 and 6.108 with 6.106. However, after like terms are combined it is clear that the numerator contains two terms that depend on the concentrations, and the denominator contains four such terms. Therefore, we need at most five constants to represent the dependence of the unidirectional flux on the concentrations. The efflux can be written in the form

$$\vec{\phi}_S = \frac{c_S^i(K_g + c_S^o)}{(K_g^2/\phi_z) + (K_g/\phi_{io})c_S^i + (K_g/\phi_{oi})c_S^o + (1/\phi_\infty)c_S^i c_S^o}, \tag{6.111}$$

where

$$K_g = \frac{a_i}{h_i} + \frac{a_i g_i}{b_i h_i} + \frac{a_o}{h_o},$$

$$\phi_z = \frac{a_i a_o}{a_i + a_o} \mathfrak{N}_{ET},$$

$$\phi_{io} = \frac{a_o b_i g_o h_i}{b_i g_o h_i + a_o g_o h_i + a_o b_o h_i + a_o b_i h_i} \mathfrak{N}_{ET}, \tag{6.112}$$

$$\phi_{oi} = \frac{a_i b_o g_i h_o}{b_o g_i h_o + a_i g_i h_o + a_i b_i h_o + a_i b_o h_o} \mathfrak{N}_{ET},$$

$$\phi_{\infty} = \frac{b_i b_o g_i g_o}{(b_i + b_o)(b_i g_o + b_o g_i + g_i g_o)} \mathfrak{N}_{ET}.$$

K_g is a sum of ratios of rate constants, and a careful examination reveals that K_g has the units of concentration. In this sense, it acts as a dissociation constant. All of the other four constants have units of flux and their physical significance, will become apparent after the following discussion of simple protocols. By manipulating Equation 6.112, it can be shown that

$$\frac{1}{\phi_z} + \frac{1}{\phi_\infty} = \frac{1}{\phi_{io}} + \frac{1}{\phi_{oi}}.$$

Hence, only four constants are required to completely characterize the relation between the unidirectional flux and the concentrations of the solute on the two sides of the membrane. Note that K_g, ϕ_z, and ϕ_∞ are identical under the transformation $i \leftrightarrow o$. The only term for which this is not obvious by inspection is the term $(a_i g_i)/(b_i h_i)$ in the expression for K_g. However, we note from Equation 6.109 that $(a_i g_i)/(b_i h_i) = (a_o g_o)/(b_o h_o)$. Therefore, this term is also invariant under the transformation $i \leftrightarrow o$. Under this transformation, $\phi_{io} \leftrightarrow \phi_{oi}$. From these symmetry properties it follows that

$$\overleftarrow{\phi}_S = \frac{c_S^o (K_g + c_S^i)}{(K_g^2/\phi_z) + (K_g/\phi_{io}) c_S^i + (K_g/\phi_{oi}) c_S^o + (1/\phi_\infty) c_S^i c_S^o}. \tag{6.113}$$

As with the simple, symmetric, four-state carrier model, the unidirectional efflux is zero if $c_S^i = 0$, and the unidirectional influx is zero if $c_S^o = 0$. This is intuitively satisfying, since if the concentration of the solute is zero on one side of the membrane there should be no flux from that side to the other.

Macroscopic and Microscopic Transport Parameters

The general, four-state carrier model has nine microscopic parameters: a_i, a_o, b_i, b_o, g_i, g_o, h_i, h_o, and $\mathfrak{N}_{ET}$. The passivity condition places one constraint

Table 6.3 Effect of simple constraints on the relation of flux to concentration for the general, four-state carrier model. For each case, the indicated flux variable, under the indicated constraint, is a rectangular hyperbolic function of the indicated concentration variable. The equivalent parameters are given in the two rightmost columns.

	Concentration		*Flux*	*Equivalent parameter*	
Protocol	constraint	variable	variable	K_{eq}	ϕ_{eq}
Zero-trans efflux	$c_S^o = 0$	c_S^i	$\vec{\phi}_S = \phi_S$	$K_g\phi_{io}/\phi_z$	ϕ_{io}
Zero-trans influx	$c_S^i = 0$	c_S^o	$\overleftarrow{\phi}_S = -\phi_S$	$K_g\phi_{oi}/\phi_z$	ϕ_{oi}
Infinite-trans efflux	$c_S^o \to \infty$	c_S^i	$\vec{\phi}_S$	$K_g\phi_\infty/\phi_{oi}$	ϕ_∞
Infinite-trans influx	$c_S^i \to \infty$	c_S^o	$\overleftarrow{\phi}_S$	$K_g\phi_\infty/\phi_{io}$	ϕ_∞
Infinite-cis efflux	$c_S^i \to \infty$	c_S^o	ϕ_S	$K_g\phi_\infty/\phi_{io}$	ϕ_{io}
Infinite-cis influx	$c_S^o \to \infty$	c_S^i	$-\phi_S$	$K_g\phi_\infty/\phi_{oi}$	ϕ_{oi}
Equilibrium exchange	$c_S^o = c_S^i$	c_S^o & c_S^i	$\vec{\phi}_S = \overleftarrow{\phi}_S$	$K_g\phi_\infty/\phi_z$	ϕ_∞

on the rate constants (Equation 6.109) so that there are eight independent microscopic parameters. However, as we have seen above, only four macroscopic parameters are required to characterize the relation of flux to concentration. Therefore, the microscopic parameters cannot be determined uniquely from measurements of flux versus concentration alone. However, the determination of the macroscopic parameters places constraints on the relations among microscopic parameters.

Simple Protocols

The relations between the fluxes and concentrations are simplified under certain constraints, called *testing protocols.* To describe these, we introduce some nomenclature. The unidirectional flux is said to flow from the *cis* side to the *trans* side. So, for an efflux, the inside of the cell is the cis side and the outside of the cell is the trans side. Conversely, for an influx, the outside of the cell is the cis side and the inside of the cell is the trans side. This nomenclature is used to describe relations between flux and concentration under different constraints on the concentrations (Table 6.3). For example, the zero-trans unidirectional efflux (which equals the total efflux for this case) can be computed from Equation 6.111 under the constraint that $c_S^o = 0$, as follows:

$$(\phi_S)_{c_S^o=0} = (\vec{\phi}_S)_{c_S^o=0} = \phi_{io}\frac{c_S^i}{(K_g\phi_{io}/\phi_z) + c_S^i},$$

which is a rectangular hyperbola with maximum equivalent flux $(\phi_{max})_{eq} = \phi_{io}$ and equivalent dissociation constant of $K_{eq} = K_g \phi_{io}/\phi_z$. The zero-trans influx and the infinite-trans efflux, and influx can be computed in a similar manner. The infinite-cis efflux and influx are computed from the total flux. For example, the efflux is obtained as

$$(\phi_S)_{c_S^i \to \infty} = (\overrightarrow{\phi_S} - \overleftarrow{\phi_S})_{c_S^i \to \infty} = \phi_{io} \frac{1}{1 + (\phi_{io} c_S^o)/(K_g \phi_\infty)}.$$

This is a decreasing rectangular hyperbolic function of c_S^o, whose parameters are given in the table. The equilibrium exchange efflux is obtained from Equation 6.111 by setting both concentrations equal: $c_S^i = c_S^o = C$. The result, after some manipulation, is another rectangular hyperbola with the parameters given in Table 6.3.

The results obtained for the simple protocols suggests the physical significance of the parameters that define the relation of flux to concentration for the general, four-state carrier model. ϕ_{io} is the maximum efflux through the membrane when $c_S^i \to \infty$ and $c_S^o = 0$; ϕ_{oi} is the maximum influx through the membrane when $c_S^i = 0$ and $c_S^o \to \infty$; ϕ_∞ is the maximum influx and efflux through the membrane when $c_S^i \to \infty$ and $c_S^o \to \infty$. ϕ_z is obtained by considering the unidirectional flux of carrier when both concentrations are zero. Under these conditions, $\phi_z = a_i \mathfrak{N}_E^i$, $a_o \mathfrak{N}_E^o = a_i \mathfrak{N}_E^i$, and $\mathfrak{N}_E^i + \mathfrak{N}_E^o = \mathfrak{N}_{ET}$. Combining these relations, we obtain the expression for ϕ_z given in Equation 6.112. Thus, ϕ_z is the unidirectional flux of unbound carrier when the solute concentration is zero. There is also a simple interpretation of the dissociation constant K_g. Consider the unidirectional efflux when both concentrations are small. Then Equation 6.111 reduces to $\overrightarrow{\phi_S} = (\phi_z/K_g)c_S^i$. A similar expression is found for the influx. Therefore,

$$\phi_S = \frac{\phi_z}{K_g}(c_S^i - c_S^o),$$

and by comparison with Equation 6.61 it is apparent that K_g acts as the effective dissociation constant for the general, four-state carrier model at low solute concentration.

Consistency with the Simple, Symmetric, Four-State Carrier

As a check of consistency, we examine the unidirectional efflux predicted by the general, four-state carrier model for the special case when the rate constants are symmetric and the interfacial binding reactions are fast. This flux should be the same as that predicted by the simple, symmetric four-state carrier model. Furthermore, the examination provides insight into the simple model. Symmetry of the rate constants implies that $a_i = b_i = \alpha$, $a_o = b_o = \beta$, $g_o = g_i$, and $h_o = h_i$. Substitution of these relations into Equation 6.107 yields

$$\vec{\phi}_S \approx \left(\frac{\alpha g_o h_i}{\alpha g_o + \beta g_i + g_i g_o}\right) c_S^i \mathfrak{N}_E^i.$$

If the interfacial reactions are much faster than the translocation reactions, then the rate constants g_i, g_o, h_i, and h_o are much larger than a_i, a_o, b_i, and b_o. Therefore,

$$\vec{\phi}_S \approx \frac{\alpha}{g_i/h_i} c_S^i \mathfrak{N}_E^i.$$

Since it is assumed that the interfacial binding reactions are fast, they can be assumed to be at equilibrium at each instant in time. Therefore, $c_S^i \mathfrak{N}_E^i / \mathfrak{N}_{ES}^i = K = g_i/h_i$; $c_S^o \mathfrak{N}_E^o / \mathfrak{N}_{ES}^o = K = g_o/h_o$; and

$$\vec{\phi}_S \approx \alpha \mathfrak{N}_{ES}^i,$$

which is the result assumed in the derivation of the flux for the simple, symmetric, four-state carrier model in Equation 6.50.

We can also determine $\mathfrak{N}_{ES}^i$ for symmetric rate constants and for fast interfacial binding reactions by substituting these constraints into Equation 6.105 to obtain

$$\frac{\mathfrak{N}_{ES}^i}{\mathfrak{N}_{ET}} \approx \left(\frac{\beta}{\alpha + \beta}\right)\left(\frac{c_S^i}{c_S^i + K}\right),$$

which is the same as Equation 6.55. This demonstrates that the simple, symmetric, four-state carrier model is a special case of the general, four-state carrier model. The simple model results if the interfacial binding reactions are assumed to be much faster than the translocation reactions and if the rate constants are assumed to be symmetric.

Similarities and Differences Between Simple, Symmetric and General, Four-State Carrier Models

We have investigated in detail two four-state mechanisms for carrier-mediated transport: the simple, symmetric and the general, four-state carrier. These two mechanisms have a number of similar kinetic properties and several significant differences.

Dependence of Unidirectional Fluxes on Concentration For the simple, symmetric four-state carrier model, the unidirectional fluxes depend only on the cis concentration (Equations 6.63 and 6.64), e.g., $\vec{\phi}_S$ depends only on c_S^i. In contrast, for the general model, the unidirectional fluxes depend upon both c_S^i and c_S^o.

Passivity of the Relation of Flux to Concentration For the simple, symmetric model, transport is inherently passive. For the general model, transport is passive provided that the condition given in Equation 6.109 is met.

Relation of Flux to Concentration for Special Protocols For all the special protocols given in Table 6.3, the relation of the indicated flux to the indicated concentration is a rectangular hyperbola for both the simple, symmetric and the general model. For the simple, symmetric model, the equivalent dissociation constant and the equivalent maximum flux are the same for each protocol and are K and $(\phi_S)_{max}$, respectively. For the general model, the equivalent dissociation constant and the equivalent maximum flux depend upon the protocol; the results are given in Table 6.3.

6.4.5 Other Carrier Models

All the models considered thus far have four states, two of which can bind solute but only one binding site is available at any instant of time. At any instant of time the binding site is available either at the cytoplasmic or extracellular face of the membrane. We wish briefly to discuss two different generalizations of this model. In the first generalization, there are still two states that can bind solute of which only one is accessible at any one time. However, there may be any number of intermediate states of either the bound or unbound carrier. Then we shall briefly describe a model in which both binding sites can be accessible simultaneously.

6.4.5.1 Models with One Accessible Binding State

The general, four-state carrier model described in the previous section is defined by eight rate constants (Figure 6.36). We call these the *microscopic parameters*. If the transport mechanism is passive, one constraint (Equation 6.109) is imposed on these rate constants. However, the steady-state kinetic equations that relate flux to concentration of solute (Equations 6.111 and 6.113) are defined by five *macroscopic parameters,* of which four are independent. These macroscopic parameters are algebraic functions of the microscopic parameters (Equations 6.112). Measurements of steady-state flux can determine the macroscopic parameters but not the microscopic parameters. The microscopic parameters are not observable directly by measurements of the steady-state flux through the membrane. Furthermore, a set of macroscopic parameters that characterize the relation between flux and concentration of a carrier system are not related uniquely to a set of rate constants. Many different carrier mechanisms can have the same set of macroscopic parameters. To make this matter clearer, we first write the carrier mechanism for the general, four-state carrier model as a reaction sequence, as follows:

$$S^i + E^i \underset{g_i}{\overset{h_i}{\rightleftharpoons}} ES^i \underset{b_o}{\overset{b_i}{\rightleftharpoons}} ES^o \underset{h_o}{\overset{g_o}{\rightleftharpoons}} E^o + S^o,$$

$$E^i \underset{a_o}{\overset{a_i}{\rightleftharpoons}} E^o.$$

The binding reactions at the two interfaces yield a bound complex that has two intermediate states, ES^i and ES^o. The unbound carrier has no intermediate states in the general, four-state carrier model. However, we could imagine a multistate carrier model in which there were N intermediate states of the bound complex and M intermediate states of the unbound complex, as follows:

$$S^i + E^i \rightleftharpoons ES^1 \rightleftharpoons ES^2 \ldots ES^{N-1} \rightleftharpoons ES^N \rightleftharpoons E^o + S^o,$$

$$E^i \rightleftharpoons E^1 \rightleftharpoons E^2 \ldots E^{M-1} \rightleftharpoons E^M \rightleftharpoons E^o.$$

It has been shown (Lieb and Stein, 1974a; Lieb, 1982) that the relation of flux to concentration is identical to that for the general, four-state model (Equations 6.111 and 6.113).[5] The only difference between such models is in the relation of the macroscopic parameters to the microscopic parameters. Therefore, from measurements of the steady-state relation of flux to concentration it is, in general, not possible to determine the kinetic mechanism uniquely.

6.4.5.2 Models with Two Accessible Binding States

Another class of carrier models relaxes the assumption that at any instant of time the carrier binds solute at one membrane face or the other, but not at both. Models in which the carrier can bind solute simultaneously at each

---

5. It is algebraically tedious to find the relation of flux to concentration for more general reaction mechanisms than the general, four-state carrier model. Graph theoretic methods have been found helpful in this regard (King and Altman, 1956; Mason and Zimmermann, 1960; Volkenstein and Goldstein, 1966; Cha, 1968; Fromm, 1970).

face of the membrane have many properties that are similar to those of the models we have developed (Carruthers, 1984). However, these models have been tested much less extensively against measurements than have the single accessible binding site models.

6.5 Hexose Transport in Cells

As discussed in Section 6.1, the carrier models described in Section 6.4 can account qualitatively for the distinguishing characteristic of hexose transport in cells. We review these points briefly. Because the flux of the hexose transport mechanism is not limited by any of the factors that limit the diffusion of a solute through the membrane, this flux can be "facilitated." The passivity of the flux follows directly from the model. The structural specificity of the flux results because the binding of the solute to the carrier is structure-specific. The saturation of the flux as a function of solute concentration occurs because each carrier has a maximum rate of transport and there are a discrete number of carriers in the membrane. Hence, when all the carriers are transporting solute at maximal rate, the flux through the membrane is a maximum. Inhibition results because more than one substance can bind to the carrier and inhibit the solute from binding and being transported.

In this section, we compare measurements on cells with predictions of carrier models in order to assess the models quantitatively. We will see that the matter is complicated by several factors. First, the kinetic characteristics of the hexose transporter are different in different cell types. Second, there are several testing protocols that can be performed on cells. While the model may be capable of fitting the results for a particular testing protocol, it may not be capable of fitting results from other testing protocols with the same model parameters. Thus, to assess the validity of the model, we must consider the results of several testing protocols.

6.5.1 Experimental Measurements and Methods for Estimating the Kinetic Parameters

6.5.1.1 The Initial Slope Method

A typical experiment designed to measure the flux for one of the testing protocols given in Table 6.3, the zero-trans efflux protocol, involves loading a single large cell (such as the giant axon of the squid) or a population of small cells (such as erythrocytes) with a known concentration of radioactive sugar and measuring the rate at which the sugar quantity in the cell declines or the rate at which the sugar appears extracellularly. The flux can be obtained from such measurements. For example, in the measurements shown in Figure 6.1, the extracellular concentration of galactose is measured as a function of time after erythrocytes have been loaded with galactose and placed in an extracellular compartment that initially contained no galactose. The parameter is the initial intracellular concentration of galactose. Conservation of galactose implies that the rate of increase of extracellular galactose quantity is proportional to the efflux of galactose,

$$\frac{1}{A}\frac{dn_S^o(t)}{dt} = \phi_{eq}\left(\frac{c_S^i(t)}{c_S^i(t) + K_{eq}}\right).$$

If the extracellular volume is $\mathcal{V}_o$, then the initial rate of change of concentration is

$$\left.\frac{dc_S^o(t)}{dt}\right|_{t=0} = \frac{A\phi_{eq}}{\mathcal{V}_o}\left(\frac{c_S^i(0)}{c_S^i(0) + K_{eq}}\right).$$

Therefore, initially the extracellular concentration changes linearly in time with a slope that is proportional to the flux, which is a rectangular hyperbolic function of the initial intracellular concentration of galactose.

The initial slope is plotted versus concentration in Figure 6.2, and the points are fit well with a rectangular hyperbola. Thus, these results are consistent with the carrier models presented in Section 6.4—both the simple, symmetric and the general, four-state models predict this result for the *zero-trans* protocol. Alternatively, the relation of flux to concentration can be plotted in double reciprocal coordinates, as is shown for another set of measurements in Figure 6.37. In these coordinates the relation of flux to concentration is well represented as a straight line, which implies that the relation of flux to concentration is a rectangular hyperbola. In addition, these measurements allow an estimate to be made of the equivalent dissociation constant and the equivalent flux.

6.5.1.2 Integrated Flux Equation

Another method to test carrier models and to estimate the transport parameters is to measure the kinetics of sugar transport as a function of time and see if the model predicts the time variation of either sugar quantity or flux. To understand these measurements, consider the schematic diagram of a cell

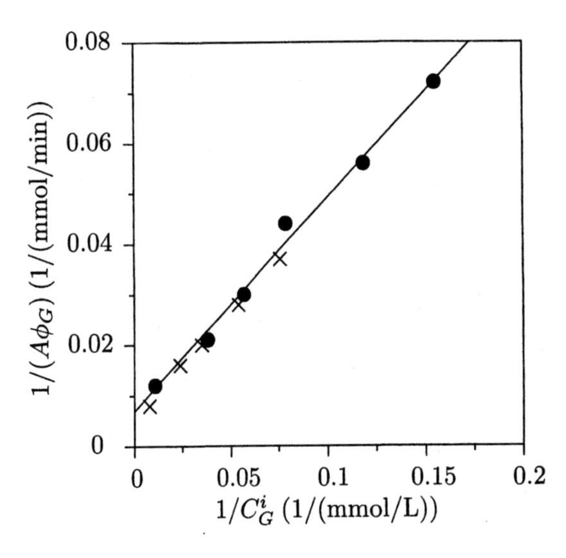

Figure 6.37 Galactose flux versus galactose concentration in reciprocal coordinates for erythrocytes (adapted from Miller, 1971, Figure 2). The total rate of transport of galactose for a population of cells is shown where A is the surface area of all the cells. The filled circles are based on zero-trans efflux measurements, the crosses on equilibrium exchange efflux measurements.

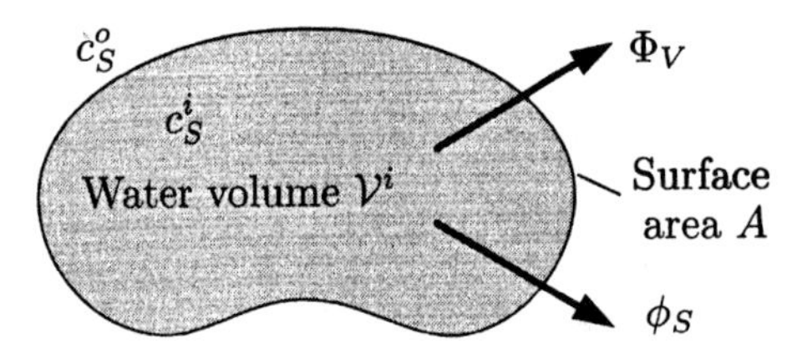

Figure 6.38 Schematic diagram of a cell showing reference directions for key transport variables.

that transports a sugar, as shown in Figure 6.38. First, we shall compute the quantity of intracellular sugar as a function of time in the zero-trans protocol, given that sugar flux is a rectangular hyperbolic function of concentration. Then we will examine the measurements.

Let us assume that the sugar is not metabolized in the cell, so that a net efflux of sugar must balance the reduction of hexose in the cell.[6] Under these conditions the rate of decrease of the quantity of intracellular sugar is proportional to the efflux of sugar:

$$-\frac{1}{A}\frac{dn_S^i(t)}{dt} = \phi_S(t) = \phi_{eq}\frac{c_S^i(t)}{c_S^i(t) + K_{eq}}. \tag{6.114}$$

6. This technical problem turns out to be an important limitation in studies of glucose transport. Glucose is rapidly metabolized in cells, and transport studies are more difficult to interpret unless the metabolism of glucose is separately characterized. An approach to circumventing this technical problem is to block the metabolism of glucose; another is to use a sugar that is not metabolized.

As the sugar leaves the cell, there is a reduction in the intracellular osmotic pressure so that water will also leave and the cell will shrink. This change in volume affects the intracellular concentration, and we must take this into account. However, the matter is simplified because sugar transport is much slower than water transport. Therefore, we assume that the cell is always in osmotic equilibrium, so that $C_\Sigma^i = C_\Sigma^o$, which can be written as

$$C_\Sigma^o = \frac{n_S^i + \tilde{N}^i}{\mathcal{V}^i}, \tag{6.115}$$

where $\tilde{N}^i = \sum_k \tilde{n}_k^i$ is the sum of all the other solutes, which are assumed to be impermeant in this analysis. This property allows us to relate the sugar quantity to the sugar concentration as follows:

$$c_S^i(t) = \frac{n_S^i(t)}{\mathcal{V}^i(t)} = C_\Sigma^o\left(\frac{n_S^i(t)}{n_S^i(t) + \tilde{N}^i}\right).$$

It is useful to express the sugar conservation equation in normalized variables, as follows:

$$\frac{dn_S(t)}{dt} = -\alpha\frac{n_S(t)}{n_S(t) + 1}, \tag{6.116}$$

where

$$n_S(t) = n_S^i(t) / \left(\frac{K_{eq}\tilde{N}^i}{C_\Sigma^o + K_{eq}}\right) \quad \text{and} \quad \alpha = \frac{A\phi_{eq}C_\Sigma^o}{K_{eq}\tilde{N}^i}.$$

Equation 6.116 can be integrated directly:

$$\int_{n_S(0)}^{n_S(t)}\left(1 + \frac{1}{n_S'(t)}\right) dn_S'(t) = -\alpha\int_0^t dt',$$

to yield

$$n_S(t) - n_S(0) + \ln\left(\frac{n_S(t)}{n_S(0)}\right) = -\alpha t.$$

If we define $n(t) = n_S(t)/n_S(0)$, then, after rearrangement of terms, the solution can be written in the form

$$\frac{1 - n(t)}{\ln n(t)} - \left(\frac{\alpha}{n_S(0)}\right)\frac{t}{\ln n(t)} = \frac{1}{n_S(0)}. \tag{6.117}$$

Thus, if $(1 - n(t))/\ln n(t)$ is plotted versus $t/\ln n(t)$, the result is a straight line and the slope and intercepts of the line determine $n_S(0)$ and α.

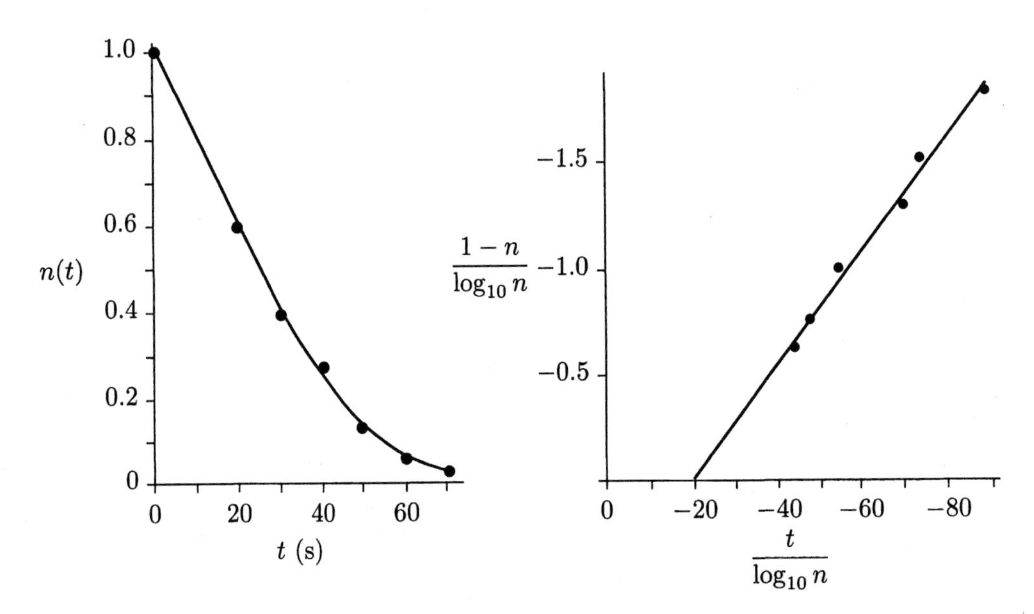

Figure 6.39 Kinetics of the quantity of intracellular glucose of human erythrocytes in linear (left panel) and normalized coordinates (right panel) under zero extracellular glucose concentration (adapted from Karlish et al., 1972, Figure 1).

ϕ_{eq} and K_{eq} can be determined from $n_S(0)$, α, the compositions of intracellular and extracellular solutions, and the cell dimensions.

Figure 6.39 shows a comparison between measurements of glucose efflux from human erythrocytes under zero-trans conditions and calculations of the integrated flux equation based on the assumption that flux is a rectangular hyperbolic function of concentration. These calculations fit the time dependence of glucose quantity quite well. The fit is shown both in linear coordinates and in the normalized logarithmic coordinates, in the latter of which the kinetics plot as a straight line. Thus, the time course of intracellular glucose is well fit if the flux is assumed to be a rectangular hyperbolic function of the glucose concentration.

6.5.2 Applicability of Carrier Models to Measurements from Cells

Both the initial slope measurements (Figures 6.1 and 6.2) and the kinetic measurements (Figure 6.39) demonstrate the validity of the rectangular hyperbolic dependence of flux on concentration and also yield estimates of the kinetic parameters ϕ_{eq} and K_{eq} that characterize the flux for a given protocol. This is a

Table 6.4 A comparison of the kinetic parameters for 3-O-methylglucose transport in barnacle muscle cells (Carruthers, 1983) and in squid giant axons (Baker and Carruthers, 1981a) by several different testing protocols. In both preparations, the intracellular solution was controlled by internal dialysis. The temperature was 10°C for barnacle (*Balanus nubilus*) muscle cells and 15°C for squid (*Loligo forbesi*) giant axons. The surface of barnacle muscle cells is irregular, making an estimate of surface area difficult. Thus, the estimated fluxes of the barnacle muscle fibers are not directly comparable to those for the squid giant axon.

	Barnacle muscle cell		*Squid giant axon*	
Protocol	K_{eq} (mmol/L)	ϕ_{eq} (pmol/cm^2·s)	K_{eq} (mmol/L)	ϕ_{eq} (pmol/cm^2·s)
Zero-trans efflux	6.0 ± 1.2	11.6 ± 0.8	5.5	8.6
Zero-trans influx	4.6 ± 0.8	10.5 ± 0.5	1.34	1.99
Infinite-trans efflux			5.5	2.1
Infinite-trans influx			1.34	1.99
Infinite-cis efflux			1.34	8.6
Infinite-cis influx			6.0	2.2
Equilibrium exchange			1.39	2.0

necessary but not a sufficient test of either the simple, symmetric four-state carrier model or the general, four-state carrier model. Both of these models make strong predictions on the kinetic parameters obtained with different testing protocols. For example, the simple, symmetric model predicts that the kinetic parameters are the same for all the testing protocols. In contrast, the general model allows for a difference in the kinetic parameters obtained with different testing protocols. However, the general model does place some constraints on these different kinetic parameters.

Measurements of the kinetic parameters for different testing protocols are now available for a variety of cell types for a number of different sugars. Examples of the kinetic parameters for marine invertebrate muscle cells and neurons are shown in Table 6.4. Because of the large dimensions of these cells, it is possible to dialyze the cells and hence to control the composition of both the intracellular and extracellular media. In addition, flux measurements can be made on individual cells. The measurements shown in Table 6.4. are for the sugar 3-O-methylglucose, which was shown not to be metabolized in these cells—which eases interpretation of the results. Interestingly, the zero-trans influx and efflux have approximately the same kinetic parameters for the barnacle muscle fibers but have very different kinetic parameters for the squid giant axon. In the squid giant axon, the maximum zero-trans efflux exceeds the maximum zero-trans influx by about a factor of 4. Note that the ratio ϕ_{eq}/K_{eq} is 16 nm/s for zero-trans efflux and 15 nm/s for zero-trans influx. Thus, these two ratios are about the same, as predicted by the general four-state carrier model (Table 6.3). Thus, the measurements in the barnacle muscle fibers are consistent with the simple, symmetric four-state carrier model but those in the giant axons of squid are not; however, the latter appear consistent with the general, four-state carrier model, with appreciable asymmetry in the kinetic constants.

The kinetics of glucose transport has been studied most extensively in human erythrocytes. These cells have the advantages that they are readily available by venous puncture and no dissection of the cells is required. However, erythrocytes are small cells, so that glucose transport cannot be studied on individual cells; rather, a population of cells is used. In addition, control of the intracellular solution is necessarily indirect. Thus, although the techniques for obtaining the cells are simple, the methods for estimating the kinetic parameters are relatively indirect, and the interpretation of the results has been controversial (Stein, 1986; Wheeler and Whelan, 1988). Furthermore, there are a large number of experimental variables that affect the kinetic parameters. These include, e.g., temperature and whether the blood is fresh, has been stored cold, or is outdated transfusion blood (no longer suitable for transfusions). Nevertheless, it is clear that the transport of glucose through the erythrocyte membrane is highly asymmetric—the simple, symmetric four-state model cannot account for the measurements. The controversy most recently has centered on whether or not the general, four-state carrier model can account for the kinetic measurements. Table 6.5 shows a particular collection of results obtained by one investigator, using outdated transfusion blood, all at the same temperature, and using the initial slope method for measuring the kinetic parameters. Thus, many of the factors that vary from one study to another are the same in this study. The asymmetry of the transport is readily seen in the kinetic parameters; for example, note the differences for zero-trans efflux and zero-trans influx. Based on the measured values of K_{eq} and ϕ_{eq} for the different protocols, a set of macroscopic parameters of the general model, K_g, ϕ_{io}, ϕ_{oi}, ϕ_z, ϕ_∞, were estimated. From these macroscopic parameters, a set of theoretical values of K_{eq} and ϕ_{eq} were computed for each protocol. The theoretically determined macroscopic parameters are roughly equivalent to the measured values. Thus, these results appear roughly consistent with the generalized, four-state carrier model. Other collections of measurements on glucose transport in human erythrocytes are not consistent with this model

Table 6.5 A comparison of kinetic parameters for D-glucose transport in erythrocytes at a temperature of 0°C obtained by several different testing protocols (Wheeler and Whelan, 1988). The initial slope method was used to estimate the parameters for outdated transfusion blood. The flux is expressed as a rate of change of concentration, as indicated in Figure 6.1. Two sets of parameters are shown. One set is based directly on measurements; the other set set was computed for the general, four-state carrier model with the parameters $K_g = 0.2$ mmol/L, $\phi_{io} = 0.062$, $\phi_{oi} = 0.003$, $\phi_z = 0.0029$, $\phi_\infty = 0.4$ mmol/L · s.

	Measurements		*Theory*	
Protocol	K_{eq} (mmol/L)	ϕ_{eq} (mmol/(L·s))	K_{eq} (mmol/L)	ϕ_{eq} (mmol/(L·s))
Zero-trans efflux	2.7	0.064	4.3	0.062
Zero-trans influx	0.52	0.0064	0.21	0.003
Infinite-trans efflux	19.3	0.61	26	0.40
Infinite-trans influx	0.8	0.39	1.3	0.4
Infinite-cis influx	39	0.0020	26	0.003
Equilibrium exchange	42	0.83	27	0.40

(Stein, 1986). However, the lack of agreement may well be due to errors in estimation of parameters for the different protocols (Wheeler and Whelan, 1988).

6.5.3 Conclusions

The general, four-state carrier model is among the simplest of a class of carrier-mediated transport models that can account for transport of sugars, amino acids, nucleotides, etc., in a large variety of cells, from bacteria to human red blood cells. These models "explain" many of the features shared by such transport systems, such as the structural specificity of the apparent permeability, the saturation of flux with increasing solute concentration, and the phenomenon of competitive inhibition. However, there appear to be a variety of transport mechanisms that differ in many details. For example, the transport of sugars varies in numerous kinetic details from one type of cell to another. In some cells transport is symmetrical, whereas in others it is grossly asymmetric. These studies can indicate in broad terms the manner in which solutes are transported across the membrane and can place constraints on the possible kinetic mechanisms. However, as we have seen, kinetic studies of steady-state fluxes cannot by themselves determine all the rate constants. Hence, they cannot uniquely determine the kinetic mechanism. This determination is more readily approached using the methods of molecular biology.

6.6 Regulation of Glucose

The blood glucose level in humans normally lies in the range of 5–10 mmol/L, a level that is required for the metabolic needs of cells. If the level drops below about 2 mmol/L, the metabolic disturbance is called *hypoglycemia*. The effect of hypoglycemia on the central nervous system can produce coma. The cells in the nervous system are particularly vulnerable to oxygen and glucose deficits and die in minutes if deprived of their sources of energy. The normal glucose level is maintained in healthy humans despite the episodic nature of both glucose intake and utilization. Glucose intake rises every four to five hours, after meals, and falls for eight hours during sleep. Glucose utilization also rises during exercise. How is blood glucose concentration maintained in the face of the variation in intake and utilization? What are the roles of the different cell types in the body in glucose regulation? The hormone *insulin* plays a key role in the answers to these questions.

6.6.1 The Discovery of the Role of Insulin: A Historical Perspective

Diabetes mellitus was a particularly devastating disease prior to 1923—its victims showed symptoms of starvation while their urine and blood sugar contained abnormally high concentration of glucose. Apparently, diabetics were not able to utilize the carbohydrates they ingested. Although the symptoms of diabetes were described in the earliest written records, the causes of the disease were unknown until the nineteenth century (Bliss, 1982; Ashcroft and Ashcroft, 1992).

It was formerly thought that the main function of the pancreas was to provide digestive juices to the small intestines via the pancreatic duct. However, in 1869 Paul Langerhans described a distinct group of cells that were different from but were found in small clusters among the more numerous acinar cells (which secrete the pancreatic juices). These constellations of cells were subsequently called the *islets of Langerhans*. In 1889, von Mering and Minkowski discovered that surgical removal of the pancreas in dogs produced the symptoms of diabetes. Furthermore, ligation of the pancreatic duct alone without removing the pancreas did not produce the symptoms of diabetes. Thus, it seemed that the pancreas produced two secretions—an external secretion of digestive juices into the small intestine, and an internal secretion whose removal caused the symptoms of diabetes to appear in animals.

At the turn of the century, the cells of the islets of Langerhans were implicated as the site of the internal secretion, but the identity of this internal secretion was sought without success. An important finding was that ligation of the pancreatic duct led to atrophy of the acinar cells, but not to atrophy of the islets. While preparing a lecture on carbohydrate metabolism for medical students, Frederick G. Banting had the idea that ligation of the pancreatic duct could be used to eliminate the acinar cells and their digestive secretions, and thus to make possible the isolation of the pancreas's internal secretion. The idea was that previous attempts to isolate the internal secretion were foiled perhaps because during the extraction procedure the digestive enzymes mixed with and inactivated the internal secretion. Elimination of the digestive enzymes allowed Banting and Charles H. Best (working in the laboratory of John J. R. MacLeod) to show that the islets of Langerhans produced an internal secretion, which they called *insulin*, which when administered to dogs whose pancreas had been removed could reverse the symptoms of diabetes. Later, insulin was purified with the assistance of James B. Collip, and tests in human diabetics showed that administration of insulin dramatically and rapidly reversed the devastating symptoms of the disease.[7] Thus, it was discovered that insulin played a key role in glucose utilization.

Today, we know that diabetes represents a family of diseases that result in pathological utilization of nutrients. Insulin-dependent diabetes mellitus (IDDM), which was formerly called juvenile diabetes or type I diabetes, is thought to be an autoimmune disease in which the immune system destroys the β cells in the pancreas. Thus, insulin is not produced endogenously but must be provided exogenously. Non-insulin-dependent diabetes mellitus (NIDDM), which was formerly called adult-onset diabetes or type II diabetes, is caused by a diminished response of cells to the presence of insulin.

Insulin has an honored role in biology. In the late 1920s it was found that insulin was a protein, and in 1953 Sanger[8] determined its amino acid sequence. Insulin was the first protein to be sequenced. As indicated in Figure 1.18, insulin is a relatively small protein. Other important benchmarks in the history of insulin are as follows: insulin was the first protein to be chemically synthesized (1963); it was one of the first proteins whose three-dimensional structure was determined (1972); it was produced commercially by recombinant DNA technology (1979); the insulin gene was cloned (1980).

---

7. The history of the discovery of insulin and the subsequent award of a Nobel Prize to Banting and MacLeod in 1923, and not to either Best or Collip, is a fascinating and controversial chapter in the history of science (Bliss, 1982).

8. Sanger was awarded the Nobel Prize in 1959 for this work.

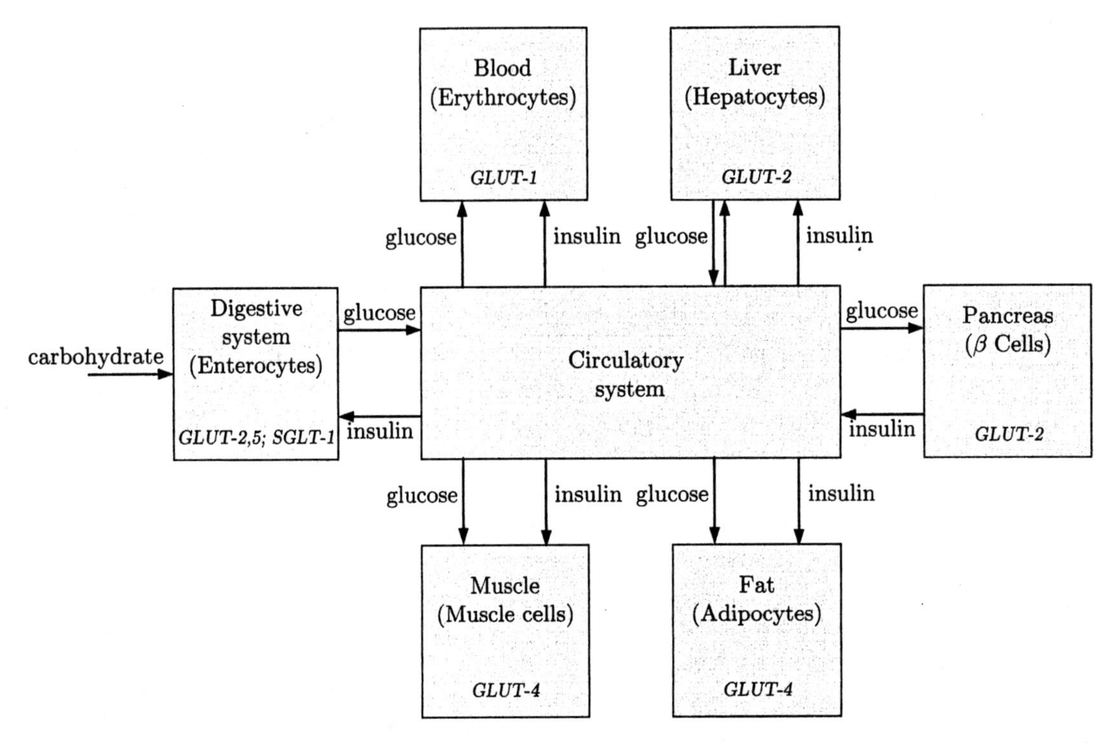

Figure 6.40 A block diagram of blood glucose control showing the interaction of blood glucose and insulin on different cell types. Arrows indicate the direction of flow of glucose and insulin. The type of glucose transporter in each cell type is also indicated.

6.6.2 Glucose Absorption, Utilization, Storage, and Control

The regulation of glucose in the body is a complex process involving several organs. This section presents a simple description of this process to indicate the distinct roles of different cell types in glucose regulation. These roles are shown schematically in Figure 6.40. A more detailed overview of this process can be found elsewhere (Vander et al., 1990).

6.6.2.1 Absorption

As discussed in Chapter 2, carbohydrates are broken down into simple sugars in the digestive system, and glucose is transported from the lumen of the gut into the circulatory system via the intestinal absorptive epithelial cells (enterocytes). This is a two-step process in which glucose is first transported from the lumen into an enterocyte via glucose-sodium cotransporters (Chapter 8) located on the mucosal surface of the enterocyte. This system can concentrate glucose in the enterocyte because the energy required to transport glucose *up* its concentration gradient is provided by the stored energy in the difference of sodium concentration. The intracellular glucose is then transported out of the enterocyte into the circulatory system by passive glucose transporters located on the serosal side of the enterocyte. A similar system is at work in the kidney tubule where glucose is reabsorbed from the urine and transported into the circulatory system. Both the absorption of glucose in the small intestine and the reabsorption of glucose in the kidneys increase the blood glucose concentration.

6.6.2.2 *Utilization*

All cells take up glucose from the blood via a variety of passive glucose transport systems. The glucose is utilized as a substrate for biosynthesis and as a source of energy. Each molecule of glucose yields 2 molecules of ATP by anaerobic glycolysis or 36 molecules of ATP by aerobic glycolysis. ATP is used to power energy-requiring cellular processes.

6.6.2.3 *Storage*

Certain cells also take up glucose and store it or convert it to other molecules for later use. These include hepatocytes (liver cells), muscle cells, and adipocytes (fat cells). Hepatocytes can synthesize glycogen from glucose (glycogenesis) and breakdown glycogen to glucose (glycogenolysis). Glycogenesis and glycogenolysis are catalyzed by the enzymes glycogen synthetase and glycogen phosphorylase, respectively. The catalytic action of these enzymes is under hormonal control. Thus, hepatocytes can store glucose by synthesizing glycogen and can release glucose by breaking down glycogen, both under hormonal control. The liver can store sufficient glucose to satisfy the body's needs for about four hours. Muscle cells can synthesize glycogen from glucose, but the glucose is primarily for the use of the muscle cell and is not a major storage site for the rest of the body. However, the products of anaerobic glycolysis in muscles (pyruvate and lactate) are recycled via the circulatory system to the liver where they are used to produce glucose, which is used to synthesize glycogen. In addition, the liver produces glycerol from glucose; glycerol is used to synthesize fatty acids, which are released into the circulatory system and stored in adipocytes. This pathway is reversible: triglycerides stored in adipocytes can be converted to glycerol, which is absorbed in the liver and used to produce glucose. Thus, fat stores in the body can be used to replenish depleted stores of glucose when the glycogen stores are depleted.

Adipocytes also absorb glucose from the circulatory system and convert the glucose to fatty acids. During fasting, most cells utilize fatty acids released by fat cells for a source of energy. The exceptions are the cells of the nervous system, which utilize glucose exclusively. Hence, the nervous system is particularly vulnerable to hypoglycemia.

6.6.2.4 Control

Glucose is also taken up by cells whose role is to control blood glucose and hence glucose utilization. The β cells in the islets of Langerhans take up glucose by a passive glucose-carrier system. By a complex mechanism, the influx of glucose leads to the release of insulin. The mechanism is as follows: the increase in intracellular glucose leads to an increase in intracellular ATP concentration; which closes a special set of potassium channels in the cell membrane;[9] which depolarizes the β cell membrane; which opens voltage-gated calcium channels in the membrane; which results in an increase in intracellular calcium concentration; which releases insulin by exocytosis into the circulatory system. The insulin in the blood is distributed in the body and binds to insulin receptors, found primarily on three cell types: hepatocytes, adipocytes, and muscle cells. The binding of insulin to the insulin receptor produces an intracellular signal that acts on intracellular processes in a variety of ways. In adipocytes and muscle cells, binding of insulin to its receptor results in an increase in glucose transport by the recruitment of glucose transporter from intracellular stores. In addition, in all three cell types, insulin binding leads to an increase in glycogen synthesis. Since all of these responses result in glucose uptake from the blood, the effect of an increase in blood insulin is to decrease blood glucose. At the same time, blood glucose is also taken up by the α cells in the pancreas. These cells secrete the hormone glucagon into the blood. Glucagon is an antagonist to insulin and is secreted into the bloodstream when blood glucose drops. Glucagon acts to release glucose from storage sites and thus to raise the blood glucose level.

6.6.3 Summary

While all cells take up glucose, there are distinct physiological differences in the characteristics of the uptake. Most cells take up glucose in a manner that is independent of insulin concentration. However, adipocytes, hepatocytes, and

---

9. Ion channels are discussed briefly in Chapters 7 and 8 and more fully elsewhere (Weiss, 1996, Chapter 6).

muscle cells contain insulin receptors that control either the uptake of glucose or its utilization or both. Enterocytes require glucose uptake mechanisms that can concentrate glucose. Not surprisingly, the different roles of glucose transporters in these different cells are served by somewhat different glucose transporters.

6.7 Molecular Biology of Glucose Transporters

Studies of the relations between macroscopic variables such as flux and concentration can place constraints on models of the molecular mechanisms that underlie carrier-mediated transport, but such studies cannot determine these mechanisms uniquely. Insight into these mechanisms has been obtained by isolating the membrane macromolecules (called carriers or transporters) that perform the transport function. Examination of the structure and the chemical properties of these molecules has shed light on the molecular mechanisms of carrier-mediated transport, on the basis for the seemingly large diversity of kinetic properties seen in glucose transport in different cell types, and on the evolution of glucose transporters.

6.7.1 Density of Glucose Transporters

Cytochalasin B inhibits sugar transport at a site located on the cytoplasmic side of the membrane. Using radioactively labeled cytochalasin B, it has been shown that the inhibitor binds with high affinity to cellular membranes that contain the glucose transporter. This property of cytochalasin B has been important in identifying the glucose transporter during isolation of the transporter and as a means for estimating the number of glucose transporters in cellular membranes. Based on cytochalasin B binding measurements, it has been estimated that human erythrocytes contain about 3×10^5 glucose transporters per cell, which implies a density of sites of about 2×10^3 sites/μm^2. Estimates for other cells range from 4×10^3 to 6×10^5 sites/μm^2 (Carruthers, 1984). According to these estimates, if the transporters are distributed uniformly in the membrane, then they are spaced about 1–20 nm apart.

6.7.2 Isolation of the Glucose Transporter

Erythrocytes can be burst osmotically to produce *ghosts*, which can continue to perform stereospecific sugar transport (e.g., they transport D-glucose but not L-glucose) and to bind cytochalasin B. If the proteins are dissolved out of the membrane of these cells and incorporated in lipid bilayers, the lipid bilayers also transport sugars in a stereospecific manner and bind cytochalasin B. Thus, this reconstituted transporter maintains some of its characteristic transport properties during these procedures. This method for assaying the presence of the transporter proved useful in the initial isolation of the glucose transporter from the membranes of human erythrocytes (Kasahara and Hinkle, 1976, 1977). Erythrocyte ghosts were fractionated to yield a preparation that was rich in membranes. This membrane preparation was washed free of peripheral proteins, placed in a membrane-dissolving detergent, and fractionated further by chromatography. Fractions were incorporated into lipid membranes and assayed for stereospecific sugar transport. A fraction that had the requisite transport properties and that bound cytochalasin B was identified. This fraction was a glycoprotein with a molecular weight of ~55 kD. Isolation of the transporter has allowed determination of its primary structure as well as its kinetic properties.

6.7.3 Structure of Glucose Transporters

The complete amino acid sequence of the glucose transporter of human hepatoma cells[10] was determined (Mueckler et al., 1985) by isolating the DNA encoding the protein, determining the sequence of nucleotides in the DNA, and translating the sequence of nucleotides into a sequence of amino acids. Isolation of the DNA was accomplished by first making an antibody to the purified glucose transporter. The antibody was used as a marker for the transporter protein. Complementary DNA (cDNA) was produced from a cell that expressed the protein. Different fragments of the cDNA were incorporated into bacteriophage genes to produce an expression library. That is, different bacterial colonies had different cDNA fragments and hence produced different messenger RNA (mRNA) transcribed from the cDNA, and expressed different proteins. The antibody raised against the glucose transporter was used to determine which colony produced the glucose transporter. The nucleotide sequence of the cDNA fragment that coded for the glucose transporter was determined, which defined the amino acid sequence of the protein. The hepatoma glucose transporter, which was shown to be almost completely homologous with the human erythrocyte glucose transporter, contains 492 amino acids in a single chain. This amino acid sequence defines the primary struc-

10. These cells are malignant hepatocytes that transport sugars in a stereo-specific manner and bind cytochalasin B.

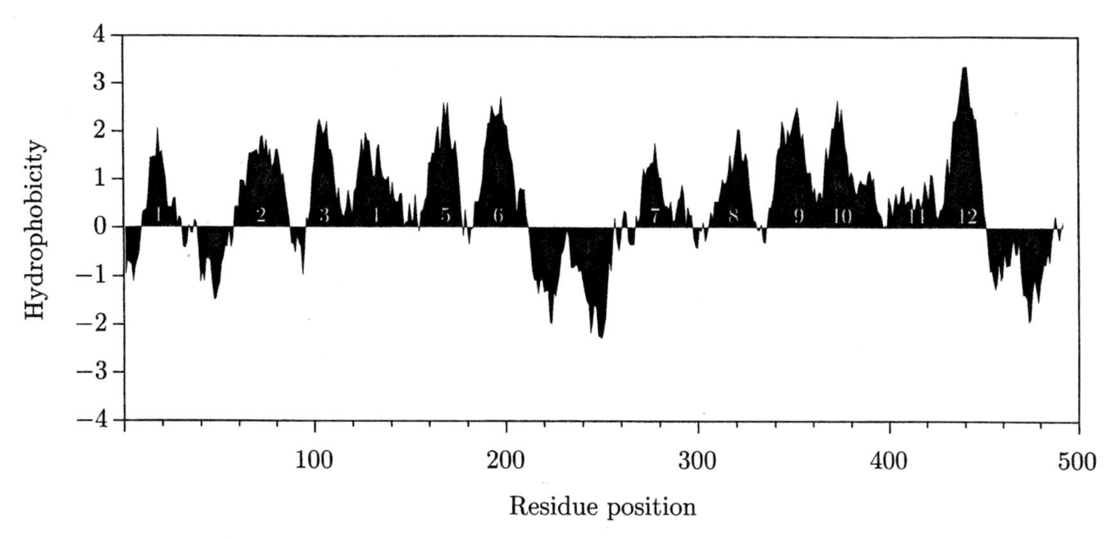

Figure 6.41 Hydrophobicity of GLUT-1. The amino acid sequence for glucose transporter GLUT-1 (Mueckler et al., 1985), was obtained from the Swiss and PIR and Translated database (accession no. P11166). The hydrophobicity for each amino acid residue was assigned (Kyte and Doolittle, 1982), and the resultant sequence of hydrophobicities was averaged with a window that was thirteen residues long to yield the averaged hydrophobicity shown. Twelve hydrophobic segments are indicated.

ture of the protein but not its three-dimensional structure, which is bound to give insights into its mechanism of operation. As with other membrane-spanning proteins, the glucose transporter sequence can be parsed into hydrophilic and hydrophobic segments (Figure 6.41). The glucose transporter contains twenty-five segments, of which thirteen consist of predominantly short (7–14 amino acids) chains of hydrophilic amino acids; two segments are distinctly longer chains of amino acids. These thirteen segments alternate with twelve segments of predominantly hydrophobic amino acids. Each of twelve hydrophobic segments consists of 21 amino acids, which are believed to form α helices of about the right dimensions to span the membrane. On this basis, it was inferred that the hydrophobic segments spanned the membrane and the hydrophilic segments were either intracellular or extracellular. Of the two hydrophilic segments, one forms an intracellular loop between hydrophobic segments 6 and 7; the other forms an extracellular loop between segments 1 and 2. Many of the features of this proposed structure have been identified by chemical means; e.g., it was determined that both end segments of the protein occur intracellularly, and the location of the major extracellular hy-

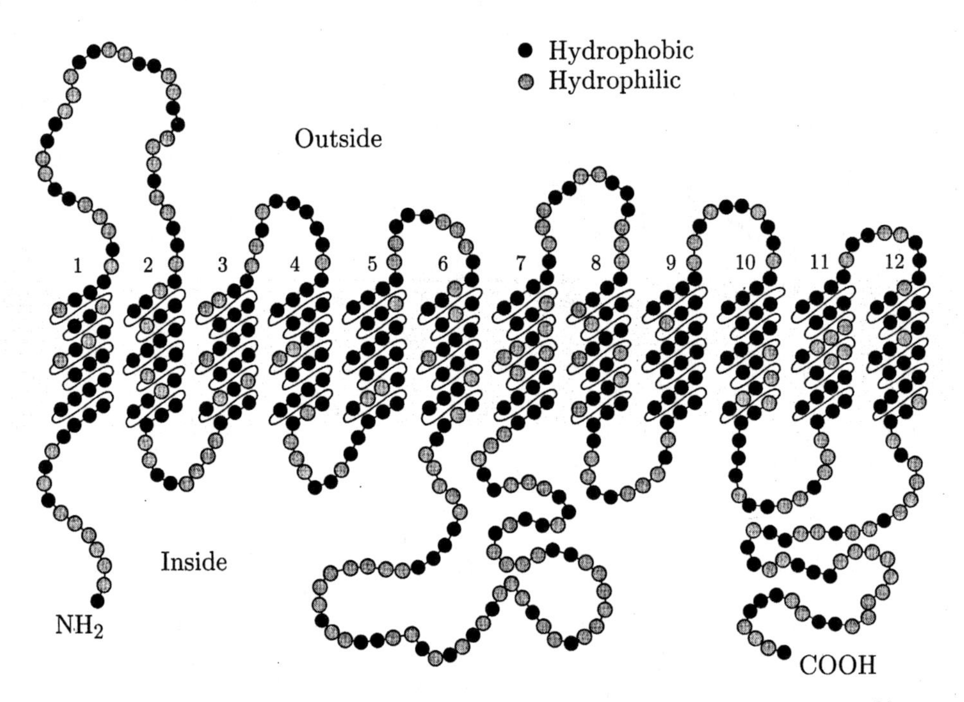

Figure 6.42 Proposed two-dimensional structure of the glucose transporter of human erythrocytes (adapted from Lienhard et al., 1992, figure on p. 88).

drophilic loop has been verified. The putative structure shown in Figure 6.42 gives a plausible two-dimensional representation of the glucose transporter in the membrane.

Examination of the pattern of amino acids in the membrane-spanning sections has suggested that five segments (3, 5, 7, 8, and 11) have hydrophilic amino acids on one side of the α helical cylinder and hydrophobic amino acids on the other side. Thus, it seems plausible that these segments could form an aqueous pore through which glucose could be transported (Figure 6.43). The lumen of the pore would be lined by the hydrophilic amino acid residues, whereas the hydrophobic residues would interface with the remainder of the transporter as well as with the surrounding lipids.

The availability of an antibody for the glucose transporter of erythrocytes allowed an inventory to be made of the occurrence of this transporter in the body. It was widely found in a number of tissues, but little was found either in the liver or in muscles, both of which are important sites of glucose transport. Thus, a search for additional glucose transporters was begun. A number of distinct glucose transporters have been found, of which five are shown in

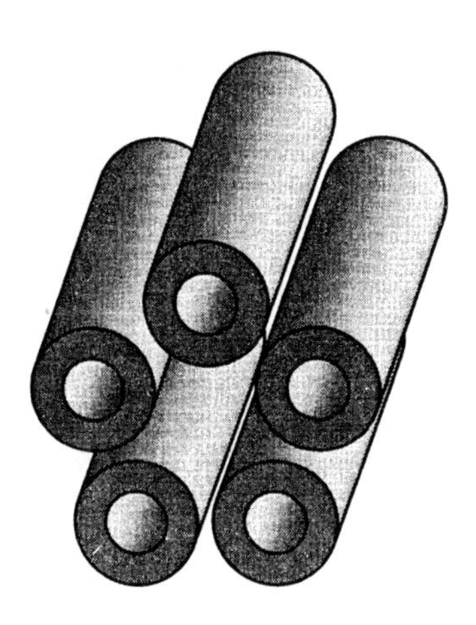

Figure 6.43 Proposed three-dimensional structure of the pore formed by the glucose transporter. Each cylinder represents a transmembrane α helix corresponding to one of the five segments 3, 5, 7, 8, and 11 (see Figure 6.42). The pentagonal arrangement of segments forms a central pore through which glucose is presumed to permeate. The remaining seven membrane-spanning segments (not shown) are assumed to be arranged around the five pore-forming segments.

Table 6.6. They are called GLUT-1, GLUT-2, GLUT-3, etc., and are numbered in the order in which they were discovered. The originally sequenced glucose transporter of human erythrocytes is GLUT-1. The glucose transporters make up a family of transporters that share many features. All contain about 500 amino acids with a similar pattern of hydrophobicity: 12 hydrophobic segments occur separated by hydrophilic segments. About 50% of the amino acid sequence is the same or similar in all five transporters. Each of these transporters has been highly conserved—e.g., for GLUT-1 there is great homology between the amino acid sequences for human, rat, rabbit, mouse, and pig. The amino acids sequence of rat GLUT-1 is 98% identical to that of human GLUT-1.

The different glucose transporters are distributed differently in different organs and show kinetic differences, as indicated in Table 6.6 and Figure 6.40. GLUT-1 is widely found in tissues and is thought to provide a basal rate of glucose uptake in cells. Glucose transport by GLUT-1 is highly asymmetric. GLUT-2 is found in tissues involved in glucose regulation—the liver and kidneys, which release glucose into the circulatory system, and the pancreas, which secretes insulin into the circulatory system in response to an increase in blood glucose. Transport of glucose via GLUT-2 is symmetric, with a value of K_{eq} of about ten times higher for these cells than for erythrocytes, suggesting that in the normal range of blood glucose these cells take up glucose from the blood roughly in proportion to the blood glucose concentration. GLUT-3 is found in cells that have a high affinity for glucose and hence a low value of K_{eq}. Thus, the influx of glucose in these cells is not limited by the normal blood glucose concentration. GLUT-4 is found in cells whose glucose transport is affected by insulin. GLUT-5 is found primarily in tissues that secrete glucose into the circulatory system. Its role is not yet clear, but it appears to be a high-

Table 6.6 Summary of glucose transporters found in different tissues (Gould and Bell, 1990; Gould and Holman, 1993). The major sites of expression of the glucose transporter is given. The abundance of the transporter is indicated as high (H) or low (L). The kinetic properties of the transporter are summarized by indicating the kinetic symmetry and affinity of each transporter.

Transporter	# AA	Major sites	Kinetic properties	Proposed function
Passive transport				
GLUT-1	492	Placenta, erythrocytes, brain, blood-tissue barrier (H); adipose tissue, muscle (L)	Asymmetric, high affinity	Basal glucose uptake in cells
GLUT-2	524	Liver, kidney proximal tubule, small intestine basolateral membrane, pancreatic β-cells (H)	Symmetric, low affinity	Glucose regulation
GLUT-3	496	Brain, nerve (H); placenta, kidney, fibroblasts, liver, cardiac muscle (L)	Symmetry unknown	High-affinity glucose uptake
GLUT-4	509	Skeletal muscle, cardiac muscle, adipose tissue	Symmetric, high affinity	Insulin-mediated glucose uptake
GLUT-5	501	Small intestines (apical membrane), adipose tissue (H); muscle, adipose tissue (L)	Symmetry unknown, high affinity	Fructose transporter
Sodium-coupled, secondary active transport				
SGLT-1	664	Small intestines (apical membranes), kidney	—	Active uptake of glucose

affinity transporter of fructose. Finally, SLGT-1 is found in the intestines and kidney and is a sodium-glucose cotransporter.

6.7.4 Recruitment of Glucose Transporters by Insulin

In muscle and adipose cells, insulin increases the maximum flux (ϕ_{eq}) without changing the dissociation constant (K_{eq}) much. This property implies that either insulin recruits (activates) transporter molecules or it increases the

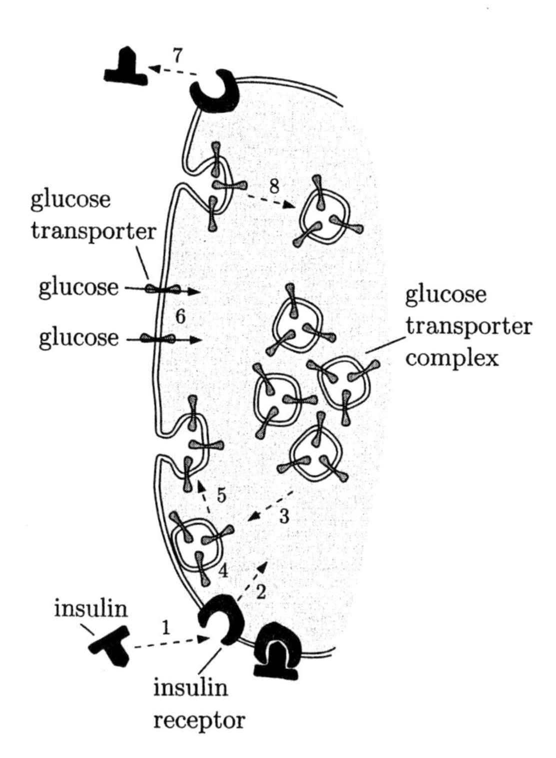

Figure 6.44 A model for the role of insulin in regulation of hexose transport (adapted from Karnieli et al., 1981, Figure 7). The steps in the regulation of glucose transport are: binding (1) of insulin to the insulin receptor, which releases (2) an intracellular signal, which recruits (3) glucose transporter complexes from an intracellular pool, which causes the binding (4) of glucose transporter complexes to the plasma membrane, where these complexes fuse (5) with the membrane, resulting in the transport of glucose (6) via the glucose transporter. Unbinding (7) of the insulin from the insulin receptor results in the invagination (8) of membrane-containing glucose transporters and storage of these complexes intracellularly.

maximum transport rates of each transporter, or both. The first possibility is favored by the results of experiments designed to estimate the location of glucose transporters in the presence and absence of insulin. In the absence of insulin, a large number of glucose transporters are found in the cytoplasm. In the presence of insulin, the number of cytoplasmic transporters is decreased and the number of membrane-bound transporters is increased. These types of experiments lead to schemes such as the one shown in Figure 6.44. In this scheme, insulin binds to the insulin receptor, which releases a signal, or *second messenger,* intracellularly. The second messenger triggers processes that induce glucose transport vesicles to be incorporated into the membrane of the cell, where they can increase the flux of glucose.

Exercises

6.1 List four properties of carrier-mediated transport mechanisms that are distinct from transport by diffusion.

6.2 Figure 6.1 shows measurements of extracellular galactose concentration as a function of time, obtained from a solution containing human erythrocytes loaded with radioactive galactose. The concentration is expressed per liter of cell water. A liter of cell water at the concentration of erythrocytes in this experiment contained about 1.4×10^{13} erythrocytes. Each erythrocyte has a surface area of about 137 μm^2. Determine the efflux of galactose in $mol/cm^2 \cdot s$ for an initial concentration of galactose of 100 mmol/L. You may ignore the change in volume of the erythrocytes that accompanies efflux of galactose.

6.3 Discuss why saturation of the dependence of measured flux on concentration difference across a membrane is consistent with the existence of a discrete number of carriers, each of which exhibits saturation in the relation of flux to concentration difference. Does saturation of the relation of the measured flux to the concentration difference prove that there are a discrete number of saturable carriers in the membrane? If not, why not?

6.4 The dissociation constant of a binding reaction of an enzyme with its substrate is $K = 5$ mmol/L. If the total quantity of enzyme is 2 mmol/L and the unbound substrate concentration is 9 mmol/L, what are the concentrations of bound and unbound enzyme at equilibrium?

6.5 Two ligands, A and B, bind to an enzyme with dissociation constants K_A and K_B, respectively, where $K_A > K_B$. The *affinity* of a ligand for an enzyme is a measure of how readily the two bind. Thus, a higher-affinity ligand binds more readily to an enzyme than does a lower-affinity ligand. For which of the ligands, A or B, does the enzyme have the greater affinity? Explain.

6.6 A chemical reaction has a rate constant α at 5°C. The reaction has a $Q_{10} = 3$. Assume that the rate constant obeys the theory of absolute reaction rates.

a. Assume that the temperature dependence of the rate factor A in Equation 6.35 is negligible. What is the rate constant for the reaction at 20°C?

b. Assess the effect on the result in part a of assuming that the temperature variation of A is negligible.

6.7 Equation 6.50 relates the flux of solute bound to carrier to the density of bound carrier in its inside and outside states. But, in the steady state,

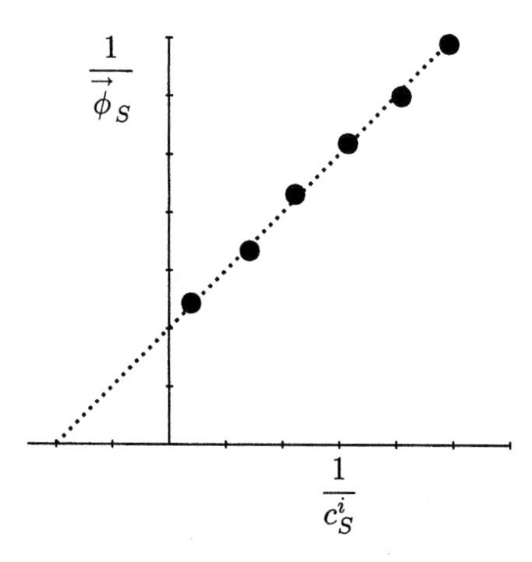

Figure 6.45 Efflux versus intracellular concentration of sugar in reciprocal coordinates (Exercises 6.10 and 6.11). The measurements are indicated by the filled circles; the dotted line has been fit to these measurements.

the density of carrier in each of these two states is a constant. How is it possible for bound carrier to move from the inside to the outside state without a change in the density of bound carrier in either state?

6.8 Consider the simple, symmetric, four-state carrier shown in Figure 6.21. For each of the following conditions, find $\mathfrak{N}_E^i$, $\mathfrak{N}_E^o$, $\mathfrak{N}_{ES}^i$, $\mathfrak{N}_{ES}^o$, and ϕ_S. Explain the physical significance of each of your answers.

a. $\alpha = 0$.

b. $\beta = 0$.

c. $K = 0$.

6.9 For the simple, symmetric, four-state carrier shown in Figure 6.21, let $c_S^i = c_S^o = 0$. Sketch the carrier density in each of its four states as a function of α/β. Give a physical interpretation of the results.

6.10 Figure 6.45 shows a plot of the efflux of a sugar from a cell as a function of the intracellular sugar concentration in reciprocal coordinates obtained with the zero-trans protocol. Assume that sugar transport across the membrane of this cell is adequately described by the simple, symmetric, four-state carrier model. Plot the influx of sugar as a function of the extracellular sugar concentration in the same reciprocal coordinates for the zero-trans protocol. Explain your plot.

6.11 Figure 6.45 shows a plot of the efflux of a sugar from a cell as a function of the intracellular sugar concentration in reciprocal coordinates obtained with the zero-trans protocol. Is this result consistent with the predictions of the general, four-state carrier model with arbitrarily asymmetric rate constants shown in Figure 6.36? Explain.

6.12 It is known that the membrane of a certain cell transports a sugar according to the simple, symmetric, four-state carrier model shown in Figure 6.21. However, the rate constants (α and β) and the dissociation constant (K) are unknown. Determine whether each of these three parameters can be estimated from measurements of the relation of steady-state flux of sugar through the membrane to the sugar concentration only. If the parameter can be so estimated, describe a procedure for determining the constant. If it cannot, explain why not.

6.13 This problem involves solute transported by a carrier mechanism. The graphs shown in Figure 6.46 were all obtained with the zero-trans protocol; that is, the unidirectional flux was measured from the cis side of the membrane of a cell into the trans side, in which the solute concentration was zero. In each graph, the flux is plotted versus concentration in double reciprocal coordinates. For each of the following descriptions, choose a graph to which it applies and state a reason for your choice. If the statement applies to none of the graphs, answer "none."

a. The graph is for the transport of a solute in the absence (curve 1) and in the presence (curve 2) of a competitive inhibitor.

b. The graph is for the transport of a solute in the absence (curve 1) and in the presence (curve 2) of a noncompetitive inhibitor.

c. The graph shows both the influx and efflux for a simple, symmetric, four-state carrier model.

d. The graph shows both the influx and efflux for a carrier model that cannot be a simple, symmetric, four-state carrier model but may be a general, four-state carrier model.

6.14 For each of the following statements indicate whether it is true or false and briefly explain your reasons.

a. Insulin regulates the transmembrane transport of glucose in erythrocytes.

b. Insulin regulates the transmembrane transport of glucose in skeletal muscle cells.

c. During fasting, the liver secretes glucose into the circulatory system.

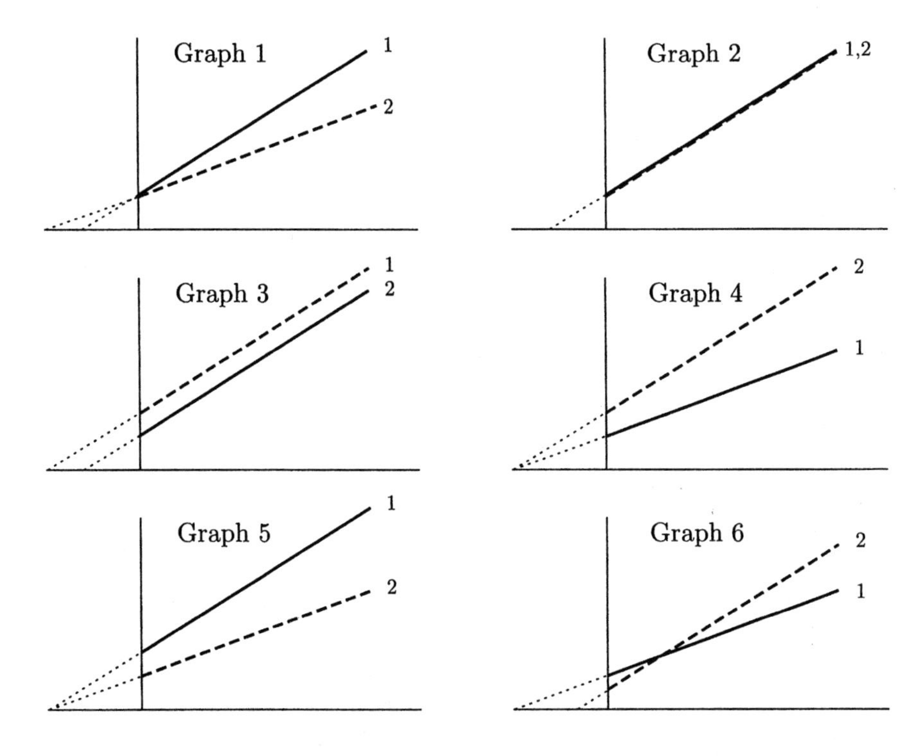

Figure 6.46 Unidirectional flux versus concentration plotted in double reciprocal coordinates for the zero-trans protocol; i.e., the reciprocal of flux is plotted versus the reciprocal of concentration (Exercise 6.13). In each panel, one trace is shown as a solid line and the other as a dashed line. In Graph 2 the solid and dashed traces overlap.

d. During fasting, muscle cells secrete glucose into the circulatory system.

e. Insulin acts on insulin-dependent cells by increasing the glucose transport through individual glucose transporters.

f. Insulin is secreted into the circulatory system by the β-cells in the islets of Langerhans of the pancreas.

6.15 At $t = 0$ a cell is placed in a large extracellular solution containing concentration C_0 of an uncharged solute n. The concentration C_0 remains constant. The intracellular concentration of solute n, $c_n^i(t) = 0$ for $t < 0$, and three different responses are shown in Figure 6.47 for $t > 0$.

a. Which of these responses could be due to a cell that transports the solute by diffusion? Explain.

b. Which of these responses could be due to a cell that transports the solute by a simple, symmetric, four-state carrier? Explain.

c. Which of these responses could be due to a cell that transports the solute by an active transport mechanism? Explain.

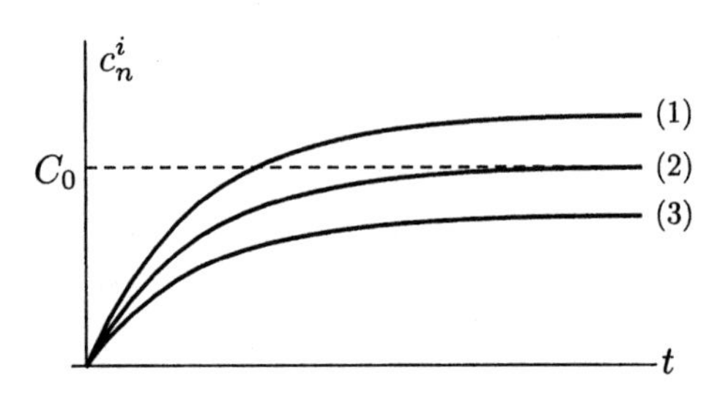

Figure 6.47 Intracellular concentration of solute n as a function of time in response to a step of concentration from 0 to C_0 applied at $t = 0$ (Exercise 6.15).

6.16 Assume that the transport of xylose across the membrane of a cell can be represented by the simple, symmetric, four-state carrier model.

a. Compute the initial rate of increase of xylose concentration in a spherical cell (radius 10 μm) that initially contains no xylose when it is placed in a bath that contains 1 mmol/L xylose. Assume that (1) the total carrier density is 10^{-10} mol/m^2, (2) the dissociation constants for all of the xylose/enzyme binding reactions are equal to 10 mmol/L, and (3) the rates of translocation of enzyme for all translocation reactions are equal to 10^5/s.

b. Can xylose be transported across the membrane if the dissociation constant for xylose/carrier binding is zero?

   i. Explain your answer by giving a mathematical argument.

   ii. Explain your answer by giving a physical argument.

6.17 Figure 6.48 shows the relation between the initial flux of sucrose across a cell membrane and sucrose concentration when the bath contains (1) sucrose with concentration c_S^o, (2) sucrose with concentration c_S^o plus 10 mmol/L of substance A, and (3) sucrose with concentration c_S^o plus 10 mmol/L of substance B.

a. When the bath contains low concentrations of sucrose (e.g., 1 mmol/L), which condition (1, 2, or 3) results in the greatest influx of sucrose? Explain.

b. When the bath contains high concentrations of sucrose (e.g., 200 mmol/L), which condition (1, 2, or 3) results in the greatest influx of sucrose? Explain.

c. Does substance A act as a competitive inhibitor? Explain.

d. Does substance B act as a noncompetitive inhibitor? Explain.

6.18 To what do the terms *GLUT-1*, *GLUT-2*, *GLUT-3*, etc., refer?

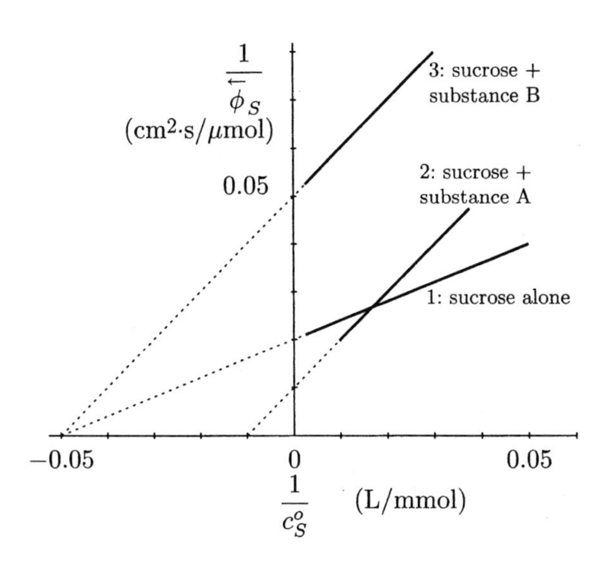

Figure 6.48 Relation between sucrose flux and sucrose concentration in the presence of other substances plotted in double reciprocal coordinates (Exercise 6.17).

6.19 What is the significance of regions numbered 1–12 in Figure 6.41?

6.20 What is the mechanism by which insulin increases glucose transport into muscle cells?

6.21 Describe the role of pancreatic β-cells in the regulation of blood glucose.

Problems

6.1 A membrane of thickness d separates two well-stirred solutions with concentrations of the permeant solute a of $c_a^1(t)$ and $c_a^2(t)$, respectively, as shown in Figure 6.49. The volumes of solution on the two sides of the membrane are $\mathcal{V}_1$ and $\mathcal{V}_2$, and the area of the membrane is A. The membrane contains fixed binding sites, b, that bind the solute a. The total concentration of binding sites is C binding sites per cm^3. The binding/unbinding reaction is defined as

$$a + b \rightleftharpoons ab.$$

This reaction is sufficiently fast, compared to the time rate of change of the concentrations, that you may assume that the reaction is at equilibrium at every instant of time and at each location in the membrane with the dissociation constant K; i.e.,

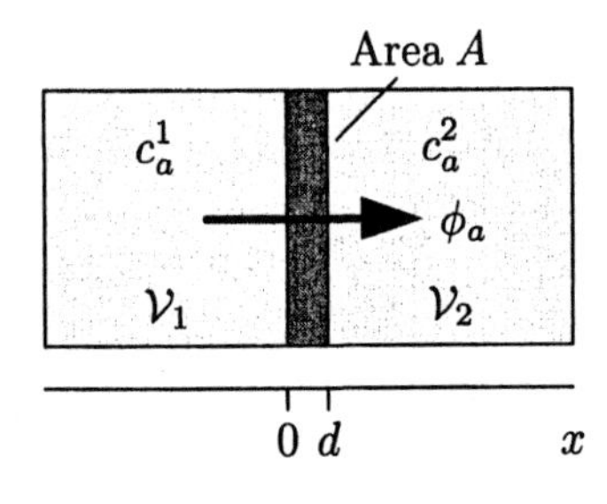

Figure 6.49 Diffusion-reaction system (Problem 6.1).

$$K = \frac{c_a(x,t)c_b(x,t)}{c_{ab}(x,t)},$$

where at each time t and location x in the membrane, $c_a(x,t)$ is the concentration of mobile or unbound solute, $c_b(x,t)$ is the concentration of empty binding sites, and $c_{ab}(x,t)$ is the concentration of sites to which solute is bound. The membrane:solution partition coefficient for a is k_a, which is defined as

$$k_a = \frac{c_a(0,t)}{c_a^1(t)} = \frac{c_a(d,t)}{c_a^2(t)}.$$

a. Determine the dependence of the concentration of bound solute, $c_{ab}(x,t)$, on the concentration of mobile solute, $c_a(x,t)$, and the constants C and K.

b. Show that for sufficiently low solute concentration $c_{ab}(x,t) = \alpha c_a(x,t)$, and find α in terms of C and K.

c. Assume that

   i. the mobile solute obeys Fick's first law so that

   $$\phi_a(x,t) = -D\frac{\partial c_a(x,t)}{\partial x},$$

   where D is a diffusion coefficient and $\phi_a(x,t)$ is the flux of a;

   ii. the solute concentration is low (as in part b).

   Show that for these assumptions,

   $$\frac{\partial c_a(x,t)}{\partial t} = D_{eff}\frac{\partial^2 c_a(x,t)}{\partial x^2},$$

   and find D_{eff} in terms of D, C, and K.

d. Determine τ_e, which is the time constant for equilibration of the concentrations in the two solutions, under the assumption that transport

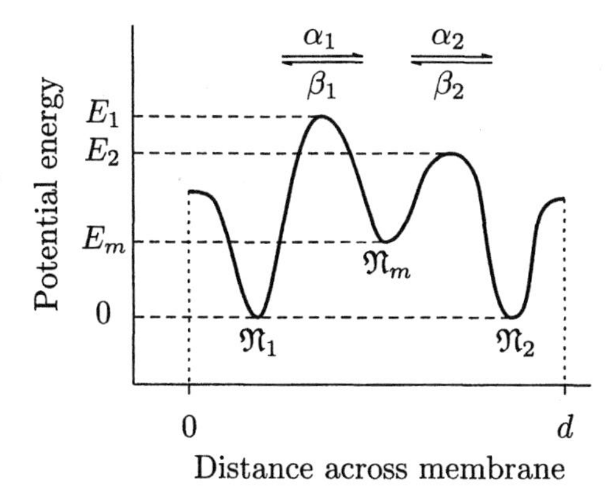

Figure 6.50 Membrane with two potential energy peaks (Problem 6.2).

in the membrane is in steady state. Sketch the dependence of τ_e on the concentration of binding sites C.

e. Estimate τ_d, which is the time for the spatial distribution of unbound solute to reach its steady-state distribution in the membrane. Sketch the dependence of τ_d on the concentration of binding sites C.

f. Explain how the concentration of binding sites affects whether transport in a membrane is in steady state.

6.2 An uncharged solute that is transported across a membrane must pass two potential energy barriers (with peak energies E_1 and E_2, respectively) in order to cross the membrane (Figure 6.50). Between these two barriers is a potential energy well (with energy E_m). $\mathfrak{N}_1$, $\mathfrak{N}_m$, and $\mathfrak{N}_2$ are the densities of solute (in mol/cm^2) in side 1 of the membrane, in the membrane potential energy well, and in side 2 of the membrane, respectively. $\mathfrak{N}_1$, $\mathfrak{N}_m$, and $\mathfrak{N}_2$ satisfy first-order reaction kinetics with rate constants α_1, β_1, α_2, and β_2, as shown in Figure 6.50. Let c_1 and c_2 be the solute concentrations (in mol/cm^3) in the solutions bathing sides 1 and 2 of the membrane. Assume $\mathfrak{N}_1 = dc_1$ and $\mathfrak{N}_2 = dc_2$, where d is a constant.

a. For the steady-state condition, $d\mathfrak{N}_m/dt = 0$, find the flux ϕ, (in mol/cm$^2 \cdot$ s) of solute from side 1 to side 2 in terms of c_1, c_2, d, and the rate constants.

b. Find a constraint on the rate constants such that Fick's law is satisfied, i.e.,

$$\phi = P\,(c_1 - c_s)\,,$$

and find an expression for P under this constraint.

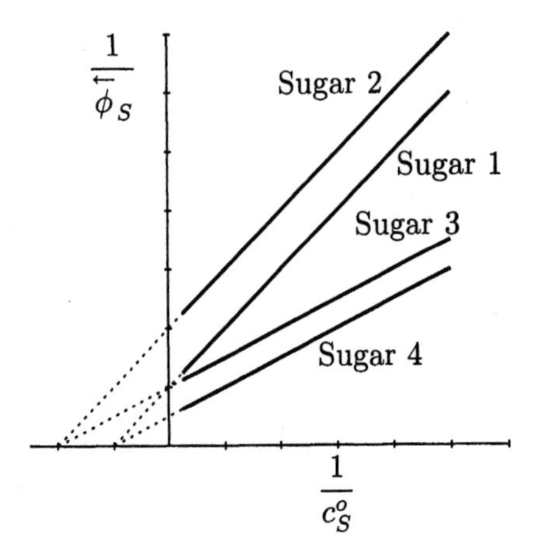

Figure 6.51 Flux versus concentration in reciprocal coordinates (Problem 6.3). The straight lines were fit to the measurements (not shown).

c. Assume that the rate constants are related to the potential energies according to the theory of absolute reaction rates, and that all the rate constants have the same value when $E_1 = E_2 = E_m = 0$. For the conditions found in b, determine the effect of varying E_m on P.

6.3 The unidirectional transport of sugar molecules into red blood cells of badgers, $\overleftarrow{\phi}_s$, is measured by means of radioactive tracers as a function of the external concentration of the sugar molecules, c_S^o. The results for four different types of sugar molecules are plotted in Figure 6.51. It is known that the transport of ten types of sugar molecules through badger red-cell membranes can be described by a simple, symmetric, four-state carrier of the type described in Section 6.4. The values for the parameters K and $(\overleftarrow{\phi}_s)_{max}$ for these sugars are given in Table 6.7, including the value for sugar 1. Determine the identity of the three measured sugars (2, 3, and 4) among those (A–I) in the table. Briefly indicate the reasons for your choices.

6.4 A monosaccharide, M, is known to be transported through a cell membrane by a carrier, so that

$$\phi_s = \phi_{max}\left(\frac{c^i}{K + c^i} - \frac{c^o}{K + c^o}\right),$$

where c^i is the intracellular concentration of M, c^o is the external concentration of M, ϕ is the outward flux of M (mol/cm^2·s), and ϕ_{max} is the maximum flux with which the carrier system is capable of transporting M. The area of the cell, A, is 10^{-6}cm^2, and K is 100 mmol/L. The following experiment is performed: The cell initially contains zero moles

Table 6.7 Parameters of carrier model (Problem 6.3).

Sugar species	K (mmol/L)	$(\overleftarrow{\phi}_S)_{max}$ (μmol/cm$^2\cdot$s)
A	20	20
B	20	80
C	40	40
D	40	80
E	40	40
F	20	10
G	10	40
H	20	5
I	10	20
1	20	40

of M, and at $t = 0$ the cell is placed in an isotonic solution containing a concentration of M equal to c^o (constant), where $c^o \ll K$. The internal concentration of M is found to be

$$c^i(t) = c^o(1 - e^{-t/\tau}), \qquad t \geq 0,$$

where $\tau = 100$ s. The volume of the cell remains roughly constant at 10^{-10} ml throughout the experiment. Determine ϕ_{max}.

6.5 The point of this problem is to investigate a carrier model that differs from the one presented in Section 6.4 but has the same macroscopic transport properties. Consider the carrier model shown in Figure 6.52. The solute, S, is assumed to be insoluble in the membrane except when combined chemically with the carrier, E. Conversely, the carrier, E, and the carrier-solute complex, ES, can diffuse in the membrane but do not leave the membrane. Assume that the diffusion coefficients for E and ES are the same and that the system is in the steady state, so that

$$\phi_S = \phi_{ES} = \alpha\left(\mathfrak{N}^i_{ES} - \mathfrak{N}^o_{ES}\right) = -\phi_E = -\alpha\left(\mathfrak{N}^i_E - \mathfrak{N}^o_E\right),$$

where α is a measure of the diffusion coefficient of the enzyme in the membrane. Assume that the reactions at the membrane surfaces take place so rapidly (compared to the carrier diffusion rates) that they are effectively at equilibrium, i.e.,

$$\frac{c^o_S \mathfrak{N}^o_E}{\mathfrak{N}^o_{ES}} = \frac{c^i_S \mathfrak{N}^i_E}{\mathfrak{N}^i_{ES}} = K,$$

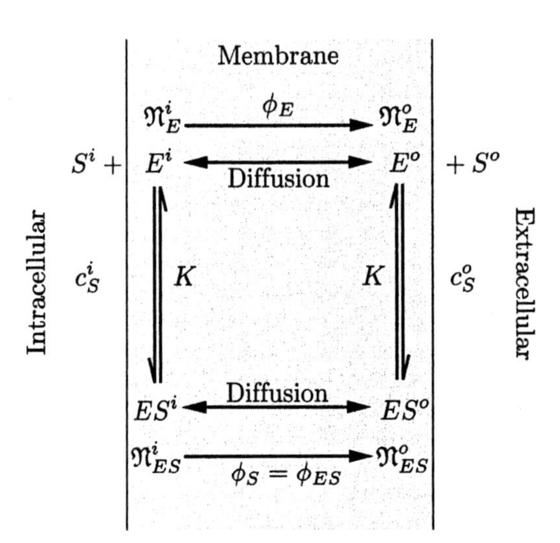

Figure 6.52 Alternative simple, symmetric, four-state carrier model (Problem 6.5).

where K is the dissociation constant and is assumed to be the same at both faces of the membrane. Assume that the total amount of carrier, bound and unbound, is constant.

a. Find and sketch the concentrations of unbound and bound enzyme, $\mathfrak{N}_E(x)$ and $\mathfrak{N}_{ES}(x)$, respectively, as a function of location in the membrane, x.

b. Show that the total concentration of ES, $\overline{\mathfrak{N}_{ES}}$, and E, $\overline{\mathfrak{N}_E}$, are

$$\overline{\mathfrak{N}_{ES}} = \frac{\mathfrak{N}_{ES}^i + \mathfrak{N}_{ES}^o}{2} \quad \text{and} \quad \overline{\mathfrak{N}_E} = \frac{\mathfrak{N}_E^i + \mathfrak{N}_E^o}{2}.$$

c. Show that the relation of solute flux to solute concentration is

$$\phi_S = (\phi_S)_{max} \left(\frac{c_S^i}{c_S^i + K} - \frac{c_S^o}{c_S^o + K} \right),$$

where

$$(\phi_S)_{max} = \alpha \mathfrak{N}_{ET}.$$

d. Assume that $c_S^i \ll K$ and that $c_S^o \ll K$. Show that

$$\phi_S = P_S(c_S^i - c_S^o),$$

and find P_S.

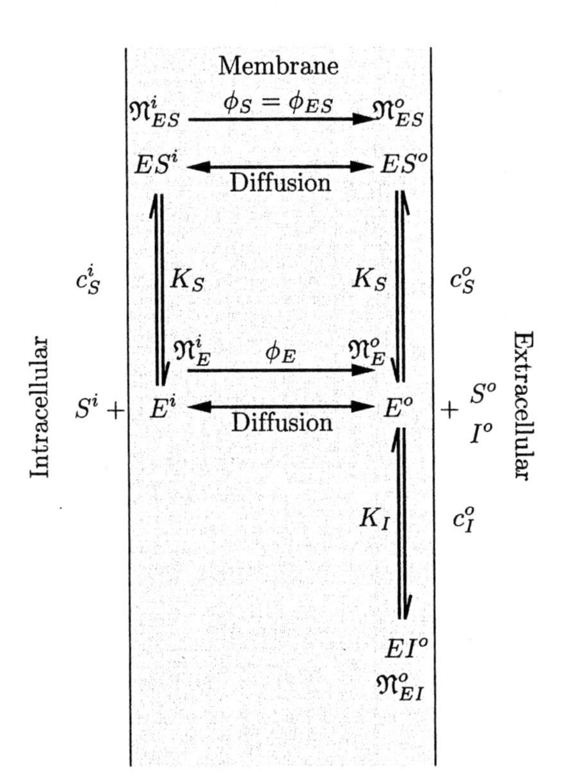

Figure 6.53 Alternative competitive inhibition (Problem 6.6).

6.6 Solute S is transported across the membrane by a carrier model whose kinetic diagram is shown in Figure 6.53. This scheme is similar to that investigated in Problem 6.5. However, note that in this problem it is assumed that $c^i_S = 0$. A second solute, an inhibitor, I, can combine with E at the outside surface of the membrane to form the complex EI, which does not diffuse through the membrane. Assume that all the chemical reactions are in the steady state.

$$\phi_S = -\phi_E = \phi_{ES} = \alpha\left(\mathfrak{N}^i_{ES} - \mathfrak{N}^o_{ES}\right),$$

$$K_S = \frac{c^o_S \mathfrak{N}^o_E}{\mathfrak{N}^o_{ES}}; \qquad K_I = \frac{c^o_I \mathfrak{N}^o_E}{\mathfrak{N}^o_{EI}}; \qquad \mathfrak{N}_{ET} = \mathfrak{N}^o_{EI} + \frac{\mathfrak{N}^o_{ES} + \mathfrak{N}^i_{ES}}{2} + \frac{\mathfrak{N}^o_E + \mathfrak{N}^i_E}{2}.$$

a. Show that

$$\phi_S = (\phi_S)_{max}\left(\frac{c^o_S}{c^o_S + K_S(1 + \frac{c^o_I}{K_I})}\right),$$

where

$$(\phi_S)_{max} = \alpha \mathfrak{N}_{ET},$$

and $\mathfrak{N}_{ET}$ is the total concentration of E in the membrane.

b. Sketch ϕ_S versus c_S^o for various values of c_I^o/K_I.

c. Explain why I acts as an inhibitor of the transport of S.

6.7 A spherical cell of radius 30 μm is loaded with glucose and immersed in a large quantity of isotonic, glucose-free solution. You may assume that the external glucose concentration remains zero and that the change in cell water volume is negligible. The cell membrane transports glucose by a simple, symmetric, four-state carrier with dissociation constant K and maximum flux ϕ_M. Two different experiments are performed to measure intracellular glucose concentration, $c_G(t)$.

Experiment 1 When the initial internal glucose concentration is low, i.e., $c_G(0) \ll K$, it is found that

$$c_G(t) = c_G(0)e^{-t/\tau},$$

where $\tau = 10$ minutes.

Experiment 2 When the initial internal glucose concentration is high, i.e., $c_G(0) \gg K$, it is found that the initial rate of change of internal glucose concentration is

$$\left(\frac{dc_G}{dt}\right)_{t=0} = -10^{-8}\,\text{mol/cm}^3 \cdot \text{s}.$$

Determine the values of K and ϕ_M.

6.8 D-glucose, D-galactose, and D-xylose are transported by a carrier resident in the membrane of a cell under study. The cell is loaded with a *low concentration* of one of these sugar molecules and is placed in an isotonic bath containing no sugar. When this experiment is repeated for each of the three sugar molecules, it is found that the intracellular sugar concentration equilibrates with the bath most rapidly when the cell is loaded with D-glucose, and least rapidly when the cell is loaded with D-xylose. D-galactose gives an intermediate equilibration time. The efflux of sugar, $\vec{\phi}_s$, is plotted versus the internal sugar concentration, c_s^i, in a Lineweaver-Burke plot, shown in Figure 6.54, for an external sugar concentration that is zero ($c_s^o = 0$). Results are shown for the three sugar molecules,

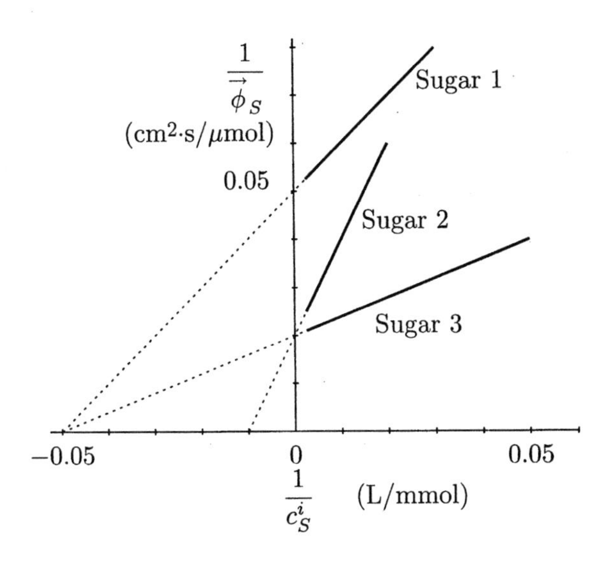

Figure 6.54 Flux versus concentration in reciprocal coordinates for sugars (Problem 6.8). Straight lines (shown) have been fit to the measurements (not shown).

Table 6.8 Transport parameters for sugars (Problem 6.8).

Solute	Sugar #	K_i (mmol/L)	$(\vec{\phi}_s)_{max}$ (μmol/cm$^2\cdot$s)
D-glucose			
D-galactose			
D-xylose			

but the correspondence of sugar 1, sugar 2, and sugar 3 to D-glucose, D-galactose, and D-xylose is not given.

Identify D-glucose, D-galactose, and D-xylose with sugar 1, 2, and 3 and determine the maximum efflux, $(\vec{\phi}_s)_{max}$, and the dissociation constant, K_i, for each sugar in Table 6.8.

6.9 A spherical cell of radius r is loaded with glucose and placed in a large bath of isotonic, glucose-free solution. The cell membrane contains a glucose carrier with dissociation constant K and maximum flux ϕ_m. The internal concentration of glucose at time t is $c_G(t)$, and the initial concentration is C.

a. Find the differential equation satisfied by $c_G(t)$ in terms of r, K, and ϕ_m.

b. Solve the differential equation to obtain a relation between $c_G(t)$ and t. You need not obtain c_G as an explicit function of t!

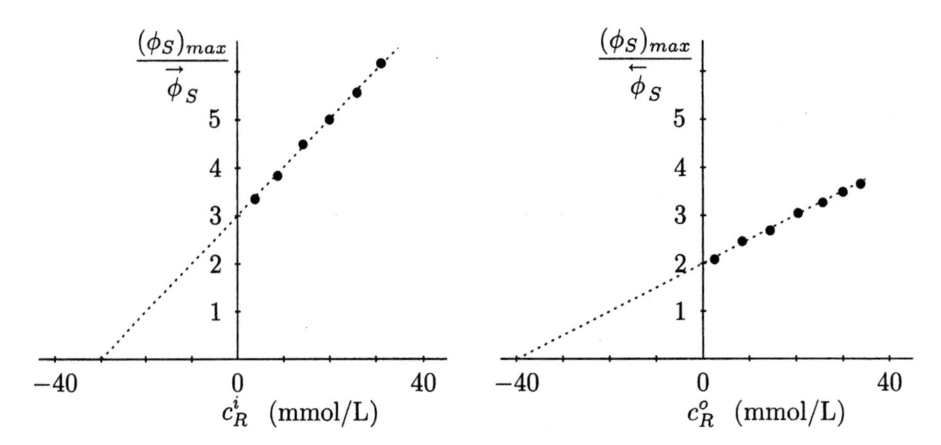

Figure 6.55 Competitive inhibition (Problem 6.10).

c. Find the solution for $K \gg c_G(t)$. Explain the significance of the form of this solution.

d. Find the solution for $K \ll c_G(t)$. Explain the significance of the form of this solution.

6.10 The transport of solute S is measured through a cellular membrane in the presence of a second solute, R, that is also transported across the membrane; the results are shown in Figure 6.55. In the left panel, the unidirectional outward flux, $\overrightarrow{\phi}_S$, is plotted as a function of c_R^i with c_S^i held constant. In the right panel, the unidirectional inward flux, $\overleftarrow{\phi}_S$, is plotted as a function of c_R^o with c_S^o held constant.

Are the results shown in Figure 6.55 consistent or inconsistent with the simple, symmetric, four-state carrier model as shown in Figure 6.27? Explain.

6.11 A cell transports the sugar S across its membrane by the carrier-mediated transport mechanism whose kinetic diagram is shown in Figure 6.56. In this mechanism, the unbound enzyme E cannot translocate across the membrane ($\phi_E = 0$).

a. For steady-state conditions, determine the flux of S across the membrane in terms of the concentrations, rate constants, and dissociation constants. Explain your answer.

b. Suppose that at $t = 0$ the internal concentration of S is C_S and the external concentration is $2C_S$. Will the concentration of S inside the cell change? Explain your reasoning.

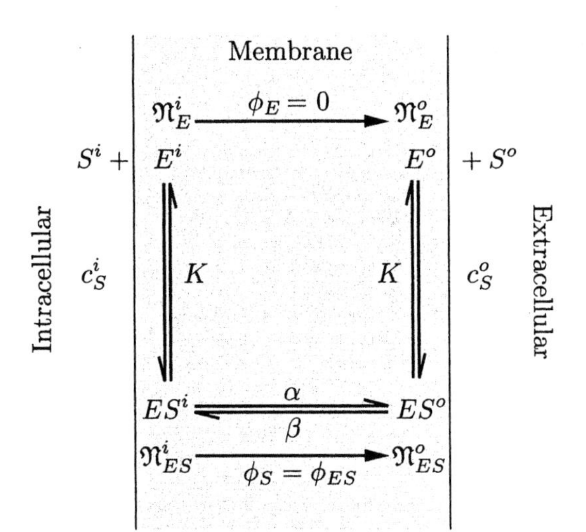

Figure 6.56 Exchange diffusion (Problem 6.11).

c. Suppose that at $t = 0$ the internal concentration of S is C_S and the external concentration is $2C_S$, but the external S is radioactive. The enzyme E cannot distinguish S from radioactive S. Will the concentration of radioactive S inside the cell change? Explain your reasoning.

d. Why is this transport mechanism called "exchange diffusion"?

6.12 The kinetics of glucose transport is studied in three types of spherical cells that differ in their membrane characteristics and in their radii; the radii are 6, 9, and 27 μm, respectively. All three types of cells transport D-glucose with simple, symmetric, four-state carrier models. The steady-state relation between the influx of D-glucose, $\overleftarrow{\phi}_G$, and the external D-glucose concentration, c_G^o, is shown in Figure 6.57 for each cell type.

At $t = 0$, initially glucose-free cells are placed in a large bath containing 0.01 mmol/L of D-glucose.

a. Find the internal concentrations of D-glucose in all three cell types after the cells have equilibrated.

b. Determine the time course of internal glucose concentration for each cell type. State and justify any approximations you make.

c. Which cell type equilibrates most rapidly; which the least rapidly?

6.13 Figure 6.58 shows measurements of the effect of cytochalasin B in the extracellular solution on the efflux of glucose from erythrocytes under zero-trans conditions. These measurements were obtained by the initial slope method.

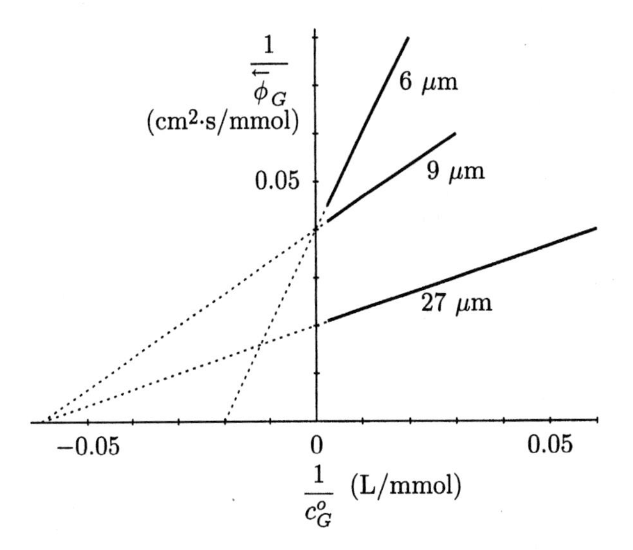

Figure 6.57 Kinetic parameters (Problem 6.12).

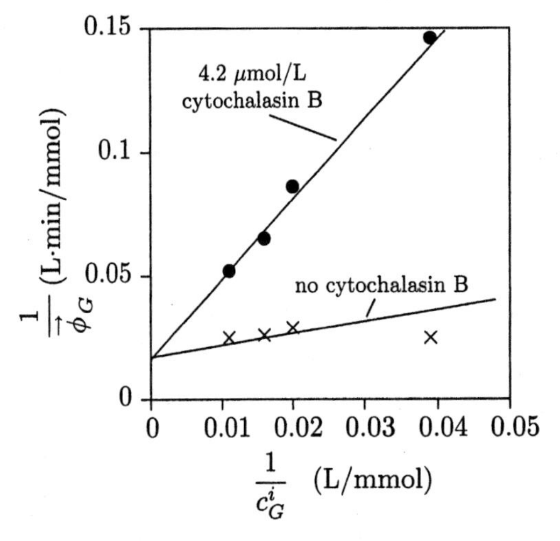

Figure 6.58 Effect of cytochalasin B on glucose efflux from erythrocytes (Devés and Krupka, 1978) (Problem 6.13). The efflux of glucose, $\overrightarrow{\phi}_G$, is plotted versus the intracellular concentration of glucose, c_G^i, in double reciprocal coordinates in the absence (×) and in the presence (filled circles) of cytochalasin B. The flux is expressed as the number of millimoles of glucose per liter of water that includes the cells per minute.

a. Determine the equivalent maximum flux, ϕ_{eq}, and dissociation constant, K_{eq}, for the measurements in the absence of cytochalasin B. Use the same units as shown for the measurements.

b. Determine the equivalent maximum flux, ϕ_{eq}, and dissociation constant, K_{eq}, for the measurements in the presence of cytochalasin B.

c. In this experiment, does cytochalasin B act as a competitive or a noncompetitive inhibitor? Explain.

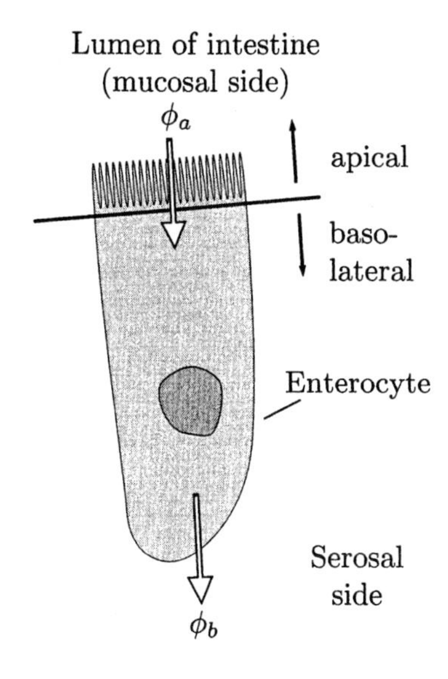

Figure 6.59 Schematic representation of an enterocyte (Problem 6.14). The cell membrane has an apical part that separates the interior of the cell from the lumen of the intestine and a basolateral part that separates the interior of the cell from extracellular space on the serosal side. ϕ_a represents the flux of glucose from the lumen of the intestine through the apical part of the cell membrane and into the cell. ϕ_b represents the flux of glucose from the cell through the basolateral part of the cell membrane and into the extracellular serosal space.

6.14 Glucose is transported into the body by enterocytes (Figure 6.59), which are absorptive epithelial cells that line the small intestine. Transport through the apical part of the cell membrane, which faces the lumen of the small intestine, is coupled to the transport of Na^+. Transport through the basolateral membrane of the cell, which faces the serosal side, is via a glucose carrier. Assume that the glucose carrier in the basolateral part of the cell can be represented by the simple, symmetric, four-state carrier model. Let K represent the dissociation constant for the binding of glucose to the carrier, and let ϕ_{max} represent the maximum flux through the carriers in the basolateral part of the membrane. Let A_a and A_b represent the areas of the apical and basolateral membranes, respectively. Let $\mathcal{V}$ represent the volume of the cell. Assume that A_a, A_b, and $\mathcal{V}$ are constant with respect to time. Assume that glucose is not produced, consumed, or bound by any intracellular mechanism.

a. In the steady state, the concentration of glucose in the cell is constant. Determine a relation that ϕ_a and ϕ_b must satisfy in the steady state.

b. Determine a relation between ϕ_b and c_s^i and c_s^o dictated by the transport properties of the basolateral membrane, where c_s^o is the extracellular concentration of glucose on the serosal side of the membrane.

c. Assume that the flux, ϕ_a, from the lumen of the intestine into the cell is constant and that the concentration of glucose on the serosal

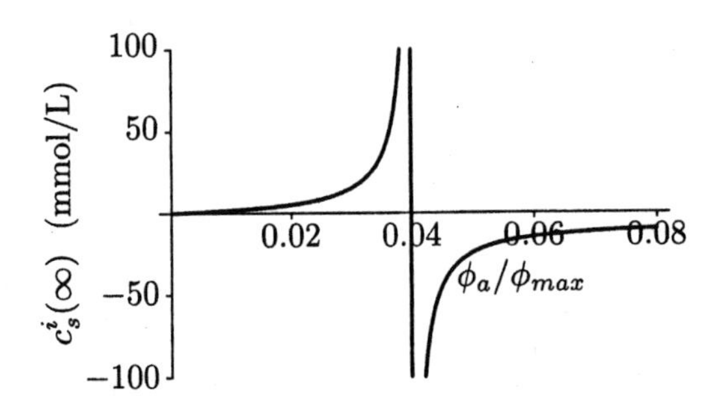

Figure 6.60 Plot of $c_s^i(\infty)$ versus ϕ_a/ϕ_{max} for $c_s^o = 0$, $K = 5$ mmol/L, and for $A_a/A_b = 25$ (Problem 6.14).

side is zero ($c_s^o = 0$). Using the results of parts a and b, determine an expression for $c_s^i(\infty)$ in terms of ϕ_a, ϕ_{max}, K, A_a, and A_b.

d. $c_s^i(\infty)$ is plotted versus ϕ_a/ϕ_{max} for $c_s^o = 0$, $K = 5$ mmol/L, and for $A_a/A_b = 25$ in Figure 6.60.

   i. Explain the *physical significance* of the value of $c_s^i(\infty)$ when $\phi_a = 0$.

   ii. Note from Figure 6.60 that $c_s^i(\infty)$ increases rapidly as ϕ_a/ϕ_{max} increases from 0 to 0.04. Give a *physical interpretation* for this result.

   iii. For $\phi_a/\phi_{max} > 0.04$, $c_s^i(\infty) < 0$. What is the *physical significance* of this result?

6.15 In equilibrium exchange measurements, a cell is allowed to equilibrate in an extracellular concentration of a sugar so that the sugar concentration, C_S, is the same on both sides of the membrane. At $t = 0$, a small quantity, N_S^* moles, of radioactively labeled sugar is added to the extracellular solution, which has a volume $\mathcal{V}^o$. The concentration of radioactive sugar on the outside of the membrane at time t, $c_S^{o*}(t)$, is so small that the sugar concentration essentially remains constant. Let the radioactively labeled sugar concentration on the inside of the cell at time t be $c_S^{i*}(t)$. The objective of the problem is to derive the kinetics of radioactively labeled intracellular sugar concentration under the following assumptions: (1) the sugar is transported by the simple, symmetric, four-state carrier model indicated in Figure 6.21; (2) the membrane cannot discriminate the labeled from the unlabeled sugar; (3) the quantity of radioactively labeled sugar is conserved; (4) the cell water volume, $\mathcal{V}^i$, remains constant because the isotonicity of both the intracellular and the extracellular medium remain constant.

a. Show that for $c_S^{i*}(t)/C_S \ll 1$ and $c_S^{o*}(t)/C_S \ll 1$,

$$\phi_S^*(t) = \gamma\left(c_S^{i*}(t) - c_S^{o*}(t)\right),$$

and find the value of γ.

b. Show that conservation of radioactive sugar leads to a differential equation of the form

$$\frac{dc_S^{i*}(t)}{dt} + \alpha c_S^{i*}(t) = \beta N_S^*, \text{ for } t > 0,$$

and find the values of α and β.

c. If $c_S^{i*}(0) = 0$, find $c_S^{i*}(t)$ for $t > 0$.

d. Explain how you would determine the value of the kinetic parameters K and $(\phi_S)_{max}$ from equilibrium exchange measurements.

e. Briefly discuss the advantages and disadvantages, for estimating the kinetic parameters, of using the equilibrium exchange measurements versus the integrated flux method based on the zero-trans protocol described in Section 6.5.

6.16 This problem deals with the effect on the interpretation of measurements of transmembrane solute diffusion of the binding of a solute to cytoplasmic macromolecules.

a. A spherical cell of radius r is immersed in a large volume of solution containing the solute S at concentration $c_S^o = C$. Both this concentration and the volume of the cell remain constant throughout the measurements. The solute is transported through the membrane by diffusion; the permeability of the membrane for S is P_S. The intracellular solute concentration has an exponential time course:

$$c_S^i(t) = C_1\left(1 - e^{-t/\tau_1}\right).$$

i. Find an expression for C_1 in terms of r, C, and P_S only.

ii. Find an expression for τ_1 in terms of r, C, and P_S only.

b. Now consider a cell that contains the macromolecule M at a *total* cytoplasmic concentration of C_T but is otherwise identical to the cell in part a. Macromolecule M cannot pass through the membrane of the cell, but it binds to the solute S according to the second-order reversible reaction

$$S + M \rightleftharpoons SM,$$

with dissociation constant K, where $K \gg C$. This binding reaction is fast, and the concentrations of reactants and products can be assumed to be at their equilibrium values. The uptake of S is again measured for $c_S^o = C$ as above. The concentration of unbound S has an exponential time dependence:

$$c_S^i(t) = C_2\left(1 - e^{-t/\tau_2}\right).$$

(*Hint:* Remember that although the total quantity of S [both unbound and bound] is conserved, it is the concentration of unbound S that determines diffusion through the membrane.)

i. Find an expression for the relation of the concentration of bound, c_{SM}^i, to unbound, c_S^i, solute concentration in terms of C_T and K.

ii. Find an expression for C_2 in terms of r, C, P_S, C_T, and K only.

iii. Find an expression for τ_2 in terms of r, C, P_S, C_T, and K only.

c. The concentration of S has an exponential time dependence with (as in b) and without (as in a) the binding of S intracellularly. From measurements of $c_S^i(t)$ and the known values of r and C *only*, can one distinguish between (1) diffusion of S across the membrane without binding; (2) diffusion of S across the membrane with binding of S to an intracellular macromolecule? Explain.

6.17 Solute S is transported through a membrane by the simple, symmetric, four-state carrier model, whose kinetic diagram is shown in Figure 6.21. The enzyme can be found in four different states: unbound to solute at either the inside or outside faces of the membrane, or bound to solute at either face. The steady-state densities of enzymes in these four states are $\mathfrak{N}_E^i$, $\mathfrak{N}_E^o$, $\mathfrak{N}_{ES}^i$, and $\mathfrak{N}_{ES}^o$ mol/cm^2; the total enzyme density is $\mathfrak{N}_{ET} = \mathfrak{N}_E^i + \mathfrak{N}_E^o + \mathfrak{N}_{ES}^i + \mathfrak{N}_{ES}^o$. The state of the enzyme system is depicted schematically for four different conditions in Figure 6.61. Answer question a–h and give brief explanations for your choices.

a. True or false: For all four conditions (1)–(4), $\phi_E = -\phi_{ES}$.

b. Which of the following statements applies to (1)?

i. $c_S^i > K$.

ii. $c_S^i = K$.

iii. $c_S^i < K$.

c. True or false: The transition from (1) to (3) can be achieved by changing c_S^i only.

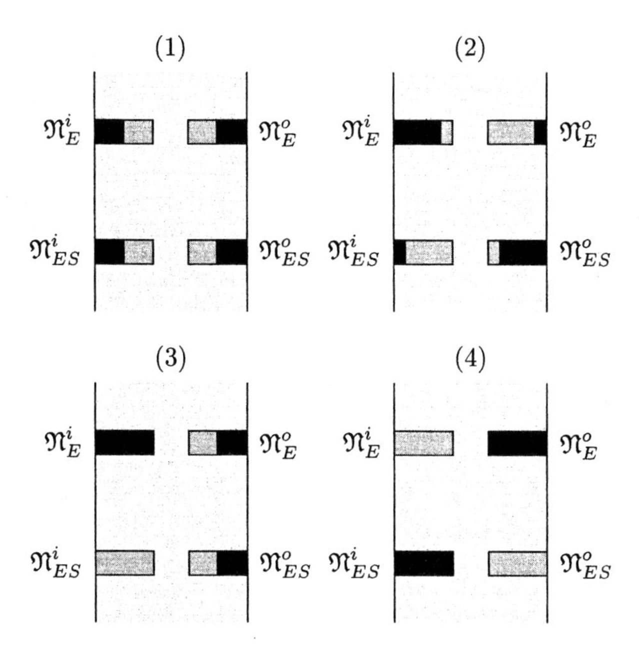

Figure 6.61 State of a simple, symmetric, four-state carrier shown in Figure 6.21 for four different conditions (Problem 6.17). Each dark bar represents one state of the enzyme; the length of each bar is proportional to the fraction of enzyme in that state. The maximum number of enzymes in each state is shown by a thin open rectangle.

d. True or false: In (2), $\phi_S > 0$.

e. True or false: In (1), $\phi_S = 0$.

f. True or false: In (3), $c_S^i = 0$.

g. True or false: The transition from (1) to (3) can be achieved by changing K only.

h. For which of the conditions, (1)-(4), is the magnitude of the flux of S equal to the maximum flux possible for any concentration of S?

6.18 The normal concentration of glucose in the blood plasma of humans is about 5-10 mmol/L. For zero-trans influx of glucose, the equivalent dissociation constant (K_{eq}) is about 20 mmol/L for β-cells in the pancreas and about 0.5 mmol/L for erythrocytes. You may assume that the intracellular concentration of glucose in both cells is zero and that the maximum glucose flux is the same for both cells.

a. For a blood glucose level of 5 mmol/L, find the ratio of glucose influx in β-cells to that in erythrocytes, ϕ_β/ϕ_e. Why is this ratio different from 1?

b. Suppose the blood glucose rises to 10 mmol/L. Compute the percentage change of glucose influx in both β-cells and erythrocytes. Why do these percentages differ?

c. For the normal range of blood glucose, describe the approximate relation between influx of glucose and blood glucose concentration for β-cells and for erythrocytes. Discuss the relevance of these relations for the functions of these two cell types.

6.19 In the simple, symmetric, four-state carrier model, the unidirectional efflux of sugar is related to the intracellular sugar concentration by

$$\vec{\phi}_S = (\phi_S)_{max} \frac{c_S^i}{c_S^i + K}.$$

We have seen that this rectangular hyperbolic relation is a straight line in reciprocal coordinates. However, there are other simple transformations that also make plotting this characteristic convenient. Two of these are explored in this problem.

a. Consider the relation of $c_S^i/\vec{\phi}_S$ versus c_S^i.

   i. Sketch this relation.

   ii. Sketch a family of different plots for different concentrations of a *noncompetitive* inhibitor.

   iii. Sketch a family of different plots for different concentrations of a *competitive* inhibitor.

b. Consider the relation $\vec{\phi}_S$ versus $\vec{\phi}_S/c_S^i$.

   i. Sketch this relation.

   ii. Sketch a family of different plots for different concentrations of a *noncompetitive* inhibitor.

   iii. Sketch a family of different plots for different concentrations of a *competitive* inhibitor.

6.20 This problem concerns the general, four-state carrier model shown in Figure 6.36.

a. Show that at sufficiently low concentrations of solute, the relation of flux to concentration approaches that of Fick's Law with a permeability

$$P_S = \frac{a_o b_i g_o h_i}{(a_i + a_o)(b_i g_o + g_i g_o + b_o g_i)} \mathfrak{N}_{ET}.$$

b. Show that this result reduces to that predicted by the simple, symmetric, four-state model shown in Equation 6.62.

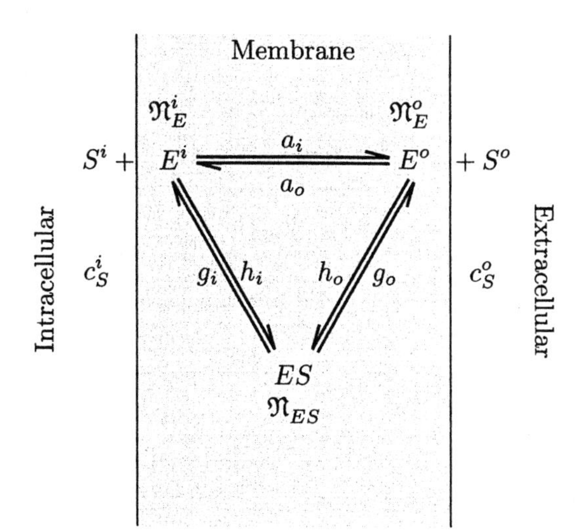

Figure 6.62 Kinetic diagram for a general, three-state carrier model (Problem 6.21).

6.21 The general, three-state kinetic model shown in Figure 6.62 differs from the general, four-state carrier model (Figure 6.36) in the number of intermediate states of the bound complex of carrier and solute.

a. Determine the steady-state distribution of state densities; i.e., find $\mathfrak{N}_E^i$, $\mathfrak{N}_E^o$, and $\mathfrak{N}_{ES}$ in the steady state.

b. Determine the unidirectional efflux and influx as a function of the rate constants and concentrations.

c. Show that the unidirectional efflux and influx can be expressed as Equations 6.111 and 6.113.

d. Determine the relation between the macroscopic parameters K_g, ϕ_z, ϕ_{io}, ϕ_{oi}, and ϕ_∞ and the microscopic parameters a_i, a_o, g_i, g_o, h_i, and h_o.

6.22 The nucleoside uridine is transported across human erythrocyte membranes by a passive carrier mechanism. A summary of transport measurements for uridine is shown in Table 6.9.

a. Explain why these measurements are inconsistent with the simple, symmetric, four-state carrier model.

b. The remaining parts of the problem involve checking whether the results are consistent with the general, four-state carrier model.

   i. Estimate ϕ_{io} and ϕ_{oi} from the zero-trans measurements.

   ii. Estimate ϕ_∞ from the equilibrium exchange data.

Table 6.9 A comparison of kinetic parameters for uridine transport in human erythrocytes at a temperature of 25°C obtained by several different testing protocols (Cabantchik and Ginsburg, 1977; Lieb, 1982) (Problem 6.22). The dissociation constant is expressed in units of concentration of millimoles per liter of cell water; the flux as the rate of change of concentration. Each parameter is given as a mean ± a standard error of the mean.

Protocol	K_{eq} (mmol/L)	ϕ_{eq} (mmol/(L·min))
Zero-trans efflux	0.40 ± 0.12	1.98 ± 0.31
Zero-trans influx	0.073 ± 0.069	0.53 ± 0.038
Infinite-cis efflux	0.252 ± 0.096	—
Infinite-cis influx	0.937 ± 0.226	—
Equilibrium exchange	1.29 ± 0.11	7.54 ± 0.45

iii. Estimate ϕ_z from the results of parts i and ii.

iv. Compute an estimate of the mean ± SE for K_g from each of the five measurements in Table 6.9.

v. Are the results obtained in ii–v consistent with the general, four-state carrier model or not? Explain.

References

Books and Reviews

Anderson, O. S. (1989). Kinetics of ion movement mediated by carriers and channels. In Fleischer, S. and Fleischer, B., eds. *Methods in Enzymology*, vol. 171, *Biomembranes*, pt. R, *Transport Theory: Cells and Model Membranes*, 62–112. Academic Press, New York.

Ashcroft, F. M. and Ashcroft, S. J. H., eds. (1992). *Insulin: Molecular Biology to Pathology*. Oxford University Press, New York.

Atkinson, M. A. and Maclaren, N. K. (1990). What causes diabetes? *Sci. Am.*, 263:62–71.

Baldwin, S. A. and Lienhard, G. E. (1981). Glucose transport across plasma membranes: Facilitated diffusion systems. *Trends Biochem. Sci.*, 6:208–211.

Bliss, M. (1982). *The Discovery of Insulin*. University of Chicago Press, Chicago.

Bull, H. B. (1964). *An Introduction to Physical Biochemistry*. F. A. Davis, Philadelphia.

Cahill, G. F. (1971). Physiology of insulin in man. *Diabetes*, 20:785–799.

Carruthers, A. (1984). Sugar transport in animal cells: The passive hexose transfer system. *Prog. Biophys. Mol. Biol.*, 43:33–69.

Cohen, J. J. and Kassirer, J. P. (1982). *Acid-Base*. Little, Brown & Co., Boston.

Davson, H. (1970). *A Textbook of General Physiology*. J. and A. Churchill, Gloucester, England.

Dowben, R. M. (1969). *General Physiology: A Molecular Approach.* Harper & Row, New York.

Ebbing, D. D. (1984). *General Chemistry*. Houghton Mifflin, Boston.

Eisenberg, D. and Crothers, D. (1979). *Physical Chemistry with Applications to the Life Sciences*. Benjamin-Cummings, Menlo Park, CA.

Eyring, H., Lin, S. H., and Lin, S. M. (1980). *Basic Chemical Kinetics*. John Wiley & Sons, New York.

Friedman, M. H. (1986). *Principles and Models of Biological Transport*. Springer-Verlag, New York.

Gould, G. W. and Bell, G. I. (1990). Facilitative glucose transporters: An expanding family. *Trends Biochem. Sci.*, 15:18–23.

Gould, G. W. and Holman, G. D. (1993). The glucose transporter family: Structure, function and tissue-specific expression. *Biochem. J.*, 295:329–341.

Hendrickson, J. B., Cram, D. J., and Hammond, G. S. (1970). *Organic Chemistry*. McGraw-Hill, New York.

Kotyk, A. and Janácěk, K. (1970). *Cell Membrane Transport: Principles and Techniques*. Plenum, New York.

LeFevre, P. G. (1961). Sugar transport in the red blood cell: Structure-activity relationships in substrates and antagonists. *Pharmacol. Rev.*, 13:39–70.

LeFevre, P. G. (1972). Transport of carbohydrates by animal cells. In Hokin, L. E., ed., *Metabolic Pathways,* vol. 6, *Metabolic Transport,* 385–454. Academic Press, New York.

Lehninger, A. H. (1970). *Biochemistry*. Worth, New York.

Lienhard, G. E., Slot, J. W., James, D. E., and Mueckler, M. M. (1992). How cells absorb glucose. *Sci. Am.*, 266:86–91.

Mason, S. J. and Zimmermann, H. J. (1960). *Electronic Circuits, Signals, and Systems*. John Wiley & Sons, New York.

Moore, W. J. (1972). *Physical Chemistry*. Prentice-Hall, Englewood Cliffs, NJ.

Neame, K. D. and Richards, T. G. (1972). *Elementary Kinetics of Membrane Carrier Transport*. Halstead, New York.

Nystrom, R. A. (1973). *Membrane Physiology*. Prentice-Hall, Englewood Cliffs, NJ.

Rothstein, A. (1980). Fashions in membranes. *Ann. N.Y. Acad. Sci.*, 358:96–102.

Schultz, S. G. (1980). *Basic Principles of Membrane Transport*. Cambridge University Press, Cambridge, England.

Segel, I. H. (1975). *Enzyme Kinetics*. John Wiley & Sons, New York.

Simpson, I. A. and Cushman, S. W. (1986). Hormonal regulation of mammalian glucose transport. *Ann. Rev. Biochem.*, 55:1059–1089.

Stein, W. D. (1967). *The Movement of Molecules across Cell Membranes*. Academic Press, New York.

Stein, W. D. (1986). *Transport and Diffusion across Cell Membranes*. Academic Press, New York.

Stein, W. D. (1990). *Channels, Carriers, and Pumps*. Academic Press, New York.

Thorens, B. (1993). Facilitated glucose transporters in epithelial cells. *Ann. Rev. Physiol.*, 55:591–608.

Vander, A. J., Sherman, J. H., and Luciano, D. S. (1990). *Human Physiology: The Mechanisms of Body Function.* McGraw-Hill, New York.

Walmsley, A. R. (1988). The dynamics of the glucose transporter. *Trends Biochem. Sci.*, 13:226–231.

Weiss, T. F. (1996). *Cellular Biophysics,* vol. 2, *Electrical Properties.* MIT Press, Cambridge, MA.

Wheeler, T. J. and Hinkle, P. C. (1985). The glucose transporter of mammalian cells. *Ann. Rev. Physiol.*, 47:503–517.

Wright, E. M. (1993). The intestinal Na^+/Glucose cotransporter. *Ann. Rev. Physiol.*, 55:575–589.

Original Articles

Ämmälä, C., Ashcroft, F. M., and Rorsman, P. (1993). Calcium-independent potentiation of insulin release by cyclic AMP in single β-cells. *Nature*, 363:356–358.

Appleman, J. R. and Lienhard, G. E. (1985). Rapid kinetics of the glucose transporter from human erythrocytes: Detection and measurement of a half-turnover of the purified transporter. *J. Biol. Chem.*, 260:4575–4578.

Appleman, J. R. and Lienhard, G. E. (1989). Kinetics of the purified glucose transporter: Direct measurement of the rates of interconversion of transporter conformers. *Biochemistry*, 28:8221–8227.

Atwater, I. and Sherman, A. (1993). Importance of islet cell synchrony for the β-cell glucose response. *Biophys. J.*, 65:565.

Baker, G. F., Basketter, D. A., and Widdas, W. F. (1978). Asymmetry of the hexose transfer system in human erythrocytes: Experiments with non-transposable inhibitors. *J. Physiol.*, 278:377–388.

Baker, G. F. and Naftalin, R. J. (1979). Evidence of multiple operational affinities for D-Glucose inside the human erythrocyte membrane. *Biochim. Biophys. Acta*, 550:474–484.

Baker, G. F. and Widdas, W. F. (1973a). The asymmetry of the facilitated transfer system for hexoses in human red cells and the simple kinetics of a two component model. *J. Physiol.*, 231:143–165.

Baker, G. F. and Widdas, W. F. (1973b). The permeation of human red cells by 4,6-*O*-Ethylidene-α-D-Glucopyranose (ethylidene glucose). *J. Physiol.*, 231:129–142.

Baker, P. F. and Carruthers, A. (1981a). 3-0-methylglucose transport in internally dialysed giant axons of *Loligo. J. Physiol.*, 316:503–525.

Baker, P. F. and Carruthers, A. (1981b). Sugar transport in giant axons of *Loligo. J. Physiol.*, 316:481–502.

Baker, P. F. and Carruthers, A. (1983). Insulin regulation of sugar transport in giant muscle fibres of the barnacle. *J. Physiol.*, 336:397–431.

Baldwin, S. A., Baldwin, J. M., and Lienhard, G. E. (1982). Monosaccharide transporter of the human erythrocyte: Characterization of an improved preparation. *Biochemistry*, 21:3836–3842.

Baldwin, S. A. and Henderson, P. J. F. (1989). Homologies between sugar transporters from eukaryotes and prokaryotes. *Ann. Rev. Physiol.*, 51:459–471.

Basketter, D. A. and Widdas, W. F. (1978). Asymmetry of the hexose transfer system in human erythrocytes: Comparison of the effects of cytochalasin B, phloretin and maltose as competetive inhibitors. *J. Physiol.*, 278:389–401.

Blok, J., Gibbs, E. M., Lienhard, G. E., Slot, J. W., and Geuze, H. J. (1988). Insulin-induced translocation of glucose transporters from post-golgi compartments to the plasma membrane of 3T3-L1 adipocytes. *J. Cell Biol.*, 106:69–76.

Brahm, J. (1983). Kinetics of glucose transport in human erythrocytes. *J. Physiol.*, 339:339–354.

Britton, H. G. (1964). Permeability of the human red cell to labelled glucose. *J. Physiol.*, 170:1–20.

Brozinick, J. T., Jr., Etgen, G. J., Jr., Yaspelkis, B. B., III, and Ivy, J. L. (1994). The effects of muscle contraction and insulin on glucose-transporter translocation in rat skeletal muscle. *Biochem. J.*, 297:539–545.

Cabantchik, Z. I. and Ginsburg, H. (1977). Transport of uridine in human red blood cells. *J. Gen. Physiol.*, 69:75–96.

Cairns, M. T., Alvarez, J., Panico, M., Gibbs, A. F., Morris, H. R., Chapman, D., and Baldwin, S. A. (1987). Investigation of the structure and function of the human erythrocyte glucose transporter by proteolytic dissection. *Biochim. Biophys. Acta*, 905:295–310.

Carruthers, A. (1983). Sugar transport in giant barnacle muscle fibres. *J. Physiol.*, 336:377–396.

Carruthers, A. and Melchior, D. L. (1983). Asymmetric or symmetric? Cytosolic modulation of human erythrocyte hexose transfer. *Biochim. Biophys. Acta*, 728:254–266.

Cha, S. (1968). A simple method for derivation of rate equations for enzyme-catalyzed reactions under the rapid equilibrium assumption or combined assumptions of equilibrium and steady state. *J. Biol. Chem.*, 243:820–825.

Charron, M. J., Brosius, F. C. I., Alper, S. L., and Lodish, H. F. (1989). A glucose transport protein expressed predominately in insulin-responsive tissues. *Proc. Natl. Acad. Sci. U.S.A.*, 86:2535–2539.

Cirillo, V. P. (1961). The transport of non-fermentable sugars across the yeast cell membrane. In Kleinzeller, A. and Kotyk, A., eds., *Membrane Transport and Metabolism*, 343–351. Academic Press, New York.

Clarke, J. F., Young, P. W., Yonezawa, K., Kasuga, M., and Holman, G. D. (1994). Inhibition of the translocation of GLUT1 and GLUT4 in 3T3-L1 cells by the phosphatidylinositol 3-kinase inhibitor, wortmannin. *Biochem. J.*, 300:631–635.

Cleland, W. W. (1963). The kinetics of enzyme-catalyzed reactions with two or more substrates or products: I. Nomenclature and rate equations. *Biochim. Biophys. Acta*, 67:104–137.

Cushman, S. W. and Wardzala, L. J. (1980). Potential mechanism of insulin action on glucose transport in the isolated rat adipose cell. *J. Biol. Chem.*, 255:4758–4762.

Devés, R. and Krupka, R. M. (1978). Cytochalasin B and the kinetics of inhibition of biological transport. *Biochim. Biophys. Acta*, 510:339–348.

Devés, R. and Krupka, R. M. (1980). Testing transport systems for competition between pairs of reversible inhibitors. *J. Biol. Chem.*, 255:11870–11874.

Efrat, S., Tal, M., and Lodish, H. F. (1994). The pancreatic β-cell glucose sensor. *Trends Biochem. Sci.*, 19:535–538.

Fischbarg, J., Cheung, M., Czegledy, F., Li, J., Iserovich, P., Kuang, K., Hubbard, J., Garner, M., Rosen, O. M., Golde, D. W., and Vera, J. C. (1993). Evidence that facilitative glucose transporters may fold as β-barrels. *Proc. Natl. Acad. Sci. U.S.A.*, 90:11658–11662.

Fischbarg, J., Kuang, K., Vera, J. C., Arant, S., Silverstein, S. C., Loike, J., and Rosen, O. M.

(1990). Glucose transporters serve as water channels. *Proc. Natl. Acad. Sci. U.S.A.*, 87:3244–3247.

Fromm, H. J. (1970). A simplified method for deriving steady-state rate equations using a modification of the "Theory of Graphs" procedure. *Biochem. Biophys. Res. Comm.*, 40:692–697.

Ginsburg, H. (1978). The galactose transport in human erythrocytes: The transport mechanism is resolved into two simple asymmetric antiparallel carriers. *Biochim. Biophys. Acta*, 506:119–135.

Ginsburg, H. and Ram, D. (1975). Zero-trans and equilibrium-exchange efflux and infinite-trans uptake of galactose by human erythrocytes. *Biochim. Biophys. Acta*, 382:369–376.

Hankin, B. L., Lieb, W. R., and Stein, W. D. (1972). Rejection criteria for the asymmetric carrier and their application to glucose transport in the human red blood cell. *Biochim. Biophys. Acta*, 288:114–126.

Harris, E. J. (1964). An analytical study of the kinetics of glucose movement in human erythrocytes. *J. Physiol.*, 173:344–353.

Heidrich, D., Kliesch, W., and Quapp, W. (1991). *Properties of Chemically Interesting Potential Energy Surfaces*, vol. 56 of *Lecture Notes in Chemistry*. Springer-Verlag, New York.

Helgerson, A. L. and Carruthers, A. (1987). Equilibrium ligand binding to the human erythrocyte sugar transporter. *J. Biol. Chem.*, 262:5464–5475.

Hill, T. L. and Kedem, O. (1966). Studies in irreversible thermodynamics. III. Models for steady state and active transport across membranes. *J. Theor. Biol.*, 10:399–441.

James, D. E., Strube, M., and Mueckler, M. (1989). Molecular cloning and characterization of an insulin-regulatable glucose transporter. *Nature*, 338:83–87.

Jarvis, S. M., Hammond, J. R., Paterson, A. R. P., and Clanachan, A. S. (1983). Nucleoside transport in human erythrocytes: A simple carrier with directional symmetry in fresh cells, but with directional asymmetry in cells from outdated blood. *Biochem. J.*, 210:457–461.

Jones, M. N. and Nickson, J. K. (1981). Monosaccharide transport proteins of the human erythrocyte membrane. *Biochim. Biophys. Acta*, 650:1–20.

Jung, C. Y. and Rampal, A. L. (1977). Cytochalasin B binding sites and glucose transport carrier in human erythrocyte ghosts. *J. Biol. Chem.*, 252:5456–5463.

Kahn, B. B., Charron, M. J., Lodish, H. F., Cushman, S. W., and Flier, J. S. (1989). Differential regulation of two glucose transporters in adipose cells from diabetic and insulin-treated diabetic rats. *J. Clin. Invest.*, 84:404–411.

Karlish, S. J. D., Lieb, W. R., Ram, D., and Stein, W. D. (1972). Kinetic parameters of glucose efflux from human red blood cells under zero-*trans* conditions. *Biochim. Biophys. Acta*, 255:126–132.

Karnieli, E., Zarnowski, M. J., Hissin, P. J., Simpson, I. A., Salans, L. B., and Cushman, S. W. (1981). Insulin-stimulated translocation of glucose transport systems in the isolated rat adipose cell. *J. Biol. Chem.*, 256:4772–4777.

Kasahara, M. and Hinkle, P. C. (1976). Reconstitution of D-glucose transport catalyzed by a protein fraction from human erythrocytes in sonicated liposomes. *Proc. Natl. Acad. Sci. U.S.A.*, 73:396–400.

Kasahara, M. and Hinkle, P. C. (1977). Reconstitution and purification of the D-glucose transporter from human erythrocytes. *J. Biol. Chem.*, 252:7384–7390.

King, E. L. and Altman, C. (1956). A schematic method of deriving the rate laws for enzyme-catalyzed reactions. *J. Phys. Chem.*, 60:1375–1378.

Krupka, R. M. and Devés, R. (1983). Kinetics of inhibition of transport systems. *Int. Rev. Cytol.*, 84:303–352.

Kyte, J. and Doolittle, R. F. (1982). A simple method for displaying the hydropathic character of a protein. *J. Mol. Biol.*, 157:105–132.

Lachaal, M., Spangler, R. A., and Jung, C. Y. (1993). High K_m of GLUT-2 glucose transporter does not explain its role in insulin secretion. *Am. J. Physiol.*, 265:E914–E919.

Lacko, L. and Wittke, B. (1982). The competetive inhibition of glucose transport in human erythrocytes by compounds of different structures. *Biochem. Pharmacol.*, 31:1925–1929.

Levi, G. and Raiteri, M. (1993). Carrier-mediated release of neurotransmitters. *Trends in Neurosci.*, 16:415–419.

Levine, M., Oxender, D. L., and Stein, W. D. (1965). The substrate facilitated transfer of the glucose carrier across the human erythrocyte membrane. *Biochim. Biophys. Acta*, 109:151–163.

Lieb, W. R. (1982). A kinetic approach to transport studies. In Ellory, J. C. and Young, J. D., eds., *Red Cell Membranes: A Methodological Approach*, 135–164. Academic Press, New York.

Lieb, W. R. and Stein, W. D. (1974a). Testing and characterizing the simple carrier. *Biochim. Biophys. Acta*, 373:178–196.

Lieb, W. R. and Stein, W. D. (1974b). Testing and characterizing the simple pore. *Biochim. Biophys. Acta*, 373:165–177.

Liu, B. (1973). *Ab initio* potential energy surface for linear H_3. *J. Chem. Phys.*, 58:1925–1937.

Lowe, A. G. and Walmsley, A. R. (1986). The kinetics of glucose transport in human red blood cells. *Biochim. Biophys. Acta*, 857:146–154.

May, J. M. and Mikulecky, D. C. (1982). The simple model of adipocyte hexose transport. *J. Biol. Chem.*, 257:11601–11608.

Miller, D. M. (1971). The kinetics of selective biological transport. V. Further data on the erythrocyte-monosaccharide transport system. *Biophys. J.*, 11:915–923.

Mueckler, M. (1990). Family of glucose-transporter genes: Implications for glucose homeostasis and diabetes. *Diabetes*, 39:6–11.

Mueckler, M., Caruso, C., Baldwin, S. A., Panico, M., Blench, I., Morris, H. R., Allard, W. J., Lienhard, G. E., and Lodish, H. F. (1985). Sequence and structure of a human glucose transporter. *Science*, 229:941–945.

Murrell, J. N., Carter, S., Farantos, S. C., Huxley, P., and Varandas, A. J. C. (1984). *Molecular Potential Energy Functions*. John Wiley & Sons, New York.

Naderi, S., Carruthers, A., and Melchior, D. L. (1989). Modulation of red blood cell sugar transport by lyso-lipid. *Biochim. Biophys. Acta*, 985:173–181.

Rauchman, M. I., Wasserman, J. C., Cohen, D. M., Perkins, D. L., Hebert, S. C., Milford, E., and Gullans, S. R. (1992). Expression of GLUT-2 cDNA in human B lymphoctyes: Analysis of glucose transport using flow cytometry. *Biochim. Biophys. Acta*, 1111:231–238.

Runyan, K. R. and Gunn, R. B. (1989). Generation of steady-state rate equations for enzyme and carrier-transport mechanisms: A microcomputer program. In Fleischer, S. and Fleischer, B., eds., *Methods in Enzymology*, vol. 171, *Biomembranes*, pt.

R, *Transport Theory: Cells and Model Membranes*, 164–190. Academic Press, New York.

Sanders, D., Hansen, U. P., Gradmann, D., and Slayman, C. L. (1984). Generalized kinetic analysis of ion-driven cotransport systems: A unified interpretation of selective ionic effects on Michaelis parameters. *J. Membr. Biol.*, 77:123–152.

Segel, L. A. (1988). On the validity of the steady state assumption of enzyme kinetics. *Bull. Math. Biol.*, 50:579–593.

Sen, A. K. and Widdas, W. F. (1962a). Determination of the temperature and pH dependence of glucose transfer across the human erythrocyte membrane measured by glucose exit. *J. Physiol.*, 160:392–403.

Sen, A. K. and Widdas, W. F. (1962b). Variations of the parameters of glucose transfer across the human erythrocyte in the presence of inhibitors of transfer. *J. Physiol.*, 160:404–416.

Siegbahn, P. and Liu, B. (1978). An accurate three-dimensional potential energy surface for H_3. *J. Chem. Phys.*, 68:2457–2465.

Simpson, I. A. and Cushman, S. W. (1985). Hexose transport regulation by insulin in the isolated rat adipose cell. In Czech, M. P., ed., *Molecular Basis of Insulin Action*, 399–422. Plenum, New York.

Slot, J. W., Geuze, H. J., Gigengack, S., Lienhard, G. E., and James, D. E. (1991). Immunolocalization of the insulin regulatable glucose transporter in brown adipose tissue of the rat. *J. Cell Biol.*, 113:123–135.

Sogin, D. C. and Hinkle, P. C. (1978). Characterization of the glucose transporter from human erythrocytes. *J. Supramol. Struct.*, 8:447–453.

Stein, W. D. (1976). An algorithm for writing down flux equations for carrier kinetics, and its application to co-transport. *J. Theor. Biol.*, 62:467–478.

Thorens, B., Sarkar, H. K., Kaback, H. R., and Lodish, H. F. (1988). Cloning and functional expression in bacteria of a novel glucose transporter present in liver, intestine, kidney, and β-pancreatic islet cells. *Cell*, 55:281–290.

Truhlar, D. G. and Horowitz, C. J. (1978). Functional representation of Liu and Siegbahn's accurate *ab initio* potential energy calculations for $H + H_2$. *J. Chem. Phys.*, 68:2466–2476.

Unger, R. H. (1991). Diabetic hyperglycemia: Link to impaired glucose transport in pancreatic β cells. *Science*, 251:1200–1205.

Varandas, A. J. C. (1979). A LEPS potential for H_3 from force field data. *J. Chem. Phys.*, 79:3786–3795.

Varandas, A. J. C., Brown, F. B., Mead, C. A., Truhlar, D. G., and Blais, N. C. (1987). A double many-body expansion of the two lowest-energy potential surfaces and nonadiabatic coupling for H_3. *J. Chem. Phys.*, 86:6258–6269.

Volkenstein, M. V. and Goldstein, B. N. (1966). A new method for solving the problems of the stationary kinetics of enzymological reactions. *Biochim. Biophys. Acta*, 115:471–477.

Wheeler, T. J. (1986). Kinetics of glucose transport in human erythrocytes: Zero-*trans* efflux and infinite-*trans* efflux at 0°C. *Biochim. Biophys. Acta*, 862:387–398.

Wheeler, T. J. (1994). Accelerated net efflux of 3-*O*-methylglucose from rat adipocytes: A reevaluation. *Biochim. Biophys. Acta*, 1190:345–354.

Wheeler, T. J. and Whelan, J. D. (1988). Infinite-Cis kinetics support the carrier model for erythrocyte glucose transport. *Biochem.*, 27:1441–1450.

Widdas, W. F. (1952). Inability of diffusion to account for placental glucose transfer in the sheep, and consideration of the kinetics of a possible carrier transfer. *J. Physiol.*, 118:23–39.

Widdas, W. F. (1954). Facilitated transfer of hexoses across the human erythrocyte membrane. *J. Physiol.*, 125:163–180.

Wilbrandt, W. and Rosenberg, T. (1961). The concept of carrier transport and its corollaries in pharmacology. *Pharmacol. Rev.*, 13:109–183.

You, G., Smith, C. P., Kanai, Y., Lee, W., Stelzner, M., and Hediger, M. A. (1993). Cloning and characterization of the vasopressin-regulated urea transporter. *Nature*, 365:844–847.

Yousef, L. W. and Macey, R. I. (1989). A method to distinguish between pore and carrier kinetics applied to urea transport across the erythrocyte membrane. *Biochim. Biophys. Acta*, 984:281–288.

Zeller, W. P., Goto, M., Parker, J., Cava, J. R., Gottschalk, M. E., Filkins, J. P., and Hofmann, C. (1994). Glucose transporters (GLUT1, 2, & 4) in fat, muscle and liver in a rat model of endotoxic shock. *Biochem. Biophys. Res. Commun.*, 198:923–927.

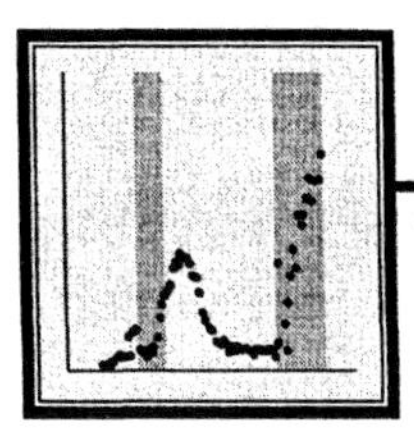

7 Ion Transport and Resting Potential

It must be admitted that the statement that, under physiological conditions, the ratio of sodium ions pumped out to potassium ions pumped in is 3:2 is sometimes made in a manner more appropriate to the recitation of a creed than to the recounting of an experimental finding.
—Glynn, 1984

7.1 Introduction

7.1.1 The Importance of Ion Transport

Previous chapters have described the transport of uncharged solutes across cellular membranes. This chapter describes the transport of charged solutes, or ions, across cellular membranes. The maintenance of ion concentrations and the flow of ions across membranes has important consequences for many cellular processes.

7.1.1.1 Dependence of Cellular Processes on Intracellular Ion Concentration

A number of cellular processes depend on the concentrations of intracellular ions. For example, the catalytic action of key cytoplasmic enzymes depends critically on the concentrations of ions such as H^+, K^+, and Mg^{++}. Many secretory processes in which cells secrete chemical compounds into the extracellular environs critically depend on the concentrations of divalent ions such as Ca^{++} and Mg^{++}. Cellular motility is dependent on the concentration of Ca^{++} in cells with motile cilia and flagella as well as in muscle cells. In the latter cells, Ca^{++} plays a key role in the coupling of electrical excitation of the muscle cell to its contraction. The coupling of cells through gap junctions is dependent on the intracellular pH as well as the intracellular concentration of calcium.

7.1.1.2 Coupling of Transport of Ions and Uncharged Solutes

Understanding ion transport mechanisms is also important for understanding the transport of uncharged molecules. First, changes in the concentration of an ion can change osmotic pressures, which may change the volume of the cell, which in turn will change the concentrations of uncharged solutes. These changes in concentrations will affect the transport of the uncharged solutes. In this sense, the transport of different solutes is linked indirectly via osmotic effects. In addition, there are more direct links. For example, in certain transporting epithelial cells the transport of glucose is linked, via a cotransport process, directly to the transport of sodium. Thus, in this instance, the flux of glucose can be affected by changes in membrane potential that affect the transport of sodium (see Chapter 8).

7.1.1.3 Ion Transport as a Substrate for Information Transmission in the Nervous System

Information is encoded into nerve messages by the sensory nervous system and processed by the central nervous system to produce outputs by the motor system. These messages are in the form of changes in electric potential across neuronal membranes caused by the flow of certain ions. Thus, the transport of ions across the nerve membrane is the substrate for information handling by these cells.

7.1.2 The Maintained Difference of Potential and Concentration Across Cellular Membranes

There is a difference of potential across the membranes of all living cells, both plant and animal cells, called the *resting potential*. This resting potential can be measured by placing one electrode in contact with the cell's cytoplasm and another in contact with the extracellular solution outside the cell (Figure 7.1). For large cells, such as the giant axon of the squid, the intracellular electrode can be a glass capillary tube (100 μm in diameter) that is filled with an electrolytic solution and inserted longitudinally into the cell (Figure 7.2). For smaller cells, micropipettes with tip diameters as small as 0.05 μm are used to impale cells (as shown schematically in Figure 7.2). Elaborate methods have been developed to ease puncture of the membranes of small cells. For

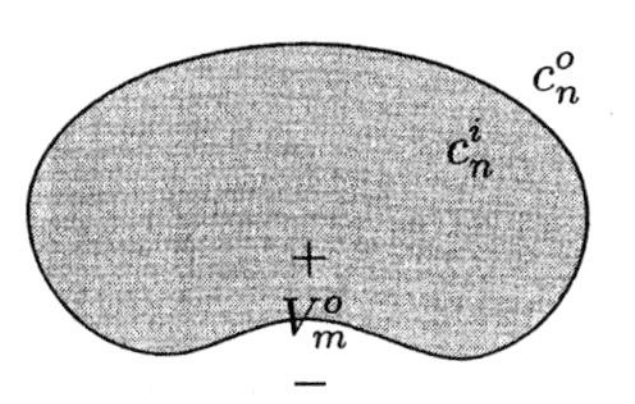

Figure 7.1 Schematic diagram of a cell showing the positive reference direction for definition of the resting membrane potential, V_m^o. The concentrations of ion n intra- and extracellularly are c_n^i and c_n^o, respectively.

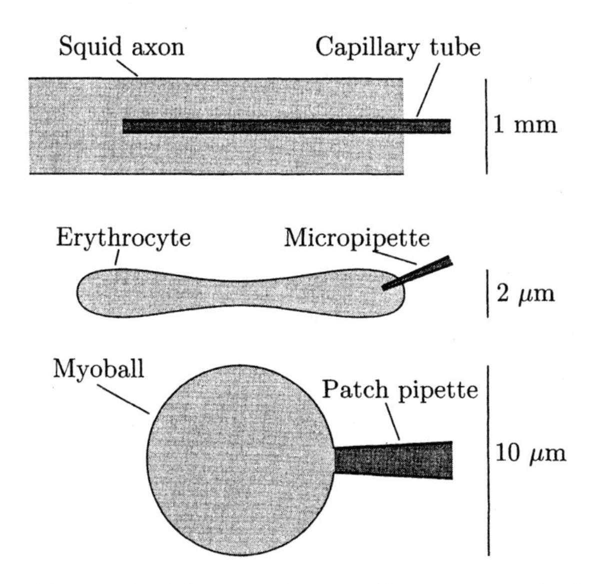

Figure 7.2 Schematic diagram illustrating methods for recording the resting potential of a cell. All these electrodes are glass tubes filled with an electrolyte, typically KCl. For large cells, such as the giant axon of the squid, a glass capillary tube can be inserted longitudinally into the cell cytoplasm. For small cells, such as a human erythrocyte, a micropipette is pulled from a glass capillary tube and used to record from the cell. Penetrating micropipettes, with tip diameters down to about 0.05 μm, are used to impale cells. Patch pipettes are somewhat larger in diameter and are sealed to the membrane of a cell. The membrane patch in the pipette lumen is ruptured by hydraulic pressure to establish contiguity between the electrolyte in the pipette and the cytoplasm.

example, the fine tips of the micropipettes have been beveled so that the tips resemble microscopic hypodermic needles (Brown and Flaming, 1974). The micropipettes have been attached to devices that deliver large accelerations to them (Brown and Flaming, 1977). Finally, special "floating" micropipettes have been used to record from cardiac muscle cells in beating hearts. More recently, somewhat larger pipettes, called *patch pipettes*, with smooth tips

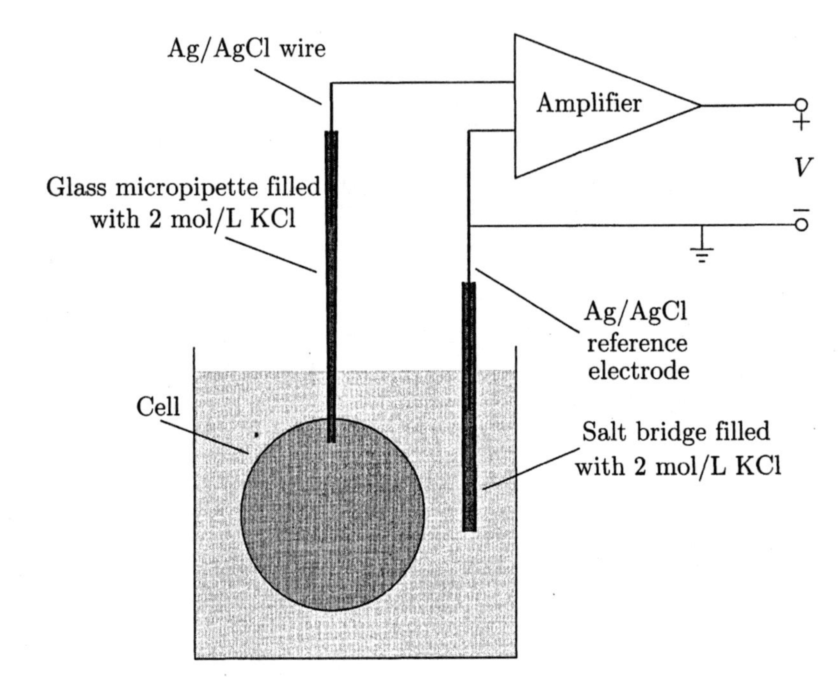

Figure 7.3 Schematic diagram of a typical recording arrangement used to measure intracellularly from a cell. A glass micropipette filled with 2 mol/L KCl is used to impale the cell, and a glass capillary tube filled with 2 mol/L KCl is used to record from the extracellular bath. This salt bridge often contains KCl in a gel (agar) to impede diffusion of KCl out of the reference electrode and into the bath. Ag/AgCl metal wires are used to connect the electrolyte electrodes to the amplifier inputs.

have been used to form seals to cellular membranes (Hamill et al., 1981). With application of suction to one of these pipettes, the membrane closing the lumen of the pipette is removed, leaving a pipette/cell combination where the interior of the pipette is in contact with the cell's cytoplasm (Figure 7.2). Patch pipettes can be used to record from cells having a range of dimensions, from the very largest to the smallest. It now appears that patch pipettes cause less damage to cells than do penetrating micropipettes.

The electric potential between two electrodes, one in contact with the cell cytoplasm and one in contact with extracellular solution, is measured with such glass electrodes, which are filled with an electrolyte. The electrolyte is typically a near-saturated solution of KCl (see Problem 7.2). A metal electrode in the pipette is connected to an amplifier to record the potential (Figure 7.3). The interpretation of such measured potentials is complicated, because each junction between dissimilar materials in the measurement circuit introduces a junction potential. Hence, the recorded potential is a sum of the potential across the membrane of the cell plus the sum of all the junction potentials in the measurement system (see Problem 7.3). With careful methodology, the ambiguities in potential can be minimized.

With such techniques, resting potentials have been measured for a large variety of plant and animal cells. It has been found that the polarity of the

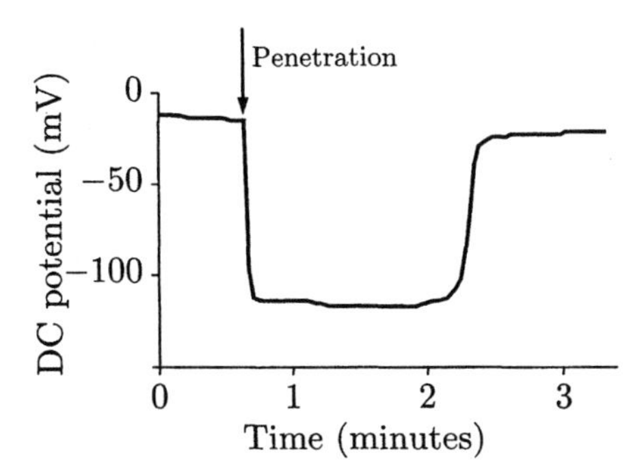

Figure 7.4 Resting potential of a cell (Mulroy et al., 1974). The DC potential recorded with a micropipette is shown as the micropipette is advanced into the sensory epithelium of the inner ear of a lizard. At the time indicated by the arrow, the membrane of the cell is punctured as signaled by a change in DC potential of about 100 mV, which represents the resting potential of this cell. Contact with the cell is lost about 1 1/2 minutes later as signaled by a loss of this negative DC potential.

Table 7.1 Ion concentrations for squid giant axon (Hodgkin, 1964; Baker, Hodgkin, et al., 1971), for frog muscle fiber (Conway, 1957), and for human erythrocytes (Davson, 1970). The salt concentrations in both intracellular and extracellular solutions are much higher in marine animals than in terrestrial animals.

	Concentration (mmol/L)						
Ion	*Squid*			*Frog*		*Human*	
	Cytoplasm	Blood	Seawater	Cytoplasm	Plasma	Cytoplasm	Plasma
K^+	400	20	10	124	2.25	150	5.35
Na^+	50	440	460	10.4	109	12-20	144
Cl^-	40-150	560	540	1.5	77.5	73.5	111
Ca^{++}	0.0001	10	10	4.9	2.1	—	6.4
Mg^{++}	10	54	53	14.0	1.25	5.6	2.14

resting potential, which we designate as V_m^o and whose reference direction is defined in Figure 7.1, is almost invariably negative. That is, the cytoplasm of the cell is at a negative potential with respect to the extracellular space, and the value of the potential lies in the range $-10^2 < V_m^o < -10^1$ mV for most cells investigated (e.g., Figure 7.4).

There is also a maintained difference in ion concentration across the membranes of living cells (Table 7.1). The cytoplasm contains a high concentration of potassium ions and a low concentration of sodium ions, whereas the reverse is true for the extracellular solution. The difference in ionic composition on the two sides of the membrane affects the electrical properties of cells. The differences in potential and ion concentration across the membrane are linked, so changes in potential can result in concentration changes and changes in ion concentration can result in membrane potential changes. The relation between the membrane potential and ion concentrations is controlled by the cellular membrane. In this chapter we shall examine the relation of resting potential to ion concentrations and explore the physical-chemical mechanisms responsible for maintaining the difference in potential and concentration across cellular membranes.

7.2 Continuum Electrodiffusion

In Chapter 3 we considered the transport of particles subject to a concentration gradient only. However, if the particles have an electric charge and if they are subjected to an electric potential gradient, transport will arise from two physical mechanisms: diffusion due to a particle concentration gradient and drift (also called migration) due to an electric potential gradient. The Nernst-Planck equation expresses the current density of a species of charged particles in terms of the concentration and potential gradients. If the species of charged particle is conserved, then the current density and concentration of that particle are also linked by a continuity equation. Finally, the concentrations of all the charged particles in electrodiffusion are linked to the electric potential via Poisson's equation. These relations are discussed and their consequences are derived in this section.

7.2.1 Electrodiffusion Equations

7.2.1.1 The Nernst-Planck Equation

As we discussed in Chapter 3, particles in a medium are scattered because their thermal kinetic energy causes collisions with particles of the medium. This collision process causes no net velocity of each particle, but does cause a net flux of particles down their concentration gradient as described by Fick's first law of diffusion (Equation 3.18),

$$\phi_{\text{diffusion}} = -D\frac{\partial c}{\partial x},$$

where $D = l^2/(2\tau)$, where l and τ are the mean free path and mean free time, respectively. Suppose that the particles also experience a body force in the positive x-direction, implying that between collisions the particles will be accelerated in the x-direction. This acceleration causes an increment in velocity in the x-direction during the interval between collisions. As shown in Section 3.3, this *drift velocity* is $v = uf$, where f is the force on a mole of particles and u is the molar mechanical mobility. The flux of particles due to their drift is

$$\phi_{\text{drift}} = cv = cuf.$$

Suppose the particles are charged with valence z and the body force is caused by an electric field with electric field intensity $\mathcal{E} = -\partial\psi/\partial x$, where ψ is the electric potential. Then the body force on a mole of particles is $f = zF\mathcal{E} = -zF\partial\psi/\partial x$, where F is Faraday's constant (the charge on a mole of univalent particles), which is about 9.65×10^4 C/mol. If we put all these factors together, then we have

$$\phi_{\text{drift}} = -cuzF\frac{\partial\psi}{\partial x}.$$

The kinetic model of the motion of a particle suggests that the instantaneous velocity of each particle is the sum of its drift and diffusion velocity. Therefore, it is reasonable that the net flux of particles is the sum of that caused by drift and diffusion.

For any ion species n, the flux due to diffusion and to drift is

$$\phi_n = \underbrace{-D_n\frac{\partial c_n(x,t)}{\partial x}}_{\text{diffusion}} \underbrace{-u_n z_n F c_n(x,t)\frac{\partial\psi(x,t)}{\partial x}}_{\text{drift}}, \tag{7.1}$$

where D_n is the diffusion coefficient, c_n is the concentration, u_n is the mobility[1], and z_n is the valence of the nth ion. The dimensions of the drift term are as follows: $\partial\psi(x,t)/\partial x$ is the potential gradient in N/C; z_nF is the charge on a mole of ion n in C/mol; $z_nF(\partial\psi(x,t)/\partial x)$ is the force on a mole of particles in N/mol; $u_nz_nF(\partial\psi(x,t)/\partial x)$ is the velocity of the particles in m/s; and the flux is the product of the velocity and the concentration.

The current density, $J_n(x,t)$, in A/cm^2 is related to the flux by

$$J_n(x,t) = z_nF\phi_n(x,t). \tag{7.2}$$

1. u_n is the molar mechanical mobility of ion n. In some fields (e.g., solid-state physics), it is customary to use the *molar electrical mobility*, $\hat{u}_n$, where $\hat{u}_n = |z_n|Fu_n$. $\hat{u}_n$ has units of (cm/s)/(V/cm). In terms of the molar electrical mobility, the Einstein relation is $D_n = (RT\hat{u}_n)/(|z_n|F)$.

Substituting Equation 7.1 into 7.2 yields

$$J_n(x,t) = -z_nF\left(D_n\frac{\partial c_n(x,t)}{\partial x} + u_nz_nFc_n(x,t)\frac{\partial \psi(x,t)}{\partial x}\right). \tag{7.3}$$

Despite the fact that Equation 7.3 contains two terms, one resulting from Fick's law and the other from Ohm's law, the equation is known as the Nernst-Planck equation of electrodiffusion. The Nernst-Planck equation can be expressed in several useful and equivalent forms. Using the Einstein relation, $D_n = u_nRT$ (Equation 3.26), the Nernst-Planck equation can be expressed as

$$J_n(x,t) = -u_nz_nFc_n(x,t)\left(\frac{RT}{c_n(x,t)}\frac{\partial c_n(x,t)}{\partial x} + z_nF\frac{\partial \psi(x,t)}{\partial x}\right), \tag{7.4}$$

which can be rearranged to yield an expression in terms of the logarithmic derivative of concentration,

$$J_n(x,t) = -u_nz_nFc_n(x,t)\frac{\partial}{\partial x}\left(RT\ln c_n(x,t) + z_nF\psi(x,t)\right). \tag{7.5}$$

The Electrochemical Potential

The Nernst-Planck equation can also be expressed as

$$J_n(x,t) = -u_nz_nFc_n(x,t)\frac{\partial \tilde{\mu}_n(x,t)}{\partial x}, \tag{7.6}$$

where

$$\tilde{\mu}_n(x,t) = \mu_n^o + RT\ln c_n(x,t) + z_nF\psi(x,t). \tag{7.7}$$

Here $\tilde{\mu}_n$ is the electrochemical potential, and μ_n^o is its reference value at unit concentration and zero potential. Note that the two terms that depend on solute concentration and potential are the stored chemical energy and the stored electrostatic energy per mole, respectively. Thus, the flux of ions and the current density carried by the ions are proportional to the electrochemical potential gradient.

7.2.1.2 Conservation of Particles and Charge

If each ionic species is conserved, each ion will satisfy its own continuity equation, which follows directly from Equation 3.6:

$$\frac{\partial J_n(x,t)}{\partial x} = -z_nF\frac{\partial c_n(x,t)}{\partial t}. \tag{7.8}$$

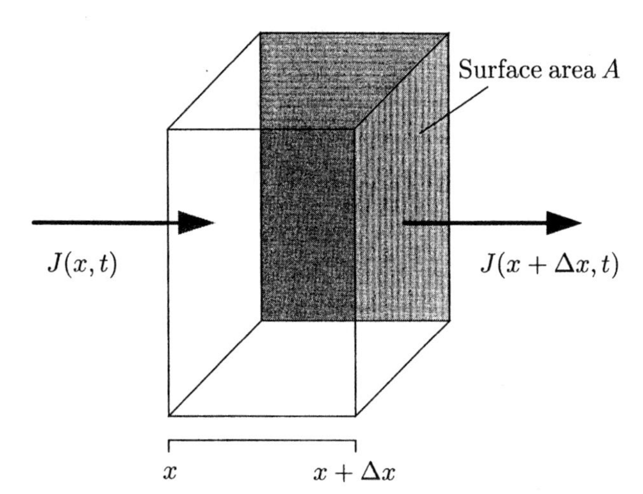

Figure 7.5 Volume element used to derive conservation of charge.

If each of the particles is conserved, the charge of a population of charged particles will also be conserved. This result can be seen by summing Equation 7.8 over all the charge species. In this section we shall give an alternative derivation that is similar to previous derivations of the continuity equation. In Section 3.1 we derived the continuity of particle relation based on the assumption that particles are conserved. In Section 4.4.2 we derived a continuity-of-mass relation based on the assumption that mass was conserved. In an analogous manner, we can derive a continuity-of-charge relation based on the assumption that charge is conserved.

Consider an incremental volume element of cross-sectional area A and width Δx with charge density ρ and current density J, as shown in Figure 7.5. Conservation of charge implies that the net charge flowing into the volume element in a time interval of duration Δt must equal the increase in charge in the volume element. Therefore,

$$J(x,t)A\Delta t - J(x+\Delta x,t)A\Delta t = \rho(x,t+\Delta t)A\Delta x - \rho(x,t)A\Delta x. \tag{7.9}$$

Rearranging the terms in Equation 7.9 and letting $\Delta x \to 0$ and $\Delta t \to 0$ yields

$$\frac{\partial J(x,t)}{\partial x} = -\frac{\partial \rho(x,t)}{\partial t}, \tag{7.10}$$

which in three dimensions can be written as

$$\nabla \cdot \mathbf{J} = -\frac{\partial \rho}{\partial t}. \tag{7.11}$$

7.2.1.3 Poisson's Equation

Poisson's equation, which is derivable from Gauss's law, links the charge density to the electric potential as follows:

$$\frac{\partial^2 \psi(x,t)}{\partial x^2} = -\frac{\rho(x,t)}{\epsilon}, \tag{7.12}$$

where ρ is the charge density (C/cm^3) and ϵ is the permittivity of the medium. The charge density consists of two terms, the mobile ionic charges and the fixed or immobile charges in the medium. Thus,

$$\rho(x,t) = F \sum_{n=1}^{N} z_n c_n(x,t) + \rho_f(x,t), \tag{7.13}$$

where ρ_f is the fixed charge density.

7.2.1.4 Conclusions

An electrodiffusion system is completely specified by Equations 7.4, 7.8, 7.12, and 7.13. Thus, there are N Nernst-Planck equations (Equation 7.4) linking the $2N+1$ variables $J_1, J_2, \ldots, J_N, c_1, c_2, \ldots, c_N, \psi$. If the ions are each conserved, then there will be an additional N continuity equations (Equation 7.8) that link these variables. The final relation linking the $2N+1$ variables is Poisson's equation (Equation 7.12). If the fixed charges are specified as a material property, and if appropriate boundary conditions are specified, then these equations constitute a complete specification of an electrodiffusion problem; that is, there are $2N+1$ variables and equations. In principle, these equations can be solved for the concentration and the current density of each of the mobile ions as well as the electric potential. However, these equations can be nonlinear because of the term $c_n(x,t)(\partial\psi(x,t)/\partial x)$ in Equation 7.4, which complicates their solution.

7.2.2 Electrodiffusive Equilibrium Condition

From the Nernst-Planck equation we can directly determine the conditions that must apply at *electrodiffusive equilibrium,* i.e., the conditions at which the flux of the nth ion is zero. If we set $J_n = 0$ in Equation 7.5 and if we recognize that at electrodiffusive equilibrium all the variables are time independent, then we obtain the condition

$$-u_n z_n^2 F^2 c_n(x,\infty)\frac{d}{dx}\left(\frac{RT}{z_n F}\ln c_n(x,\infty)+\psi(x,\infty)\right)=0. \tag{7.14}$$

There are several ways to satisfy this electrodiffusive equilibrium condition. Three of these are trivial: (1) if $u_n = 0$, which occurs if the particles are fixed and cannot drift or diffuse; (2) if $z_n = 0$, which implies that the particles are uncharged and do not carry a current even if they can diffuse; and (3) if $c_n = 0$, which occurs if there are no particles.

The more interesting case occurs if $u_n \neq 0$, $z_n \neq 0$, and $c_n \neq 0$. Then we can divide these terms out of Equation 7.14 so that this condition is

$$\frac{d\left(\ln c_n(x,\infty)\right)}{dx} = -z_n\beta\frac{d\psi(x,\infty)}{dx}, \tag{7.15}$$

where $\beta = F/RT$. This equation can be integrated to yield

$$c_n(x,\infty) = c_n(x_o)e^{-z_n\beta(\psi(x,\infty)-\psi(x_o))}, \tag{7.16}$$

where location x_o provides a point of reference for the potential. Equation 7.16 gives the spatial distribution of charged particles in an electrostatic field at electrodiffusive equilibrium. Examination of Equation 7.16 reveals that for the special case in which $\psi(x,\infty) - \psi(x_o) = 0$ and/or $z_n = 0$, the concentration at electrodiffusive equilibrium is uniform in space, just as we found for the equilibrium of an uncharged solute in Chapter 3—i.e, the condition for *diffusive equilibrium*. Equation 7.16 can be solved for the potential difference to yield

$$\psi(x,\infty) - \psi(x_o) = \frac{RT}{z_n F}\ln\left(\frac{c_n(x_o)}{c_n(x,\infty)}\right). \tag{7.17}$$

Hence, at electrodiffusive equilibrium the spatial distribution of potential is proportional to the logarithm of the solute concentration.[2]

The exponent in Equation 7.16 can be written as $-z_nF\left(\psi(x,\infty) - \psi(x_o)\right)/RT$ and hence equals the stored electrical energy per mole, $z_nF\psi$, divided by the thermal energy per mole, RT. This result makes intuitive sense, since the spatial distribution of molecules will depend on two competing factors: the tendency to drift due to their electric field and the tendency to diffuse due to their thermal energy. If the potential distribution is sharply lo-

2. An alternative derivation of Equation 7.17 can be obtained if we set $J_n = 0$ in Equation 7.6. Then at equilibrium the electrochemical potential is a constant, i.e., $\mu(x,\infty)$ is constant, and Equation 7.17 follows directly from Equation 7.7

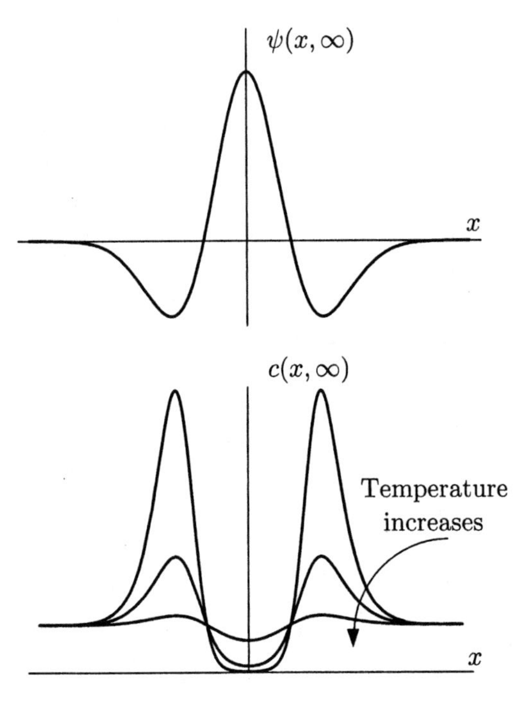

Figure 7.6 The spatial distribution of electric potential and ion concentration at electrodiffusive equilibrium for different temperatures.

calized (Figure 7.6) and the electrical stored energy is large compared to the thermal energy, then the concentration profile will also be sharply localized, with a large concentration of particles at the potential minima and few particles at the potential maxima. As the temperature is raised, the thermal energy will increase and the spatial distribution of concentration will become less peaked. In the limit as the temperature becomes arbitrarily large, the spatial distribution of particles will approach a uniform distribution at equilibrium. This result is similar in form to one derived in Section 6.2.2.

A more intuitive interpretation of electrodiffusive equilibrium is obtained from a mechanical analog. Imagine a three-dimensional rigid surface whose peaks and valleys represent local maxima and minima in gravitational potential energy. This spatial distribution of potential energy is analogous to the electric potential. Now place a collection of balls on the surface, and shake the surface to transmit some kinetic energy to the balls. The kinetic energy of the balls, delivered mechanically in this analogy, is analogous to the thermal kinetic energy. The balls will tend to roll into the valleys and remain there. Gentle vibration of the surface will tend to distribute the balls among the valleys, and some will be found between valleys at any time. However, if the surface is shaken sufficiently, we will find that the balls are equally likely to be found at any location on the surface. The balls are in constant motion, and just as soon as they fall into a valley they will be jolted out of it.

7.2.3 Electroneutrality

The strong forces of attraction between oppositely charged ions in solution tend to neutralize the net charge in a solution. As we shall see, there is a strong tendency for this charge neutralization to occur, and the resultant property of the solution is called *electroneutrality*. In this section we shall explore the temporal and spatial properties of electroneutrality. We shall deal with two questions: (1) If charge is deposited in an electrolytic solution, how long will it take for electroneutrality to be established? This question leads to the definition of the *charge relaxation time* as the measure of the time scale for establishing electroneutrality. (2) If a fixed charge is placed in an electrolytic solution, how far away from the fixed charge is the solution electrically neutral? This question leads to the definition of the *Debye length* as the spatial measure of charge distribution in the electrolyte.

7.2.3.1 Charge Relaxation Time

Consider an electrolytic solution with uniform composition. Now imagine that a charge density $\rho(0)$ is applied uniformly to this solution at time zero. What will happen? For this uniform solution, the current density is related to the electric field (negative gradient of the electric potential) according to Ohm's law (see Equation 7.45):

$$\mathbf{J} = \sigma_e \mathcal{E} \quad \text{where } \mathcal{E} = -\nabla\psi,$$

where σ_e is the electric conductivity of the medium. In a homogeneous conductor,

$$\nabla \cdot \mathbf{J} = \sigma_e \nabla \cdot \mathcal{E}.$$

But Gauss's law implies that $\nabla \cdot (\epsilon\mathcal{E}) = \rho(t)$, where $\rho(t)$ is the charge density (C/cm^3) at time t. Therefore,

$$\nabla \cdot \mathbf{J} = \sigma_e \nabla \cdot \mathcal{E} = \frac{\sigma_e}{\epsilon}\rho. \tag{7.18}$$

If we substitute the equation for the continuity of charge, Equation 7.10, into Equation 7.18 under the assumption that the charge density is increased uniformly in the solution, we obtain

$$\frac{d\rho(t)}{dt} + \frac{\sigma_e}{\epsilon}\rho(t) = 0, \tag{7.19}$$

which has the solution

$$\rho(t) = \rho(0)e^{-t/\tau_r}, \tag{7.20}$$

where τ_r is the charge relaxation time

$$\tau_r = \frac{\epsilon}{\sigma_e}. \tag{7.21}$$

Thus, for a spatially uniform distribution of charge (so that no diffusion takes place), the charge density relaxes exponentially to zero and the charges move to the boundaries of the solution, i.e., they get as far away from each other as possible. The time constant for this relaxation can be computed for physiological saline solutions. For such solutions $\sigma_e \approx 0.01$ S/cm and $\epsilon \approx 80\epsilon_o \approx 80 \times 8.85 \times 10^{-14}$F/cm. Thus, $\tau_r \approx 0.7$ nanoseconds!

7.2.3.2 The Debye Length

Consider a plate located at $x = 0$ that contains a fixed surface charge density Q_f and is bathed (for $x > 0$) by an electrolyte whose cation and anion concentrations are $c_+(x)$ and $c_-(x)$, respectively (Figure 7.7). Near this plate, mobile ions whose charge is opposite to that of the fixed charge, or counterions, will be attracted, whereas those with the same charge will be repelled. Thus, near the plate there is a space charge region where electroneutrality does not hold. We wish to determine the dimensions of this region. For locations far from this plate, we assume that the solution is electrically neutral, and hence, that $c_+(\infty) = c_-(\infty) = C$. For simplicity, we shall assume that the electrolyte is symmetrical, i.e., that the valences are equal in magnitude or that $-z_- = z_+ = z$. We wish to find the equilibrium distribution of electric potential and ion concentration in the solution. The electric potential and ion concentrations satisfy two fundamental relations in the region $x > 0$, Poisson's equation and the equilibrium condition derived in Section 7.2.2. In one dimension, Poisson's equation is

$$\frac{d^2\psi(x)}{dx^2} = -\frac{\rho}{\epsilon}, \tag{7.22}$$

where ρ is the charge density and ϵ is the permittivity. For each ion, the equilibrium condition derived in Section 7.2.2 is

$$c_+(x) = Ce^{-zF\psi(x)/RT} \quad \text{and} \quad c_-(x) = Ce^{zF\psi(x)/RT}, \tag{7.23}$$

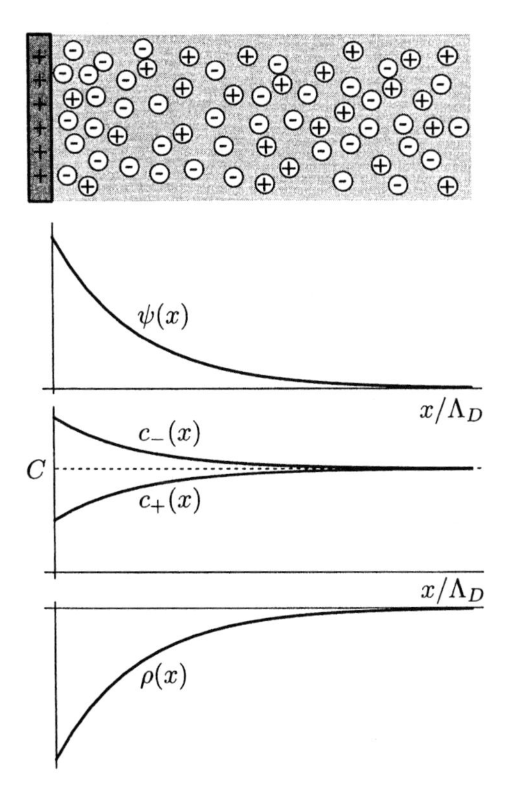

Figure 7.7 The spatial distribution of charge near a plate containing positive fixed charges. The counterions are anions and are in higher concentration near the plate than far from the plate. The cations are at a lower concentration near the plate than far from the plate. The spatial distributions of both mobile ions are exponential, with space constant equal to the Debye length.

where we have assumed that $\psi(\infty) = 0$, which is used as a reference for the potential. Equations 7.22 and 7.23 are linked by the equation

$$\rho(x) = z_+ F c_+(x) + z_- F c_-(x). \tag{7.24}$$

Now substitute Equation 7.23 into 7.24 and the result into Equation 7.22 to obtain the Poisson-Boltzmann equation,

$$\begin{aligned} \frac{d^2\psi(x)}{dx^2} &= -\frac{1}{\epsilon}\left(z_+ F C e^{-zF\psi(x)/RT} + z_- F C e^{zF\psi(x)/RT}\right) \\ &= \frac{2zFC}{\epsilon}\sinh\left(\frac{zF\psi(x)}{RT}\right). \end{aligned} \tag{7.25}$$

The Poisson-Boltzmann equation can also be expressed in normalized coordinates

$$\frac{d^2\Psi(X)}{dX^2} = \sinh\Psi(x), \tag{7.26}$$

where

$$\Psi(x) = \frac{\psi(x)}{RT/zF} \tag{7.27}$$

is the normalized potential,

$$X = \frac{x}{\Lambda_D}$$

is the normalized distance, and Λ_D is called the *Debye length* and is expressed as

$$\Lambda_D = \sqrt{\frac{\epsilon RT}{2z^2F^2C}}. \tag{7.28}$$

Therefore, the Debye length is a measure of the spatial extent of the potential distribution and, as we shall see, a measure of the distance over which electroneutrality is violated.

A particularly simple and informative solution of the Poisson-Boltzmann equation is about obtained for $zF\psi(x)/RT \ll 1$.[3] Equation 7.25 becomes

$$\frac{d^2\psi(x)}{dx^2} = \frac{\psi(x)}{\Lambda_D^2}. \tag{7.29}$$

Given the assumption that $\psi(x)$ is bounded for $x > 0$, the solution to Equation 7.29 is

$$\psi(x) = \psi(0)e^{-x/\Lambda_D}. \tag{7.30}$$

The potential at $x = 0$ can be found by matching the boundary condition at the plate, where the normal component of the electric field must be proportional to the surface charge density, i.e.,

$$-\epsilon\left(\frac{d\psi(x)}{dx}\right)_{x=0} = Q_f. \tag{7.31}$$

Substitution of Equation 7.30 into 7.31 yields $\psi(0) = Q_f\Lambda_D/\epsilon$, so that the solution for the potential is

$$\psi(x) = \frac{Q_f\Lambda_D}{\epsilon}e^{-x/\Lambda_D}. \tag{7.32}$$

3. For a univalent ion RT/F is about 26 mV, so the condition is satisfied when the potential difference is much less than 26 mV.

If we use the condition that $zF\psi(x)/RT \ll 1$, then Equation 7.23 becomes

$$c_+(x) = C\left(1 - \frac{zF}{RT}\psi(x)\right) \quad \text{and} \quad c_-(x) = C\left(1 + \frac{zF}{RT}\psi(x)\right). \tag{7.33}$$

Substitution of Equation 7.32 into Equation 7.33 yields the spatial distribution of concentration of the cations and the anions,

$$c_+(x) = C\left(1 - \frac{zFQ_f\Lambda_D}{RT\epsilon}e^{-x/\Lambda_D}\right) \quad \text{and} \quad c_-(x) = C\left(1 + \frac{zFQ_f\Lambda_D}{RT\epsilon}e^{-x/\Lambda_D}\right). \tag{7.34}$$

From Equations 7.32 and 7.34, we see that the potential decreases to zero for distances that are several Debye lengths from the plate. In addition, for distances several Debye lengths from the plate, the cation and anion concentrations approach their values at $x = \infty$. The space charge region can also be computed by substituting Equation 7.34 into 7.24 and combining terms to yield

$$\rho(x) = -\frac{Q_f}{\Lambda_D}e^{-x/\Lambda_D}. \tag{7.35}$$

Hence, the net charge in the space charge layer has a sign that is opposite that of the charge on the plate and a magnitude that is proportional to the charge on the plate. The space charge also decays according to the Debye length.

The Debye length has an interesting relation to other characteristics of an ionic solution. As we shall see (Equation 7.46), the conductivity of a symmetrical ionic solution, assuming that the mobilities of both the anion and cation are the same, is

$$\sigma_e = 2z^2F^2uC = \frac{2z^2F^2}{RT}DC, \tag{7.36}$$

where u is the mechanical mobility and D is the diffusion coefficient. Combining Equations 7.28 and 7.36, we obtain

$$\frac{2z^2F^2C}{RT} = \frac{\epsilon}{\Lambda_D^2} = \frac{\sigma_e}{D}.$$

Using the definition of the relaxation time from Equation 7.21, we obtain the result

$$D = \frac{\Lambda_D^2}{\tau_r}. \tag{7.37}$$

In words, the diffusion coefficient of an ion is the characteristic or Debye length squared divided by the characteristic or relaxation time. Since diffusion coefficients of ions in water are on the order of 10^{-5}cm/sec^2 and the relaxation time is on the order of 1 ns, the Debye length is on the order of 10 Å.

7.2.3.3 Conclusion

The relaxation time and the Debye length for small ions in water show that electroneutrality holds over observation durations that exceed 1 ns and over distances that exceed 10 Å. Provided that these conditions are met, ionic solutions obey electroneutrality; that is, the net charge density of the solution is zero, i.e.,

$$\sum_n z_n F c_n = 0, \tag{7.38}$$

where the summation is taken over all the charged particles.

7.2.4 Steady-State Conditions

By definition, for steady-state conditions none of the electrodiffusion variables depend upon time—but they may have nonzero values that depend upon position. If c_n is independent of time, the continuity equation for the nth ion (Equation 7.8) shows that J_n is independent of x. Hence, J_n is a constant, i.e., independent of x and t. Therefore, the Nernst-Planck equation for ion n is

$$J_n = -\frac{z_n F^2 u_n}{\beta}\left(\frac{dc_n(x)}{dx} + z_n \beta c_n(x)\frac{d\psi(x)}{dx}\right), \tag{7.39}$$

and Poisson's equation is

$$\frac{d^2\psi(x)}{dx^2} = -\frac{\rho(x)}{\epsilon}. \tag{7.40}$$

Although simpler than the time-varying equations for electrodiffusion, the time-invariant or steady-state equations for electrodiffusion are also difficult to solve. However, Equation 7.39 can be put into a somewhat more tractable form by multiplying both sides of the equation by the integrating factor $e^{z_n\beta\psi(x)}$ and rearranging the terms to yield

$$J_n e^{z_n\beta\psi(x)} = -\frac{z_n F^2 u_n}{\beta}\frac{d}{dx}\left(c_n(x)e^{z_n\beta\psi(x)}\right). \tag{7.41}$$

Integration of Equation 7.41 over the limits x_o to x yields

$$J_n = -\frac{z_n F^2 u_n}{\beta}\left(\frac{c_n(x)e^{z_n\beta\psi(x)} - c_n(x_o)e^{z_n\beta\psi(x_o)}}{\int_{x_o}^{x} e^{z_n\beta\psi(y)}dy}\right). \tag{7.42}$$

Thus, in the steady state, the potential must satisfy both Poisson's equation and the transcendental integral relation shown in Equation 7.42. The solution will depend upon a specification of the fixed charge distribution in the medium. Two simple solutions to the steady-state electrodiffusion equations follow if additional physically plausible, simplifying assumptions are made. These are the Henderson liquid junction potential solution (see Problem 7.2) and the Goldman constant field potential solution, which is described in some detail in Appendix 7.1.

7.2.4.1 Ohm's Law for Electrolytes

In a homogeneous electrolyte with uniform concentration of ion n, the flux due to diffusion is zero and the Nernst-Planck equation becomes

$$J_n(x,t) = -u_n z_n^2 F^2 c_n \frac{\partial\psi(x,t)}{\partial x}. \tag{7.43}$$

If several species are present, then the total current density is

$$J(x,t) = \sum_n J_n(x,t) = -\left(\sum_n u_n z_n^2 F^2 c_n\right)\frac{\partial\psi(x,t)}{\partial x} \tag{7.44}$$

and Equation 7.44 can be written as

$$J(x,t) = -\sigma_e \frac{\partial\psi(x,t)}{\partial x}, \tag{7.45}$$

where σ_e is called the *electric conductivity of the solution* and equals

$$\sigma_e = \sum_n u_n z_n^2 F^2 c_n. \tag{7.46}$$

The conductivity is the reciprocal of the resistivity, $\sigma_e = 1/\rho_e$.

Suppose we have a uniform current density in the x-direction in the solution, so that the total current flowing through the area A (Figure 7.8) is

$$I = JA = -\frac{A}{\rho_e}\frac{\partial\psi}{\partial x}. \tag{7.47}$$

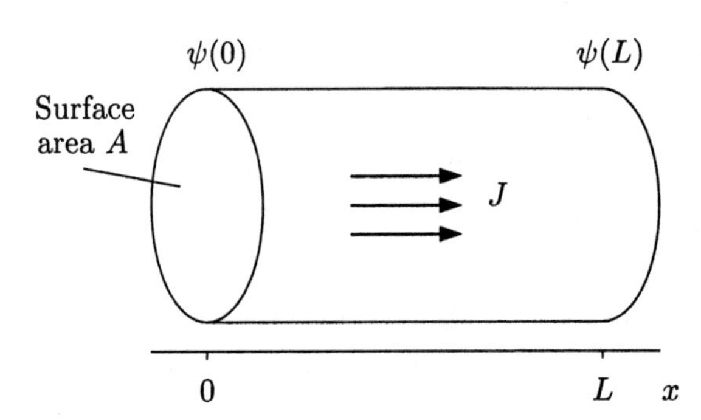

Figure 7.8 Illustration of geometry for defining Ohm's law for an electrolyte.

Integrating Equation 7.47 over the length L, we obtain

$$\int_0^L \frac{I\rho_e}{A} dx = -\int_{\psi(0)}^{\psi(L)} d\psi. \tag{7.48}$$

Therefore,

$$V = \mathcal{R}I, \quad \text{or} \quad I = \mathcal{G}V, \tag{7.49}$$

which is Ohm's law, where $V = \psi(0) - \psi(L)$, which is the potential difference across the volume element. The resistance of the volume element $\mathcal{R} = \rho_e L/A$, and the conductance of the volume element $\mathcal{G} = 1/\mathcal{R}$. The units of resistance are called *ohms*; one ohm is the ratio of a potential difference of one volt when one ampere flows through the resistance. The units of conductance are *siemens*[4]; one siemen is the reciprocal of one ohm. The resistance (conductance) of a volume element of an electrolytic solution can be calculated from the geometry and the resistivity (or conductivity), which can be computed from the concentrations and mobilities of the ions comprising the solution (Equation 7.46).

7.2.4.2 Molar Conductivity

Equation 7.46 shows that the conductivity of an electrolytic solution is a sum of contributions of the individual ions, which suggests that we can define the conductivity contributed by individual ions as

$$\sigma_n = u_n z_n^2 F^2 c_n, \tag{7.50}$$

4. The unit of conductance is also called the *mho*.

which is a function of the concentration of the ion. We can define the molar conductivity, Λ_n, as

$$\Lambda_n = \frac{\sigma_n}{c_n} = u_n z_n^2 F^2 = \frac{D_n}{RT} z_n^2 F^2. \tag{7.51}$$

The molar conductivity, molar mechanical mobility, and diffusion coefficient are related by physical constants (as well as the valence of the ion), and hence, a measurement of one of these quantities can lead to an estimate of the other two. Since electrical measurements of conductivity are relatively easy to make, this is a particularly simple way to estimate the diffusion coefficient.

Different disciplines define the molar mobility and molar conductivity in a variety of ways. To clarify these relations, let us define the *molar equivalent concentration*, $\hat{c}_n$, as

$$\hat{c}_n = |z_n| c_n. \tag{7.52}$$

We can think of the molar equivalent concentration as follows: a molar concentration c_n of an ion whose valence is z_n can combine chemically with a concentration of univalent ions that equals $|z_n|c_n$ to form a neutral salt. The latter is the molar equivalent concentration. The *molar mechanical mobility*, u_n, is the velocity acquired by a mole of particles subject to a unit mechanical force. The molar *electrical mobility*, $\hat{u}_n$, is the velocity acquired by a mole of charged particles subject to a unit electric field. Therefore, the relation between the molar electrical mobility and the molar mechanical mobility is

$$\hat{u}_n = |z_n| F u_n. \tag{7.53}$$

By substitution of Equations 7.52 and 7.53 into Equation 7.50, the conductivity can be expressed in terms of the simple expression

$$\sigma_n = F \hat{c}_n \hat{u}_n, \tag{7.54}$$

and the molar equivalent conductivity can be defined as

$$\hat{\Lambda}_n = \frac{\sigma_n}{\hat{c}_n} = F \hat{u}_n. \tag{7.55}$$

Table 7.2 shows values of the molar equivalent conductivity obtained by electrical measurements of the conductivity of ion solutions, along with derived values of the molar electrical and mechanical mobilities and the diffusion coefficient. Of course, all of the relations between these quantities are valid for

Table 7.2 Molar equivalent conductivity ($\hat{\Lambda}_n$), molar electrical mobility ($\hat{u}_n$), molar mechanical mobility (u_n), and diffusion coefficient (D_n) at 25°C and at infinite dilution. These quantities are related according to Equations 7.51 through 7.55. The molar conductivity was obtained from Robinson and Stokes (1965, Appendix 6.1).

Ion	$\hat{\Lambda}_n$	$\hat{u}_n = \hat{\Lambda}_n/F$	$u_n = \hat{u}_n/\lvert z_n\rvert F$	$D_n = u_n RT$
species n	$\left(\frac{\text{S/m}}{\text{equiv/m}^3}\right)$	$\left(\frac{\text{m/s}}{\text{V/m}}\right)$	$\left(\frac{\text{m/s}}{\text{N/mol}}\right)$	$\left(\frac{\text{m}^2}{\text{s}}\right)$
H^+	$3.50\ 10^{-2}$	$3.63\ 10^{-7}$	$3.76\ 10^{-12}$	$9.31\ 10^{-9}$
Li^+	$3.87\ 10^{-3}$	$4.01\ 10^{-8}$	$4.16\ 10^{-13}$	$1.03\ 10^{-9}$
Na^+	$5.01\ 10^{-3}$	$5.19\ 10^{-8}$	$5.38\ 10^{-13}$	$1.33\ 10^{-9}$
K^+	$7.35\ 10^{-3}$	$7.62\ 10^{-8}$	$7.89\ 10^{-13}$	$1.96\ 10^{-9}$
Rb^+	$7.78\ 10^{-3}$	$8.06\ 10^{-8}$	$8.35\ 10^{-13}$	$2.07\ 10^{-9}$
Cs^+	$7.73\ 10^{-3}$	$8.01\ 10^{-8}$	$8.30\ 10^{-13}$	$2.06\ 10^{-9}$
Tl^+	$7.47\ 10^{-3}$	$7.74\ 10^{-8}$	$8.02\ 10^{-13}$	$1.98\ 10^{-9}$
NH_4^+	$7.36\ 10^{-3}$	$7.52\ 10^{-8}$	$7.79\ 10^{-13}$	$1.96\ 10^{-9}$
$CH_3NH_3^+$	$5.87\ 10^{-3}$	$6.08\ 10^{-8}$	$6.30\ 10^{-13}$	$1.56\ 10^{-9}$
TMA^+	$4.49\ 10^{-3}$	$4.65\ 10^{-8}$	$4.82\ 10^{-13}$	$1.19\ 10^{-9}$
TEA^+	$3.27\ 10^{-3}$	$3.39\ 10^{-8}$	$3.51\ 10^{-13}$	$0.87\ 10^{-9}$
Mg^{++}	$5.30\ 10^{-3}$	$2.75\ 10^{-8}$	$1.42\ 10^{-13}$	$0.35\ 10^{-9}$
Ca^{++}	$5.95\ 10^{-3}$	$3.08\ 10^{-8}$	$1.60\ 10^{-13}$	$0.40\ 10^{-9}$
Sr^{++}	$5.94\ 10^{-3}$	$3.08\ 10^{-8}$	$1.60\ 10^{-13}$	$0.40\ 10^{-9}$
Ba^{++}	$6.36\ 10^{-3}$	$3.30\ 10^{-8}$	$1.71\ 10^{-13}$	$0.42\ 10^{-9}$
F^-	$5.54\ 10^{-3}$	$5.74\ 10^{-8}$	$5.95\ 10^{-13}$	$1.47\ 10^{-9}$
Cl^-	$7.64\ 10^{-3}$	$7.92\ 10^{-8}$	$8.21\ 10^{-13}$	$2.03\ 10^{-9}$
Br^-	$7.81\ 10^{-3}$	$8.09\ 10^{-8}$	$8.38\ 10^{-13}$	$2.08\ 10^{-9}$
I^-	$7.68\ 10^{-3}$	$7.96\ 10^{-8}$	$8.25\ 10^{-13}$	$2.04\ 10^{-9}$
NO_3^-	$7.15\ 10^{-3}$	$7.41\ 10^{-8}$	$7.68\ 10^{-13}$	$1.90\ 10^{-9}$
Acetate$^-$	$4.09\ 10^{-3}$	$4.24\ 10^{-8}$	$4.39\ 10^{-13}$	$1.09\ 10^{-9}$
$SO_4^=$	$8.00\ 10^{-3}$	$4.15\ 10^{-8}$	$4.30\ 10^{-13}$	$1.06\ 10^{-9}$

(infinitely) dilute solutions of electrolytes. For appreciable concentrations of ions, the relations among the various quantities deviate from those defined in Equations 7.51 through 7.55.

7.3 The State of Intracellular Ions

Are the ions in the cytoplasm free to diffuse and drift as they are in an aqueous ionic solution, or are they retarded somehow by the cytoplasmic components? Alternatively, can we regard the cytoplasm as an aqueous ionic bath? We shall deal with these questions in this section. Comprehensive measurements of the mobility and the diffusion coefficient of potassium ions in cytoplasm were obtained (Hodgkin and Keynes, 1953) in experiments performed on cuttlefish nerve axons (which have diameters of about 200 μm). The measurements and their interpretation are of inherent interest and make use of the theory of electrodiffusion, which has been developed in the previous sections. In these experiments, a drop of artificial seawater containing labeled potassium ions ($^{42}K^+$, which has a half-life of $12\frac{1}{2}$ hours) was applied to the axon and allowed to exchange with the intracellular potassium for some time. The axon was then washed to eliminate any adherent radioactive seawater and was placed in a measuring cell in which longitudinal electrodes were applied to create a constant potential gradient in the longitudinal direction in the cytoplasm. A Geiger counter was used to measure the longitudinal spatial distribution of the concentration of radioactive potassium ions at various times in the presence and absence of the potential gradient.

In order to evaluate the results of this experiment, we need to know the space-time evolution of the concentration of a population of particles subject to both drift and diffusion in a constant electric potential gradient. To obtain the differential equation satisfied by the concentration, we combine Equation 7.1 with the continuity equation (Equation 3.6) for ion n to yield

$$\frac{\partial c_n(x,t)}{\partial t} = D_n \frac{\partial^2 c_n(x,t)}{\partial x^2} + u_n z_n F c_n(x,t) \frac{\partial^2 \psi(x,t)}{\partial x^2} + u_n z_n F \frac{\partial \psi}{\partial x} \frac{\partial c_n(x,t)}{\partial x}. \tag{7.56}$$

If the potential gradient is constant, then the second term on the right-hand side of Equation 7.56 is zero, and the equation becomes

$$\frac{\partial c_n(x,t)}{\partial t} = D_n \frac{\partial^2 c_n(x,t)}{\partial x^2} + u_n z_n F \frac{\partial \psi}{\partial x} \frac{\partial c_n(x,t)}{\partial x}. \tag{7.57}$$

One method of finding the solution to Equation 7.57 in response to an initial placement of n_o mol/cm^2 at $x = 0$ at $t = 0$ is to guess that the solution has a form that is similar to that of the solution of the diffusion equation except that the presence of the potential gradient causes a drift in the concentration of charged particles. We try a solution of the form

$$c_n(x,t) = \frac{n_o}{\sqrt{4\pi D_n t}} e^{-(x-vt)^2/4D_n t}, \tag{7.58}$$

where v is the drift velocity. Substitution of Equation 7.58 into Equation 7.57 yields

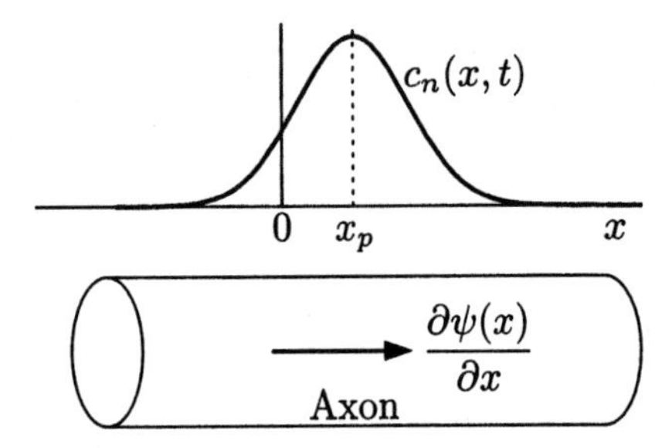

Figure 7.9 Schematic diagram of an arrangement for measuring the mobility and the diffusion coefficient of ions in the cytoplasm of a nerve fiber. The concentration is drawn for $\partial\psi(x)/\partial x < 0$.

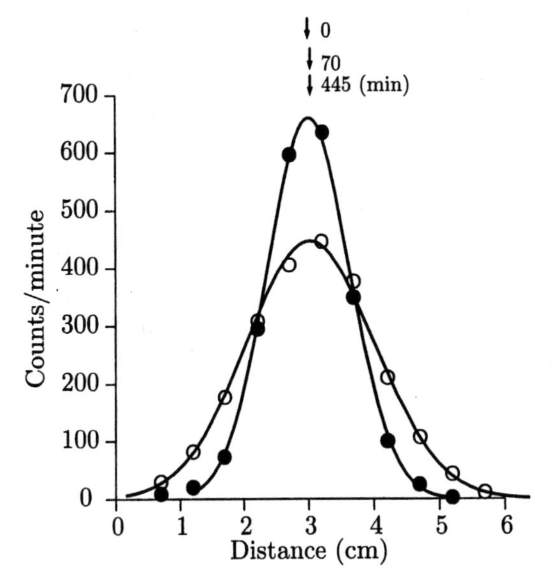

Figure 7.10 Distribution of radioactive potassium along a nerve fiber in the absence of a longitudinal electric field (adapted from Hodgkin and Keynes, 1953, Figure 4). The filled circles give the count rate at the initial time 0, and the open circles give the count rate 445 minutes later. The arrows indicate the peak of the count rate pattern according to a fit of the data with a Gaussian function. These peaks are shown for three times and do not differ appreciably. The curves are Gaussian functions that have parameters estimated from the measurements.

$$v = -u_n z_n F \frac{\partial\psi}{\partial x}. \tag{7.59}$$

Thus, the solution is a Gaussian distribution of concentration whose width spreads out as $\sqrt{2D_n t}$ and whose peak value moves at a constant velocity equal to $-u_n z_n F \partial\psi/\partial x$ (Figure 7.9). Therefore, by measuring the spatial distribution of the concentration at various times, we can estimate both the diffusion constant and the mobility independently.

The measurements of Hodgkin and Keynes (1953) are compared with the solutions given in Equations 7.58 and 7.59 in Figures 7.10 and 7.11. In the absence of an electric field, the spatial distribution of radioactive potassium, which is proportional to the count rate, spreads out in space, but the position of the peak of the distribution does not vary appreciably with time (Figure 7.10). The diffusion coefficient can be estimated from the change in the

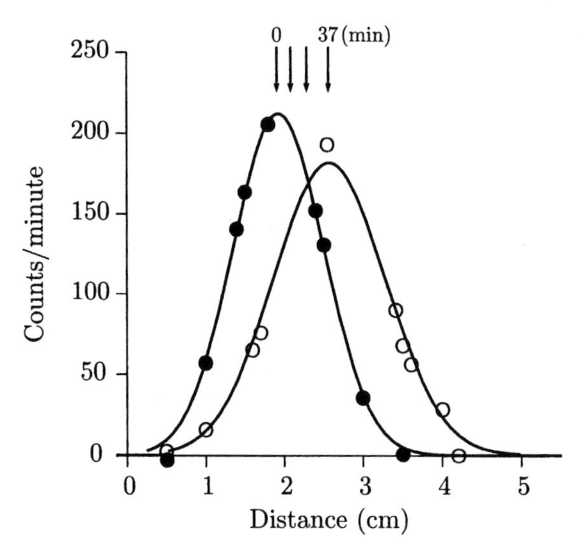

Figure 7.11 Distribution of radioactive potassium along a nerve fiber in the presence of an electric field (adapted from Hodgkin and Keynes, 1953, Figure 3). The filled circles represent the spatial distribution of counts at time 0, and the open circles represent the spatial distribution of counts 37 minutes later. The measurements were used to estimate the parameters of Gaussian functions that are fitted to the data. The arrows mark the maximum of the fitted functions at times 0 and 37 minutes and at two intermediate times (11 and 23 minutes).

width of the distribution with time. Upon application of a longitudinal electric field, the spatial distribution of potassium spreads out and migrates along the axon (Figure 7.11). The mobility of potassium can be estimated from the migration of the peak of the distribution, and the diffusion coefficient can be estimated from the change in the width of the distribution. The theoretical curves fit these measurements to within their variability. Furthermore, the diffusion coefficient has been found to agree with the mobility (according to the Einstein relation, Equation 3.26) and with the diffusion coefficient of potassium in an aqueous solution at the concentration found in the cytoplasm.[5] In these respects, potassium ions in cytoplasm behave similarly to those in an aqueous solution. The results obtained with potassium cannot necessarily be extrapolated to other ions. For example, calcium ions are not free to diffuse

5. Measurements of the diffusion coefficient and mobility of potassium in 0.5 mol/L of KCl (near the concentration in the cytoplasm of the squid giant axons) at 18°C (approximately the temperature of the measurements in squid axons) are 1.5×10^{-9} m^2/s and 5.4×10^{-13} (m/s)/(N/mol), respectively (Hodgkin and Keynes, 1953).

intracellularly (Hodgkin and Keynes, 1957; Zhou and Neher, 1993; Allbritton et al., 1992), but appear to be bound to a variety of intracellular organelles and cytoplasmic calcium-binding proteins (e.g., calmodulin).

7.4 Macroscopic Model of Passive Ion Transport

7.4.1 Derivation from Microscopic Models

The macroscopic model of ion transport that we shall obtain in this section can represent transport for many different microscopic models, two of which we shall illustrate.

7.4.1.1 Steady-State Electrodiffusion Through a Homogeneous Membrane

We shall consider steady-state electrodiffusion across a homogeneous membrane in a manner analogous to our treatment of steady-state diffusion across a membrane (Section 3.6). From this microscopic electrodiffusion model we shall derive a macroscopic model that relates the membrane potential to the membrane current density.

We shall assume that a membrane separates two ionic solutions whose concentrations for ion n are c_n^i and c_n^o, respectively (Figure 7.12). The current density carried by this ion is J_n, which is constant in the steady state. The concentration and the electric potential in the membrane are $c_n(x)$ and $\psi(x)$, which are independent of t in the steady state. In the steady state, if we use the Einstein relation, the Nernst-Planck equation can be written as

$$J_n = -u_n z_n^2 F^2 c_n(x) \frac{d}{dx}\left(\frac{RT}{z_n F}\ln c_n(x) + \psi(x)\right). \tag{7.60}$$

By separating variables and integrating Equation 7.60, we obtain

$$J_n \int_0^d \frac{dx}{u_n z_n^2 F^2 c_n(x)} = -\int_{x=0}^{x=d} d\left(\frac{RT}{z_n F}\ln c_n(x) + \psi(x)\right),$$

which yields

$$J_n = \frac{\psi(0) - \psi(d) - \dfrac{RT}{z_n F}\ln\left(\dfrac{c_n(d)}{c_n(0)}\right)}{\displaystyle\int_0^d \frac{dx}{u_n z_n^2 F^2 c_n(x)}}. \tag{7.61}$$

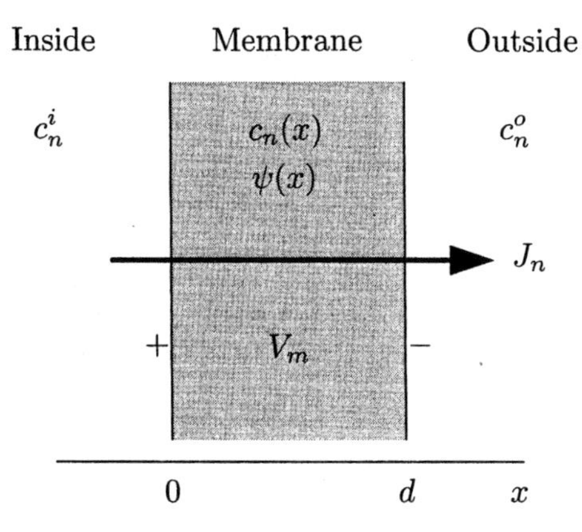

Figure 7.12 Schematic diagram showing variables important in describing electrodiffusion through a membrane.

We shall assume that the ion concentrations at the boundaries partition themselves according to the partition coefficient, as follows:

$$\frac{c_n(0)}{c_n^i} = \frac{c_n(d)}{c_n^o} = k_n. \tag{7.62}$$

Therefore, Equation 7.61 becomes

$$J_n = G_n(V_m - V_n), \tag{7.63}$$

where V_n is the Nernst equilibrium potential, which is

$$V_n = \frac{RT}{z_n F} \ln\left(\frac{c_n^o}{c_n^i}\right); \tag{7.64}$$

the membrane potential, V_m, is defined as

$$V_m = \psi(0) - \psi(d); \tag{7.65}$$

and the membrane conductance is defined as

$$G_n = \frac{1}{\displaystyle\int_0^d \frac{dx}{u_n z_n^2 F^2 c_n(x)}}. \tag{7.66}$$

The relation between membrane potential and current density given by Equation 7.63 can be represented by an equivalent electrical network (Figure 7.13). This relation defines a macroscopic model for ion transport through a membrane. The current density results from two mechanisms: the presence of an electric potential across the membrane V_m, which results in drift of ions, and the presence of a concentration difference across the membrane,

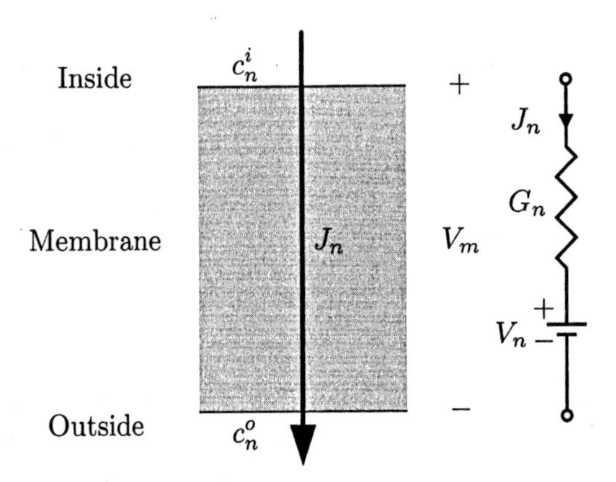

Figure 7.13 Electric network model of passive ion transport through a membrane.

which results in diffusion of ions. The chemical potential energy difference across the membrane is manifested as an equivalent potential V_n in the network model.

7.4.1.2 *Steady-State Electrodiffusion Through a Porous Membrane*

Suppose the membrane consists of an impermeant matrix and water-filled channels through which ions are transported by electrodiffusion. With this assumption, the Nernst-Planck equation can be integrated over the length of such a channel. Since the problem is identical, except for some dimensions, to that for electrodiffusion in a homogeneous membrane, the current voltage relation must have the same form as Equation 7.63. We express the current voltage relation for a single open channel as

$$\mathcal{I} = \gamma(V_m - V_n), \tag{7.67}$$

where $\mathcal{I}$ is the current through an open channel and γ is the conductance of the open channel. As is discussed elsewhere (Weiss, 1996, Chapter 6), the passive transport of ions through membranes is via microscopic ionic channels, but these channels are not always open. The state of conduction of the channel is gated by some environmental variable (e.g., the membrane potential, the binding of a ligand to the membrane, etc.). These channels are either open, or closed at any given time. When a channel is open, an open channel current $\mathcal{I}$ flows through the channel. When a channel is closed, no current flows through the channel. The state of conduction of the channel switches randomly from the open to the closed state. Thus, the average current through the membrane is $\mathcal{I}x$, where x is the probability that the channel is open. If there are $\mathcal{N}$ channels per cm^2 of membrane, then the average current density of ion n through the membrane is

$$J_n = \gamma x \mathcal{N}(V_m - V_n) = G_n(V_m - V_n), \tag{7.68}$$

where $G_n = \gamma x \mathcal{N}$ is the conductance of these channels per unit area of membrane. Therefore, this simple channel model yields the same macroscopic relation between membrane current density and membrane potential as that for a homogeneous membrane model. The only difference is the form of the conductance G_n.

7.4.1.3 Summary

Both integration of the Nernst-Planck equation for continuous electrodiffusion through a membrane and a simple model for membrane channels lead to the same macroscopic relation between current density and membrane potential (Equation 7.63) and to the same macroscopic network model. We shall now explore the properties of this model.

7.4.2 Properties of the Macroscopic Model

7.4.2.1 The Meaning of the Nernst Equilibrium Potential

Equation 7.63 shows that if $V_m = V_n$, then $J_n = 0$, and there is no net flow of the nth ion through the membrane. That is, the nth ion is in electrodiffusive equilibrium across the membrane (provided that there are no differences in temperature or hydrostatic pressure across the membrane). Therefore, V_n is called an *equilibrium potential*. If $V_m \neq V_n$, then $J_n \neq 0$ (provided that $G_n \neq 0$), and there is a net flow of the nth ion.

7.4.2.2 Energetic Interpretation of the Nernst Equilibrium Potential

By cross multiplying the terms in Equation 7.64, expressing the membrane potential as the difference in inside minus outside potential, i.e., $V_n = \psi_i - \psi_o$, and rearranging terms, we can interpret the electrodiffusive equilibrium condition in energetic terms as follows:

$$z_n F(\psi_i - \psi_o) + RT(\ln c_n^i - \ln c_n^o) = 0. \tag{7.69}$$

The first term on the left-hand side of Equation 7.69 is the stored electrostatic energy per mole at the inside of the membrane minus that at the outside.

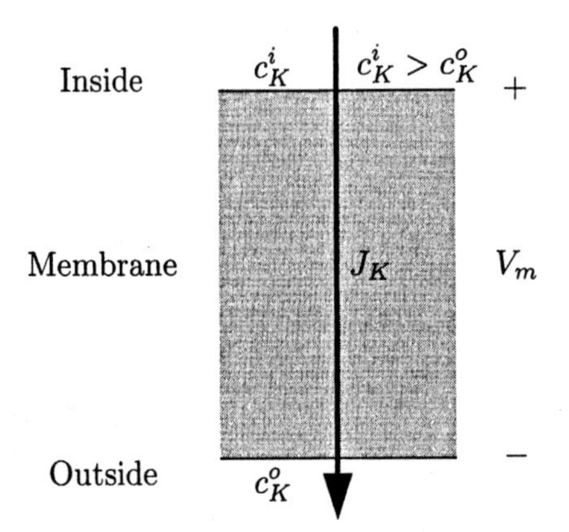

Figure 7.14 Illustration of the polarity of the Nernst equilibrium potential and its relation to ion concentration.

The second term is the stored chemical energy per mole at the inside of the membrane minus that at the outside. Thus, the sum of these differences is the electrochemical energy gained in transferring a mole of ions from the outside to the inside of the membrane. At electrodiffusive equilibrium, this difference must be zero. By shuffling the terms in Equation 7.69 and adding the reference electrochemical potential to each side, we obtain

$$\mu_n^o + RT \ln c_n^i + z_n F \psi_i = \mu_n^o + RT \ln c_n^o + z_n F \psi_o,$$

which states that at electrodiffusive equilibrium the electrochemical potentials on the two sides of the membrane must be the same.

7.4.2.3 The Polarity of the Nernst Equilibrium Potential

Suppose that we have a membrane that is permeable only to potassium ions and that there is a difference in the concentration of potassium on the two sides of the membrane as indicated in Figure 7.14. Because of the difference in concentration, there will be a tendency for potassium to diffuse from the high-concentration side to the low-concentration side of the membrane. However, because of the presence of the membrane potential there will also be a tendency for the ions to drift across the membrane. For which polarity of V_m will the tendency to diffuse just cancel the tendency to drift and prevent the net flow of potassium ions across the membrane? Since potassium ions have a positive charge, if the low-concentration side is at a higher electric potential than the high-concentration side, the tendency to drift can cancel the tendency to diffuse. This physical argument is consistent with Equation 7.64, which indicates that if $c_n^i > c_n^o$ and if $z_n > 0$, then $V_n < 0$. From Table 7.1, we see that the polarities of the Nernst equilibrium potentials for key ions are $V_K < 0$, $V_{Na} > 0$, and $V_{Cl} < 0$.

7.4.2.4 The Magnitude of the Nernst Equilibrium Potential

The magnitude of the Nernst equilibrium potential depends on the product of two factors, RT/F and the logarithm of the concentration ratio. From Equation 7.64 we can see that RT/F equals the Nernst equilibrium potential for an e-fold ratio of concentrations of ion n. That is, at electrodiffusive equilibrium an e-fold ratio of concentrations is just balanced by a potential that equals RT/F. It is convenient to express the Nernst equilibrium potential in terms of the logarithm of the concentration to the base 10, so that

$$V_n = \frac{RT}{z_nF}\ln\left(\frac{c_n^o}{c_n^i}\right) = \frac{RT}{z_nF\log_{10}e}\log_{10}\left(\frac{c_n^o}{c_n^i}\right).$$

To determine the value of the Nernst potential, we substitute the values of the molar gas constant, $R = 8.31\ \mathrm{J\cdot K^{-1}\cdot mol^{-1}}$; Faraday's constant, $F = 9.65\times10^4$ C/mol; and the absolute temperature, $T = 273.15 + t_c$, where t_c is the temperature in centrigrade, into the expression for RT/F. At room temperature, $t_c = 24°\mathrm{C}$, $RT/F \approx 26$ mV, and $RT/(F\log_{10}e) \approx 59$ mV. Thus, a 10-fold concentration ratio gives a Nernst equilibrium potential of 59 mV. The dependence of $RT/(F\log_{10}e)$ on temperature is shown in Figure 7.15 and indicates that for most of the physiological temperature range this quantity is in the range of 54–62 mV. At a temperature of 24°C, we have

$$V_n = \frac{26}{z_n}\ln\left(\frac{c_n^o}{c_n^i}\right) = \frac{59}{z_n}\log_{10}\left(\frac{c_n^o}{c_n^i}\right)\ [\mathrm{mV}]. \tag{7.70}$$

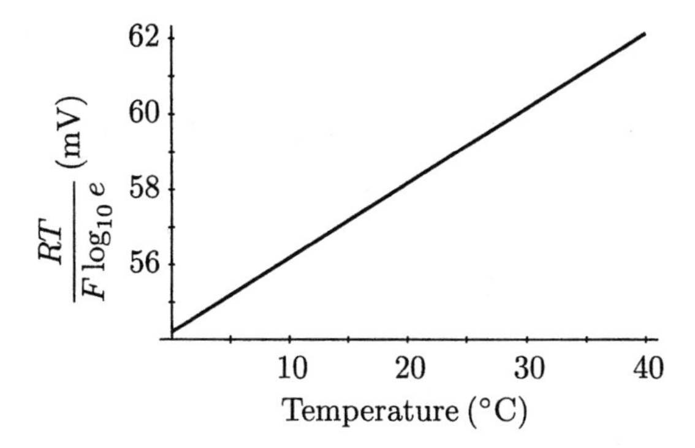

Figure 7.15 Variation with temperature of the factor $RT/(F\log_{10}e)$.

Thus, we note that for univalent ions, for which $1/100 < c_n^o/c_n^i < 100$ (which is commonly found for cells), $-118 < V_n < +118$ mV. As we shall see, the Nernst potentials bound the membrane potential.[6] This constraint explains why membrane potentials generated by individual cells tend to be in the range of ± 100 mV.

7.4.2.5 How Is the Nernst Equilibrium Potential Generated?

To see how the Nernst equilibrium potential could be generated, consider the situation shown in Figure 7.16. Imagine a bath separated into two compartments by a membrane permeable only to ion n, which for simplicity we shall assume has a positive valence and has a higher concentration on side 2 than on side 1. For $t < 0$, the solutions are separated by impermeable partitions so that they cannot mix through the membrane. At $t = 0$ these partitions are removed and the two ionic solutions come in contact with the two sides of the membrane, but only the nth ion can pass through the membrane. Since there is a higher concentration of these ions on side 2 than on side 1, there will be an initial tendency for ions to diffuse from side 2 to side 1. Since these ions are positively charged, there will be a net transfer of positive charge across the membrane from side 2 to side 1. Since both solutions were initially electrically neutral, this transfer of positive charge to side 1 leaves an equal negative charge on side 2. Since the solutions on both sides of the membrane are good electrical conductors, the charges will distribute themselves on the two sides of the membrane in a very short amount of time given by the relaxation time (see Section 7.2.3). This separation of charge across the membrane gives rise to an electric potential difference across the membrane, with side 1 at a positive potential with respect to side 2. This electric potential difference retards the flow of positive ions from side 2 to side 1. When this potential reaches the Nernst equilibrium potential, then the net flow becomes zero and ion n is in equilibrium across the membrane.

In the process of generation of this potential, some quantity of ion n has been transferred from side 2 to side 1 across the membrane. Is this quantity appreciable? Let us estimate the number of ions per unit area that need to be displaced across the membrane to charge that membrane to a Nernst

---

6. This constraint is strictly true for components of membrane potential produced by passive transport processes. Active transport processes, to be discussed later, allow the membrane potential to exceed the bounds imposed by the Nernst potentials.

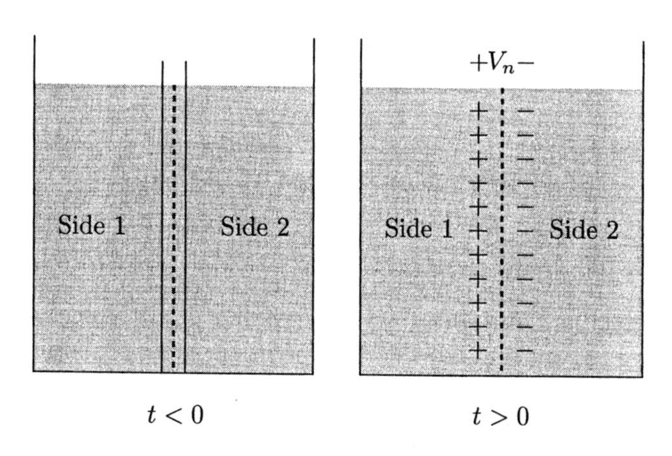

Figure 7.16 Illustration of the generation of the Nernst equilibrium potential. A bath is separated into two compartments by a membrane permeable only to ion n.

equilibrium potential of V_n. The charge per unit area of membrane is simply $C_m V_n$, where C_m is the electrostatic capacitance of a unit area of membrane. Therefore, the number of moles of ion n per unit area of membrane displaced across the membrane is $C_m V_n / z_n F$.

To get an indication of how many ions this involves, we shall compute the thickness, δ, of a unit area of solution that contains this number of ions. Therefore, we wish to compute δ such that $\delta c_n = C_m V_n / z_n F$. For the following typical values, $C_m \approx 1\mu\text{F/cm}^2$ (Weiss, 1996, Chapter 3), $V_n \approx 100$ mV, $z_n = 1$, $F \approx 10^5$ C/mol, and $c_n \approx 10^{-4}\text{mol/cm}^3$, we find that $\delta \approx 1$ Å. Thus, to charge one unit area of membrane to a potential of 100 mV when it is bathed in a solution of concentration of about 10^{-4}mol/cm^3 requires the number of ions contained in a slab of solution of unit area and of thickness equal to 1 Å. Thus, if the bath thicknesses were greater than 1 μm, the concentration would change by less than 0.01%. Thus, for the membranes of cells, charging the membrane will not change the concentration of the solutions appreciably.

7.4.2.6 Dependence of the Nernst Equilibrium Potential on the Valence

Equation 7.64 demonstrates that the Nernst equilibrium potential depends inversely on the valence. The physical basis of this dependence can be seen from Equation 7.1, which shows that the flux due to the electric potential gradient is proportional to the valence, whereas the flux due to diffusion is independent of the valence. This result occurs because the force on a charged particle in an electric field is proportional to the particle's charge. One consequence of this property is that, for example, a divalent ion requires only half the potential gradient to balance a given concentration gradient.

7.4.2.7 Consequences of Assuming the Ionic Conductances Are Non-Negative Quantities

Equation 7.66 indicates that the ionic conductances are non-negative quantities. Therefore, Equation 7.63 suggests that for $G_n > 0$,

$$\text{if } V_m - V_n > 0, \text{ then } J_n > 0$$
$$\text{if } V_m - V_n < 0, \text{ then } J_n < 0$$

Thus, the assumption that the ionic conductances are non-negative quantities is equivalent to assuming that the current carried by the nth ion flows down the electrochemical potential gradient. This type of transport is called *passive transport*. In later sections we shall see that there are mechanisms in the membrane that are capable of transporting ions up the electrochemical potential gradient. This type of transport is called *active transport*.

7.4.2.8 Significance of the Ionic Conductances

The magnitude of the current density through the membrane for a given value of $V_m - V_n$ is proportional to G_n (Equation 7.63). If G_n is a large quantity, then we say that the membrane is highly permeable to ion n. If $G_n = 0$, then $J_n = 0$ for all values of $V_m - V_n$ and we say that the membrane is impermeable to ion n. Thus, G_n is an electrical measure of the permeability of the membrane to ion n.

7.4.2.9 Concurrent Electrodiffusive Equilibrium of Several Ions

If several ions are concurrently at electrodiffusive equilibrium across a membrane, then their Nernst equilibrium potentials must all be equal, or

$$\frac{RT}{z_1F}\ln\left(\frac{c_1^o}{c_1^i}\right) = \frac{RT}{z_2F}\ln\left(\frac{c_2^o}{c_2^i}\right) = \cdots = \frac{RT}{z_nF}\ln\left(\frac{c_n^o}{c_n^i}\right), \tag{7.71}$$

which leads to constraints on the concentrations of these ions

$$\left(\frac{c_1^o}{c_1^i}\right)^{1/z_1} = \left(\frac{c_2^o}{c_2^i}\right)^{1/z_2} = \cdots = \left(\frac{c_n^o}{c_n^i}\right)^{1/z_n}. \tag{7.72}$$

For example, suppose that a membrane is permeable only to potassium and chloride ions. Then the condition that both ions are in equilibrium across the membrane is that their Nernst equilibrium potentials are equal, which implies (see Equation 7.72) that $c_K^o/c_K^i = c_{Cl}^i/c_{Cl}^o$ or that $c_K^o c_{Cl}^o = c_K^i c_{Cl}^i$. This condition is called the *Donnan relation.*

7.5 Resting Potential of Uniform Isolated Cells

7.5.1 Model 1: A Single Permeant Ion (the Bernstein Model)

By the beginning of the twentieth century, a number of observations had been made that were relevant to understanding the basis of cellular resting potentials. For example, measurements with string galvanometers had shown that the potential difference between electrodes placed on the cut end and the intact end of a muscle was negative (Figure 7.17). This potential was initially called the *demarcation potential* and later called the *injury potential*. Since an electrode placed on the cut end of a muscle was clearly in closer electrical contact with the cell interiors than was an electrode on the intact portion, the polarity of the demarcation potential suggested that the insides of cells were at a negative potential with respect to the cell exteriors. It was found that the magnitude of the demarcation potential could be reduced by increasing the extracellular potassium concentration, whereas this potential was relatively insensitive to changes in external sodium concentration. Furthermore, it was known that cells contained a much higher concentration of potassium and a much lower concentration of sodium than did the extracellular solutions. These observations, together with the development of the theory of thermodynamic equilibrium across semipermeable membranes (by Nernst and others) led Bernstein (1902) to propose the hypothesis that the resting potential of cells (V_m^o) was equal to the Nernst equilibrium potential for potassium

$$V_m^o = V_K = \frac{RT}{F \log_{10} e} \log_{10}\left(\frac{c_K^o}{c_K^i}\right). \tag{7.73}$$

The idea was that cellular membranes at rest were permeable to potassium ions and impermeable to *all* other ions. This idea can be represented in an

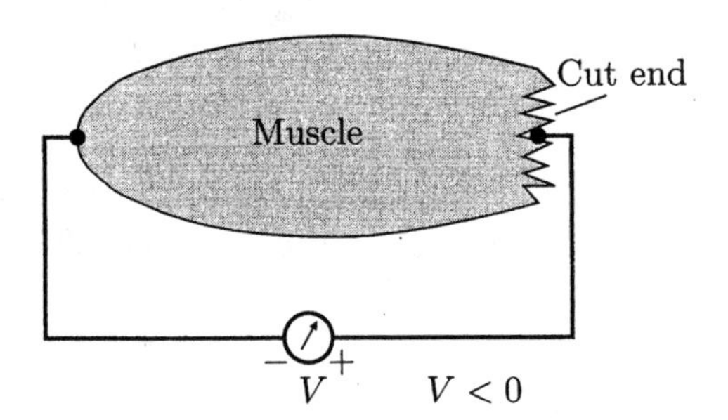

Figure 7.17 Recording the demarcation potential from the cut end of a muscle.

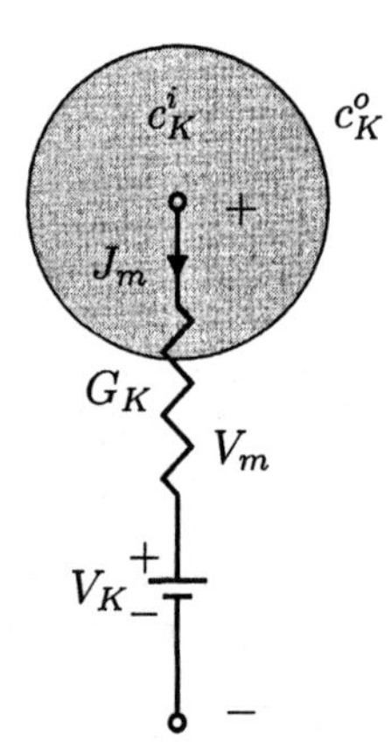

Figure 7.18 Network model of a cell according to the Bernstein model.

equivalent network model of a cell (Figure 7.18). At rest, by definition the total membrane current is zero, so there is no net charge transfer across the membrane. But for a single permeant ion, like potassium, the membrane current equals the current carried by potassium. Hence, at rest potassium is in electrodiffusive equilibrium across the membrane. Therefore, the resting potential across the membrane must be the Nernst equilibrium potential for potassium. This hypothesis fitted the rather indirect evidence that was then available. It explained the polarity of the demarcation potential, the sensitivity of this potential to changes in potassium ion concentration, and its insensitivity to changes in sodium concentration. The measurements of the demarcation potential were not sufficiently accurate to test this hypothesis directly, and the Bernstein model endured for over four decades as the accepted explanation of the existence of the inferred cellular resting potential. Because this hypothesis is close to explaining contemporary measurements, this model of resting potentials persists to this day as the simplest explanation of resting potentials. However, this model is fundamentally incorrect, as we shall see.

What does this model predict about the relation of the resting potential to changes in ion concentration? As indicated in Figure 7.19, the Bernstein model (Equation 7.73) predicts that if the resting membrane potential is plotted versus the logarithm of the external potassium concentration, the result will be a straight line with a slope of 59 mV/decade at a temperature of 24°C. The model also predicts that the resting potential should be independent of the concentration of any other ion. Any device that produces a potential that depends on concentration according to Equation 7.73 is called a *perfect potassium electrode.* Thus, the Bernstein model represents the cellular membrane as a potassium electrode.

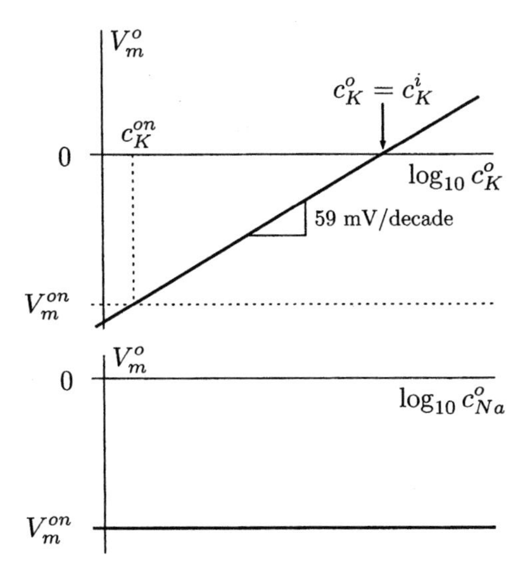

Figure 7.19 The dependence of resting potential on external potassium (upper panel) and external sodium (lower panel) concentration at 24°C as predicted by the Bernstein model. At the normal extracellular concentration of potassium, c_K^{on}, the membrane potential is at its normal value, V_m^{on}. The membrane potential changes in proportion to the logarithm of the potassium concentration and has the value zero when the external and internal potassium concentrations are equal. The membrane potential is independent of the external sodium concentration.

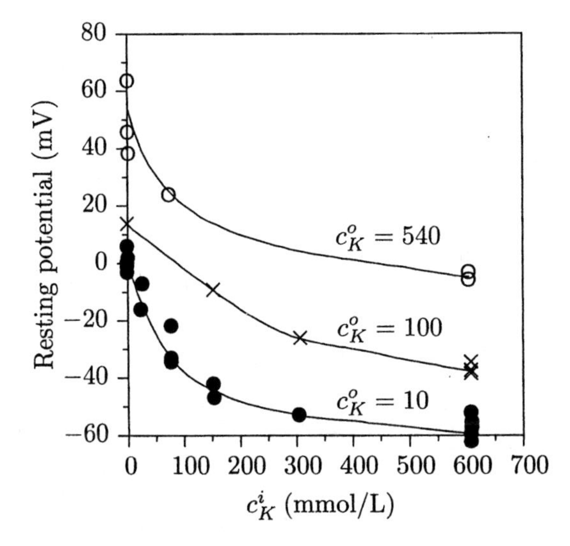

Figure 7.20 Dependence of resting potential on intracellular and extracellular potassium concentration in the squid giant axon (adapted from Baker et al., 1962a, Figure 2). The resting potential is plotted versus the intracellular potassium concentration; the parameter is the extracellular potassium concentration in units of mmol/L. The internal potassium concentration was changed by replacing isotonic KCl with NaCl. The external solution was artificial seawater with the indicated potassium concentration.

7.5.2 Dependence of Resting Potential on Ion Concentration

The strong dependence of resting potential on both the intracellular and extracellular potassium concentration, predicted by the Bernstein model, is indicated in Figure 7.20. In these measurements, squid giant axons were perfused with artificial solutions and immersed in a bath whose composition was controlled. Thus, the dependence of the resting potential on both intracellular

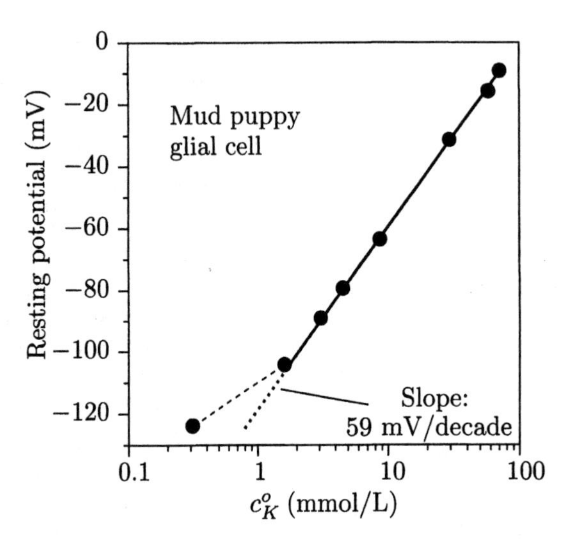

Figure 7.21 Dependence of the resting potential of glial cells in the optic nerve of the mud puppy (*Necturus maculosus*) on external concentration of potassium (adapted from Kuffler et al., 1966, Figure 8). The mean value of the resting potential in a solution with a normal potassium concentration was −89 mV. The points represent averages of forty-two measurements whose standard deviations were less than the point width.

and extracellular ion concentration was investigated. At all extracellular concentrations of potassium, increasing the intracellular potassium concentration made the resting potential more negative. For any value of the intracellular potassium concentration, increasing the extracellular potassium concentration made the resting potential more positive. If the intracellular concentration and extracellular concentrations of sodium and potassium were reversed so that the extracellular concentration of potassium was high and the intracellular concentration of potassium was low, then the resting potential reversed in sign and became positive. If intracellular and extracellular concentrations of potassium were approximately equal, then the resting potential was near zero. All these predictions are qualitatively consistent with the Bernstein model (Equation 7.73).

A direct quantitative test of the Bernstein model is to compare such measurements of the resting potential with Equation 7.73 in the manner illustrated schematically in Figure 7.19. As shown in Figure 7.21, measurements from some cells (in this case, glial cells in the optic nerve of an amphibian) suggest that the membrane acts as a potassium electrode for a large range of potassium concentrations. However, the data deviate from this prediction at the lowest external potassium concentration. Results for a number of other cell types are shown in Figure 7.22. In several of the cell types, the data approach that predicted for a potassium electrode, but only at the highest external potassium concentrations. For several cell types, even at high external potassium concentrations V_m^o depends linearly on $\log c_K^o$, but the slope is less than that predicted for a potassium electrode. For low external potassium

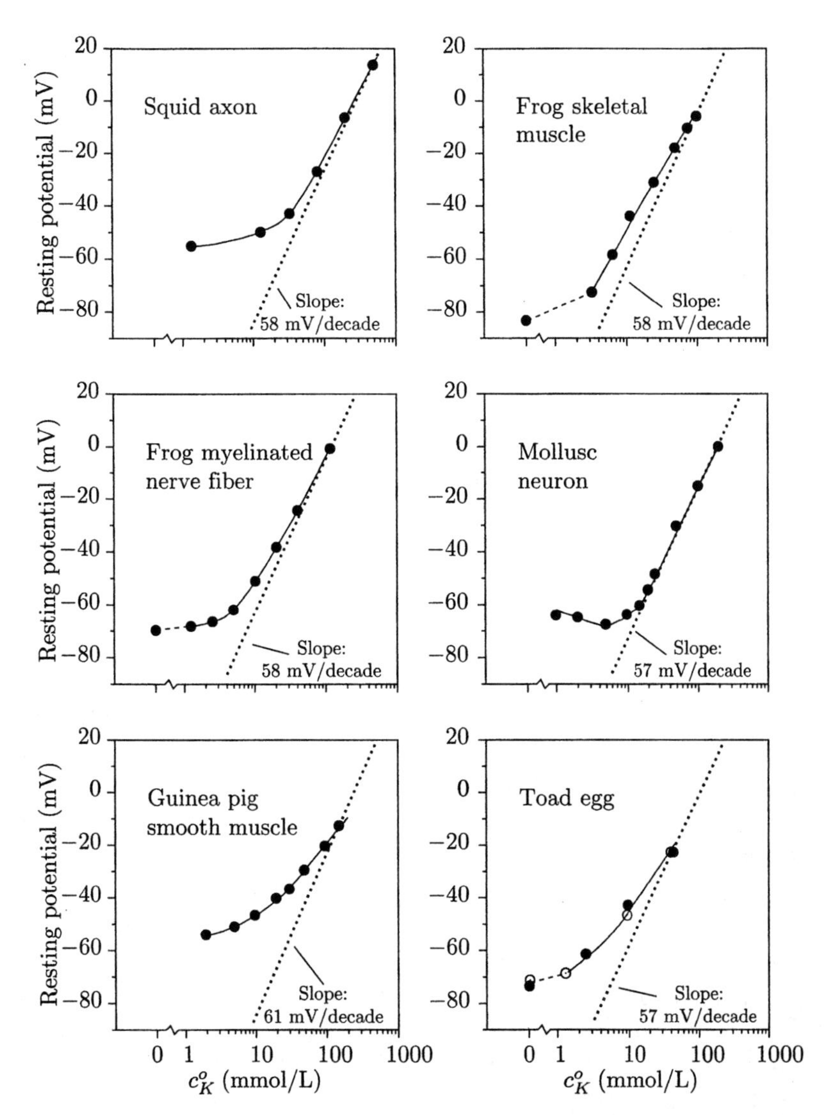

Figure 7.22 Dependence of resting potential on external potassium concentration for several cell types: squid (*Loligo pealii*) axon (adapted from Curtis and Cole, 1942, Figure 2); average of results from six frog skeletal (sartorius) muscle fibers (adapted from Ling and Gerard, 1950); average of results from six frog (*Rana esculenta*) myelinated sciatic nerve fibers (adapted from Huxley and Stämpfli, 1951, Figure 3); mollusc (*Anisodoris nobilis*) neuron (adapted from Gorman and Marmor, 1970, Figure 3); smooth muscle fiber in the *taenia coli* of the guinea pig (adapted from Holman, 1958, Figure 6); average results from egg cells of the toad (*Bufo vulgaris formosus* Boulenger) (adapted from Maéno, 1959, Figure 2). The dependence of potential on potassium concentration for a potassium electrode at the temperature of the measurements is shown as a dotted line in each panel.

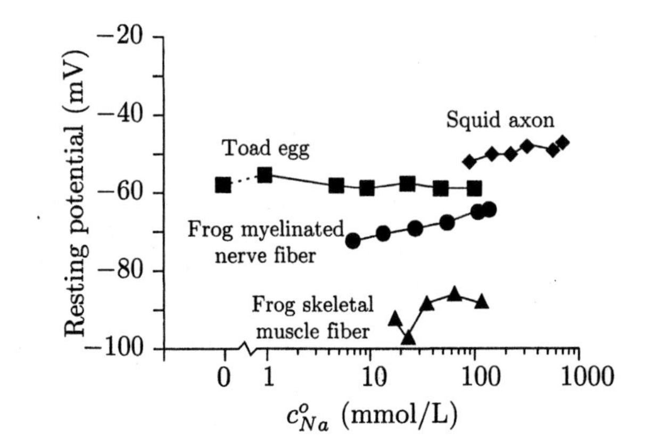

Figure 7.23 Dependence of resting potential on external sodium concentration for several cell types: squid (*Loligo pealii*) axon (adapted from Hodgkin and Katz, 1949, Table 4); frog (*Rana temporaria*) skeletal (sartorius) muscle fibers (adapted from Nastuk and Hodgkin, 1950, Figure 10); frog (*Rana esculenta*) myelinated sciatic nerve fibers (adapted from Huxley and Stämpfli, 1951, Figure 4); average results from egg cells of the toad (*Bufo vulgaris formosus* Boulenger) (adapted from Maéno, 1959, Figure 1).

concentration near the normal physiological concentration, V^o_m is relatively independent of c^o_K for all these cells. Thus, we conclude that the dependence of the membrane potential of these cells on potassium concentration does not behave as would a potassium electrode.

A further prediction of the Bernstein model is that the resting potential should be independent of the sodium concentration. Figure 7.23 shows that the resting potential is not independent of the sodium concentration as the Bernstein model predicts, although the effect of changes in sodium concentration is much smaller than the effect of potassium concentration changes. These results, together with direct measurements of the flux of radioactive ions, demonstrate that membranes are permeable to ions in addition to potassium. Thus, the resting potential of a cell does not equal the Nernst equilibrium potential for potassium.

7.5.3 Model 2: Multiple Permeant Ions

Since cellular membranes are permeable to several ions, we shall represent the relation between membrane potential and membrane current density by a parallel circuit in which each branch represents the current carried by a particular ion species through the membrane (Figure 7.24). We represent each ionic branch by a macroscopic model for permeation of an ion through the

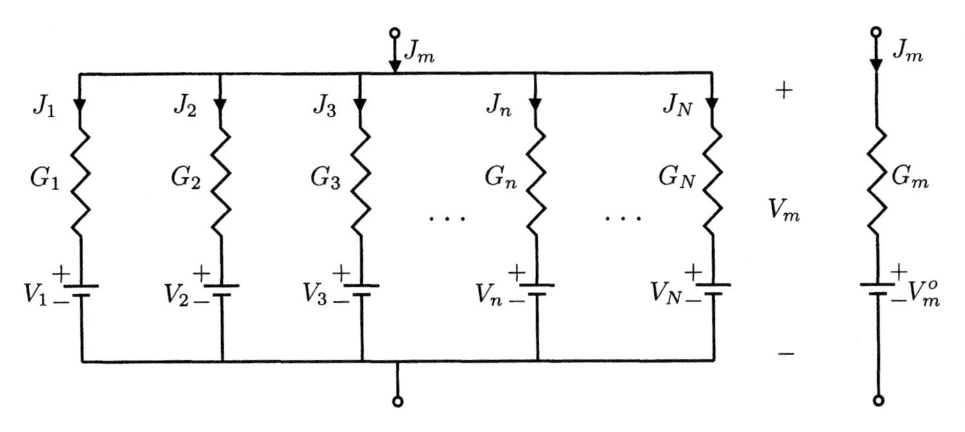

Figure 7.24 Network model of membrane showing a parallel combination of independent branches, one for each ion transported. A Thévenin's equivalent network is shown on the right side.

membrane (Figure 7.13). Thus, the total membrane current carried by all these branches is

$$J_m = \sum_n J_n,$$

and from Equation 7.63

$$J_m = \sum_n G_n(V_m - V_n). \tag{7.74}$$

Now assume that we have a cell with a membrane that has uniform electrical properties, and that the cell is isolated electrically from other cells, i.e., not in electrical contact with other cells. For such a cell, the total current flowing out of the cell is simply $I_m = AJ_m$, where A is the surface area of the cell (Figure 7.25). We define the resting state as the state for which the membrane potential is constant. To ensure that the membrane potential is constant at rest, the net charge entering the cell must be zero. Therefore, at rest there is no net flow of current out of the cell, i.e., $I_m = 0$; therefore, $J_m = 0$, and the membrane potential is, by definition, equal to the resting potential. Note that the multiple permeant ion model (model 2) differs fundamentally from the single permeant ion model (model 1). For multiple permeant ions, the resting condition does not guarantee ionic equilibrium for any of the ions.

The resting state for the multiple-ion model implies that

$$\sum_n G_n(V_m^o - V_n) = 0, \tag{7.75}$$

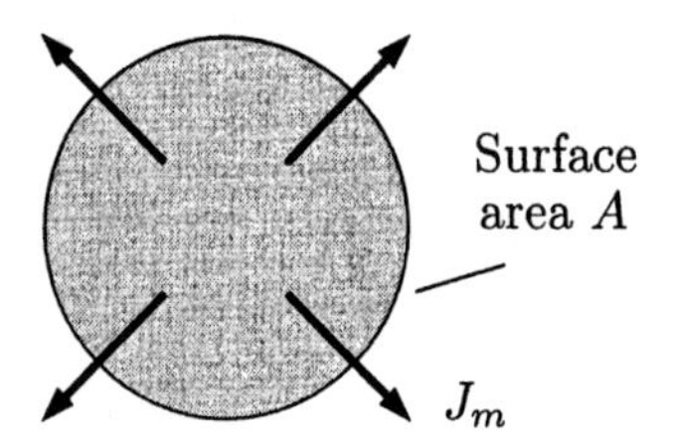

Figure 7.25 Schematic diagram of a uniform isolated cell.

and solving Equation 7.75 for V_m^o, we obtain

$$V_m^o = \sum_n \frac{G_n}{G_m} V_n, \tag{7.76}$$

where

$$G_m = \sum_n G_n, \tag{7.77}$$

where G_m is the total conductance of a unit area of membrane. As indicated in Equation 7.76, the resting membrane potential is a weighted sum of the Nernst equilibrium potentials of all the ions. The weighting factor for each ion, G_n/G_m, is the ratio of the ionic conductance for that ion to the total membrane ionic conductance, G_m. Thus, if the membrane is highly permeable to ion j, then G_j will be an appreciable fraction of G_m, and V_m^o will be near V_j. Because the weighting factors are numbers between zero and one, it can be shown that V_m^o is bounded by the set of Nernst equilibrium potentials.

Equation 7.74 can also be rewritten in terms of V_m^o and G_m, as follows:

$$J_m = \left(\sum_n G_n\right)V_m - \sum_n G_n V_n = G_m\left(V_m - \sum_n \frac{G_n}{G_m} V_n\right) = G_m(V_m - V_m^o). \tag{7.78}$$

Thus, the parallel network of channels has a Thévenin's equivalent network as shown on the right side of Figure 7.24. This equivalent network of the membrane expresses graphically the voltage-current characteristic embodied in Equation 7.78.

For different cell types, different ionic species take on more or less importance in determining the resting potential. However, one ion, potassium, is ubiquitously important in determining the resting potential. Sodium usually has a small effect, and other ions may or may not make an appreciable contribution. Thus, in most cases we shall represent the relation between the resting potential and the ionic currents by an equivalent network with three branches (Figure 7.26). One branch represents the passive transport of potassium ions, another the passive transport of sodium ions, and the third the

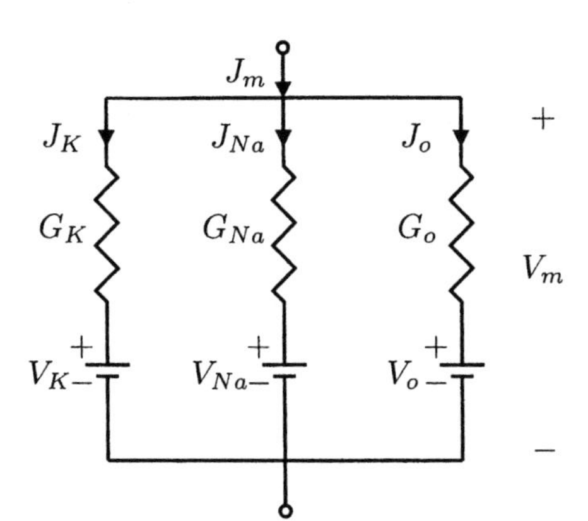

Figure 7.26 Equivalent network model of a cell that has three independent ion channels.

Table 7.3 Resting values of ionic conductances and concentrations in the giant axon of the squid.

Ion	G_n (S/cm^2)	G_n/G_m	c_n^o/c_n^i	V_n (mV)
K^+	3.7×10^{-4}	0.55	0.05	−72
Na^+	1×10^{-5}	0.016	9.8	+55
Leakage	3.0×10^{-4}	0.44	—	−49

passive transport of all other ions. This *other* branch may represent Cl^- current in some cells, Ca^{++} current in other cells, etc. In isolated squid giant axons, this other branch is called *leakage* and represents not only transport of ions other than sodium and potassium, but also ion transport through leakage pathways caused by the isolation of the axon. In general, the *other* channel can be regarded as a Thévenin equivalent network for all passive transport that contributes ionic membrane current in addition to sodium and potassium.

It is instructive to examine the relation of the resting potential to the membrane ionic conductances and the Nernst equilibrium potentials for the different ions. Table 7.3 shows typical values for the giant axon of the squid in normal seawater. The resting potential consistent with the parameters given in the table is −68 mV. The relation between the membrane potential and the Nernst equilibrium potentials is depicted in the diagram of potential shown in Figure 7.27. The membrane potential lies near but somewhat above the Nernst equilibrium potential for potassium. This property is typical of most cells investigated. We summarize the fact that there is a potential across the membrane at rest by saying the membrane is *polarized* at rest. If the membrane potential exceeds its normal resting value, we say the membrane is *depolar-*

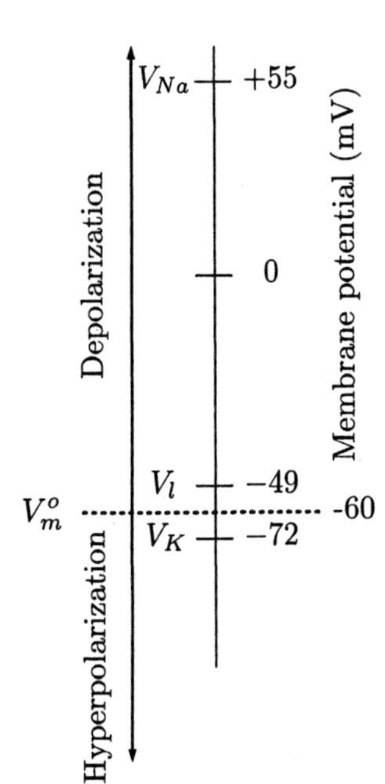

Figure 7.27 Diagram of potential for the squid giant axon.

ized. If the membrane potential is below its normal resting value, we say the membrane is *hyperpolarized.*

Model 2 explicitly allows for the observation that membranes are permeable to more than one ion. What does this model predict about the relation between the resting membrane potential and the concentrations of ions? Let us consider the dependence of the resting membrane potential on the external potassium concentration. For this case, we substitute the expression for the Nernst equilibrium potential for potassium into Equation 7.76 to obtain

$$V_m^o = \frac{RT}{F}\left(\frac{G_K}{G_m}\right)\ln\left(\frac{c_K^o}{c_K^i}\right) + \sum_{n\neq K}\frac{G_n}{G_m}V_n.$$

We isolate the effect of c_K^o on V_m^o by expressing the resting potential as follows:

$$V_m^o = 59\frac{G_K}{G_m}\log_{10} c_K^o + A\ [\text{mV}],$$

where A depends upon c_K^i and both the internal and the external concentrations of all ions except potassium, but is independent of c_K^o. By isolating the dependence of the resting potential on the external potassium concentration, we can see that this model predicts that the resting membrane potential depends linearly on the logarithm of the concentration of external potassium and that the slope of the line is less than or equal to that of a potassium

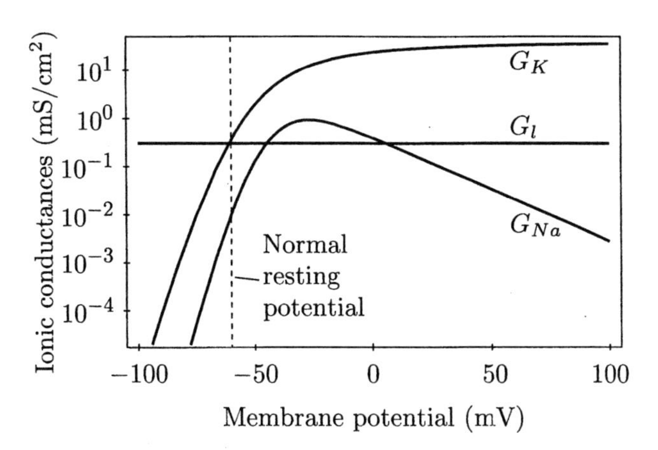

Figure 7.28 Dependence of the steady-state values of ionic conductances on the membrane potential for the giant axon of the squid according to the Hodgkin-Huxley model. The normal value of the resting potential, i.e., the membrane potential at the normal value of the extracellular potassium concentration, is indicated with a dashed line.

electrode (59 mV/decade at 24°C). Thus, this model can account for the measurements at fairly high concentrations of external potassium, but not for the change of slope of the dependence of V_m^o on c_K^o that occurs at low concentrations (Figure 7.22).

7.5.4 Model 3: Independent Passive Voltage-Gated Ion Channels

The saturation in the resting potential as the external potassium concentration is decreased results primarily from the voltage-dependent properties of the ionic conductances, which are discussed elsewhere (Weiss, 1996, Chapter 4). The ionic conductances to both sodium and potassium, and in some cells to other ions as well, are not constant, but explicitly depend on the value of the membrane potential.[7] The dependence of ionic conductances on membrane potential is indicated in Figure 7.28 for the Hodgkin-Huxley model, which is discussed elsewhere (Weiss, 1996, Chapter 4), based on direct measurements of the conductances in squid axons. Note that when the membrane potential is near its normal resting value, the potassium conductance of the squid giant axon is approximately equal to the leakage conductance, which is not voltage dependent; the sodium conductance is more than a factor of ten smaller than either the potassium conductance or the leakage conductance. As the membrane potential increases, the potassium conductance increases

7. The ionic conductances have also been found to depend explicitly on the ion concentrations, but this effect is less important in explaining the saturation of the resting potential on concentration than is the effect of the membrane potential on the ionic conductances. The explicit effect of ion concentration on ion conductances is omitted from the treatment in this section.

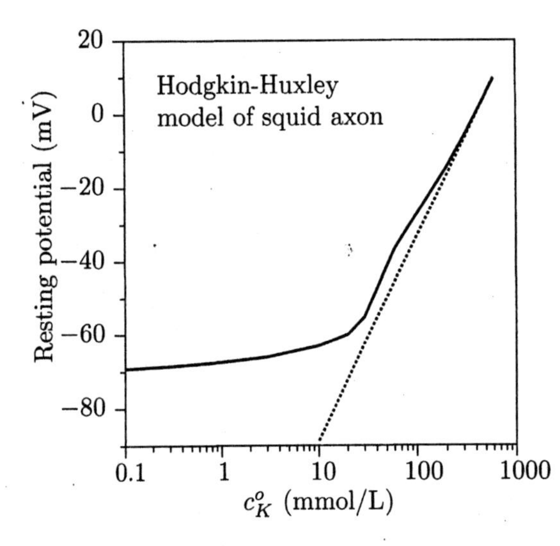

Figure 7.29 Dependence of resting potential on the external potassium concentration according to the Hodgkin-Huxley model. The dashed line shows the Nernst equilibrium potential for potassium

monotonically and the sodium conductance increases to a maximum value and then decreases, but the leakage conductance remains constant. For large membrane potentials, the potassium conductance is appreciably larger than either the leakage conductance or the sodium conductance. Hence, for such large values, the membrane potential of the squid axon should approach that of a potassium electrode.

Thus, if the voltage dependence of each conductance is known, we can compute the resting potential from the following version of Equation 7.75:

$$\sum_n G_n(V_m^o)(V_m^o - V_n) = 0, \tag{7.79}$$

which takes the voltage dependence into account. Because the conductances depend upon membrane potential in complex manners (Weiss, 1996, Chapter 4), this equation cannot be solved algebraically, but it can be solved numerically. For each value of ion concentration, the appropriate value of the Nernst equilibrium potential is used in Equation 7.79 and V_m^o is found iteratively such that the left-hand side of the equation equals 0. The dependence of the resting potential on external potassium concentration, computed in this manner, is shown in Figure 7.29 for the voltage-dependent conductances shown in Figure 7.28. Although the conductances are not explicit functions of concentration in this model, they are implicit functions of concentration, because as the concentration is changed, the membrane potential changes. The values of the conductances, normalized to the total membrane conductance, are shown as a function of concentration in Figure 7.30. At a high concentration of potassium, the membrane potential is large (more positive) and the potas-

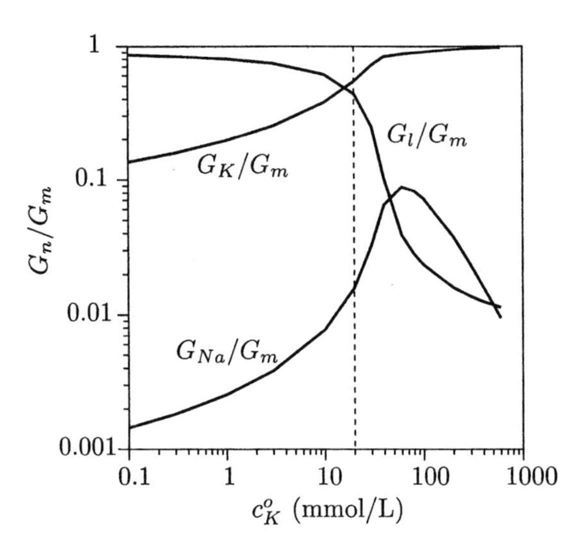

Figure 7.30 Dependence of ionic conductances on the external potassium concentration according to the Hodgkin-Huxley model. The ionic conductances are normalized to the total membrane conductance. In this model, the conductances are explicit functions of membrane potential, but not of concentration. The concentration dependence shown here results because the membrane potential depends upon concentration and the conductances depend upon membrane potential.

sium conductance is a large fraction of the membrane conductance. Therefore, the membrane potential is near the potassium equilibrium potential. At lower potassium concentration, the membrane potential is lower and the potassium conductance is a smaller fraction of the total membrane conductance. Hence, the resting membrane potential departs from that of a potassium electrode. The slope of the relation of membrane potential to the logarithm of the potassium concentration becomes very small.

As indicated in Figure 7.31, with the same voltage-dependent conductances, computation of the dependence of resting potential on sodium concentration indicates that there is a small increase in the resting potential as the external sodium concentration is raised. Thus, the dependence of resting potential on ion concentrations can be explained by Model 3, shown in Figure 7.32, which has conductances that depend on V_m^o. A model of this type can, in principle, account for data such as those shown in Figures 7.22 and 7.23.

7.5.5 Molecular Basis of Passive Ion Transport Through Channels

The type of passive ion transport described thus far in this chapter is mediated by ion channels. Ion channels are formed by macromolecules that span the lipid bilayer. These macromolecules have multiple discrete states, some of which allow ion transport, while others do not. Techniques have been developed to record the ionic currents through individual ion channels, and the ion channel macromolecules have been isolated for several types of channels. The electrophysiological behavior of these channels and their molecular structure

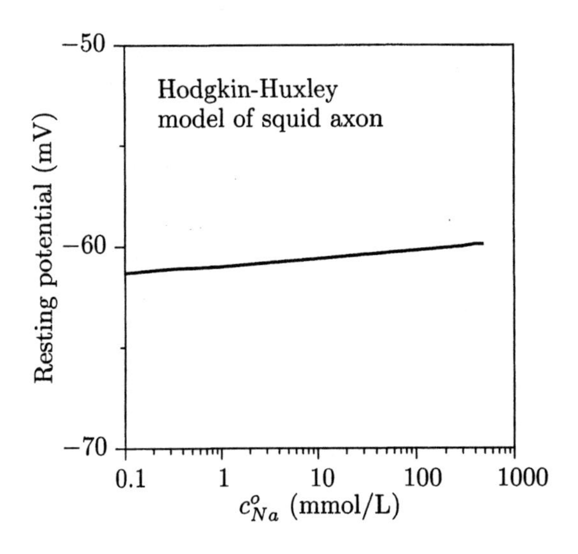

Figure 7.31 Dependence of resting potential on the external sodium concentration according to the Hodgkin-Huxley model.

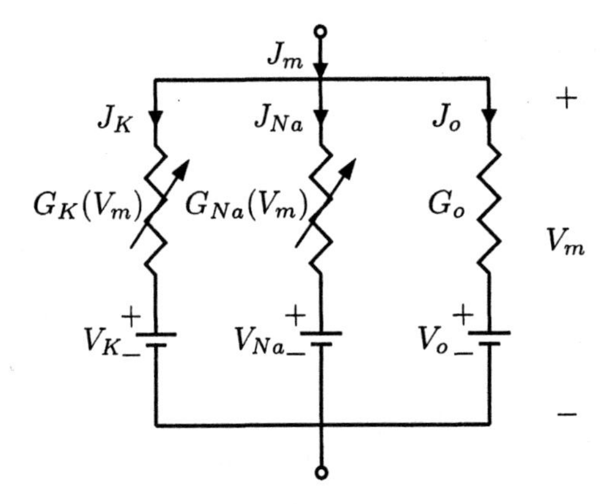

Figure 7.32 Equivalent network model of a cell that has three independent ion channels, two of which have voltage-dependent conductances.

are described elsewhere (Weiss, 1996, Chapter 6). A further analysis of resting potential requires characterization of individual ionic channels to determine their contribution to the resting potential.

7.6 Inadequacy of Passive Ion Transport Models

7.6.1 Instability of the Resting Potential

There is an inherent difficulty with passive ion transport models of the type discussed thus far. Since the Nernst equilibrium potentials for different ions differ, there exist ions for which V_m^o does not equal the Nernst equilibrium potential. Thus, all the ions cannot simultaneously be in electrodiffusive equilibrium. Any ion for which $G_n \neq 0$ and for which $V_n \neq V_m^o$ will not be in electrodiffusive equilibrium, and there will be a net passive flux of that ion down its electrochemical potential gradient. This flux is in the direction to reduce the difference in electrochemical potential. Thus, the concentration difference and the potential difference across the membrane will decrease. But is the magnitude of this flux appreciable in the lifetime of the cell? Perhaps it is only a trickle and does not influence cell concentrations much over a lifetime. In this scheme, a cell is created with a mature concentration of ions that trickles away during the cell's lifetime. This scenario raises the puzzling question of how the concentration gradients were established in the first place.

We need a quantitative analysis of flux magnitudes to determine whether the flux of ions is appreciable in the lifetime of a cell. We can estimate this flux for the squid giant axon using the values given in Table 7.3. The currents carried by potassium and sodium down their electrochemical potential gradients are

$$J_K = G_K(V_m^o - V_K) = 0.37 \times 10^{-3}(-60 + 72) \times 10^{-3} \approx +4\,\mu\text{A/cm}^2 \qquad (7.80)$$

and

$$J_{Na} = G_{Na}(V_m^o - V_{Na}) = 1 \times 10^{-5}(-60 - 55) \times 10^{-3} \approx -1\,\mu\text{A/cm}^2. \qquad (7.81)$$

These estimates give the order of magnitude and the direction of flow of potassium and sodium ions at rest. There is a net outflow (or efflux) of potassium and a net inflow (or influx) of sodium. Despite the fact that the potassium conductance exceeds the sodium conductance at rest by almost three orders of magnitude, sodium is further from equilibrium than potassium (Figure 7.27); therefore, the currents carried by these two ions differ by less than an order of magnitude for the parameters given here. From the current densities we can estimate the fluxes for each of these ions as follows:

$$\phi_K = \frac{J_K}{F} \approx 10^{-5} \times 4 \times 10^{-6} = 40\,\text{pmol/(cm}^2 \cdot \text{s)}, \qquad (7.82)$$

and

$$\phi_{Na} = \frac{J_{Na}}{F} \approx -10^{-5} \times 1 \times 10^{-6} = -10\,\text{pmol/(cm}^2 \cdot \text{s)}. \qquad (7.83)$$

Thus, at rest, on the order of 10 picomoles of sodium and potassium flow down their electrochemical potential gradients through a square centimeter of membrane each second. Is this an appreciable flux, or is it merely a trickle in the life of a squid? In other words, at this flux magnitude, how long would

Table 7.4 Net flux of ions across the membranes of nerve axons during a propagated action potential (Cohen and De Weer, 1977). The ion fluxes are given per action potential.

Preparation	K^+ efflux	Na^+ influx
	(pmol/cm^2)	
Loligo forbesi axon	3.0	3.5
Loligo pealei axon	3.7	—
Sepia officinalis axon	3.6	3.8
Homarus nerve	4.1	5.2
Carcinus nerve	1.7–20	—
Rabbit vagus nerve	1	—

it take for the intracellular concentrations of ions to change appreciably? To answer this question, we compute the number of moles of an ion contained in a cell per unit surface area of the cell. This number is simply $c_n^i \mathcal{V}/A$, where $\mathcal{V}$ is the cell volume and A is the surface area of the cell. This result shows that if the flux is the same for cells of different size, small cells (for which the ratio $\mathcal{V}/A$ is small) will incur a larger change in concentration per unit time than will large cells.

For the sake of making a quantitative assessment of this effect, we shall make an estimate of the time it takes to make an appreciable change in concentration in a large animal cell, the giant axon of the squid. The squid giant axon is a cylindrical cell with a radius of about 200 μm. Therefore, the number of moles of potassium per unit of surface area is $c_K^i a/2$, where a is the radius. Using the data in Table 7.1, we find that the number of moles of potassium per square centimeter of membrane is about 2×10^{-5}. Thus, the concentration will change by about 10% in $10^6/20$ seconds or about half a day. Clearly this would be a problem for the squid at rest. But the squid uses his giant axon to conduct action potentials that trigger contractions of the mantle muscles that propel the squid through water.

As indicated in Table 7.4 and as discussed elsewhere (Weiss, 1996, Chapter 4), during the passage of each action potential there are additional leakages of sodium and potassium down their electrochemical potential gradients. Thus, the concentrations will tend to equilibrate more rapidly when the squid giant axon is carrying out its normal physiological function. The situation is even more precipitous in small cells, which have a smaller volume of stored ions and in which the time to change concentrations at rest is likely to be more that an order of magnitude shorter than in the squid axon. Thus, passive transport models of the type described earlier have a fatal flaw: they are not stable.[8] The passive leakage of ions reduces the concentration gradients, which in turn reduces the magnitudes of the Nernst potentials, which in turn reduces the membrane resting potential.

7.6.2 Instability of the Cell Volume

We have seen that if all the ions to which a membrane is permeable are not at equilibrium and if ion transport is passive, then the membrane potential will not be stable because the ion concentration gradients will run down. But suppose the membrane is permeable to only two ions and they obey the conditions for a Donnan equilibrium. Can this simple situation account for cellular resting potentials? The general answer to this question is no! We shall defer a detailed analysis of this question to Chapter 8, where the general question of cellular homeostasis is discussed. Suffice it to say that due to the presence of fixed intracellular charges, it is in general not possible to simultaneously satisfy conditions of electroneutrality, ionic equilibrium (in which transport is by passive mechanisms), and osmotic equilibrium for an animal cell whose membrane cannot sustain an appreciable hydraulic pressure difference.

7.7 Active Ion Transport

The intrinsic problem with the passive models described thus far is that ions are not necessarily in electrodiffusive equilibrium; that is, there is a net flow of ions so as to reduce the ion concentration gradients across the membrane. However, membranes do not contain just the type of passive transport mechanisms we have discussed thus far. They may contain several mechanisms that transport the same ion. For example, sodium is transported by several types of transport mechanisms that are functionally and structurally distinct. These include passive and active transport mechanisms described in this chapter as well as carrier-based exchange mechanisms in which the electrochemical potential gradient for sodium is used to transport other solutes (Chapter 8). No matter what the identity of the transport mechanisms, the condition that

8. In principle, a model such as the Bernstein model in which a membrane is permeable to only one ion and for which that ion is in equilibrium avoids this logical inconsistency. However, as we have seen, measurements show that most cellular membranes are permeable to more than one ion.

guarantees the constancy of the concentration of ion n is simply that when summed over all mechanisms, there is no net transport of ion n; that is

$$\sum_k \phi_n^k = 0 \quad \text{or} \quad \sum_k J_n^k = 0, \tag{7.84}$$

where ϕ_n^k and J_n^k are the flux and current density, respectively, for the transport of ion n by transport mechanism k.

The condition in Equation 7.84 guarantees that the concentrations in the cell will be constant. We call this *cellular quasi-equilibrium,* because the cell is in equilibrium in the sense that its solute concentrations are constant. However, a particular transport mechanism may have a nonzero flux even though the net flux of all mechanisms transporting an ion is zero. For example, we can have cellular quasi-equilibrium so that there is no net flux of any ion. However, a particular ion may not be in electrodiffusive equilibrium, so that there is a net flux due to electrodiffusion that must be compensated by an equal and opposite flux by some other mechanism.

Passive transport is down the electrochemical potential gradient. Therefore, to satisfy Equation 7.84 membranes must contain mechanisms that transport ions in a direction opposite to their electrochemical potential gradients. Such transport is called *active transport*. A variety of distinct mechanisms satisfying this definition of active transport are known to occur. In this section, we shall be concerned with *primary active transport* mechanisms in which the energy for actively transporting the ions comes from coupling the transport directly to an energy-yielding chemical reaction involving an energy-storage molecule, i.e., adenosine triphosphate or ATP. We shall consider one such active transport system, the so-called *sodium-potassium pump* ubiquitously found in the plasma membranes of animal cells. Other active transport mechanisms are described in Chapter 8. Before we consider the properties and molecular basis of active ion transport, we shall first describe model 4 of the resting potential of a uniform isolated cell in which active transport is assumed to maintain the ion concentrations.

7.7.1 Model 4: Model of Resting Potential, Including Both Active and Passive Transport

Model 4 of the voltage-current characteristic of a patch of membrane (Figure 7.33) contains two mechanisms of ion transport, one passive ion transport and the other active ion transport. Passive transport is represented in a manner identical to the representation in model 2. Passive transport of ion n has

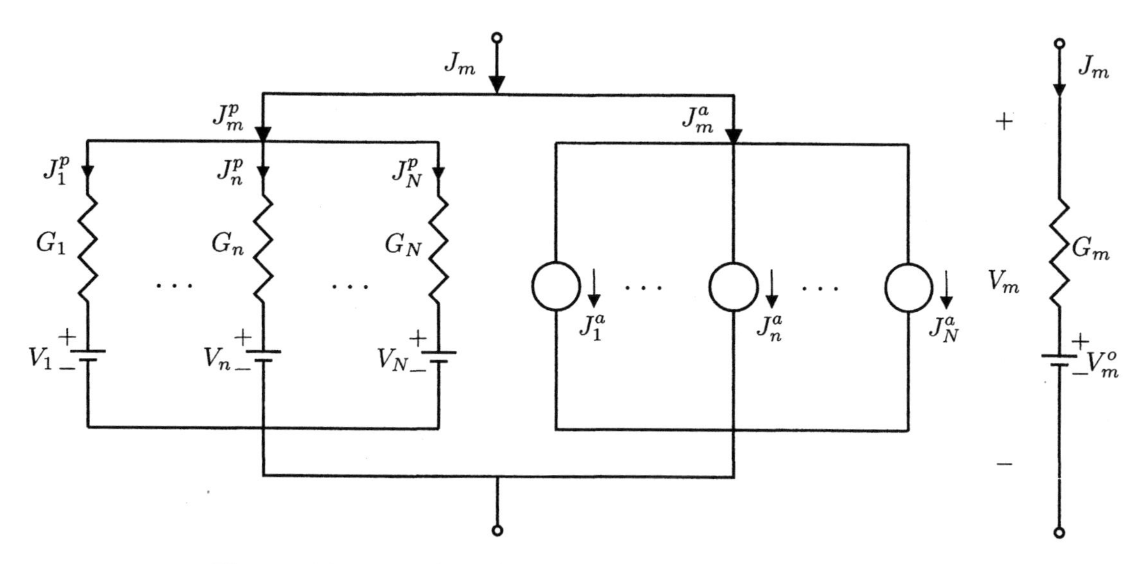

Figure 7.33 Network model of a membrane with passive and active ion transport.

a current density J_n^p; active transport of ion n is represented by a current density source J_n^a. The purpose of J_n^a is to maintain the concentration of ion n stable. If the concentration of ion n is constant, then there is no net transport of that ion, i.e.,

$$J_n^p + J_n^a = 0. \tag{7.85}$$

The condition given in Equation 7.85 guarantees the maintenance of ion concentration and, consequently, potential gradients across the membrane. There are several caveats. First, to guarantee constancy of ion concentrations, the summation of ionic currents must be taken over all mechanisms that transport that ion. In this section we shall consider only passive ion transport mediated by channels and active ion transport mediated by the sodium-potassium pump. Second, Equation 7.85 implies that there must be a nonzero active transport current density for each ion whose Nernst equilibrium potential does not equal the resting membrane potential. In this section we shall consider primarily sodium and potassium ions, but Equation 7.85 must hold for all permeant ions. Third, Equation 7.85 is an equilibrium condition. Under nonequilibrium conditions, the net flux ion n will not be zero and the cellular content of ion n will change. For example, we shall find that active transport is governed by a sequence of chemical reactions that take time to respond to concentration changes. So if the concentration in the cell is changed suddenly, these mechanisms will show a transient response in active current during which the equilibrium condition (Equation 7.85) will not necessarily hold.

For a uniform isolated cell at rest, the total membrane current density, J_m, is zero, which for model 4 implies that

$$J_m = J_m^p + J_m^a = 0$$

or

$$\sum_n G_n(V_m^o - V_n) + \sum_n J_n^a = 0,$$

from which we can solve for V_m^o and get[9]

$$V_m^o = \sum_n \frac{G_n}{G_m} V_n - \frac{1}{G_m} \sum_n J_n^a. \tag{7.86}$$

Using these relations we can once again write the voltage-current relation of the membrane as follows:

$$J_m = \sum_n G_n(V_m - V_n) + \sum_n J_n^a = G_m \left(V_m - \left(\sum_n \frac{G_n}{G_m} V_n - \frac{1}{G_m} \sum_n J_n^a \right) \right),$$

which can be written in the form

$$J_m = G_m(V_m - V_m^o), \tag{7.87}$$

which has the same form as that of model 2 except that V_m^o contains an additional term due to active transport.

The first term on the right-hand side of Equation 7.86 is the contribution of passive transport to the resting potential, while the second term is the contribution of active transport. We refer to the $\{J_n^a\}$ as *ion pumps*. It is important to distinguish two effects of ion pumps on the resting membrane potential: an indirect effect and a direct effect. First, the pumps maintain the concentration ratios, which in turn maintain the Nernst potentials, which maintain the resting potential. Therefore, the pumps have an indirect effect on the membrane potential. However, if $\sum_n J_n^a \neq 0$, then the net current flow due to active transport contributes an additional direct effect on the resting membrane potential.

---

9. If the conductances depend upon membrane potential, then this equation cannot be solved explicitly, but can be solved numerically if all the parameters are known. The voltage dependence of the conductances is an algebraically complicating factor in describing the effect of active transport on the resting potential. For the sake of conceptual simplicity, we shall ignore this voltage dependence in this section.

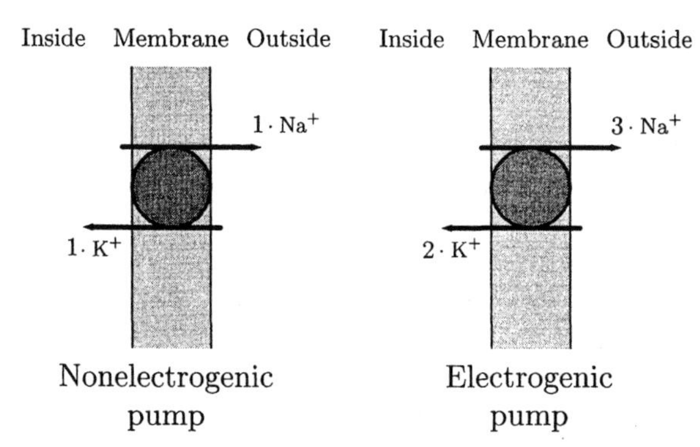

Figure 7.34 Schematic diagrams of nonelectrogenic and electrogenic pumps.

A pump for which $\sum_n J_n^a \neq 0$ is called an *electrogenic pump*. A *nonelectrogenic pump* is one for which $\sum_n J_n^a = 0$. A nonelectrogenic pump makes no direct contribution to the resting potential, although it does make an indirect contribution. An example of a nonelectrogenic pump is one for which the transport of two ion species is coupled so that no net charge is transferred across the membrane in one cycle of the pump (Figure 7.34). For example, every time one sodium ion is transported outward through the membrane, one potassium ion is transported inward. Thus, no net current flows through the membrane. An electrogenic pump transfers a net charge across the membrane. For example, a pump that transfers three sodium ions out for every two potassium ions transferred into a cell is electrogenic. As is made clear by Equation 7.86, the magnitude of the electrogenic effect depends not only on the net membrane current due to the pumps, but also on the membrane conductance.

7.7.2 Properties of Active Transport of Ions by the Sodium-Potassium Pump

Since individual ions are not labeled to indicate their manner of transport and since some ions are normally transported passively (e.g., they diffuse) in both directions across the membrane, separating active from passive transport posed some initial experimental difficulties. However, as indicated schematically in Figure 7.35, active transport exceeds passive transport for both the efflux of sodium and the influx of potassium, whereas passive transport exceeds active transport for both the influx of sodium and the efflux of potassium. Thus, an initial, although imperfect, basis for comparing properties of active transport with properties of passive transport was to examine, for example, the behavior of sodium efflux (predominantly transported actively) and to compare this with sodium influx (predominantly transported passively). As

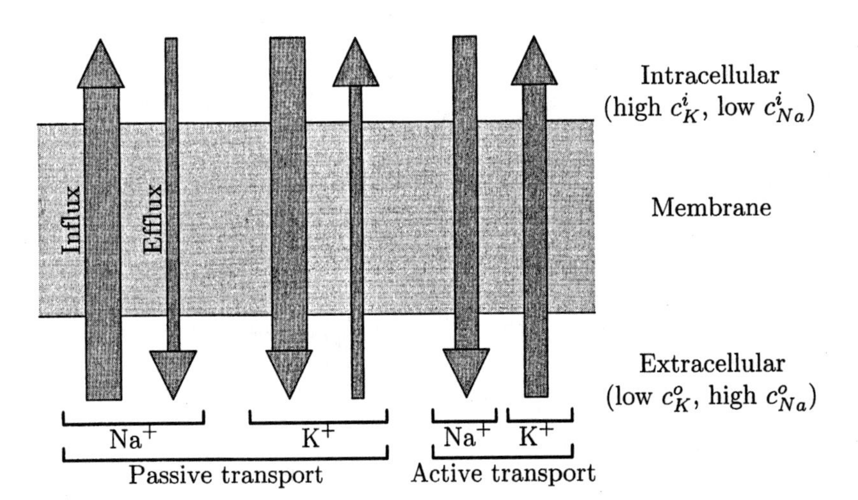

Figure 7.35 Active and passive ion traffic across a membrane. The width of each arrow indicates the flux magnitude in a qualitative manner.

properties of active transport via the sodium-potassium pump became better understood, methods for separation of active and passive flux components were improved.

7.7.2.1 Dependence on Metabolism

Studies of ion fluxes in invertebrate axons revealed that the efflux of sodium was dependent upon the existence of an intracellular store of ATP that is continually produced by cellular metabolism. For example, if axons are subjected to metabolic poisons that block the production of ATP, such as cyanide, dinitrophenol, or azide, the efflux of sodium is greatly reduced. Examples of such results are shown in Figures 7.36 and 7.37. In these experiments, cuttlefish nerve axons were loaded with radioactive sodium ($^{24}Na^+$) by placing the axons in radioactive sea water. The axons were washed with fresh seawater that was collected and counted at regular time intervals. The change in count rate with time was proportional to the efflux of radioactive sodium. Figure 7.36 shows that if the axon is bathed in seawater containing cyanide, which inhibits oxidative phosphorylation in the cell and thus decreases the production of ATP, the efflux of sodium is reversibly reduced. Similarly (Figure 7.37), if an axon is bathed in seawater containing dinitrophenol (DNP), which decouples respiration from phosphorylation of ADP and thus also reduces the production of ATP in the cell, the efflux of sodium is also reduced.

A particularly revealing experiment on the effects of metabolic inhibitors on active and passive sodium transport is summarized in Figure 7.38. In this experiment, the axon is placed in seawater containing radioactive sodium ($^{24}Na^+$) and then placed in a measurement cell and washed continually to re-

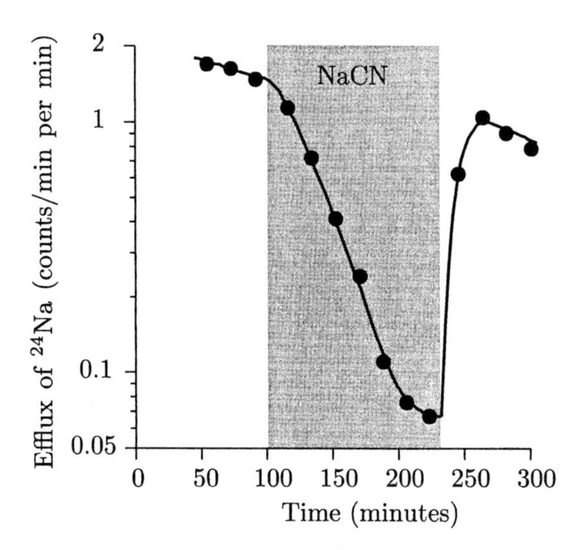

Figure 7.36 Effect of a metabolic poison (cyanide) on the efflux of sodium from a cuttlefish axon (adapted from Hodgkin and Keynes, 1955, Figure 4). The axon was in artificial seawater; during the interval shown, the artificial seawater contained 1 mmol/L of sodium cyanide (NaCN). The ordinate scale is logarithmic, so that an exponential reduction of the flux with time will plot as a straight line.

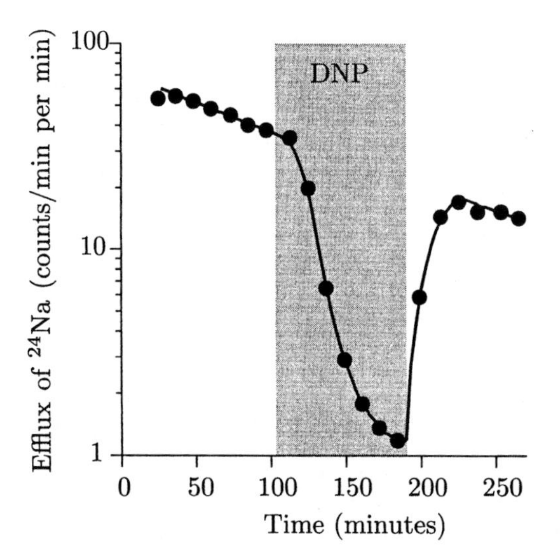

Figure 7.37 Effect of a metabolic poison, dinitrophenol (DNP) on the efflux of sodium from a cuttlefish axon (adapted from Hodgkin and Keynes, 1955, Figure 3). The axon was bathed in artificial seawater to which DNP, at a concentration of 0.2 mmol/L, was added in the indicated interval.

move the radioactive sodium of the effluent. The axon itself is counted, the count rate being proportional to the intracellular concentration of sodium. During interval 1 the axon is placed for 10 minutes in radioactive seawater. During interval 2 the axon is in normal seawater and the count rate is above the background rate, indicating that some radioactive sodium entered the axon during interval 1. The count rate decreases during interval 2, indicating that some sodium is being transported out of the axon. During interval 3 the axon is placed in radioactive seawater for 5 minutes and stimulated elec-

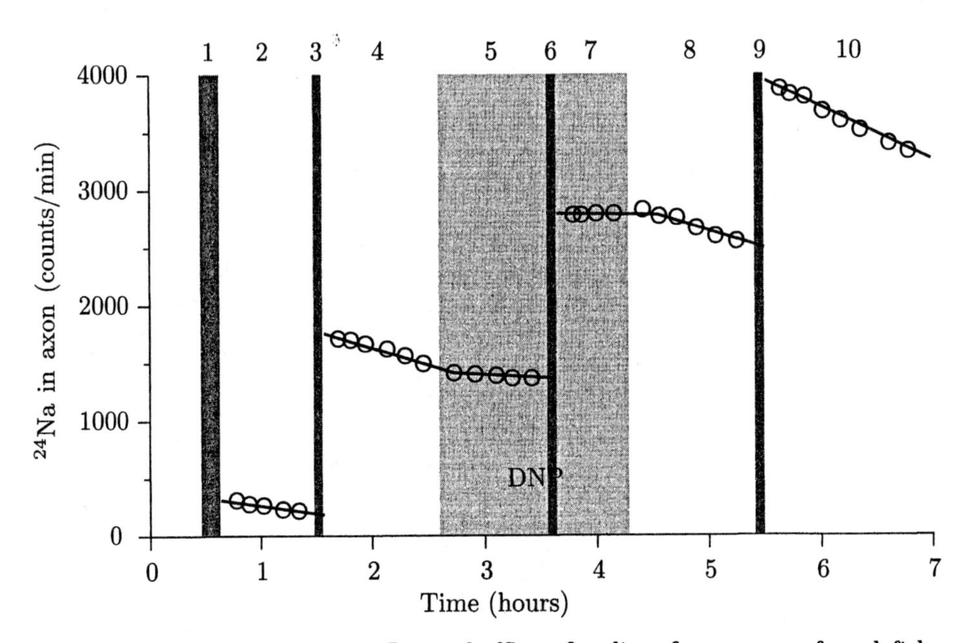

Figure 7.38 Effect of DNP on influx and efflux of sodium from axons of cuttlefish (adapted from Hodgkin and Keynes, 1955, Figure 9). During intervals 1, 3, 6, and 9 the axon is placed in a bath containing radioactive sodium. Interval 1 is 10 minutes long. The other three intervals are 5 minutes long, and the axon is stimulated electrically at fifty shocks per second to produce action potentials. During intervals 2, 4, 5, 7, 8, and 10 the axon is in a measurement chamber through which normal seawater (without radioactive sodium) is perfused. DNP (at a concentration of 0.2 mmol/L) is added to the seawater during intervals 5–7.

trically to produce action potentials. Each action potential is known to result in the net influx of sodium and efflux of potassium. At the end of interval 3 the count rate has increased substantially, indicating that there has been a substantial influx of sodium during this interval. During interval 4 the count rate decreases, indicating that sodium is being transported out of the axon.

At the onset of interval 5, DNP is added to the seawater. The slope of the count rate is decreased during interval 5, indicating a decrease in the efflux of sodium in the presence of DNP. During interval 6 the axon is placed in radioactive seawater containing DNP and stimulated electrically. The increase in the count rate at the end of interval 6 is about the same as the increase at the end of interval 3, demonstrating that the DNP has not affected the influx of sodium. During interval 7 the axon is in normal seawater containing DNP and the count rate is relatively constant, indicating that the efflux of sodium is low. When normal seawater (not containing DNP) is restored at the onset of interval 8, the count rate starts to decrease, indicating that sodium is being

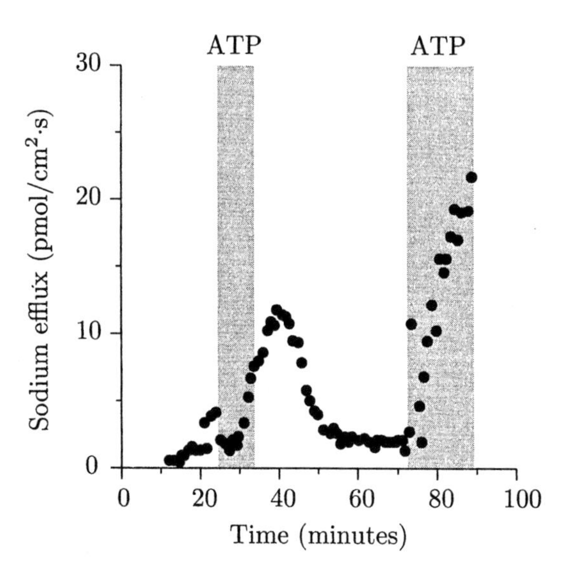

Figure 7.39 The efflux of sodium is plotted versus time while a solution is perfused through dialysis tubing inserted intracellularly into a giant axon of the squid (adapted from Mullins and Brinley Jr., 1967, Figure 3). Except during the shaded intervals shown, the dialysis solution contains no ATP. During the shaded intervals 5 mmol/L of ATP is added to the dialysis solution.

transported out of the axon. The cycle of measurements is repeated during intervals 8–10. This experiment demonstrates that DNP reduces sodium efflux appreciably without affecting sodium influx much.

Measurements of the concentration of ATP in invertebrate axons revealed that soaking the axons in these metabolic poisons indeed reduced the intracellular concentration of ATP (Caldwell, 1960). Furthermore, in axons that were perfused intracellularly by means of dialysis tubing, the intracellular concentration of ions and of ATP could be controlled experimentally. Perfusion with ATP-free solutions for extended periods of time reduced the sodium efflux dramatically. Reintroduction of ATP to the perfusion fluid increased sodium efflux, as shown in Figure 7.39. Thus, ATP is required for normal efflux of sodium. As indicated in Figure 7.40, the flux of sodium is a saturating function of the concentration of ATP in the cytoplasm of the axon. In these measurements, activation of the sodium-potassium pump for ATP is half maximal at an ATP concentration of about 0.175 mmol/L in this case. Since at rest the concentration of ATP in the squid giant axon is in the range of 3–5 mmol/L, these results suggest that the pump is essentially fully activated by ATP at rest.

7.7.2.2 Effect of Temperature

Blood is routinely withdrawn from human blood donors, stored for some time, and then transfused to an acceptor patient. In the process of blood management, it has been found that if blood is stored at low temperature, the erythrocytes tend to accumulate sodium and to lose potassium. If the blood

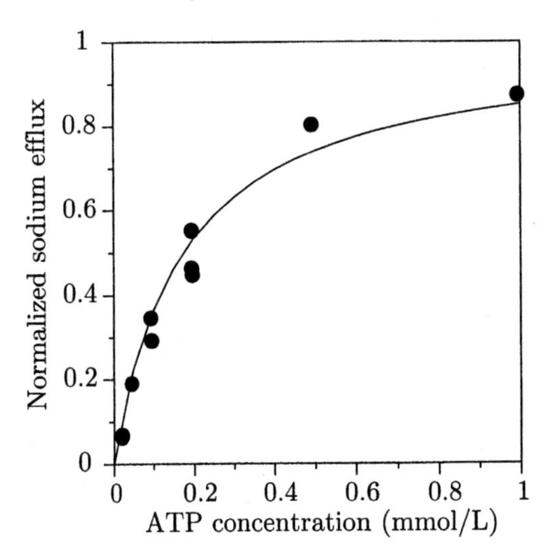

Figure 7.40 Dependence of sodium efflux on internal ATP concentration in squid giant axons (adapted from Beaugé and DiPolo, 1981b, Figure 3). The measured effluxes are normalized to the value of the efflux at an ATP concentration of 3 mmol/L. The curve is a rectangular hyperbola with an asymptotic value of 1 and a concentration at a normalized flux of 0.5 that equals 0.175 mmol/L. The extracellular solution contained in mmol/L: K^+ 10 and Na^+ 440. The dialysis solution contained in mmol/L: K^+ 310 and Na^+ 70.

is warmed to body temperature and if the plasma contains a metabolizable substrate such as glucose, the normal intracellular composition of the erythrocytes is restored (Maizels, 1949). In experiments on single-cell preparations, it has been shown that active transport is much more sensitive to changes in temperature than is passive ion transport. Typically, the Q_{10} for active transport of sodium and potassium is 3.3, whereas that for passive transport of ions is 1.1–1.4 (Hodgkin and Keynes, 1955). This difference explains, in a qualitative manner, the problems incurred with blood storage. At low temperature, active transport of sodium and potassium is inhibited more than is passive transport. Therefore, ion concentration differences run down as sodium leaks into the cells from the surrounding plasma and potassium leaks out. When the cells are warmed and supplied with a source of energy, active transport is restored, which restores the original concentration differences.

7.7.2.3 Cardiac Glycosides: Specific Blockers of the Sodium-Potassium Pump

A group of steroids called *cardiac glycosides* have been used for over two hundred years in the treatment of congestive heart failure. Administration of these drugs to patients with failing hearts has been known to increase the force of contraction of the heart. It was not until the 1950s (Schatzmann, 1953) that the site of action of these drugs was determined. The cardiac glycosides are specific inhibitors of the sodium-potassium pump. The different members of this family of drugs, which includes digitalis, ouabain, strophan-

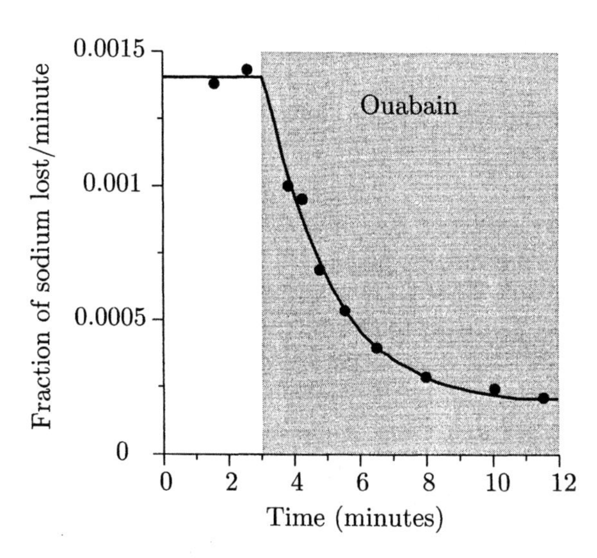

Figure 7.41 Effect of ouabain on sodium efflux from squid giant axons (adapted from Baker and Willis, 1972b, Figure 2). The concentration of ouabain in the solution was 10 μmol/L.

thidin, and others, have different potencies. Active efflux of sodium is blocked in a matter of minutes by micromolar concentrations of ouabain applied extracellularly (Figure 7.41); intracellular injection of ouabain at much higher concentrations produces no appreciable inhibition. Because of their specificity of action, these drugs have been enormously useful in studies of the sodium-potassium pump. The blockage of active transport of these ions allows assessment of the contribution of active transport to total flux across membranes. In fact, blockage of a flux component by cardiac glycosides is taken as strong evidence that that component is transported by the sodium-potassium pump. Furthermore, changes in flux caused by ouabain are used to measure the component of ion flux that is carried by the sodium-potassium pump.

7.7.2.4 Estimation of the Density of Pump Sites

That the action of ouabain extracellularly is very long lasting—i.e., the inhibition persists for tens of minutes after the drug is applied—has suggested that the drug binds to an extracellular site on the membrane. Therefore, examination of the kinetics of drug binding has been of interest, and the steady-state binding has been used to determine the density of pump-inhibiting binding sites. The amount of radioactive ouabain bound to either a collection of cells or to single cells has been measured in several cell types. By estimating the total surface area of cell membrane in such preparations and assuming that one molecule of ouabain binds to one pump site, estimates of pump densities, in units of pump sites/μm^2, have been found to be in the range of 500–5000 in nerve, muscle, kidney, and cultured heart cells, but less than 1 in erythrocytes (Landowne and Ritchie, 1970; Baker and Willis, 1972a; Venosa and Horowicz, 1981).

7.7.2.5 Dependence on Ion Concentration

The purpose of the sodium-potassium pump is to pump sodium out and potassium into cells so as to maintain the differences in concentrations of these ions across the membrane in the face of passive transport of these ions, which is in a direction so as to dissipate the concentration differences. Thus, we might expect that sodium efflux would depend upon intracellular sodium concentration and that potassium influx would depend upon extracellular potassium concentration. However, in principle the active transport of sodium and potassium could result from two separate mechanisms, one dependent on intracellular sodium that pumps sodium out, and the other dependent on extracellular potassium that pumps potassium in; both would depend upon the presence of ATP. However, this is not the case; i.e., the efflux of sodium and the influx of potassium are linked. Furthermore, both the influx of potassium and the efflux of sodium depend upon the concentration of both intracellular sodium and extracellular potassium.

The effect of extracellular potassium on the efflux of sodium and the influx of potassium is illustrated in the experiment shown in Figure 7.42. Samples of human erythrocytes were stored at a temperature of 2°C in a sodium solution. The low temperature was intended to inhibit the action of the sodium-potassium pump and to allow the cells to accumulate sodium and to lose potassium. The cells were washed, burst osmotically, and the Na^+, K^+, and hemoglobin concentrations measured. The Na^+ and K^+ concentrations were normalized to 5 mmol/L of hemoglobin, which is the normal concentration of hemoglobin in these cells. Since different samples contain different numbers of cells and since the Na^+, K^+, and hemoglobin concentrations should be proportional to the number of cells, this normalization should reduce the variability in measurements of Na^+ and K^+ concentrations due to the different numbers of cells in different samples. After measuring the composition for a sample of cold-stored cells, another cold-stored cell sample was placed in a sodium solution at a temperature of 37°C for one hour and the concentrations measured. The concentrations had not changed appreciably. The next sample of cold-stored cells was placed in a low-sodium solution for two hours, and then the population was split in two. Half the cells were placed in a test solution that consisted of the original sodium solution adjusted so

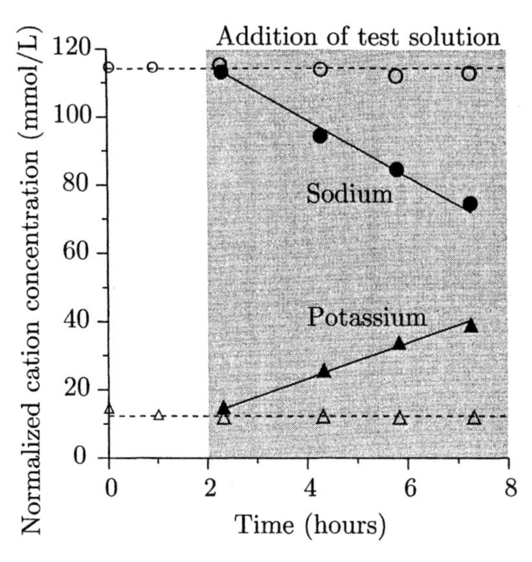

Figure 7.42 Linkage between sodium and potassium flux in human erythrocytes (adapted from Post and Jolly, 1957, Figure 1). After storage at 2°C in a sodium solution for nine weeks, the Na^+ and K^+ contents of the cell population were measured at time 0. The cells were then stored for various time intervals (indicated on the abscissa) at 37°C. The circles represent measurements of Na^+ concentration, and the triangles represent measurements of K^+ concentration. The small open symbols are measurements in the low-sodium solution. The large filled symbols are measurements with potassium in the extracellular solution; the large open symbols are measurements in control solutions that contained no potassium.

that KCl contributed 21 mmol/L. The other half, that was a control group, was placed in a test solution that did not contain KCl. The measurements were repeated with several samples for several hours. The results show that when the test solution contained only Na^+ and no K^+ ions, the internal Na^+ and K^+ content remained the same: intracellular Na^+ was high, and K^+ was low. If the test solution contained potassium, the cells lost sodium and accumulated potassium. Thus, the influx of potassium and the efflux of sodium were linked and did not occur in the absence of extracellular potassium. A similar experiment showed that both fluxes were also dependent on the presence of intracellular sodium.

Measurements such as those illustrated in Figure 7.42 give important insights into the qualitative nature of the fluxes generated by the sodium-potassium pump and their dependence on ion concentrations. However, the quantitative characterization of pump transport that is needed to develop mathematical models of pump kinetics requires systematic measurements with stringent control of electrical and chemical variables. Special attention

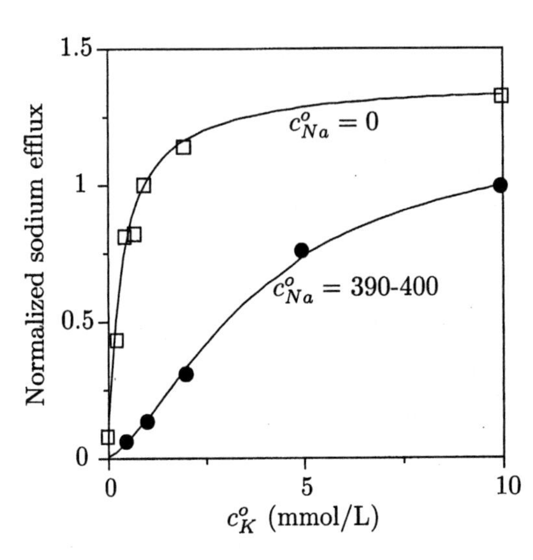

Figure 7.43 Dependence of sodium efflux on external potassium concentration in squid giant axons (adapted from Rakowski et al., 1989, Figure 6A). Ion channel blockers were used to reduce passive transport, and the increment of sodium efflux caused by the cardiac glycoside dihydrodigitoxigenin was measured. Results are shown for external sodium concentrations of 0 and 390–400 mmol/L. The data are plotted on a normalized flux scale; the normalization is the same for both curves.

must be paid to the elimination of contaminating fluxes due to other transport mechanisms.

Such controlled measurements have led to several generalizations about the role of ion concentration in pump activation. For example, sodium efflux depends on the extracellular concentration of potassium as illustrated by the measurements from squid giant axons in Figure 7.43. If extracellular potassium is removed, the coupled efflux of sodium and influx of potassium are reduced to near zero. An increase in external potassium concentration increases the sodium efflux until near 10 mmol/L, where the effect of sodium efflux on external potassium concentration saturates. Since this is the normal range of external potassium concentration in squid axon, the sodium-potassium pump is fully activated, i.e., shows the maximum flux, at the normal extracellular concentration of potassium. The results also show that reducing extracellular sodium concentration from near 400 mmol/L, which is near the normal extracellular concentration, increases the sodium efflux. In general, it has been found that potassium has a number of extracellular agonists with which it competes for activation of the sodium pump. A variety of evidence suggests that the effectiveness of different extracellular cations to activate the pump satisfies the sequence $Tl^+ > K^+ > Rb^+ > NH_4^+ > Cs^+ > Li^+ > Na^+$.

Similarly, the sodium pump is activated by intracellular sodium; in the absence of intracellular sodium, the coupled efflux of sodium and influx of potassium are near zero. In contrast to the relatively nonspecific dependence of sodium efflux on extracellular cations, the sodium-potassium pump is activated rather specifically by intracellular sodium; that is, sodium cannot be readily replaced by other monovalent cations. These results show that both the accumulation of potassium extracellularly and the accumulation of sodium intracellularly, which reduce the differences in concentration of these two ions across the membrane, increase the efflux of sodium and the influx of potassium. Both of these changes are in the direction to increase the concentration difference of these ions across the membrane.

7.7.2.6 A Phenomenological Model of the Dependence of Active Ion Fluxes on Ion Concentrations and on the Concentration of ATP

For each mole of ATP split, the pump normally transports ν_{Na} moles of sodium outward and ν_K moles of potassium inward. As we shall see, the pump kinetics are complex. However, we know that the pump rate increases with an increase in the intracellular sodium, intracellular ATP (Figure 7.40), and extracellular potassium (Figure 7.43) concentrations. A simple phenomenological model (Lew and Bookchin, 1986) that captures this behavior in an approximate manner is to assume that each molecule of $(Na^+\text{-}K^+)$-ATPase binds 1 ATP molecule, ν_{Na} molecules of Na^+, and ν_K molecules of K^+ independently and that the pump rate α_{ATP} has the following dependence on these concentrations:

$$\alpha_{ATP}(t) = \alpha_{max}\left(\frac{c^i_{ATP}(t)}{c^i_{ATP}(t) + K^i_{ATP}}\right)\left(\frac{c^i_{Na}(t)}{c^i_{Na}(t) + K^i_{Na}}\right)^{\nu_{Na}}\left(\frac{c^o_K(t)}{c^o_K(t) + K^o_K}\right)^{\nu_K}, \quad (7.88)$$

where $\alpha_{ATP}(t)$ is the pump rate in units of moles of ATP split per unit time per unit area of membrane; α_{max} is the maximum pump rate; K^i_{ATP}, K^i_{Na}, and K^o_K are dissociation constants; and $c^i_{ATP}(t)$ is the intracellular concentration of ATP. Since it is known that the pump rate is reduced by an increase of the intracellular concentration of potassium and by an increase in the extracellular concentration of sodium, Equation 7.88 can be modified to reflect this behavior by replacing K^i_{Na} with $K^i_{Na}(1 + C^i_K/K^i_K)$ and K^o_K by $K^o_K(1 + C^o_{Na}/K^o_{Na})$. The fluxes of sodium and potassium carried by the pump are

$$\phi^a_{Na}(t) = \nu_{Na}\alpha_{ATP}(t) \quad \text{and} \quad \phi^a_K(t) = -\nu_K\alpha_{ATP}(t). \quad (7.89)$$

7.7.2.7 Electrogenicity of the Sodium-Potassium Pump

Measurements have consistently shown that the active efflux of sodium exceeds the active influx of potassium, suggesting that the sodium-potassium pump is electrogenic. We can estimate the order of magnitude of the direct contribution of active transport to the resting potential of the squid giant axon using measurements of the active transport of sodium and potassium at rest. Although simultaneous measurements of the active transport of sodium and potassium in squid axon are not available, we can get some indication of the order of magnitude of this potential by using average measurements (Baker, Blaustein, Keynes, et al., 1969), which suggest that the fluxes of sodium and potassium by active transport are about $+36.3 \pm 5.1$ and -17.9 ± 1.9 pmol/cm$^2 \cdot$ s, so that using Faraday's constant, the current densities of sodium and potassium are about $+3.6$ and $-1.8\ \mu$A/cm^2, respectively. From Equation 7.86 we can see that the contribution of active transport to the resting potential is $(V_m^o)_a = -\sum_n J_n^a/G_m$, which, using Table 7.3, gives $(V_m^o)_a \approx -(3.6 - 1.8\,\mu\text{A/cm}^2)/(0.68\,\text{mS/cm}^2) = -2.6\,\text{mV}$. This rough calculation suggests that active transport in the squid axon might account for 2–3 millivolts of *hyperpolarization* at rest.

Figure 7.44 illustrates the results of an experiment to test this notion. Two experimental variables are used to see if inhibition of the sodium-potassium pump causes a depolarization. First, addition of the cardiac glycoside strophanthidin causes a depolarization which, in this study, averages 1.4 mV. Concurrently the rate of efflux of sodium is reduced by an order of magnitude. Both these results are consistent with an inhibition of the sodium-potassium pump. The depolarization results because the addition of strophanthidin eliminates the hyperpolarization that is caused by active transport. The effect of a change in external potassium concentration is more complex, since such a change will affect the direct contributions to the resting potential of both passive (see Section 7.5) and active (see Figure 7.43) transport. Furthermore, at low extracellular potassium concentrations both of these contributions are expected to be small, but to have opposite polarities. Reduction of the extracellular concentration of potassium makes the Nernst equilibrium potential more negative, which produces a hyperpolarizing component of the resting potential by passive transport mechanisms. In contrast, reduction of potassium concentration inhibits the sodium-potassium pump, which removes the pump-induced hyperpolarization and thus depolarizes the membrane. The measurements show that in the absence of strophanthidin a reduction in extracellular potassium decreases the efflux of sodium, but causes a net hyperpolarization. However, during a strophanthidin-induced inhibition of sodium efflux, a reduction of the extracellular potassium causes an even larger hyperpolarization and a return to 10 mmol/L potassium causes a net depolarization above that in the absence of strophanthidin. Thus, in the absence of strophanthidin the small hyperpolarization caused by potassium reduction results from a hyperpolarization due to passive transport that is partly offset by a depolarization caused by reduction of active transport. This experiment demonstrates the complexity of interpreting experiments in which both active

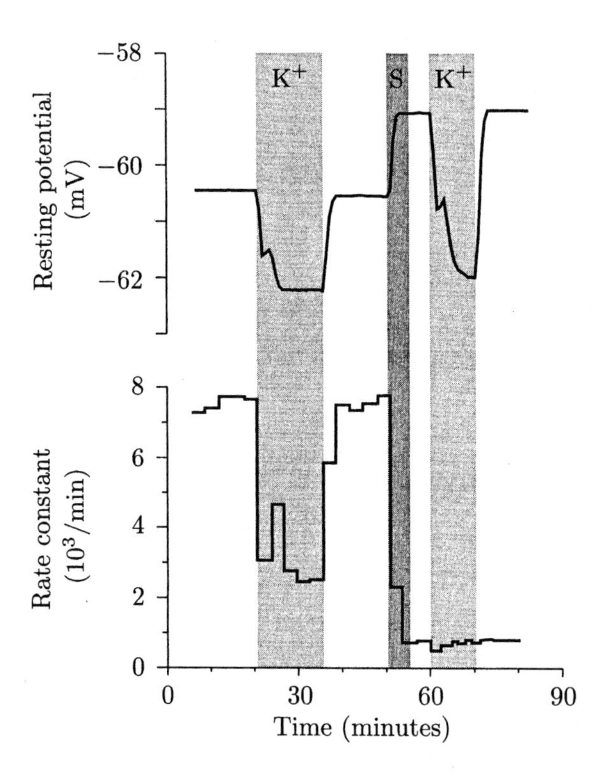

Figure 7.44 The effect of extracellular potassium concentration and of strophanthidin on sodium efflux and resting potential of the squid giant axon (adapted from De Weer and Geduldig, 1973, Figure 1). The axon was injected with radioactive ^{22}Na and immersed in seawater containing 10 mmol/L potassium except during the intervals noted, when either the potassium concentration was reduced to 1 mmol/L (K^+) or the solution contained 10^{-5} mmol/L of strophanthidin (S). The upper trace shows the resting potential, and the lower trace shows the rate constant for sodium efflux.

and passive transport are present and demonstrates that there is a small electrogenic component of the resting potential in squid axon at rest.

Equation 7.86 suggests that if the pump were stimulated to increase its pump rate, a larger electrogenic effect could be produced. This effect is illustrated in the experiments described in Figure 7.45. In these experiments, large neurons (200 μm in diameter) in the snail, *Helix asperasa,* were impaled with three micropipettes. One pair of micropipettes was used to inject ions into the cell cytoplasm. By passing current between these two micropipettes, ions were injected into the cell without passing current through the cell membrane, which would cause a change in membrane potential. The two current-carrying micropipettes were filled with different cations so that the direction of current flow could be used to control which cation was injected. For example, one micropipette was filled with sodium acetate and the other with lithium acetate. When the sodium acetate micropipette was the anode, sodium was injected. When the lithium acetate micropipette was the anode, lithium was injected. A third micropipette was used to measure the membrane potential with respect to an extracellular reference electrode. The results show that injection of Na^+ causes a large hyperpolarization of the membrane that does not oc-

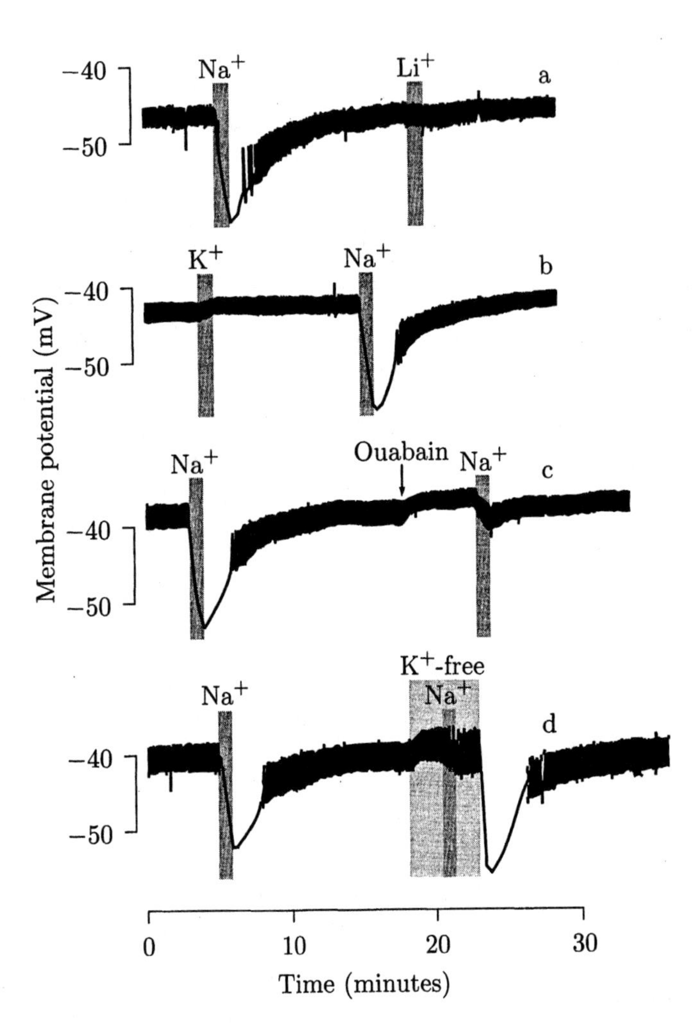

Figure 7.45 Effect on the membrane potential of the injection of ions and of other treatments known to affect active transport (adapted from Thomas, 1969, Figures 3–5). The membrane potential has been recorded using low bandwidth, which attenuated the action potentials and accounts for the width of the traces. Traces a and b show that injections of Na^+ produce a hyperpolarization of about 10 mV, but neither Li^+ nor K^+ does. Trace c shows that the hyperpolarization caused by Na^+ injection is blocked by ouabain application. Trace d shows that removal of extracellular K^+ blocks the hyperpolarization caused by Na^+ injection, but restoration of extracellular K^+ produces the hyperpolarization.

cur with either Li^+or K^+ injection. Furthermore, either application of ouabain extracellularly or removal of extracellular potassium blocks the hyperpolarization response to sodium injection. Thus, the effect of sodium injection into these neurons is consistent with the hypothesis that sodium injection increases the intracellular concentration of sodium (with an estimated increase of 6 mmol/L for the results shown), which stimulates the sodium pump to increase the coupled transport of sodium and potassium across the membrane in an electrogenic manner. The net current flow resulting from the fluxes of sodium and potassium must be outward so as to cause a hyperpolarization of the membrane.

To test the latter notion, additional electrodes were used to measure the membrane current and the intracellular sodium concentration, while the po-

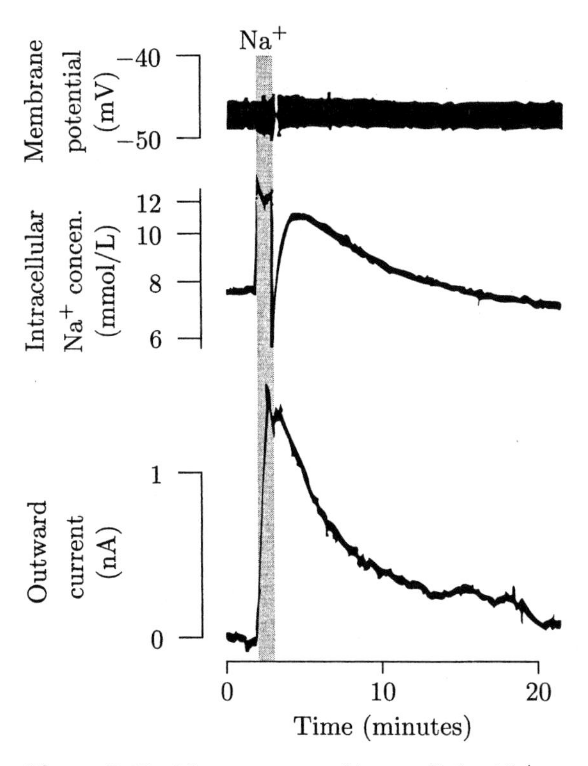

Figure 7.46 Measurement of intracellular Na^+ concentration and pump current that results from an intracellular injection of Na^+(adapted from Thomas, 1969, Figure 11). The Na^+ was injected during the time interval that is shown shaded by passing 31.3 nA for 1 minute between two intracellular micropipettes. The membrane potential was maintained constant by a feedback system that maintained low-frequency components of the potential constant; the width of the trace indicates the presence of action potentials that were not prevented by the feedback system. The intracellular sodium concentration was measured with an ion-selective electrode. During the injection of Na^+, the output of the sodium electrode (which had a very large input impedance) and the amplifier produced an artifact, obviating the measurement of concentration.

tential across the membrane was held near its resting value (Figure 7.46). Injection of Na^+ led to a transient increase in intracellular sodium concentration and an outward current through the membrane. The current rose linearly during the injection and decayed approximately exponentially after termination of the injection. Both the changes in sodium concentration from the resting value and the pump current were proportional and exponential functions of time. Thus, the pump current is proportional to the increment in sodium concentration above its resting value. These results are a convincing demonstration that the sodium-potassium pump in these cells is electrogenic.

7.7.2.8 Stoichiometry of the Pump

The sodium-potassium pump transports sodium out and potassium into the cell and hydrolyzes ATP. Thus, we can imagine that the pump mechanism has three substrates: sodium, potassium, and ATP. In order to determine the kinetic mechanism of this pump, it is important to determine the stoichiometry of the relations among these three substrates, i.e., the number of molecules of sodium and potassium transported per molecule of ATP hydrolyzed. This objective has proved to be challenging. First, it is somewhat difficult to separate the transport of sodium and potassium by the sodium-potassium pump from other transport mechanisms. Second, determining the stoichiometric ratios requires measurement of three quantities, which is even more difficult to do accurately in a single cell. Third, because the transport mechanism is dependent on so many variables (e.g., intracellular and extracellular concentrations of sodium and potassium, intracellular concentrations of magnesium and ATP, etc.), it has been difficult to measure the fluxes and ATP consumption in different cells with sufficient accuracy to allow meaningful ratios to be taken across measurements. Nevertheless, measurement of active transport of sodium and potassium in erythrocytes showed that the two fluxes were linked and that the ratio of sodium efflux to potassium influx was about 3:2 (Post and Jolly, 1957). Later measurements showed that the ratio of the number of sodium ions pumped out to the number of ATP molecules hydrolyzed was about 3:1 (Garrahan and Glynn, 1967e). This result has led to the notion that the pump transports three sodium ions out and two potassium ions in for every molecule of ATP hydrolyzed. While this stoichiometry is supported moderately well by measurements in erythrocytes (see quote at beginning of chapter), measurements in single cells, such as squid giant axons, have given somewhat variable results.

The 3:2:1 stoichiometry predicts that a simple relation exists between the pump current and the fluxes. Suppose that the rate of depletion of ATP by the pump is α mol/cm$^2 \cdot$ s and that for each ATP molecule split ν_{Na} molecules of Na^+ are transported outward and ν_K molecules of K^+ are transported inward. Then the fluxes of Na^+ and K^+ are $\phi_{Na} = \nu_{Na}\alpha$ and $\phi_K = -\nu_K\alpha$, respectively, and the current densities are

$$J_{Na}^a = \nu_{Na} F\alpha \quad \text{and} \quad J_K^a = -\nu_K F\alpha. \tag{7.90}$$

Therefore, the total membrane current density due to the active transport of Na^+ and K^+, or the pump current, is

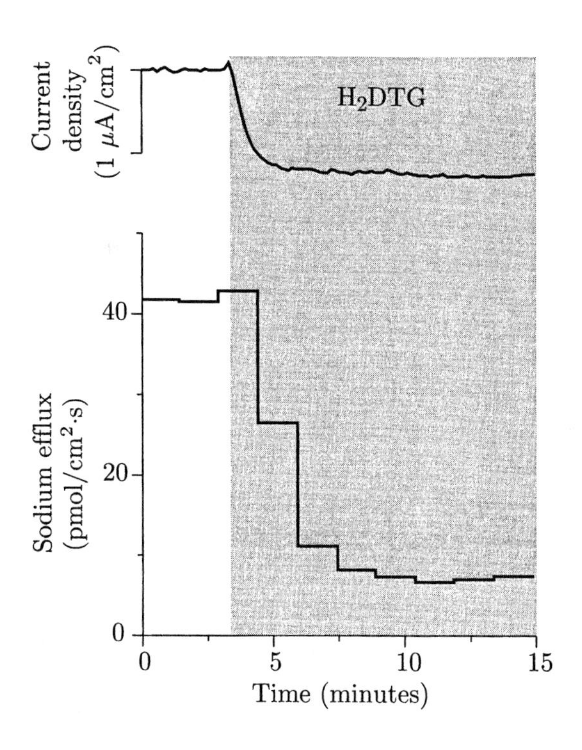

Figure 7.47 Simultaneous measurements of sodium efflux and pump current in the same squid giant axon (adapted from Rakowski et al., 1989, Figure 8D). The membrane potential was maintained at 0 mV, and the current through the axon was measured as a function of time. Radioactive Na^+ was added to the dialysis solution and its efflux was measured by sampling the extracellular solution. During the time interval represented by the shaded region, H_2DTG was added to the extracellular solution. Both the current and the efflux of sodium decreased. The scales have been chosen so that $F\phi$ and J are equal, which allows a direct comparison of flux to current density.

$$J_{\text{pump}} = (\nu_{Na} - \nu_K)F\alpha. \tag{7.91}$$

Thus, the ratio of sodium current to pump current, $\nu_{Na}/(\nu_{Na} - \nu_K)$, determines the stoichiometric ratios of sodium and potassium. If the pump stoichiometry were 3 : 2 : 1 then the ratio of sodium current to pump current should be 3 : 1.

An experiment to test this notion is described in Figure 7.47. A dialyzed squid axon was perfused with an ionic solution that contained both ATP and blockers of passive transport channels, while the external ionic solution also contained channel blockers. The current and efflux of sodium were measured in response to an application of the cardiac glycoside dihydrodigitoxigenin (H_2DTG) at constant membrane potential. The efflux of sodium was measured by adding ^{22}Na to the perfusion medium and counting samples of the extracellular solution. The application of H_2DTG resulted in a reduction in the efflux of sodium and a reduction in the current through the membrane. These reductions were presumably consequences of the inhibition of the sodium pump by this cardiac glycoside. The ratio of the change in sodium current to the change in membrane current caused by H_2DTG averaged 2.87 ± 0.07 (for twenty-five pairs of measurements), which did not differ appreciably from 3 as predicted by the 3:2:1 stoichiometry for the simple model of the sodium-potassium pump. Furthermore, this ratio did not depend appreciably on the membrane potential. Thus, these well-controlled measurements to estimate pump stoichiometry support the 3:2:1 hypothesis.

7.7.2.9 *Is the Energy Released in Hydrolysis of ATP Sufficient to Pump Sodium and Potassium?*

The sodium-potassium pump mechanism sketched thus far requires that the energy liberated in the hydrolysis of ATP exceed the energy required to pump Na^+ and K^+ against their electrochemical potential gradients. Is the energy enough? The change in free energy required to pump ν_K moles of potassium into a cell is simply $\nu_K(FV_m^o + RT\ln(c_K^i/c_K^o))$, and the energy required to pump ν_{Na} moles of sodium out of a cell is $\nu_{Na}(-FV_m^o + RT\ln(c_{Na}^o/c_{Na}^i))$ (see Section 7.4). Thus, the total energy per mole required to pump the two ions, which equals the increase in their free energy on exchange across the membrane, $\Delta\mathcal{G}_{\text{ions}}$, is

$$\Delta\mathcal{G}_{\text{ions}} = \underbrace{FV_m^o(\nu_K - \nu_{Na})}_{\text{electrical work}} + \underbrace{\nu_K RT\ln\left(\frac{c_K^i}{c_K^o}\right) + \nu_{Na}RT\ln\left(\frac{c_{Na}^o}{c_{Na}^i}\right)}_{\text{chemical work}}. \tag{7.92}$$

The first term on the right-hand side of Equation 7.92 represents the increase in electrostatic energy or the *electrical work* required to move the charged particles through an electric potential gradient. The second term represents the increase in chemical energy or the *chemical work* required to move particles against a concentration gradient. The reaction for the hydrolysis of ATP is

$$\text{ATP} \rightleftharpoons \text{ADP} + \text{P}, \tag{7.93}$$

and the change in free energy for this reaction, $\Delta\mathcal{G}_{\text{ATP}}$, is the free energy of the products minus the free energy of the reactants, which is

$$\Delta\mathcal{G}_{\text{ATP}} = \Delta\mathcal{G}_{\text{ATP}}^o + RT\ln\left(\frac{c_{ADP}\cdot c_P}{c_{ATP}}\right), \tag{7.94}$$

where $\Delta\mathcal{G}_{\text{ATP}}^o$ is the standard change in free energy.

Thus, the total free energy change for the sodium-potassium pump is the sum of the free energies for transporting the ions and hydrolyzing ATP and is

$$\Delta\mathcal{G}_{\text{pump}} = \Delta\mathcal{G}_{\text{ions}} + \Delta\mathcal{G}_{\text{ATP}}. \tag{7.95}$$

$\Delta\mathcal{G}_{\text{pump}}$ represents the difference in free energy per mole for the products minus that for the reactants. Thus, if this free energy difference is negative, the products have less free energy than the reactants and the reaction will proceed in the direction toward the products.

We shall evaluate these energies for typical values of the variables. We assume a sodium-potassium pump with a stoichiometry of 3:2:1, a temperature of 300 K, a cell with a resting potential of −70 mV, and concentration ratios of $c_K^i/c_K^o = 20$ and $c_{Na}^o/c_{Na}^i = 10$. Then the chemical work term has the value $300 \cdot 8.31(2\ln 20 + 3\ln 10) \approx 32$ kJ/mol, and the electrical work is $9.65 \times 10^4(-70 \times 10^{-3})(-1) \approx 7$ kJ/mol. Thus, $\Delta\mathcal{G}_{\text{ions}} \approx 39$ kJ/mol. Estimates of $\Delta\mathcal{G}_{\text{ATP}}$ at normal intracellular concentrations of the products and reactants are approximately −60 kJ/mol(De Weer, 1984). Thus, about 40 kJ/mol are required to transport both sodium and potassium actively, and about 60 kJ/mol are available from the hydrolysis of ATP. Thus, enough energy is available in ATP to power the pump.

7.7.2.10 Voltage Dependence of Pump Current

Another interpretation can be given to Equation 7.95 (De Weer et al., 1988). At equilibrium the total change in free energy of the pump is zero, i.e., the free energy of the reactants equals that of the products, and the pump flux goes to zero. We call the membrane potential at this equilibrium $V_m^o = V_{\text{pump}}$. Substitution of $\Delta\mathcal{G}_{\text{pump}} = 0$ in Equation 7.95 and rearrangement of terms yields

$$V_{\text{pump}} = \frac{\Delta\mathcal{G}_{\text{ATP}}}{F(\nu_{Na} - \nu_K)} + \frac{\nu_{Na}}{\nu_{Na} - \nu_K}V_{Na} + \frac{\nu_K}{\nu_{Na} - \nu_K}V_K, \tag{7.96}$$

where V_{Na} and V_K are the Nernst equilibrium potentials for sodium and potassium, respectively. The value of V_{pump} can be estimated as follows using the same values used in the previous section. The first term is $-60 \times 10^3/96.5 \times 10^3 \approx -0.62$ V or −620 mV. The two other terms are $3 \cdot 60 - 2 \cdot (-77) \approx +330$ mV. Thus, $V_{\text{pump}} \approx -290$ mV. The importance of this result is that we expect the pump current to be zero for a large hyperpolarization. A corollary to this result is that we do not expect the pump current to be constant as a function of membrane potential. Since the pump apparently displaces charged particles across the membrane, this result is not really surprising.

The voltage dependence of steady-state pump current has been measured in several different cell types (e.g., Figure 7.48) using methods similar to those used to measure pump stoichiometry illustrated in Figure 7.47. These measurements illustrate a decrease in pump current of 50% for a membrane potential change of −80 mV. The results show that the pump current is not

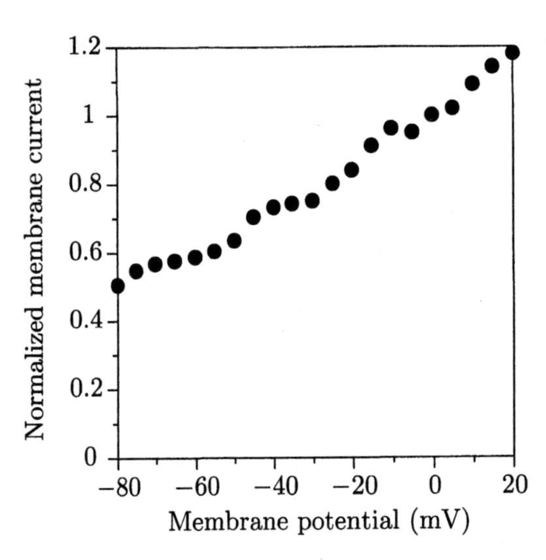

Figure 7.48 Measurements of the voltage dependence of pump current in squid giant axons(adapted from Rakowski et al., 1989, Figure 15B). The pump current is the average current measured for several cells and normalized to its value at 0 mV. The pump current was measured by using pharmacological blockers of sodium and potassium currents caused by voltage-gated ion channels and measuring the component of membrane current blocked by the cardiac glycoside dihydrodigitoxin.

constant and does decline as the cell membrane is hyperpolarized. However, demonstration of a decline of the pump current to zero at a very large hyperpolarization has not been feasible, since such large hyperpolarizations cause irreversible damage to the membranes of most cells.

These results suggest that models of sodium pumps of the type shown in Figure 7.33 are not accurate, since they ignore the effective internal conductance of the pump. A more accurate model would include this conductance as shown in Figure 7.49. How large is this conductance? In the squid giant axon, the sodium pump current density is typically less than 1 $\mu A/cm^2$ near a membrane potential of 0 mV (e.g., Figure 7.47). If this current declined linearly to $V_{pump} \approx -290$ mV, then this would imply that $G_{pump} \approx 3$ $\mu S/cm^2$, which is at least two orders of magnitude less than the membrane conductance of passive transport mechanisms (De Weer et al., 1988). Therefore, Figure 7.33 provides a reasonable approximation of the electrical effect of the sodium-potassium pump on the membrane potential.

7.7.2.11 Other Modes of Operation of the Sodium-Potassium Pump

The sodium-potassium pump normally operates by transporting three molecules of sodium out and two molecules of potassium in for each molecule of ATP split. However, by changing the compositions of the products and reactants from their normal ranges, it is possible to stimulate additional modes of operation of the pump. For example, if the ratio of intracellular concentration of ATP to that of the product of ADP and P is reduced drastically and if both

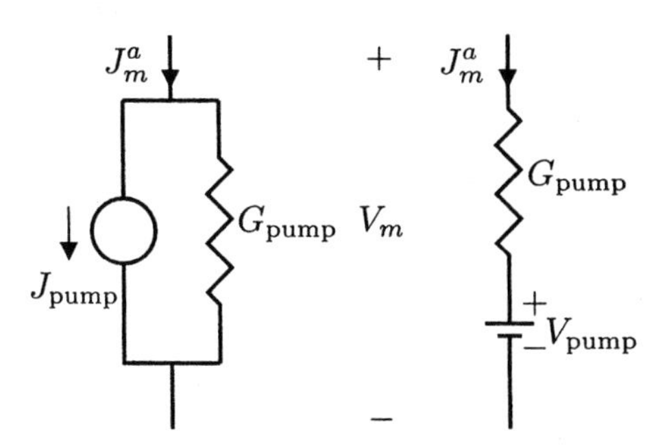

Figure 7.49 Norton (left) and Thévenin (right) equivalent circuits for the sodium-potassium pump. G_{pump} is the equivalent conductance of the pump, V_{pump} is the equilibrium potential of the pump, and $J_{pump} = -G_{pump}V_{pump}$. J_m^a is the total membrane current density transported by the sodium-potassium pump.

extracellular potassium and intracellular sodium, both of which activate the pump in the forward direction of operation, are removed, the pump operates in a backward direction. The pump synthesizes ATP from ADP and P, pumps sodium in and potassium out, and causes an inward pump current (Garrahan and Glynn, 1967c; De Weer and Rakowski, 1984). In addition, the pump results in a 1:1 exchange of either sodium or potassium in the absence of potassium or sodium, respectively. These exchanges depend upon ATP and are blocked by ouabain.

7.7.2.12 Summary of Properties of Active Transport

The purpose of the sodium-potassium pump is to create and to maintain ion concentration differences across the membrane, which then form the energy source that is used by cells for a variety of tasks, including electrical signaling by electrically excitable cells, volume control, and coupled transport of metabolites. In steady state, the passive flow of sodium and potassium down their concentration gradients is just balanced by active transport of these ions. Experiments on ion transport have led to a simple diagram (Figure 7.50) that summarizes, in a qualitative manner, many of the physiological properties of the sodium-potassium pump. First, the pump mechanism is highly asymmetric. It is activated by the intracellular presence of sodium, magnesium, and ATP and by the extracellular presence of potassium (or its agonists). The pump mechanism is blocked by cardiac glycosides, such as ouabain, applied extracellularly but not intracellularly. The stoichiometry of the pump appears to be that for each molecule of ATP hydrolyzed to ADP and phosphate, three molecules of sodium are transported outward and two molecules of potassium are transported inward. Under these circumstances, the sodium-potassium pump is electrogenic, producing a net outward current that normally results in a few millivolts of membrane hyperpolarization.

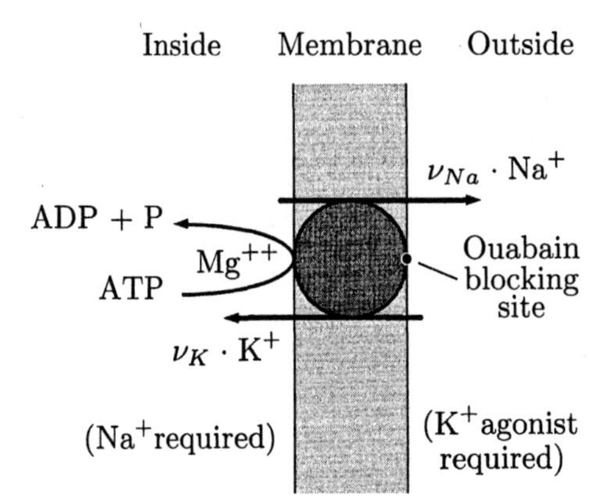

Figure 7.50 Summary of properties of active sodium transport.

7.7.3 ($Na^+ - K^+$)-ATPase

In a seminal study, Skou (1957) found that homogenized crab leg nerves contained a fraction that could catalyze the hydrolysis of ATP (as given in Equation 7.93); that is, the fraction contained an adenosine triphosphatase, or ATPase. The efficacy of the ATPase in catalyzing the reaction, i.e., in accelerating the progress of the reaction, was assessed by measuring the quantity of one of the products of the reaction, inorganic phosphate, as a function of time and as a function of the composition of the medium. The following results were found: (1) No appreciable inorganic phosphorus was produced when the solution contained ADP but no ATP. (2) In the presence of ATP, the quantity of phosphorus produced was consistent with the reaction shown in Equation 7.93 in that the reaction came to completion when one phosphate was released per ATP molecule; other phosphates in ADP were not released. (3) Magnesium was an obligatory cofactor for the reaction; the reaction was not catalyzed in the absence of magnesium. (4) In the presence of both ATP and Mg^{++} the enzyme activity was small, but with the addition of *both* Na^+ and K^+ the activity increased appreciably. (5) The enzyme activity was a complex function of the concentrations of ATP, Na^+, K^+, and Mg^{++}. These observations suggested that the ATPase found in the tissue fraction was responsible for the transport properties associated with the sodium-potassium pump. Since this substance requires both Na^+ and K^+ and catalyzes the hydrolysis of ATP, it has been called (Na^+-K^+)-ATPase.

Subsequently, (Na^+-K^+)-ATPase has been found ubiquitously in animal cell membranes, but it is especially rich in tissues with large energy requirements, such as brain and transporting epithelia such as kidney (Schuurmans

Stekhoven and Bonting, 1981). Further studies indicated that the catalytic activity of (Na^+-K^+)-ATPase is blocked by cardiac glycosides at concentrations similar to those at which ion transport by the sodium-potassium pump is blocked. Thus, there is little doubt left that (Na^+-K^+)-ATPase is the molecular basis of the sodium-potassium pump. The dependence of the catalytic action of (Na^+-K^+)-ATPase and of the transport properties of the sodium-potassium pump on ion composition and on the presence of ouabain are very similar. Furthermore, incorporation of (Na^+-K^+)-ATPase into artificial lipid bilayers has been shown to lead to the appropriate coupled transport of Na^+ and K^+. More recently, studies of (Na^+-K^+)-ATPase have had two major objectives: to define the chemical reaction sequence by which Na^+ and K^+ are linked to hydrolysis of ATP and to determine the structure of (Na^+-K^+)-ATPase.

7.7.3.1 Kinetics of ($Na^+ - K^+$)-ATPase

(Na^+-K^+)-ATPase apparently has two primary conformations called E_1 and E_2. These two conformations have different affinities for ATP, ouabain, Na^+, and K^+. The two conformations can be detected, e.g., they have a small difference in intrinsic fluorescence and produce large differences in fluorescence when extrinsic fluorescent probes are bound to the molecule. Thus, it has been possible to measure the rates of transition between states as a function of composition of the bathing medium and to relate this to the composition-dependent enzymatic activity. After several decades of work, it is clear that the chemical kinetics of (Na^+-K^+)-ATPase is complex largely because there are so many substrates and because the enzyme is normally membrane bound.

It is generally agreed that many, but not all, of the properties of the enzymatic action and the transport properties of (Na^+-K^+)-ATPase can be explained by a relatively simple chemical cycle. This cycle, which is called the *Post-Albers model* and shown schematically in Figure 7.51, is thought to be responsible for the activity of the sodium-potassium pump at normal physiological concentrations. In the cyclic reaction sequence, E_1 and E_2 alternately bind and release sodium and potassium ions and catalyze the hydrolysis of ATP. The E_1 conformation has an ion-binding site that faces cytoplasm and binds sodium ions, whereas E_2 has an ion-binding site that faces extracellularly and binds potassium. The reaction sequence, beginning with state $E_1 \cdot$ ATP, is as follows:

- With ATP bound to E_1, the enzyme binds three sodium ions on the cytoplasmic face of the membrane.

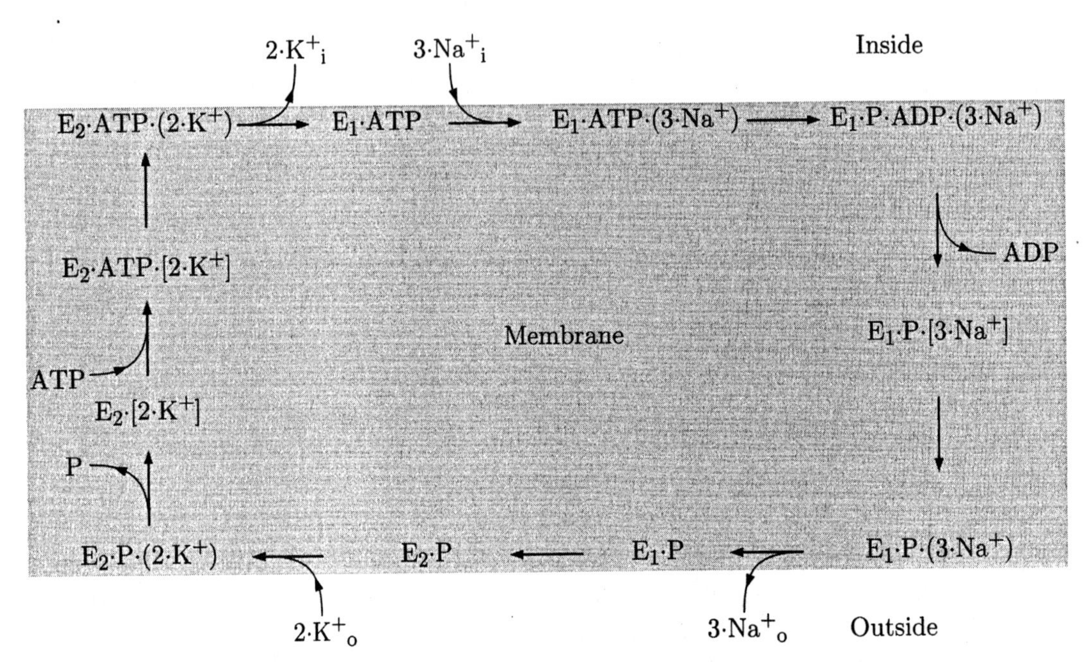

Figure 7.51 Reaction scheme for kinetic cycle of $(Na^+\text{-}K^+)$-ATPase (adapted from Skou, 1988, Figure 2). Each reaction is reversible; the arrows indicate the normal forward direction of the cycle. The Es represent the $(Na^+\text{-}K^+)$-ATPase in its different states. The square brackets, e.g. $[3 \cdot Na^+]$, indicate that the ions are "occluded"; that is, they are trapped within the $(Na^+\text{-}K^+)$-ATPase and cannot be released into solution in this state.

- Hydrolysis of ATP to ADP, then release of ADP with resultant phosphorylation of the enzyme, results in an occluded state, $E_1 \cdot P \cdot [3 \cdot Na^+]$, from which the sodium ions cannot be released into solution without a change in conformation of the $(Na^+\text{-}K^+)$-ATPase.
- The occluded state converts to the nonoccluded state, from which three sodium ions are released extracellularly to yield the phosphorylated enzyme $E_1 \cdot P$. The release of sodium ions may be a staged release of individual ions through intermediate reactions (not shown).
- The conformation $E_1 \cdot P$ converts to the conformation $E_2 \cdot P$
- The conformation $E_2 \cdot P$ binds two potassium ions on the extracellular face of the membrane.
- Dephosphorylation of $E_2 \cdot P \cdot (2 \cdot K^+)$ leads to a complex in which potassium ions are occluded.

- Binding of ATP leads, in two stages, to a conformation from which two potassium ions are released on the cytoplasmic face of the membrane and conversion of the enzyme to its $E_1 \cdot ATP$ conformation.

In one cycle a molecule of ATP is split, three sodium ions are transported from the cytoplasm to the extracellular space, and two potassium ions are transported in the opposite direction.

7.7.3.2 Characterization of ($Na^+ - K^+$)-ATPase

(Na^+-K^+)-ATPase is an integral membrane protein that has been extracted from membranes by use of detergents. Purification of these extracts has shown that (Na^+-K^+)-ATPase consists of two subunits: the α subunit is a protein with a molecular weight of about 100,000 daltons, and the β subunit is a glycoprotein with a molecular weight of about 40,000 daltons. In the membrane, (Na^+-K^+)-ATPase most likely consists of a tetramer, $\alpha_2\beta_2$, plus the lipids necessary for its activity. Although the β subunit is required for activity of the enzyme, its function is unknown. The α subunit has been shown to contain both the ATP- and the ouabain-binding sites.

The amino acid sequences of both the α and β subunits have been determined from characterization of complementary DNA and consist of 1,016 and 302 amino acids, respectively. By analyzing the hydrophobicity of the amino acid sequences, it has been inferred that the α subunit has eight hydrophobic regions and the β subunit has only one. Thus, consistent with other integral membrane proteins, the α subunit probably has eight transmembrane regions and the β subunit probably has just one.

7.8 Comparison of Active and Passive Transport

As we have seen, active and passive ion transport are distinctly different. A comparison of properties of passive and active transport of sodium and potassium is summarized in Table 7.5. In this chapter we have discussed the molecular basis of the coupled active transport of sodium and potassium via (Na^+-K^+)-ATPase, but we have not discussed the molecular basis of passive transport of sodium and potassium. As shown elsewhere (Weiss, 1996, Chapter 6), a major pathway for passive ion transport through membranes is via gated ion channels that are distinct from the pumps described in this chapter and from the carriers described in Chapters 6 and 8.

Table 7.5 Distinctions between properties of active and passive transport of ions (Keynes, 1975, adapted from Table 1). The passive transport is via voltage-gated sodium and potassium channels and the active transport via (Na^+-K^+)-ATPase. TTX is tetrodotoxin, a specific blocker of the voltage-gated sodium channel, and TEA is tetraethylammonium, a blocker of voltage-gated potassium channels.

Property	Passive transport	Active transport
Flux direction	Down electrochemical potential gradient	Up electrochemical potential gradient
Source of energy	Ion concentration difference across membrane	ATP
Voltage dependence	Ion conductances depend strongly on membrane potential	Weak dependence on membrane potential
Transport blockers	TTX blocks Na^+; TEA blocks K^+; cardiac glycosides have no effect	TTX and TEA have no effect; cardiac glycosides block
External calcium	Affects ionic conductances	No effect
Selectivity	Li^+ can substitute for Na^+	Li^+ substitutes poorly for Na^+
Effect of temperature	Flux is weakly temperature dependent	Flux is strongly temperature dependent
Site density	27–500 sites/μm^2	500–5000 sites/μm^2
Maximum flux of Na^+	10^4 pmol/cm^2	60 pmol/cm^2
Metabolic inhibitors	No effect until ion concentrations run down	CN and DNP block ATP production which blocks transport

Appendix 7.1 The Goldman Constant Field Model

As described in Section 7.2, the equations for electrodiffusion are nonlinear, and general solutions are not available for either time-varying or time-invariant cases. However, solutions are available under certain simplifying assumptions. Among the best known of such solutions is the Goldman solution described in this section. The Goldman solution was proposed first as a microscopic model for ion transport through biological membranes (Goldman,

1943; Hodgkin and Katz, 1949). In this model the membrane is regarded as a homogeneous medium in which ions are free to move according to the laws of electrodiffusion. The details of this model for ion transport in biological membranes are no longer valid. However, the Goldman model is of interest for several reasons. First, the Goldman model gives an instructive solution to an electrodiffusion problem. Second, the equations derived from this model are widely used as empirical bases for relating membrane potentials to concentrations and permeabilities. Third, some of the results derived from the Goldman model are much more robust than they would appear from the simple derivation given here (Problem 7.4).

Derivation of the Voltage-Current Characteristic

Consider a membrane that separates two solutions as shown in Figure 7.12. The two solution phases are assumed to be well mixed; that is, their compositions are assumed to be uniform. Under steady-state conditions, the current density of ion n is constant, but in general the concentrations and the electric potential are functions of position in the membrane x. In the Goldman model it is assumed that the net charge density in the membrane, due to fixed and mobile charges, is zero. That is, electroneutrality is imposed on a microscopic scale. With this additional assumption, which overspecifies the equations of electrodiffusion, the steady-state electrodiffusion equations are readily solved. If this charge is zero, then from Equation 7.40 we have

$$\frac{d^2\psi(x)}{dx^2} = 0, \tag{7.97}$$

which implies that the potential gradient, which is the electric field, is a constant in the membrane. Hence, the assumption has been called the *constant field assumption* (Goldman, 1943; Hodgkin and Katz, 1949). If the membrane has thickness d and the potential across the membrane is V_m, then

$$\frac{d\psi(x)}{dx} = -\frac{V_m}{d}, \tag{7.98}$$

so that $\psi(x) = \psi(0) - V_m(x/d)$ and

$$V_m = \psi(0) - \psi(d). \tag{7.99}$$

Under steady-state conditions and with a constant electric field, Equation 7.42, which expresses the relation of current carried by the nth ion to the concentration of that ion and to the electric potential, can be evaluated for $x_o = 0$ to $x = d$ to yield

$$J_n = -\frac{z_n F^2 u_n}{\beta} \left(\frac{c_n(d) e^{z_n \beta \psi(d)} - c_n(0) e^{z_n \beta \psi(0)}}{\int_0^d e^{z_n \beta (\psi(0) - V_m (x/d))} \, dx} \right). \tag{7.100}$$

We assume that at the membrane solution interfaces the membrane potential is continuous and that ion n distributes itself according to the partition coefficient k_n, i.e.,

$$\frac{c_n(0)}{c_n^i} = k_n \text{ and } \frac{c_n(d)}{c_n^o} = k_n. \tag{7.101}$$

Evaluation of the integral in Equation 7.100 and substitution of Equations 7.99 and 7.101 into 7.100 yields

$$J_n = \frac{k_n z_n u_n F^2}{\beta d} z_n \beta V_m \left(\frac{c_n^o - c_n^i e^{z_n \beta V_m}}{1 - e^{z_n \beta V_m}} \right), \tag{7.102}$$

which can also be written as

$$J_n = z_n F P_n (z_n \beta V_m) \left(\frac{c_n^o - c_n^i e^{z_n \beta V_m}}{1 - e^{z_n \beta V_m}} \right), \tag{7.103}$$

where $P_n = k_n u_n RT/d = k_n D_n/d$ is the permeability of the membrane to ion n.

Properties of the Voltage-Current Characteristic

Equation 7.103 relates the current density carried by an ion to its concentration and to the potential across the membrane. Let us examine properties of this relation (Figure 7.52). The current is zero if and only if the numerator of Equation 7.103 is zero, which occurs if

$$V_m = V_n \equiv \frac{1}{z_n \beta} \ln \left(\frac{c_n^o}{c_n^i} \right) = \frac{RT}{z_n F} \ln \left(\frac{c_n^o}{c_n^i} \right), \tag{7.104}$$

where V_n is the Nernst equilibrium potential, which is discussed in detail in Section 7.4. The asymptotic behavior of J_n is

$$\lim_{V_m \to \infty} J_n = (z_n^2 F P_n c_n^i) V_m \text{ and } \lim_{V_m \to -\infty} J_n = (z_n^2 F P_n c_n^o) V_m.$$

Thus, the asymptotic slopes differ, and the J_n-V_m characteristic shows rectification that depends on the ratio of concentrations; that is, the ratio of

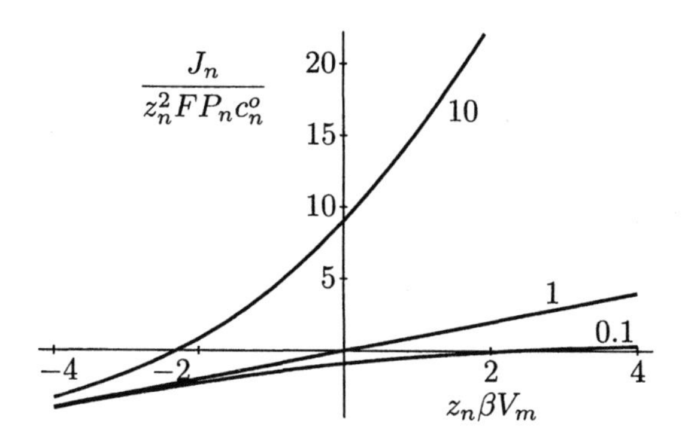

Figure 7.52 Normalized current-voltage relation for the Goldman constant field model of steady-state electrodiffusion in a membrane.

asymptotic slopes equals the ratio of concentrations. Note that for $c_n^i = c_n^o$ the relation between the current and the membrane potential is that of a linear resistor.

The Unidirectional Flux Ratio

Equation 7.103 can also be written as the difference between unidirectional current densities as follows:

$$J_n = \overrightarrow{J_n} - \overleftarrow{J_n},$$

where $\overrightarrow{J_n}$ is the outward unidirectional current density and can be obtained from Equation 7.103 by setting $c_n^o = 0$, and $\overleftarrow{J_n}$ is the inward unidirectional current density and can be obtained from Equation 7.103 by setting $c_n^i = 0$. Therefore,

$$\overleftarrow{J_n} = z_n^2 F P_n V_m \left(\frac{c_n^o}{1 - e^{z_n \beta V_m}} \right), \quad \text{and} \quad \overrightarrow{J_n} = z_n^2 F P_n V_m \left(\frac{c_n^i e^{z_n \beta V_m}}{1 - e^{z_n \beta V_m}} \right). \tag{7.105}$$

The ratio of unidirectional current densities (and hence, the ratio of unidirectional fluxes) is obtained from Equation 7.105 as follows:

$$\frac{\overrightarrow{\phi_n}}{\overleftarrow{\phi_n}} = \frac{\overrightarrow{J_n}}{\overleftarrow{J_n}} = \frac{c_n^i}{c_n^o} e^{z_n \beta V_m} = e^{z_n \beta (V_m - V_n)}, \tag{7.106}$$

where V_n is the Nernst equilibrium potential for ion n. At equilibrium, the current density, J_n, is zero (as is the net flux of ion n) and the unidirectional current densities (and unidirectional fluxes) must be equal. Hence, at equilibrium the flux ratio is one, and this occurs when $V_m = V_n$.

The Goldman Equation for the Resting Potential

Equation 7.103 is a nonlinear relation between ionic current and membrane potential that has zero ionic current when the membrane potential equals the Nernst equilibrium potential. Hence, the relation between ionic current and membrane potential can be represented by a network model (Figure 7.24). The conductance of the membrane for ion n, G_n.

If several ions can permeate the membrane, then the total current density through the membrane is

$$J_m = \sum_n J_n = \sum_n z_n^2 F P_n V_m \left(\frac{c_n^o - c_n^i e^{z_n \beta V_m}}{1 - e^{z_n \beta V_m}} \right).$$

Next we will consider univalent electrolytes for which $z_n = \pm 1$. Let V_m^o be the value of V_m for which the membrane current density is zero, or the *resting membrane potential*. Then

$$\sum_{z_n=+1} F P_n V_m^o \left(\frac{c_n^o - c_n^i e^{\beta V_m^o}}{1 - e^{\beta V_m^o}} \right) + \sum_{z_n=-1} F P_n V_m^o \left(\frac{c_n^o - c_n^i e^{-\beta V_m^o}}{1 - e^{-\beta V_m^o}} \right) = 0,$$

which can be written as

$$\sum_{z_n=+1} F P_n V_m^o \left(\frac{c_n^o - c_n^i e^{\beta V_m^o}}{1 - e^{\beta V_m^o}} \right) + \sum_{z_n=-1} F P_n V_m^o \left(\frac{c_n^i - c_n^o e^{\beta V_m^o}}{1 - e^{\beta V_m^o}} \right) = 0$$

or

$$\frac{F V_m^o}{1 - e^{\beta V_m^o}} \left(\left(\sum_{z_n=+1} P_n c_n^o + \sum_{z_n=-1} P_n c_n^i \right) - e^{\beta V_m^o} \left(\sum_{z_n=+1} P_n c_n^i + \sum_{z_n=-1} P_n c_n^o \right) \right) = 0.$$

The coefficient of the bracketed term is nonzero for all finite values of V_m^o. Therefore, the bracketed term must be zero, and the value of V_m^o for which this term is zero is

$$V_m^o = \frac{RT}{F} \ln \left(\frac{\sum_{z_n=+1} P_n c_n^o + \sum_{z_n=-1} P_n c_n^i}{\sum_{z_n=+1} P_n c_n^i + \sum_{z_n=-1} P_n c_n^o} \right). \tag{7.107}$$

Suppose the membrane is permeable to only three ions, e.g., K^+, Na^+, and Cl^-. Then Equation 7.107 can be expressed as follows:

$$V_m^o = \frac{RT}{F} \ln \left(\frac{c_K^o + (P_{Na}/P_K) c_{Na}^o + (P_{Cl}/P_K) c_{Cl}^i}{c_K^i + (P_{Na}/P_K) c_{Na}^i + (P_{Cl}/P_K) c_{Cl}^o} \right). \tag{7.108}$$

Equation 7.108 gives the relation between the resting potential predicted by the Goldman theory and the ion concentrations in terms of two parameters, the ratios of permeabilities of the membrane, P_{Na}/P_K and P_{Cl}/P_K.

Exercises

7.1 Let ψ represent electric potential, ϕ_n represent solute flux, c_n concentration, z_n valence, F Faraday's constant, u_n molar mechanical mobility, R the molar gas constant, D_n the diffusion coefficient, and T absolute temperature. Identify the units of the expressions in parts a–g as one of the following: A, moles/s; B, moles/cm · s; C, moles/cm^2 · s; D, moles/cm^3 · s; E, amps/cm^2; F, amps; G, coul/mol; H, coul/cm^3; I, volts/cm; J, volts; X, none of the above. If the answer to any part is X, provide proper units.

a. $\dfrac{\partial \psi(x,t)}{\partial x}$

b. $z_n F c_n(x,t)$

c. $u_n z_n F c_n(x,t)$

d. $D_n \dfrac{\partial c_n(x,t)}{\partial x}$

e. F

f. $z_n F \phi_n$

g. $\dfrac{RT}{z_n F}$

7.2 As shown in Figure 7.53, compartments 1 and 2 contain well-stirred solutions of potassium chloride and are separated by a membrane that is permeable to only potassium. The potential between compartment 1 and 2 is V_m. The concentrations of KCl in compartments 1 and 2 are 100 mmol/L and 10 mmol/L, respectively.

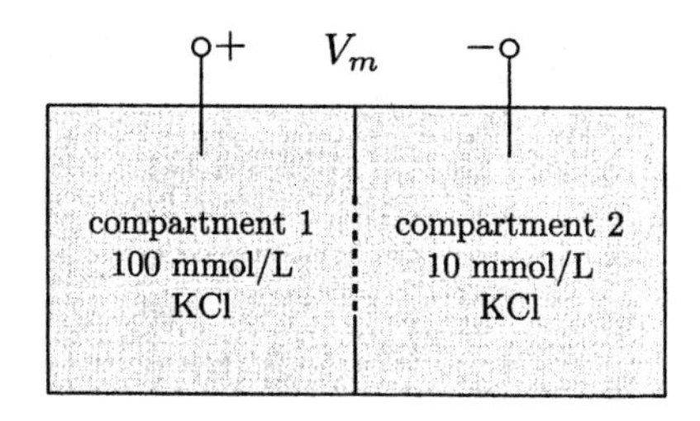

Figure 7.53 Two compartments separated by a membrane that is permeable to potassium only (Exercise 7.2).

a. Determine the equilibrium value of V_m, and give a physical explanation of the sign of the potential.

b. A battery is now connected to the solutions, so that $V_m = -30$ mV. In which direction will current flow through the membrane? Explain.

c. Draw an equivalent electrical network for the condition indicated in part b. Label the nodes that represent compartments 1 and 2, V_m, and label I_m, defined as the current that flows through the membrane in the direction from compartment 1 to compartment 2.

7.3 Define *electroneutrality*, and briefly explain its physical basis.

7.4 Define the *Nernst equilibrium potential*, and briefly explain its physical basis.

7.5 Describe the distinctions between the following terms that refer to ion transport across a cellular membrane: electrodiffusive equilibrium, steady state, resting conditions, and cellular quasi-equilibrium.

7.6 The following is a discussion of electroneutrality (Nicholls et al., 1992):

The intracellular and extracellular solutions must each be electrically neutral. For example, a solution of chloride ions alone cannot exist; their charges must be balanced by an equal number of positive charges on cations such as sodium or potassium (otherwise electrical repulsion would literally blow the solution apart).

Briefly critique this discussion of electroneutrality.

7.7 A beaker of solution contains a concentration of chloride ions of 100 mmol/L.

a. Determine which of the following statements is correct, and explain your selection.

i. There are no other ions in the solution but chloride.

ii. There may be other ions in the solution besides chloride.

iii. There must be other ions in the solution besides chloride.

b. Suppose it is known that the solution contains sodium ions in addition to the chloride ions, but does not contain any other ion species. Can the sodium concentration be determined from the information given? If so, what is it? If not, why not?

7.8 Figure 7.54 shows a schematic diagram of two compartments with rigid walls separated by a rigid membrane that is permeable only to a positive ion. The anions are not shown.

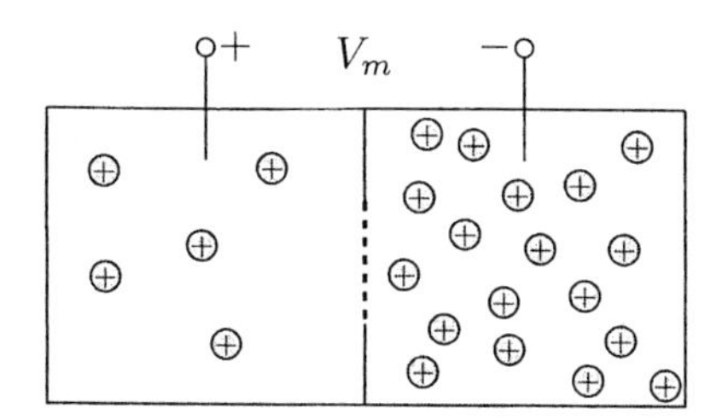

Figure 7.54 Two compartments separated by a membrane that is permeable to only the positive ion (Exercise 7.8). The schematic diagram indicates that the concentration of positive ions is larger in the right compartment than in the left.

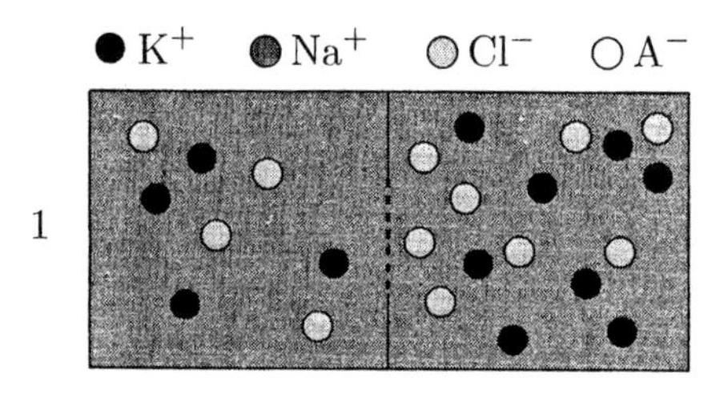

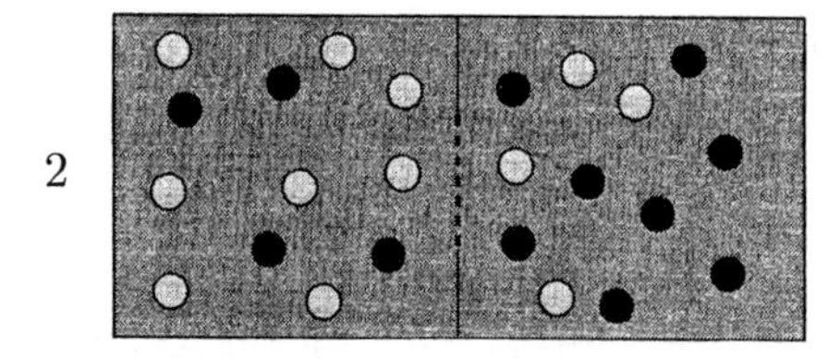

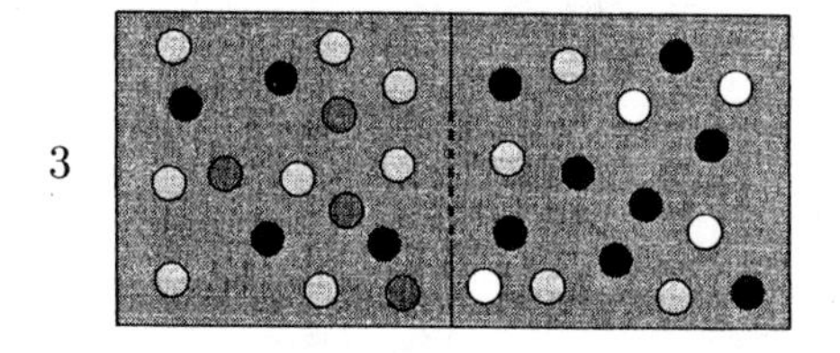

Figure 7.55 Two compartments separated by a membrane with different ions on the two sides of the membrane (Exercise 7.9). The three panels show different distributions of ions. The diagrams represent the compositions of the bulk solutions on the two sides of the membrane.

a. At electrodiffusive equilibrium, what is the sign of V_m?

b. Make a rough sketch like Figure 7.54 that includes both the anions and the cations. Explain your sketch.

c. On a scale of expanded length, make a rough sketch that indicates the concentration of cations and anions on either side of the membrane. Explain your sketch.

7.9 Figure 7.55 shows three panels in which the distributions of Na^+, K^+, Cl^-, and an impermeant anion A^- are shown schematically in two rigid compartments separated by a rigid semipermeable membrane. The problem

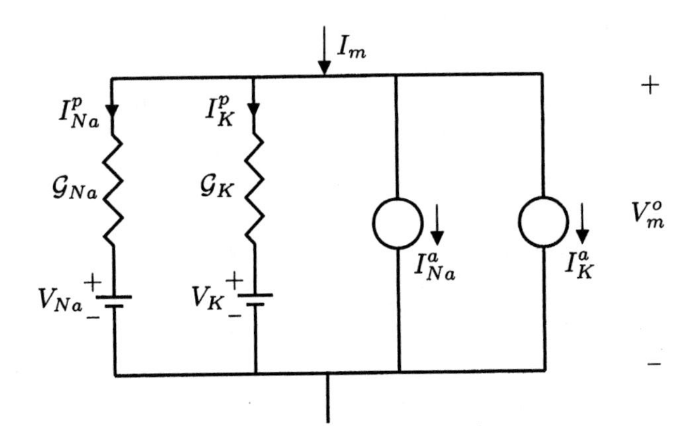

Figure 7.56 Network model of a glial cell (Exercise 7.10).

concerns the distributions at electrodiffusive equilibrium for different membrane characteristics.

a. If the membrane were permeable only to K^+, which of the three distributions would be possible at electrodiffusive equilibrium?

b. If the membrane were permeable to both K^+ and Cl^-, which of the three distributions would be possible at electrodiffusive equilibrium?

7.10 In this exercise, the measurements of resting potential of a glial cell as a function of extracellular potassium concentration shown in Figure 7.21 are to be interpreted in terms of the network model shown in Figure 7.56. Assume that $c^o_{Na} = 150$ mmol/L and $c^i_{Na} = 15$ mmol/L and that the external solution is maintained isotonic with the cytoplasm by controlling impermeant solutes. Assume that sodium and potassium concentrations are constant, except for c^o_K, and that the pump system, which consists of I^a_{Na} and I^a_K, is nonelectrogenic.

a. Consider only the region for which the data are well fitted by the straight line of slope 59 mV/decade. Indicate whether the following statements are true or false, and give a brief reason for each answer.

   i. $I_m = 0$.
   ii. $V^o_m \approx V_K$.
   iii. $g_{Na} \gg g_K$.
   iv. $V_{Na} > V_K$.
   v. $c^i_K = 100$ mmol/L.
   vi. $I^a_K = -I^a_{Na}$.

Table 7.6 Potassium and sodium concentrations (Exercise 7.11).

	Concentration (mmol/L)	
	Inside	Outside
Potassium	150	15
Sodium	15	150

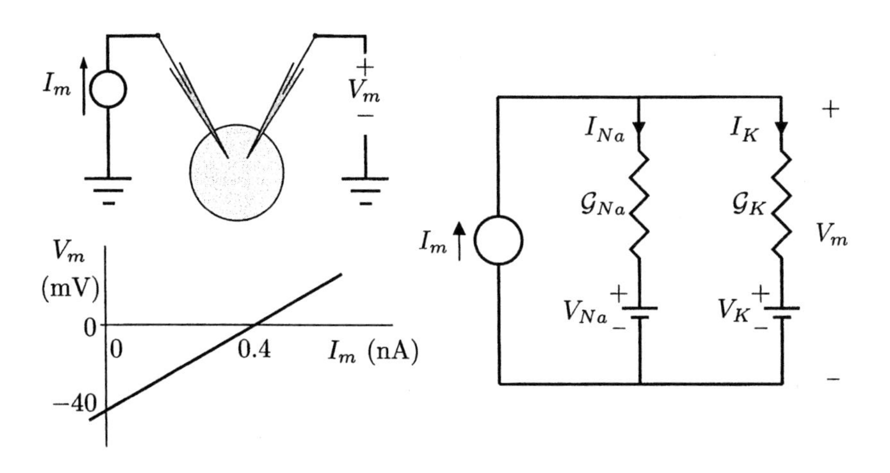

Figure 7.57 Relation between membrane potential and membrane current through an isolated cell (Exercise 7.11).

vii. $I_K^p \gg I_{Na}^p$.

viii. $I_{Na}^a = -\mathcal{G}_{Na}(V_m^o - V_{Na})$.

b. It is proposed that deviation of the data from the straight line for the lowest c_K^o is a result of a change in $\mathcal{G}_K$ that occurs when $V_m^o < -110$ mV. For the data shown, is this a reasonable hypothesis? Does it require that $\mathcal{G}_K$ for $V_m^o = -125$ mV be larger or smaller than $\mathcal{G}_K$ for $V_m^o > -100$ mV? Explain.

7.11 The ionic concentrations of a uniform isolated cell are given in Table 7.6. An electrode is inserted into the cell and connected to a current source so that the current through the cell membrane is I_m. The steady-state voltage across the cell membrane, V_m, is determined as a function of the current as shown in Figure 7.57. Assume that (1) the cell membrane is permeable to only K^+ and Na^+ ions, (2) the Nernst equilibrium potentials are $V_n = (60/z_n)\log_{10}(c_n^o/c_n^i)$ (mV), (3) the ion concentrations are constant, (4) the active transport processes make no contribution to these measurements.

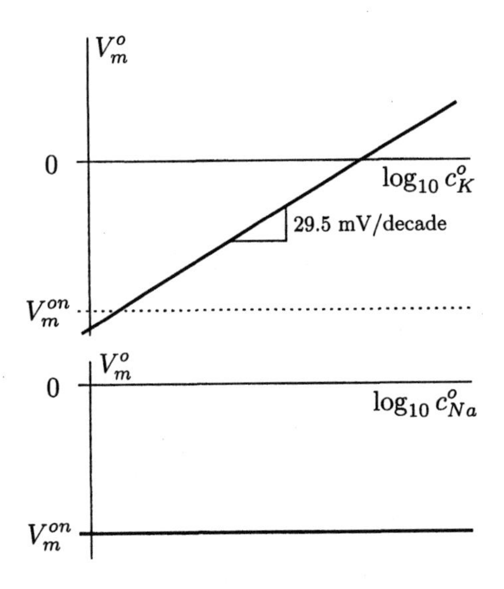

Figure 7.58 Dependence of resting potential on sodium and potassium concentration at 24°C (Exercise 7.12).

a. Determine the equilibrium potentials for sodium and potassium ions, V_{Na} and V_K.

b. What is the resting potential of the cell with these ionic concentrations?

c. With the current I_m adjusted so that $V_m = V_K$, what is the ratio of the sodium current to the total membrane current, I_{Na}/I_m?

d. What is the total conductance of the cell membrane $\mathcal{G}_m = \mathcal{G}_{Na} + \mathcal{G}_K$?

e. Determine $\mathcal{G}_{Na}$ and $\mathcal{G}_K$.

7.12 I.M. Fumbler has published a paper in which he reported measurements of membrane potential in a cell for a range of extracellular potassium and sodium concentrations as shown in Figure 7.58. Remarkably, the membrane potential is proportional to the logarithm of the potassium concentration, but the slope is 29.5 mV/decade. The membrane potential is independent of sodium concentration. Fumbler claims that the results are a proof that Bernstein's theory, which states that the resting potential of a cell is the potassium equilibrium potential, is incorrect. In a letter to the editor, Sharp Wan claims that Fumbler's results can be explained by the Bernstein theory if Fumbler's experimental technique of piercing the cell membrane with a measuring electrode introduced a "leakage resistance" between the inside and outside of the cell.

In a paragraph with an accompanying diagram or two, explain the quantitative basis of Wan's argument, and determine how big the

Table 7.7 The table shows the compositions of intracellular and extracellular ions in four different cells and also indicates which ions are permeant through the cell membranes (Exercise 7.14). Ion A is an impermeant anion.

Cell number	Permeant ions	Composition (mmol/L) Intracellular KCl	Intracellular NaCl	Intracellular KA	Extracellular KCl	Extracellular NaCl	Extracellular KA
1	K^+	150	10	0	10	150	0
2	Cl^-	150	10	0	10	150	0
3	K^+ & Cl^-	150	10	0	10	150	0
4	K^+ & Cl^-	0	10	150	0	150	10

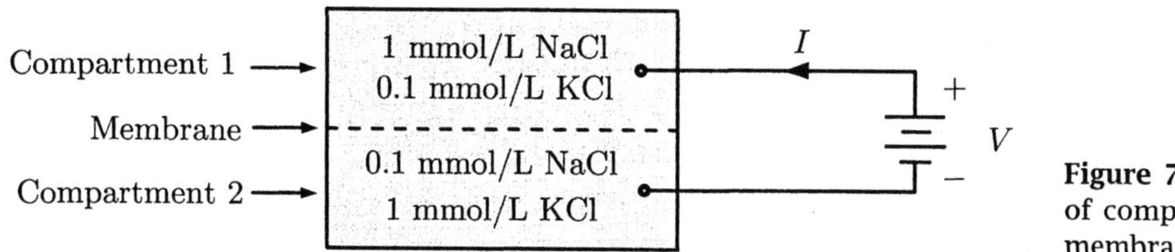

Figure 7.59 Relation of compartments and membrane (Problem 7.15).

leakage resistance must be relative to the net resistance of the cell membrane. Assume that the equilibrium potential for potassium is $V_K = 59\log(c_K^o/c_K^i)$ (mV).

7.13 Active ion transport is said to have a direct and an indirect effect on the resting potential of a cell. Define both effects, and discuss the distinction between them.

7.14 The intracellular and extracellular ion concentrations are given in Table 7.7 for four cells. The permeant ions are also listed for each cell. You may assume that the permeant ions are transported passively only and that there are no mechanisms of active ion transport in these cells. The temperature is 24°C. For each cell, determine whether the solutions are in electrodiffusive equilibrium across the membrane, and if they are, determine the resting membrane potential.

7.15 Two compartments of a fluid-filled chamber are separated by a membrane as shown in Figure 7.59. The area of the membrane is 100 cm^2, and the volume of each compartment is 1000 cm^3. The solution in compartment 1 contains 1 mmol/L NaCl and 0.1 mmol/L KCL. The solution in compartment 2 contains 0.1 mmol/L NaCl and 1 mmol/L KCL. The temperatures of the solutions are 24°C. The membrane is known to be permeable to a single ion, but it is not known if that ion is sodium, potassium, or chloride. Electrodes connect the solutions in the compartments to a battery. The current, I, was measured with the battery voltage $V = 0$ and was found to be $I = -1$ mA.

a. Identify the permeant ion species. Explain your reasoning.

b. Draw an equivalent circuit for the entire system, including the battery. Indicate values for those components whose values can be determined.

c. Determine the current I that would result if the battery voltage were set to 1 volt. Explain your reasoning.

7.16 A cell has been arranged so that solutions with known compositions can be perfused either intracellularly or extracellularly. In each part of this problem, the perfusion sequence has three separate intervals. During the first interval a control solution is perfused, and the cell is allowed to equilibrate. During the second interval a test solution is perfused, and once again the cell is allowed to come to equilibrium. During the third and final interval the cell is again perfused by the control solution that was used during the first interval. The problem is to determine the effect of the test solution on the flux of ions through the membrane and on the resting potential of the cell. Assume the following:

- The only important fluxes are the passive and active fluxes of sodium and potassium, ϕ^p_{Na}, ϕ^a_{Na}, ϕ^p_K, and ϕ^a_K. As the reference direction for all fluxes, outward fluxes are considered as positive.
- The active transport is due to the sodium potassium ATPase, so that three sodium ions are transported out of the cell and two potassium ions are transported into the cell for each molecule of ATP that is split.
- The difference between the test and control solutions is adequate to produce a physiological response.
- The concentrations of intracellular and extracellular solutions not controlled by the perfusion do not change appreciably over the duration of the test interval.

a. The cell is perfused extracellularly with a control solution that approximates the cell's normal extracellular solution. The test solution is similar to the control solution, but also contains ouabain. For each of the fluxes (ϕ^p_{Na}, ϕ^a_{Na}, ϕ^p_K, and ϕ^a_K) and for the resting potential across the membrane (V_m^o), determine the effect of the test solution by indicating whether the quantity increases, decreases, or stays the same. Explain your reasoning.

b. The cell is perfused intracellularly with a control solution that approximates the cell's normal intracellular solution except that it contains no ATP. The test solution is the same as the control solution, but contains ATP. For each of the fluxes (ϕ_{Na}^p, ϕ_{Na}^a, ϕ_K^p, and ϕ_K^a) and for the resting potential across the membrane (V_m^o), determine the effect of the test solution by indicating whether the quantity increases, decreases, or stays the same. Explain your reasoning.

Problems

7.1 This problem deals with (among other things) the question, Do the intracellular ions in a cell behave as if they were in an aqueous solution with the same ionic concentration as the intracellular medium?

a. The measured resistivity of squid axoplasm (the cytoplasm of squid axons) is $\rho_e = 30\ \Omega$-cm at 18° C. Estimate the resistivity of squid axoplasm by assuming that when an electric field is applied to axoplasm, current is carried by only Na^+, K^+, and Cl^- ions and that each of these ions has a diffusion coefficient and mobility in axoplasm equal to that found in aqueous solution. Concentrations of these ions in squid axoplasm are provided in Table 7.8. Compare your estimate with the measured value of resistivity, and briefly discuss the factors that might be responsible for any differences you found.

b. The measured resistivity of the external fluids around the squid axon is $\rho_e = 22\ \Omega$-cm, which is lower than that of axoplasm. Katz (1966, p.47) states, "The difference may be due to such factors as immobility

Table 7.8 Composition of axoplasm (Problem 7.1).

Ion	*Concentration (mmol/L)* Axoplasm	External	D (cm^2/s)	Ionic crystal radius (Å)
Sodium	50	460	1.33×10^{-5}	0.95
Potassium	400	10	1.96×10^{-5}	1.33
Chloride	100	540	2.03×10^{-5}	1.81

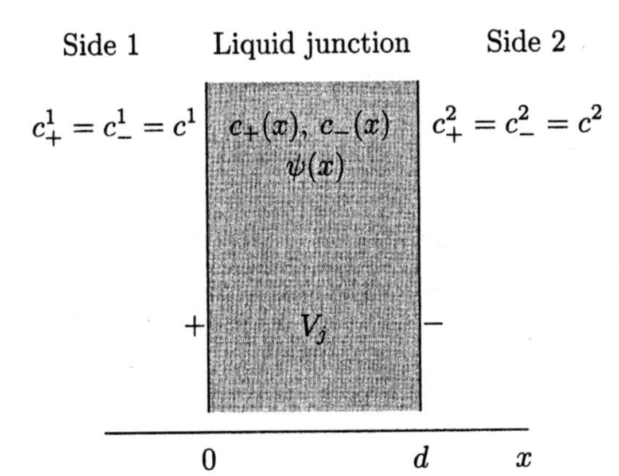

Figure 7.60 Liquid junction potential (Problem 7.2).

or low mobility of some of the internal anions and the presence of nonsolvent or membrane enclosed intracellular particles that do not contribute to current flow and may present obstacles to it." Discuss this suggestion from a quantitative perspective.

c. Assume that the Stokes-Einstein relation (see Section 3.3) holds for Na^+, K^+, and Cl^- in aqueous solution. Compare the apparent radius of each ion based on measurements of the diffusion constant in water with the radius of each ion obtained from measurements of ionic crystals. What factors could cause the differences? The viscosity of water at 25°C is $\eta \approx 0.9$ mPa · s.

7.2 A liquid junction potential exists across the boundary between two salt solutions. In this problem we explore a simple theory of such a liquid junction between two binary salt solutions that have the same ions, but at different concentrations. Let the positive ion (cation) be designated with a positive subscript and the negative ion (anion) be designated with a negative subscript. Both the cation and the anion are assumed to have unit valence. Assume that the flow of cations and anions in the boundary can be modeled by simple steady-state electrodiffusion and that the Nernst-Planck equation describes the current carried by each ion. The situation is shown in Figure 7.60. The boundary extends from $x = 0$ to $x = d$. Within the boundary assume that electroneutrality guarantees that $c_+(x) = c_-(x) = c(x)$ for $0 \le x \le d$. Also assume that particles do not accumulate at the interfaces of the two solutions, so that the concentrations are continuous at these interfaces. Assume that steady-state conditions guarantee that the total current through the boundary is zero, i.e., $J = J_+ + J_- = 0$.

a. Show that

$$\frac{d\psi(x)}{dx} = -\left(\frac{u_+ - u_-}{u_+ + u_-}\right)\frac{RT}{Fc(x)}\frac{dc(x)}{dx}.$$

b. Show that the potential across the boundary, called the *liquid junction potential*, V_j, is expressed by

$$V_j = \left(\frac{u_+ - u_-}{u_+ + u_-}\right)\frac{RT}{F}\ln\frac{c^2}{c^1}.$$

c. Let ϕ_+ and ϕ_- be the flux of cations and anions, respectively. Then define the total flux of salt as $\phi_t = \phi_+ + \phi_-$. Show that

$$\phi_t = -u_t RT\frac{dc(x)}{dx},$$

where the effective mobility of the salt solution is

$$u_t = \frac{4u_+u_-}{u_+ + u_-}.$$

d. Sketch u_t and $V_j/\left((RT/F)\ln(c^2/c^1)\right)$ versus u_+/u_-.

e. By examining the results obtained in the previous parts, answer the following questions precisely, and explain your answer.

   i. Is the salt solution in electrodiffusive equilibrium?

   ii. What is the relation between the mobility of the cation and the anion in uniform solutions and through the liquid junction? Why do they differ? Is the faster-moving ion (the one with the higher mobility) sped up or retarded in the junction? Why?

f. Two micropipettes filled with 2 mol/L of KCl, in contact with the intracellular and extracellular solutions, respectively, are used to measure the potential across the membrane of a cell whose intracellular solution is perfused with 150 mmol/L KCl and whose extracellular bath is perfused with 5 mmol/L KCl, with sucrose added to maintain osmotic equilibrium. Estimate the effect of the liquid junction potentials at the two electrodes on the measurement of the potential between the two electrodes.

g. Suppose you wish to measure the resting potential of a cell using a glass micropipette filled with a binary salt solution (such as shown in Figure 7.2). You are free to compose a binary salt from any cation/anion pair shown in Table 7.2. Which choice will minimize the liquid junctional potential? Explain.

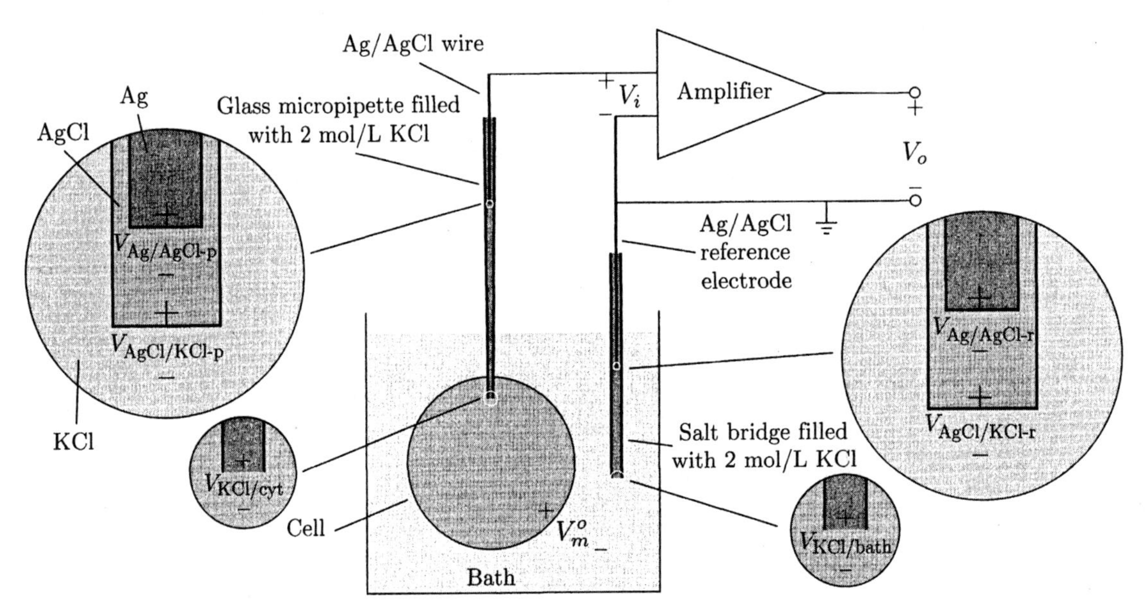

Figure 7.61 Arrangement of electrodes for measuring the resting potential of a cell (Problem 7.3). The circular insets indicate portions of the electrodes at higher magnification, so that the junctions between dissimilar materials can be defined.

7.3 A typical arrangement of electrodes and an amplifier used to measure the resting potential of a cell, V_m^o, are shown in Figure 7.61. This problem concerns some of the technical problems incurred in interpreting measurements of the resting potential. Starting with the positive terminal of the amplifier, the connections include a silver (Ag) wire with a coating of AgCl that is inserted into the micropipette containing a solution of 2 mol/L KCl. This pipette penetrates the cellular membrane, and the KCl in the lumen of the pipette is in contact with the cell cytoplasm, forming a liquid junction. The negative terminal of the amplifier is in contact with a similar electrode arrangement that is immersed in a bath, forming the extracellular solution of the cell. At each junction between dissimilar materials, there is a junctional potential that is defined in a figure insert. The potentials between the silver and the silver chloride in the micropipette and the reference electrodes are $V_{Ag/AgCl-p}$ and $V_{Ag/AgCl-r}$, respectively; the potentials between the silver chloride and the KCl solutions are $V_{AgCl/KCl-p}$ and $V_{AgCl/KCl-r}$; and the potentials between the KCl in the electrodes and between the cytoplasm and the bath are $V_{KCl/cyt}$ and $V_{KCl/bath}$.

a. Express the input voltage to the amplifier, V_i, in terms of the resting potential of the cell and the junctional potentials.

b. If the potential across the junction between the silver and the silver chloride layer, $V_{Ag/AgCl}$, is constant, how does that affect the results you found in part a)?

c. If a metal electrode is placed in an electrolyte, an electrolytic chemical reaction will take place between the metal and the ions in solution. For an arbitrary metal this chemical reaction may not be reversible, and the consequent difference in potential between the metal and the electrolyte will not be stable. Such an electrode is called *polarizable.* Most reactive metals such as copper form polarizable electrodes in salt solutions. The Ag/AgCl electrode is an example of a *nonpolarizable* electrode that is widely used to measure cellular potentials. At the AgCl/KCl boundary of a Ag/AgCl electrode, the reaction is reversible, since there are common charge carriers in both the metal and the electrolyte. This result can be seen from the reactions that occur at the junction, which are

$$AgCl \rightleftharpoons Ag^+ + Cl^-$$

and

$$Ag^+ + e \rightleftharpoons Ag,$$

where e denotes an electron. In the first reaction, silver chloride ionizes to produce silver and chloride ions; in the second, silver ions combine with electrons to form metallic silver. If these two reactions are combined, they yield

$$AgCl + e \rightleftharpoons Ag + Cl^-.$$

Clearly the equilibrium state of this reaction depends upon both the chloride concentration and the electric potential. It can be shown (Bull, 1964) that $V_{AgCl/KCl} = V^o_{AgCl/KCl} - (RT/F)\ln c_{Cl}$, where $V_{AgCl/KCl} \approx 222$ mV for a 1 mol/L solution of KCl at 25°C.

i. What is the net contribution of both $V_{Ag/AgCl-p}$ and $V_{Ag/AgCl-r}$ to the voltage at the input of the amplifier, V_i?

ii. Now suppose the salt bridge is removed from the reference electrode, so that the Ag/AgCl wire is immersed directly in the bath solution. Now what is the net contribution of $V_{Ag/AgCl-p}$ and

$V_{Ag/AgCl-r}$ to the voltage at the input of the amplifier, V_i? What happens to this contribution as the bath solution is changed?

7.4 The Goldman constant-field model yields a number of conclusions that are valid under more general conditions than those assumed in the derivation given in Appendix 7.1. This problem deals with an alternate derivation of some of the results given in Appendix 7.1.

Let $\vec{\phi}_n$ and $\overleftarrow{\phi}_n$ be outward and inward unidirectional fluxes, respectively, of ion n through a membrane. Therefore, the net flux of n is $\phi_n = \vec{\phi}_n - \overleftarrow{\phi}_n$. Let c_n^o and c_n^i be the outside and inside concentrations of ion n.

a. Assume that the outward flux is proportional to the inside concentration and the inward flux to the outside concentration of ion n. That is, $\vec{\phi}_n = P_n^i c_n^i$ and $\overleftarrow{\phi}_n = P_n^o c_n^o$, where P_n^i and P_n^o are independent of concentration. Show that

$$\frac{\vec{\phi}_n}{\overleftarrow{\phi}_n} = e^{(z_n F/RT)(V_m - V_n)} = \frac{c_n^i}{c_n^o} e^{(z_n F/RT)V_m},$$

where V_n is the Nernst equilibrium potential for the nth ion. All potentials are defined to be positive when the inside of the membrane is more positive than the outside. *Hint*: Note that when $V_m = V_n$, $\vec{\phi}_n / \overleftarrow{\phi}_n = 1$ and $\vec{\phi}_n - \overleftarrow{\phi}_n = 0$.

b. Now assume that all the ions are univalent ($|z_n| = 1$). From the results in part a, show that under steady-state conditions

$$V_m^o = \frac{RT}{F} \ln \left(\frac{\sum_{z_n=+1} P_n^o c_n^o + \sum_{z_n=-1} P_n^i c_n^i}{\sum_{z_n=+1} P_n^o c_n^i + \sum_{z_n=-1} P_n^i c_n^o} \right).$$

7.5 *Hint:* You may need to consult a handbook (such as the *Handbook of Chemistry and Physics*) to complete all parts of this problem.

Consider a uniform, electrically isolated spherical cell 50 μm in diameter with a membrane thickness of 70 Å. The membrane capacitance of the cell is $C_m = 1\,\mu\text{F/cm}^2$, and the resting potential of the cell is $V_m^o = -70$ mV. The cell membrane is permeable only to potassium ions and the membrane conductance for potassium is $G_K = 10^{-3}\,\text{S/cm}^2$ and is constant. The only positively charged ion in the cytoplasm is potassium, which is maintained at a constant concentration of $c_K^i = 0.2$mol/L. The extracellular concentration of potassium is maintained constant at 0.002 mol/L. The temperature is 24°C.

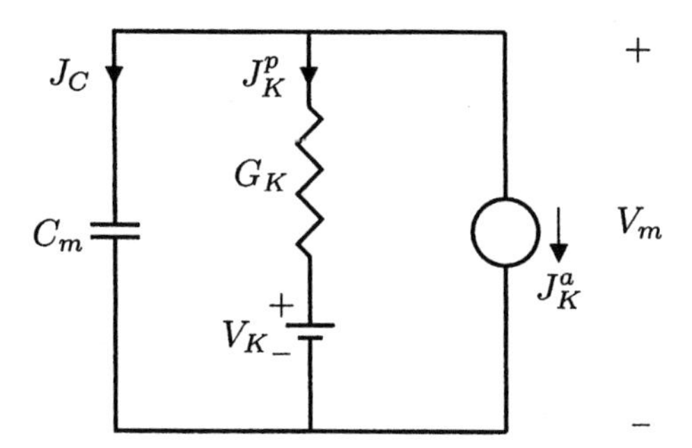

Figure 7.62 Network model of passive and active transport (Problem 7.5).

a. Compute the equivalent dielectric constant of the membrane, and compare this value to that found for typical electrical insulators (glass, oil).

b. Compute the electric field strength in the membrane (assuming it is a uniform insulator), and compare this value to the dielectric breakdown field strength of typical electrical insulators (glass, oil).

c. Assume that the presence of the potential across the membrane in the absence of a net current through the membrane implies a separation of charge across the membrane. Compute the surface charge per unit area at the membrane-solution interface. What fraction of the ions inside the cell is required to produce this surface charge?

d. Assume that the membrane is a homogeneous conductor, and compute the bulk resistivity (ρ_m) of the membrane material, then compare this value to that found for typical good conductors (copper), ionic solutions, semiconductors (germanium, silicon), and insulators (glass, oil).

Since the membrane is permeable only to potassium, we shall assume that the equivalent circuit shown in Figure 7.62 represents the electrical properties of a unit area of membrane. J_c is the capacitance current density, and J_n^p and J_n^a are the passively and actively transported potassium current densities, respectively

e. What is the value of the potassium equilibrium potential, V_K?

f. Under resting conditions, how much current per unit area is supplied by active transport, i.e., what is J_K^a?

Suppose that the cell is poisoned by a substance whose only effect is to set $J_K^a = 0$. You may assume that c_K^o remains constant.

g. Estimate the value of the resting potential of the poisoned cell. State all your assumptions explicitly. Estimate the change in c_K^i that results from the poisoning.

7.6 The experiments of Hodgkin and Keynes described in Section 7.3 showed that both the mobility and the diffusion coefficient of potassium ions in squid cytoplasm did not differ appreciably from their values in aqueous solutions. This problem involves interpretation of the results of these experiments as shown in Figures 7.10 and 7.11.

a. Draw a sketch of an axon, and indicate the direction of the electric field consistent with the measurements.

b. From the measurements in the absence of an electric field, estimate the diffusion coefficient of potassium in the axon.

c. From the measurements in the presence of an electric field, estimate both the molar mechanical mobility and the diffusion coefficient of potassium in the axon.

d. Compare the results obtained in part c with those predicted by the Einstein relation.

7.7 The following experiment is performed to indicate the presence of a membrane potential component attributable to active transport mechanisms. Sodium ions are injected intracellularly into a neuron without passing a net current through the membrane. The effects of changes in concentration of ions other than sodium are demonstrated to be negligible. Figure 7.63 depicts a record of the time course of the injection rate of sodium and a record of the corresponding change in membrane potential. Sodium is injected at a constant rate of 48×10^{-12}mol/min for 0.5 minutes. The membrane potential, V_m, starts at its resting value of -60 mV and hyperpolarizes (approximately linearly) to -76 mV during the injection of sodium. When the sodium injection ceases, the membrane potential returns to its resting value with an exponential time course whose time constant is 5 minutes. A model for the transport of ions across the membrane of the neuron is shown in Figure 7.64. $\mathcal{G}_{Na}$, $\mathcal{G}_K$, and C can be assumed to be constants. The current sources, I_{Na}^a and I_K^a, represent the active transport of ions across the membrane.

a. Show that passive transport mechanisms cannot account for the data observed by calculating the expected change in membrane potential in response to the injection of sodium ions. Assume that I_{Na}^a and I_K^a are

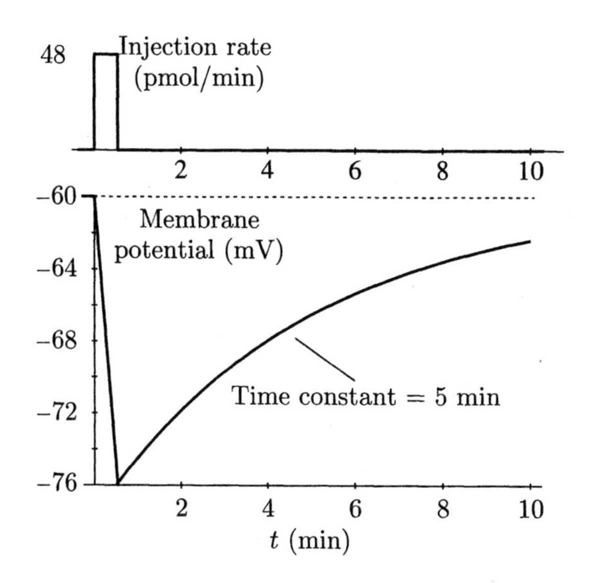

Figure 7.63 Effect of injecting sodium ions into a nerve cell (Problem 7.7).

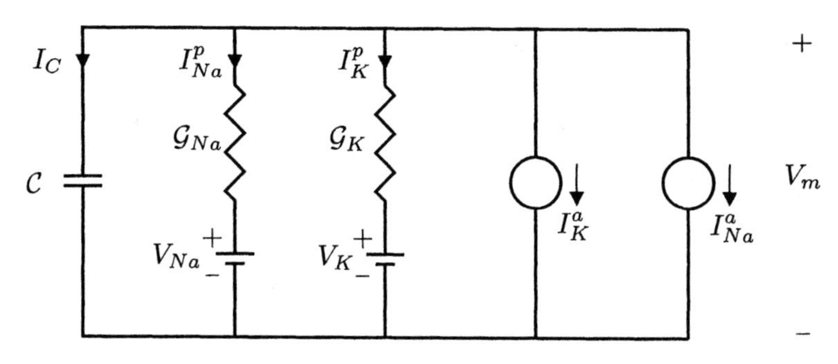

$\mathcal{G}_{Na} = 5$ nS Volume of cell, $\mathcal{V}_c = 4 \times 10^{-6}$ cm^3
$\mathcal{G}_K = 0.5\ \mu$S Normal intracellular concentration
$C = 10^{-9}$ F of sodium $c_{Na}(0) = 6 \times 10^{-6}$ mol/cm^3

Figure 7.64 Model of a neuron (Problem 7.7).

constant and that under resting conditions these currents maintain constant ion concentrations.

b. An active transport (sodium pump) mechanism is proposed to account for the data. Assume that

$$I^a_{Na}(t) = I^a_{Na}(0) + \Delta I^a_{Na}(t)$$

and

$$c_{Na}(t) = c^i_{Na}(0) + \Delta c^i_{Na}(t),$$

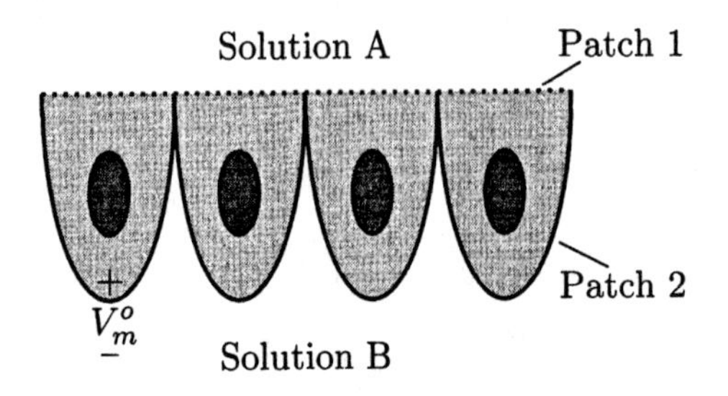

Figure 7.65 An epithelium is shown that consists of a layer of cells each of which contains two membrane patches (Problem 7.8).

Table 7.9 Cell with two membrane patches (Problem 7.8).

Ion	*Concentration (mmol/L)* Cytoplasm	*Solution* A	B
Sodium	15	15	150
Potassium	150	150	15

where $\Delta I_{Na}^a(t)$ is the deviation of the sodium pump current from its normal resting value, $I_{Na}^a(0)$, and $\Delta c_{Na}(t)$ is the deviation of the intracellular sodium concentration from its normal resting value, $c_{Na}^i(0)$. Then

$$\Delta I_{Na}^a(t) = \frac{F\mathcal{V}_c}{\tau_a}\Delta c_{Na}^i(t),$$

where Δc_{Na}^i is in mol/cm^3, F is Faraday's constant (96,500 C/mol), and τ_a is the time constant (in seconds). All other parameters are the same as in part a. Prove that the model predicts the measured change in membrane potential.

7.8 Each small cell shown in Figure 7.65 consists of two uniform patches of membrane, 1 and 2, that differ in permeability; patch 1 is permeable to Na^+ and K^+, while patch 2 is permeable to K^+ only. The potassium conductance of each patch is the same and equals G siemens, and the sodium conductance of patch 1 is the same as the potassium conductance. This cell is part of an epithelium that separates two solutions, A and B. Because of the locations of tight junctions in the epithelium, patch 1 is in contact with solution A, and patch 2 is in contact with solution B. The Na^+ and K^+ compositions of the cell's cytoplasm and of solutions A and B are given in Table 7.9. You may assume that there are no electrical junctions between this cell and its neighbors, that the ionic

Table 7.10 Cell that transports sodium and potassium by active and passive transport (Problem 7.9).

Ion	*Concentration (mmol/L)*	
	Internal	External
Sodium	15	106
Potassium	150	3

pumps that maintain the cell's composition are nonelectrogenic, and that the electric potential difference and the electrical resistance between solutions A and B are both zero.

a. Determine an equivalent electrical network model of this cell that is appropriate for determining its resting potential.

b. Determine the numerical value of the resting potential, V_m^o.

7.9 A uniform isolated small cell has a membrane that is permeable only to sodium and potassium ions and contains an active transport mechanism that transports three sodium ions outward and two potassium ions inward for every molecule of ATP split into ADP and phosphate. Summed over the entire membrane of this cell, the active transport system splits 10^{-17} moles of ATP per second. Assume that the cell is at quasi-equilibrium, so that the concentrations of all ions are constant. The cell has a total membrane conductance of 10^{-10} siemens. The temperature is 24°C. The ionic concentrations of sodium and potassium across the membrane are given in Table 7.10. The potassium conductance exceeds the sodium conductance of this cell.

a. Determine the value of the component of the resting membrane potential, V_m^o, that is directly attributable to active transport.

b. Determine the value of the resting membrane potential, V_m^o.

c. Determine the values of the sodium, $\mathcal{G}_{Na}$, and potassium, $\mathcal{G}_K$, conductances of the membrane.

7.10 A plant cell whose intracellular (sap) concentration for sodium, potassium, and chloride is shown in Table 7.11 is immersed in an external solution whose temperature is 24°C and whose composition is given in Table 7.11. The composition of the sap is constant in time, as is the difference of potential between the inside and the outside of this cell; the inside of the cell is at a potential of −86 mV with respect to the outside. The passive transport of each ion through the membrane is independent

Table 7.11 Passive and active transport in a plant cell (Problem 7.10).

Ion	*Concentration (mmol/L)*	
	Sap	External
Sodium	15	0.1
Potassium	28.7	1.0
Chloride	38	1.3

of the concentrations of all other solutes to which the membrane is permeable. For each of the ions shown in Table 7.11, determine if that ion is transported across the membrane by an active transport mechanism and, if so, in which direction across the membrane (inward or outward) it is transported.

7.11 The resting membrane potential, V_m^o, of two uniform isolated cells is measured as a function of the external concentration of potassium, c_K^o, with the sodium concentration held fixed at its normal value, c_{Na}^{on}, and then as a function of the external sodium concentration, c_{Na}^o, with the potassium concentration held fixed at its normal value, c_K^{on}. The results for these two cells are shown in Figure 7.66. You may assume that for each cell: (1) the external solutions are isotonic, (2) the membranes are impermeable to ions other than potassium and sodium, (3) the internal concentrations of potassium and sodium are maintained constant by a nonelectrogenic active transport mechanism, (4) the total membrane conductance is 10 nS, (5) the normal resting potential is −60 mV, (6) the internal concentration of sodium is 20 mmol/L.

a. For each cell, determine the total conductance of the membrane to potassium and to sodium, $\mathcal{G}_K$ and $\mathcal{G}_{Na}$, respectively. Enter the results into Table 7.12. If the value is indeterminate from the information given, put an × in the appropriate space.

b. For each cell, determine the internal and external concentrations of potassium and the external concentration of sodium at the normal resting potential (−60 mV) and enter the results into Table 7.12. If the value is indeterminate from the information given, put an × in the appropriate space.

c. For each cell, determine the simplest equivalent electrical network model that relates the dependence of the resting potential of the cell on the ion concentrations. Indicate the values of all elements in the network.

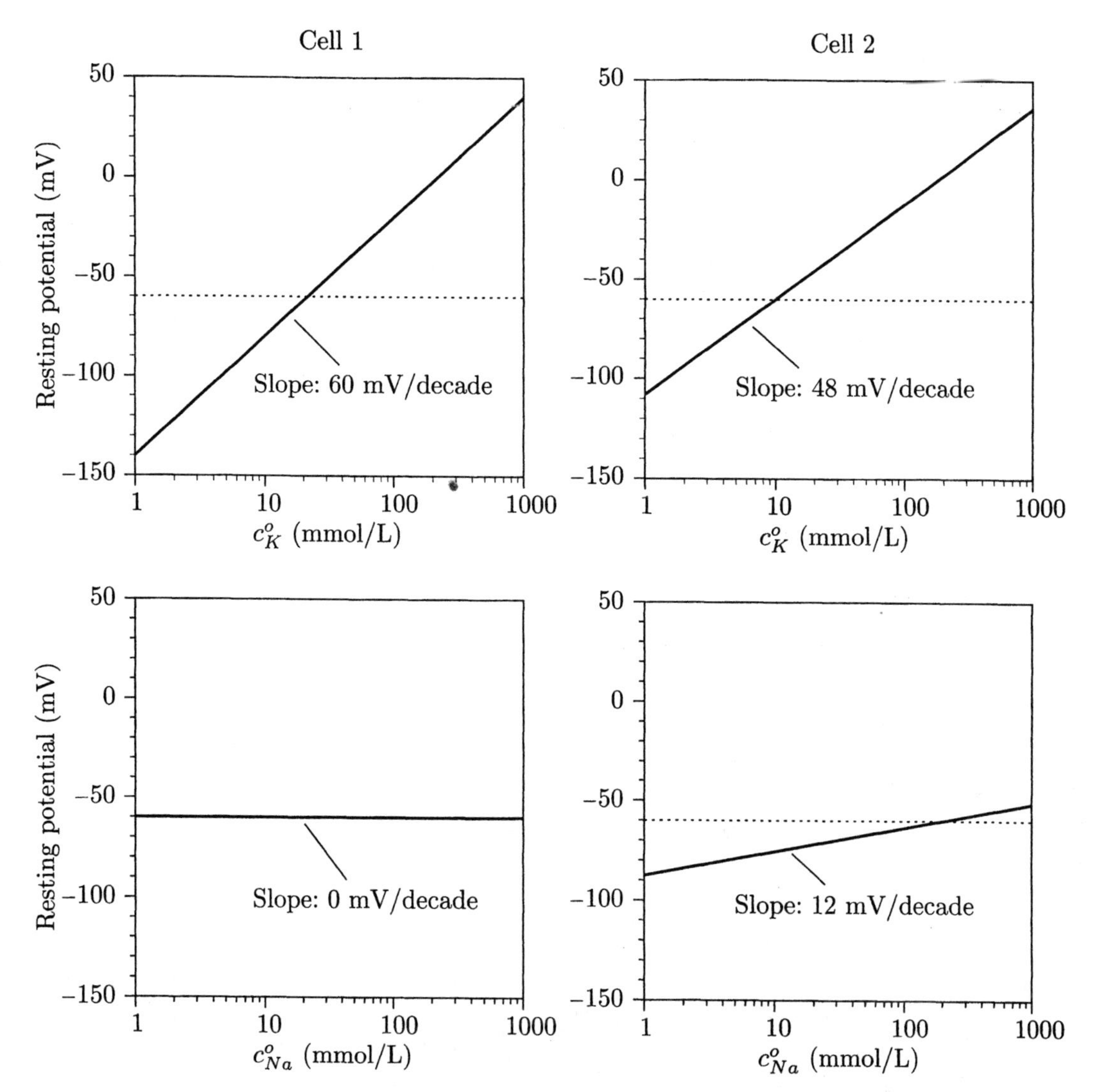

Figure 7.66 Effect of changes in ion concentration on resting potential (Problem 7.11).

7.12 A micropipette is used to puncture the membrane of a cell so that the membrane potential can be measured as shown schematically in Figure 7.67. Assume that the cell membrane behaves as a potassium electrode and that the normal concentrations of potassium are 150 mmol/L inside and 15 mmol/L outside the cell. The membrane conductance of the cell is 10^{-3} S/cm^2, and the cell has a diameter of 30 μm. The puncture of the membrane introduces a small leakage path between the inside

Table 7.12 Effect of changes in ion concentration on resting potential (Problem 7.11).

	Cell 1	Cell 2
$\mathcal{G}_K$ (nS)		
$\mathcal{G}_{Na}$ (nS)		
c_K^{on} (mmol/L)		
c_K^{in} (mmol/L)		
c_{Na}^{on} (mmol/L)		
c_{Na}^{in} (mmol/L)	20	20

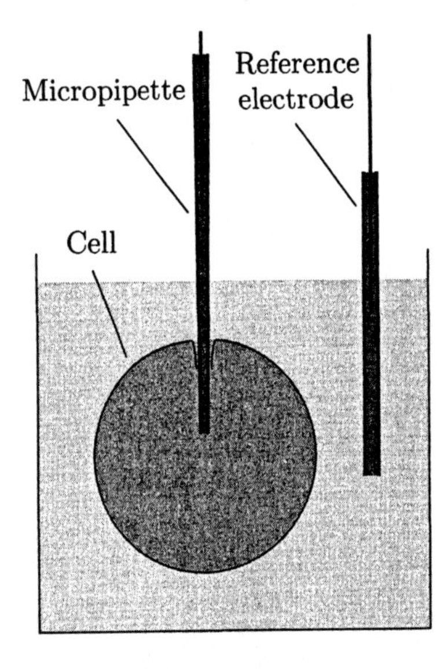

Figure 7.67 Effect on resting potential of the shunt resistance introduced by puncturing a cellular membrane with a micropipette (Problem 7.12).

and the outside of the cell. This leakage pathway can be represented by a constant resistance whose value is 100 MΩ. You may assume that the intracellular concentration remains constant and that the temperature is 24°C.

a. Determine the value of the resting potential of the cell before the cell is punctured by the micropipette.

b. Determine an equivalent circuit of the cell with the micropipette in place. You may assume that the micropipette resistance as seen from the tip is negligibly high.

c. Determine the value of the resting potential of the cell after the cell is punctured by the micropipette.

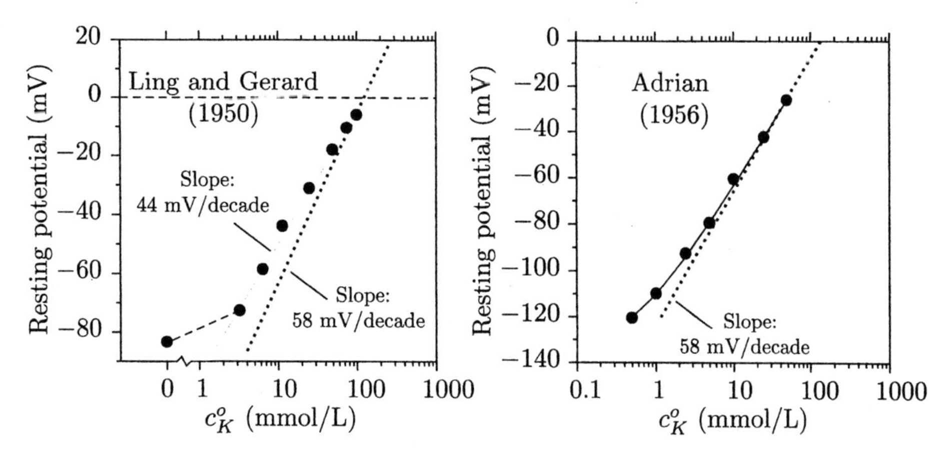

Figure 7.68 Two sets of measurements (Ling and Gerard, 1950; Adrian, 1956) of the dependence of the resting potential of frog sartorius muscle fibers on external potassium concentration c_K^o (Problem 7.13). The dotted lines have slopes predicted for a perfect potassium electrode.

d. Determine the dependence of the resting potential on external potassium concentration with the micropipette in place.

e. Discuss the results shown in Figure 7.22 for the frog skeletal muscle in light of the conclusions reached in this problem.

7.13 Figure 7.68 shows two sets of measurements of the resting potential of frog sartorius muscle fibers as a function of the potassium concentration in an extracellular Ringer's solution. One set of measurements (Ling and Gerard, 1950) were among the first measurements obtained with intracellular micropipettes, while the other measurements (Adrian, 1956) were obtained later. In both experiments, the potassium concentration was varied by substituting equimolar quantities of KCl for NaCl in the Ringer's solution bathing the muscle. While the two sets of measurements are qualitatively similar, there are quantitative differences. For example, the slopes of the measurements at large c_K^o differ; the data of Ling and Gerard fall on a line with a slope of 44 mV/decade, whereas those of Adrian approach a line with a slope of 58 mV/decade, as would be expected for a perfect potassium electrode.

Assume that the total membrane conductance of the sartorius muscle fiber is $\mathcal{G}_m = 2 \times 10^{-6}$ S and that the intracellular concentration of ions did not change during the measurements shown in Figure 7.68.

a. For Adrian's measurements:

   i. Estimate the potassium conductance, $\mathcal{G}_K$, for large values of c_K^o.

   ii. Estimate the intracellular concentration of potassium, c_K^i.

b. For the purposes of this part of the problem, assume that the measurements of Adrian accurately represent the relation between resting potential and extracellular potassium concentration for frog sartorius muscle fibers. The purpose of this part is to evaluate the hypothesis that in the measurements of Ling and Gerard penetration of the muscle cell membrane by a micropipette introduced an appreciable leakage path between the inside and the outside of the muscle cell. This leakage path was assumed to be caused by the hole in the membrane created by, but not filled by, the micropipette.

   i. For the measurements of Ling and Gerard, construct an electrical network model that consists of two conductances and two batteries and that takes into account the possibility of the leakage path. Explain what each element represents.

   ii. Determine all four elements from the measurements at high c_K^o. For all elements whose values are constant, evaluate the constant. For all elements whose values are not constant, simplify the expression as much as possible.

   iii. Discuss whether your calculations support or contradict the hypothesis about the Ling and Gerard measurements.

   iv. Which features of the measurements, if any, will your model not explain?

7.14 The inner ear of a mammal, which contains the organs of hearing and equilibrium, consists of two fluid-filled compartments separated by a multicellular membrane and enclosed in the temporal bone of the skull (Figure 7.69). The two compartments contain solutions, *perilymph* and *endolymph*, whose sodium and potassium concentrations differ (Table 7.13). There is also a constant difference of potential between these compartments called the *endolymphatic potential*, $V_{EP} = 100$ mV (Figure 7.69). You may assume that the inner ear temperature in mammals is 38°C.

a. In this part of the problem, assume that the endolymphatic potential results from passive transport of sodium and potassium across the membrane. To which of these ions is the membrane more permeable? Support your conclusion with a quantitative analysis.

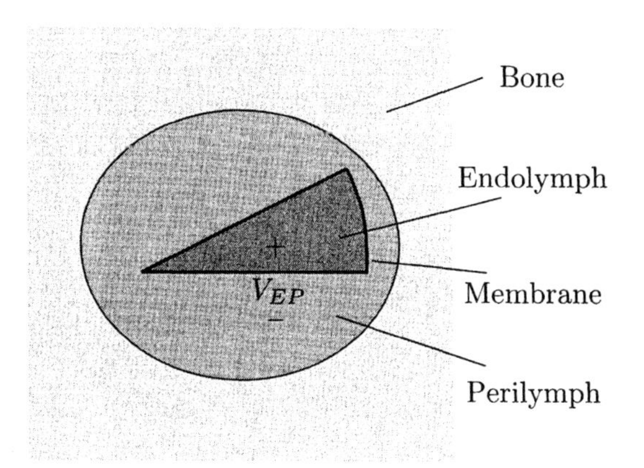

Figure 7.69 Schematic diagram of the inner ear of a mammal (Problem 7.14). An epithelium (membrane) separates two liquid-filled compartments that are filled with two fluids of very different compositions—endolymph and perilymph.

Table 7.13 Composition of inner ear fluids (Problem 7.14).

Ion	*Concentration (mmol/L)*	
	Perilymph	Endolymph
Sodium	140	2
Potassium	5	150

b. Experiments are performed to test the hypothesis that passive transport of sodium and/or potassium accounts for the endolymphatic potential. In separate experiments, the endolymphatic potential is measured as the concentration of sodium, and the concentration of potassium in perilymph is changed (Figure 7.70). Are these results consistent with the hypothesis? Support your conclusion with a quantitative analysis.

c. Describe a mechanism, involving the transport of sodium and potassium only, that would account for the data on the endolymphatic potential that are given in Table 7.13 and Figure 7.70. Give an equivalent circuit for this mechanism, and support your conclusions with a quantitative analysis that demonstrates how your model is consistent with these data.

7.15 The membrane of a cell is known to be permeable to only Na^+ and K^+, with passive ionic conductances of $G_{Na} = 10^{-6}\,S/cm^2$ and $G_K = 10^{-3}\,S/cm^2$, and contains a (Na^+- K^+)-ATPase active transport mechanism that transports three molecules of Na^+ outward and two molecules of K^+ inward through the membrane for each molecule of ATP split into

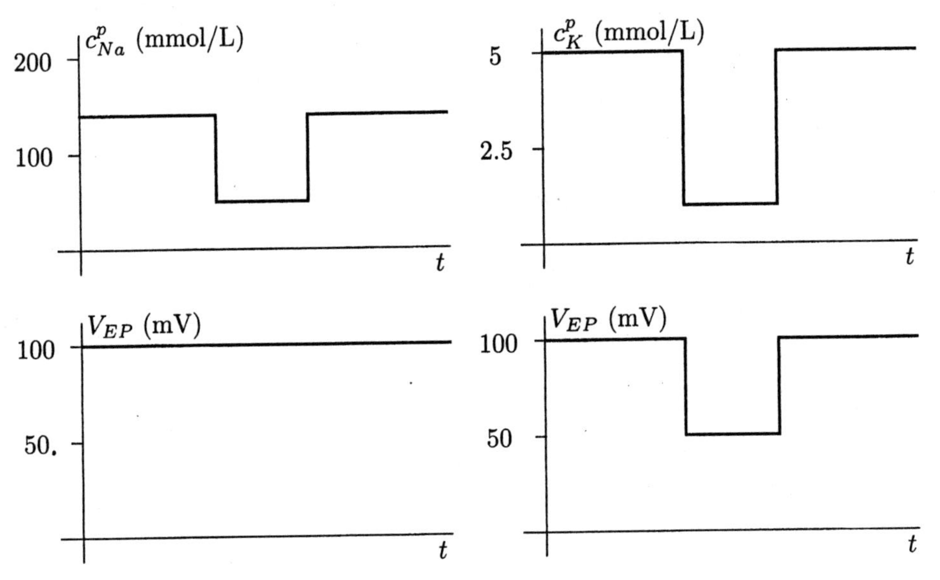

Figure 7.70 Effect of changes in perilymphatic sodium (left-hand side) and potassium (right-hand side) concentration on the endolymphatic potential (Problem 7.14).

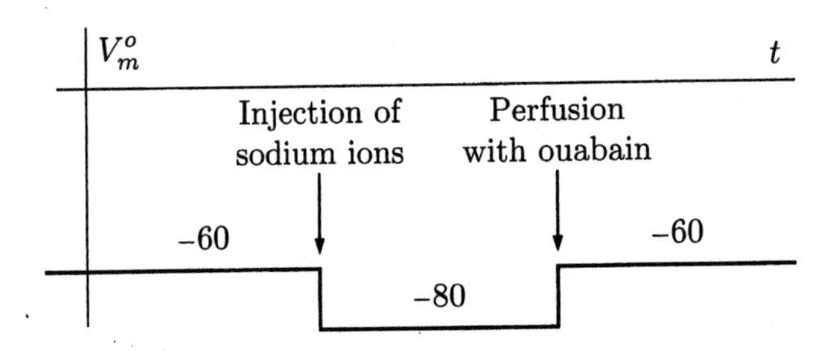

Figure 7.71 Effect of sodium injection and ouabain perfusion on resting potential (Problem 7.15).

ADP and phosphate. In the experiment shown in Figure 7.71, the cell is kept at a temperature of 24°C. For $t < t_1$, the cell is in its normal resting state, with resting potential $V^o_m \approx -60$ mV and the resting ion concentrations shown in Table 7.14. At $t = t_1$ Na^+ is injected rapidly into the cell interior, without the passage of any current through the membrane, to double the intracellular sodium concentration. When the injection is completed, the membrane potential hyperpolarizes to approximately −80 mV. At $t = t_2$ ouabain, a blocker of the active transport mechanism, is applied to the cell, and the membrane potential returns to approximately its normal resting value.

Table 7.14 Concentrations of ions (Problem 7.15).

Ion	Concentration (mmol/L)	
	Internal	External
Sodium	15	135
Potassium	156	15

a. Is the active transport mechanism electrogenic or nonelectrogenic? Explain.

b. Does the active transport mechanism contribute appreciably to the resting potential for $t < t_1$? Explain.

c. How many moles of ATP are split per second by a cm^2 of membrane during the interval (t_1, t_2)?

7.16 The pH of extracellular solutions is typically 7.3. Consider a cell with a resting potential of -70 mV.

a. If H^+ were transported passively across the membrane, what would be the value of the intracellular pH?

b. Measurements indicate that the intracellular pH is typically in the range of 7.0–7.2. What can you conclude about the transmembrane transport of H^+?

7.17 Dr. Tropsnart Evitca has found a cell with a resting potential of -70 mV and a sodium ion concentration $c^o_{Na}/c^i_{na} = 8$. On the basis of measurements using radioactive tracers, the good doctor has proposed that Na^+ is transported actively across the membrane of this cell and that four molecules of Na^+ are extruded from the cell for each molecule of ATP hydrolyzed. At the intracellular concentration of ATP, you may assume that hydrolyzing one mole of ATP liberates 30 kJ of energy. Is the doctor's model energetically feasible?

7.18 This problem concerns estimates of the diffusion coefficient and mobility of calcium in the axoplasm of squid giant axons with methods such as those used for potassium (Section 7.3). Figure 7.72 shows the count rate of radioactivity (which is proportional to the calcium concentration) along a squid giant axon 159 minutes after injecting $^{45}CaCl_2$ intracellularly along the axon.

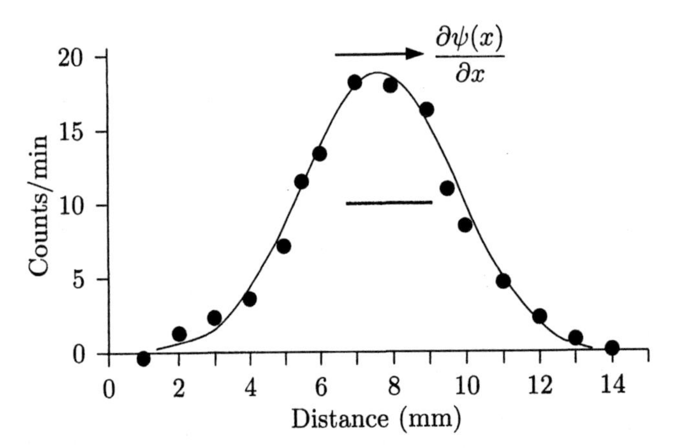

Figure 7.72 Measurements of count rate as a function of distance along a squid axon (Hodgkin and Keynes, 1957) (Problem 7.18). The initially injected region is shown with a horizontal line. The measurements (filled circles) were obtained 159 minutes later. For 110 minutes a current was passed through the axoplasm with a voltage gradient of 0.51 V/cm. The solid line is a calculation that was fitted to the measurements (see text).

a. This part deals with estimates of the molar mechanical mobility.

   i. Estimate the displacement of the center of gravity of the count rate.

   ii. From the estimate in part i, estimate the molar mechanical mobility of calcium.

   iii. Compare this estimate of the molar mechanical mobility of calcium in the axoplasm with the molar mechanical mobility of calcium in water.

b. This part deals with estimates of the diffusion coefficient. The diffusion of solute from a pulse of concentration of width $2a$ (see Section 3.5) can be expressed in the form

$$c(x,t) = \frac{C_0}{2}\left(\text{erf}\left(\frac{x+a}{2\sqrt{Dt}}\right) - \text{erf}\left(\frac{x-a}{2\sqrt{Dt}}\right)\right),$$

where

$$\text{erf}(x) = \frac{2}{\sqrt{\pi}}\int_0^x e^{-y^2}\,dy.$$

The presence of a potential gradient shifts this spatial distribution a fixed distance that is proportional to elapsed time.

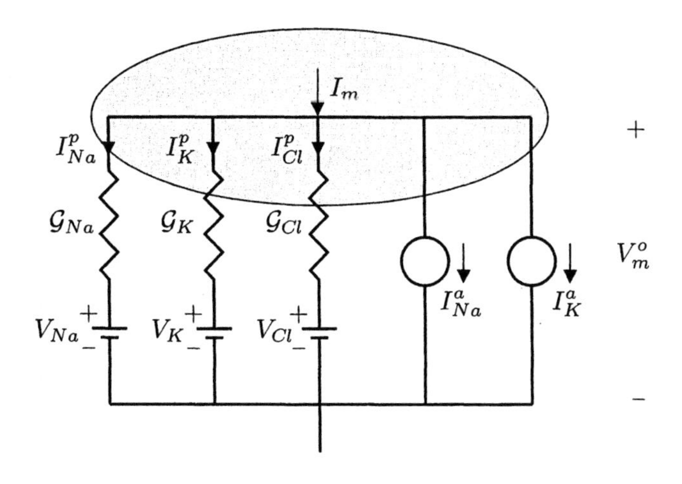

	c^i_n	c^o_n	V_n	g_n/g_K
	(mmol/L)		(mV)	
Na^+	10	140	+68	0.1
K^+	140	10	−68	1
Cl^-		150		1

Figure 7.73 Model of a cell whose membrane contains both active and passive transport of sodium and potassium, but only passive transport of chloride, (Problem 7.19). The intracellular and extracellular concentrations, Nernst equilibrium potentials, and conductance ratios are given for sodium and potassium. Some information is also given for chloride; blank entries represent unknown quantities.

i. The curve shown in Figure 7.72 was drawn for $a = 0.116$ cm and $2\sqrt{Dt} = 0.262$ cm. Estimate the diffusion coefficient for these measurements.

ii. Compare this estimate of the diffusion coefficient of calcium in the axoplasm with the diffusion coefficient of calcium in water.

c. What do you conclude from these measurements?

7.19 Consider the model of a cell shown in Figure 7.73. The cell has channels for the passive transport of sodium, potassium, and chloride as well as a pump that actively transports sodium out of the cell and potassium into the cell. The pump ratio is $I^a_{Na}/I^a_K = -1.5$. The intracellular and extracellular concentrations, Nernst potentials, and conductance ratios are given in Figure 7.73. The cell also contains impermeant intracellular ions. Assume that the cell is in equilibrium at $t = 0$, i.e., assume that at $t = 0$ the cell has reached a condition for which all solute concentrations, the cell volume, and the membrane potential are constant.

a. Choose one of the following statements, and explain why it is true.
   i. The cell resting potential depends on $\mathcal{G}_{Cl}$.
   ii. The cell resting potential depends on V_{Cl}.
   iii. The cell resting potential depends on both $\mathcal{G}_{Cl}$ and V_{Cl}.
   iv. The cell resting potential does not depend on $\mathcal{G}_{Cl}$.
   v. The cell resting potential does not depend on V_{Cl}.
   vi. The cell resting potential does not depend on either $\mathcal{G}_{Cl}$ or V_{Cl}.

b. Determine V_m^o.

c. At $t = 0$, the external concentration of chloride is reduced from 150 mmol/L to 50 mmol/L by substituting an isosmotic quantity of an impermeant anion for chloride. Assume that the concentrations of sodium and potassium both inside and outside the cell remain the same and that the volume of the cell does not change.
   i. Determine $V_m^o(0+)$, the value of the membrane potential immediately after the change in solution. You may ignore the effect of the membrane capacitance.
   ii. Determine $V_m^o(\infty)$, the value of the membrane potential after the cell has equilibrated.
   iii. Determine $c_{Cl}^i(\infty)$, the intracellular chloride concentration after the cell has equilibrated.
   iv. Give a physical interpretation of your results in parts i, ii, and iii.
   v. Discuss the validity of the assumptions that the sodium and potassium concentrations in the cell are constant and that the volume does not change.

References

Books and Reviews

Blostein, R. (1989). Ion pumps. *Curr. Opinion Cell Biol.*, 1:746–752.

Bull, H. B. (1964). *An Introduction to Physical Biochemistry*. F. A. Davis, Philadelphia.

Caldwell, P. C. (1968). Factors governing movement and distribution of inorganic ions in nerve and muscle. *Physiol. Rev.*, 48:1–64.

Cohen, L. B. and De Weer, P. (1977). Structural and metabolic processes directly related to action potential propagation. In Brookhart, J. M. and Mountcastle, V. B., eds., *Handbook of Physiology,* sec. 1, *The Nervous System,* vol. 1, pt. 1. Williams & Wilkins, Baltimore, MD.

Davson, H. (1970). *A Textbook of General Physiology*. J. and A. Churchill, Gloucester, England.

Dean, R. B. (1941). Theories of electrolyte equilibrium in muscle. In Cattell, J., ed., *Biological Symposia*, vol. 3, *Muscle*, 331–348. Jaques Cattell, Lancaster, PA.

De Weer, P. (1984). Electrogenic pumps: Theoretical and practical considerations. In Blaustein, M. P. and Lieberman, M., eds., *Electrogenic Transport: Fundamental Principles and Physiological Implications*, 1–15. Raven, New York.

De Weer, P. (1986). The electrogenic sodium pump: Thermodynamics and kinetics. *Fortsch. Zool.*, 33:387–399.

De Weer, P. (1992). Cellular sodium-potassium transport. In Seldin, D. W. and Giebisch, G., eds., *The Kidney: Physiology and Pathophysiology*, 93–112. Raven, New York.

De Weer, P., Gadsby, D. C., and Rakowski, R. F. (1988). Voltage dependence of the Na-K pump. *Ann. Rev. Physiol.*, 50:225–241.

Glynn, I. M. (1968). Membrane adenosine triphosphatase and cation transport. *Br. Med. Bull.*, 24:165–169.

Glynn, I. M. (1984). The electrogenic sodium pump. In Blaustein, M. P. and Lieberman, M., eds., *Electrogenic Transport: Fundamental Principles and Physiological Implications*, 33–48. Raven, New York.

Glynn, I. M. and Ellory, C., eds. (1985). *The Sodium Pump*. Company of Biologists, Cambridge, England.

Glynn, I. M. and Karlish, S. J. D. (1990). Occluded cations in active transport. *Ann. Rev. Biochem.*, 59:171–205.

Grodzinsky, A. J. and Melcher, J. R. (1975). Fields, forces and flows: Background for physiology. Notes for subject 6.561, Department of Electrical Engineering and Computer Science, Massachusetts Institute of Technology.

Hodgkin, A. L. (1951). The ionic basis of electrical activity in nerve and muscle. *Biol. Rev.*, 26:339–409.

Hodgkin, A. L. (1964). *The Conduction of the Nervous Impulse.* Charles C. Thomas, Springfield, IL.

Jørgensen, P. L. (1975a). Isolation and characterization of the components of the sodium pump. *Quart. Rev. Biophys.*, 7:239–274.

Katz, B. (1966). *Nerve, Muscle, and Synapse.* McGraw-Hill, New York.

Keynes, R. D. (1975). Organization of the ionic channels in nerve membranes. In Tower, D. B., ed., *The Nervous System*, vol. 1, *The Basic Neurosciences*. Raven, New York.

Kuffler, S. W., Nicholls, J. C., and Martin, A. R. (1984). *From Neuron to Brain: A Cellular Approach to the Function of the Nervous System*, 2d ed. Sinauer, Sunderland, MA.

Läuger, P. (1991). *Electrogenic Ion Pumps*. Sinauer, Sunderland, MA.

Ling, G. N. (1962). *A Physical Theory of the Living State: The Association-Induction Hypothesis*. Blaisdell, New York.

MacInnes, D. A. (1961). *The Principles of Electrochemistry*. Dover, New York.

Nicholls, J. C., Martin, A. R., and Wallace, B. G. (1992). *From Neuron to Brain: A Cellular and Molecular Approach to the Function of the Nervous System*, 3d ed. Sinauer, Sunderland, MA.

Post, R. L. (1981). The sodium and potassium ion pump. In Singer, T. P. and Ondarza, R. N., eds., *Molecular Basis of Drug Action*, 299–331. Elsevier, New York.

Ritchie, J. M. (1971). Electrogenic ion pumping in nervous tissue. *Curr. Topics Bioenerg.*, 4:327–356.

Robinson, R. A. and Stokes, R. H. (1959). *Electrolyte Solutions*. Butterworths, London.
Schuurmans Stekhoven, F. and Bonting, S. L. (1981). Transport adenosine triphosphatases: Properties and functions. *Physiol. Rev.*, 61:1-76.
Skou, J. C. (1965). Enzymatic basis for active transport of Na^+ and K^+ across cell membrane. *Physiol. Rev.*, 45:596-616.
Skou, J. C. (1975). The (Na^+ + K^+) activated enzyme system and its relationship to transport of sodium potassium. *Quart. Rev. Biophys.*, 7:401-434.
Skou, J. C. (1988). Overview: The Na,K-pump. In Fleisher, S. and Fleisher, B., eds., *Biomembranes*, pt. P, *ATP-Driven Pumps and Related Transport: The Na,K-Pump*, 1-25. Academic Press, New York.
Thomas, R. C. (1972a). Electrogenic sodium pump in nerve and muscle cells. *Physiol. Rev.*, 52:563-594.
Weiss, T. F. (1996). *Cellular Biophysics*, vol. 2, *Electrical Properties*. MIT Press, Cambridge, MA.

Original Articles

Adrian, R. H. (1956). The effect of internal and external potassium concentration on the membrane potential of frog muscle. *J. Physiol.*, 133:631-658.
Adrian, R. H. and Slayman, C. L. (1966). Membrane potential and conductance during transport of sodium, potassium and rubidium in frog muscle. *J. Physiol.*, 184:970-1014.
Allbritton, N. L., Meyer, T., and Stryer, L. (1992). Range of messenger action of calcium ion and inositol 1,4,5-triphosphate. *Science*, 258:1812-1815.
Apell, H. (1989). Electrogenic properties of the Na,K pump. *J. Membr. Biol.*, 110:103-114.
Baker, P. F., Blaustein, M. P., Hodgkin, A. L., and Steinhardt, R. A. (1969). The influence of calcium on sodium efflux in squid axons. *J. Physiol.*, 200:431-458.
Baker, P. F., Blaustein, M. P., Keynes, R. D., Manil, J., Shaw, T. I., and Steinhardt, R. A. (1969). The ouabain-sensitive fluxes of sodium and potassium in squid giant axons. *J. Physiol.*, 200:459-496.
Baker, P. F., Foster, R. F., Gilbert, D. S., and Shaw, T. I. (1971). Sodium transport by perfused giant axons of *loligo*. *J. Physiol.*, 219:487-506.
Baker, P. F., Hodgkin, A. L., and Ridgeway, E. B. (1971). Depolarization and calcium entry in squid giant axons. *J. Physiol.*, 218:709-755.
Baker, P. F., Hodgkin, A. L., and Shaw, T. I. (1961). Replacement of the protoplasm of a giant nerve fibre with artificial solutions. *Nature*, 190:885-887.
Baker, P. F., Hodgkin, A. L., and Shaw, T. I. (1962a). The effects of changes in internal ionic concentrations on the electrical properties of perfused giant axons. *J. Physiol.*, 164:355-374.
Baker, P. F., Hodgkin, A. L., and Shaw, T. I. (1962b). Replacement of the axoplasm of giant nerve fibres with artificial solutions. *J. Physiol.*, 164:330-354.
Baker, P. F. and Willis, J. S. (1972a). Binding of the cardiac glycoside ouabain to intact cells. *J. Physiol.*, 224:441-462.
Baker, P. F. and Willis, J. S. (1972b). Inhibition of the sodium pump in squid giant axons by cardiac glycosides: Dependence on extracellular ions and metabolism. *J. Physiol.*, 224:463-475.
Beaugé, L. A. and DiPolo, R. (1979a). Sidedness of the ATP-Na^+-K^+ interactions with the Na^+ pump in squid axons. *Biochim. Biophys. Acta*, 553:495-500.

Beaugé, L. A. and DiPolo, R. (1979b). Vanadate selectivity inhibits the K_o^+-activated Na^+ efflux in squid axons. *Biochim. Biophys. Acta*, 551:220–223.
Beaugé, L. A. and DiPolo, R. (1981a). An ATP-dependent sodium-sodium exchange in strophanthidin poisoned dialysed squid giant axons. *J. Physiol.*, 315:447–460.
Beaugé, L. A. and DiPolo, R. (1981b). The effects of ATP on the interactions between monovalent cations and the sodium pump in dialysed squid axons. *J. Physiol.*, 314:457–480.
Bernstein, J. (1902). Untersuchungen zur thermodynamik der bioelectrischen strome. *Pflügers Arch. Ges. Physiol.*, 92:521–562.
Blaustein, M. P. and Hodgkin, A. L. (1969). The effect of cyanide on the efflux of calcium from squid axons. *J. Physiol.*, 200:497–527.
Blostein, R. and Polvani, C. (1992). Altered stoichiometry of the Na,K-ATPase. *Acta Physiol. Scand.*, 146:105–110.
Brinley, F. J., Jr. (1968). Sodium and potassium fluxes in isolated barnacle muscle fibers. *J. Gen. Physiol.*, 51:445–477.
Brinley, F. J., Jr. and Mullins, L. J. (1967). Sodium extrusion by internally dialyzed squid axons. *J. Gen. Physiol.*, 50:2303–2331.
Brinley, F. J., Jr. and Mullins, L. J. (1968). Sodium fluxes in internally dialyzed squid axons. *J. Gen. Physiol.*, 52:181–211.
Brown, K. T. and Flaming, D. G. (1974). Beveling of fine micropipets by a rapid precision method. *Science*, 185:693–695.
Brown, K. T. and Flaming, D. G. (1977). New microelectrode techniques for intracellular work on small cells. *Neuroscience*, 2:813–827.
Bühler, R., Stürmer, W., Apell, H., and Läuger, P. (1991). Charge translocation by the Na,K-pump: I. Kinetics of local field changes studied by time-resolved fluorescence measurements. *J. Membr. Biol.*, 121:141–161.
Caldwell, P. C. (1960). The phosphorus metabolism of squid axons and its relationship to the active transport of sodium. *J. Physiol.*, 152:545–560.
Caldwell, P. C., Hodgkin, A. L., Keynes, R. D., and Shaw, T. I. (1960a). The effects of injecting "energy-rich" phosphate compounds on the active transport of ions in the giant axons of *Loligo*. *J. Physiol.*, 152:561–590.
Caldwell, P. C., Hodgkin, A. L., Keynes, R. D., and Shaw, T. I. (1960b). Partial inhibition of the active transport of cations in the giant axons of *Loligo*. *J. Physiol.*, 152:591–600.
Caldwell, P. C., Hodgkin, A. L., Keynes, R. D., and Shaw, T. I. (1964). The rate of formation and turnover of phosphorus compounds in squid giant axons. *J. Physiol.*, 171:119–131.
Chapman, J. B. (1973). On the reversibility of the sodium pump in dialyzed squid axons. *J. Gen. Physiol.*, 62:643–646.
Chapman, K. M. (1982). The Na-K pump as a current source. *J. Gen. Physiol.*, 80:473–479.
Conway, E. J. (1957). Nature and significance of concentration relations of potassium and sodium ions in skeletal muscle. *Physiol. Rev.*, 37:84–132.
Cooper, K., Jakobsson, E., and Wolynes, P. (1985). The theory of ion transport through membrane channels. *Prog. Biophys. Mol. Biol.*, 46:51–96.
Curtis, H. J. and Cole, K. S. (1942). Membrane resting and action potentials from the squid giant axon. *J. Cell. Comp. Physiol.*, 19:135–144.
De Weer, P. and Geduldig, D. (1973). Electrogenic sodium pump to resting potential in squid giant axon. *Science*, 179:1326–1328.

De Weer, P. and Geduldig, D. (1978). Contribution of sodium pump to resting potential of squid giant axon. *Am. J. Physiol.*, 235:C55–C62.

De Weer, P. and Rakowski, R. F. (1984). Current generated by backward-running electrogenic Na pump in squid giant axons. *Nature*, 309:450–452.

DiPolo, R. (1973). Calcium efflux from internally dialyzed squid giant axons. *J. Gen. Physiol.*, 62:575–589.

DiPolo, R. (1979). Calcium influx in internally dialyzed squid giant axons. *J. Gen. Physiol.*, 73:91–113.

DiPolo, R. and Beaugé, L. (1980). Mechanisms of calcium transport in the giant axon of the squid and their physiological role. *Cell Calcium*, 1:147–169.

Draper, M. H. and Weidmann, S. (1951). Cardiac resting and action potentials recorded with an intracellular electrode. *J. Physiol.*, 115:74–94.

Finkelstein, A. and Mauro, A. (1963). Equivalent circuits as related to ionic systems. *Biophys. J.*, 3:215–237.

Flynn, F. and Maizels, M. (1950). Cation control of human erythrocytes. *J. Physiol.*, 110:301–318.

Forbush, B., III. (1979). Sodium and potassium fluxes across the dialyzed giant axon of *Myxicola*. *J. Membr. Biol.*, 46:185–212.

Gadsby, D. C., Rakowski, R. F., and De Weer, P. (1993). Extracellular access to the Na,K pump: Pathway similar to ion channel. *Science*, 260:100–103.

Garrahan, P. J. and Glynn, I. M. (1967a). The behavior of the sodium pump in red cells in the absence of external potassium. *J. Physiol.*, 192:159–174.

Garrahan, P. J. and Glynn, I. M. (1967b). Factors affecting the relative magnitudes of the sodium:potassium and sodium:sodium exchanges catalysed by the sodium pump. *J. Physiol.*, 192:189–216.

Garrahan, P. J. and Glynn, I. M. (1967c). The incorporation of inorganic phosphate into adenosine triphosphate by reversal of the sodium pump. *J. Physiol.*, 192:237–256.

Garrahan, P. J. and Glynn, I. M. (1967d). The sensitivity of the sodium pump to external sodium. *J. Physiol.*, 192:175–188.

Garrahan, P. J. and Glynn, I. M. (1967e). The stoichiometry of the sodium pump. *J. Physiol.*, 192:217–235.

Goldman, D. E. (1943). Potential, impedance, and rectification in membranes. *J. Gen. Physiol.*, 27:37–60.

Gorman, A. L. F. and Marmor, M. F. (1970). Contributions of the sodium pump and ionic gradients to the membrane potential of a molluscan neurone. *J. Physiol.*, 210:897–917.

Hamill, O. P., Marty, A., Neher, E., Sakmann, B., and Sigworth, F. J. (1981). Improved patch-clamp techniques for high-resolution current recording from cells and cell-free membrane patches. *Pflügers Arch.*, 391:85–100.

Hansen, U. P., Gradmann, D., Sanders, D., and Slayman, C. L. (1981). Interpretation of current-voltage relationships for "active" ion transport systems: I. Steady-state reaction-kinetic analysis of class-I mechanisms. *J. Membr. Biol.*, 63:165–190.

Hilgemann, D. W. (1994). Channel-like function of the Na,K pump probed at microsecond resolution in giant membrane patches. *Science*, 263:1429–1432.

Hobbs, A. S. and Albers, R. W. (1980). The structure of proteins involved in active membrane transport. *Ann. Rev. Biophys. Bioeng.*, 9:259–291.

Hodgkin, A. L. and Horowicz, P. (1959). The influence of potassium and chloride on the membrane potential of single muscle fibres. *J. Physiol.*, 148:127–160.

Hodgkin, A. L. and Katz, B. (1949). The effect of sodium ions on the electrical activity of the giant axon of the squid. *J. Physiol.*, 108:37–77.

Hodgkin, A. L. and Keynes, R. D. (1953). The mobility and diffusion coefficient of potassium in giant axons from *Sepia*. *J. Physiol.*, 119:513–528.

Hodgkin, A. L. and Keynes, R. D. (1955). Active transport of cations in giant axons from *Sepia* and *Loligo*. *J. Physiol.*, 128:28–60.

Hodgkin, A. L. and Keynes, R. D. (1957). Movements of labelled calcium in squid giant axons. *J. Physiol.*, 138:253–281.

Holman, M. E. (1958). Membrane potentials recorded with high-resistance microelectrodes; and the effects of changes in ionic environment on the electrical and mechanical activity of the smooth muscle of the *taenia coli* of the guinea-pig. *J. Physiol.*, 141:464–488.

Huxley, A. F. and Stämpfli, R. (1951). Effect of potassium and sodium on resting and action potentials of single myelinated nerve fibres. *J. Physiol.*, 112:496–508.

Jørgensen, P. L. (1975b). Purification and characterization of Na^+,K^+-ATPase V. Conformational changes in the enzyme transitions between the Na-form and the K-form studied with tryptic digestion as a tool. *Biochim. Biophys. Acta*, 401:399–415.

Jørgensen, P. L. (1982). Mechanism of the Na^+,K^+ pump protein structure and conformations of the pure (Na^+ + K^+)-ATPase. *Biochim. Biophys. Acta*, 694:27–68.

Jørgensen, P. L. (1985). Conformation E_1-E_2 transitions in $\alpha\beta$-units related to cation transport by pure Na,K-ATPase. In *The Sodium Pump*, 83–96. International Conference on Na,K-ATPase 4. Company of Biologists, Cambridge, England.

Jørgensen, P. L. (1992a). Functional domains of Na,K-ATPase: Conformational transitions in the α-subunit and ion occlusion. *Acta Physiol. Scand.*, 146:89–94.

Jørgensen, P. L. (1992b). Na,K-ATPase, structure and transport mechanism. In de Pont, J. J. H. H. M., ed., *Molecular Aspects of Transport Proteins*, 1–26. Elsevier, New York.

Jørgensen, P. L. and Anderson, J. P. (1988). Structural basis for E_1-E_2 conformational transitions in Na,K-pump and Ca-pump proteins. *J. Membr. Biol.*, 103:95–120.

Karlish, S. J. D., Goldshleger, R., Tal, D. M., Capasso, J. M., Hoving, S., and Stein, W. D. (1992). Identification of the cation binding domain of Na/K-ATPase. *Acta Physiol. Scand.*, 146:69–76.

Kawakami, K., Noguchi, S., Noda, M., Takahashi, H., Ohta, T., Kawamura, M., Nojima, H., Nagano, K., Hirose, T., Inayama, S., Hayashida, H., Miyata, T., and Numa, S. (1985). Primary structure of the α-subunit of *Torpedo californica* (Na^+ + K^+) ATPase deduced from cDNA sequence. *Nature*, 316:733–736.

Krupka, R. M. (1993a). Coupling mechanisms in active transport. *Biochim. Biophys. Acta*, 1183:105–113.

Krupka, R. M. (1993b). Coupling mechanisms in ATP-driven pumps. *Biochim. Biophys. Acta*, 1183:114–122.

Kuffler, S. W., Nicholls, J. G., and Orkand, R. K. (1966). Physiological properties of glial cells in the central nervous system of amphibia. *J. Neurophys.*, 29:768–787.

Landowne, D. and Ritchie, J. M. (1970). The binding of tritiated ouabain to mammalian non-myelinated nerve fibres. *J. Physiol.*, 207:529–537.

Lassen, U. V. and Sten-Knudsen, O. (1968). Direct measurements of membrane potential and membrane resistance of human red cells. *J. Physiol.*, 195:681–696.

Läuger, P. (1984). Thermodynamic and kinetic properties of electrogenic ion pumps. *Biochim. Biophys. Acta*, 779:307–341.

Läuger, P. and Apell, H. J. (1988). Voltage dependence of partial reactions of the Na^+/K^+ pump: Predictions from microscopic models. *Biochim. Biophys. Acta*, 945:1–10.

Lederer, W. J. and Nelson, M. T. (1984). Sodium pump stoichiometry determined by simultaneous measurements of sodium efflux and membrane current in barnacle. *J. Physiol.*, 348:665–677.

Levitt, D. G. (1986). Interpretation of biological ion channel flux data: Reaction-rate versus continuum theory. *Ann. Rev. Biophys. Biophys. Chem.*, 15:29–57.

Lew, V. L. and Bookchin, R. M. (1986). Volume, pH, and ion-content regulation in human red cells: Analysis of transient behavior with an integrated model. *J. Membr. Biol.*, 92:57–74.

Ling, G. and Gerard, R. W. (1950). External potassium and membrane potential of single muscle fibres. *Nature*, 165:113–114.

Maéno, T. (1959). Electrical characteristics and activation potential of *Bufo* eggs. *J. Gen. Physiol.*, 43:139–157.

Maizels, M. (1949). Cation control in human erythrocytes. *J. Physiol.*, 108:247–263.

Maizels, M. (1951). Factors in the active transport of cations. *J. Physiol.*, 112:59–83.

Maizels, M. (1954). Active cation transport in erythrocytes. *Symp. Soc. Exp. Biol.*, 8:202–227.

Mullins, L. J., Adelman, W. J., Jr., and Sjodin, R. A. (1962). Sodium and potassium ion effluxes from squid axons under voltage clamp conditions. *Biophys. J.*, 2:257–274.

Mullins, L. J. and Brinley, F. J., Jr. (1967). Some factors influencing sodium extrusion by internally dialyzed squid axons. *J. Gen. Physiol.*, 50:2333–2355.

Mullins, L. J. and Brinley, F. J., Jr. (1969). Potassium fluxes in dialyzed squid axons. *J. Gen. Physiol.*, 53:704–740.

Mullins, L. J. and Noda, K. (1963). The influence of sodium-free solutions on the membrane potential of frog muscle fibers. *J. Gen. Physiol.*, 47:117–132.

Mullins, L. J. and Requena, J. (1979). Calcium measurement in the periphery of an axon. *J. Gen. Physiol.*, 74:393–413.

Mullins, L. J. and Requena, J. (1981). The "late" Ca channel in squid axons. *J. Gen. Physiol.*, 78:683–700.

Mulroy, M. J., Altmann, D. W., Weiss, T. F., and Peake, W. T. (1974). Intracellular electric responses to sound in a vertebrate cochlea. *Nature*, 249:482–485.

Nastuk, W. L. and Hodgkin, A. L. (1950). The electrical activity of single muscle fibers. *J. Cell. Comp. Physiol.*, 35:39–72.

Nieto-Frausto, J., Läuger, P., and Apell, H. (1992). Electrostatic coupling of ion pumps. *Biophys. J.*, 61:83–95.

Overbeek, J. T. G. (1956). The Donnan equilibrium. *Prog. Biophys. Biophys. Chem.*, 6:58–84.

Pedemonte, C. H. and Kaplan, J. H. (1990). Chemical modification as an approach to elucidation of sodium pump structure-function relations. *Am. J. Physiol.*, 258:C1–C23.

Pedersen, P. L. and Carafoli, E. (1987). Ion motive ATPases: II. Energy coupling and work output. *Trends Biochem. Sci.*, 12:186–189.

Post, R. L. and Jolly, P. C. (1957). The linkage of sodium, potassium, and ammonium active transport across the human erythrocyte membrane. *Biochim. Biophys. Acta*, 25:118–128.

Post, R. L., Kume, S., Tobin, T., Orcutt, B., and Sen, A. K. (1969). Flexibility of an active center in sodium plus-potassium adenosine triphosphatase. *J. Gen. Physiol.*, 54:306–326.

Rakowski, R. F. (1989). Simultaneous measurement of changes in current and tracer flux in voltage-clamped squid giant axon. *Biophys. J.*, 55:663–671.

Rakowski, R. F., Gadsby, D. C., and De Weer, P. (1989). Stoichiometry and voltage dependence of the sodium pump in voltage-clamped, internally dialyzed squid giant axon. *J. Gen. Physiol.*, 93:903–941.

Rakowski, R. F. and Paxson, C. L. (1988). Voltage dependence of Na/K pump current in *Xenopus* oocytes. *J. Membr. Biol.*, 106:173–182.

Schatzmann, V. H. J. (1953). Herzglykoside als hemmstoffe für den aktiven kalium- und natriumtransport durch die erythrocytenmembran. *Helv. Physiol. Acta*, 11:346–354.

Shull, G. E., Schwartz, A., and Lingrel, J. B. (1985). Amino-acid sequence of the catalytic subunit of the (Na^+ + K^+)ATPase deduced from a complementary DNA. *Nature*, 316:691–695.

Sjodin, R. A. (1966). Long duration responses in squid giant axons injected with ^{134}cesium sulfate solutions. *J. Gen. Physiol.*, 50:269–278.

Sjodin, R. A. and Beaugé, L. A. (1969). The influence of potassium- and sodium-free solutions on sodium efflux from squid giant axons. *J. Gen. Physiol.*, 54:664–674.

Sjodin, R. A. and Beaugé, L. A. (1973). An analysis of the leakages of sodium ions into and potassium ions out of striated muscle cells. *J. Gen. Physiol.*, 61:222–250.

Sjodin, R. A. and Henderson, E. G. (1964). Tracer and non-tracer potassium fluxes in frog sartorius muscle and the kinetics of net potassium movement. *J. Gen. Physiol.*, 47:605–638.

Sjodin, R. A. and Medici, A. (1975). Inhibitory effect of external sodium ions on the potassium pump in striated muscle. *Nature*, 255:632–633.

Sjodin, R. A. and Mullins, L. J. (1967). Tracer and nontracer potassium fluxes in squid giant axons and the effects of changes in external potassium concentration and membrane potential. *J. Gen. Physiol.*, 50:533–549.

Sjodin, R. A. and Ortiz, O. (1975). Resolution of the potassium ion pump in muscle fibers using barium ions. *J. Gen. Physiol.*, 66:269–286.

Skou, J. C. (1957). The influence of some cations on an adenosine triphosphatase from peripheral nerve. *Biochim. Biophys. Acta*, 23:394–401.

Skou, J. C., Nørby, J. G., Maunsbach, A. B., and Esmann, M., eds. (1988a). *Progress in Clinical and Biological Research,* vol. 268A, *The* Na^+,K^+*-Pump,* pt. A, *Molecular Aspects*. Alan R. Liss, New York.

Skou, J. C., Nørby, J. G., Maunsbach, A. B., and Esmann, M., eds. (1988b). *Progress in Clinical and Biological Research,* vol. 268B, *The* Na^+,K^+*-Pump,* pt. B, *Cellular Aspects*. Alan R. Liss, New York.

Stanton, M. G. (1983). Origin and magnitude of transmembrane resting potential in living cells. *Philos. Trans. R. Soc. London, Ser. B*, 301:85–141.

Stürmer, W., Bühler, R., Apell, H., and Läuger, P. (1991). Charge translocation by the Na,K-pump: II. Ion binding and release at the extracellular face. *J. Membr. Biol.*, 121:163–176.

Thomas, R. C. (1969). Membrane current and intracellular sodium changes in a snail neurone during extrusion of injected sodium. *J. Physiol.*, 201:495–514.

Thomas, R. C. (1972b). Intracellular sodium activity and the sodium pump in snail neurones. *J. Physiol.*, 220:55–71.

Venosa, R. A. and Horowicz, P. (1981). Density and apparent location of the sodium pump in frog sartorius muscle. *J. Membr. Biol.*, 59:225–232.

Wood, D. W. (1957). The effect of ions upon neuromuscular transmission in a herbivorous insect. *J. Physiol.*, 138:119–139.

Zahler, R., Gilmore-Hebert, M., Baldwin, J. C., Franco, K., and Benz, E. J., Jr. (1993). Expression of α isoforms of the Na,K-ATPase in human heart. *Biochim. Biophys. Acta*, 1149:189–194.

Zhou, Z. and Neher, E. (1993). Mobile and immobile calcium buffers in bovine adrenal chromaffin cells. *J. Physiol.*, 469:245–273.

8 Cellular Homeostasis

One of the remarkable properties of living cells is their capacity to maintain a relatively constant volume throughout life. Since cells generally contain a large number of charged macromolecules which cannot pass through the plasma membrane, osmotic forces produce a constant tendency to swell. In general, such systems can avoid swelling only if the hydrostatic pressure is sufficiently higher inside than outside the cell, or if the cell surface is impermeable to a large fraction of the solutes in the external solution. In some bacterial and plant cells the former mechanism appears to be responsible for volume stability. . . . For many years, it was believed that animal cells in general and red cells in particular (which lack a tough wall to support a pressure gradient) maintained a stable volume because they were impermeable to cations, or at least to the major extracellular cation, sodium. . . . " It was "pointed out that loss of cation impermeability would lead to colloid-osmotic swelling and hemolysis. With the use of isotopic tracers, it became clear that the surface of red cells (and animal cells in general) is permeable to sodium and potassium. It became necessary, therefore, to discard the idea that red cells maintain a stable volume because they are impermeable to cations and also to find another explanation for the fact that they do not swell and hemolyze.

—Tosteson and Hoffman, 1960

8.1 Introduction

As indicated in the quote at the beginning of Chapter 2, by the end of the nineteenth century it was recognized that a higher organism has the capacity to regulate its internal environment. The composition of extracellular solutions, the body temperature, etc., are all highly regulated in the face of changes in the environment composition and temperature. This regulation of the internal environment of an organism has been called *homeostasis*. More recently, it has become equally clear that an individual cell also has the capacity to regulate its intracellular environment, a function referred to as *cellular homeostasis*. Homeostatic mechanisms are required to maintain the intracellular composition if cells are to carry out their physiological functions. This chapter deals with cellular homeostatic mechanisms. By *homeostatic mechanisms* we mean those mechanisms that maintain cellular volume, membrane potential, and intracellular composition when the environmental variables are invariant and the cell is at rest, as well as those mechanisms that regulate these variables when an environmental variable is changed. This topic is much more complex than the investigation of individual transport mechanisms in Chapters 3–7, since homeostasis inherently involves the integration of numerous transport mechanisms as well as cytoplasmic reactions that are coupled to the transport mechanisms.

It is difficult to study homeostatic mechanisms in a quiescent cell that may be subject to only minute changes in environmental variables. Hence, measurements have been made of cell responses to relatively large perturbations in environmental variables. We will examine such homeostatic responses first. Next we will examine the homeostatic equations that all cells must satisfy and then explore solutions to these equations for simple cell models. These simple cell models will be based on the material already developed in Chapters 3–7. This analysis will establish the roles of impermeant ions, electrical constraints, and active transport mechanisms in homeostasis. However, cells contain a plethora of mechanisms that affect homeostasis. We will then inventory different transport mechanisms and illustrate their representation in a few selected cells that are relatively well characterized. We will end with qualitative comments on some of the mechanisms thought to be responsible for homeostasis.

8.2 Volume Regulation

8.2.1 Background

If human erythrocytes are suddenly placed in distilled water, they will swell rapidly and then burst (lyse). This is indeed an extreme primary response to a large and rapidly applied osmotic stress. Most cells will not show such an extreme response to a less extreme osmotic stimulus. Most cells will, in response to a rapid change in osmotic pressure, exhibit a rapid change in volume consistent with the primary response we discussed in Chapter 4. However, the volume of such a cell will then slowly return to near its normal value. This property is called *volume regulation*. It is an example of a homeostatic response.

Table 8.1 The composition of seawater and fresh water (adapted from Schmidt-Nielsen, 1990, Tables 8.1 and 8.2). The compositions of seawater and fresh water vary with locale. The composition of fresh water is based on the mean composition of water in North American rivers.

Ion	*Osmolality (mmol/kg)*	
	Seawater	Fresh water
Sodium	475.4	0.39
Magnesium	54.17	0.21
Calcium	10.34	0.52
Potassium	10.07	0.04
Chloride	554.4	0.23
Sulfate	28.56	0.21
Bicarbonate	2.37	1.11

To gain some insight into the usefulness of volume regulation, consider those organisms that live in an osmotically stressful environment. For example, consider migratory fish, such as salmon that are born in fresh water, migrate to the ocean, and return to fresh water to spawn. The difference in composition between fresh water and seawater is large (Table 8.1). Nevertheless, animals that live in an estuary, where fresh water and seawater mix, have adapted to life in an environment whose salinity changes every twelve hours with the tides between that of fresh water and that of seawater. The salinity of seawater varies in the oceans, but is typically about 34%, which means that 34% of the weight of 1 kg of water is due to salts. In contrast, the salinity of fresh water is less than 0.5%. Brackish water (such as occurs in estuaries, salt marshes, and some land-locked seas) has a salinity of between 0.5 and 30%. Now imagine an animal in an estuary before the tide has come in, where the water has a salinity near that of fresh water. When the tide comes in, seawater enters the estuary and its salinity increases about two orders of magnitude. This osmotic shock is enormous. Since the osmotic pressure of seawater is about that of a 1 mol/L solution, the osmotic shock is about 25 atmospheres (see Section 4.3). The theory described thus far implies that such a large osmotic change will cause a primary response that will suck the water out of the organism; the organism should shrink. However, these animals do not shrink much. Furthermore, humans do not shrink appreciably if they take a dip in the ocean.

Which mechanisms allow organisms to survive such enormous osmotic pressure changes? As shown in Figure 8.1, an organism can be considered to

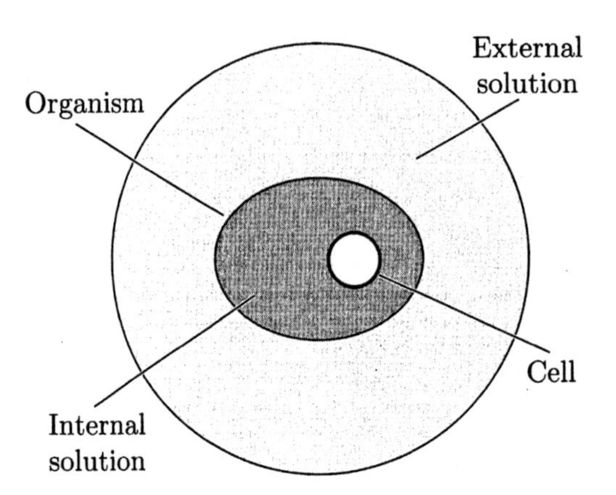

Figure 8.1 Relation between external and internal solutions of an aquatic organism. The external solution represents the solution of the medium bathing the organism, the internal solution the body solutions, including blood, interstitial solutions, etc.

have three compartments: an intracellular compartment, an extracellular or internal compartment, and an external compartment (for example, the water in the estuary). Aquatic animals that live in environments that have a narrow range of salinities are called *stenohaline*. Aquatic animals that live in a wide range of environmental salinities, such as occurs in an estuary, are called *euryhaline*. In euryhaline aquatic animals, there are two extreme possibilities for adjustment to a change in external salinity. Either the internal solution takes on the composition of the external solution (as occurs in animals called *osmoconformers*) or the internal solution is regulated by some organ system (as occurs in animals called *osmoregulators*). In Figure 8.2, the relation between osmolarity of the internal and external solutions is shown for two marine invertebrates, one an osmoconformer, the other an osmoregulator. In the osmoregulator the internal osmolarity is relatively independent of the external osmolarity. In contrast, in the osmoconformer the internal osmolarity approaches that of the external solution. Vertebrates in general, and mammals in particular, are very good osmoregulators.

In osmoconformers the internal solution osmolarity varies widely as the external osmolarity varies. Therefore, the fact that an osmoconformer can exist in an estuary without large changes in the volumes of its constituent cells suggests that its cells have mechanisms that allow the regulation of cell volume, but on a time scale that is much slower than that of the primary response. In contrast, osmoregulators have specialized organ systems that regulate the osmolarity of internal body fluids. However, even in good osmoregulators, such as mammals, individual cells may be subjected to large osmotic transients. For example, after a person drinks a glass of water the solution in the alimentary canal becomes temporarily hyposmotic, thus subjecting the cells of the epithelium to a hyposmotic solution. Also, blood cells

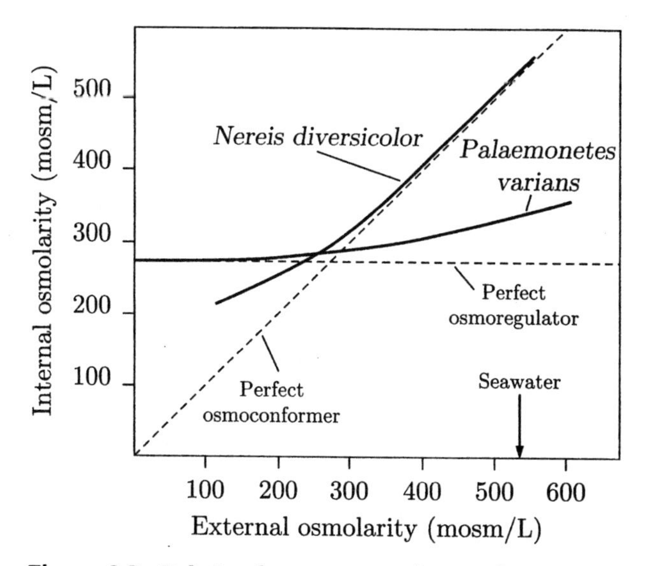

Figure 8.2 Relation between osmolarity of internal and external solutions for an osmoconformer (the marine worm *Nereis diversicolor*) and an osmoregulator (the shrimp *Palaemonetes varians*) (adapted from Schmidt-Nielsen, 1990, Figure 8.2). The horizontal dashed line shows the relation of osmolarities for a perfect osmoregulator. The internal osmolarity is constant independent of the external osmolarity. The diagonal dashed line shows that for a perfect osmoconformer the internal and external osmolarities are identical.

traversing the capillary system are subjected to large changes in osmolarity as they traverse the peripheral circulation of organs such as the kidney. Thus, cells must have mechanisms to regulate their volumes and hence the concentrations of cytoplasmic substances in the face of osmotic challenges.

8.2.2 Volume Regulatory Responses

Figure 8.3 shows the volume response of lymphocytes on a relatively long time scale—much longer than that of the primary response we studied in Chapter 4. In response to a hypotonic solution, the cells swell rapidly as water enters to restore osmotic equilibrium. This response is the primary response of the cell, which acts approximately as an osmometer. However, on a time scale of tens of minutes the volume of the cell decreases to near its isotonic value. This response is called a *regulatory volume decrease* (RVD). When the isotonic solution is restored, water rapidly leaves the cell to restore osmotic equilibrium—another indication of the primary osmometric response of the cell. However, after tens of minutes the volume increases to near its isotonic

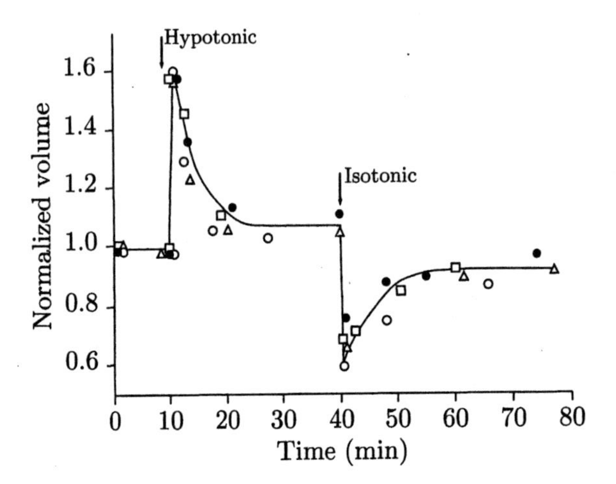

Figure 8.3 Volume regulatory responses of lymphocytes (adapted from Stein, 1990, Figure 7.2). Lymphocytes were initially in an isotonic saline solution. Then the extracellular solution was made hypotonic. After 30 minutes in this hypotonic solution, the solution was made isotonic again.

value. This later response is called a *regulatory volume increase* (RVI). These types of regulatory responses to changes in osmolarity are seen in many, but not all cells.

8.2.3 Conclusions

In response to a change in an environmental variable, cells exhibit primary responses that result in a change in cell volume. However, regulatory responses that are slower than the primary responses occur and restore each cell's volume to its unperturbed value. Presumably such mechanisms are at work when the cell is relatively quiescent and contribute to cellular homeostasis by correcting volume changes that result from small perturbations in environmental variables.

8.3 General Equations for Homeostasis

Cellular homeostatic equations are the set of equations that describes all the cellular homeostatic variables, which include the cell solute composition, the cell volume, the membrane potential, etc. There are two types of homeostatic equations: those that follow from conservation laws and the constitutive relations that define specific membrane transport mechanisms and specific cytoplasmic binding reactions. We will discuss the former in this section.

As we have repeatedly seen in Chapters 3–7, conservation of mass leads to continuity equations that link intracellular solute quantity to solute flux and cell water volume to volume flux. In addition, conservation of charge links the concentrations of ions. These conservation laws follow from first principles and are robust constraints on cellular variables. That is, they apply to all cells. In previous chapters, we have already seen these conservation principles applied in isolation when considering one or another particular transport mechanism. In this chapter we will generalize these equations to apply to homeostasis of a cell.

8.3.1 Kinetic Equations

8.3.1.1 Conservation of Solute

If we assume that each solute to which a membrane is permeant is conserved, then the rate of decrease of unbound intracellular solute quantity must equal the sum of the efflux of the solute plus the rate at which that solute is bound in the cytoplasm. Since there may be a number of different transport mechanisms that transport a solute, the total solute flux is the sum of the fluxes over all transport mechanisms. Similarly, there may be several mechanisms that bind the solute intracellularly. Therefore, the conservation of solute can be expressed as

$$-\frac{1}{A(t)}\frac{dn_n^i(t)}{dt} = \sum_j \phi_n^j(t) + \sum_j \alpha_n^j(t), \tag{8.1}$$

where $A(t)$ is the surface area of the cell, $n_n^i(t)$ is the total unbound intracellular quantity of solute n (in mol), ϕ_n^j is the outward flux of solute n for transport mechanism j (in mol/(cm^2 · s)), and α_n^j is the rate of binding of solute by intracellular mechanism j per unit surface area of membrane (in mol/(cm^2 · s)). The concentrations and quantities are related by $c_n^i(t) = n_n^i(t)/\mathcal{V}(t)$, where $\mathcal{V}(t)$ is the cell water volume.

8.3.1.2 Conservation of Water Volume

Conservation of water volume requires that any efflux of water through the membrane must result in a reduction of water volume in the cell. In the presence of both hydraulic and osmotic pressure difference across the membrane, conservation of water volume requires that

$$-\frac{1}{A(t)}\frac{d\mathcal{V}(t)}{dt} = \Phi_V(t) = \mathcal{L}_V\left((p_i - p_o) - RT\left(\sum_n \sigma_n c_n^i(t) - \sum_n \sigma_n c_n^o(t)\right)\right), \tag{8.2}$$

where $p_i - p_o$ is the difference in the hydraulic pressures on the two sides of the membrane. The second term is the difference in osmotic pressure on the two sides of the membrane, where R is the molar gas constant, T is the absolute temperature, σ_n is the reflection coefficient for solute n, $c_n^o(t)$ and $c_n^i(t)$ are the extracellular and intracellular concentrations of solute n (in mol/(cm^3), and $\mathcal{L}_V$ is the hydraulic conductivity of the membrane. If all solutes have reflection coefficients that equal 1 and if the hydraulic pressure difference across the membrane is negligible (as is usually the case in animal cells), water volume conservation can be expressed as

$$-\frac{1}{A(t)}\frac{d\mathcal{V}(t)}{dt} = \Phi_V(t) = \mathcal{L}_V RT\left(\sum_n c_n^o(t) - \sum_n c_n^i(t)\right). \tag{8.3}$$

8.3.1.3 *Conservation of Charge: Electroneutrality*

On a temporal scale that is greater than the charge relaxation time and a spatial scale that exceeds the Debye length, both the intracellular and extracellular media must be electrically neutral (see Section 7.2.3). Therefore, the sum of the charge densities of all solutes must be zero, which gives

$$\sum_n z_n F c_n^i(t) = 0 \quad \text{and} \quad \sum_n z_n F c_n^o(t) = 0,$$

where z_n is the valence of solute n and F is Faraday's constant. These equations can be simplified by removing the common factor F to yield

$$\sum_n z_n c_n^i(t) = 0 \quad \text{and} \quad \sum_n z_n c_n^o(t) = 0. \tag{8.4}$$

8.3.2 Quasi-Equilibrium Equations

At cellular quasi-equilibrium, all the homeostatic variables have reached their final values, i.e., all the derivatives in Equations 8.1 and 8.3 are zero, and if $\sigma_n = 1$ and the hydraulic pressure difference across the membrane is zero, then the set of homeostatic equations is as follows:

$$\begin{aligned}
\sum_j \phi_n^j(\infty) + \sum_j \alpha_n^j(\infty) &= 0,\\
\sum_n c_n^o(\infty) - \sum_n c_n^i(\infty) &= 0,\\
\sum_n z_n c_n^i(\infty) &= 0,\\
\sum_n z_n c_n^o(\infty) &= 0.
\end{aligned} \tag{8.5}$$

We call these equations quasi-equilibrium equations because the *net* flux of each solute is zero. However, this condition may result from the balance of two or more transport processes, each of which has a nonzero flux. Thus, there may be transport mechanisms that produce a net flux that are balanced by other mechanisms that produce an equal and oppositely directed flux.

8.3.3 Solutions of the Equations for Homeostasis

Both the time-varying equations (Equations 8.1, 8.3, and 8.4) and the quasi-equilibrium equations (Equations 8.5) make up a complete set of equations that can, in principle, be solved. Let us assume that all the extracellular concentrations are known, as are the intracellular quantities of any impermeant solutes. Therefore, the unknowns or dependent variables are the quantities of all the permeant solutes, as well as the volume of the cell and the membrane potential if any of the solutes are charged.

To make this more precise, suppose there are N solutes of which M are permeant. Then there are M flux equations that are differential equations in the time-varying case and algebraic equations in the case of quasi-equilibrium. There is also one equation for electroneutrality of the intracellular solutes and one equation for water volume conservation. Thus, there are $M + 2$ equations that involve the $M + 2$ independent variables — M permeant solute quantities, the volume, and the membrane potential. Thus, the number of equations and the number of unknowns are equal, and the number of equations is sufficient to obtain a solution. The concentrations of all the intracellular solutes can be obtained from the quantities and the volume.

8.4 Homeostasis for Simple Cell Models

As we shall see later in this chapter, in a typical cell homeostasis involves a large number of mechanisms that transport and bind solutes, all operating simultaneously. In order to understand homeostasis in such a cell, all these homeostatic mechanisms need to be fully characterized. This type of analysis has been approached for only a few cell types, and in these cases the equations for cellular homeostasis are complex and need to be solved numerically (Lew et al., 1979; Jakobsson, 1980; Wolf, 1980; Werner and Heinrich, 1985; Lew and Bookchin, 1986; Mintz et al., 1986; Moroz et al., 1989; Weinstein, 1992). It is difficult to obtain insight into the homeostatic mechanism by examining these solutions without understanding simpler systems. Therefore, in this section we consider a sequence of models of cellular homeostasis that are of increasing complexity, but sufficient simplicity that they can be understood thoroughly. These simple models point out the difficulty that cells face in maintaining their compositions, membrane potentials, and volumes. These models also point out those factors that stabilize and those that destabilize these homeostatic mechanisms. While even the most complex of these models pales in complexity when compared to even the simplest cells, these simple models must be understood thoroughly if the more complex homeostatic mechanisms of real cells are to be understood.

In all these simple models, we shall make the following assumptions:

- The cell membrane is freely distensible, so that no hydraulic pressure difference occurs across the cell membrane.
- The cytoplasmic and extracellular solutions are well mixed, so that the concentrations are uniform in space in each solution.
- The extracellular solution is extensive, so that its composition is unaffected by the flux of solute through the cell membrane.
- Unless it is stated otherwise, intracellular solutes are free to diffuse and are not bound by intracellular macromolecules.
- Transport of water is much more rapid than that of any solute, so that the cell is always in osmotic equilibrium.

These assumptions imply that these simple models will not apply to cells with cell walls such as plant cells and certain bacteria, but may be expected to yield insights into the homeostatic mechanisms at work in animal cells.

8.4.1 Solute Flux Equations

Neither the kinetic nor the quasi-equilibrium equations of cell homeostasis can be solved without a specification of the equations of solute flux. In this section, we shall use the simple transport mechanisms studied in Chapters 3–7 to examine homeostasis in simple cell models. We shall review these transport mechanisms briefly.

8.4.1.1 Diffusion of a Nonelectrolyte

According to the dissolve-diffuse mechanism described in Chapter 3, the flux of a nonelectrolyte solute is given by Fick's first law for membranes as

$$\phi_n^d(t) = P_n\left(c_n^i(t) - c_n^o(t)\right), \tag{8.6}$$

where ϕ_n^d is the flux of solute n by diffusion and P_n is the permeability of the membrane to solute n.

8.4.1.2 Carrier-Mediated Transport of a Nonelectrolyte

According to the simple symmetric, four-state carrier model for carrier-mediated transport described in Chapter 6,

$$\phi_n^c(t) = (\phi_n)_{max}\left(\frac{c_n^i(t)}{c_n^i(t)+K} - \frac{c_n^o(t)}{c_n^o(t)+K}\right), \tag{8.7}$$

where ϕ_n^c is the flux of solute n by this simple carrier model, $(\phi_n)_{max}$ is the maximum flux of solute n, and K is the dissociation constant.

8.4.1.3 Passive Ion Transport Through Ion Channels

According to the macroscopic model of ion transport described in Chapter 7,

$$J_n^p(t) = G_n\left(V_m^o(t) - V_n(t)\right), \tag{8.8}$$

where $J_n^p(t)$ is the current density carried by ion n by passive transport, G_n is the specific conductance of the membrane for ion n, $V_m^o(t)$ is the resting membrane potential, and $V_n(t)$ is the Nernst equilibrium potential for ion n, which can be a function of time if the ion concentrations are varying in time.[1] Therefore, the flux due to passive transport of ion n is

$$\phi_n^p(t) = \frac{G_n}{z_n F}\left(V_m^o(t) - V_n(t)\right). \tag{8.9}$$

8.4.1.4 Active Transport for a Hypothetical Generic Pump

We shall find it useful to define a simple hypothetical generic pump with which we can explore the role of active transport in homeostasis in the simplest possible circumstances. We shall call this pump mechanism a *generic pump* and define its flux equation as

$$\phi_n^a(t) = k_a c_n^i(t), \tag{8.10}$$

1. This assumption implies that the ion crosses the membrane through a channel that is perfectly ion-selective to ion n. If it does not, then V_n can represent the channel equilibrium (reversal) potential and G_n must represent the fraction of the channel conductance due to ion n.

where ϕ_n^a is the flux of solute n by active transport, c_n^i is the intracellular concentration of the pumped solute, and k_a is a proportionality constant that has units of permeability.

8.4.1.5 *Ion Flux Due to the Sodium-Potassium Pump*

For each mole of ATP split by (Na^+-K^+)-ATPase, the pump transports ν_{Na} moles of sodium outward and ν_K moles of potassium inward. The simple phenomenological model, presented in Section 7.7, that approximates the normal behavior of the pump is to assume that each molecule of (Na^+-K^+)-ATPase binds 1 ATP molecule, ν_{Na} molecules of Na^+, and ν_K molecules of K^+ independently and that the pump rate α_{ATP} has the following dependence on these concentrations:

$$\alpha_{ATP}(t) = \alpha_{max}\left(\frac{c_{ATP}^i(t)}{c_{ATP}^i(t) + K_{ATP}^i}\right)\left(\frac{c_{Na}^i(t)}{c_{Na}^i(t) + K_{Na}^i}\right)^{\nu_{Na}}\left(\frac{c_K^o(t)}{c_K^o(t) + K_K^o}\right)^{\nu_K}, \quad (8.11)$$

where $\alpha_{ATP}(t)$ is the pump rate in units of moles of ATP split per unit time per unit area of membrane; α_{max} is the maximum pump rate; K_{ATP}^i, K_{Na}^i, and K_K^o are dissociation constants; and $c_{ATP}^i(t)$ is the intracellular concentration of ATP. A further refinement of this model takes into account the effects of the intracellular concentration of potassium and the extracellular concentration of sodium. This refinement can be obtained by modifying Equation 8.11 by replacing K_{Na}^i with $K_{Na}^i(1 + C_K^i/K_K^i)$ and K_K^o by $K_K^o(1 + C_{Na}^o/K_{Na}^o)$. The fluxes of sodium and potassium carried by the pump are

$$\phi_{Na}^a(t) = \nu_{Na}\alpha_{ATP}(t) \quad \text{and} \quad \phi_K^a(t) = -\nu_K\alpha_{ATP}(t). \quad (8.12)$$

8.4.2 Nonelectrolyte Solutes

The osmotically active components of cytoplasm and of extracellular solutions are largely ions. Hence, models of cells that include only nonelectrolyte solutes are highly unrealistic. Nevertheless, such models do reveal the role of impermeant cytoplasmic solutes in the control of cellular volume in the simplest circumstances possible. Therefore, it is valuable to explore this role in a nonelectrolyte context before we explore the more realistic models that include ionic solutes.

8.4.2.1 *Impermeant Solutes*

If all solutes are assumed to be impermeant and if they do not bind to the cell cytoplasm, then both $\phi_n^j = 0$ and $\alpha_n^j = 0$. Therefore, Equation 8.1 implies that

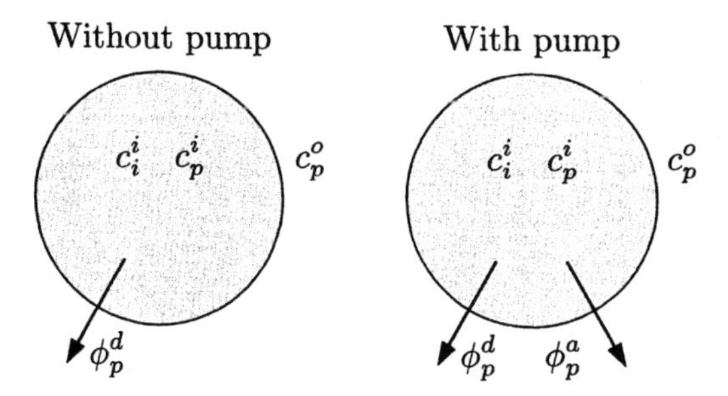

Figure 8.4 Models of transport of nonelectrolytes in simple cell models. Both cell models have intracellular concentrations of a permeant and an impermeant solute of c_p^i and c_i^i, respectively, and an extracellular concentration of the permeant solute of c_p^o. In the model shown in the left panel, the only flux of permeant solute is diffusive, ϕ_p^d, whereas the model in the right panel also contains an active flux, ϕ_p^a.

n_n^i is constant, and Equation 8.3 implies that the equilibrium volume satisfies

$$0 = C_\Sigma^o - \frac{N_\Sigma^i}{\mathcal{V}(\infty)}, \tag{8.13}$$

where $N_\Sigma^i = \sum_n n_n^i$ is the total solute quantity in the cell and $C_\Sigma^o = \sum_n c_n^o$ is the total extracellular solute concentration. Equation 8.13 is the equation of a perfect osmometer and implies that the volume of the cell adjusts itself to any change in external concentration of any solute by a change in volume so as to reach an equilibrium volume of $\mathcal{V}(\infty) = N_\Sigma^i / C_\Sigma^o$. This type of response was analyzed in some detail in Chapter 4.

8.4.2.2 *One Permeant and One Impermeant Solute: The Simple Pump and Leak*

While the simple model with one impermeant solute leads to a stable solution to the homeostatic equations, the addition of a permeant solute leads to no stable solution to the homeostatic equations if the permeant solute is transported by passive mechanisms only. In the simplest model that exhibits both the problem and its solution (Post and Jolly, 1957), consider that the cell contains an impermeant solute of quantity n_i^i and a permeant solute whose concentration is $c_p^i(t)$ intracellularly, as shown in Figure 8.4 (left panel). The extracellular solution contains only the permeant solute at concentration c_p^o. Since the water transport is assumed to be rapid and the cell is assumed to be in osmotic equilibrium at all times, the osmolarities of the cytoplasm and of the extracellular medium are the same at each instant of time. Therefore,

$$c_i^i(t) + c_p^i(t) = c_p^o. \tag{8.14}$$

Passive Transport Only: Cell Flooding

Let us assume that the permeant solute is transported through the membrane by diffusion, so that the flux is

$$\phi_p^d(t) = P_p\left(c_p^i(t) - c_p^o\right),$$

where P_p is the permeability of the membrane for the permeant solute. Conservation of solute p (Equation 8.1) implies that

$$-\frac{1}{A(t)}\frac{dn_p^i(t)}{dt} = \phi_p^d(t) = P_p\left(c_p^i(t) - c_p^o\right). \tag{8.15}$$

Examination of Equations 8.14 and 8.15 reveals that both equations cannot be satisfied at equilibrium. At diffusive equilibrium, Equation 8.15 implies that $c_p^i(\infty) = c_p^o(\infty)$, which contradicts osmotic equilibrium, described by Equation 8.14. To examine what will happen, we combine Equations 8.14 and 8.15 to obtain

$$\frac{1}{A(t)}\frac{dn_p^i(t)}{dt} = P_p c_i^i(t).$$

We can express $c_i^i(t) = n_p^i(t)/\mathcal{V}(t)$ and $\mathcal{V}(t) = (n_p^i(t) + n_i^i)/c_p^o$. Therefore, $c_p^i(t) = (n_p^i(t)/(n_p^i(t) + n_i^i))c_p^o$, so that

$$\frac{1}{A(t)}\frac{dn_p^i(t)}{dt} = P_p c_p^o\left(\frac{n_i^i}{n_i^i + n_p^i(t)}\right). \tag{8.16}$$

To solve this equation requires expressing the cell surface area, $A(t)$, in terms of the volume, $\mathcal{V}(t)$, a relation that is dictated by the geometry of the cell. Since the volume can be expressed in terms of the solute quantity $n_p^i(t)$, Equation 8.16 can be solved explicitly. However, the qualitative behavior of the solution can be discerned directly by examining the differential equation. Since all the quantities in Equation 8.16 are positive, the quantity of intracellular permeant solute, $n_p^i(t)$, will continually increase. Therefore, the volume, $\mathcal{V}(t)$, will continually increase due to the influx of water; that is, the cell will flood! We can see, physically, that if at some instant in time the cell is in osmotic equilibrium, the intracellular concentration of permeant solute will be less than that present extracellularly. Therefore, permeant solute will flow into the cell in such a direction as to establish diffusive equilibrium. This solute influx will require inflow of water to restore osmotic equilibrium. The only way to satisfy both diffusive and osmotic equilibrium is when $c_i^i \to 0$, which occurs when the cell volume is arbitrarily large. Thus, the presence of impermeant intracellular solutes guarantees that the cell cannot be in equilibrium in the presence of passive transport alone.

Pump and Leak

Inclusion of a simple active transport mechanism removes the problem that occurs with just passive transport. Let us assume that the permeant solute is transported actively out of the cell at a rate that is proportional to the intracellular concentration of the solute, i.e.,

$$\phi_p^a(t) = k_a c_p^i(t),$$

where k_a defines the pump rate for a given concentration of permeant solute and has the units of a permeability. Therefore, the total flux of permeant solute now includes both a passive and an active transport component. Conservation of solute p (Equation 8.1) now implies that

$$-\frac{1}{A(t)}\frac{dn_p^i(t)}{dt} = \phi_p^d(t) + \phi_p^a(t) = P_p\left(c_p^i(t) - c_p^o\right) + k_a c_p^i(t). \tag{8.17}$$

Quasi-Equilibrium Solution

At quasi-equilibrium, the solution to Equation 8.17 implies that $P_p(c_p^i(\infty) - c_p^o) + k_a c_p^i(\infty) = 0$, which no longer conflicts with osmotic equilibrium (Equation 8.14). We can combine the two relations,

$$c_i^i(\infty) + c_p^i(\infty) = c_p^o,$$

$$P_p\left(c_p^i(\infty) - c_p^o\right) + k_a c_p^i(\infty) = 0,$$

to solve for all the variables of interest. The quasi-equilibrium intracellular concentrations are

$$c_p^i(\infty) = \left(\frac{P_p/k_a}{1 + (P_p/k_a)}\right) c_p^o,$$

$$c_i^i(\infty) = \left(\frac{1}{1 + (P_p/k_a)}\right) c_p^o,$$

and the intracellular quantity of permeant solute and the cell water volume are

$$n_p^i(\infty) = (P_p/k_a) n_i^i,$$

$$\mathcal{V}(\infty) = \left(1 + (P_p/k_a)\right) \frac{n_i^i}{c_p^o}.$$

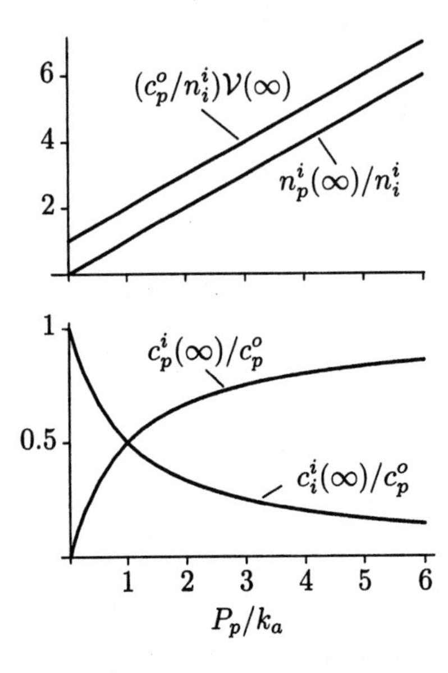

Figure 8.5 Quasi-equilibrium solutions of the simple pump and leak model for nonelectrolytes are plotted versus P_p/k_a, which is the ratio of the permeability of the membrane for the permeant solute and the rate at which it is pumped out of the cell. The variables are plotted in normalized ordinate scales of volume, $(c_p^o/n_i^i)\mathcal{V}(\infty)$, and intracellular quantity of permeant solute, $n_p^i(\infty)/n_i^i$, and the concentrations are normalized to the extracellular concentration.

These solutions are plotted in Figure 8.5. The normalized variables $c_p^i(\infty)/c_p^o$, $c_i^i(\infty)/c_p^o$, $n_p^i(\infty)/n_i^i$, and $\mathcal{V}(\infty)c_p^o/n_i^i$ depend upon a single parameter: the leak-to-pump ratio P_p/k_a.

The solutions have the following simple interpretation. When the leak-to-pump ratio is zero, i.e., when there is no leakage of the permeant solute through the membrane, then the concentration of permeant solute intracellularly is zero, i.e., $c_p^i(\infty) = 0$. That is, the pump maintains the intracellular concentration at zero no matter what the initial concentration of permeant solute. Also, the concentration of impermeant solute is $c_i^i(\infty) = c_p^o$, so that osmotic equilibrium is maintained. The volume of the cell is $\mathcal{V}(\infty) = n_i^i/c_p^o$, which is just the value that occurs with no permeant solute. As the leak-to-pump ratio is increased, the quantity of permeant solute in the cell increases linearly with the ratio P_p/k_a, and the concentration of permeant solute increases hyperbolically. As the concentration of intracellular permeant solute increases, water flows into the cell to increase its volume. Since the volume of the cell water increases, the concentration of impermeant solute decreases even though the quantity of impermeant solute remains constant. Clearly, as the ratio P_p/k_a becomes arbitrarily large, i.e., the leakage greatly exceeds the pump rate, the cell quantity of permeant solute and the cell water volume become arbitrarily large. Under these circumstances, the solution approaches that derived when there is only passive transport of the permeant solute.

Kinetic Solution

As we found in previous chapters, the kinetic equation depends upon the shape of the cell which, in turn, forces a relation between the cell surface area and the volume. For the sake of simplicity, we shall assume that the cell surface area A does not change as the cell volume changes. The solution for this case differs in detail from that obtained for other assumptions about cell shape, but the qualitative behavior is similar.

From Equation 8.14 we find that

$$\mathcal{V}(t) = \frac{n_i^i + n_p^i(t)}{c_p^o},$$

which can be combined with Equation 8.17 to eliminate the volume. With the assumption that the surface area is constant, the kinetic equation is

$$-\frac{dn_p^i(t)}{dt} = (P_p + k_a)Ac_p^o\left(\frac{n_p^i(t)}{n_p^i(t) + n_i^i}\right) - AP_pc_p^o.$$

It is helpful to define normalized coordinates as follows: $\hat{n}(\tau) = n_p^i(t)/n_i^i$, $\hat{n}_\infty = P_p/k_a$, $\hat{\mathcal{V}}(\tau) = \mathcal{V}(t)c_p^o/n_i^i$, and $\tau = \alpha t$, where $\alpha = (P_p + k_a)c_p^oA/n_i^i$. Then the equation in normalized coordinates is

$$\frac{d\hat{n}(\tau)}{d\tau} + \frac{\hat{n}(\tau)}{1 + \hat{n}(\tau)} = \frac{\hat{n}_\infty}{1 + \hat{n}_\infty}.$$

The normalized volume, $\hat{\mathcal{V}}(\tau) = 1 + \hat{n}(\tau)$, has the same time course as the quantity of permeant solute, so there is no need to explore the volume as well. The equation in normalized coordinates depends upon a single parameter, making it relatively easy to explore the behavior of the solution. The solutions to this differential equation (shown in Figure 8.6) indicate that if the cell initially contains no permeant solute, then the quantity of permeant solute will increase monotonically to reach its quasi-equilibrium value. As P_p/k_a increases, the quasi-equilibrium value of the intracellular quantity of permeant solute increases, as we have seen previously, but the time course for reaching that quasi-equilibrium value gets longer even in these normalized coordinates. This response property indicates the nonlinear nature of the kinetic equation.

8.4.2.3 Summary

The simple cell models with nonelectrolyte solutes clearly show that the addition of impermeant solutes in the presence of permeant solutes causes a cell to swell if only passive transport mechanisms are present. Active transport

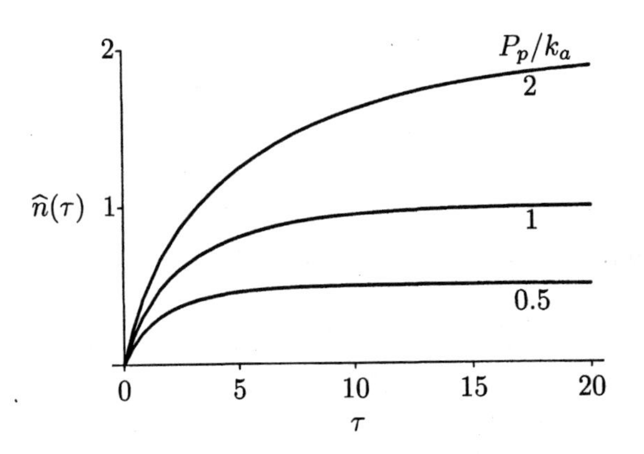

Figure 8.6 The quantity of intracellular permeant solute as a function of time predicted by the simple pump and leak model for nonelectrolytes is plotted for different values of the ratio P_p/k_a. The variables are plotted in normalized coordinates. The initial quantity of the permeant solute in the cell was zero.

mechanisms can counter the effects of solute leakage and can stabilize the cell volume. When stabilized by an active transport pump, the cell responses, such as the intracellular quantity of permeant solute or the cell water volume, in response to a change in composition of the extracellular solution, change monotonically in time and saturate. The saturated values of the cell responses depend on the ratio of the permeability of the membrane for the permeant solute to the rate constant of the active transport mechanism.

8.4.3 Ionic Solutes

Analysis of even simple models with ionic solutes is more complex than analysis of those with nonelectrolytic solutes. However, these models give important insights into the role of electrostatic forces in cellular homeostasis.

8.4.3.1 A Single Ion Transported Passively

With the Same Binary Electrolyte on Both Sides of the Membrane

The simplest cell model we can consider in discussing ionic solutes is one in which the membrane is permeant to only one ion, arbitrarily assumed to be the cation, as indicated in Figure 8.7 (upper left panel). We assume that the electrolytes are binary, with cation concentrations c_+^i and c_+^o intracellularly and extracellularly, respectively, and anion concentrations c_-^i and c_-^o. By definition of a binary electrolyte, the valences are $z_+ = +1$ and $z_- = -1$. Passive flux of the cation is assumed to be given by $\phi_+ = G_+(V_m^o - V_+)/F$, where V_+ is the Nernst equilibrium potential for the cation.

Now we apply the general equations for homeostasis and seek a solution. Electroneutrality of the intracellular and extracellular solutions requires that

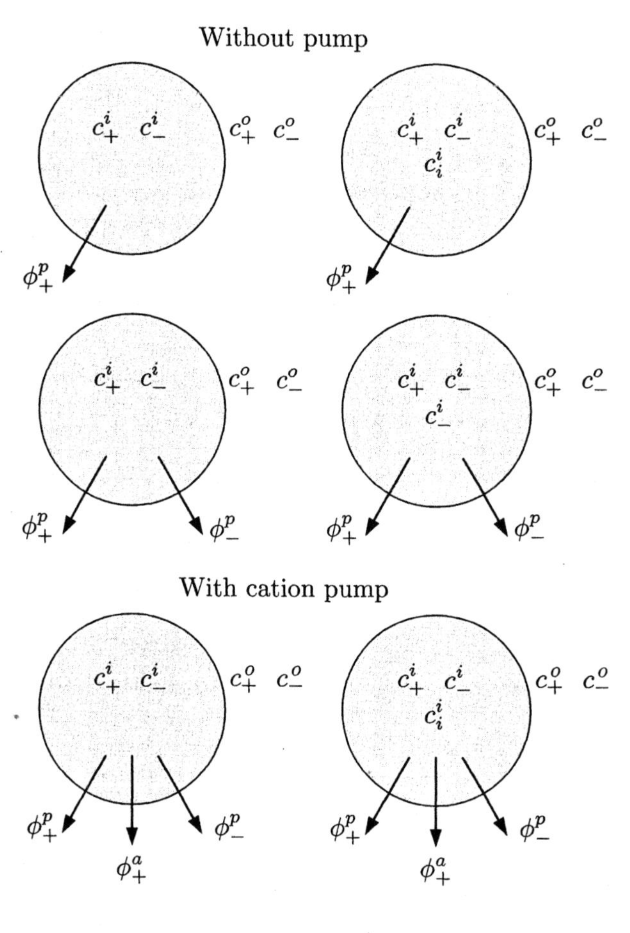

Figure 8.7 Models of transport of electrolytes in simple cell models. The upper four cell models contain only passive transport mechanisms; the two lower cell models also contain active cation transport. In each cell model there is a binary electrolyte with cation and anion concentrations of c^i_+ and c^i_- inside and c^o_+ and c^o_- outside. In the two models in the upper panels, the membrane is assumed to be permeable to only the cation. In the lower four models, the membrane is assumed to be permeable to both the cation and the anion. Three of the models contain an impermeant solute intracellularly with concentration c^i_i.

$$c^i_+(t) = c^i_-(t) \text{ and } c^o_+ = c^o_-,$$

respectively. Osmotic equilibrium requires that

$$c^i_+(t) + c^i_-(t) = c^o_+ + c^o_-.$$

Combining electroneutrality with osmotic equilibrium shows that

$$c^i_+(t) = c^o_+ = c^i_-(t) = c^o_- \doteq C.$$

Thus, a solution to these equations is possible only if the concentrations of all ions have the same constant value both intracellularly and extracellularly. Since the cation is permeant, at equilibrium the passive transport of cation must be zero, so that $G_+(V^o_m - V_+)/F = 0$. Therefore, $V^o_m = V_+$, but since all the concentrations are the same, $V_+ = 0$. Thus, for this case the only possible solution is the trivial solution for which all the concentrations are the same and the membrane potential is zero. This solution applies to a cell of any volume.

With an Additional Impermeant Intracellular Ion

We shall next consider that the intracellular solution contains a binary electrolyte plus an impermeant solute with concentration c_i^i and valence z_i (upper right panel of Figure 8.7). The membrane is assumed to be permeable to only the cation.

Derivation Electroneutrality of the extracellular solutions requires that $c_+^o = c_-^o \doteq C$, and for the intracellular solution it implies that

$$c_+^i(t) - c_-^i(t) + z_i c_i^i(t) = 0. \tag{8.18}$$

Osmotic equilibrium requires that

$$c_+^i(t) + c_-^i(t) + c_i^i(t) = c_+^o + c_-^o = 2C. \tag{8.19}$$

These two equations can be combined so as to express the intracellular cation and anion concentrations in terms of the other variables as follows:

$$c_+^i(t) = C - \left(\frac{1+z_i}{2}\right) c_i^i(t),$$

$$c_-^i(t) = C - \left(\frac{1-z_i}{2}\right) c_i^i(t). \tag{8.20}$$

But the membrane is impermeable to the anion. Thus, the quantity of intracellular anion must be constant. Thus,

$$\frac{n_-^i}{\mathcal{V}(t)} = C - \left(\frac{1-z_i}{2}\right) \frac{n_i^i}{\mathcal{V}(t)},$$

which can be solved for $\mathcal{V}(t)$ to yield

$$\mathcal{V}(t) = \frac{n_-^i + (1-z_i)n_i^i/2}{C}.$$

While the intracellular quantity of impermeant ion is constant, the concentration will not be if the volume changes. We can find this concentration from the volume with the relation $c_i^i(t) = n_i^i/\mathcal{V}(t)$, which yields

$$c_i^i(t) = C\left(\frac{n_i^i}{n_-^i + (1-z_i)n_i^i/2}\right).$$

Substitution of $c_i^i(t)$ into Equation 8.20 yields

$$c_+^i(t) = C\left(1 - \left(\frac{1+z_i}{2}\right)\frac{n_i^i}{n_-^i + (1-z_i)n_i^i/2}\right).$$

Similarly, we find that

$$c_-^i(t) = C\left(\frac{n_-^i}{n_-^i + (1-z_i)n_i^i/2}\right).$$

Finally, the quantity of intracellular cation can change because the membrane is permeant to this ion. We can find an expression for the intracellular quantity of cation from the relation $n_+^i(t) = c_+^i(t)\mathcal{V}(t)$, which yields

$$n_+^i(t) = n_-^i - z_i n_i^i.$$

The flux of the cations must be zero at equilibrium, so that $V_m^o(\infty) = V_+(\infty)$, which implies that

$$V_m^o(\infty) = V_+(\infty) = \frac{RT}{F}\ln\left(\frac{c_+^o}{c_+^i(\infty)}\right) = -\frac{RT}{F}\ln\left(1 - \left(\frac{1+z_i}{2}\right)\frac{n_i^i}{n_-^i + (1-z_i)n_i^i/2}\right).$$

Thus, we have a complete solution to all of the equilibrium values of the dependent variables $c_+^i(\infty)$, $c_-^i(\infty)$, $c_i^i(\infty)$, $n_+^i(\infty)$, $\mathcal{V}(\infty)$, and $V_m^o(\infty)$ in terms of the independent variables C, n_-^i, n_i^i, and z_i.

Dependence on C To examine the dependence of the dependent variables on the independent variables, we shall first consider the dependence on C with the other independent variables fixed. Note that the concentration of the extracellular ions C affects all the variables in a simple manner; all the intracellular concentrations are scaled by C, the intracellular quantities are independent of C, the volume varies inversely with C, and the membrane potential is independent of C. The fact that $n_+^i(\infty)$ is independent of C implies that the cation, to which the membrane is permeable, is not transported across the membrane when C is varied. This property occurs because the membrane is impermeant to the anion and any transport of cation would violate electroneutrality of the intracellular solution. Therefore, in the absence of a permeant anion this hypothetical model cell acts as if it were impermeant to the cation. This result explains why the volume varies inversely with extracellular concentration of the cation in a manner identical to that of a perfect osmometer. This variation of the volume causes the concentration of all ions to be proportional to the extracellular cation concentration. Because the intracellular concentration of the cation is proportional to the extracellular concentration, the Nernst equilibrium potential for the cation is independent of the cation concentration. Therefore, the membrane potential, which equals the Nernst equilibrium potential for the cation, is also independent of extracellular cation concentration.

Dependence on Characteristics of Impermeant Solutes Next we shall examine the dependence of the dependent variables on the impermeant solute, whose parameters are z_i and n_i^i, under the assumption that n_-^i is a positive constant. To examine this dependence, it is helpful to normalize the dependent variables so as to remove their dependence on C and on n_-^i. Therefore, we define normalized variables as $\hat{c}_+ = c_+^i(\infty)/C$, $\hat{c}_- = c_-^i(\infty)/C$, $\hat{c}_i = c_i^i(\infty)/C$, $\hat{n}_i = n_i^i/n_-^i$, $\hat{\mathcal{V}} = \mathcal{V}(\infty)C/n_-^i$, and $\hat{V}_m^o = V_m^o/(RT/F)$. We can now express the equilibrium, normalized volume, concentrations, quantities, and membrane potential as

$$\hat{n}_+ = 1 - z_i\hat{n}_i,$$

$$\hat{\mathcal{V}} = 1 + \left(\frac{1 - z_i}{2}\right)\hat{n}_i,$$

$$\hat{c}_+ = \frac{1 - z_i\hat{n}_i}{\hat{\mathcal{V}}},$$

$$\hat{c}_- = \frac{1}{\hat{\mathcal{V}}},$$

$$\hat{c}_i = \frac{\hat{n}_i}{\hat{\mathcal{V}}},$$

$$\hat{V}_m^o = -\ln(\hat{c}_+).$$

Note, that all the normalized dependent variables depend upon two parameters only, $\hat{n}_i$ and z_i. It is much easier to explore the dependence of the normalized variables on these two parameters than the dependence of unnormalized variables on the four unnormalized parameters. In examining this dependence, we do not restrict z_i to being an integer for the following reason. Suppose there were N different intracellular impermeant ions, each with concentration $c_{in}^i(t)$ and valence z_{in}. Then the total concentration of these impermeant ions would be

$$c_i^i(t) = \sum_{n=1}^{N} c_{in}^i(t),$$

and the average valence of the total concentration would be the ratio of the charge density of the total concentration to the charge density if all the constituents had a valence of +1, or

$$z_i = \frac{\sum_{n=1}^{N} z_{in} F c_{in}^i(t)}{\sum_{n=1}^{N} F c_{in}^i(t)} = \frac{\sum_{n=1}^{N} z_{in} c_{in}^i(t)}{\sum_{n=1}^{N} c_{in}^i(t)}.$$

Thus, a mixture of impermeant solutes would have an average valence that need not be an integer. Because the concentrations and quantities and the volume cannot be negative, the parameter values for which a physically plausible solution exists are limited. Examination of the equations reveals that if both $\hat{n}_i z_i \leq 1$ and $(1 - z_i)\hat{n}_i \geq -2$, then all the quantities are non-negative. But since $\hat{n}_i$ is non-negative, both inequalities are satisfied when $\hat{n}_i z_i \leq 1$.

The dependence of all the normalized variables is plotted versus $\hat{n}_i$ with z_i as a parameter in Figure 8.8. First we check to see if the solutions reduce to those without the additional impermeant intracellular solute described in the previous section by examining the solutions with $n_i^i = 0$. The solutions for the unnormalized variables can be obtained from the unnormalized equations, and the normalized variables can be obtained either from the normalized equations or from Figure 8.8. They are $\mathcal{V}(\infty) = n_-^i/C$ or $\hat{\mathcal{V}} = 1$; $c_i^i(\infty) = 0$ or $\hat{c}_i = 0$; $c_+^i(\infty) = c_-^i(\infty) = C$ or $\hat{c}_+ = \hat{c}_- = 1$; $n_+^i(\infty) = n_-^i$ or $\hat{n}_+ = 1$; $V_m^o(\infty) = 0$ or $\hat{V}_m^o = 0$. These solutions are the same as those derived in the previous section, where the intracellular solution contained no extra impermeant solute.

Next consider the case for which the impermeable intracellular solute is a nonelectrolyte, i.e., for $z_i = 0$. Note that increasing $\hat{n}_i$ increases $\hat{\mathcal{V}}$, because the intracellular osmolarity is increased. Since the solute is uncharged, this increase in $\hat{n}_i$ is not accompanied by a change in $\hat{n}_+$, because electroneutrality is unaffected by this change. However, because an increase in $\hat{n}_i$ causes an increase in $\hat{\mathcal{V}}$ and since $\hat{n}_+$ is constant, $\hat{c}_+$ decreases. A similar argument shows that $\hat{c}_-$ decreases. However, because $\hat{c}_i$ is the ratio of $\hat{n}_i$ to $\hat{\mathcal{V}}$, when both increase the ratio depends upon the manner in which they increase. The results show that $\hat{c}_i$ decreases. Because $\hat{c}_+$ decreases from a value of 1 to 0, $\hat{V}_m^o$ is positive and increases as $\hat{n}_i$ increases.

Consider the dependence of the normalized dependent variables on the valence of the impermeant solute. For negative valences, an increase in $\hat{n}_i$ leads to an increase in $\hat{n}_+$ to maintain electroneutrality. If the valence is -1, then $\hat{n}_+$ grows linearly with $\hat{n}_i$ with a slope of 1, because for every molecule of impermeant solute added the intracellular cation quantity increases by one molecule to maintain electroneutrality. If the valence is -2, then $\hat{n}_+$ grows linearly with $\hat{n}_i$ with a slope of 2, because for every molecule of impermeant solute added the intracellular cation quantity increases by two molecules to maintain electroneutrality. A similar pattern holds true for valences of -3 and -4. For positive valences the addition of impermeant solute reduces the intracellular quantity of cation with an increment that is proportional to the

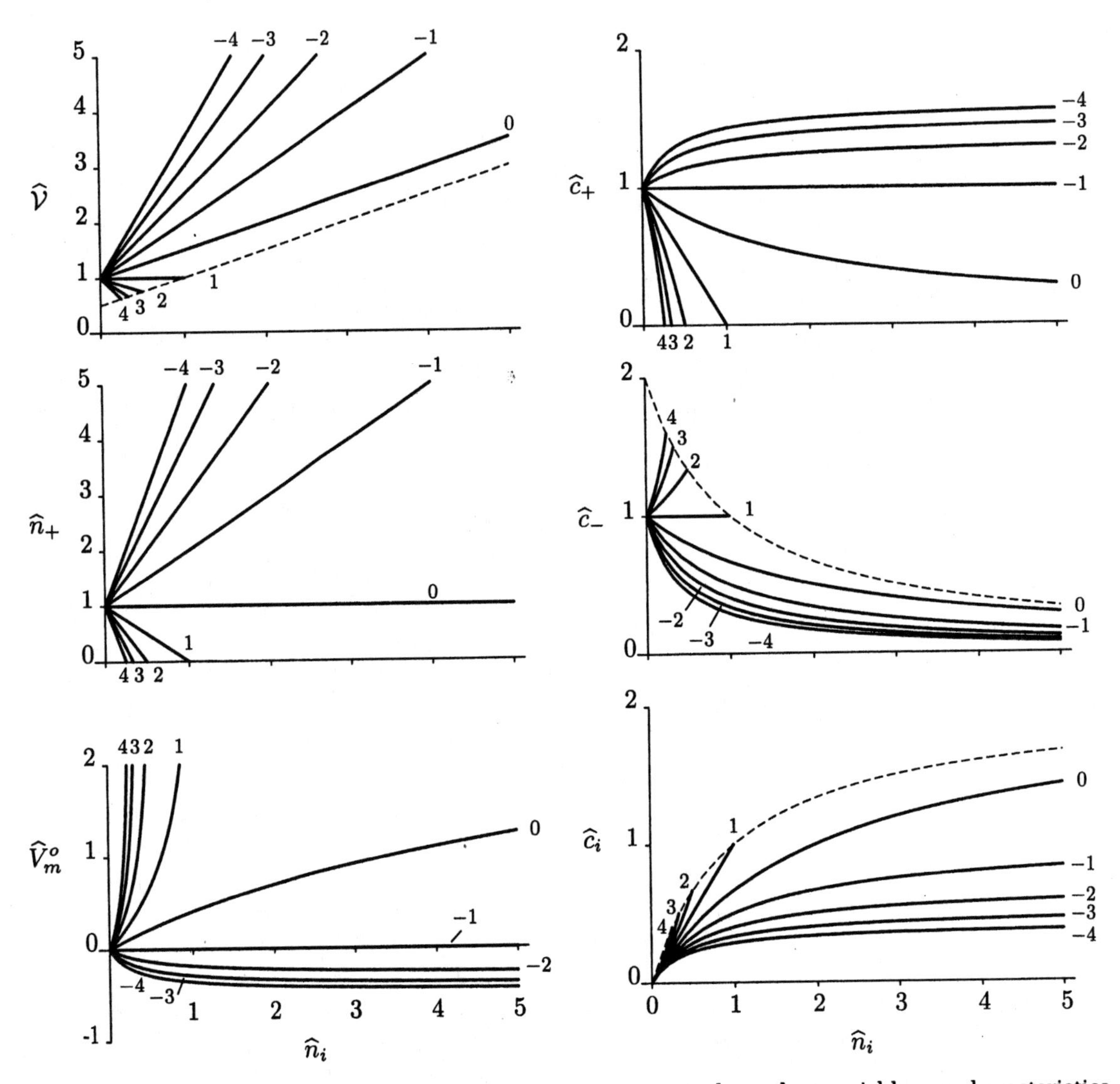

Figure 8.8 Dependence of all the homeostatic dependent variables on characteristics of the impermeant ions for a simple cell model in which there is one permeant cation transported passively. The variables are plotted versus $\widehat{n}_i$ with parameter z_i. The dashed lines define the locus of points for which $z_i \widehat{n}_i = 1$. Physically plausible solutions occur only for $z_i \widehat{n}_i \leq 1$.

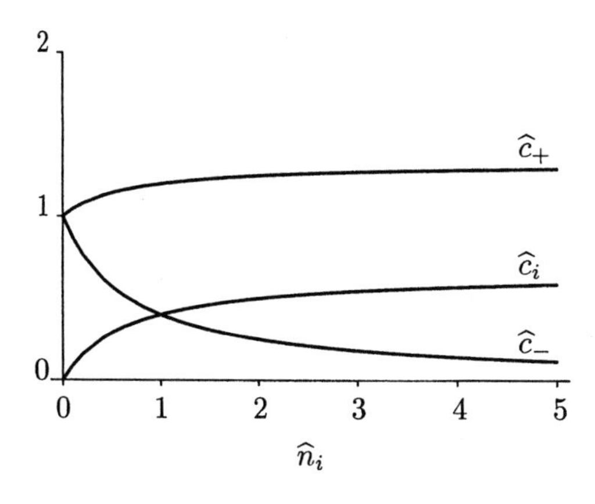

Figure 8.9 Dependence of intracellular concentrations on characteristics of the impermeant ions for a simple cell model. The concentrations are plotted versus $\widehat{n}_i$ for $z_i = -2$.

valence. This dependence of the cation quantity is reflected in the normalized volume. For negative valences an increase in impermeant solute quantity results in an increase in cation. The increase in both these quantities results in an increase in volume. For a valence of +1, an increase in impermeant solute quantity causes an equal decrease in cation quantity. Therefore, there is no net increase in intracellular quantity, and the volume does not change. For larger valences, an increase in impermeable solute quantity causes a larger increment in cation reduction. Therefore, the net intracellular quantity decreases and the volume decreases. The concentrations are ratios of quantities to the volume. Hence, their behavior can be discerned directly from these two variables.

The relation among the concentrations is shown for one value of the valence of the impermeant solute, $z_i = -2$, in Figure 8.9. Note that as $\widehat{n}_i$ increases, both $\widehat{c}_+$ and $\widehat{c}_i$ increase, but $\widehat{c}_-$ decreases because, while $\widehat{n}_-$ is constant, $\widehat{V}$ decreases.

Summary This model, in which the cell membrane is permeable to only the cation, illustrates the role of electroneutrality, which prevents transport of the permeant cation. As a result, the cell acts as a perfect osmometer and has a membrane potential that is independent of cation concentration. This model illustrates the important constraint that electroneutrality can have on cell homeostasis.

8.4.3.2 Two Ions Transported Passively

With No Impermeant Intracellular Solute

The analysis for this case (for which the model is shown in the left central panel in Figure 8.7) is identical to the case for a single permeant solute. The only viable solution, which is a trivial solution, is found if all the ion concentrations have the same constant value and the membrane potential is zero.

With an Impermeant Intracellular Solute

When an impermeant solute is added to the intracellular solution (right central panel of Figure 8.7), the behavior of the cell model changes in an important way. Once again electroneutrality of the extracellular solutions requires that $c_+^o = c_-^o \doteq C$ and that the intracellular concentrations satisfy both Equations 8.18 and 8.19 to give the relations in Equation 8.20. At equilibrium, the fluxes of both the cation and anion must be zero. Hence, the membrane potential must equal the Nernst equilibrium potential of each ion,

$$V_m^o = V_+ = V_- = \frac{RT}{F}\ln\left(\frac{C}{c_+^i(\infty)}\right) = -\frac{RT}{F}\ln\left(\frac{C}{c_-^i(\infty)}\right),$$

which leads to the relation

$$\frac{C}{c_+^i(\infty)} = \frac{c_-^i(\infty)}{C} \text{ or } c_+^i(\infty)c_-^i(\infty) = C^2. \tag{8.21}$$

Substitution of Equation 8.20 into Equation 8.21 yields

$$\left(C - \left(\frac{1+z_i}{2}\right)c_i^i(\infty)\right)\left(C - \left(\frac{1-z_i}{2}\right)c_i^i(\infty)\right) = C^2,$$

which can be factored to give

$$\left(\left(\frac{1-z_i^2}{4}\right)c_i^i(\infty) - C\right)c_i^i(\infty) = 0.$$

Therefore, there are two solutions possible for $c_i^i(\infty)$,

$$c_i^i(\infty) = \begin{cases} 0 & \text{for } z_i = \pm 1, \\ 4C/(1-z_i^2) & \text{for } z_i \neq \pm 1. \end{cases}$$

Each solution can be substituted back into Equation 8.20 to find both $c_+^i(\infty)$ and $c_-^i(\infty)$, which after simplification yields the following two sets of solutions to these equations:

$$c_i^i(\infty) = 0, \qquad c_+^i(\infty) = C, \qquad c_-^i(\infty) = C;$$

$$c_i^i(\infty) = \left(\frac{4}{1-z_i^2}\right)C, \qquad c_+^i(\infty) = \left(\frac{z_i+1}{z_i-1}\right)C, \qquad c_-^i(\infty) = \left(\frac{z_i-1}{z_i+1}\right)C.$$

The second set is not physically plausible, because there are no values of z_i for which all three concentrations are positive. Note that $4/(1-z_i^2)$ is positive when $|z_i| < 1$, while $(z_i + 1)/(z_i - 1)$ is positive when $|z_i| > 1$. Therefore, the only physically plausible solution is for $c_i^i(\infty) = 0$. Since $c_i^i(\infty) = n_i^i/\mathcal{V}(\infty)$, if $n_i^i \neq 0$ this solution can only occur if $\mathcal{V}(\infty) \to \infty$. Thus, if there is a nonzero quantity of impermeant intracellular solute, water flows into the cell and the volume diverges.[2] Just as we found for nonelectrolytes, the presence of an impermeant solute makes the equations of cellular homeostasis unstable. With passive ion transport only, osmotic equilibrium, electroneutrality, and electrodiffusive equilibrium cannot be satisfied simultaneously in the presence of an impermeant solute, no matter what its valence.

8.4.3.3 Two Ions Transported Passively, One Transported Actively

We shall next consider a simple cell model in which both the cation and the anion are transported passively, but in addition the cation is transported actively, as indicated schematically in the lower panels of Figure 8.7. We represent active cation transport by the simple generic pump model defined in Equation 8.10. Therefore, the flux for each ion is

$$\phi_+(t) = \phi_+^p(t) + \phi_+^a(t) = \frac{G_+}{F}\left(V_m^o(t) - V_+(t)\right) + k_a c_+^i(t), \tag{8.22}$$

$$\phi_-(t) = -\frac{G_-}{F}\left(V_m^o(t) - V_-(t)\right). \tag{8.23}$$

The flux of the cation and of the anion contains a term due to passive ion transport. In addition, the flux of the cation contains a term due to the simple generic pump model of active transport.

In quasi-equilibrium, each flux is zero. A zero anion flux implies that the membrane potential equals the Nernst equilibrium potential for the anion,

$$V_m^o = V_- = -\frac{RT}{F}\ln\left(\frac{C}{c_-^i(\infty)}\right) = \frac{RT}{F}\ln\left(\frac{c_-^i(\infty)}{C}\right).$$

Thus, the anion is in electro-diffusive equilibrium as well as in steady state. A zero cation flux results in a quasi-equilibrium, because the active cation transport just cancels the passive cation transport. Thus, the cation is not in electro-diffusive equilibrium. Quasi-equilibrium of both the anion and the cation constrains the quasi-equilibrium anion and cation concentrations to obey the equation

2. For an alternate derivation of this result, see Problem 8.6.

$$0 = \frac{G_+}{F}\left(\frac{RT}{F}\ln\left(\frac{c_-^i(\infty)}{C}\right) - \frac{RT}{F}\ln\left(\frac{C}{c_+^i(\infty)}\right)\right) + k_a c_+^i(\infty), \tag{8.24}$$

which can be written in the compact form

$$\left(\frac{RTG_+}{k_a F^2}\right)\ln\left(\frac{c_+^i(\infty)c_-^i(\infty)}{C^2}\right) + c_+^i(\infty) = 0. \tag{8.25}$$

We shall first consider the case in which there is no impermeant intracellular solute, and then we shall consider the case in which there is.

With No Impermeant Intracellular Solute

With no impermeant intracellular solute, electroneutrality and osmotic equilibrium guarantee (Equations 8.18–8.20) that $c_+^i(\infty) = c_-^i(\infty) = C$. Therefore, the Nernst equilibrium potential for each ion is zero. Since the resting potential of the cell equals the Nernst equilibrium potential for the anion, the resting potential is also equal to zero. Therefore, Equation 8.24 implies that $c_+^i(\infty) = 0$, which contradicts the conclusion based on osmotic equilibrium and electroneutrality, for which $c_+^i(\infty) = C$. Thus, provided that $k_a \neq 0$—i.e., the case in which there is no pump—there is no nontrivial stable solution in the absence of an impermeant intracellular solute and in the presence of a generic cation pump. As the analysis in the next section indicates, with no impermeant intracellular solute the cell volume goes to zero. This condition results because, as the concentrations of cations equilibrate across the membrane, the passive cation flux decreases, but the active cation flux remains. As the cation is pumped out of the cell, an equal amount of anion also leaves to satisfy electroneutrality. Thus, the osmolarity in the cell will decrease and water will leave the cell. This efflux will occur until the cell water volume goes to zero.

With an Impermeant Intracellular Solute

With the addition of an impermeant intracellular solute, electroneutrality and osmotic equilibrium imply that the intracellular anion and cation concentrations satisfy Equation 8.20. The analysis is simplified somewhat by using normalized variables. Let $\hat{c}_+ = c_+^i(\infty)/C$, $\hat{c}_- = c_-^i(\infty)/C$, $\hat{c}_i = c_i^i(\infty)/C$, $\hat{\mathcal{V}} = \mathcal{V}(\infty)C/n_i^i = 1/\hat{c}_i$, and $\hat{V}_m^o = V_m^o(F/RT)$.[3] With these normalized variables,

3. There is an important caveat to these normalized variables. The normalizations used to analyze different problems in this chapter are not necessarily identical. They are chosen to be useful in each individual analysis. So, for example, in the analysis in this section we define $\hat{\mathcal{V}} = \mathcal{V}(\infty)C/n_i^i$, whereas in an earlier section we defined $\hat{\mathcal{V}} = \mathcal{V}(\infty)C/n_-^i$.

Equations 8.20 become

$$\hat{c}_+ = 1 - \left(\frac{1+z_i}{2}\right)\left(\frac{1}{\hat{\mathcal{V}}}\right), \tag{8.26}$$

$$\hat{c}_- = 1 - \left(\frac{1-z_i}{2}\right)\left(\frac{1}{\hat{\mathcal{V}}}\right). \tag{8.27}$$

In terms of these normalized variables, the quasi-equilibrium cation flux equation (Equation 8.25) becomes

$$\kappa \ln(\hat{c}_+\hat{c}_-) + \hat{c}_+ = 0. \tag{8.28}$$

The quantity $\kappa = RTG_+/(k_a F^2 C)$ is proportional to the ratio of the passive conductance of the membrane for the cation (G_+) to the rate at which the cation is pumped out of the cell (k_a). Therefore, we call κ the *leak-to-pump ratio*. When κ is large, the cell has a large relative flux of cations due to passive transport, i.e., a large cation leak. When κ is small, the cell has a large relative flux of cations due to active transport.

Thus, there are three equations in the normalized variables $\hat{c}_+$, $\hat{c}_-$, and $\hat{\mathcal{V}}$. Since the general equations are transcendental, we cannot obtain an analytic solution. However, we can examine solutions for limiting cases, and we can examine numerical solutions. To simplify further analysis, let us determine $\hat{c}_+$. We need to consider two different cases. For $z_i = -1$, Equation 8.26 shows that $\hat{c}_+ = 1$. For $z_i \neq -1$, combining Equations 8.27–8.28 defines a single transcendental equation for $\hat{c}_+$,

$$\hat{c}_+ \frac{2z_i + (1-z_i)\hat{c}_+}{1+z_i} = e^{-\hat{c}_+/\kappa}, \tag{8.29}$$

which can, in principle, be solved for $\hat{c}_+$ for any value of z_i and κ. Once $\hat{c}_+$ is determined, then with the use of Equations 8.28–8.27 all the other variables can be determined as follows:

$$\hat{c}_- = \begin{cases} \frac{2z_i+(1-z_i)\hat{c}_+}{1+z_i} & \text{for } z_i \neq -1 \\ e^{-1/\kappa} & \text{for } z_i = -1 \end{cases}, \tag{8.30}$$

$$\hat{\mathcal{V}} = \begin{cases} \frac{1+z_i}{2(1-\hat{c}_+)} & \text{for } z_i \neq -1 \\ \frac{1}{1-e^{-1/\kappa}} & \text{for } z_i = -1 \end{cases}, \tag{8.31}$$

$$\hat{c}_i = \frac{1}{\hat{\mathcal{V}}}, \tag{8.32}$$

$$\hat{V}_m^o = \begin{cases} \ln \hat{c}_- & \text{for } z_i \neq -1 \\ -\frac{1}{\kappa} & \text{for } z_i = -1 \end{cases}. \tag{8.33}$$

Physical Constraints The solutions to these homeostatic equations are also subject to the physical constraints that all the concentrations and the volume must be positive quantities. We examine the effect of these constraints on $\hat{c}_+$. The constraints that $\hat{c}_- > 0$ can be found from Equation 8.30 and are

$$\hat{c}_+ > \frac{2z_i}{z_i - 1} \quad \text{for } |z_i| < 1,$$

$$\hat{c}_+ < \frac{2z_i}{z_i - 1} \quad \text{for } |z_i| > 1.$$

The constraint that $\hat{V} > 0$ can be found from Equation 8.31, which shows that when $z_i < -1$, then $\hat{c}_+ > 1$, and when $z_i = -1$, then $\hat{c}_+ = 1$ (from Equation 8.26), provided that $\kappa > 0$, and when $z_i > -1$, then $\hat{c}_+ < 1$. All the constraints are satisfied for certain values of $\hat{c}_+$ and z_i, which are illustrated in Figure 8.10. The constraints that guarantee that $\hat{c}_- > 0$ and $\hat{V} > 0$ are that $\hat{c}_+ > 0$ and the following: if $z_i < -1$, then $1 < \hat{c}_+ < 2z_i/(z_i - 1)$; if $|z_i| < 1$, then $2z_i/(z_i - 1) < \hat{c}_+ < 1$; and if $z_i > 1$, then $\hat{c}_+ < 1$;

Dependence on Leak-to-Pump Ratio The general solution can be obtained by solving the transcendental equation (Equation 8.29). Note that this equation contains two terms that depend on $\hat{c}_+$. Both terms are plotted as a function of $\hat{c}_+$ in Figure 8.11. The left-hand term depends parabolically on $\hat{c}_+$ with a single parameter, the valence of the impermeant solute z_i. The right-hand term depends exponentially on $\hat{c}_+$ with a single parameter, the leak-to-pump ratio κ. Mathematical solutions to the transcendental equation occur at the intersection of the parabolas and the exponentials. Of these solutions, only those subject to the physical constraints given in Figure 8.10 are physically plausible solutions. Inspection of Figures 8.10 and 8.11 gives insight into the dependence of the solution of the transcendental equation on κ and on z_i. Note

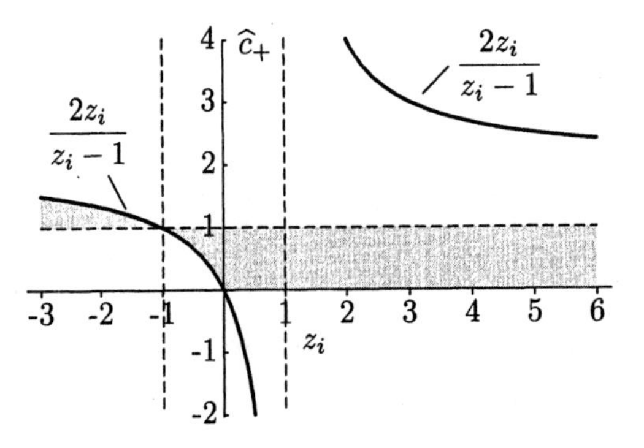

Figure 8.10 Illustration of physical constraints for the existence of solutions. Here $\hat{c}_+$ is plotted versus z_i. Solutions that satisfy the physical constraints that all the concentrations and the volume are nonnegative quantities occur in the shaded region of the $\hat{c}_+$-z_i plane.

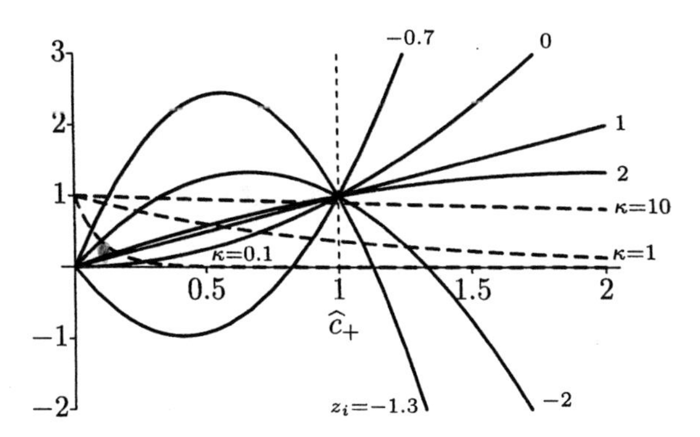

Figure 8.11 Illustration of mathematical conditions for the existence of solutions to the transcendental equation (Equation 8.29). The solid lines are parabolas that depend upon z_i. The dashed lines are exponentials that depend upon κ. Intersections of the solid and dashed lines are solutions to the transcendental equation.

that for $z_i > -1$, and for $\hat{c}_+ < 1$, increasing κ makes $\hat{c}_+$ increase toward 1. For $z_i < -1$, and for $\hat{c}_+ > 1$, increasing κ makes $\hat{c}_+$ decrease toward 1. Thus, for both conditions as the cell gets leakier, the intracellular cation concentration approaches its extracellular concentration. Numerical solutions have been obtained for the homeostatic equations using Newton's method to solve the transcendental equation with the appropriate physical constraints. The dependence of $\hat{c}_+$, $\hat{c}_-$, $\hat{c}_i$, $\hat{\mathcal{V}}$, and $\hat{V}_m^o$ on κ and z_i is shown in Figures 8.12 and 8.13.

The asymptotic behavior of the variables at arbitrarily large and small values of κ can be found directly from the equations and checked against the numerical solutions. The asymptotic values of the variables for large and small κ can be found directly from Equation 8.29. First consider a very leaky cell for which the leak-to-pump ratio is high, i.e., $\kappa \to \infty$. Under this condition, Equation 8.29 yields

$$\hat{c}_+ \frac{2z_i + (1 - z_i)\hat{c}_+}{1 + z_i} \to 1,$$

which yields a quadratic equation in $\hat{c}_+$ whose solution is

$$\hat{c}_+ \to 1 \text{ or } \frac{z_i + 1}{z_i - 1}.$$

Using Equation 8.30, we find that

$$\hat{c}_- \to 1 \text{ or } \frac{z_i - 1}{z_i + 1}.$$

Using Equation 8.31, we find that

$$\hat{\mathcal{V}} \to \infty \text{ or } \frac{1 - (z_i)^2}{4}.$$

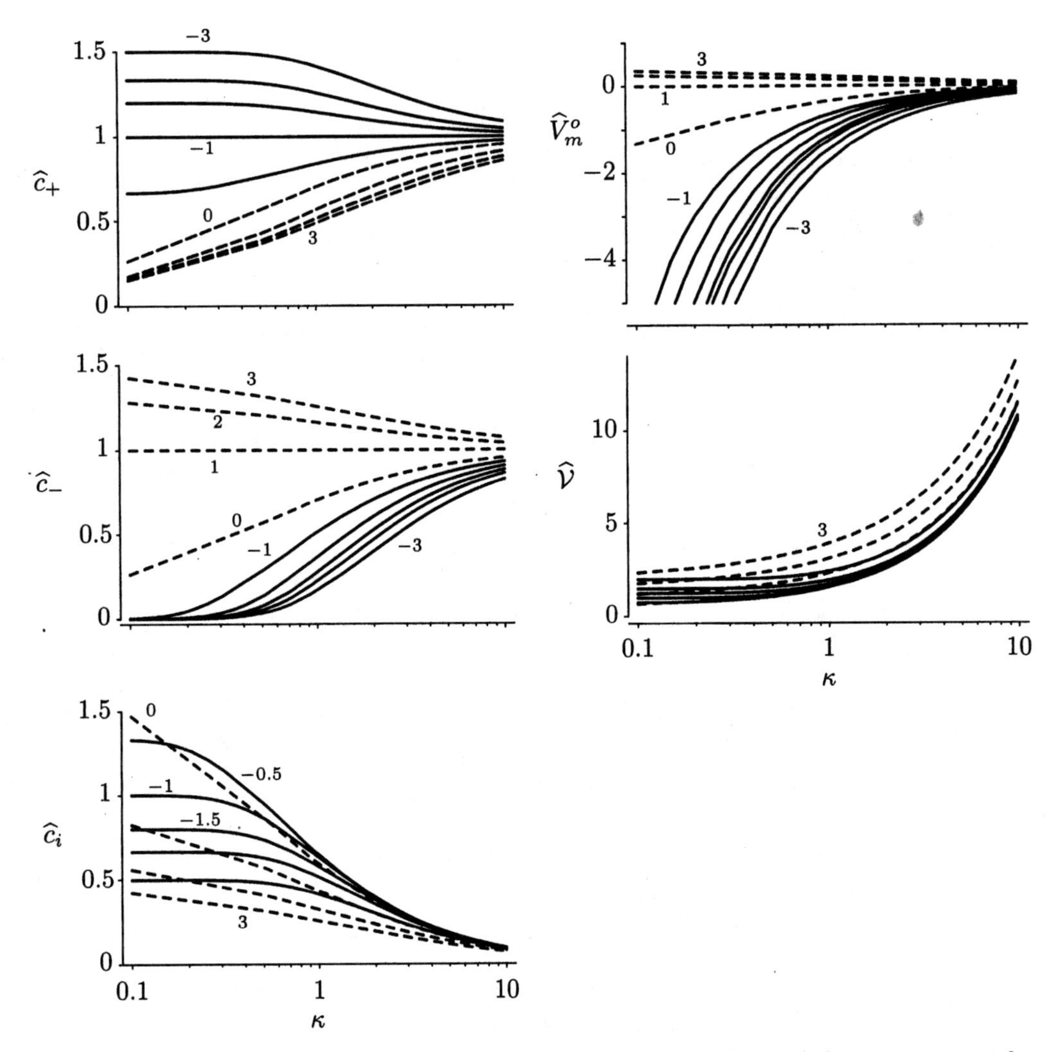

Figure 8.12 Dependence of concentrations, volume, and resting potential on the leak-to-pump ratio, κ. The results are shown for the valences of the impermeant solute $z_i = -3, -2, -1.5, -1\ -0.5, 0, 1, 2$, and 3. Solid lines are used for $z_i < 0$, and dashed lines for $z_i \geq 0$.

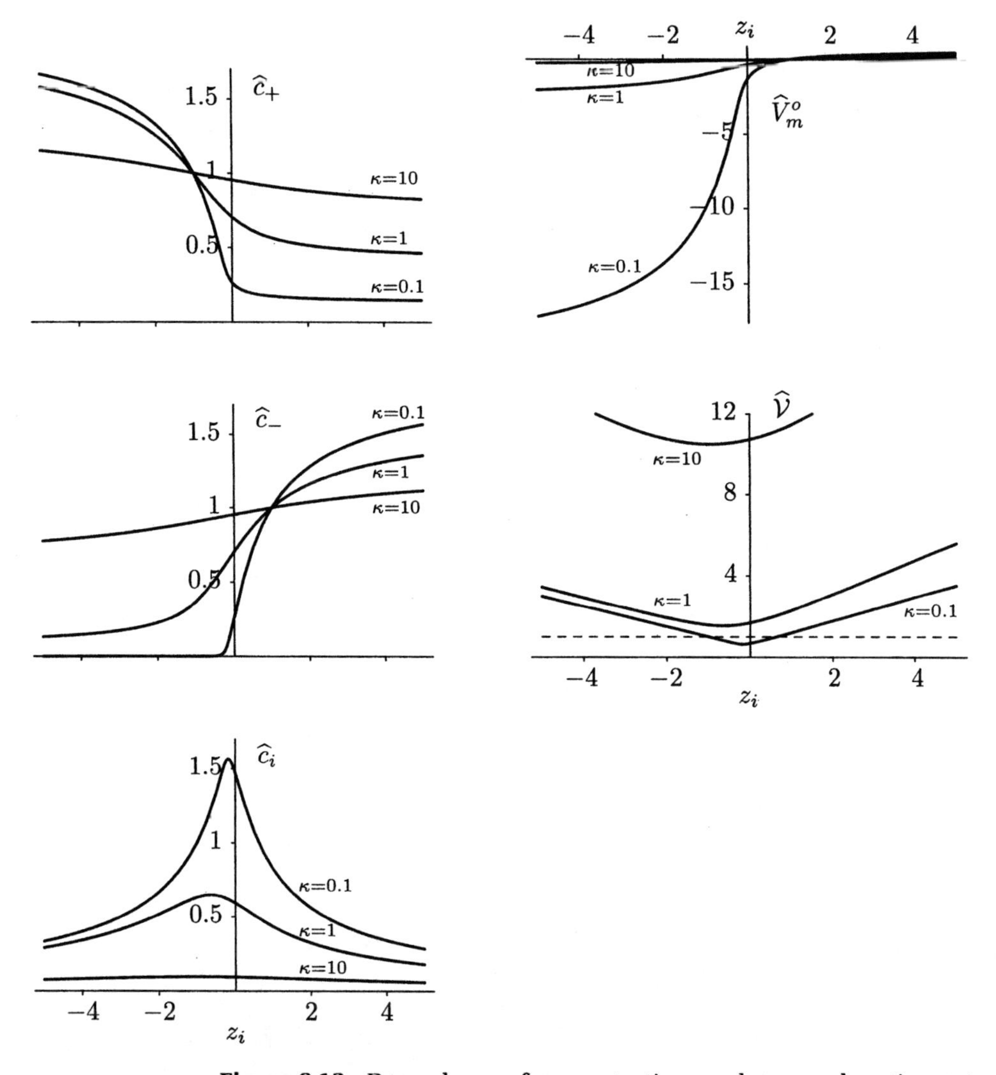

Figure 8.13 Dependence of concentrations, volume, and resting potential on the valence of the impermeant solute for leak-to-pump ratios $\kappa = 0.1$, 1, and 10.

Thus, the first solution listed for each variable gives $\hat{c}_+ = \hat{c}_- \to 1$ and $\hat{V} \to \infty$. For the second solution, the volume is negative in the range of $|z_i| > 1$ and positive in the range of $|z_i| < 1$. However, the concentrations are negative in the range of $|z_i| < 1$. Thus, the second solution is physically implausible, whereas the first solution is physically plausible.

Second, consider a cell that is not leaky and has a very small leak-to-pump ratio, i.e., $\kappa \to 0$. Under this condition, Equation 8.29 yields

$$\hat{c}_+ \frac{2z_i + (1 - z_i)\hat{c}_+}{1 + z_i} \to 0,$$

which yields a quadratic equation in $\hat{c}_+$. After imposing the condition that the concentrations and volumes are nonnegative quantities, the solutions are

$$\hat{c}_+ \to \begin{cases} \frac{2z_i}{z_i - 1} & \text{for } z_i < -1 \\ 1 & \text{for } z_i = -1 \\ 0 & \text{for } z_i > -1 \end{cases},$$

$$\hat{c}_- \to \begin{cases} 0 & \text{for } z_i \leq -1 \\ \frac{2z_i}{z_i + 1} & \text{for } z_i > -1 \end{cases},$$

$$\hat{V} \to \begin{cases} \frac{1 - z_i}{2} & \text{for } z_i \leq -1 \\ \frac{1 + z_i}{2} & \text{for } z_i > -1 \end{cases},$$

$$\hat{c}_i \to \begin{cases} \frac{2}{1 - z_i} & \text{for } z_i \leq -1 \\ \frac{2}{1 + z_i} & \text{for } z_i > -1 \end{cases},$$

$$\hat{V}_m^o \to \begin{cases} -\infty & \text{for } z_i \leq -1 \\ \ln\left(\frac{2z_i}{z_i + 1}\right) & \text{for } z_i > -1 \end{cases}.$$

Examination of the asymptotic solutions and the numerical solutions (Figure 8.12) leads to several general conclusions. As the cell becomes leakier (κ increasing), the anion and cation concentrations approach their extracellular concentrations, which equal C, and the volume of the cell increases. The intracellular concentration of impermeant solute goes to zero, because the quantity of intracellular impermeant solute is fixed and the volume becomes infinite. The membrane potential goes to zero, because the anion concentration ratio across the membrane approaches 1. As the strength of the pump is increased (κ decreasing), the concentration of the cation and anion deviate from C in a manner that depends upon the valence of the intracellular impermeant solute. For $z_i < -1$, the cation concentration increases and the anion concentration decreases. For $z_i > -1$, the cation concentration decreases and the anion concentration increases. For all z_i, the volume decreases and the intracellular concentration of impermeant solute increases. The membrane potential becomes more negative if $z_i < 1$ and slightly positive if $z_i > 1$.

Dependence on Impermeant Solute First we check the solution for an uncharged impermeant solute, i.e., $z_i = 0$. Under this condition, Equation 8.30 shows that $\hat{c}_- = \hat{c}_+$, which must be true to satisfy electroneutrality. Figure 8.13 shows the dependence of the solution on z_i for different values of κ. As the valence of the impermeant solute is increased, the concentration of the cation is decreased and the concentration of the anion is increased. This behavior is a direct consequence of electroneutrality. The volume increases as $|z_i|$ increases, since an increase in the valence of the impermeant solute will attract the counter ion to maintain electroneutrality. That is, for z_i large and positive, anions will enter the cell to maintain electroneutrality, and for z_i large and negative, cations will enter the cell to maintain electroneutrality. From Figure 8.13 it is apparent that in either case there is a net influx of counterions over coions and thus an increase in osmotic pressure. The increase in osmotic pressure leads to an increase in cell volume to maintain osmotic equilibrium. Since the resting potential is equal to the anion equilibrium potential, the shape of the dependence of the resting potential to the valence must be a logarithmic function of the dependence of the anion concentration on the valence.

The effect of the solution on the quantity of impermeant intracellular solute (n_i^i) is particularly simple. Previously we showed that for given values of z_i and κ, the solutions of the normalized variables are determined. Thus, none of these variables is affected by n_i^i. The only way in which n_i^i enters into the equations is in the relation $\hat{\mathcal{V}} = \mathcal{V}(\infty)C/n_i^i$. Therefore, for given values of z_i and κ, $\hat{\mathcal{V}}$ is fixed and so is C. Therefore, a change in n_i^i results in a change in the volume of the cell $\mathcal{V}(\infty)$ and in nothing else. With respect to a change in n_i^i, the cell behaves as a perfect osmometer.

8.4.4 Summary

An important general conclusion from these simple models is that the transports of different solutes are linked by constraints imposed by both electroneutrality and osmotic equilibrium. Hence, a change in even an impermeant intracellular solute can affect the transport of some ion. Furthermore, the transports of cations and anions are not independent, but are linked by electroneutrality. In addition to these general conclusions that apply to all cells, there are a number of specific conclusions that apply to the simple cell models. These are summarized in Table 8.2. A key specific result is that in the

Table 8.2 Qualitative summary of solutions for simple cell models.

Intracellular solutes	Transport of permeant solutes	Solution
	Nonelectrolyte solutes	
Impermeant	None	Perfect osmometer
Impermeant and permeant	Passive only	No stable solution
Impermeant and permeant	Passive and active	Stable solution
	Ionic solutes	
Binary electrolyte	Passive transport of either or both ions	Trivial solution
Binary electrolyte plus impermeant ion	Passive transport of cation	Perfect osmometer
Binary electrolyte plus impermeant ion	Passive transport of cation and anion	No stable solution
Binary electrolyte	Passive transport of cation and anion plus active cation transport	No stable solution
Binary electrolyte plus impermeant ion	Passive transport of cation and anion plus active cation transport	Stable solution

absence of active transport, the presence of an impermeant intracellular solute destabilizes the homeostatic equations for both nonelectrolytes and ions. For ions this destabilization is manifested provided that the permeant solutes are not prevented from being transported by constraints of electroneutrality. For example, if the membrane is permeant to only one ion species, electroneutrality prevents the transport of that species. Active transport pumps can stabilize the instability in the homeostatic equations caused by the presence of impermeant intracellular solutes.

8.5 Inventory of Homeostatic Mechanisms

8.5.1 Transport Mechanisms

In previous chapters we considered a number of distinct types of cellular transport mechanisms. There are only a few types, and these are shown schematically in Figure 8.14 and summarized in this section.

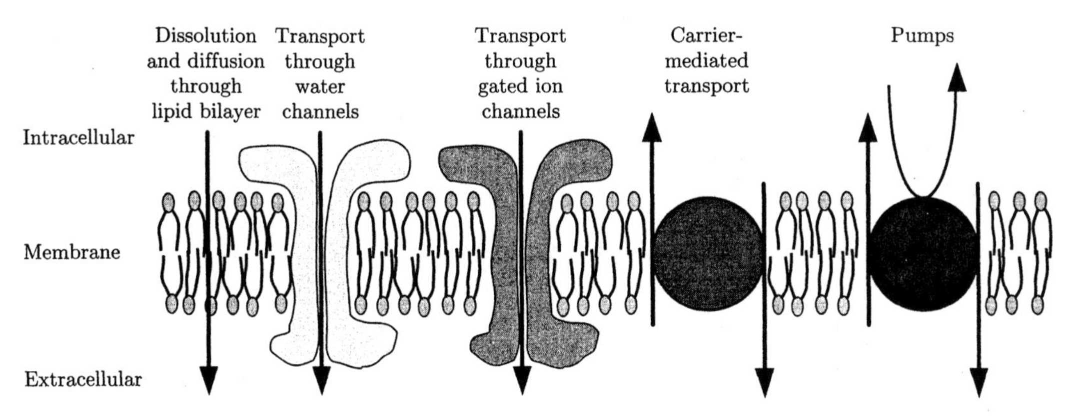

Figure 8.14 Schematic diagram of different types of membrane transport mechanisms.

8.5.1.1 Dissolution and Diffusion in the Lipid Bilayer

The notion of the dissolve-diffuse mechanism of transport through the lipid phase of the cellular membrane was introduced in Chapter 3. This relatively nonselective mechanism transports solutes according to their lipid solubility and molecular radius; highly lipid-soluble solutes of small dimensions are highly permeant, whereas poorly lipid-soluble solutes and large solutes are poorly permeant via this mechanism. Water is biologically the most important substance that is transported by the dissolve-diffuse mechanism through the lipid bilayer in which it acts as a solute. In addition, lipid-soluble hormones, the steroids, are transported into cells, and local anesthetics are transported into the lipid bilayer by this mechanism.

8.5.1.2 Transport Through Water Channels

Water is also transported into and out of cells via water channels that traverse the membrane (Chapter 4). It appears that these channels are highly selective to water, and it is not clear that solutes other than water permeate this pathway. These channels are not gated, but are normally open. Water flows convectively through these channels in response to hydraulic and osmotic pressure differences across the membrane.

8.5.1.3 Passive Transport of Ions Through Channels

As introduced in Chapter 7 and discussed more fully elsewhere (Weiss, 1996, Chapter 6), ion channels are membrane-bound macromolecules that allow the passage of ions between the extracellular space and the cytoplasm. Ions on both sides of the membrane have simultaneous access to an open ion channel. This property is distinct from that for carriers (described below), whose transport site is accessible at any instant in time to one side of the membrane only. Transport through open ion channels is driven by the electrochemical potential difference across the membrane. Ion channels are selectively permeable to particular ions. Generally speaking, transport through ion channels is rapid when compared to other ion-transporting mechanisms such as carriers and pumps. Many different types of ion channels have been identified based on their kinetic properties, ion selectivity, gating variables, pharmacology, and molecular structure. Table 8.3 gives a partial list of some of the different types of ion channels that have been identified. A more complete listing is given elsewhere (Watson and Girdlestone, 1994), and a thorough treatise on ion channels is also available (Hille, 1992).

Most ion channels are *gated* (either opened or closed) by some physical-chemical variable. These ion channels are categorized by this gating variable. In voltage-gated channels, which are found ubiquitously and abundantly in electrically excitable cells as well as in other cells, the potential across the membrane is the gating variable. That is, a change in the membrane potential either opens or closes the channel. Voltage-gated ion channels are critically involved in nerve signaling, muscle contraction, and chemical secretion. Another category of gated ion channels is the ligand-gated channels, which are gated by the binding of a ligand to a receptor site on either the cytoplasmic face or the extracellular face of the membrane.

The receptor is an integral part of the channel macromolecule. For example, the calcium-activated potassium channel is opened by an increase in the intracellular calcium concentration, and the ATP-activated potassium channel is closed by an increase in the intracellular concentration of ATP. The remaining ligand-gated channels listed in Table 8.3 are opened by the binding of a neurotransmitter (e.g., acetylcholine, the excitatory amino acids, serotonin, glycine, GABA) to the appropriate receptor on the extracellular surface of the cell. These channels are involved in nerve signaling as well as in intercellular communication in the nervous and endocrine systems.

Each type of gated ion channel can also be categorized according to the ion(s) that are normally transported. Each listing in the table represents a family of such channels that share certain characteristics, e.g., gating variable, ion species transported, pharmacological inhibitors, or channel kinetics. The inhibitors are listed when they are common to the family, and there is no listing when different members of the family have different pharmacological inhibitors. Some of the families are relatively homogeneous, such as the voltage-gated sodium channels, most members of which are inhibited with high affinity by tetrodotoxin and saxitoxin, while other listed channel families are heterogeneous. For example, four voltage-gated calcium channels have been identified, each with its own pharmacology and channel kinetics. There are dozens of different types of potassium channels.

8.5.1.4 Solute Transport by Carriers

Carriers all have the property that they transport solutes in a highly structure-specific manner, as described in Chapter 6. Carriers are distinct from channels in that the transport site is accessible to only one side of the membrane at each instant in time. The accessible site shuttles back and forth across the membrane so that the transport site can be accessible at alternate times to the two sides of the membrane. Carriers transport solutes at much lower rates than do ion channels and transport ions as well as uncharged solutes. Carriers can be further categorized by the number and direction of transport of the transported solutes. Simple carriers, such as the glucose (hexose) transporter discussed in some detail in Chapter 6, transport only a single solute. Other carriers transport two or three solutes simultaneously, as shown in Figure 8.15. If the solutes are transported in opposite directions, then the carrier is called an *exchanger*, a *countertransporter*, or an *antiporter*. If the solutes are transported in the same direction, it is called a *cotransporter* or a *symporter*.

Carriers may transport one solute actively by deriving the energy from the electrochemical potential gradient across the membrane for another of the transported solutes, a mechanism called *secondary active transport* (Section 6.4). Thus, the energy required to transport each transported solute comes only from the stored electrochemical potential energy difference across the membrane of those solutes that are transported. No additional energy is required from metabolic processes in the cell. A number of cotransporters couple the transport of some solute to the transport of sodium. Chapters 2 and 6 discussed the Na^+-glucose transporter, which transports glucose into the mucosal surface of intestinal epithelial cells. The energy for transporting glucose comes from the electrochemical potential energy of sodium across the membrane. There are a number of such sodium-coupled cotransporters, including the sodium-coupled amino acid transporter and a number that transport neurotransmitters. Among the exchangers are the anion exchanger that exchanges Cl^- for HCO_3^-, which is important in pH regulation and in volume control.

Table 8.3 Selective list of transport mechanisms, the solutes they normally transport, and examples of specific pharmacological inhibitors. Receptors that act through second messengers to open channels or that modulate channel transport have not been included. The following abbreviations are used: AA, amino acid; GABA, γ-aminobutyric acid; NMDA, *N*-methyl-D-aspartate.

Name	Solute transported	Inhibitors
	Channels	
	Voltage-gated channels	
Sodium	Sodium	Tetrodotoxin, saxitoxin
Delayed rectifier	Potassium	Tetraethylammonium
Transient potassium (A-type)	Potassium	4-aminopyridine
Inward rectifier	Potassium	Tetraethylammonium
Calcium	Calcium	
Chloride	Chloride	
	Ligand-gated channels	
Calcium-activated potassium	Potassium	Tetraethylammonium
ATP-activated potassium	Potassium	
Nicotinic acetylcholine receptor	Sodium, potassium, calcium	d-tubocurarine, α-bungarotoxin
Excitatory AA (NMDA) receptor	Sodium, potassium, calcium	Dizocilipine, phencyclidine
Excitatory AA (non-NMDA) receptor	Sodium, potassium	
Serotonin receptor	Potassium	
Glycine receptor	Chloride	Strychnine
$GABA_A$ receptor	Chloride	Picrotoxin, bicuculine
	Mechanically gated channels	
Hair-cell mechanical transducer	Cations	Aminoglycosides
Stress-gated	Cations	
	Other channels	
Water	Water	Mercury compounds
Gap-junctional	Small solutes	
Amiloride-sensitive	Sodium	Amiloride
	Carriers	
	Single substrate carriers	
Glucose transporter	Glucose	Cytochalasin B
Amino acid transporter	Amino acid	
Neurotransmitter transporter	Neurotransmitter	

Table 8.3 *(continued)*

Name	Solute transported	Inhibitors
	Multiple substrate carriers	
Anion exchanger	Chloride, bicarbonate	Stilbenedisulfonates
Na^+-Cl^- cotransporter	Sodium, chloride	Furosemide
K^+-Cl^- cotransporter	Potassium, chloride	
K^+-H^+ cotransporter	Potassium, hydrogen	
Na^+-H^+ exchanger	Sodium, hydrogen	Amiloride
Na^+-Ca^{++} exchanger	Sodium, calcium	
Na^+-K^+-Cl^- cotransporter	Sodium, potassium, chloride	Furosemide, bumetanide
Na^+-Glucose cotransporter	Sodium, glucose	
Na^+-AA cotransporter	Sodium, amino acid	
Na^+-Neurotransmitter cotransporter	Neurotransmitter	
H^+-Glucose cotransporter	Proton, glucose	
H^+-AA cotransporter	Proton, amino acid	
	Pumps	
(Na^+-K^+)-ATPase		
(Ca^{++})-ATPase		
(H^+)-ATPase		
(H^+-K^+)-ATPase		

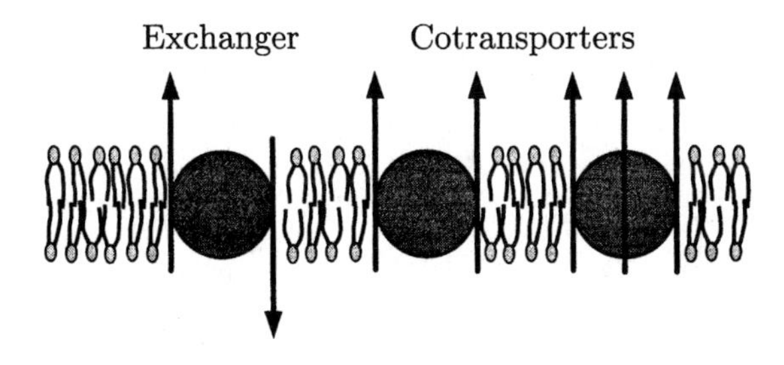

Figure 8.15 Schematic diagram of different types of carrier transport mechanisms.

8.5.1.5 *Ion Transport by Pumps*

Pumps share many features with carriers, but in addition they transport solutes by *primary active transport*. The transport of solutes by pumps is coupled to the hydrolysis of ATP. Therefore, the macromolecules responsible for these mechanisms are called ATPases. These pumps can transport ions against an electrochemical potential energy difference of the solute, i.e., uphill. Pumps are described in some detail elsewhere (Läuger, 1991). Among the most widely found pumps is (Na^+-K^+)-ATPase, which is important in maintaining the difference in sodium and potassium concentration across the membranes of cells. The proton pump is key to pH control. Proton pumps can pump protons against large proton concentration differences, for example, those in the membrane of the parietal cell lining the stomach, which produces the large quantity of acid in the lumen of the stomach.

8.5.2 Intracellular Solute-Binding/Release Mechanisms

In addition to the transport mechanisms in the cellular membrane, other mechanisms are important in cellular homeostasis. Cytoplasmic substances, and cytoplasmic organelles either release or bind a variety of substances, which will influence cellular homeostatic mechanisms. For example, there are numerous intracellular mechanisms that bind/release calcium. There are special calcium-binding proteins, such as calmodulin and troponin, that bind calcium with high affinity. Cellular organelles can sequester calcium ions via (Ca^{++})-ATPase pumps located in the organelle membranes. Similarly, protons are bound/released by numerous intracellular mechanisms in cells. For example, the hemoglobin in erythrocytes binds H^+. Solutes such as glucose are substrates for metabolic pathways. Hence, glucose entering a cell is rapidly used up. These intracellular mechanisms can be key elements in homeostatic mechanisms in cells. Any mechanism that either increases or decreases the intracellular concentration of some solute will affect the transport of that solute across the membrane.

8.5.3 Transporter Regulatory Mechanisms

Transport can be enhanced by mechanisms that either mobilize preexisting transporters or stimulate their synthesis *de novo*. In Chapter 4 we saw that the hormone vasopressin stimulates the mobilization of water channels from intracellular stores, and in Chapter 6 we saw that the hormone insulin similarly stimulates the mobilization of glucose transporters from intracellular stores.

In each case, transport is enhanced by the increase in the number of transporters. When the hormone is removed, the transporters move from the cellular membrane back to the intracellular stores. In addition to the modulation of the number of transporters in the membrane, existing membrane-bound transporters are subject to activation or inhibition by intracellular signals such as those of calcium and cyclic AMP and by phosphorylation. For example, a transporter may be inactive when not phosphorylated, but activated when phosphorylated. Thus, transport activity can be modulated by control of phosphorylation of transporters.

8.6 Transport Mechanisms in Selected Cell Types

The inventory of transport mechanisms shown in Table 8.3 gives an indication of the types of transport mechanisms that have been identified in different cells. However, all these mechanisms are not found in every cell. Different cells express different mechanisms consistent with their different requirements for transport. In this section we shall examine the complement of transport mechanisms that have been identified in certain well-characterized cells.

8.6.1 Uniform Isolated Cells

From the point of view of studying transport mechanisms, the simplest cells have a uniform membrane, are bathed in a uniform solution, and are found in isolation, i.e., they are not in direct contact with other cells in the body. Blood cells, such as erythrocytes and lymphocytes, are examples of cells that approach this ideal, and they have been studied extensively in numerous species. Figure 8.16 shows the typical complement of transport mechanisms found in human erythrocytes. There are a number of channels for passive ion transport; several carriers, including both an anion exchanger and a cotransporter; and two pumps. Some of the mechanisms are nonelectrogenic or *electrically silent*. That is, these mechanisms produce no net charge transport, and therefore, they produce no direct change in the membrane potential. For example, both the Na^+-K^+-Cl^- cotransporter and the anion exchanger, which exchanges Cl^- for HCO_3^-, are electrically silent.

Figure 8.16 represents a current view of the transporters found in erythrocytes—more may yet be discovered—but it it gives no indication of the importance of one or another type of transporter. Each icon represents a type of transport mechanism; a cell will contain many copies of each mechanism.

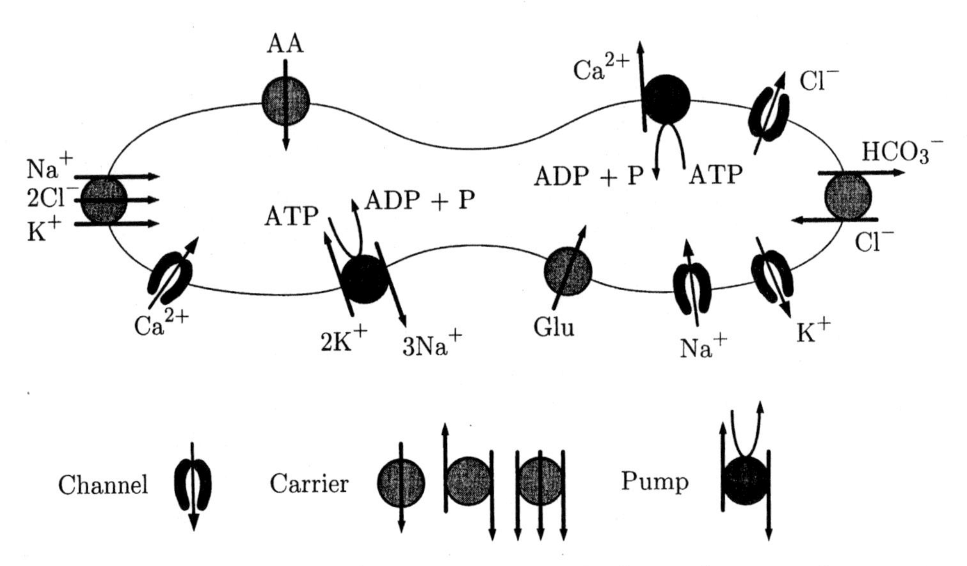

Figure 8.16 Schematic diagram of an erythrocyte displaying the types of transport mechanisms present in the membrane. Transport mechanisms include channels for Na^+, K^+, Cl^-, and Ca^{++}; carriers for glucose (Glu) and amino acids (AA); an anion-exchanger mechanism that exchanges chloride for bicarbonate and a Na^+-K^+-Cl^- cotransporter; and two pumps, a (Na^+-K^+)-ATPase and a (Ca^{++})-ATPase.

For example, it has been estimated that human erythrocytes contain 10^6 anion exchangers that exchange chloride for bicarbonate ions, 3×10^5 glucose transporters, and 10^3 (Na^+-K^+)-ATPase transporters. We already know the function of the glucose transporters as well as the (Na^+-K^+)-ATPase transporters. But why are there so many anion transporters, and what do they do?

The main function of the erythrocyte is to transport oxygen from the lungs to the body tissues and to transport carbon dioxide in the opposite direction. The anion exchanger is an important element of this process. The reaction sequence for oxygen and carbon dioxide transport is shown schematically in Figure 8.17. Cells in the body release CO_2, which is an end product of carbohydrate metabolism, by transmembrane diffusion of CO_2 dissolved in water. The dissolved CO_2 enters the plasma by diffusion through the endothelial cells that make up the walls of capillaries. The dissolved CO_2 diffuses through the membranes of erythrocytes into the cytoplasm, where it enters into the reaction $CO_2 + H_2O \rightleftharpoons HCO_3^- + H^+$, which is rapidly catalyzed by the enzyme carbonic anhydrase. This reaction is driven in the direction of production of protons by the increase in CO_2 concentration and by the removal of HCO_3^- by the anion exchanger, which exchanges HCO_3^- for Cl^- when the concentra-

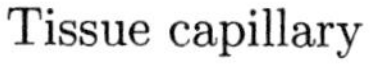

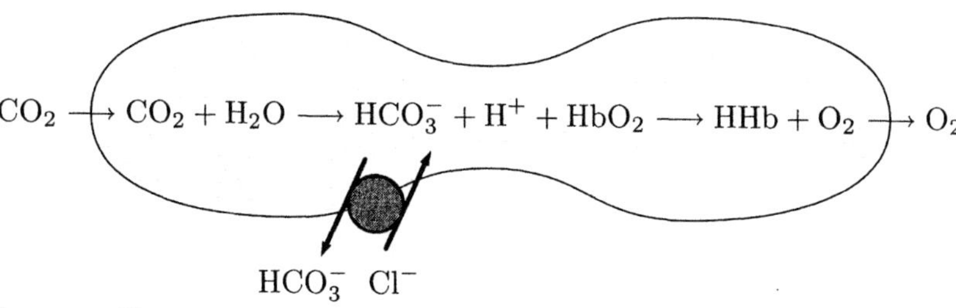

Lung capillary

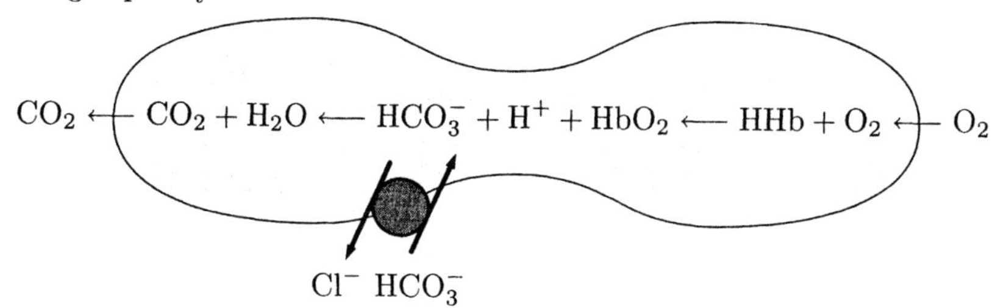

Figure 8.17 Schematic diagram of the role of the anion exchanger in oxygen and carbon dioxide transport in erythrocytes.

tion ratio of bicarbonate (intracellular to plasma) exceeds that for chloride. The hydrogen released in this reaction reacts with oxyhemoglobin by the reaction $H^+ + HbO_2 \rightleftharpoons HHb + O_2$ to form deoxyhemoglobin. Two factors influence this reaction. First, the affinity of hemoglobin for oxygen is reduced at lower pH, and deoxyhemoglobin binds protons with higher affinity than does oxyhemoglobin. The dissolved oxygen then diffuses through the membrane of the erythrocyte into the plasma and out of the plasma into the interstitial fluids, where it is taken up by cells. Because the tissue concentration of oxygen is low, diffusion of oxygen out of erythrocytes and into the tissues also drives the intracellular reactions in the directions indicated in the upper part of Figure 8.17. As indicated in the lower part of Figure 8.17, the entire process is reversed in the capillaries of the lung, in which there is a high interstitial concentration of oxygen and a low concentration of carbon dioxide.

The anion exhanger plays a critical role in the function of the erythrocyte. In the interstitial tissue capillaries, the anion exchanger removes bicarbonate, and therefore, the reaction continues in the direction of absorbing carbon dioxide and releasing oxygen. However, one consequence of this reaction, if it goes to completion, is that each molecule of carbon dioxide entering the erythrocyte results in the increase of one molecule of chloride intracellularly, the so-called *chloride shift* or *Hamburger shift*. This increase in intracellular chloride can be reversed by the action of other transporters. For example, the $Na^+/2Cl^-/K^+$ cotransporter can compensate for the increased chloride concentration by transporting chloride out of the erythrocyte. This transport will affect the sodium and potassium concentration, which can be compensated for by the (Na^+-K^+)-ATPase. The point is that all these factors act in concert to affect the homeostasis of the erythrocyte. To understand homeostasis in erythrocytes, one must characterize all of the transport processes as well as all the intracellular reactions that affect the concentrations of transported solutes. Discussion of such integrated models of whole cells is beyond our scope; the interested reader is referred to the literature (Werner and Heinrich, 1985; Lew and Bookchin, 1986; Moroz et al., 1989).

8.6.2 Cells in an Epithelium

Body compartments are bounded by layers of epithelial cells that provide a degree of isolation of one body compartment from another (Chapter 2). In addition, epithelial cells are involved in transport of numerous materials between compartments. Examination of homeostasis in epithelial cells reveals additional mechanisms at work in cells that do not occur in uniform isolated cells such as erythrocytes.

First, epithelial cells are asymmetric—at morphological, physiological, and molecular levels. The asymmetry of the transport mechanisms of epithelial cells is illustrated in Figure 8.18 for the intestinal epithelial cell (enterocyte), which lines the small intestine. This cell plays a key role in digestion of simple sugars and amino acids and the transport of these building block molecules to the vascular systems (Chapter 2). The apical (also called the mucosal or lumenal) end of the intestinal epithelial cell faces the lumen of the gut. The apical region contains a multitude of microvilli that serve to greatly expand the surface of the apical membrane to enhance absorption. The membrane of the microvilli contains sodium-coupled cotransporters for both amino acids and simple sugars. These transport amino acids and simple sugars into the enterocyte against the concentration gradients of these solutes, but down the electrochemical potential gradient for sodium. The amino acids and sugars are then transported out of the enterocyte through the basolateral (also called the serosal) region, by simple carriers uncoupled to sodium, into the interstitial space of the intestine, where they are absorbed into the circulation. In the process, there is an increase in intracellular sodium concentration. This concentration change is restored by several transporters working in concert, including (Na^+-K^+)-ATPase and the $Na^+/2Cl^-/K^+$ cotransporter, which also maintains cellular volume. As with the erythrocyte, a number of transport mechanisms and intracellular mechanisms must act in concert to support the

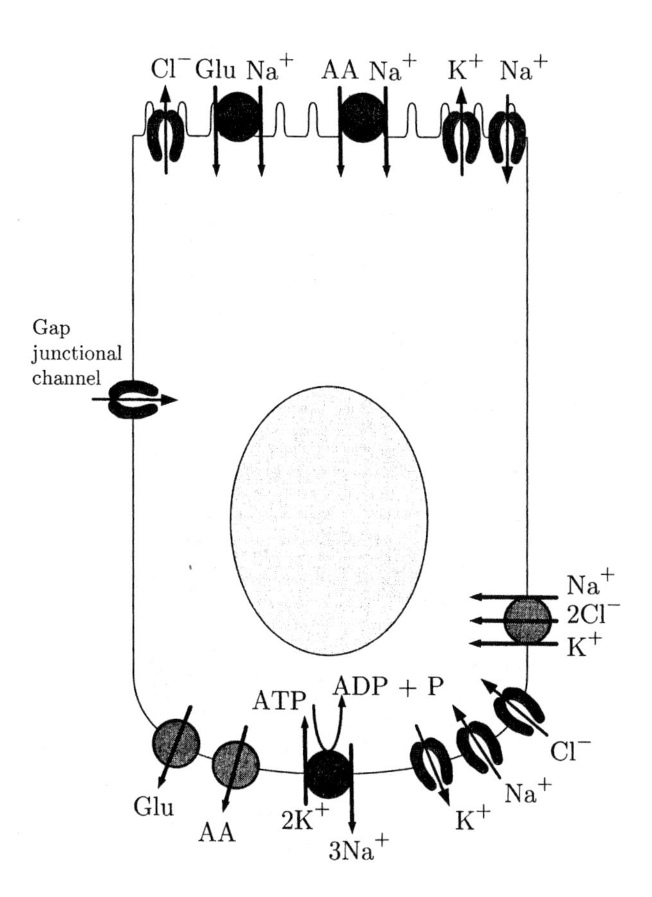

Figure 8.18 Schematic diagram of an enterocyte displaying the types of transport mechanisms present in the membrane. Transport mechanisms in the apical membrane include channels for Na^+, K^+, and Cl^- and both glucose-Na^+ and amino acid-Na^+ cotransporters. Transport mechanisms in the basolateral membrane include channels for Na^+, K^+, and Cl^-; gap junctional channels; carriers for glucose (Glu) and amino acids (AA); a Na^+-K^+-Cl^- cotransporter; and a (Na^+-K^+)-ATPase pump.

important special function of enterocytes, which is to transport amino acids and simple sugars across the epithelium and to perform general homeostatic functions.

Another characteristic property of epithelial cells is that, by definition, they form an epithelium that separates two media of different composition on the apical and basolateral sides. Tight junctions that link the individual cells (Figure 2.10) serve to isolate the compartments and to reduce the paracellular transport of matter between compartments (Figure 2.13). Thus, another asymmetry of the epithelial cell is in the composition of the solutions that bathe the apical and basolateral membranes. These solutions can differ quite radically. For example, in the inner ear of vertebrates, the sensory epithelium separates two solutions of radically different composition, endolymph on the apical side and perilymph on the basolateral side of the epithelial cells (see Problem 7.14).

Whereas perilymph has an ion composition typical of other extracellular solutions, the ion composition of endolymph resembles that of cytoplasm. In addition, electrogenic secretory cells produce a large (+100 mV) extracellular potential (endolymphatic potential) across the epithelium. Thus, the cellular homeostatic processes in the sensory epithelium of the inner ear operate in the presence of a large electrochemical potential gradient across the epithelium.

Cells in an epithelium are also coupled by gap junctions through which molecules of low molecular weight are transported. These gap junctions are found in the lateral regions of the basolateral membrane and are channels formed in both membranes of neighboring cells. Thus, cellular homeostasis in epithelial cells must take into account the communication of small solutes such as ions between cells. Thus, homeostasis in epithelial cells is more complex than in erythrocytes. Nevertheless, some progress has been made on formulating integrated models of homeostatic processes in epithelial cells (Lew et al., 1979, Ferreira and Ferreira, 1983; Mintz et al., 1986; Strieter et al., 1990; Weinstein, 1992)

8.6.3 Electrically Excitable Cells

Electrically excitable cells have still another constellation of transporters consistent with their function. For example, consider the constellation of transporters contained in the membrane of sympathetic ganglion cell neurons in the bullfrog as shown schematically in Figure 8.19. These cells are involved in regulating the heart rate. These are particularly simple neurons, having a cell body and an axon, but no dendritic tree (Weiss, 1996, Chapter 1). Inputs of other neurons occur only on the cell body. Typical of neurons, these cells have a large variety of different types of ion channels, some voltage gated, and others ligand gated. These channels are involved in providing inputs to this cell from other cells, as well as providing mechanisms that shape the electrical responses of the ganglion cell. All these channels allow Na^+, K^+, or $Ca^{++}$to flow through the membrane. Thus, during the normal electrical activity of this neuron there is a net flow of several ions down their electrochemical potential gradients. In order for homeostasis to occur, there must be compensatory transport processes that restore concentrations. Only two pumps are shown as almost certainly present in the ganglion cell, but it is probable that a number of other transporters are present. In addition, these cells undoubtedly contain calcium-binding proteins that can buffer the intracellular calcium concentration.

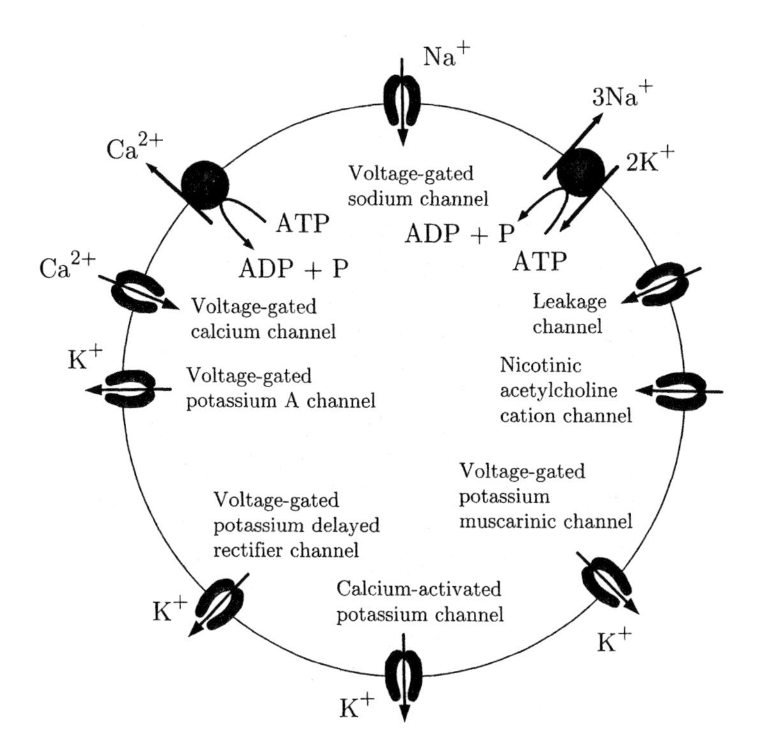

Figure 8.19 Schematic diagram of transporters in bullfrog sympathetic ganglion (Yamada et al., 1989).

8.6.4 General Comments on the Mechanisms of Volume Regulation

We have seen that different cell types have different constellations of transporters, but each type of cell must have homeostatic mechanisms to maintain concentrations, volume, and membrane potential. The homeostatic mechanisms have not been thoroughly unveiled for any cell, although there is great promise in recent progress. What can be said in general terms about homeostatic processes?

The regulatory volume responses are much slower than the primary responses of cells, suggesting that the regulatory responses result from a change in solute concentration that is followed almost simultaneously by an osmotic transport of water. There are several different strategies that cells could use to produce changes in volume by changing intracellular solute concentration. The concentration of intracellular solutes could be changed by intracellular chemical processes that either synthesize or break down intracellular solutes. Alternatively, the cell membrane could change the intracellular solute concentration in the cell by changing some mechanism that transports solutes from one side of the membrane to the other. There is evidence that both types of mechanisms occur. However, we shall focus on the transport mechanisms, about which more is known and which appear to occur quite widely in different cell types.

Which ions are likely to be involved in regulatory responses of cells resulting from transport mechanisms? During the regulatory volume decrease (RVD) response, the cell volume decreases; this would occur if water and solute were transported out of the cell. Since in many cells the primary intracellular solutes are potassium and chloride, they are prime candidates for carrying the solute flux during an RVD response. Thus, one mechanism for the RVD response is that an increase in cell volume increases the transport of potassium and chloride out of the cell. During the regulatory volume increase (RVI), cell volume increases; this would occur if water and solute were transported into the cell. Since the primary extracellular solutes in many cells are sodium and chloride, they are prime candidates for carrying the solute flux that leads to RVI. Thus, a mechanism for RVI is that an increase in cell volume results in an increase in the transport of sodium and chloride into cells, which then draws water into the cell to increase the volume. In some cells, transport of organic solutes is used to regulate cell volume.

Judging from the inventory of transport mechanisms available to cells (Table 8.3), a number of different transport mechanisms are potentially available for volume regulatory responses. Not so surprisingly, different cells appear to use different transport mechanisms to achieve volume regulation, as illustrated in Figure 8.20. Furthermore, some cells have RVD responses and not RVI responses. Thus, for example, RVD can be mediated by a volume-dependent permeability of membrane channels that transport potassium, chloride, or both ions. Such mechanisms are thought to underlie RVD responses in the epithelial cells in frog skin, B-lymphocytes, the frog urinary bladder, and the *Necturus* choroid plexus.

In all these mechanisms, it is apparent that the cell must be able to sense some signal that accompanies the change in volume in order to initiate a regulatory response. The mechanism(s) for such a sensor has not been identified unambiguously. However, there are several mechanisms for which there is evidence. First, there exist special gated ion channels, *stress-gated channels,* that open in response to the application of mechanical stress to the membrane of the cell. The idea is that, as the volume of a cell changes, the stress on the membrane changes so as to modulate ion flow through these stress-gated channels, which can affect intracellular osmolarity and therefore water flow through the membrane. These channels are permeable to cations, but not very selective to different cations. Stress-gated channels have been reported in a great diversity of cell types and implicated in volume regulation. Second, sev-

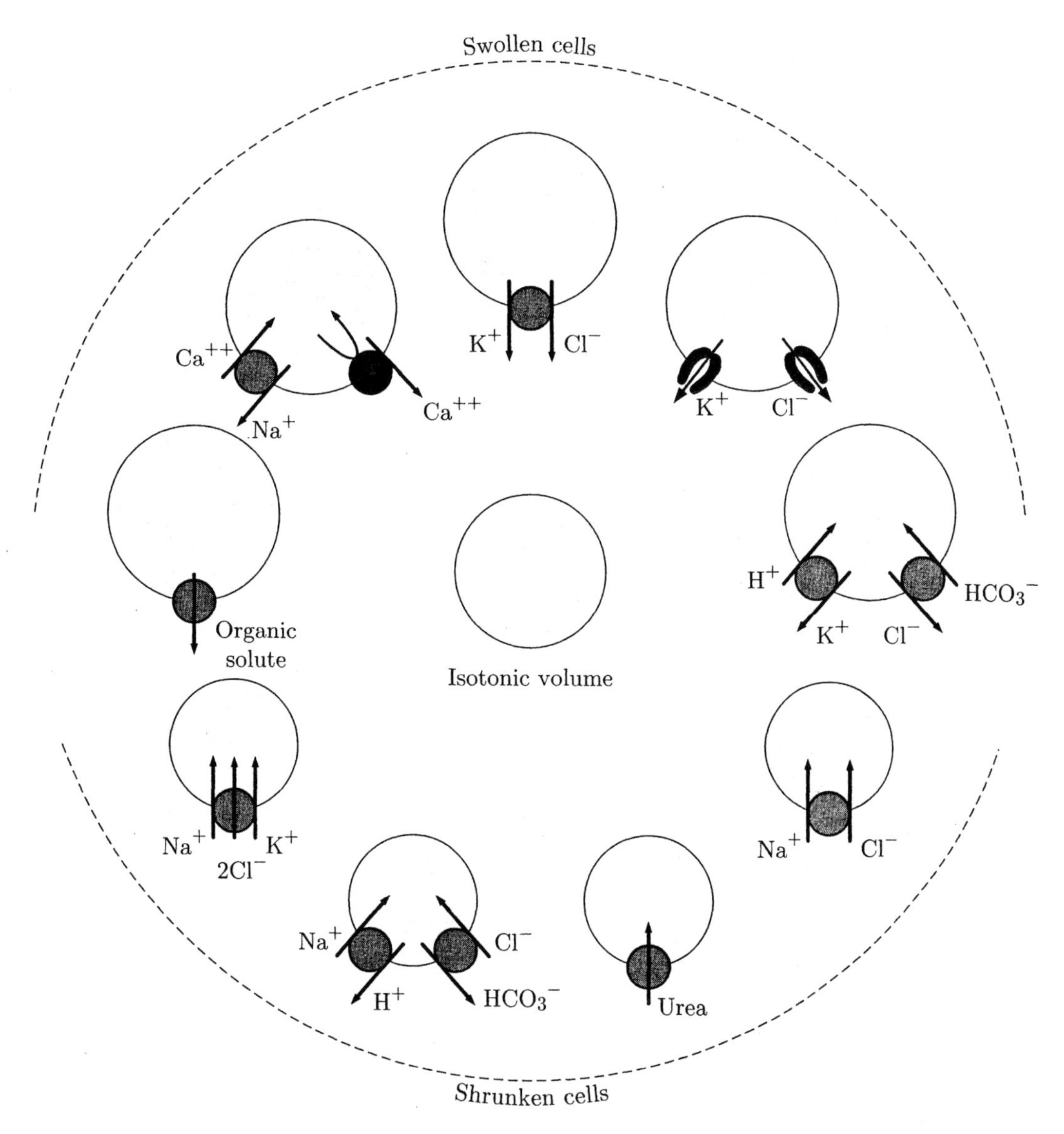

Figure 8.20 Schematic diagram showing a survey of volume regulatory mechanisms in different types of cells in which solute transport mechanisms are activated in response to a volume change (Lang et al., 1990; Lewis and Donaldson, 1990). The upper five panels show a cell whose volume has swollen to larger than its isotonic volume (shown in the center); the swelling has activated different transport mechanisms that give rise to an RVD response. The lower four panels show a cell that has shrunk to a volume that is less than its isotonic volume; this shrinkage has activated the indicated mechanisms that give rise to an RVI response.

eral of the ion transport mechanisms have kinetic properties that are sensitive to membrane stress. Thus, ion transport via these mechanisms could be modulated by membrane stress induced by changes in cell volume. Finally, when the cell volume changes, so does the concentration of all intracellular solutes. Thus, a change in some intracellular solute could signal the change in volume and act on one or another transport mechanism to induce a compensatory or regulatory response. At the moment it appears likely that, just as different cells use somewhat different strategies for volume regulation, they may use somewhat different sensors for initiating a volume regulatory response.

Exercises

8.1 A rigid rectangular volume with rigid impermeable walls is divided in two by a rigid fixed selectively permeable membrane as shown in Figure 8.21. The membrane is permeable to water. Each side of the chamber contains well-stirred solutions of solutes described in Table 8.4. The concentrations of these solutes at time $t = 0$ are indicated in the figure.

a. Is this system in equilibrium at time $t = 0$? Explain.

b. What are the sign and magnitude of the voltage V_m at $t = 0$? Explain.

c. Consider what would happen if a battery were connected between the two sides so that the voltage V_m were forced to −40 mV. Describe

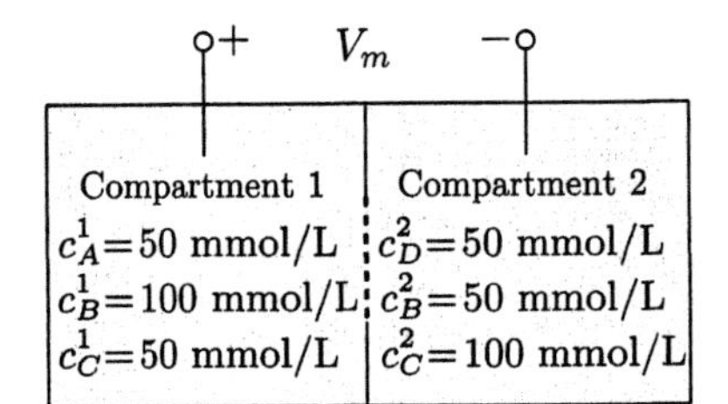

Figure 8.21 Two compartments separated by a permeable membrane (Exercise 8.1).

Table 8.4 Compositions of solutions (Exercise 8.1).

Solute	Valence	Permeability	Transport mechanism
A	−1	Impermeant	—
B	+1	Permeant	Passive electrodiffusion
C	−1	Permeant	Passive electrodiffusion
D	+1	Impermeant	—

the effect of the battery during the instant of time just after it was connected.

d. Consider the effect of adding an uncharged impermeant protein to side 1 so that its concentration would be 10 mmol/L. What would happen? Give a quantitative answer.

8.2 For each of the terms in the left column, find the most relevant description from the items in the right column.

a. Permeability	1. A condition that prevails in homogeneous electrical conductors
b. Resting potential	2. Membrane-bound transport protein
c. Collander plot	3. Hormone that modulates transmembrane glucose transport
d. Osmosis	4. Grassy area in Forest Lawn Cemetery
e. Osmole	5. A measure of the effective osmotic pressure
f. $(Na^+$-$K^+)$-ATPase	6. Osmotic pressure expressed as a molar concentration
g. Equilibrium potential	7. A blocker of $(Na^+$-$K^+)$-ATPase
h. Insulin	8. Relates membrane permeability to lipid solubility
i. Reflection coefficient, σ	9. A weighted sum of equilibrium potentials
j. Electroneutrality	10. A measure of the discrimination of a membrane for a solute versus a solvent
k. Ouabain	11. The capacity to maintain the body erect
	12. The potential across a membrane at zero membrane current
	13. The flux of solute through a membrane per unit difference in solute concentration across the membrane
	14. A position taken by Swiss utilities
	15. The potential across a membrane at which the ionic current carried by a permeant ion is zero
	16. Flow of solvent resulting from a solute concentration gradient
	17. A small burrowing mammal

8.3 Vasopressin and insulin are hormones that have important effects on transport.

a. Vasopressin affects the transport of which substance through the membrane?

b. What is the effect of an increase in systemic vasopressin concentration on the transmembrane flux of the substance given in part a?

c. Insulin affects the transport of which substance through the membrane?

d. What is the effect of an increase in systemic insulin concentration on the transmembrane flux of the substance given in part c?

e. Compare the molecular mechanisms by which vasopressin and insulin affect the transmembrane flux.

Problems

8.1 A membrane separates solution 1, which contains potassium, K^+, and chloride, Cl^-, plus an additional anion, A^-, from solution 2, which contains potassium and chloride ions only (Figure 8.22). Each solution is electrically neutral. The membrane is permeable to water and Cl^- and impermeable to K^+ and A^-. However, the permeability to Cl^- is much lower than to water, and hence, the reflection coefficient for Cl^- is approximately one. The membrane potential is V_m. The temperature is 300 K. There is no hydrostatic pressure gradient across the membrane. For $c_A^1 =$ 135 mmol/L and $c_{Cl}^2 = 150$ mmol/L, find the values of c_{Cl}^1, c_K^1, c_K^2, and V_m such that the flux of volume and the flux of ions across the membrane is zero.

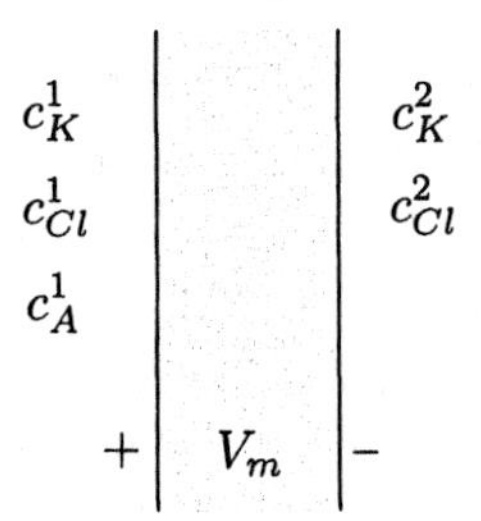

Figure 8.22 Donnan equilibrium (Problem 8.1).

8.2 The capillary wall that separates blood plasma from interstitial fluid can be characterized, for the purposes of this problem, as a membrane that is permeable to water as well as to sodium and chloride ions and impermeable to all other solutes. The concentrations of sodium and chloride ions in the interstitial fluids are both equal to 155 mmol/L. The blood plasma is known to contain a protein with concentration 1 mmol/L. The effects of all other constituents of plasma and interstitial fluids can be ignored in this problem. You may also ignore the effect of hydraulic pressure on electro-diffusive equilibrium across the membrane. The temperature is 38°C.

a. Assume that the protein is uncharged. At equilibrium, determine the following:

   i. The concentration of sodium and chloride ions in plasma.

   ii. The electric potential difference across the capillary wall (including its polarity).

   iii. The difference in osmotic pressure across the capillary wall (including its polarity).

   iv. The difference in hydraulic pressure across the capillary wall (including its polarity).

b. Assume that the protein has a valence of -50. At equilibrium, determine the following:

   i. The concentration of sodium and chloride ions in plasma.

   ii. The electric potential difference across the capillary wall (including its polarity).

   iii. The difference in osmotic pressure across the capillary wall (including its polarity).

   iv. The difference in hydraulic pressure across the capillary wall (including its polarity).

8.3 This problem concerns homeostasis in a hypothetical cell that is impermeant to solutes 1, 2, 3, 4, and 5 (all nonelectrolytes), but permeant to water. For $t < 0$, the cell is bathed in an extracellular solution having the composition shown in Table 8.5. The equilibrium intracellular composition of these solutes for $t < 0$ is also shown in the table, and the volume of water in the cell is 100 μm^3. At $t = 0$, the extracellular solution is changed to the composition shown in the table.

a. Find the equilibrium value of the volume of water in the cell.

Table 8.5 Composition of solutes for a cell that is impermeable to solutes, but permeable to water (Problem 8.3).

n	Concentration (mmol/L)		
	$t < 0$		$t > 0$
	$c_n^o(t)$	$c_n^i(t)$	$c_n^o(t)$
1	20	0	0
2	0	180	360
3	30	0	0
4	120	0	0
5	130	120	0

b. Find the equilibrium values of the intracellular concentrations of solutes 1–5.

8.4 In Section 8.4, the homeostasis of a hypothetical cell was investigated in which solutions on both sides of the membrane were the same binary electrolyte and for which the membrane was permeable only to the cation, which was transported by passive means only. It was found that the only solution possible was that all the ions had the same concentration both inside and outside the membrane and the membrane potential was zero. Now suppose that the cation has valence z_+ and the anion has valence z_-.

a. Determine the equations of electroneutrality for the intracellular and extracellular solutions.

b. Determine the equation of osmotic equilibrium across the membrane.

c. For each of the following quantities, state whether that quantity is an independent or a dependent variable, and give the reason for your choice.

i. The cation concentration inside c_+^i.

ii. The anion concentration inside c_-^i.

iii. The cation concentration outside c_+^o.

iv. The anion concentration outside c_-^o.

v. The cation quantity inside n_+^i.

vi. The anion quantity inside n_-^i.

vii. The cation valence z_+^i.

viii. The anion valence z_-^i.

ix. The cell water volume $\mathcal{V}$.

x. The membrane potential V_m^o.

d. Solve the equations for the intracellular concentrations for both the cation and the anion. Express the concentrations in terms of the independent variables only.

e. Determine the volume of the cell in terms of the independent variables.

f. Determine the membrane potential in terms of the independent variables.

g. Compare the results you have obtained for a z_+-z_- electrolyte with the results obtained for a binary electrolyte in Section 8.4.

8.5 In Section 8.4, the homeostasis of a hypothetical cell was investigated in which solutions on both sides of the membrane were the same binary electrolyte and for which the membrane was permeable only to the cation, which was transported by passive means only. It was found that the only solution possible was that all the ions had the same concentration both inside and outside the membrane and the membrane potential was zero. Now suppose that the intracellular cation has valence +1 and the intracellular anion has valence z_-. The extracellular solution is a binary electrolyte with the same cation as the intracellular solution. The concentration of both cation and anion in the extracellular solution is C, which automatically satisfies electroneutrality of the extracellular solution.

a. Determine the equation of electroneutrality for the intracellular solution.

b. Determine the equation of osmotic equilibrium across the membrane.

c. For each of the following quantities, state whether that quantity is an independent or a dependent variable, and give the reason for your choice.

i. The cation concentration inside c_+^i.

ii. The anion concentration inside c_-^i.

iii. The cation concentration outside C.

iv. The anion concentration outside C.

v. The cation quantity inside n_+^i.

vi. The anion quantity inside n_-^i.

vii. The cation valence z_+^i.

viii. The anion valence z_-^i.

ix. The cell water volume $\mathcal{V}$.

x. The membrane potential V_m^o.

d. Determine the intracellular concentrations for both the cation and the anion. Express the concentrations in terms of the independent variables only.

e. Determine the volume of the cell in terms of the independent variables.

f. Determine the membrane potential in terms of the independent variables.

g. Compare the results you have obtained for a z_+-z_- electrolyte with the results obtained for a binary electrolyte in Section 8.4.

8.6 An animal cell has an intracellular medium consisting of a permeant binary cation and an anion whose equilibrium concentrations are c_+^i and c_-^i, respectively, as well as an impermeant ion whose concentration is c_i^i and whose valence is z_i. The extracellular solution is a binary electrolyte whose concentrations are $c_+^o = c_-^o = C$. Transport through the membrane is passive, with the fluxes obeying the equations $\phi_+^p = G_+(V_m^o - V_+)/F$ and $\phi_-^p = G_-(V_m^o - V_-)/F$, where G_+ and G_- are the conductances of the membrane for the cation and anion, V_+ and V_- are the Nernst equilibrium potentials for the the cation and anion, V_m^o is the membrane resting potential, and F is Faraday's constant. There is no active transport of either the cation or the anion.

a. Find an equation that expresses electroneutrality of the intracellular solution.

b. Show that at equilibrium $c_+^i c_-^i = C^2$.

c. Using the results in parts a and b, determine both c_+ and c_- in terms of C, z_i, and c_i^i.

d. Let the osmotic pressures of the intracellular and extracellular solutions be π^i and π^o, respectively. Compute the difference in osmotic pressure between the intracellular and extracellular solution, $\pi^i - \pi^o$.

e. Determine the condition on C, z_i, and c_i^i for which the osmotic pressure difference is zero.

8.7 Figure 8.23 is reproduced directly from a portion of Figure 8.8 and shows the dependence of the normalized concentrations of the cation, the anion, and the extra impermeant solute on the normalized quantity of the

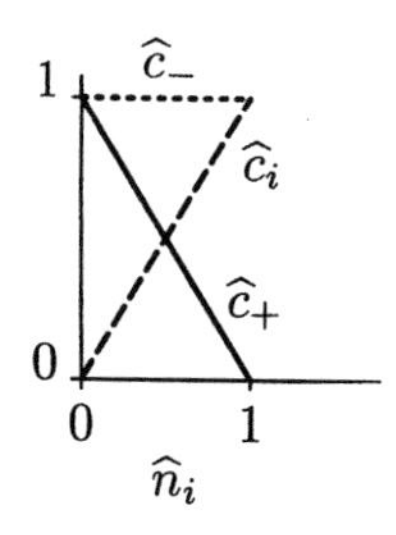

Figure 8.23 Dependence of intracellular concentrations ($\hat{c}_+$, $\hat{c}_-$, and $\hat{c}_i$) on the quantity of impermeant intracellular solute ($\hat{n}_i$) for a solute valence $z_i = +1$ (Problem 8.7). The computations are taken from Figure 8.8 and correspond to the solutions to the homeostatic equation for a simple cell model in which the membrane is permeable only to the cation.

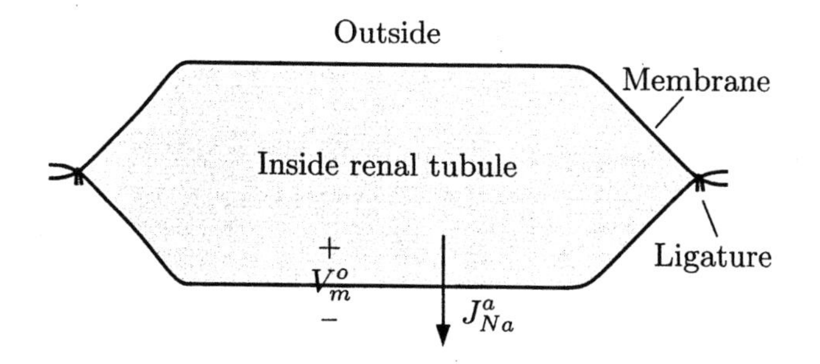

Figure 8.24 Schematic diagram of a segment of a renal tubule (Problem 8.8). Reference directions are shown for the resting potential difference across the membrane, V_m^o, and for the current density due to active sodium transport, J_{Na}^a.

extra impermeant solute intracellularly. Using the results shown in Figure 8.23:

a. Show that the intracellular solution is isotonic with the extracellular solution.

b. Show that the intracellular solution satisfies electroneutrality.

c. Explain why the solution is shown for $\hat{n}_i \leq 1$ only.

8.8 In the kidney proximal tubule, sodium ions are actively transported out of the tubules into the surrounding fluid. The transport of other ions and of water is linked to the active transport of sodium ions. This problem explores this linkage in a simple situation.

A segment of the proximal tubule, ligated at both ends as shown in Figure 8.24, encloses a solution consisting of sodium and chloride ions plus an uncharged macromolecule M and is bathed in a solution of sodium chloride. The tubular epithelium separates the inside and outside of the tubule and is labeled as the membrane in Figure 8.24. The concentrations of sodium and chloride are c_{Na}^i and c_{Cl}^i inside and c_{Na}^o and c_{Cl}^o outside. The tubule segment has volume $\mathcal{V}$ and contains n_M moles of M. The membrane is permeable to sodium and chloride ions, with ionic conductances G_{Na} and G_{Cl} (S/cm^2), but is *not* permeable to M. The membrane also contains an active transport mechanism that pro-

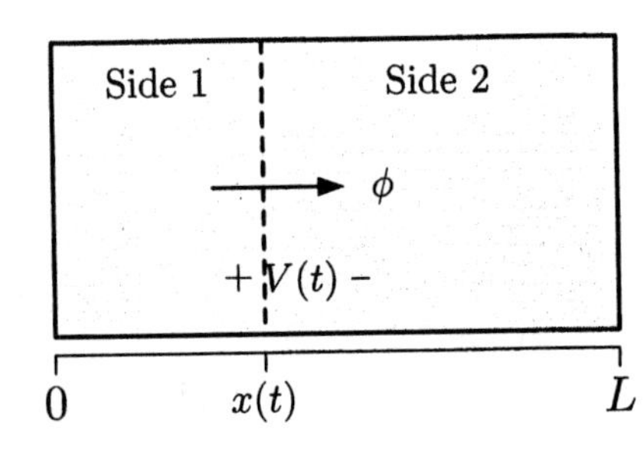

Figure 8.25 Volume element with rigid walls and movable partition (Problem 8.9). The potential across the partition is $V(t)$, and its position is $x(t)$ at time t. The positive reference direction for both volume and solute fluxes is indicated by the arrow. The cross-sectional area of the volume element is A, so that its total volume is AL.

vides a constant current density, J^a_{Na}. The membrane is deformable and cannot sustain an appreciable hydraulic pressure difference.

a. Let V_m be the potential difference between the inside and outside of the tubule. Using only the variables c^i_{Na}, c^o_{Na}, c^i_{Cl}, c^o_{Cl}, V_m, $\mathcal{V}$, n_M, G_{Na}, G_{Cl}, and J^a_{Na} and physical constants, list equations that guarantee the following:
   i. Electroneutrality of the outside solution.
   ii. Electroneutrality of the inside solution.
   iii. Osmotic equilibrium.
   iv. Equilibrium of chloride ions.
   v. Equilibrium of sodium ions.

b. Determine an equivalent network that relates the sodium and chloride current densities through the membrane to the membrane potential.

c. At equilibrium, let the membrane potential be V^o_m. Determine c^i_{Na}, c^i_{Cl}, V^o_m, and $\mathcal{V}$ at equilibrium as a function of c^o_{Na}, c^o_{Cl}, n_M, G_{Na}, G_{Cl}, J^a_{Na}, and physical constants *only*.

d. Sketch $\mathcal{V}$ as a function of J^a_{Na}, and discuss the physical meaning of the relation between these two variables in fifty words or less.

8.9 A rectangular parallelepiped volume element of cross-sectional area A and length L has rigid impermeable walls and is divided into two compartments by an infinitesimal thin, rigid, semipermeable partition that is mounted on frictionless bearings so that the partition is free to move in the x-direction as shown in Figure 8.25. The position of the partition at time t is $x(t)$, and the electric potential across the partition is $V(t)$. The partition is permeable to water and to some solutes. Assume that Φ_V is the flux of volume, and ϕ_n is the molar flux of solute, n, and c^1_n and c^2_n are the concentrations of solute n on sides 1 and 2, respectively. Reference directions for voltage and flux are shown in Figure 8.25. Table 8.6 shows characteristics of the solutes in this problem.

Table 8.6 Solute characteristics (Problem 8.9).

Solute	Valence	Solute permeability	Transport mechanism
A	0	Impermeable	—
B	0	Permeable	Diffusion
C	+1	Impermeable	—
D	−1	Impermeable	—
E	+1	Permeable	Passive electrodiffusion
F	−1	Permeable	Passive electrodiffusion

Table 8.7 Table of initial solute concentrations (Problem 8.9).

Part	*Given: Initial solute concentrations (mmol/L)* Side 1	Side 2	Find:
(i)	$c_A^1(0) = 10$	$c_A^2(0) = 5$	$x(\infty)$, $c_A^1(\infty)$, $c_A^2(\infty)$
(ii)	$c_B^1(0) = 10$	$c_B^2(0) = 5$	$c_B^1(\infty)$, $c_B^2(\infty)$
(iii)	$c_D^1(0) = 5$	$c_C^2(0) = 5$	$x(\infty)$, $V(\infty)$, $c_C^1(\infty)$, $c_C^2(\infty)$,
	$c_E^1(0) = 10$	$c_E^2(0) = 5$	$c_D^1(\infty)$, $c_D^2(\infty)$, $c_E^1(\infty)$, $c_E^2(\infty)$,
	$c_F^1(0) = 5$	$c_F^2(0) = 10$	$c_F^1(\infty)$, $c_F^2(\infty)$

a. For parts i–iii of this problem, the position of the partition at $t = 0$ is $x(0) = L/2$. For each of the initial solute concentrations given in Table 8.7, find the final (equilibrium) values of the indicated variables. Give reasons for your answers. *Note:* You may assume that $RT/F = 26$ mV for this problem.

b. Assume that the solution in the volume element contains only solute B and that there are N_B moles of B in the whole volume element. The flux of solute B across the partition is

$$\phi_B = P_B(c_B^1 - c_B^2),$$

where P_B is the permeability of the partition for solute B. Show that the differential equation that expresses conservation of solute B can be written in terms of two variables $n_B^1(t)$, which is the total number of moles of solute B in compartment 1 at time t, and $x(t)$ in the following form:

$$\frac{dn_B^1(t)}{dt} = \alpha_n \frac{\beta_n n_B^1(t) - \gamma_n x(t)}{x(L - x)},$$

and find the constants α_n, β_n, and γ_n in terms of geometry of the volume element (A and L), membrane characteristics ($\mathcal{L}_V$ and P_B), solution properties (N_B and T), and physical constants (R) only.

c. Once again, assume that the volume element contains only solute B and that there are a total N_B moles of solute B. The flux of volume in the presence of solute B is

$$\Phi_V = \mathcal{L}_V RT(c_B^2 - c_B^1),$$

where $\mathcal{L}_V$ is the hydraulic conductivity of the membrane for water. Show that the differential equation that expresses conservation of water volume can be written in terms of the same two variables, $n_B^1(t)$ and $x(t)$, in the following form:

$$\frac{dx(t)}{dt} = \alpha_x \frac{\beta_x n_B^1(t) - \gamma_x x(t)}{x(L-x)},$$

and find the constants α_x, β_x, and γ_x in terms of the same quantities specified in part b.

d. You may have noted that in part ii of part a of this problem (Table 8.7) it was not required that you find the value of $x(\infty)$ for solute B. The reason this value was not required is because, for a permeant solute, such as B, it can be shown that $x(\infty)$ depends upon the value of $\delta = (ALP_B)/(N_B RT\mathcal{L}_V)$, i.e., on the permeability of the solute, as expressed by P_B, relative to that of water, as expressed by $\mathcal{L}_V$. Thus, larger values of δ occur for highly permeable solutes. To investigate motion of the partition and the number of moles of solute B on side 1, the coupled nonlinear equations derived in parts b and c were first normalized and then solved numerically. The solutions are shown for values of δ of 0.1, 1, and 10 in Figure 8.26. Which of the curves n_1, n_2, or n_3 correspond to $\delta = 10$? Which of the curves y_1, y_2, or y_3 correspond to $\delta = 10$? Explain your reasoning.

8.10 Figure 8.27 contains six hypothetical cells, each containing a different constellation of transport mechanisms. In each of these cells, the intracellular solution consists of Na^+, K^+, Cl^-, and an impermeant anion and the extracellular solution contains Na^+, K^+, and Cl^-. Three types of transport mechanisms occur in these cells.

- For channels transport is passive and the flux of ion n is $\phi_n^p = G_n(V_m^o - V_n)/(z_n F)$, where G_n is the specific membrane conductance for ion n, V_m^o

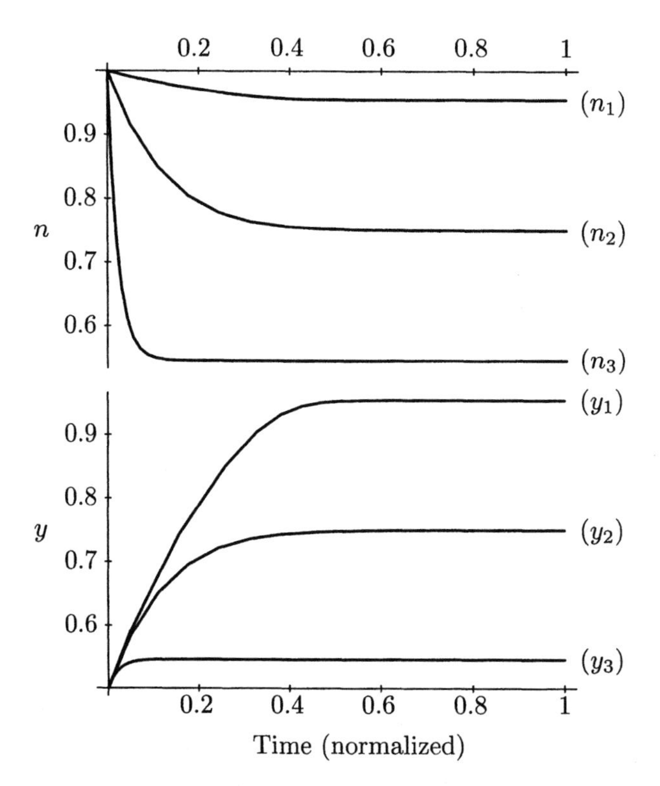

Figure 8.26 Dependence of $y = x/L$ and $n = n_B^1/N_B$ on normalized time t/τ_x for three values of δ (Problem 8.9). The time axis is normalized to $\tau_x = AL^2/(N_B RT\mathcal{L}_V)$.

is the resting membrane potential, V_n is the Nernst equilibrium potential of ion n, z_n is the valence of ion n, and F is Faraday's constant.

- The sodium-potassium pump has the characteristic that the rate of ATP hydrolysis, α_{ATP}, is a function of the concentrations of sodium and potassium both inside and outside the cell. The details of this relation are not important in this problem. The active fluxes of sodium and potassium due to the pump are $\phi_{Na}^a = \nu_{Na}\alpha_{ATP}$ and $\phi_K^a = -\nu_K\alpha_{ATP}$, where ν_{Na} and ν_K are the stoichiometric coefficients for Na^+ and K^+, respectively.
- Finally, the Na^+-K^+-Cl^- cotransporter has a rate of turnover of ions, α_{co}, that is a function of the concentrations of Na^+, K^+, and Cl^- both inside and outside the membrane. The details of this relation are not important in this problem. The cotransporter has Na^+, K^+, and Cl^- fluxes that are $\phi_{Na}^{co} = \phi_K^o = \alpha_{co}$ and $\phi_{Cl}^{co} = 2\alpha_{co}$.

For each part given below, determine if each of the statements is true, false, or does not apply, and explain each answer. To determine

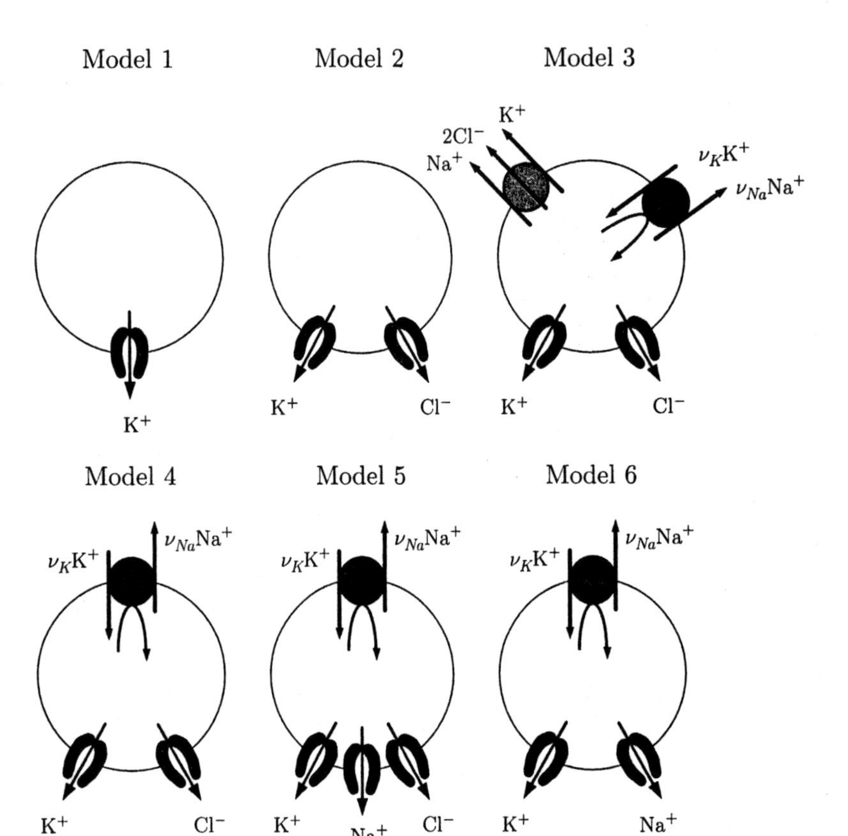

Figure 8.27 Six cell models (Problem 8.10).

the validity of most of these statements, you will need to analyze the relevant equations carefully.

a. For cell model 1:
   i. At equilibrium, $V_m^o = V_{Cl}$.
   ii. At equilibrium, $V_m^o = V_K$.
   iii. The cell volume is inherently unstable.
   iv. At equilibrium, $c_{Na}^i = 0$.
b. For cell model 2:
   i. At equilibrium, $V_m^o = V_{Cl}$.
   ii. At equilibrium, $V_m^o = V_K$.
   iii. At equilibrium, $\alpha_{ATP} = 0$.

c. For cell model 3:

  i. At equilibrium, $V_m^o = V_{Cl}$.

  ii. At equilibrium, $V_m^o = V_K$.

  iii. At equilibrium, $\alpha_{ATP} = 0$.

  iv. A change in the rate of ATP hydrolysis, α_{ATP}, will not affect the quantity of the potential differences, $V_m^o - V_{Cl}$, which drives chloride ions through the membrane.

d. For cell model 4:

  i. At equilibrium, $V_m^o = V_{Cl}$.

  ii. At equilibrium, $V_m^o = V_K$.

  iii. At equilibrium, $\alpha_{ATP} = 0$.

e. For cell model 5:

  i. At equilibrium, $V_m^o = V_{Cl}$.

  ii. At equilibrium, $V_m^o = V_K$.

  iii. At equilibrium, $\alpha_{ATP} = 0$.

  iv. $\phi_K^p / \nu_K + \phi_{Na}^p / \nu_{Na} = 0$.

f. For cell model 6:

  i. At equilibrium, $V_m^o = V_{Cl}$.

  ii. At equilibrium, $V_m^o = V_K$.

  iii. At equilibrium, the membrane potential can be written as

$$V_m^o = \frac{G_K / \nu_K}{G_K / \nu_K + G_{Na} / \nu_{Na}} V_K + \frac{G_{Na} / \nu_{Na}}{G_K / \nu_K + G_{Na} / \nu_{Na}} V_{Na}.$$

References

Books and Reviews

Amara, S. G. and Kuhar, M. J. (1993). Neurotransmitter transporters: Recent progress. *Ann. Rev. Neurosci.*, 16:73–93.

Beadle, L. C. (1943). Osmotic regulation and the faunas of inland waters. *Biol. Rev.*, 18:172–183.

Benos, D. J., Cunningham, S., Baker, R. R., Beason, K. B., Oh, Y., and Smith, P. R. (1992). Molecular characteristics of amiloride-sensitive sodium channels. *Rev. Physiol. Biochem. Pharmacol.*, 120:31–113.

Boron, W. F. (1986). Intracellular pH regulation in epithelial cells. *Ann. Rev. Physiol.*, 48:377–388.

Grinstein, S. and Foskett, J. K. (1990). Ionic mechanisms of cell volume regulation in leukocytes. *Ann. Rev. Physiol.*, 52:399–414.
Haas, M. (1989). Properties and diversity of (Na-K-Cl) cotransporters. *Ann. Rev. Physiol.*, 51:443–457.
Haas, M. (1994). The (Na-K-Cl) cotransporters. *Am. J. Physiol.*, 267:C869–C885.
Hamill, O. P., Lane, J. W., and McBride, D. W., Jr. (1992). Amiloride: A molecular probe for mechanosensitive channels. *Trends Physiol. Sci.*, 13:373–376.
Hediger, M. A. and Rhoads, D. B. (1994). Molecular physiology of sodium-glucose cotransporters. *Physiol. Rev.*, 74:993–1026.
Hille, B. (1992). *Ionic Channels of Excitable Membranes*. Sinauer, Sunderland, MA.
Hoffmann, E. K. (1977). Control of cell volume. In Gupta, B. L., Moreton, R. B., Oschman, J. L., and Wall, B. J., eds., *Transport of Ions and Water in Animals*, 285–332. Academic Press, New York.
Hoffmann, E. K. and Simonsen, L. O. (1989). Membrane mechanisms in volume and pH regulation in vertebrate cells. *Physiol. Rev.*, 69:315–382.
Kleinzeller, A. and Ziyadeh, F. N. (1990). Cell volume regulation in epithelia: With emphasis on the role of osmolytes and the cytoskeleton. In Raess, B. U. and Tunnicliff, G., eds., *Cell Volume Regulation*, 59–86. Karger, New York.
Koepsell, H. and Spangenberg, J. (1994). Topical review: Function and presumed molecular structure of Na^+-D-Glucose cotransport systems. *J. Membr. Biol.*, 138:1–11.
Lang, F., Völkl, H., and Häussinger, D. (1990). General principles in cell volume regulation. *Comp. Physiol.*, 4:1–25.
Läuger, P. (1991). *Electrogenic Ion Pumps*. Sinauer, Inc., Sunderland, MA.
Lechene, C. (1988). Physiological role of the Na-K pump. In *The Na^+, K^+-Pump*, pt. B, *Cellular Aspects*, 171–194. Alan R. Liss, New York.
Lentz, T. L. (1971). *Cell Fine Structure*. W. B. Saunders, Philadelphia, PA.
Lewis, S. A. and Donaldson, P. (1990). Ion channels and cell volume regulation: Chaos in an organized system. *News Physiol. Sci.*, 5:112–119.
Macknight, A. D. C. (1985). The role of anions in cellular volume regulation. *Pflügers Arch.*, 405:S12–S16.
Macknight, A. D. C., Grantham, J., and Leaf, A. (1993). Physiologic responses to changes in extracellular osmolality. In Seldin, D. W. and Giebisch, G., eds., *Clinical Disturbances of Water Metabolism*, 31–49. Raven, New York.
Macknight, A. D. C. and Leaf, A. (1977). Regulation of cellular volume. *Physiol. Rev.*, 57(3): 510–573.
Marger, M. D. and Saier, M. H., Jr. (1993). A major superfamily of transmembrane facilitators that catalyse uniport, symport and antiport. *Trends Biochem. Sci.*, 18:13–20.
McCarty, N. A. and O'Neil, R. G. (1992). Calcium signaling in cell volume regulation. *Physiol. Rev.*, 72:1037–1061.
Parker, J. C. (1993). In defense of cell volume? *Am. J. Physiol.*, 265:C1191–C1200.
Passow, H. (1986). Molecular aspects of band 3 protein-mediated anion transport across the red blood cell membrane. *Rev. Physiol. Biochem. Pharmacol.*, 103:61–203.
Rothstein, A. (1989). The Na^+/H^+ exchange system in cell pH and volume control. *Rev. Physiol. Biochem. Pharmacol.*, 112:235–257.
Sachs, F. (1988). Mechanical transduction in biological systems. *CRC Crit. Rev. Biomed. Eng.*, 16:141–169.

Scharff, O. and Foder, B. (1993). Regulation of cytosolic calcium in blood cells. *Physiol. Rev.*, 73:547–582.

Schmidt-Nielsen, K. (1990). *Animal Physiology: Adaptation and Environment*. Cambridge University Press, New York.

Siebens, A. W. (1985). Cellular volume control. In Seldin, D. W. and Giebisch, G., eds., *The Kidney: Physiology and Pathophysiolgy*. Raven, New York.

Spring, K. R. (1985). Determinants of epithelial cell volume. *Fed. Proc.*, 44:2526–2529.

Spring, K. R. and Hoffman, E. K. (1992). Cellular volume control. In Seldin, D. W. and Giebisch, G., eds., *The Kidney: Physiology and Pathophysiology*, 147–169. Raven, New York.

Stein, W. D. (1989). Kinetics of transport: Analyzing, testing, and characterizing models using kinetic approaches. In Fleischer, S. and Fleisher, B., eds., *Methods in Enzymology*, vol. 171, *Biomembranes*, pt. R, *Transport theory: Cells and Model Membranes,* 23–62. Academic Press, New York.

Stein, W. D. (1990). *Channels, Carriers, and Pumps*. Academic Press, New York.

Tosteson, D. C. (1964). Regulation of cell volume by sodium and potassium transport. In Hoffman, J. F., ed., *The Cellular Functions of Membrane Transport*. Prentice-Hall, Englewood Cliffs, NJ.

Treherne, J. E. (1987). Neural adaptations to osmotic and ionic stress in aquatic and terrestrial invertebrates. In Dejours, P., Bolis, L., Taylor, C. R., and Weibel, E. R., eds., *Comparative Physiology: Life in Water and on Land*, 523–535. Liviana Press, Padova, Italy.

Ussing, H. H. (1990). Volume regulation of frog skin epithelium. In Raess, B. U. and Tunnicliff, G., eds., *Cell Volume Regulation*, 87–113. Karger, New York.

Watson, S. and Girdlestone, D. (1994). 1994 Receptor and Ion Channel Nomenclature Supplement. *Trends Pharmacol. Sci.,* 5th ed.

Weiss, T. F. (1996). *Cellular Biophysics,* vol. 2, *Electrical Properties*. MIT Press, Cambridge, MA.

Original Articles

Abraham, E. H., Breslow, J. L., Epstein, J., Chang-Sing, P., and Lechene, C. (1985). Preparation of individual human diploid fibroblasts and study of ion transport. *Am. J. Physiol.*, 248:C154–C164.

Alexandre, J. and Lassalles, J. (1991). Hydrostatic and osmotic pressure activated channel in plant vacuole. *Biophys. J.*, 60:1326–1336.

Bergh, C., Kelley, S. J., and Dunham, P. B. (1990). K-Cl cotransport in LK sheep erythrocytes: Kinetics of stimulation by cell swelling. *J. Membr. Biol.*, 117:177–188.

Bookchin, R. M., Ortiz, O. E., and Lew, V. L. (1989). Mechnanisms of red cell dehydration in sickle cell anemia. In Raess, B. U. and Tunnicliff, G., eds., *The Red Cell Membrane,* 443–461. Humana Press, Clifton, NJ.

Boyle, P. J. and Conway, E. J. (1941). Potassium accumulation in muscle and associated changes. *J. Physiol.*, 100:1–63.

Brewster, J. L., de Valoir, T., Dwyer, N. D., Winter, E., and Gustin, M. C. (1993). An osmosensing signal transduction pathway in yeast. *Science*, 259:1760–1764.

Brinley, F. J., Jr. (1980). Regulation of intracellular calcium in squid axons. *Fed. Proc.*, 39:2778–2782.

Burton, R. F. (1973). The significance of ionic concentrations in the internal media of animals. *Biol. Rev.*, 48:195–231.

Burton, R. F. (1983a). Cell composition as assessed from osmolality and concentrations of sodium, potassium and chloride: Total contributions of other substances to osmolality and charge balance. *Comp. Biochem. Physiol.*, 76A:161–165.

Burton, R. F. (1983b). The composition of animal cells: Solutes contributing to osmotic pressure and charge balance. *Comp. Biochem. Physiol.*, 76B:663–671.

Burton, R. F. (1986). Internal reference standards in ionic regulation and the predictability of ionic concentrations in animals. *Comp. Biochem. Physiol.*, 83A:607–611.

Burton, R. F. (1987). On calculating concentrations of "HCO_3" from pH and pCO_2. *Comp. Biochem. Physiol.*, 87A:417–422.

Burton, R. F. (1988). The protein content of extracellular fluids and its relevance to the study of ionic regulation: Net charge and colloid osmotic pressure. *Comp. Biochem. Physiol.*, 90A:11–16.

Cala, P. M. (1983). Volume regulation by red blood cells: Mechanisms of ion transport between cells and mechanisms. *Mol. Physiol.*, 4:33–52.

Christensen, O. (1987). Mediation of cell volume regulation by Ca^{2+} influx through stretch-activated channels. *Nature*, 330:66–68.

Christensen, O. and Hoffman, E. K. (1992). Cell swelling activities K^+ and Cl^- channels as well as nonselective, stretch-activated cation channels in Ehrlich ascites tumor cells. *J. Membr. Biol.*, 129:13–36.

Civan, M. M. and Bookman, R. J. (1982). Transepithelial Na^+ transport and the epithelial fluids: A computer study. *J. Membr. Biol.*, 65:63–80.

Cohen, B. J. and Lechene, C. (1989). (Na,K)-pump: Cellular role and regulation in nonexcitable cells. *Biol. Cell*, 66:191–195.

Colclasure, G. C. and Parker, J. C. (1992). Cytosolic protein concentration is the primary volume signal for swelling-induced K-Cl cotransport in dog red cells. *J. Gen. Physiol.*, 100:1–10.

Collins, J. F., Honda, T., Knobel, S., Bulus, N. M., Conary, J., DuBois, R., and Ghishan, F. K. (1993). Molecular cloning, sequencing, tissue distribution, and functional expression of a Na^+/H^+ exchanger (NHE-2). *Proc. Natl. Acad. Sci. U.S.A.*, 90:3938–3942.

Davidson, R. M. (1993). Membrane stretch activates a high-conductance K^+ channel in G292 osteoblastic-like cells. *J. Membr. Biol.*, 131:81–92.

Deutsch, C. and Chen, L. (1993). Heterologous expression of specific K^+ channels in T lymphocytes: Functional consequences for volume regulation. *Proc. Natl. Acad. Sci. U.S.A.*, 90:10036–10040.

DiPolo, R. and Beaugé, L. (1990). Asymmetrical properties of the Na-Ca exchanger in voltage-clamped, internally dialyzed squid axons under symmetrical ionic conditions. *J. Gen. Physiol.*, 95:819–835.

DiPolo, R. and Beaugé, L. (1993a). Effects of some metal-ATP complexes on Na^+-Ca^{2+} exchange in internally dialysed squid axons. *J. Physiol.*, 462:71–86.

DiPolo, R. and Beaugé, L. (1993b). In squid axons the Ca_i^{2+} regulatory site of the Na^+/Ca^{2+} exchanger is drastically modified by sulfhydryl blocking agents. Evidences that intracellular Ca_i^{2+} regulatory and transport sites are different. *Biochim. Biophys. Acta*, 1145:75–84.

Dunham, P. B., Jessen, F., and Hoffmann, E. K. (1990). Inhibition of Na-K-Cl cotransport in Ehrlich ascites cells by antiserum against purified proteins of the cotransporter. *Proc. Natl. Acad. Sci. U.S.A.*, 87:6828–6832.

Dunham, P. B., Klimczak, J., and Logue, P. J. (1993). Swelling activation of K-Cl cotransport in LK sheep erythrocytes: A three-state process. *J. Gen. Physiol.*, 101:733–766.

Erxleben, C. (1989). Stretch-activated current through single ion channels in the abdominal stretch receptor organ of the crayfish. *J. Gen. Physiol.*, 94:1071–1083.

Eveloff, J. L. and Warnock, D. G. (1987). Activation of ion transport systems during cell volume regulation. *Am. J. Physiol.*, 252:F1–F10.

Farber, J. L., Holowecky, O. O., Serroni, A., and Van Rossum, G. (1989). Effects of ouabain on potassium transport and cell volume regulation in rat and rabbit liver. *J. Physiol.*, 417:389–402.

Ferreira, H. G. and Ferreira, K. T. G. (1983). Epithelial transport parameters: An analysis of experimental strategies. *Proc. R. Soc. London, Ser. B*, 218:309–329.

Finn, A. L. (1985). Symposium: Volume dependent pathways in animal cells. *Fed. Proc.*, 44:2499–2529.

Fitzsimons, E. J. and Sendroy, J., Jr. (1961). Distribution of electrolytes in human blood. *J. Biol. Chem.*, 236:1595–1601.

Fliegel, L. and Fröhlich, O. (1993). The Na^+/H^+ exchanger: An update on structure, regulation and cardiac physiology. *Biochem. J.*, 296:273–285.

Franciolini, F. and Petris, A. (1990). Chloride channels of biological membranes. *Biochim. Biophys. Acta*, 1031:247–259.

Frey, N., Büchner, K. H., and Zimmermann, U. (1988). Water transport parameters and regulatory processes in *Eremosphaera viridis*. *J. Membr. Biol.*, 101:151–163.

Funder, J. and Wieth, J. O. (1966). Chloride and hydrogen ion distribution between human red cells and plasma. *Acta Physiol. Scan.*, 68:234–245.

Gasbjerg, P. K. and Brahm, J. (1991). Kinetics of bicarbonate and chloride transport in human red cell membranes. *J. Gen. Physiol.*, 97:321–349.

Geck, P., Pietrzyk, C., Burckhardt, B. C., Pfeiffer, B., and Heinz, E. (1980). Electrically silent cotransport of Na^+, K^+, Cl^- in Ehrlich cells. *Biochim. Biophys. Acta*, 600:432–447.

Gilbertson, T. A., Avenet, P., Kinnamon, S. C., and Roper, S. D. (1992). Proton currents through amiloride-sensitive Na channels in hamster taste cells. *J. Gen. Physiol.*, 100:803–824.

Grinstein, S., Cohen, S., Goetz, J. D., and Rothstein, A. (1985). Na^+/H^+ exchange in volume regulation and cytoplasmic pH homeostasis in lymphocytes. *Fed. Proc.*, 44:2508–2512.

Hall, A. C. and Ellory, J. C. (1986). Evidence for the presence of volume-sensitive KCl transport in 'young' human red cells. *Biochim. Biophys. Acta*, 858:317–320.

Hamill, O. P. and McBride, D. W., Jr. (1992). Rapid adaptation of single mechanosensitive channels in *xenopus* oocytes. *Proc. Natl. Acad. Sci. U.S.A.*, 89:7462–7466.

Hoffman, P. G. and Tosteson, D. C. (1971). Active sodium and potassium transport in high potassium and low potassium sheep red cells. *J. Gen. Physiol.*, 58:438–466.

Hoffmann, E. K. (1985). Role of separate K^+ and Cl^- channels and of Na^+/Cl^- cotransport in volume regulation in Ehrlich cells. *Fed. Proc.*, 44:2513–2519.

Hokin, L. E. (1981). Reconstitution of "carriers" in artificial membranes. *J. Membr. Biol.*, 60:77–93.

Hurst, A. M. and Hunter, M. (1990). Stretch-activated channels in single early distal tubule cells of the frog. *J. Physiol.*, 430:13–24.

Jacobs, M. H. and Stewart, D. R. (1942). The role of carbonic anhydrase in certain ionic exchanges involving the erythrocyte. *J. Gen. Physiol.*, 25:539–552.

Jacobs, M. H. and Stewart, D. R. (1947). Osmotic properties of the erythrocyte. *J. Cell. Comp. Physiol.*, 30:79–103.

Jakobsson, E. (1980). Interactions of cell volume, membrane potential, and membrane transport parameters. *Am. J. Physiol.*, 238:C196–C206.

Jennings, M. L. and Al-Rohil, N. (1990). Kinetics of activation and inactivation of swelling-stimulated K^+/Cl^- transport. *J. Gen. Physiol.*, 95:1021–1040.

Kim, D. (1992). A mechanosensitive K^+ channel in heart cells. *J. Gen. Physiol.*, 100:1021–1040.

Kirber, M. T., Walsh, J. V., Jr., and Singer, J. J. (1988). Stretch-activated ion channels in smooth muscle: A mechanism for the initiation of stretch-induced contraction. *Pflügers Arch.*, 412:339–345.

Kracke, G. R. and Dunham, P. B. (1990). Volume-sensitive K-Cl cotransport in inside-out vesicles made from erythrocyte membranes from sheep of low-K phenotype. *Proc. Natl. Acad. Sci. U.S.A.*, 87:8575–8579.

Kregenow, F. M. (1973). The response of duck erythrocytes to norepinephrine and an elevated extracellular potassium: Volume regulation in isotonic media. *J. Gen. Physiol.*, 61:509–527.

Kubalski, A., Martinac, B., Ling, K., Adler, J., and Kung, C. (1993). Activities of a mechanosensitive ion channel in an *E. coli* mutant lacking the major lipoprotein. *J. Membr. Biol.*, 131:151–160.

Larsen, E. H. and Rasmussen, B. E. (1985). A mathematical model of amphibian skin epithelium with two types of transporting cellular units. *Pflügers Arch.*, 405, Suppl. 1:S50–S58.

Larsson, L., Aperia, A., and Lechene, C. (1986). Ionic transport in individual renal epithelial cells from adult and young rats. *Acta Physiol. Scand.*, 126:321–332.

Leaf, A. (1956). On the mechanism of fluid exchange of tissues *in vitro*. *Biophys. J.*, 62:241–248.

Lew, V. L. and Bookchin, R. M. (1986). Volume, pH, and ion-content regulation in human red cells: Analysis of transient behavior with an integrated model. *J. Membr. Biol.*, 92:57–74.

Lew, V. L. and Bookchin, R. M. (1991). Osmotic effects of protein polymerization: Analysis of volume changes in sickle cell anemia red cells following deoxy-hemoglobin S polymerization. *J. Membr. Biol.*, 122:55–67.

Lew, V. L., Ferreira, H. G., and Moura, T. (1979). The behaviour of transporting epithelial cells: I. Computer analysis of a basic model. *Proc. R. Soc. London, Ser. B*, 206:53–83.

Lew, V. L., Freeman, C. J., Ortiz, O. E., and Bookchin, R. M. (1989). Need and applications of integrated red cell models. In Raess, B. U. and Tunnicliff, G., eds., *The Red Cell Membrane*, 19–34. Humana Press, Clifton, NJ.

Lew, V. L., Freeman, C. J., Ortiz, O. E., and Bookchin, R. M. (1990). Application of integrated cell models: Novel predictions on the behavior of red cells and reticulocytes. *Biomed. Biochim Acta*, 49:737–741.

Lew, V. L., Freeman, C. J., Ortiz, O. E., and Bookchin, R. M. (1991). A mathematical model of the volume, pH, and ion content regulation in reticulocytes. *J. Clin. Invest.*, 87:100–112.

Lewis, R. S., Ross, P. E., and Cahalan, M. D. (1993). Chloride channels activated by osmotic stress in *T* lymphocytes. *J. Gen. Physiol.*, 101:801–826.

Mandel, L. J., Bacallao, R., and Zampighi, G. (1993). Uncoupling of the molecular "fence" and paracellular "gate" functions in epithelial tight junctions. *Nature*, 361:552–555.

Martinac, B., Adler, J., and Kung, C. (1990). Mechanosensitive ion channels of *E. coli* activated by amphipaths. *Nature*, 348:261–263.

Mathias, R. T. (1985). Steady-state voltages, ion fluxes, and volume regulation in syncytial tissues. *Biophys. J.*, 48:435–448.

McBride,, D. W., Jr., and Hamill, O. P. (1993). Pressure-clamp technique for measurement of the relaxation kinetics of mechanosensitive channels. *Trends Neurosci.*, 16:341–345.

Medina, I. R. and Bregestovski, P. D. (1988). Stretch-activated ion channels modulate the resting membrane potential during early embryogenesis. *Proc. R. Soc. London, Ser. B*, 235:95–102.

Milgram, J. H. and Solomon, A. K. (1977). Membrane permeability equations and their solutions for red cells. *J. Membr. Biol.*, 34:103–144.

Miller, R. J. (1991). The control of neuronal Ca^{2+} homeostasis. *Prog. Neurobiol.*, 37:255–285.

Minton, A. P., Colclasure, G. C., and Parker, J. C. (1992). Model for the role of macromolecular crowding in regulation of cellular volume. *Proc. Natl. Acad. Sci. U.S.A.*, 89:10504–10506.

Mintz, E., Thomas, S. R., and Mikulecky, D. C. (1986). Exploration of apical sodium transport mechanisms in an epithelial model by network thermodynamic simulation of the effect of mucosal sodium depletion: I. Comparison of three different apical sodium permeability expressions. *J. Theor. Biol.*, 123:1–19.

Miseta, A., Bogner, P., Berényi, E., Kellermayer, M., Galambos, C., Wheatley, D. N., and Cameron, I. L. (1993). Relationship between cellular ATP, potassium, sodium and magnesium concentrations in mammalian and avian erythrocytes. *Biochim. Biophys. Acta*, 1175:133–139.

Moody, W. J., Jr. (1981). The ionic mechanism of intracellular pH regulation in crayfish neurones. *J. Physiol.*, 316:293–308.

Moore, E. D. W., Etter, E. F., Philipson, K. D., Carrington, W. A., Fogarty, K. E., Lifshitz, L. M., and Fay, F. S. (1993). Coupling of the Na^+/Ca^{2+} exchanger, Na^+/K^+ pump and sarcoplasmic reticulum in smooth muscle. *Nature*, 365:657–660.

Moroz, P. A., Amaullakhanov, F. I., Kiyatkin, A. B., Pichugin, A. V., and Viteitsky, V. M. (1989). A mathematical model for the stabilization of cell volumes of erythrocytes. *Biologicheskie membranyi*, 6:409–419.

Morris, C. E. (1990). Mechanosensitive ion channels. *J. Membr. Biol.*, 113:93–107.

Morris, C. E. and Horn, R. (1991). Failure to elicit neuronal macroscopic mechanosensitive currents anticipated by single-channel studies. *Science*, 251:1246–1249.

Morris, C. E. and Sigurdson, W. J. (1989). Stretch-inactivated ion channels coexist with stretch-activated ion channels. *Science*, 243:807–809.

Motais, R., Guizouarn, H., and Garcia-Romeu, F. (1991). Red cell volume regulation: The pivotal role of ionic strength in controlling swelling-dependent transport systems. *Biochim. Biophys. Acta*, 1075:169–180.

Oliet, S. H. R. and Bourque, C. W. (1993). Mechanosensitive channels transduce osmosensitivity in supraoptic neurons. *Nature*, 364:341–343.

Orlov, S. N., Pokudin, N. I., Kotelevtsev, Y. V., and Gulak, P. V. (1989). Volume-dependent regulation of ion transport and membrane phosphorylation in human and rat erythrocytes. *J. Membr. Biol.*, 107:105–117.

Overbeek, J. T. G. (1956). The Donnan equilibrium. *Prog. Biophys. Biophys. Chem.*, 6:58–84.

Payne, J. A., Lytle, C., and McManus, T. J. (1990). Foreign anion substitution for chloride in human red blood cells: Effect on ionic and osmotic equilibria. *Am. J. Physiol.*, 259:C819–C827.

Peters, T. (1988). Basic mechanisms of cellular calcium homeostasis. *Acta Otolaryngol.*, 460:7–12.

Post, R. L. and Jolly, P. C. (1957). The linkage of sodium, potassium, and ammonium active transport across the human erythrocyte membrane. *Biochim. Biophys. Acta*, 25:118–128.

Reuben, J. P., Girardier, L., and Grundfest, H. (1964). Water transfer and cell structure in isolated crayfish muscle fibers. *J. Gen. Physiol.*, 47:1141–1174.

Rink, T. J. (1984). Aspects of the regulation of cell volume. *J. Physiol. (Paris)*, 79:388–394.

Sachs, G. and Fleischer, S. (1989). Transport machinery: An overview. In Fleischer, S. and Fleischer, B., eds., *Methods in Enzymology*, vol. 171, *Biomembranes*, 3–12. Academic Press, New York.

Sackin, H. (1989). A stretch-activated K^+ channel sensitive to cell volume. *Proc. Natl. Acad. Sci. U.S.A.*, 86:1731–1735.

Sadoshima, J., Takahashi, T., Jahn, L., and Izumo, S. (1992). Roles of mechano-sensitive ion channels, cytoskeleton, and contractile activity in stretch-induced immediate-early gene expression and hypertrophy of cardiac myocytes. *Proc. Natl. Acad. Sci. U.S.A.*, 89:9905–9909.

Schwiening, C. J., Kennedy, H. J., and Thomas, R. C. (1993). Calcium-hydrogen exchange by the plasma membrane Ca-ATPase of voltage-clamped snail neurons. *Proc. R. Soc. London, Ser. B*, 253:285–289.

Strieter, J., Stephenson, J. L., Palmer, L. G., and Weinstein, A. M. (1990). Volume-activated chloride permeability can mediate cell volume regulation in a mathematical model of a tight epithelium. *J. Gen. Physiol.*, 96:319–344.

Sukharev, S. I., Blount, P., Martinac, B., Blattner, F. R., and Kung, C. (1994). A large-conductance mechanosensitive channel in *E. coli* encoded by *mscL* alone. *Nature*, 368:265–268.

Taglietti, V. and Toselli, M. (1988). A study of stretch-activated channels in the membrane of frog oocytes: Interactions with Ca^{2+} ions. *J. Physiol.*, 407:311–328.

Thomas, R. C. (1974). Intracellular pH of snail neurones measured with a new pH-sensitive glass micro-electrode. *J. Physiol.*, 238:159–180.

Thomas, R. C. (1977). The role of bicarbonate, chloride and sodium ions in the regulation of intracellular pH in snail neurones. *J. Physiol.*, 273:317–338.

Thomas, R. C. (1989). Bicarbonate and pH_i response. *Nature*, 337:601.

Tosteson, D. C. and Hoffman, J. F. (1960). Regulation of cell volume by active cation transport in high and low potassium sheep red cells. *J. Gen. Physiol.*, 44:169–194.

Treherne, J. E. and Pichon, Y. (1978). Long-term adaptations of *sabella* giant axons to hyposmotic stress. *J. Exp. Biol.*, 75:253–263.

Turk, E., Martín, M. G., and Wright, E. M. (1994). Structure of the human Na^+/glucose cotransporter gene *SGLT1*. *J. of Biolog. Chem.*, 269:15204–15209.

Turner, R. J. (1983). Quantitative studies of cotransport systems: Models and vesicles. *J. Membr. Biol.*, 76:1–15.

Ubl, J., Murer, H., and Kolb, H. (1988a). Hypotonic shock evokes opening of Ca^{2+}-activated K channels in opossum kidney cells. *Pflügers Arch.*, 412:551–553.

Ubl, J., Murer, H., and Kolb, H. (1988b). Ion channels activated by osmotic and mechanical stress in membranes of opossum kidney cells. *J. Membr. Biol.*, 104:223–232.

Voilley, N., Lingueglia, E., Champigny, G., Mattei, M., Waldmann, R., Lazdunski, M., and Barbry, P. (1994). The lung amiloride-sensitive Na^+ channel: Biophysical properties, pharmacology, ontogenesis, and molecular cloning. *Proc. Natl. Acad. Sci. U.S.A.*, 91:247–251.

Wang, K. and Wondergem, R. (1993). Hepatocyte water volume and potassium activity during hypotonic stress. *J. Membr. Biol.*, 135:137–144.

Weinstein, A. M. (1992). Analysis of volume regulation in an epithelial model. *J. Membr. Biol.*, 54:537–561.

Werner, A. and Heinrich, R. (1985). A kinetic model for the interaction of energy metabolism and osmotic states of human erythrocytes: Analysis of the stationary "in vivo" state and of time dependent variations under blood preservation conditions. *Biomed. Biochim. Acta*, 44:185–212.

Wilson, T. H. (1954). Ionic permeability and osmotic swelling of cells. *Science*, 120:104–105.

Wolf, M. B. (1980). A simulation study of the anomalous osmotic behavior of red cells. *J. Theor. Biol.*, 83:687–700.

Worrall, D. M. and Williams, D. C. (1994). Sodium ion-dependent transporters for neurotransmitters: A review of recent developments. *Biochem. J.*, 297:425–436.

Yamada, W. M., Koch, C., and Adams, P. R. (1989). Multiple channels and calcium dynamics. In Koch, C. and Segev, I., eds., *Methods in Neuronal Modeling*, 97–133. MIT Press, Cambridge, MA.

Yang, X. and Sachs, F. (1993). Mechanically sensitive, nonselective cation channels. In Siemen, D. and Hescheler, J., eds., *Nonselective Cation Channels: Pharmacology, Physiology and Biophysics*, 79–92. Birkhäuser Verlag, Switzerland.

List of Figures

Chapter 1

Chapter 2

Chapter 3

Chapter 4

Chapter 5

Chapter 6

Chapter 7

Chapter 8

List of Tables

Chapter 1

Chapter 2

Chapter 3

Chapter 4

Chapter 6

Chapter 7

Chapter 8

Contents of Volume 2

Index

Page numbers in italic indicate figures or tables; *n* indicates footnotes.